二氧化碳驱油与埋存技术文集

胡永乐　主编

石 油 工 业 出 版 社

内 容 提 要

本文集为国家科技重大专项"大型油气田及煤层气开发"支持项目"二氧化碳驱油与埋存关键技术（2011ZX05016）"在国家"十二五"攻关期间取得的部分理论技术研究成果，内容涵盖：二氧化碳驱油与埋存机理，二氧化碳驱油藏工程方法及优化设计技术，二氧化碳驱油井筒与地面工艺设计、高效防腐及安全性评价技术，二氧化碳驱油与埋存动态监测及潜力评价技术等，较为全面地反映了我国二氧化碳驱油与埋存理论技术水平和应用进展。

本书可供从事二氧化碳驱油提高采收率、二氧化碳地质埋存等相关工作的科研人员、工程技术人员和高等院校师生参考。

图书在版编目（CIP）数据

二氧化碳驱油与埋存技术文集/胡永乐主编 .

北京：石油工业出版社，2016.5

ISBN 978-7-5183-1215-3

Ⅰ. 二…

Ⅱ. 胡…

Ⅲ. 二氧化碳-驱油-文集

Ⅳ. TE357.5-53

中国版本图书馆 CIP 数据核字（2016）第 066521 号

出版发行：石油工业出版社

    （北京安定门外安华里 2 区 1 号　100011）

    网　址：www. petropub. com

    编辑部：（010）64523712

    图书营销中心：（010）64523633

经　销：全国新华书店

印　刷：北京中石油彩色印刷有限责任公司

2016 年 5 月第 1 版　2016 年 5 月第 1 次印刷

787×1092 毫米　开本：1/16　印张：36

字数：919 千字

定价：288.00 元

前　言

二氧化碳驱油与埋存技术越来越受到人们的重视。一方面，二氧化碳驱油可提高油气藏采收率和动用率，增加油气可采储量；另一方面，二氧化碳注入储层可实现地质埋存，达到温室气体减排的目的。我国早在 20 世纪 60 年代初就开始了二氧化碳驱油提高原油采收率的室内研究和矿场试验，但由于气源条件限制，油层中气窜严重及技术不配套等原因，直到 2005 年之前这项技术在我国一直发展缓慢。但是，二氧化碳驱油与埋存技术在我国有着十分广阔的发展前景，一是由于我国拥有大量的低渗透、特低渗透油藏和一批难动用储量，适宜采用注气开发方式来提高油藏采收率和动用率；二是近年来我国二氧化碳排放量急剧增长，已成为全球碳排放主要国家之一，二氧化碳减排压力非常大。

"十一五"期间，出于对低渗透油田提高采收率和二氧化碳减排的考虑，加上松辽盆地高含二氧化碳气田投入开发需要解决伴生二氧化碳气的出路问题，二氧化碳驱油与埋存技术被重新提上日程。并在吉林油田开辟了技术应用示范区。通过技术研发与应用，初步形成技术成果，示范区试验也初见成效。"十二五"期间，国家继续大力支持这项技术的发展，在国家科技重大专项"大型油气田及煤层气开发"中设立"二氧化碳驱油与埋存关键技术"项目，针对制约我国陆相油藏二氧化碳驱油与埋存技术发展的瓶颈问题开展持续攻关。通过几年的研究，项目在二氧化碳驱油与埋存基础理论、关键技术、现场应用等方面取得了一系列重要进展。

为了系统展示本项目的研究成果，我们从攻关过程中所撰写的 111 篇学术论文中筛选了 64 篇，分专业、分领域编纂成册，以论文集的形式出版发行。这些论文内容涵盖二氧化碳驱油与埋存理论认识、关键技术及应用等。全书共分四篇，第一篇包括 18 篇论文，内容涉及二氧化碳在油藏和盐水层中的驱油与埋存机理，二氧化碳—地层油体系相态特征和表征方法，泡沫驱油提高采收率机理等；第二篇包括 16 篇论文，内容涉及不同类型油藏、不同驱替方式油藏工程优化设计技术，二氧化碳驱油藏工程方法和数值模拟技术，二氧化碳驱油开发规律认识等；第三篇包括 16 篇论文，内容涉及二氧化碳驱油地面工程优化设计技术、井筒和地面防腐技术、安全风险评价技术等；第四篇包括 13 篇论文，内容涉及二氧化碳驱油试井分析方法和动态监测技术，二氧化碳驱油与埋存潜力评价技术和典型实例分析，二氧化碳驱油适应性评价与筛选技术等。希望本文集的出版，能够对我国二氧化碳驱油与埋存技术的发展和规模化应用起到积极推动作用。

由于文集涉及的学科较多，加之作者水平有限，错误和不当之处在所难免，敬请广大读者批评指正。

目　　录

第三篇　二氧化碳驱油井筒与地面工艺设计、高效防腐及安全性评价技术

第四篇　二氧化碳驱油与埋存动态监测及潜力评价技术

CO_2 驱油与埋存关键技术进展

胡永乐　　郝明强

（中国石油勘探开发研究院）

0　引言

近年来，我国碳排放量增长迅速，成为世界上主要的温室气体排放国家之一。我国政府高度重视 CO_2 减排工作，2015 年 11 月 30 日，习近平主席在法国巴黎气候变化大会上发表题为《携手构建合作共赢、公平合理的气候变化治理机制》的重要讲话，进一步重申：我国将于 2030 年左右使二氧化碳排放达到峰值并争取早日实现，2030 年单位国内生产总值二氧化碳排放比 2005 年下降 60%~65%。我国 CO_2 减排的责任和压力很大。

从油气田开发角度来看，目前，我国新发现和已发现未动用原油储量 70% 以上均为低渗透油田和难采储量，这类油田经济有效开发难度大，原油采收率低。国内外现场实践均已证明，CO_2 驱油能够大幅度提高低渗透油田的采收率和难采储量的动用率，其中混相驱油能够提高原油采收率 10% 以上，非混相驱油能够提高原油采收率 7% 以上。根据估算，我国有 $70×10^8t$ 以上的石油地质储量适合 CO_2 驱油，若按提高原油采收率 10% 计算，可多采原油 $7×10^8t$，相当于新发现一个特大型油田，经济效益相当可观。另外，"十五"期间，我国在松辽盆地发现大量高含 CO_2 天然气藏，要实现这类气田的安全环保开发，需要解决气田开发伴生 CO_2 气的出路问题。

CO_2 驱油与埋存技术是一项同时实现"增油增气"经济效益与"减排 CO_2"社会效益双赢的技术，并且适逢我国 CO_2 减排、原油天然气增产等多重重大需求的驱动，国家科技部、财政部、发展和改革委员会在国家科技重大专项"十二五"05 专项"大型油气田及煤层气开发"中专门设立"CO_2 驱油与埋存关键技术"项目，针对我国陆相油藏 CO_2 驱油与埋存技术的瓶颈问题开展攻关，经过近 5 年的持续研究，项目在 CO_2 驱油与埋存的基础理论、关键技术和现场应用等方面取得了重要进展。

1　项目简介

"CO_2 驱油与埋存关键技术"的总体目标是：针对陆相油藏 CO_2 驱油与埋存规模化应用的技术瓶颈，通过开展实验机理、油藏工程、注采工艺、集输处理、矿场试验配套技术研究，完善 CO_2 驱油机理实验，确定 CO_2 驱油与埋存工程中关键基础参数，阐明 CO_2 驱油的开发规律，形成 CO_2 驱油藏工程与方案优化技术、CO_2 驱油高效注采工艺技术、CO_2 长距离输送及超临界注入技术、CO_2 驱油采出流体处理及循环注气技术、CO_2 驱油动态监测技术、CO_2 驱油与埋存潜力评价技术等，并制订 CO_2 驱油与埋存战略规划方案。为 CO_2 驱油采收率提高 10% 以上、实现示范区产量目标提供技术支撑，为我国大规模推广应用 CO_2 驱油与埋存技术奠定基础。

围绕这一整体目标，项目分解为 5 个层次、6 个课题开展攻关研究。5 个层次分别为应用基础研究、关键工程技术研发与应用、矿场试验、推广应用和技术集成等，具体技术路线如图 1 所示。6 个课题分别为 CO_2—地层油体系关键参数应用方法，CO_2 驱油与埋存油藏工程技术及应用，CO_2 驱油高效注采与埋存工艺技术，CO_2 超临界输送及循环注气技术，CO_2 驱油藏动态监测与方案优化技术，CO_2 驱油与埋存潜力评价及战略规划等。

项目由中国石油集团科学技术研究院牵头，联合中国石油吉林油田分公司、大庆油田有限责任公司、中国石油大学、中国地质大学、西南石油大学、东北石油大学、吉林大学等 10 余家单位，组成一支"产-学-研-用"一体化联合攻关团队。经过全体人员的共同努力，项目获得一大批有形化成果和知识产权。项目有力支撑了我国 CO_2 驱油与埋存技术研发中心建设和试验示范基地建设，现场试验取得了较好的经济效益和社会效益，促进了 CO_2 驱

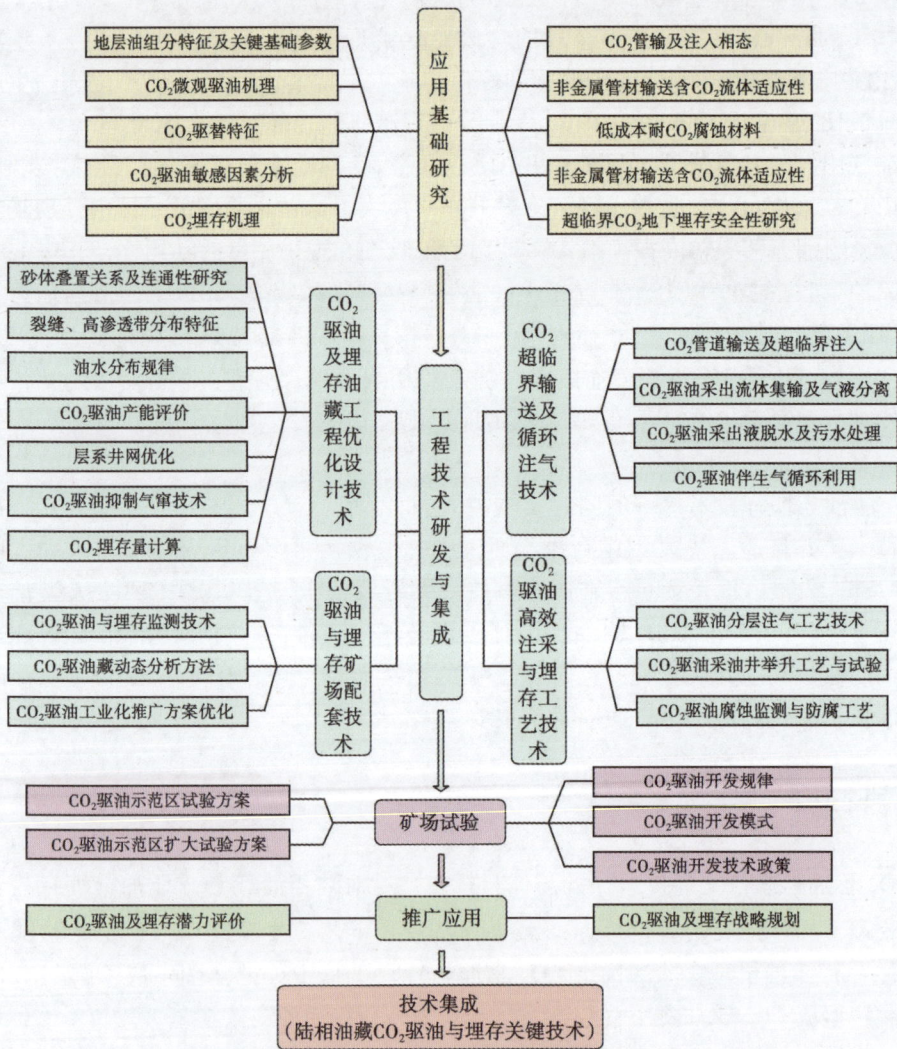

图 1　项目技术路线图

油与埋存技术在长庆、新疆等其他油区的推广应用，提升了我国在 CO_2 捕集、驱油与埋存（CCUS）领域的国际影响力和 CO_2 减排的话语权。

2 理论与技术进展

通过几年的持续攻关，在 CO_2 驱油与埋存基础理论方面取得新认识，在油藏工程、注采工艺、地面工程等核心主体技术方面有重要突破，在 CO_2 驱油与埋存动态监测、潜力评价、战略规划等配套技术方面取得新进展。现从以下 6 个方面简要介绍。

2.1 CO_2—地层油体系相态特征、关键参数及应用方法

围绕 CO_2 驱油提高采收率的基础理论问题，通过深化 CO_2—地层油相态实验研究，对原油组分与 CO_2 混相能力关系取得新认识；完善了特低渗透非均质油藏 CO_2 驱油物理模拟实验方法；建立了 CO_2—地层油体系相态基础参数数据库，并将室内实验研究与工程应用进行有机结合，满足现场生产需求。

（1）深化了 CO_2—地层油相态实验研究，揭示了 CO_2—地层油体系相间传质机理。

通过对国内外 8 个油田 22 个低渗透区块 30 口井的地层油组成特征研究，发现不同油区烃组分分布规律基本相似，即 C_1+N_2 组分含量高（摩尔分数在 10%~80% 之间）、C_2—C_6 分布分散（摩尔分数在 0.2%~10% 之间）、C_7—C_{15} 含量相对较高（摩尔分数在 2%~10% 之间）、C_8 以上组分含量随碳数增加而逐渐减少。但陆相地层原油与海相地层原油相比，其 C_2—C_6 组分含量明显偏低，C_{11+} 和胶质、沥青质含量较高，增加了 CO_2 与原油的混相难度。

针对陆相原油组成特点，将原油组成分为 6 个特征组分段：①强挥发性组分 C_1+N_2。②轻烃组分 C_2—C_6。③中间烃（液烃）组分 C_7—C_{15}。④重烃（固烃）组分 C_{16}—C_{29}。⑤超重烃组分 C_{30+}。⑥极重质组分（胶质、沥青质）。

通过研究超临界 CO_2—原油体系相间传质的动态过程，分析了不同烃组分的传质能力。揭示了升压过程中相间传质机理：传质（低压），CO_2 溶解于原油并萃取轻烃，气相富化；传质增强，富气相萃取原油中间烃形成富烃过渡相；传质剧烈，富烃相进一步萃取原油中较重烃组分；混相（高压），重质组分参与相间传质直至混相。基于 CO_2—纯烃体系相态、界面张力等实验结果，表征了各组分段与 CO_2 相间传质的能力，即烃组分传质能力随碳数增加而减弱，且各种组分对混相过程贡献程度不一。

通过大量的物理模拟实验，深化了非均质油藏 CO_2 驱油过程中油气传质特征：①CO_2 驱油时，油气两相间存在一个临界界面张力值 IFTc，当界面张力小于 IFTc 时，相间传质作用增强，才能实现动态混相驱，高效驱动孔隙中的原油。②非均质模型中 CO_2 驱油时，不同级别孔隙中的 IFTc 存在差异，IFTc 随孔隙半径减小而升高，并在 1.5mN/m 处存在拐点，相对于大孔隙，小孔隙中原油的有效动用需要进一步降低界面张力。③提高体系压力，可以降低临界界面张力，有效动用小孔隙中的原油，扩大低渗透非均质油藏中 CO_2 的微观波及体积。

（2）完善了特低渗透非均质油藏 CO_2 驱油物理模拟实验方法，深化了驱替特征认识。

针对微观模型与真实孔隙介质间的尺度和表面性质差异、一维长岩心中需要考虑非均质性、二/三维大物模中层间非均质特征的表征等问题，改进了物理模拟实验方法，制作了低渗透填砂/岩石微观可视模型、特低渗透非均质长岩心模型、低渗透多层非均质大岩石模型，

更好地模拟了真实孔隙、喉道内的作用机理及储层中的非均质性，为机理认识、驱替特征分析和波及状况识别等提供了基础手段。

通过非均质微观可视模型驱替实验，研究了不同相态条件（非混相/混相）、不同驱替方式（水平/垂直）时 CO_2 驱油机理，发现：①非均质模型水平驱替时，非均质非混相驱替过程中毛细管压力影响大于重力分异作用影响，超覆作用减弱，注入气沿优势通道指进，波及体积低；混相驱过程中消除气—液界面，均相驱替，大幅度提高原油采收率。②非均质模型垂直驱替时，非混相驱替过程中初期注入气沿优势通道指进，随后由于重力分异作用，前缘得到调整，波及体积逐渐扩大；混相驱替过程中消除气—液界面，均相驱替，大幅度提高原油采收率。并且，采用重力驱替方式能够有效扩大 CO_2 波及体积。

利用低渗透非均质长岩心模型、三维大尺度非均质模型，在不同维度下模拟 CO_2 驱替过程，探索非均质油藏中 CO_2—地层油不同混相状态时的驱替机理，揭示了不同驱替阶段的驱油特征。①水驱：沿优势通道指进，采出程度低。②CO_2 非混相驱：降低界面张力和毛细管压力，膨胀降黏改善相对渗透率，采收率显著提高，CO_2 突破后压差降低，流度比升高，气窜明显。③水气交替驱：有效提高压差，改善流度比，扩大微观波及体积，从而提高采收率。④CO_2 混相驱：消除界面张力，驱替微小孔隙中的原油，大幅度提高采收率。

（3）建立了 CO_2—地层油体系关键参数数据库，形成了关键基础参数的工程应用方法。

以影响 CO_2—原油体系混相的关键参数为重点，收集、测试了多套油气藏流体基础相态数据、CO_2—纯组分/地层油体系膨胀数据、油气藏流体组分临界参数数据、CO_2—纯组分/原油体系界面张力数据、CO_2—地层油体系最小混相压力数据、原油—CO_2—水共存时溶解度数据，形成了涵盖国内外 6 类油藏的 CO_2—地层油体系的基础参数数据库。

另外，基于大量实验数据，还建立了一系列 CO_2—地层油体系关键参数预测关系式和图版，包括：地层油黏度通用公式、新型最小混相压力预测关系式、C_{45} 以上临界参数预测公式、改进 W-K 型界面张力预测公式、CO_2—纯组分体积膨胀系数预测公式。

关键参数数据库和应用图版为 CO_2 驱油与埋存的室内研究成果向现场应用转化提供了基础，满足了现场生产需求。

（4）提出了改善 CO_2—地层油混相能力的新方法，研制出有效的化学助剂。

基于"油—气相间传质先形成过渡相、进而实现混相"的新认识，提出在 CO_2 和地层油之间预置一个低界面张力的易混段塞，从而降低 CO_2 驱油最小混相压力。应用该原理，研制出降低最小混相压力的化学试剂 CAE，该试剂是在合成烃链末端有两个甲基的低相对分子质量碳氢表面活性剂，较高压力时与 CO_2 互溶性好，易溶于原油。细管实验评价结果表明，通过前置 0.2%PV 的 CAE 段塞，F48 油藏的地层油最小混相压力从 27.3MPa 降低到 21.2MPa，有效改善了 CO_2—原油体系的混相能力。

2.2 CO_2 埋存机理与安全性评价

CO_2 埋存是目前最经济、最可靠、也是最直接有效的 CO_2 减排方式。通过开展油藏储层和盐水层中的 CO_2 埋存机理研究，在一些基础理论认识上取得新进展，同时对试验示范区进行了 CO_2 埋存安全性评价。

（1）揭示了油藏储层中 CO_2 埋存机制及规律，为潜力评价提供了理论基础。

构造埋存、溶解埋存、游离埋存和矿化埋存是 CO_2 在油藏储层中的主要埋存形式。通过建立 CO_2 在油藏中埋存的物理模拟实验方法，明确了 CO_2 在油藏中不同埋存形式的埋存机理和主控因素。

以相态实验为基础，基于不同原油体系中的 CO_2 溶解实验，开展了 CO_2 在地层原油中溶解规律及原油—CO_2 体系的物性变化特征研究，结果表明：随着 CO_2 注入量的增大，原油性质不断发生变化，其饱和压力和体积系数随着 CO_2 浓度的增加而不断增大。CO_2 在原油中的溶解能力与原始溶解气油比、原油体积系数均具有良好的线性相关性。

通过不同地层水样品中 CO_2 溶解度实验，揭示了 CO_2 在地层水中的溶解度变化规律，即溶解度随压力增加而增大，低压时增加幅度大，高压时增加幅度小，溶解度曲线在 CO_2 临界压力附近出现拐点；温度越低，CO_2 在地层水中的溶解度越高，当温度大于 100℃、压力大于 22MPa 时，CO_2 的溶解度随温度升高略有增加。同时，根据 CO_2 在地层水中的溶解度实验结果，对比了多种计算方法的适应性，并修正了 PR-GE 型状态方程模型，应用该模型建立了不同矿化度地层水在不同温度、不同压力条件下的溶解度图版。

通过油—水—CO_2 三相相态实验，分析了高温高压条件下 CO_2 在油水共存体系中相态变化及溶解度变化特征。并通过一维多孔介质中长岩心 CO_2 埋存实验，揭示了渗透率、地层倾角、驱油方式等对 CO_2 埋存量的影响规律。

（2）量化分析了盐水层中 CO_2 埋存的主控因素，深化了 CO_2 扩散规律认识。

盐水层是 CO_2 地质埋存的主要场所。国际能源署（IEA）对全球 CO_2 地质埋存总量的评估为 $10850×10^{12}t$，其中盐水层封存占埋存总量的 92%。因此，开展盐水层中 CO_2 埋存机理研究意义重大。

通过盐水层 CO_2 埋存的机理模拟（地层水矿化度为 5000mg/L，温度为 60℃），量化分析了毛细管压力、盐水矿化度、岩石压缩系数等因素对 CO_2 埋存量的影响规律。结果表明：构造埋存量随毛细管压力的增大而增加，溶解埋存量随盐水矿化度的增大而减少，游离相 CO_2 埋存量随岩石压缩系数的增大而略有减少。

利用全直径岩心驱替系统测试了 CO_2 在盐水层及油藏剩余油中的扩散速率，深化了对其扩散规律的认识。

（3）认识了 CO_2—油—水—岩石反应规律，评价了示范区 CO_2 埋存的安全性。

松辽盆地南部主要矿物组成是石英和长石，经研究认为与 CO_2 反应的主要敏感矿物由强到弱依次为：方解石→铁白云石→钠长石→钾长石，其中铁白云石和片钠铝石是注 CO_2 后主要的固碳矿物。建立了 CO_2—油—水—岩石的化学反应实验方法，测试了盐水层、常规水驱油藏、废弃油藏条件下与 CO_2 反应敏感矿物的溶蚀速率，定量评价了油藏流体的存在对 CO_2 与地层敏感矿物反应强度的影响规律。通过 X 射线衍射及扫描电镜实验，分析了 CO_2—地层水—岩石相互作用对岩石矿物成分及孔隙结构的影响，综合考虑 CO_2 在地层水中溶解和矿化反应，模拟研究了 CO_2 埋存潜力及运移规律。并基于实验测试数据，建立了示范区矿物反应模型，通过 CO_2 驱油与埋存地球化学反应模拟器，研究了 CO_2 埋存过程中的矿物捕获变化规律。

利用地球化学反应模拟和天然含 CO_2 油气藏自演类比法，开展了短期及中长期 CO_2 埋存安全性评价研究。其中，地球化学反应模拟结果表明，油藏及盐水层中的 CO_2 均可以矿物形式捕获；天然含 CO_2 油气藏自演类比表明，天然含 CO_2 油气藏后期充注的 CO_2 与油藏砂岩的长时间相互作用，形成了片钠铝石等"固碳矿物"，可在一定程度上保障 CO_2 中长期埋存的安全性。

2.3 CO₂驱油与埋存油藏工程技术

油藏工程技术是 CO_2 驱油与埋存三大主体技术之一。围绕油藏工程优化设计、注采调控、抑制气窜等瓶颈问题开展攻关，建立了低渗透油藏 CO_2 驱油藏工程调整方法，探索 CO_2 驱油扩大波及体积技术，初步形成注 CO_2 驱油提高低渗透油田采收率油藏工程技术体系，为 CO_2 驱油与埋存扩大化应用提供了技术支撑。

（1）深化 CO_2 驱油开发规律认识，形成了提高 CO_2 驱油开发效果的油藏工程调整方法。

以 CO_2 驱油微观渗流及波及规律研究、试验区 CO_2 驱油开发动态分析、国内外 CO_2 驱油藏开发特征类比为基础，确定了以含水率、气油比、换油率等指标参数作为试验区 CO_2 驱油开发阶段划分的依据，按不同油藏类型进行了开发阶段划分，明确了不同水驱阶段油藏转 CO_2 驱油主要指标变化规律和阶段工作任务，并提出了阶段调整对策。

建立了湖泊三角洲前缘相沉积储层 CO_2 驱油注采调控模式，提出了"气驱注采调控单元"和"气驱渗流各向异性综合指数"的概念和表征方法，通过对同一注采单元区带的注采联调、联控，有效防控气窜，实现均衡波及。建立了适合低渗透油藏 CO_2 水气交替驱的油藏工程设计方法，并且以现有水驱井网为基础，通过岩心驱替实验、油藏数值模拟研究，提出了以瞬时换油率为依据确定水气交替时机、气水比、段塞大小等开发技术政策界限的方法，同时兼顾不同采油工艺及地面产出气处理能力，进行油藏工程设计及方案指标预测，提高了预测精度和开发效果。

（2）形成了 CO_2 驱油动态分析、调控及开发效果评价技术。

建立了以混相分析为核心的动态分析方法，内容包括：①注气状况分析。与水驱油相比，特低渗透油藏注气具有"强、低、高"的特点，即吸气能力强，可进入渗透率更低的储层，地层压力保持水平高。②混相和见效特征分析。通过分析地层压力、生产动态反映、井流物组分和物性特征，并经过试井和油藏数值模拟研究，对区块混相状况做出综合评价。③生产动态特征分析。与水驱油相比，低渗透油藏 CO_2 驱油具有"高、异、窜"的特点，即产液产油能力提高，见效差异更明显，部分井发生气窜。

建立了 CO_2 驱油定量化注采设计与调控方法，形成了防窜—调采—控套相结合的注采调控技术。对现场可操控的生产井井底流压和注入井注入压力进行优化设计，提高了 CO_2 驱油的调控精度和油藏管理水平。吉林油田黑79北小井距试验区采取水气交替+周期采油+控流压+控套工艺个性化组合措施，增产效果显著。

针对 CO_2 驱油开发的特点，参照聚合物驱油开发效果评价方法，并与水驱油开发效果做对比，建立了 CO_2 驱油开发效果评价方法及指标体系。内容包括技术、经济和安全环保3个类别15项指标，其中主要评价指标8项，辅助评价指标7项，创新提出了地混压力系数、温室气体减排效益和环境监测异常率3个新指标。

（3）研发了抑制气窜的泡沫体系、注入工艺与成泡控制装置。

研发出适合试验区油藏条件的3种泡沫配方体系，测试了不同温度和压力条件下 CO_2—泡沫体系的各相体积、黏度、密度等相态参数，通过 CO_2—泡沫体系的相互作用关系，完善了 CO_2—泡沫体系相态表征技术，为工程应用提供了依据。

设计出稳定成泡的注入工艺及控制装置，通过检测泡沫体系密度获知泡沫实际生成量，调控注入方式使 CO_2—泡沫体系平均密度处于稳定成泡范围，确保泡沫稳定进入油藏，解决了现场作业中无法明确井口成泡状态的问题，为 CO_2 泡沫驱油现场施工提供了技术支撑。

2.4 CO₂ 驱油高效注采与埋存工艺技术

针对因层间矛盾突出导致储层吸气不均、高气液比条件下泵效下降、井筒腐蚀等问题，开展了 CO_2 驱油与埋存分层注气工艺技术、采油井高效举升工艺技术、采油井腐蚀监测及防腐技术等攻关，形成了一套 CO_2 驱油与埋存高效注采新工艺、新工具，满足了现场生产需求。

（1）完善了 3 种分层注气工艺和配套工具，在现场实现 2~3 层段分层注气。

根据不同的工艺原理，完善了 3 种分层注气工艺技术，即同心双管分层注气工艺、井下选择性分层注气工艺、单管分层注气工艺，通过调整注入剖面改善了层间矛盾。针对 CO_2 超临界注入技术要求，设计了由多级文丘里管串联气嘴、防气密封封隔器、气密封油管、CO_2 防返吐配注器以及防返吐底阀组成的井下配注器分层注气管柱；管柱整体耐温 120℃、耐压 25MPa，克服了超临界 CO_2 气体黏度小、节流压差建立困难、密度大、对气嘴冲蚀大、影响气嘴使用寿命等问题。同时建立了精度较高的 CO_2 注入井筒温度压力场耦合计算模型，研究了五参数法分层测试及调配工艺，解决了测试技术难题。

优选黑 59-10-4 井、黑 59-4-2 井等 7 口井开展同心双管分层注气、井下选择性分层注气现场试验，分层注气效果良好，吸气剖面测试也已取得成功。

（2）完善了 3 种采油井高效举升工艺技术，提高抽油泵泵效 10% 以上。

CO_2 突破以后，油井气液比大幅度上升，造成泵效下降。针对 CO_2 驱高气液比生产井，完善了气举—助抽—控套一体化举升工艺、防气泵高效防气举升工艺和高效防气装置举升工艺 3 种举升技术。其中，气举—助抽—控套一体化举升工艺通过在采油井井下安装控套阀，实现了油井控制套压生产，该工艺在泵下实现气液分离，减少了进入泵筒内的气体量，从而提高举升效率，利用气体能量实现气举助抽；适用于多种高气液比生产井，降低了安全隐患，且不影响地面正常生产、加药等工艺。防气泵高效防气举升工艺通过增加中空管结构，在泵筒运动时将泵内气体排入油管，降低泵内的气液比，从而有效提高泵效；该工艺通过调整冲程和中空管体积，来满足不同机械采油参数和气油比；适用于气液比小于 300m³/m³ 的生产井，也可与控套阀和分离装置结合使用。高效防气装置举升工艺通过自主研发的高效防气装置，实现油套环空分离、入口进液分离、外管和中心管环空分离、旋转离心分离四重分离。该工艺将抽油泵安装在高效防气装置内部，以有效控制流压和沉没度；适用于气液比大于 300m³/m³ 的生产井，可与控套阀和防气泵结合使用。

通过 30 余井次的现场试验，应用防气泵举升效果良好，满足气油比达 500m³/m³ 的油井正常生产，动液面升高，气油比明显降低。平均日增液 18.63t，日增油 6.61t，平均每口井的泵效提高 11.5%。

（3）研发了井下在线腐蚀监测技术，优选了防腐剂并优化了加注工艺。

根据管柱腐蚀状况，只能定性判断且无法知道井下腐蚀速率随井深变化趋势的问题，创新研发了井下电化学阵列腐蚀速率在线监测系统。该系统基于电化学噪声技术，通过直接测量金属腐蚀时产生的微电流计量腐蚀速率，可以定量地确定局部或不均匀腐蚀的速率。地面试验证明，该系统能够在 45MPa、120℃ 情况下长期使用，腐蚀速率测量精度满足 0.076mm/a 的国家测量标准。

针对不同管材优选了低成本缓蚀剂：N80 管材适合采用 IMC871GH1 型缓蚀剂、20 钢管材适合采用 ZK789C 型缓蚀剂、Q235 管材适合采用 IMC80BH 型缓蚀剂、P110 管材适合采用

ZK682 型缓蚀剂。并对缓蚀剂加注方式、加注点、加注量等进行了定量研究，结果表明，注入井和采出井，采用连续注入或者连续注入和间歇注入相结合的方式；采出井加注点在油井井口向油套环空或者直接加注到油井底部，注入井加注点在注入泵入口端 1~3m 内；对于井下管系，建立了缓蚀剂用量公式。

2.5 CO_2 超临界输送及循环注气技术

围绕 CO_2 输送、注入与集输处理等地面工程方面的瓶颈问题，针对吉林油田示范区，开展了 CO_2 驱油与埋存地面工程技术应用基础研发、关键技术攻关和现场应用试验，形成了 CO_2 捕集、输送、注入、采出流体集输处理和产出气循环注入等关键技术系列，并完善了相关标准和规范。

（1）形成了 CO_2 捕集、长距离管道输送和注入技术，在矿场实践中形成 3 种 CO_2 注入模式。

在 CO_2 捕集方面，验证和改进了含 CO_2 天然气胺法脱碳工艺。天然气 CO_2 含量 30% 以下时，胺法脱碳装置运行平稳，各项运行指标优于设计；局部优化了胺法脱碳工艺，减少了胺液损失和循环量。

在 CO_2 相态机理方面，扩大温度压力范围，完善了 CO_2 及含 CO_2 天然气物性参数工程图版，进一步深化了高含 CO_2 天然气相态特征及物性认识。在温度为 40~60℃、压力为 1~33MPa 条件下，完成了 5 种 CO_2 浓度的含 CO_2 天然气物性参数和相态测试，得到 CO_2 密度随相态变化的"突变"和"渐变"规律，为 CO_2 输送注入设计提供了依据。

在 CO_2 长距离管道输送方面，明确了 CO_2 管输特性，形成了管输相态控制方法，建立了管输优化设计和运行模型，明确了 CO_2 最优输送方式。并针对吉林大情字井油田 CO_2 驱工业化试验区，研究了 CO_2 气态、液态和超临界状态等多种管道输送方案，以各循环注入站需求为依据，通过技术经济对比论证，优选了管输设计方案。

在 CO_2 注入方面，明确了 CO_2 超临界注入需要重点考虑的因素，提出了 CO_2 相态、水合物及腐蚀控制方法，设计了 CO_2 超临界注入流程，完善了 CO_2 液相注入技术和超临界注入技术。建成并运行黑 59 试验区 CO_2 超临界注入中试装置，最高试验压力达到 26MPa。同时，通过对比分析国外 CO_2 捕集与埋存（CCS）项目中 CO_2 气体回注系统设计特点，明确了 CO_2 压缩机优化方向。

在矿场试验方面，形成了 3 种 CO_2 注入模式。一是满足先导试验的小站橇装注入模式：黑 59 试验区早期采用罐车拉运液态注入，后期采用单井采出 CO_2 气源，单井集气脱水、气相短距离管输、氨冷液化、液相高压注入，CO_2 驱油产出气不分离超临界混合注入；该模式适用于规模较小的零散区块的 CO_2 驱油先导试验。二是满足扩大试验集中注入模式：采用 CO_2 液态注入、CO_2 驱油采出流体气液分离和产出气分离后液相循环注入；该模式适用于试验井数较多、规模较大且邻近 CO_2 气源的区块，如黑 79 试验区毗邻净化厂 CO_2 捕集和储存系统。三是满足工业化应用超临界注入模式：采用气相管道输送、超临界注入、CO_2 驱油产出气不分离全部超临界回注，流程简化、经济性好；该模式适用于规模和距离范围广的试验区，可用于工业化推广，如大情字井油田黑 46 试验区注气模式，以长岭气田产出的 CO_2 为气源。

（2）完善了 CO_2 驱油采出流体集输处理和产出气循环注入技术，支撑建成 2 套循环注气中试装置。

通过模拟和现场取样实验分析，得到 CO_2 驱油采出流体密度、黏度和 CO_2 溶解度变化规律。通过 CO_2 驱油采出流体集输流程、材质和计量现场试验，提出了 CO_2 驱油采出流体最佳集输流程、防腐形式和计量方法，改进了气液分离器结构。通过优化集输和分离工艺，形成了两种计量方法，优选脱水及污水处理剂，形成了 CO_2 驱油采出液脱水及污水处理技术。

提出了直接回注、混合回注和分离提纯后回注 3 种 CO_2 驱油产出气循环注入技术路线，研究了胺法、变压吸附法及精馏与低温提馏耦合分离技术等相关分离工艺，采用变压吸附法可满足 CO_2 含量从 3% 到 90% 变化需要。建成黑 59 试验区超临界混合回注中试装置和黑 79 试验区变压吸附分离提纯中试装置。其中，黑 59 试验区超临界混合回注中试装置采用国产压缩机、四级压缩，实现 CO_2 及伴生气混注，现场最高运行压力为 26MPa，规模达 $5 \times 10^4 m^3/d$；黑 79 试验区变压吸附分离提纯中试装置可实现黑 79 南试验区产出气（CO_2 含量为 5%~90%）、长岭气田营城组天然气（CO_2 含量 27%）及其混合气的分离，规模达 $8 \times 10^4 m^3/d$。

（3）完善了 2 项 CO_2 驱油地面工程标准及控制体系。

CO_2 驱油地面工程腐蚀风险控制体系：建立了 CO_2 驱油腐蚀模拟试验装置，应用了在线腐蚀监测和缓蚀剂加注工艺，优选适合 CO_2 驱油采出流体的玻璃钢管材，完善了应力对金属管道 CO_2 腐蚀影响理论。

CO_2 驱油地面工程 HSE 风险控制体系：认识了 CO_2 及 CO_2 驱油产出气在管道中泄漏扩散规律，分析了 CO_2 净化、集输、输送、注入、分离、回注等环节的风险，提出了措施预案，完善了 HSE 体系。

提出了 CO_2 气态输送、超临界注入的 CO_2 驱油地面工程主体技术路线，建成了 CO_2 捕集、输送和循环注入系统，实现了 CO_2 驱油产出气循环注入和 CO_2 零排放。

2.6 CO_2 驱油与埋存配套技术

除了重点攻关 CO_2 驱油与埋存的基础理论，以及油藏工程、采油工程和地面工程三大主体技术以外，还开展了 CO_2 驱油与埋存油藏监测技术、潜力评价技术和战略规划等方面的研究，为我国 CO_2 驱油与埋存安全和科学发展提供技术支撑和决策依据。

（1）形成了 CO_2 驱油与埋存油藏监测技术，为 CO_2 驱油与埋存技术评价提供了手段。

第一，形成了 CO_2 与地层油混相状况监测技术。一是通过三项监测判断混相状况，即通过地层压力监测，确定是否满足达到混相的压力条件；通过采油井生产动态监测，从动态反映上分析混相状况；通过地面油气组分和物性监测、地层油高压物性监测，取得混相的直接证据。二是完善了 CO_2 混相驱油藏监测 4 个系列、14 项监测技术，其中 4 个系列分别是：CO_2 驱油注气井监测技术系列、CO_2 驱油采油井监测技术系列、CO_2 驱油注采井间监测技术系列和 CO_2 驱油混相状态监测技术系列。三是确定了 CO_2 驱油藏监测项目的优化原则。

第二，形成了 CO_2 运移及前缘监测技术。通过前缘形态分析和时间推移监测结果的对比分析，明确了试验区气驱前缘的推进特征，如对于黑 79 南试验区反九点井网，注入气仅波及边井，很难达到角井。

第三，形成了 CO_2 驱油项目环境监测技术。环境监测项目分常态监测项目和应急监测项目。监测阶段划分为注气前阶段、注气阶段、项目关闭阶段和项目关闭后阶段。2012—2014 年，分别对黑 59 和黑 79 两个试验区的大气、土壤和地表水中 CO_2 含量进行了监测，结果显示为各项参数正常，对环境没有影响。

（2）完善了 CO_2 驱油与埋存潜力评价技术，并开发了数据库软件，评价了中国石油典型油区的 CO_2 驱油与埋存潜力。

基于大量最小混相压力实验结果，建立了 CO_2 驱油最小混相压力预测模型，并开发了预测软件，与新疆、吐哈、长庆和吉林等油田的实际资料对比，软件预测精度可达 90% 以上。根据 CO_2 埋存机理，建立了 1 套 CO_2 驱油理论埋存量和有效埋存量的计算方法。利用油藏数值模拟方法，研究了不同驱替方式下地层参数、流体参数、注入参数等对 CO_2 驱油和埋存的影响规律，并采用响应曲面法，建立了纯气注入和 WAG 注入混相驱油、纯气注入和 WAG 注入非混相驱油 4 种不同驱替方式下 CO_2 驱油提高采收率和埋存系数的快速评价模型。

在最小混相压力和埋存影响因素分析的基础上，建立了 CO_2 驱油提高采收率和地质埋存潜力评价计算模型，并提出了潜力评价工作流程。在此基础上，开发了相应的潜力评价数据库软件，应用该软件可以计算 CO_2 驱油的埋存系数、埋存量、增油量和采收率等，并且可以实现 CO_2 驱油和地质埋存评价数据库的图形编辑，油藏分布图展示以及 CO_2 驱油提高采收率和地质埋存评价等功能。

利用该项研究成果，对中国石油典型油区的 CO_2 驱油与埋存潜力进行了评价，初步明确了我国利用油气藏埋存 CO_2 的理论埋存量和有效埋存量，以及这项技术的应用前景。

（3）建立了 CO_2 捕集、驱油与埋存战略规划方法，制订了规划方案。

建立了中国石油重点油区附近 8 个主要行业 1000 多个大型 CO_2 集中排放源的数据库，该数据库包含各 CO_2 排放企业的产能、位置以及排放气中 CO_2 气体的浓度、排放因子等信息。在统一空间地理信息系统中录入 CO_2 排放源的位置信息和油田的位置信息，建立了我国主要 CO_2 排放企业和主要油田的位置关系图册。

通过借鉴国际权威机构的 CO_2 减排成本估算方法，建立了 CO_2 驱油经济评价方法，并进行了影响因素分析。以实际 CO_2 驱油项目为对象，对多个油田区块的 CO_2 驱油承受成本和 CO_2 来源成本进行了测算，结果表明，许多项目 CO_2 供给成本大于 CO_2 驱油提高采收率（CO_2-EOR）所能承受的成本；指出 CCS 技术本身的发展、激励政策—政府投资补贴、碳定价机制的发育完善是填补 CO_2 来源成本与驱油和埋存承受成本之间空白的最有效途径。

基于经济评价和超结构建模优化方法，利用 GAMS 软件对典型油区的潜力区块与 CO_2 排放源进行了 CCS-EOR 整体系统优化，获取了不同油价、不同产量规模等条件下油田潜力区与 CO_2 排放源之间的静态、动态最优匹配方案。通过分析 CO_2 驱油产业发展战略规划情景要素，构建了 CO_2-EOR 战略规划情景框架，制订了公司 CCUS 近期发展规划方案和中长期的发展战略设想。

通过集成应用以上 6 个方面的研究成果，吉林油田建成了国际一流的 CCUS 示范工程，形成了具有中国石油特色的 CCS-EOR 减排增效一体化模式和技术系列，为我国规模实施 CO_2 驱油提供了技术示范，也为国家制定可持续发展相关政策法规提供了参考。

3 存在问题及下步攻关方向

尽管我国近几年 CO_2 驱油与埋存技术取得飞速发展，但要进行大规模的工业化推广和应用还有很长的路要走，还存在很多技术瓶颈有待深入研究和解决。根据目前该项技术的发展现状和应用水平，结合我国的油藏地质特征和在矿场实践过程中存在的问题，笔者认为在今后的攻关过程中需要重点关注以下问题。

（1）我国陆相原油重质组分含量高，混相困难，驱油效率低，而且物理模拟实验技术尚不完善。需要深入分析不同类型原油组成与 CO_2 驱油混相能力的相关性，提出改善混相条件的新方法；进一步完善实验方法，更好地模拟地下油藏条件，为 CO_2 驱油特征、改善混相条件、扩大波及体积等研究奠定基础，形成陆相油藏 CO_2 驱油与埋存理论方法和实验技术。

（2）陆相油藏非均质性强、气窜更为严重，造成波及体积小、采收率低；并且目前我国 CO_2 驱油技术应用领域还比较局限，主要用于常规低渗、特低渗透油藏。应加大扩大气驱波及体积技术研究力度，以进一步提高 CO_2 驱油采收率；开展不同类型油藏 CO_2 驱油藏工程设计方法与技术研究，如裂缝性油藏的水动力调整及化学辅助结合 CO_2 驱油技术，致密油藏水平井开发 CO_2 吞吐技术等，以扩大 CO_2 驱油与埋存技术的应用范围。

（3）目前 CO_2 注入井的测试一直存在施工难度大、测试效果不连续、无法长期监测等问题，而目前分层注水工艺正在由传统的机械投捞测试方式向边测试边配调或全自动调配方向发展，建议创新理念开发 CO_2 自动注入技术；针对矿化度较高的油藏开展防腐防垢一体化工艺技术研究，研发不同气液比阶段采油井举升技术。

（4）未来的 CO_2 驱油地面工程技术领域，需要基于高效低耗、安全环保的原则，优化简化 CO_2 驱油采出流体集输处理工艺，降低 CO_2 驱油采出流体集输处理系统运行能耗、推广 CO_2 超临界管道输送技术、实践 CO_2 驱油产出气混合回注，形成 CO_2 密相增压注入技术、不同 CO_2 流体计量方式检定方法。

（5）CO_2 注入和埋存的安全性关乎人类的生存环境和生命财产，需要进行安全监测、科学防控风险，保证项目安全运行。目前的监测手段、安全评价方法以及风险控制技术等还不完善，急需加强 CO_2 驱油与埋存安全风险评价方法，以及注采工程、地面工程风险控制技术研究，制定 CO_2 驱油与埋存生产运行管理 HSE 规范；发展新的监测工具和手段，建立 CO_2 地质埋存综合监测技术体系，对 CO_2 地质埋存项目进行实时连续监测，确保 CO_2 的安全永久埋存。

（6）CO_2 驱油与埋存项目经济风险高、跨部门跨行业组织困难是这项技术在我国推广缓慢的一个重要原因。需要研究我国 CO_2 驱油与埋存技术的可持续发展模式，科学制定 CCS-EOR 产业的相关政策法规和技术路线图，做好 CCS-EOR 产业中长期发展规划，积极促进 CO_2 驱油与埋存技术的规模化推广应用。

第一篇
二氧化碳驱油与埋存机理

CO_2 在链状烷烃中的溶解性能及膨胀效应

韩海水[1,2] 袁士义[3] 李　实[1,2] 刘晓蕾[4] 陈兴隆[1,2]

(1. 中国石油勘探开发研究院；2. 提高石油采收率国家重点实验室；
3. 中国石油天然气集团公司科技管理部；4. 中国科学院渗流流体力学研究所)

摘　要： 选取原油中含量普遍较高的 5 种链状正构烷烃与不同比例的 CO_2 组成油—气体系。通过体系在不同温度下的恒质膨胀实验，研究 CO_2 在 5 种链状正构烷烃中的溶解性能及体系膨胀效应。研究表明，链状正构烷烃—CO_2 体系的压力—体积关系曲线并非是严格意义的两段直线，弯曲程度受温度、压力、CO_2 含量、正构烷烃类别等因素影响。体系的泡点压力随温度升高呈直线增大趋势，随 CO_2 含量的增加大幅度增大。当压力较低时，CO_2 在不同链状正构烷烃中的溶解度近似相同，而高压时溶解度随烷烃碳原子数增大而减小。CO_2 溶于链状正构烷烃中可造成体系不同程度的体积膨胀，膨胀系数的大小受温度、压力影响不大，随 CO_2 含量的增加快速增大，随烷烃碳数的增大直线下降。体积膨胀作用对油井增产有重要意义。

关键词： 链状正构烷烃；二氧化碳；恒质膨胀实验；溶解度；体积膨胀

0　引言

向各类油层中注入 CO_2 可不同程度地提高石油采收率，原因在于 CO_2 和原油接触，不仅可以使原油黏度降低，油水、油—气界面张力降低，还可使原油体积发生一定程度的膨胀，这种膨胀作用既增加了油藏的含油饱和度，又补充了地层能量，对油田开发具有积极意义[1-3]。针对 CO_2—原油（组分）体系体积变化特征的研究主要分为两类：（1）以典型油藏原油—CO_2 体系为研究对象，研究体系的体积变化特征。文献［4-7］报道了国内外部分油田原油—CO_2 体系的体积变化特征实验研究成果，为油田实施注 CO_2 采油提供了室内实验和理论支撑，但成果适用性较差，且多是一些表面的现象和规律，不能深层次揭示原油中各类烃物质与 CO_2 接触时温度、压力、体积的变化规律。（2）针对物理化学及热力学的理论研究需要而进行的某种烃类物质—CO_2 体系实验，研究结果主要用来验证和校正相应理论计算模型。文献［8-12］研究了 CO_2 与苯、甲苯、正己烷、正庚烷、正癸烷、正十六烷、水等纯物质组成的二元体系在相对较低温度（小于 40 ℃）下的部分状态参数变化特征，主要包括相平衡、泡点压力、膨胀系数等，同时提出或修正了相平衡计算理论模型和状态方程，但其结果主要用来完善热力学理论，且测试样品不够系统，没有给出较为完整的 CO_2—烃类体系 PVT（压力—体积—温度）变化规律。

目前，对引起原油膨胀的关键组分认识尚不清晰，CO_2 在不同原油组分中的溶解膨胀规律亦无系统报道。前人对中国 5 个代表性油区原油样品的组分进行了测试和分析[12]，发现链状烃类含量较高，达 56.4%~72.5%，因此研究 CO_2—链状正构烷烃体系的体积变化特征有助于深入认识和理解 CO_2 使原油膨胀的原理。

本研究选取正己烷、正壬烷、正十二烷、正十四烷和正十六烷 5 种典型链状正构烷烃，通过恒质膨胀实验，较系统地研究了 CO_2 在链状正构烷烃中的溶解性能、CO_2—链状正构烷烃体系压力—体积关系特点以及膨胀规律，并结合相关物理化学及分子热力学理论给出解释，最后分析了体积膨胀作用对石油开发的重要意义。

1 恒质膨胀实验

恒质膨胀实验在恒温恒组成条件下，将系统从单相状态变换至油气两相状态，根据压力和体积的关系来研究相态变化及部分状态参数变化。恒质膨胀实验是研究高压地层流体体系体积变化特征的主要方法。

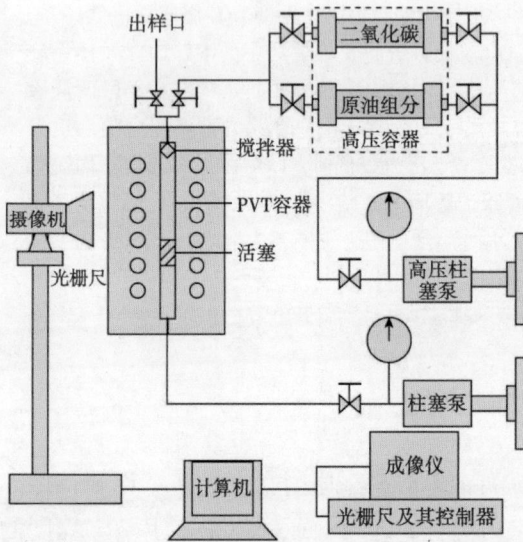

图 1　无汞高压 JEFRI PVT 装置

1.1 实验装置及药品

实验选用加拿大 Donald Baker Robinson 公司生产的无汞高压 JEFRI PVT 装置。该套装置主要由 PVT 容器、摄像机、成像仪、光栅尺及其控制器、计算机、柱塞泵等构成（图 1）。其最大工作压力为 103.4MPa，最高操作温度为 180 ℃，样品筒体积为 130mL。PVT 容器具有前后视窗，可观察容器内的相态变化，容器内样品的体积可通过摄像机和光栅尺进行精确计量。此外，实验还需要两个高压容器及一台高压柱塞泵计量泵。

实验用链状烷烃主要有正己烷、正壬烷、正十二烷、正十四烷和正十六烷，均为百灵威科技有限公司的产品，分析纯。

实验用 CO_2 为北京兆格气体科技有限公司生产的产品，纯度为 99.996%。清洗实验装置所用石油醚为北京化工厂生产的分析纯试剂。

1.2 实验方案

将 CO_2 分别与正己烷、正壬烷、正十二烷、正十四烷和正十六烷组成 CO_2 摩尔分数为 25%，50% 和 75% 的 15 个二元体系。将各个体系分别在温度为 50℃，70℃，90℃ 和 110℃ 下进行恒质膨胀实验。为了数据的完整性和对比分析，设置了纯正己烷、正壬烷、正十二烷、正十四烷和正十六烷在以上 4 个温度下的恒质膨胀实验。

1.3 实验步骤

实验操作参考相应国家标准[13]，其主要步骤如下：

（1）实验样品的配制。5 种链状正构烷烃和不同含量 CO_2 组成的所有二元体系样品均按如下方法在 50℃ 下配制完成。首先向 PVT 釜内充入一定量的 CO_2，稳定 4h 后读取 PVT 釜中样品的压力和体积，再向釜中泵入设定样品所需体积的原油组分。

（2）PVT釜内达到设定温度后，打开搅拌器、升高压力使流体形成单相。稳定4h，保证流体均匀稳定。

（3）以0.5mL/min的速度退泵，观察釜内压力变化及相态变化，当发现压力不再下降甚至升高或者有气泡冒出时，停泵、静置30min，记录压力即为粗测的饱和压力。

（4）升压至30MPa，恒压稳定30min记录压力、活塞位置（光栅尺示数）。

（5）恒压模式逐级退泵，当接近粗测的泡点压力时，改为恒速定体积退泵，分别稳定30min计数。

（6）当气液两相体积大致相当时，停止这一组实验。将压力升高至泡点压力以上，温度升高到下一个温度点，准备下一组实验。

2 实验结果和分析

2.1 压力—体积关系曲线特征

2.1.1 正构烷烃—CO_2体系典型压力—体积关系曲线

按照上述方法进行实验，将所测得的数据点做成压力—体积关系曲线。图2为50℃时CO_2—正己烷体系（CO_2摩尔分数75%）压力—体积关系曲线。

一般认为曲线上单相段和两相段的交点即为该体系的泡点，对应的压力为泡点压力。然而，事实上压力—体积曲线泡点两侧并非是此前认为的两条直线，而是呈近似抛物线趋势（图2）。但在泡点附近可认为是两条直线，进而求得泡点压力（7.36MPa）。

不同组实验样品的绝对体积不同，为了便于比较分析，以泡点压力时体积为基准，定义相对体积为绝对体积与泡点处体积之比，即泡点压力处的相对体积为1。

2.1.2 温度的影响

图3为不同温度下正己烷—CO_2体系（CO_2摩尔分数75%）压力—相对体积关系曲线。单相区内，50℃时压力—相对体积变化稍偏离直线，这种偏离趋势随温度的升高而增大，即温度越高，曲线越弯曲。当温度达110℃时变为一条光滑的曲线（无拐点），其形态类似于凝析气。气液两相区域内曲线形态随温度变化不大。

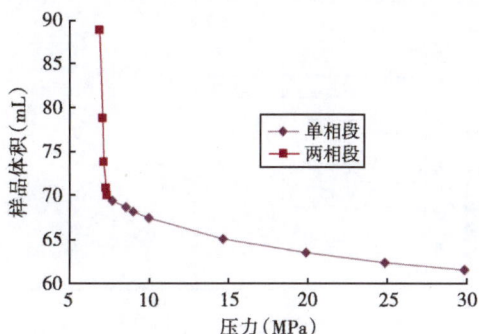

图2 50℃时正己烷—CO_2体系压力—体积
关系曲线（CO_2摩尔分数75%）

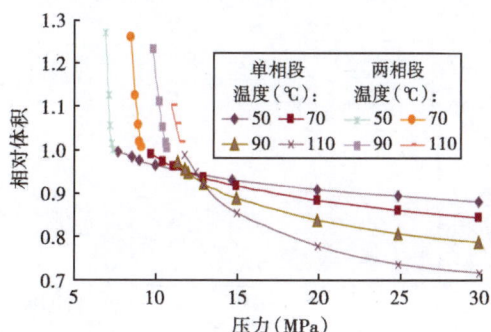

图3 不同温度下正己烷—CO_2体系压力—相对
体积关系曲线（CO_2摩尔分数75%）

2.1.3 CO_2 含量的影响

同一种原油组分内含有不同量的 CO_2 气体，在相同温度下，单相段的曲线形态有着较大差别。图 4 为不同 CO_2 含量的正己烷—CO_2 体系在 110℃ 时压力—相对体积关系曲线。体系中 CO_2 摩尔分数为 25% 和 50% 时曲线的单相段变化不大，但达到 75% 时，单相段明显弯曲。气液两相段曲线形态随 CO_2 摩尔分数增大变化不大。

2.1.4 不同烃组分间的影响

在相同温度、相同 CO_2 摩尔分数条件下，不同的原油烃组分也会对压力—相对体积曲线造成较大的影响。图 5 为 CO_2 摩尔分数 75%、温度 110℃ 时的 CO_2—链状正构烷烃体系压力—相对体积关系曲线。含正十二烷、正十四烷和正十六烷 3 种烷烃体系的单相段接近直线，正壬烷则稍微偏离了直线，正己烷则完全偏离直线成为一条曲线。在两相段亦有类似特征。

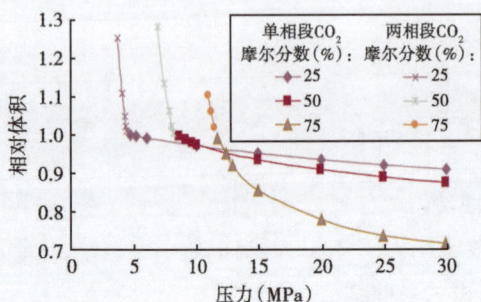

图 4　不同 CO_2 摩尔分数的正己烷—CO_2 体系在 110℃ 时压力—相对体积关系曲线

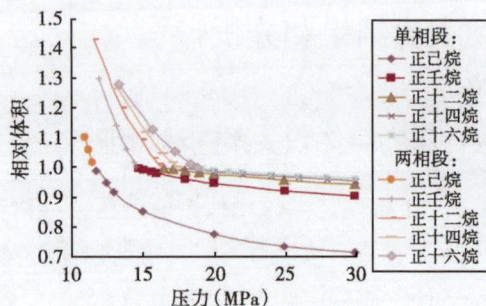

图 5　链状正构烷烃—CO_2 体系在 110℃ 时压力—相对体积关系曲线（CO_2 摩尔分数 75%）

2.2　泡点压力及 CO_2 溶解性能

所测量体系中泡点压力也可称作饱和压力，即为体系维持单相需要的最小压力，其大小主要由体系组成和温度决定。

2.2.1 温度对泡点压力的影响

同一烷烃—CO_2 体系中，随温度的升高，饱和压力呈现较好的线性变化（图 6）。

2.2.2 CO_2 含量的影响

相同温度下，体系的泡点压力随 CO_2 摩尔分数的增加而增加（图 7）。当体系中 CO_2 摩

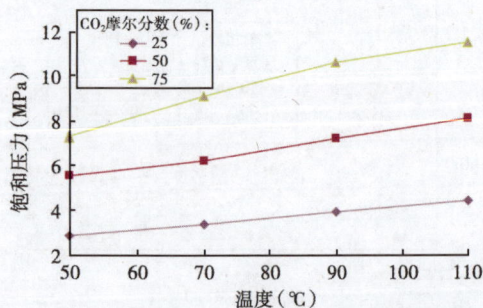

图 6　不同 CO_2 含量下正己烷—CO_2 体系饱和压力随温度的变化

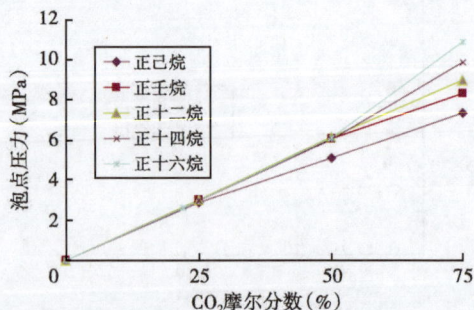

图 7　50℃ 时泡点压力随 CO_2 摩尔分数的变化

尔分数较小时，泡点压力呈现出直线增加的趋势；当体系中 CO_2 摩尔分数较高时，则泡点压力根据烃组分的不同呈现不同的增加趋势。此外，还可以看出组分越重（碳数越大），其泡点压力受 CO_2 摩尔分数影响越大，增加幅度越大。

2.2.3 泡点压力受烃组分影响

对于正己烷、正壬烷、正十二烷、正十四烷和正十六烷 5 种组分，当 CO_2 含量为 25% 时，体系泡点压力几乎相同；当 CO_2 含量为 50% 时，正己烷—CO_2 体系泡点压力略低，而其余 4 种组分体系泡点压力几乎相同；当 CO_2 含量为 75% 时，链状正构烷烃—CO_2 体系泡点压力随碳数的增加呈现近直线增加趋势（图 8）。

2.2.4 CO_2 在链状正构烷烃组分中的溶解度

CO_2 在链状正构烷烃组分中的溶解度曲线见图 9。当压力小于 6MPa，CO_2 在除正己烷外的 4 种链状正构烷烃组分中溶解度近乎相同，而高压条件下，则随碳数增加 CO_2 溶解度降低。

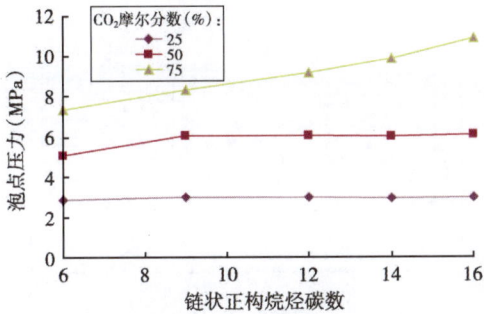

图 8 50℃时正构烷烃—CO_2
体系泡点压力随碳数变化

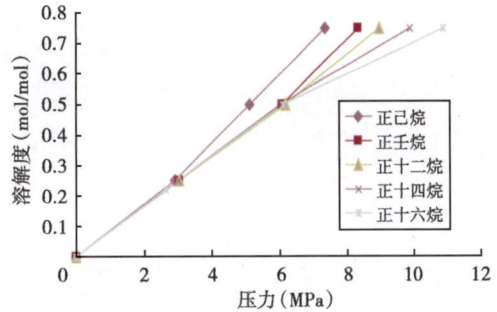

图 9 50℃时 CO_2 在 5 种链状
正构烷烃组分中的溶解度

2.3 体积膨胀

CO_2 溶于原油（烃组分）中，可造成一定程度的体积膨胀。为定量评价膨胀程度，定义膨胀系数为同等温度、压力条件下原油（烃组分）溶解 CO_2 后体积与溶解 CO_2 前体积之比。

2.3.1 温度、压力影响

研究发现，温度和压力都会不同程度地影响原油—CO_2 体系的膨胀系数。图 10 和图 11

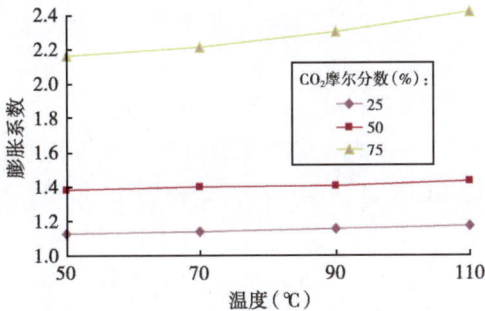

图 10 30MPa、不同温度下正己烷—CO_2
体系的膨胀系数

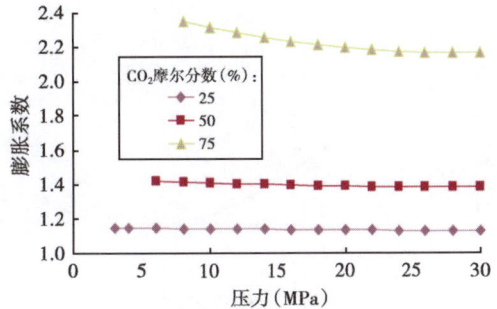

图 11 50℃、不同压力下正己烷—CO_2
体系的膨胀系数

分别为正己烷—CO_2 体系膨胀系数随温度和压力的变化关系。由图 10 和图 11 可见，当正己烷中 CO_2 摩尔分数小于 50% 时，其膨胀系数随温度和压力的变化并不大，但当 CO_2 含量达到 75% 时，体系膨胀系数随温度和压力的变化很明显。

2.3.2 CO_2 含量影响

相对其他气体，CO_2 在一定的压力和温度下与链状正构烷烃具有较好的互溶性。随着烃组分中溶解 CO_2 量的增加，其体积亦会不同程度地增大。从膨胀系数来看，烃组分体积膨胀程度和 CO_2 摩尔分数呈现凹形上升关系（图 12）。

2.3.3 烃组分影响

不同碳数的链状正构烷烃，CO_2 对其膨胀作用的整体规律为随碳数增加，膨胀作用减弱。当体系 CO_2 摩尔分数小于 50% 时，膨胀作用随碳数变化不大，但 CO_2 摩尔分数达到 75% 时，膨胀作用随碳数增加而大幅度降低。50 ℃ 时，若体系 CO_2 摩尔分数为 75%，可使正己烷体积膨胀系数达 2.16，正十六烷体积膨胀系数达 1.50（图 13）。

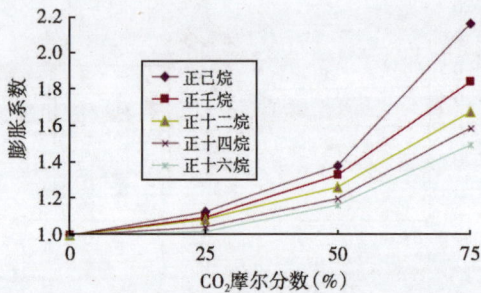

图 12 50℃，30MPa 条件下 CO_2 含量对链状正构烷烃组分的体积膨胀作用

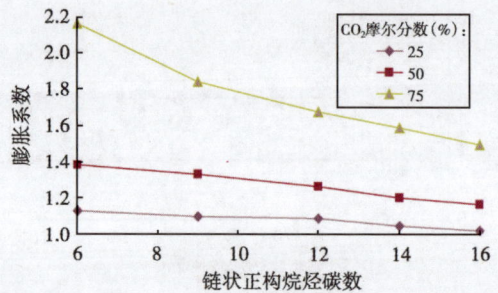

图 13 膨胀系数随烃组分的变化（50℃，30MPa）

3 理论分析

3.1 压缩性

压力—体积关系曲线形态主要受测量样品压缩性的影响，用压缩系数来衡量，其定义为体积随压力的变化率。测量样品压缩性越差，则压力—体积曲线越接近直线；反之则压力—体积曲线越弯曲。

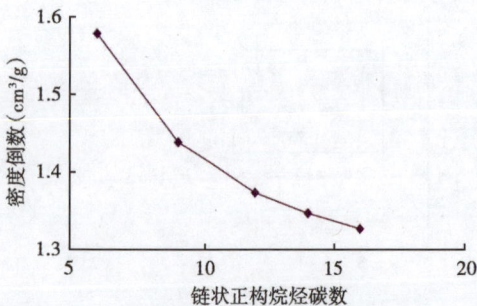

由于链状正构烷烃的结构相似，在特定压力条件下，可以用其密度的倒数来度量链状正构烷烃—CO_2 体系的压缩性。由图 14 可见，50 ℃、常压下，链状正构烷烃密度的倒数随碳数的增大迅速减小。可以证实，随碳数增加，体系压缩性快速减小。

图 14 50℃、常压下正构烷烃密度倒数与碳数关系

3.2 溶解过程及机理

由图8可见，当体系中 CO_2 摩尔分数为25%时，5种正构烷烃泡点压力几乎相同；当 CO_2 摩尔分数为50%时，只有正己烷—CO_2 体系泡点压力略低，其余4个体系泡点压力几乎相同；而当 CO_2 摩尔分数为75%时，泡点压力随碳数的增加呈近直线增加趋势。由图12可见，正构烷烃—CO_2 体系的膨胀系数与 CO_2 含量并非呈均匀的直线增加趋势，而是斜率递增的增加趋势。

以上现象可以从 CO_2 溶入正构烷烃过程的角度加以解释。当向烃组分中加入少量 CO_2 时（如摩尔分数为25%），体积相对较小的 CO_2 分子首先对烃组分分子间的空隙进行填充，分散于烃分子间。此时体系的总体积几乎不变（图15），需要很小的压力即可确保 CO_2 完全溶解于烃组分的空隙中。不同碳数的烃分子间均会存在一定的空隙，因此正构烷烃—CO_2 体系的饱和压力也相差不大。

继续向烃组分中加入 CO_2 并使其摩尔分数达到50%，此时烃组分和 CO_2 两种分子数量相当，烃分子间空隙已无法容纳这些 CO_2，导致总体积发生一定程度的膨胀，CO_2 分子与烃分子相互分散。同时需要较大的压力才能使两种分子相互溶解，维持系统的单相。

继续向烃组分中加入 CO_2 使其摩尔分数达到75%，此时在分子数量上 CO_2 绝对占

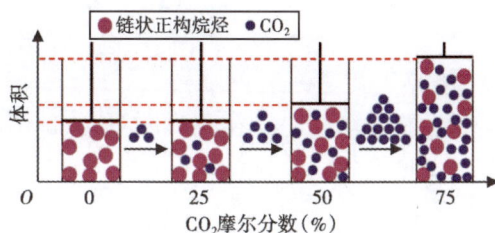

图15 CO_2 溶解于链状正构烷烃过程示意图

优，烃分子分散在 CO_2 分子间，CO_2 分子极易聚集分离出来形成单独的气相，体系需要更大的压力才能维持单相。由于体系中两种分子的总数量急剧增加，体系的总体积也大幅度增加。不同碳数正构烷烃分子间空隙和分子性质的不同导致体系总体积的增大倍数和泡点压力均有较大不同，碳链越长、分子越大的烃越难与 CO_2 互溶，因此需要更大的泡点压力。

3.3 溶解度和膨胀程度

由图9、图10可知，相同温度条件下 CO_2 在不同碳数的链状正构烷烃中的溶解度和膨胀程度都有所不同。依据分子学理论及 CO_2 在烃组分中溶解过程理论进行分析，可做出如下解释。

当 CO_2 含量较低时，CO_2 分子主要用于填充烃分子间隙，而5种链状正构烷烃结构相似，只是分子链长度存在差异，分子间隙差别并不大，因此促使体系饱和所需压力不大。外观表现为不同碳数烃对 CO_2 溶解度差别不大，体系的总体积也基本不发生膨胀效应。

当 CO_2 含量较高时，CO_2 分子数目的增多使得烃分子间隙无法容纳，必须施加较大的压力迫使烃分子彼此远离来维持 CO_2—烃体系的单相状态。而碳数较多的正构烷烃分子体积和质量较大，分子间吸引力较大，需施加更大的压力来扩大分子间隙进而维持体系的单相状态。外观表现为 CO_2 在碳数较多、碳链较长的烃组分中溶解度较小，膨胀性能较弱。

3.4 石油开发意义

实验证明，温度为50℃条件下，链状正构烷烃中混入的 CO_2 摩尔分数达75%时，最多可使总体积膨胀系数达2.16，当 CO_2 摩尔分数为50%时，总体积膨胀系数达 1.20～1.40。

因此在注CO_2开发过程中，较易形成CO_2—原油分散体系，原油体积的膨胀对提高石油采收率具有重要意义。

若假设原油体积不变，向未注水开发的油藏中注入CO_2，则体积膨胀作用将转化为油藏压力的升高，根据状态方程可预测注入后地层压力，再根据产量公式，可计算出原油体积膨胀带来的油井产量增量。

若假设油藏压力不变，向高含水油藏中注入CO_2，则由此发生的原油体积膨胀将导致油藏中含油饱和度增加，这将使油水相对渗透率曲线中油相的相对渗透率大幅度升高，进而提高原油在地层孔隙中的流动能力，从而达到油井增产的目的。

4　结论

（1）链状正构烷烃—CO_2体系的压力—体积关系曲线并非是严格意义的两条直线，其弯曲程度主要受温度、压力、CO_2含量、正构烷烃类别等因素影响。

（2）链状正构烷烃—CO_2体系泡点压力随温度升高呈直线增大趋势，随CO_2含量的增加大幅度增大。

（3）低压时CO_2在不同链状正构烷烃中的溶解度近似相同，高压时溶解度随烷烃碳数增大而减小。

（4）向正构烷烃中加入的CO_2首先填充分子间空隙，分散在烃分子间；进而与烃分子相互分散，直至烃分子分散在CO_2分子间。

（5）CO_2溶于链状正构烷烃中可使体系不同程度地膨胀，膨胀系数最高可达2.16。膨胀系数的大小受温度、压力影响不大，随CO_2含量的增大快速增大，随烷烃碳原子数的增大快速下降。

（6）CO_2溶解于原油中的体积膨胀作用对油田开发过程的油井增产具有重要意义。

参 考 文 献

[1] 杨承志，岳清山，沈平平．混相驱提高石油采收率：上册［M］．北京：石油工业出版社，1991：26-102.

[2] 沈平平，廖新维．二氧化碳地质埋存与提高石油采收率技术［M］．北京：石油工业出版社，2009：128-161.

[3] Whorton L P, Brownscombe E R, Dyes A B. Method for producing oil by means of carbon dioxide：US，2623596［P］．1952-12-30.

[4] Dong M, Huang S S, Srivastava R. A laboratory study on near-miscible CO_2 injection in Steelman reservoir［J］. JCPT, 2001, 40（2）：53-61.

[5] 谢尚贤．大庆油田CO_2驱室内实验研究［J］．大庆石油地质与开发，1991，10（4）：32-35.

[6] 王利生，郭天民．江汉油藏油及其注二氧化碳体系高压粘度的实验测定［J］．石油大学学报：自然科学版，1994，18（4）：125-130.

[7] 杨胜来，王亮，何建军，等．CO_2吞吐增油机理及矿场应用效果［J］．西安石油大学学报：自然科学版，2004，19（6）：23-26.

[8] Nematilay E, Taghikhani V, Ghotbi C. Measurement and correlation of CO_2 solubility in the systems of CO_2/toluene, CO_2/benzene, and CO_2/n-hexane at near-critical and supercritical conditions［J］. Journal of Chemical & Engineering Data, 2006, 51（6）：2197-2200.

［9］ Wagner Z, Wichterle I. High-pressure vapor-liquid equilibrium in systems containing carbon dioxide, 1-hexene, and n-hexane ［J］. Fluid Phase Equilibria, 1987, 33 （1）: 109-123.

［10］ Lay E N. Measurement and correlation of bubble point pressure in （$CO_2 + C_6H_6$）, （$CO_2 + CH_3C_6H_5$）, （$CO_2 + C_6H_{14}$）, and （$CO_2 + C_7H_{16}$） at temperatures from 293. 15 to 313. 15 K ［J］. Journal of Chemical & Engineering Data, 2010, 55 （1）: 223-227.

［11］ Eduardo Pereira Siqueira Campos C, Gomes D' Amato Villardi H, Luiz Pellegrini Pessoa F, et al. Solubility of carbon dioxide in water and hexadecane: Experimental measurement and thermodynamic modeling ［J］. Journal of Chemical & Engineering Data, 2009, 54 （10）: 2881-2886.

［12］ 沈平平, 秦积舜, 李实, 等. 注 CO_2 提高原油采收率的多相多组分相态理论 ［R］. 北京: 中国石油勘探开发研究院, 2013.

［13］ 郑希谭, 孙文悦, 李实, 等. GB/T 26981—2011 油气藏流体物性分析方法 ［S］. 北京: 中国标准出版社, 2012.

CO$_2$ 在地层水中溶解对驱油过程的影响

汤 勇[1]　杜志敏[1]　孙 雷[1]　刘 伟[2]　陈祖华[2]

(1. 西南石油大学油气藏地质及开发工程国家重点实验室；2. 中国石化华东分公司)

摘 要： 利用 CO$_2$—烃—地层水相平衡热力学模型模拟计算了 CO$_2$ 在地层水中的溶解规律。建立了考虑 CO$_2$ 在地层水中溶解的一维长岩心数值模拟模型，模拟计算了注 CO$_2$ 驱替过程中原油采出程度、气油比、油气水饱和度剖面、CO$_2$ 在地层油和地层水中摩尔分数剖面的变化规律。研究表明：CO$_2$ 在地层水中的溶解量随着压力的升高而增加，随着温度的升高而降低；当温度达到100℃以上或压力达到20MPa以上时，压力和温度对 CO$_2$ 在水中溶解量影响变小。注气初期，考虑 CO$_2$ 溶解时采出程度比不考虑溶解时低，注气突破时间更迟，油墙向生产井端推进速度更慢。含水饱和度越高，影响程度越大。当含水饱和度为 0.67、注入 1.0 倍烃孔隙体积 CO$_2$ 时，考虑 CO$_2$ 溶解采出程度比不考虑 CO$_2$ 溶解低约6%。CO$_2$ 在地层水中溶解可导致 CO$_2$ 损失，使得 CO$_2$ 驱油见效时间滞后。

关键词： 注气；混相驱；CO$_2$ 驱替；地层水；溶解模型

0 引言

用温室气体 CO$_2$ 注入油藏实现 CO$_2$ 的地质埋存和提高石油采收率，是温室气体减排和资源化利用的结合点。室内实验和现场试验表明，CO$_2$ 是一种有效的驱油剂[1-7]，CO$_2$ 在油藏储层中溶解于水的能力比常规烃类物质大一个数量级[8-9]。由于油藏储层中总是存在束缚水、地层共存水或边底水，且水体体积往往比油藏体积大很多倍，这样当 CO$_2$ 注入油藏中时，可能存在大量的 CO$_2$ 溶解于地层水中，难以和地层油有效接触。同时，地层水的存在可能对 CO$_2$—地层油相态变化和混相驱油过程造成影响。目前国内外注 CO$_2$ 混相驱机理研究和注 CO$_2$ 方案设计均忽略了 CO$_2$ 在地层水中的溶解[10-11]。这使得 CO$_2$ 方案设计或注 CO$_2$ 方案预测可能存在一定的不确定性。因此，研究地层水存在对注 CO$_2$ 混相驱 CO$_2$—地层原油间的相态特征和地层水对 CO$_2$ 混相驱效率影响规律，对于注 CO$_2$ 提高采收率矿场试验具有重要意义[4]。

1 CO$_2$—烃—地层水相平衡热力学理论模型

1.1 油气水三相平衡理论模型

设一个由 n_c 个组分构成的含 CO$_2$ 油气水体系，取 1mol 的质量数作为分析单元，则当体系处于气—液—液三相平衡时，结合组分在平衡时的气相、液相Ⅰ和液相Ⅱ各相中平衡常数的定义，可以导出气—液—液三相平衡数值模型方程组[7,10]为：

$$\sum_{i=1}^{n_c} y_i = \sum_{i=1}^{n_c} \frac{k_{i,1} z_i}{1 + L_2(k_{i,2} - 1) + V(k_{i,1} - 1)} = 1 \tag{1}$$

$$\sum_{i=1}^{n_c} x_{i,1} = \sum_{i=1}^{n_c} \frac{z_i}{1 + L_2(k_{i,2} - 1) + V(k_{i,1} - 1)} = 1 \tag{2}$$

$$\sum_{i=1}^{n_c} x_{i,2} = \sum_{i=1}^{n_c} \frac{k_{i,2} z_i}{1 + L_2(k_{i,2} - 1) + V(k_{i,1} - 1)} = 1 \tag{3}$$

$$V + L_1 + L_2 = 1 \tag{4}$$

$$k_{i,1} = y_i / x_{i,1} \tag{5}$$

$$k_{i,2} = x_{i,2} / x_{i,1} \tag{6}$$

$$V_{yi} + L_1 x_{i,1} + L_2 x_{i,2} = z_i \tag{7}$$

式中，z_i 代表油—气—水体系中第 i 个组分的总摩尔分数；y_i，$x_{i,1}$ 和 $x_{i,2}$ 分别代表平衡时气相、液相Ⅰ、液相Ⅱ各相中第 i 个组分的摩尔分数；$k_{i,1}$ 和 $k_{i,2}$ 分别为气相与液相Ⅰ、液相Ⅱ与液相Ⅰ之间 i 组分平衡常数；V，L_1 和 L_2 分别代表平衡时气相、液相Ⅰ和液相Ⅱ的摩尔分数。

式（1）至式（7）是一个高度非线性方程组，联立求解方程组，可以算出气相、液相Ⅰ和液相Ⅱ各相的平衡摩尔分数 V，L_1，L_2 和各相摩尔分数 y_i，$x_{i,1}$，$x_{i,2}$。结合 PR 状态方程，方程组可以采用 Newton-Raphson 迭代法求解[4]。

1.2 CO_2 在水中的溶解模型

Henry 定律[12]是指气体在液体内的溶解量与其分压成正比。混合气体在液体内的溶解量与其中各个气体的分压有关，而与混合气体的总压力无关：

$$f_i^w = x_{wi} H_i \tag{8}$$

式中，x_{wi} 为水相中 i 组分的摩尔分数；f_i^w 为 i 组分的逸度；H_i 为 Henry 系数；$i = 1$，2，$\cdots$，n。

$$H_i = H_i^* \exp\left[\frac{V_i^\infty (p - p^*)}{RT}\right] \tag{9}$$

在已知 H_i^* 和 V_i^∞ 的前提下，利用式（9）可计算出任何压力下的 Henry 系数 H_i。

参考压力 p^* 下的 H_i^* 通常是根据组分 i 在纯水中溶解度的实验数据确定。笔者使用了 Nghiem 的拟合结果[10]，将 Nghiem 结果表中的数据作为参考点进行内插或外插，可得出任何压力、温度下的 H_i 值。

2 CO_2 在地层水中的溶解规律

以国内某油田注 CO_2 混相驱 PVT 相态实验数据为基础，利用油—气—水三相平衡理论模型，计算了不同温度和压力条件下 CO_2 在地层水中的溶解规律以及 CO_2 含量对溶解度的影响。模拟计算的 CO_2—烃—水体系组成见表 1，其中 CO_2 在烃体系的摩尔分数分别为 10%，30%，50% 和 70%。

表 1　模拟计算样品组成

样品号	H_2O	N_2	C_1	CO_2	C_2—iC_4	nC_4—C_6	C_7—C_{18}	C_{19}—C_{32}
样品 1	50.00	0.15	1.54	35.05	0.13	0.52	6.53	6.08
样品 2	55.59	0.16	1.72	27.79	0.15	0.58	7.26	6.76
样品 3	62.54	0.18	1.93	18.76	0.17	0.65	8.16	7.60
样品 4	71.48	0.21	2.21	7.15	0.19	0.75	9.33	8.69

以样品 1 为基础模拟计算的 CO_2 在水中的溶解度随温度和压力的变化曲线如图 1 所示。由图 1 可见：同一压力下，溶解度随着温度的升高而降低，当温度达到 100℃ 以后，温度对溶解度影响减小；同一温度下，随着压力的升高，CO_2 在水中溶解量增加；当压力大于 20MPa 以后，压力升高对 CO_2 在水中溶解度影响减小。当温度在 100℃ 左右、压力为 20~42MPa 条件下，CO_2 在地层水中的溶解质量分数约为 5.5%。因此，在油藏地层水体积较大的情况下，CO_2 在水中的溶解量是不能忽略的。

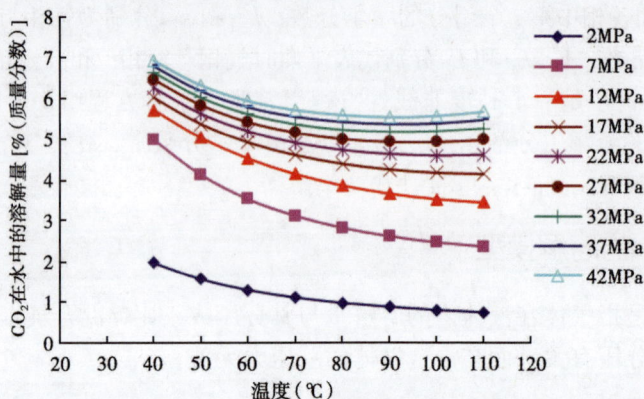

图 1　CO_2 在地层水体系中溶解量随压力、温度的变化

在地层压力 32MPa 下，不同 CO_2 含量油—气—水体系中 CO_2 在地层水中溶解量的模拟结果如图 2 所示。由图 2 可见，在压力和温度保持不变的条件下，CO_2 在水中的溶解量随注

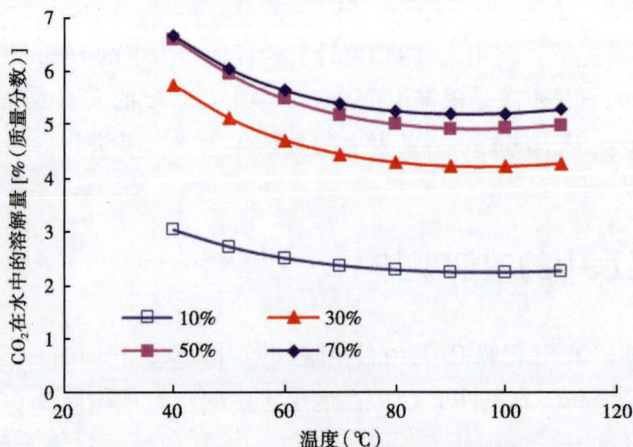

图 2　CO_2 含量对 CO_2 在地层水中溶解量的影响

入 CO_2 含量的增加而增加，当 CO_2 在油—气体系中的摩尔分数达到50%后，CO_2 含量的增加对 CO_2 在水中的溶解影响减小。

3 CO_2 在地层水中的溶解对驱油效率的影响

3.1 建立模型

岩心模拟计算所用地层原油的PVT参数来自该油田地层原油高压PVT实验数据。储层参数和模拟条件均来自实际的长岩心注气驱替实验数据，其中岩心组总长度为95.67cm，直径为2.54cm，岩心有效孔隙度为13.65%～15.94%，气测渗透率为7.72～44.46mD，调和平均渗透率为14.59mD。运用CMG的GEM组分模型器，建立一维长岩心模拟模型，其中网格总数为20个，一端设置一口注入井，另一端设置一口采油井。注入气为纯 CO_2 气。模拟温度为地层温度110℃，压力为地层压力32MPa，该压力条件下注 CO_2 驱已经达到多级接触混相。相对渗透率曲线水相临界流动饱和度端点为0.47。

当原始含水饱和度为0.47和0.67，考虑 CO_2 在水相中溶解和不考虑 CO_2 在水相中溶解时，模拟计算注 CO_2 驱油采出程度、气突破时间、油气水饱和度分布剖面、CO_2 组分在油气水相中的分布剖面以及 CO_2 的损失，从而认识 CO_2 溶解对驱油过程的影响，其中 CO_2 在水中的溶解量通过Henry定律计算。

3.2 结果分析

图3（a）为含水饱和度为0.47时考虑 CO_2 溶解和不考虑 CO_2 溶解的长岩心注 CO_2 驱原油采出程度和气油比变化计算结果曲线。由图3可见，当 CO_2 注气量为0.5HCPV（烃孔隙体积）时，考虑 CO_2 在地层水溶解的原油采出程度比不考虑地层水溶解的低2.5%。同时，当注入气量超过0.8HCPV后采出程度的影响降低。

图3（b）为含水饱和度为0.67时考虑 CO_2 溶解和不考虑 CO_2 溶解的原油采出程度和气油比变化计算结果。由图3（b）可见，考虑 CO_2 在地层水中溶解时，当注气为0.6HCPV和1.0HCPV时，原油采出程度比不考虑溶解时分别低3.7%和6%。

由图3可见，考虑 CO_2 在地层水中溶解时，气油比曲线升高的时间比考虑 CO_2 溶解的

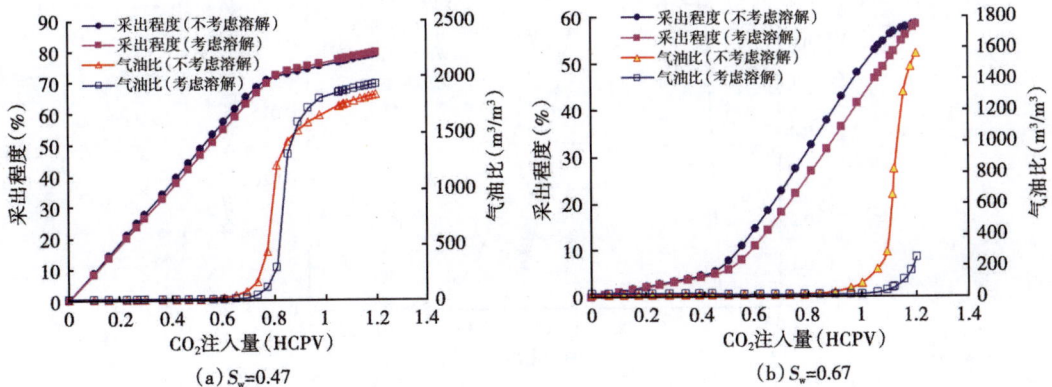

图3 CO_2 驱采出程度和气油比与注入量的关系

滞后。这说明考虑 CO_2 溶解时，损失部分 CO_2 在地层水中，与地层油有效接触的 CO_2 量减少，总体上导致 CO_2 驱气突破时间推迟。

由图 3 可见，初始含水饱和度 $S_w = 0.67$ 时，考虑 CO_2 在水中溶解比不考虑溶解采出程度和气油比的差别比初始含水饱和度 0.47 更显著。这说明初始含水饱和度越高，CO_2 在地层水中溶解对 CO_2 驱油过程影响越显著。其主要原因是含水饱和度高使得 CO_2 在地层水中的溶解量更大，CO_2 损失更大。

图 4 和图 5 分别为初始含水饱和度为 0.67，注气 0.6HCPV 时油气水饱和度分布和 CO_2 在油相和水相中的摩尔分数变化。由图 4 可见，考虑 CO_2 溶解时含油饱和度和含气饱和度较不考虑 CO_2 溶解时更靠近注入端网格，出现滞后现象。这说明，由于 CO_2 在地层水中的溶解损失，形成的油墙需要更长的时间才能推进到生产井一端。

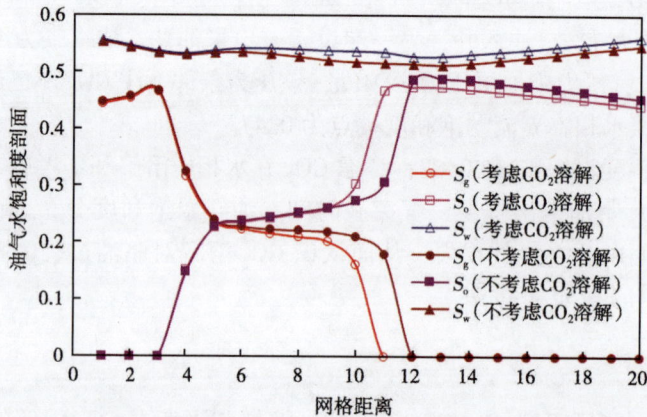

图 4　考虑 CO_2 溶解和不考虑 CO_2 溶解时油气水饱和度剖面对比图

由图 5 可见，CO_2 在油中溶解的摩尔分数约为 70%。考虑 CO_2 溶解时，CO_2 在地层水中溶解的摩尔分数约为 2%，这时 CO_2 在地层油中溶解的过渡段相对更短。这说明，考虑 CO_2 在地层水中溶解时形成的混相带比不考虑 CO_2 溶解时更短。

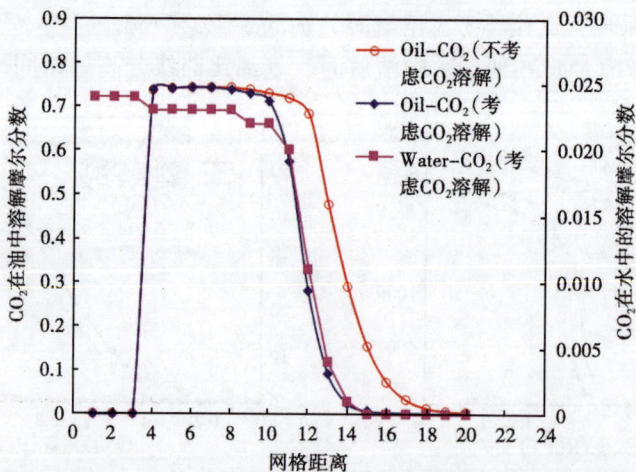

图 5　考虑 CO_2 溶解和不考虑 CO_2 溶解时 CO_2 在油水相中的摩尔分数剖面对比图

4 结论

（1）CO_2 在地层水中的溶解量随着温度的升高而降低，随着压力的升高而升高。当温度达到 100℃ 以上或压力达到 20MPa 以上时，压力和温度对 CO_2 在水中的溶解量影响减小。在通常的油藏温度压力条件下，CO_2 在地层水中的溶解质量分数约为 5.5%。

（2）注气初期，考虑 CO_2 溶解时的原油采出程度比不考虑 CO_2 溶解时更低，气突破时间更晚。含水饱和度越高，CO_2 在地层水中溶解对采出程度和气油比影响越大，其采出程度可相差约 6%。

（3）考虑 CO_2 在地层水中溶解时，CO_2 在地层油中溶解的过渡段更短，油墙向生产井端推进速度相对更慢。

（4）注气初期，由于注入 CO_2 在地层水中的溶解而导致有效 CO_2 的损失，可能使得注 CO_2 驱油效果出现滞后。CO_2 注入油藏驱油及埋存方案设计时不能忽略 CO_2 在地层水中的溶解。

参 考 文 献

［1］李孟涛，单文文，刘先贵，等．超临界二氧化碳混相驱油机理实验研究［J］．石油学报，2006，27（3）：80-83.

［2］熊钰，孙良田，孙雷，等．基于模糊层次分析法的注 CO_2 混相驱油藏综合评价方法［J］．石油学报，2002，23（6）：60-62.

［3］Nutakki R，Firoozabadi A，Wong T W，et al. Calculation of multiphase equilibrium for water-hydrocarbon systems at high temperature［R］. SPE 17390, 1988.

［4］Chang Y B，Coats B K，Nolen J S. A compositional model for CO_2 floods including CO_2 solubility in water［R］. SPE 35164, 1998.

［5］Zuluaga E，Monsalve J C. Water vaporization in gas reservoirs［R］. SPE 84829, 2003.

［6］沈平平，袁士义，韩冬，等．中国陆上油田提高采收率潜力评价及发展战略研究［J］．石油学报，2001，22（1）：45-48.

［7］Egermann P，Chalbaud C，Duquerroix J P，et al. An integrated approach to parameterize reservoir models for CO_2 injection in aquifers［R］. SPE 102308, 2006.

［8］Martin F D. Carbon dioxide flooding［J］J. Pet. Tech. , 1992, 44（4）：396-400.

［9］李士伦，张正卿，冉新权，等．注气提高石油采收率技术［M］．成都：四川科学技术出版社，2001：22-24.

［10］沈平平，黄磊．二氧化碳—原油多相多组分渗流机理研究［J］．石油学报，2009，30（2）：247-251.

［11］汤勇，孙雷，周涌沂，等．注富烃气凝析/蒸发混相驱机理评价［J］．石油勘探与开发，2005，32（2）：133-136.

［12］施文，桓冠仁，郭尚平．二氧化碳—烃—水系统相平衡闪蒸计算方法研究［J］．石油勘探与开发，1992，19（3）：48-56.

地层油关键组分与 CO_2 混相能力的相关性研究

李 实 张 可 马德胜 秦积舜 陈兴隆

(中国石油勘探开发研究院提高石油采收率国家重点实验室)

摘 要：利用高温高压可视流体相态仪，"肉眼"直接观察到煤油、凝析油和地层油这3种典型油品与 CO_2 混相的动态过程，进行了对比分析。并对地层油与 CO_2 混相过程中的气相分层现象中的每层气体组成做出了定量、定性分析。研究还发现：经过对20个区块23组样品的最小混相压力（MMP）测试结果分析，提出"最小混相密度"的概念（MMD）作为确定MMP的关键参数，并给出了包含胶质、沥青质影响因素在内的MMD确定公式，为MMP的确定提出了新的准则。

关键词：地层油；关键组分；混相；相关性

0 引言

针对当前日益扩大的 CO_2 驱油现场试验的开展，石油工作者、领域专家对 CO_2 混相驱的"真实混相过程"一直存在困惑，并未真正地、直观地研究过混相的动态过程，而现有的研究方法主要是物理模拟的驱替实验，并结合油气界面特征分析手段，虽然这些实验方法模拟了多孔介质中的驱替动态特征，但不能直观地展现 CO_2 对原油的萃取或抽提过程以及抽提组分组成[1-6]。在研究PVT筒中的死油与 CO_2 的接触实验中发现，在加压过程中，气相中分出了清晰的4个轻组分层（图1），这对解释 CO_2 混相驱的前缘驱替特征有着重要的指导作用[7-9]。

因此，笔者选出3种具有代表性的油品（煤油、凝析油和地层油）模拟 CO_2 混相实验过程，并定量地分析了地层油在混相过程中产生的分层特征，对积累多年的30个原油样品的全组分数据与对应的MMP进行分析，提出了最小混相密度作为确定MMP的新准则[10-14]。该研究使 CO_2 混相过程清晰地呈现在面前，有助于分析实际油藏条件下的 CO_2 混相驱过程。

图1 CO_2 萃取死油分层现象

1 实验设备及材料

实验设备：法国ST公司生产的油藏流体相态仪，该装置可实现全可视、高温（200℃）、高压（150MPa）、拍摄记录实验过程及数据；色谱分析仪，安捷伦公司生产气相色谱仪。

实验材料：航空煤油、新疆某井凝析油、大庆某井地层油样品；CO_2 为北京中国科学院气体厂生产，工业纯 99.9%。

2 结果分析

使用国际领先的实验技术手段，观测超临界 CO_2 在不同类型的原油体系相间传质的动态过程，并定量地分析了关键组分对油气混相的影响。以下实验中，对系统加压方式均采用不搅拌、恒速进泵方式，速度梯度为 0.5mL/min。

2.1 煤油动态混相特征

经色谱分析，煤油组成：C_7—C_{12} 含量为 99.7%（摩尔分数），密度为 $0.78g/cm^3$。从图 2 可以看出，初始条件下煤油与 CO_2 截然分为两层，界面清晰可见 [图 2（a）]；随着压力的增加，煤油层体积开始膨胀，此时以 CO_2 溶解为主，煤油中的少量气体组分被萃取[15] [图 2（b）]；继续增加压力，此时以煤油组分大量被萃取为主，CO_2 溶解为辅 [图 2（c）]；继续增加压力，油—气界面剧烈传质、界面混沌现象出现 [图 2（d）]；再增加压力，CO_2—煤油完全混相，油—气界面完全消失 [图 2（e）]。由于煤油中中间烃较多，混相过程持续时间较短。

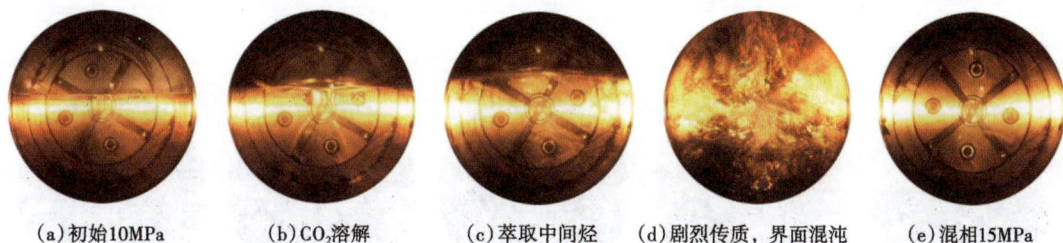

(a)初始10MPa　　(b)CO_2溶解　　(c)萃取中间烃　　(d)剧烈传质，界面混沌　　(e)混相15MPa

图 2　CO_2 与煤油动态混相过程

2.2 凝析油动态混相特征

经色谱分析，凝析油组成：C_2—C_6 为 10.7%（摩尔分数），C_7—C_{15} 为 75.8%（摩尔分数），C_{16}—C_{29} 为 7.3%（摩尔分数），C_{30+} 为 6.1%（摩尔分数），油密度为 $0.79g/cm^3$。从图 3 可以看出，初始条件下凝析油与 CO_2 分为两层，界面清晰可见 [图 3（a）]；随着压力的

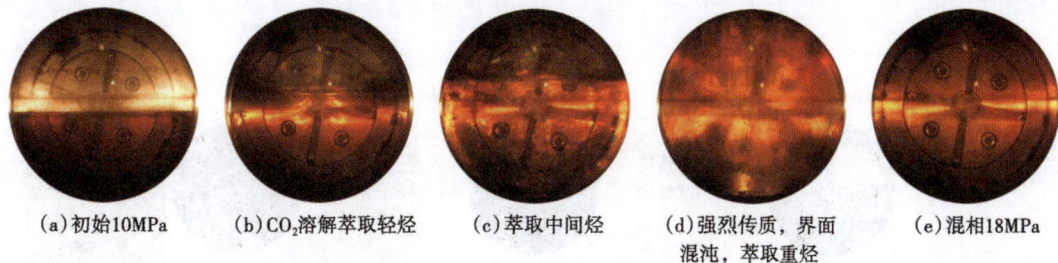

(a)初始10MPa　　(b)CO_2溶解萃取轻烃　　(c)萃取中间烃　　(d)强烈传质，界面混沌，萃取重烃　　(e)混相18MPa

图 3　CO_2 与凝析油动态混相过程

增加，凝析油体积开始膨胀，此时以 CO_2 溶解为主，煤油中的 C_2—C_6 组分被萃取 ［图 3 (b)］；继续增加压力，此时凝析油中 C_7—C_{15} 一部分组分被萃取，C_2—C_6 几乎被完全萃取 ［图 3 (c)］；继续增加压力，油—气界面剧烈传质，界面混沌现象出现，C_7—C_{15} 被大量萃取，C_{16}—C_{29} 中的一部分也挥发至气相中 ［图 3 (d)］；再增加压力，C_{30+} 组分也被溶解，CO_2—凝析油完全混相，油气界面完全消失 ［图 3 (e)］。由于凝析油中含有少量重质组分，因此混相过程持续时间长于煤油。

2.3　地层油动态混相特征

经色谱分析，地层油组成：C_2—C_6 为 10.5%（摩尔分数），C_7—C_{15} 为 28.5%（摩尔分数），C_{16}—C_{29} 为 17.9%（摩尔分数），C_{30+} 为 12.2%（摩尔分数），油密度为 0.85g/cm^3。从图 4 可以看出，初始条件下 CO_2—地层油溶解现象与煤油和凝析油类似，油气界面十分清晰 ［图 4 (a)］；随着压力的增加，地层油体积开始膨胀，此时以 CO_2 溶解为主，少量的 C_2—C_6 组分被萃取，极少量 C_7—C_{15} 被抽提，形成少量薄雾区域 ［图 4 (b)］；压力继续增加，地层油中间烃组分被大量萃取，形成富烃带，该区域在油气混相过程中起着重要的作用 ［图 4 (c)］；继续增加压力，油—气界面传质加剧，界面混沌现象出现，大量重质组分参与混相，油—气界面比较模糊 ［图 4 (d)］；再增加压力，大量的 C_{30+} 组分也被溶解，CO_2—地层油完全混相，油—气界面完全消失，形成单一相，但是其中可能包含着少量未溶解的重质组分 ［图 4 (e)］。由于地层油中含有大量重质组分，混相过程持续时间大于凝析油。

(a) 初始10MPa　　(b) CO_2溶解萃取轻烃　(c) 萃取中间烃，　(d) 传质加剧，界面混沌，　(e) 混相25MPa
　　　　　　　　　　　　　　　　　　　形成富烃相　　　重烃参与传质

图 4　CO_2 与地层油动态混相过程

2.4　CO_2—陆相地层油体系的相间传质机理

图 5 为对不同阶段的 CO_2—地层油混相过程中的每个挥发层进行分层取样后，利用色谱进行气相组分分析（红色虚线为取样分析区域）。烃组分传质能力随碳数增加而减弱：（1）轻烃组分（C_2—C_6）强传质，极易混相；（2）中间组分（液态烃 C_7—C_{15}）较强传质，易

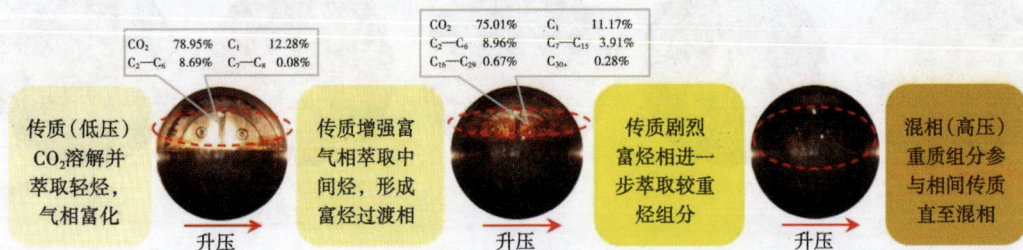

| CO_2 | 78.95% | C_1 | 12.28% |
| C_2—C_6 | 8.69% | C_7—C_8 | 0.08% |

CO_2	75.01%	C_1	11.17%
C_2—C_6	8.96%	C_7—C_{15}	3.91%
C_{16}—C_{29}	0.67%	C_{30+}	0.28%

传质（低压）CO_2溶解并萃取轻烃，气相富化　　升压　　传质增强富气相萃取中间烃，形成富烃过渡相　　升压　　传质剧烈富烃相进一步萃取较重烃组分　　升压　　混相（高压）重质组分参与相间传质直至混相

图 5　CO_2—陆相地层油体系的相间传质

混相；（3）重组分（固烃 C_{16+}）弱传质，不易混相；（4）极性重组分（胶质沥青）极弱传质，易沉积。

3 最小混相密度确定 CO_2—地层油最小混相压力

依据 25 个区块 23 组实验，分析地层油组分对 MMP 的影响[16-17]，如图 6 所示。通过对数据点的拟合并附带胶质、沥青质含量得出烃组分系数与最小混相密度间的计算公式。

（1）低胶质沥青［小于 6%（质量分数）］地层油采用：$D_{MMP} = -0.188X_f + 0.732$ （1）

（2）高胶质沥青［大于 6%（质量分数）］地层油采用：$D_{MMP} = -0.352X_f + 0.988$ （2）

其中：$X_f = (C_2-C_{15}) / (C_1+N_2+C_{16+})$，表征组分与 MMP 的相关性。

针对上式，在已知地层油组分和油藏温度的前提下计算出 X_f，根据胶质、沥青质含量选用式（1）或式（2）确定 D_{MMP}，再根据油藏温度和 D_{MMP} 通过查 CO_2 密度表，便可以得出所对应的 CO_2 压力，即为 CO_2—地层油体系的 MMP。

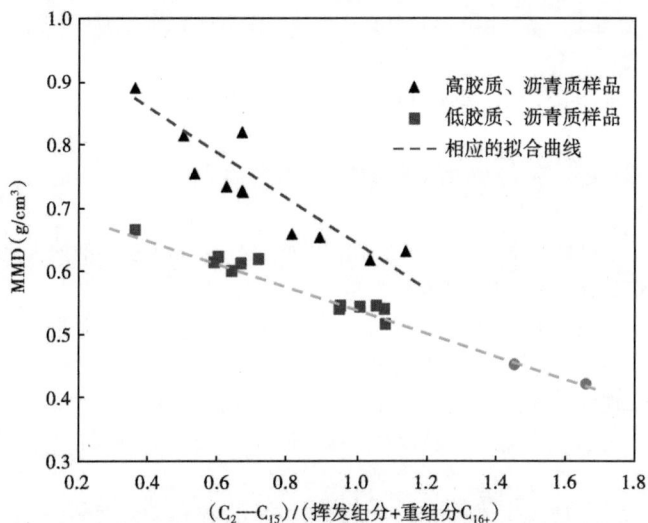

图 6　地层油组分与 CO_2 最小混相密度曲线

4 结论

（1）煤油—CO_2 混相过程中，油气界面剧烈传质，界面混沌现象显著，基本无分层现象；凝析油—CO_2 混相过程中，由于凝析油中含有少量重质组分，导致混相过程持续时间长于煤油。

（2）地层油—CO_2 混相过程中，随着压力的缓慢增加，地层油中组分抽提顺序不同，导致气相中分出显著的多层现象，该现象有助于今后研究地层多孔介质中的 CO_2 驱替过程。

（3）提出最小混相密度确定地层油—CO_2 最小混相压力的概念，并通过对数据点的拟合，建立了胶质、沥青质含量，最小混相密度与地层油—CO_2 体系最小混相压力间关系，为 CO_2 混相驱技术推广提供了一种有效的预测方法。

参 考 文 献

[1] 凌建军, 黄鹏. 非混相 CO_2 开采稠油 [J]. 油气采收率技术, 1996, 3 (2): 13-18.

[2] 毛振强, 陈凤莲. CO_2 混相驱最小混相压力确定方法研究 [J]. 成都理工大学学报: 自然科学版, 2005, 32 (1): 61-64.

[3] 周伦先, 褚小兵. CO_2 气井相态特征 [J]. 油气井测试, 2006, 15 (5): 35-37.

[4] 李孟涛, 张英芝, 杨志宏, 等. 低渗透油藏 CO_2 混相驱提高采收率试验 [J]. 石油钻采工艺, 2005, 27 (12): 43-46.

[5] 黄廷章, 刘福海, 杨正明. 复杂化学流体在多孔介质中的传质 [J]. 力学学报, 2002, 34 (2): 256-261.

[6] 李振泉. 油藏条件下溶解 CO_2 的稀油相特性实验研究 [J]. 石油大学学报: 自然科学版, 2004, 28 (3): 43-45.

[7] 吴文有, 张丽华, 陈文彬. CO_2 吞吐改善低渗透油田开发效果可行性研究 [J]. 大庆石油地质与开发, 2001, 20 (6): 52-54.

[8] 童敏, 李相方, 胡永乐, 等. 多孔介质对凝析气相态的影响 [J]. 石油大学学报: 自然科学版, 2004, 28 (5): 62-64.

[9] 李士伦, 周守信, 杜建芬, 等. 国内外注气提高石油采收率技术回顾与展望 [J]. 油气地质与采收率, 2002, 9 (2): 1-5.

[10] 孙业恒, 吕广忠, 王延芳, 等. 确定 CO_2 最小混相压力的状态方程法 [J]. 油气地质与采收率, 2006, 13 (1): 82-84.

[11] 彭宝仔, 罗虎, 陈光进, 等. 用界面张力法测定 CO_2 与原油的最小混相压力 [J]. 石油学报, 2007, 28 (3): 93-95.

[12] 李向良, 王庆奎, 李振泉, 等. CO_2 多次抽提作用对地层油析蜡温度影响的实验研究 [J]. 大庆石油地质与开发, 2007, 26 (3): 106-110.

[13] 刘炳官, 朱平, 雍志强. 江苏油田 CO_2 混相驱现场试验研究 [J]. 石油学报, 2002, 23 (4): 56-60.

[14] 李孟涛, 单文文, 刘先贵, 等. CO_2 混相驱油机理实验研究 [J]. 石油学报, 2006, 27 (3): 80-83.

[15] 蒋明, 赫恩杰. 二连低渗透砂岩油藏开发中的几点认识 [J]. 石油学报, 2000, 21 (4): 58-64.

[16] 苏畅, 孙雷, 李士伦. CO_2 混相驱多级接触过程机理研究 [J]. 西南石油学院学报, 2001, 23 (2): 33-36.

[17] 郝永卯, 薄启炜, 陈月明. CO_2 驱油实验研究 [J]. 石油勘探与开发, 2005, 32 (2): 110-112.

超临界 CO_2 动态混相驱过程机理研究

陈 文[1]　汤 勇[1]　梁 涛[3]　孙 雷[1]　刘 伟[2]　陈祖华[2]

(1. 西南石油大学油气藏地质及开发工程国家重点实验室;
2. 中国石化华东油田分公司; 3. 中国石油勘探开发研究院)

摘　要: 以国内 CS 油田注 CO_2 混相驱典型实例为基础,在油藏地层流体注 CO_2 驱膨胀实验和细管最小混相压力实验拟合的基础上,建立一维组分注气驱细管模型。应用所建立的模型,模拟研究 CO_2 注入过程中油气两相组成、油气两相黏度、密度和界面张力等动态特性参数沿注气井到生产井距离的变化规律,以及注气量和注气压力对动态特性参数的影响规律。研究结果显示: CO_2 在原油中的溶解能力强,工程混相条件下,摩尔分数达到 0.7。注入 CO_2 抽提原油中的中间烃,甚至 C_{19+} 的重烃,与地层油在前缘达到混相。 CO_2 注入量增加,混相带增长, CO_2 波及区域增加,有利于驱油效率增加。随着注入压力的提高,从非混相到混相, CO_2 在地层油中的溶解量增加,界面张力降低,油的黏度降低。达到混相后,继续增加压力对驱油影响变小。

关键词: CO_2 混相驱;混相机理;细管模拟;注气

0　引言

注 CO_2 混相驱是近年来 CO_2 气体减排和注气提高采收率研究的热点[1]。用 CO_2 作为油藏提高采收率的驱油剂已研究多年,室内实验和现场试验都表明 CO_2 是一种有效的驱油剂[1-10]。 CO_2 与原油的混相压力较低,注 CO_2 的效果非常明显,注 CO_2 技术在全球已得到推广运用[2-10]。我国大庆油田、吉林油田、江苏油田和草舍油田先后实施了注 CO_2 驱的先导实验研究,已经取得了一定的现场效果。那么在实施 CO_2 混相驱过程, CO_2 与地层油到底发生了怎样的传质过程, CO_2 注入后对地层油的性质在动态上有怎样的影响,混相驱过程的工程注入参数,注入压力和注气比例怎样影响混相过程,这些都是在实施混相驱矿场试验中必须认识的机理。国内外关于注气机理的论述很多,总体上可分为一次接触混相、多次接触混相和非混相驱 3 种,而多次接触混相又分为蒸发气驱混相和凝析气驱混相两种[5]。这主要是从静态没有考虑流动的情况下对混相机理的评价[8-9],关于 CO_2 动态混相过程的评价相对较少。细管实验为地层油和注入气提供了一个在多孔介质中连续接触的环境,可以动态模拟 CO_2 与地层油的多次接触混相过程,但是细管实验无法实现多级接触过程各种物性参数的测量,难以分析注 CO_2 过程地层油与注入 CO_2 气之间发生的传质作用。因此,通过实际油藏注 CO_2 驱相态实验和细管实验的模拟,在组分模型中再现 CO_2 与地层油的动态传质过程,认识超临界 CO_2 动态混相驱过程,从而在实践中更好地利用 CO_2 驱技术,达到环保和提高采收率的目的。本文以 CS 油田注 CO_2 混相驱实验为基础,在注气相态模拟和细管实验模拟的基础上,研究 CO_2 驱替过程的相态传质和地层流体物性变化。

1　CS 油藏注 CO_2 相态实验模拟

CS 油藏地层流体组成见表 1。由表 1 可以看出，该流体 C_{7+} 组分摩尔分数为 84.03%，重质组分含量很高，属于较重的黑油。为了优化组分模型中状态方程参数，改善对原油性质的预测精度，在不影响模拟结果的前提下和考虑注 CO_2 气混相驱情形下，按组分性质相近的原则[11]，把地层原油组成划分为 7 个拟组分，分别为 CO_2，N_2，C_1，C_2—iC_4，nC_4—C_6，C_7—C_{18}，C_{19}—C_{33}。地层流体 PVT 相态实验拟合结果见表 2，总体上拟合效果好，满足进一步模拟精度需要。

表 1　CS 油藏地层流体组成表

组分	N_2	CO_2	C_1	C_2	C_3	iC_4	nC_4	iC_5	nC_5	C_6	C_{7+}
组成［%(摩尔分数)］	0.97	0.34	10.29	0.36	0.45	0.08	0.16	0.04	1.8	1.48	84.03

表 2　饱和压力与单次闪蒸数据拟合结果表

项　　目	实验值	拟合值	绝对误差	相对误差（%）
饱和压力（MPa）	4.0000	3.9800	0.0200	0.50
单次闪蒸气油比（m^3/m^3）	11.4300	12.0300	0.600	5.25
地层原油密度（g/cm^3）	0.8219	0.8200	0.0019	0.23
地面脱气油密度（g/cm^3）	0.8860	0.8470	0.0390	4.40
地层油黏度（$mPa \cdot s$）	7.0200	6.9970	0.0230	0.33

地层温度 110℃下地层油注 CO_2 气膨胀实验拟合结果如图 1 所示，其中注入气中 CO_2 摩尔分数为 99.93%。由图 1 可见，膨胀实验拟合值和实验值基本吻合。

2　拟三元相图确定 CS 油藏注 CO_2 混相压力

在地层流体 PVT 相态实验和注气膨胀实验拟合的基础上，模拟研究 CS 地层流体在地层温度 110℃和 32MPa 下的注 CO_2 拟三元相图，结果如图 2 所示。由图 2 可见，注入 CO_2 不断抽提地层油的中间组分，CO_2 气被加富，富化的 CO_2 混合气不断向前继续与地层油接触，

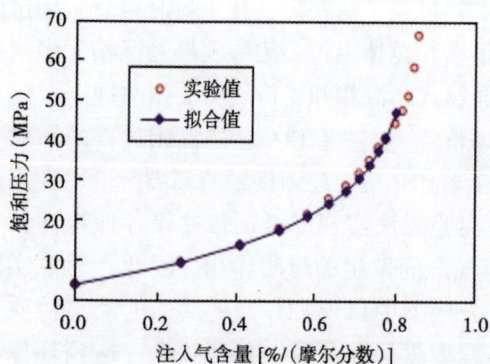

图 1　CS 油藏地层油注 CO_2 气膨胀
实验饱和压力拟合（110℃）图

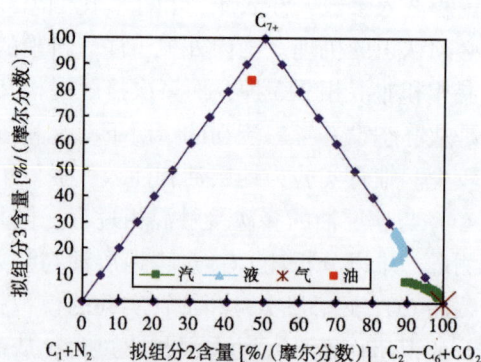

图 2　CS 地层油注 CO_2 气拟
三元相图（110℃）

不断被加富，最后形成富含中间烃的 CO_2 气，与地层油达到多级接触混相（向前接触混相或蒸发混相）。而后缘的地层油被 CO_2 抽替后，又不断与新鲜的 CO_2 接触，不断被抽替，形成高含重质烃的油相。本次模拟结果显示，CO_2 与 S198 井地层流体在 32MPa 下通过多级接触达到混相。由此，通过相态确定的多级接触最小混相压力（MMP）为 32MPa。

3 CS 油藏注 CO_2 细管实验最小混相压力模拟

在注气 PVT 相态模拟的基础上，进行了注 CO_2 气细管实验最小混相压力拟合。细管模型尺寸设计为细管长 8.3m，横截面为正方形，边长 0.00386m。平均孔隙度为 0.39，渗透率为 3.98D。网格划分 X 方向为 40 个，Y 方向和 Z 方向各为 1 个，网格步长为 $X = 0.4575$m，$DY = 0.00386$m，$DZ = 0.00386$m。在初始端设置一口生产井，另一段设置一口注入井。

注入溶剂为 CO_2 气。整个驱替过程在恒定温度110℃（地层温度）条件下进行。模拟计算注入 1.2 倍烃孔隙体积（HCPV）后的原油采收率。模拟结果见表 3 和图 3。由表 3 和图 3 可见，相同压力下模拟计算采收率和实验采收率，细管模拟的最小混相压力为 30.18MPa，比实验值 29.34MPa 略高。而细管模拟结果比拟三元相图确定的混相压力低，这是因为拟三元相图模拟的是严格意义上的混相状态（界面张力为 0），而细管实验的混相压力属于工程意义上的混相压力（界面张力没达到 0）。

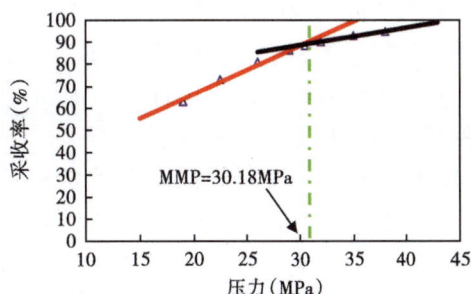

图 3　CO_2 驱细管实验最小混相压力拟合结果（110℃）图

表 3　不同压力下注 CO_2 气细管实验采收率拟合结果表

驱替压力（Pa）		22.5	26	29	30.5	32	35	38
采收率（%）	实验	59.75	75.84	90.21	92.6	93.81	96.16	—
	模拟	73.05	81.01	86.19	88.27	90.05	92.81	94.71

4 CO_2 混相驱机理分析

4.1 组分变化规律

选取达到混相的 30.18 MPa 时注入 1.2HCPV 的 CO_2 气，油气两相中组分 CO_2，C_2—C_6 和 C_{19}—C_{32} 随网格的变化，如图 4 和图 5 所示。由图可见，网格 10 到网格 28 这个区间属于油气混相过渡带，在该区域地层油相中溶解大量的 CO_2，摩尔分数达到 0.6。C_{19+} 含量逐渐降低，说明地层油相变得更轻质，而且过渡带前缘的 C_{19+} 含量低于后缘，C_2—C_6 含量后缘高于前缘，说明注气井一端油相相对较重。由图 5 可见，气相中的 C_2—C_6 和 C_{19+} 含量增加，说明地层条件下过渡带的气相从油相中抽提了部分中间烃和少量重烃，总体上富化。在过渡带的前缘，油气两相的组成，特别是 C_2—C_6+CO_2 的含量基本接近，达到向前接触混相的目的。

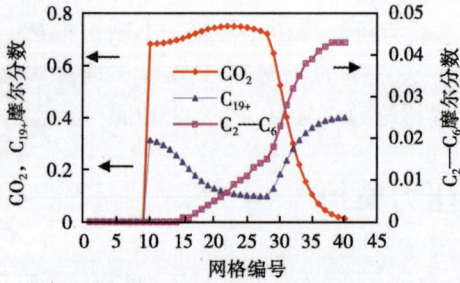

图4 油相中各组分含量随网格变化图

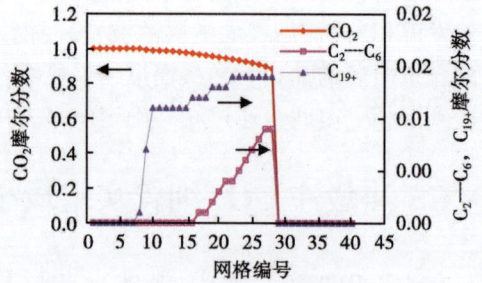

图5 气相中各组分含量随网格变化图

4.2 注气量的影响

模拟研究了30.18 MPa下不同注气量（0.1PV，0.3PV，0.6PV，0.9PV和1.2PV）条件下，油气过渡带、油中溶解CO_2含量和油相黏度的变化，如图6和图7所示。由图6、图7可见，随着注气量从0.1PV增加到1.2PV，混相过渡带的长度逐渐增加，由于压力始终保持不变。因此，油相中溶解的CO_2摩尔分数基本保持在0.7，前缘低后缘略高。在30.18MPa下注气，油相黏度降低在混相带前缘有最低值，0.64~0.67mPa·s，远低于原始地层油的黏度。而在混相过渡带的后缘，油相黏度增加，这与图4的分析结果一致（后缘的油相组成变重）。因此，注CO_2主要通过前缘混相，属于蒸发式混相过程。

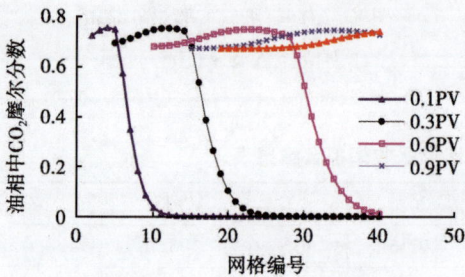

图6 油相中CO_2含量随注气量的变化图

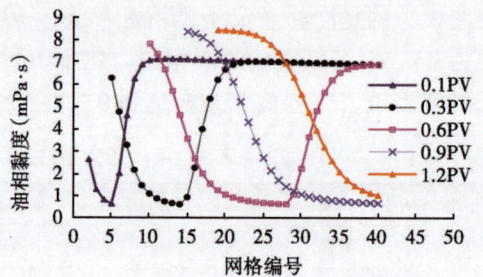

图7 油相黏度随注气量的变化图

4.3 注入压力对混相过程的影响

图8、图9和图10分别显示了当注气量为0.6PV时，不同压力下油气界面张力、油相中CO_2含量和地层油黏度随网格的变化情况。

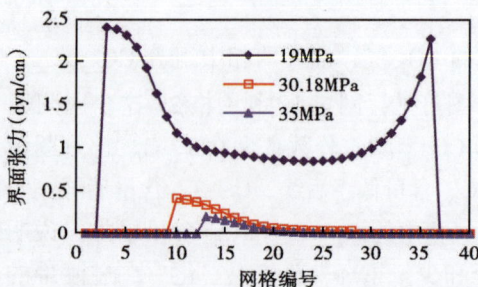

图8 不同注入压力下界面张力随网格变化图

由图8可见，随着注入压力的增加，油气界面张力降低，压力越高，界面张力越低，当压力增加到30.18 MPa后油气界面张力近似为0，而压力从30.18MPa升高到35 MPa，界面张力变化幅度很小。由图9可见，压力越高，CO_2在原油中溶解量增加，当压力从30.18MPa增加到35MPa时，CO_2在原油中的溶解比例基本不再增加。由图10可见，随着注气压力的增加，油相黏度降低，压力达到 30.18MPa 后，黏度随着压力升高降低幅度非常小。当压力较低时（19MPa），在注气端地层油黏度反而高于原始地层油黏度，这说明注气后在过渡带的后缘油的组成变重，同时也注气达到混相或近混相后，再增加压力对改善驱油效率意义已经不大。

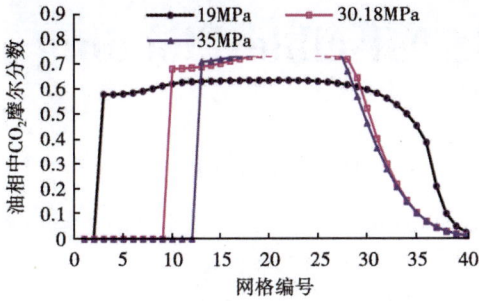

图 9 不同注入压力下油相各组分
含量随网格变化图

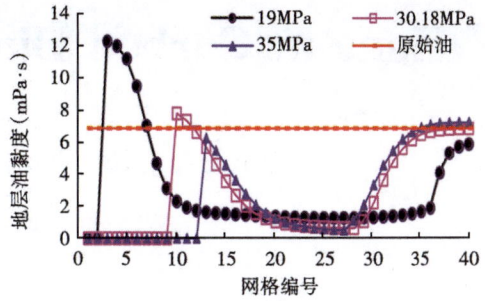

图 10 不同注入压力下地层
油黏度随网格变化图

5 结论

（1）细管实验模拟得到 CO_2 驱混相压力为工程混相，比严格意义的界面张力为 0 的混相压力低，但在实际油藏注气驱过程中往往已经足够。

（2）CO_2 在原油中的溶解能力强，工程混相条件下，摩尔分数达到 0.7。注入 CO_2 抽提原油中的中间烃，甚至 C_{19+} 的重烃，与地层油在前缘达到混相。

（3）CO_2 注入量增加，混相带增长，CO_2 波及区域增加，有利于驱油效率增加。

（4）随着注入压力的提高，从非混相到混相，CO_2 在地层油中的溶解量增加，界面张力降低，油的黏度降低。达到工程混相后，继续增加压力对驱油影响变小。

参 考 文 献

[1] 谷丽冰，李治平，侯秀林 . 二氧化碳地质埋存研究进展 [J]. 地质科技情报，2008，27（4）：23-26.

[2] 郭万奎，廖广志 . 注气提高采收率技术 [M]. 北京：石油工业出版社，2003.

[3] 缪明富，彭子成，钟国利 . 利用二氧化碳资源 提高气田开发效益 [J]. 石油与天然气化工，2005，34（6）：470-481.

[4] Moritis G . Special report enhanced oil recovery [J]. Oil Gas Journal, 2004 (15)：53-65.

[5] 马涛，汤达祯，蒋平，等 . 注 CO_2 提高采收率技术现状 [J]. 油田化学，2007，24（4）：23-27.

[6] 郭平，李士伦，杜志敏，等 . 低渗透油藏注气提高采收率评价 [J]. 西南石油学院学报，2002，24（5）：1-6.

[7] 汤勇，孙雷，周涌沂，等 . 注气混相驱机理评价方法 [J]. 新疆石油地质，2004，25（4）：414-416.

[8] 汤勇，孙雷，周涌沂，等 . 注富烃气凝析/蒸发混相驱机理评价 [J]. 石油勘探与开发，2005，4（2）：133-137.

[9] 李士伦，郭平，戴磊，等 . 发展注气提高采收率技术 [J]. 西南石油学院学报，2000，22（3）：42-46.

[10] 李福恺 . 黑油和组分模型的应用 [M]. 北京：科学出版社，1996.

Characteristic Analysis of the Transition Zone in Carbon Dioxide Miscible Flooding

Liao Changlin   Liao Xinwei   Li Ju   Lu Ning

(China University of Petroleum)

Abstract: Study on the characteristic of the transition zone in CO_2-oil system has important meaning for the research of CO_2 miscible flooding. In this paper, one of crude oil samples in Xinjiang oilfield was taken as an example. The minimum miscible pressure (MMP) of CO_2-oilsystem was confirmed through laboratory experiment and numerical simulation separately. And the characteristic of the transition zone was analyzed. The transition zone size and interfacial tension in miscible process were quantified. And their variation tendencies along with the change of the pressure and CO_2 injection volume were studied. The results show that it is easier to reach miscible state in higher pressure and CO_2 injection volume. This work provides a reference for the further research of CO_2 miscible flooding.

Keywords: transition zone; MMP; CO_2 miscible flooding

0   Introduction

Research results[1-3] show that the technology of CO_2 flooding to enhance oil recovery can make up for the deficiency of water flooding and enhance oil production effectively. CO_2 flooding can enhance oil recovery 7%~23% according to the statistics[4]. It is divided into miscible flooding and immiscible flooding on basis of the principle of CO_2 flooding, and it is proved that the recovery of miscible flooding is better than the latter's. In the miscible process, the variable characteristic of the transition zone reflects the miscibility in CO_2-oil system greatly. For this reason, the characteristic of the transition zone in CO_2-oil system was studied through slim tube experiment and numerical simulation separately. It has vital significance for CO_2 miscible flooding.

1   Slim Tube Experiment

The 2328-86 type slim tube miscible instrument which was produced by RUSKA company was used in slim tube experiment. Its gas permeability and porosity was $5.43\mu m^2$ and 0.2 separately. And the injection rate was $15cm^3/h$. The oil sample was compounded by crude oil and gas from one of the oil reservoirs in Xinjiang oilfield by PVT instrument. The experimental procedure were presented as follows: (1) Compounding oil sample; (2) Saturating oil into the slim tube; (3) Injecting CO_2 into the slim tube with constant pressure; (4) Increasing pressure to repeat the process of flooding step by step until the inflection point in the curve of recovery vs. pressure appears.

Pressure automatic tracking system, three-phase automatic separating system and computer soft were used to monitor the experimental process. The volume of gas and liquid must be noticed when gas breaks through. The oil recovery is increasing with pressure according to the data analysis. But the range of increase is very small after the inflection point. In general, the condition that the recovery reach 80% when gas breaks through or its value reach 90%–95% in the inflection point is conducted as the criterion of miscible flooding[4]. So the MMP of the reservoir is 19.1MPa which is obtained from the change rule of oil recovery vs. pressure, see Fig. 1.

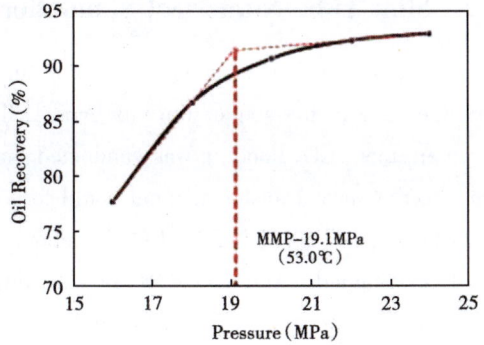

Fig. 1   Oil Recovery vs. Pressure in Slim Tube Experiment

Although slim tube experiment is the most widely used method to determine MMP currently, it need slim tube numerical simulation to verify whether it would achieve a good connection with numerical simulation. Therefore, in the process of slim tube numerical simulation, the experiment data should be considered as the reference value to fit the relevant parameters of the model to achieve the simulation object.

2   Numerical Simulation

2.1   Crude Oil PVT Matching

Reliable PVT data in the process of numerical simulation of CO_2 flooding plays an important role. The crude oil components were regrouped into 10 pseudo-components such as N_2, CO_2, C_1, C_2, C_3, C_4, C_5, C_6, C_7–C_{27}, C_{28+}, Peng-Robinson 3 state equation was used to achieve flash calculation to match experiment data such as the data of constant composition expansion, constant volume depletion and differential liberation. The state equation parameters were obtained (Table 1) and used to calculate the properties of oil and gas.

Table 1   The State Equation Parameters of Crude Oil

Comp	Mol-f(%)	MW(g/mol)	P-crit(MPa)	T-crit(K)	OmegaA	OmegaB	S-shift	V-crit	Z-crit
N_2	0.858	28.01	33.944	126.2	0.089	0.151	−0.13	0.09	0.29
CO_2	0.031	44.01	69.596	299.1	0.421	0.071	−0.05	0.09	0.26
C_1	42.09	16.04	26.197	150.7	0.210	0.014	−0.14	0.10	0.20
C_2	3.741	30.07	27.476	239.7	1.209	0.152	−0.10	0.15	0.20
C_3	1.696	44.10	24.138	286.9	0.647	0.109	−0.09	0.20	0.20
C_4	1.450	58.12	37.188	409.9	0.409	0.122	−0.08	0.26	0.28
C_5	1.501	72.15	38.195	439.5	0.686	0.089	−0.04	0.31	0.32
C_6	2.6	84	34.521	467.3	0.797	0.068	0.01	0.35	0.31
C_7–C_{27}	32.7	177.8	21.447	673.1	0.722	0.114	0.33	0.67	0.26
C_{28+}	13.333	564.2	4.796	975.4	1.363	0.146	0.76	2.26	0.13

2.2　Slim Tube Numerical Simulation

The cross sectional dimensions of slim tube were 3.86mm×3.86mm, and the other parameters were the same as the actual slim tube model. The result of the PVT matching was used as the crude oil parameters. CO_2 flooding was conducted under different pressure point. Some studies[4] showed that effective mass transfer channel could form after 0.7 PV CO_2 was injected, and the simulation could be stopped after 1.2PV CO_2 injection. Then the oil recoveries in different pressure points could be obtained. And the MMP of slim tube numerical simulation was achieved by miscibility judgment criterion. The value of the MMP is 18.6 MPa. Comparing to the experiment data, the relative error is 2.62%. It indicates that the numerical model could be used to simulate the process of slim tube experiment correctly.

2.3　Characteristic Analysis of Transition Zone

One-dimension numerical model was adopted to analyze the characteristic of transition zone. And the mixing process of CO_2-oil system was studied on basis of the variation tendency of interfacial tension. The molar density, molar fraction and parachor of the oil and gas were adopted to calculate the interfacial tension in the numerical model, see Eq.(1):

$$\sigma = \Big[\sum_{i=1}^{N_c} [P]_i (b_L^m x_i - b_v^m y_i) \Big]^4 \tag{1}$$

Here, σ is interfacial tension, $[P]_i$ is the parachor of the component i, x_i, y_i is the molar fraction of oil and gas respectively, is the molar density of oil and gas respectively.

The CO_2-oil system was achieved mixed-phase when the simulation pressure is larger than MMP. And then the residual oil saturation is very low in the area which is swept by CO_2. The transition zone can be observed when oil and gas contacted each other, as shown in Fig. 2. The interfacial tension in the front edge of the transition zone is the largest and reduces gradually from the front to the back. It indicates that CO_2 and crude oil has just started to contact and cannot reach miscibility immediately which lead higher value of interfacial tension. CO_2 extract the light components from the crude oil and then mixed. The mixture is transmitted from the front to the back with flooding. And it contacts with CO_2 ceaselessly. The interfacial tension of transition zone is reduced gradually and reached miscible phase finally. It can be observed from Fig. 2 that the transition zone moves forward with the flooding and disappears in the outlet.

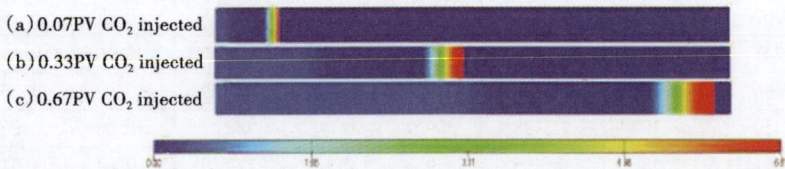

(a) 0.07PV CO_2 injected
(b) 0.33PV CO_2 injected
(c) 0.67PV CO_2 injected

Fig. 2　The Change Process of the Interfacial Tension of the Transition Zone (20 MPa)

As it is shown in Fig. 3, the width of the transition zone is increasing with the increase of the CO_2 injection volume before the mass transfer channel is built. And the transition zone disappears after the mass transfer channel is built. Meanwhile, the width of the transition zone is reducing with the increasing of displacement pressure. It indicates that the higher the displacement pressure is, the faster the CO_2-oil system reaches miscible phase. As it is shown in Fig. 4, the mass transfer channel is built when CO_2 injection volume is among 0. 7PV and 0. 8PV. And the interfacial tension in the front edge of the transition zone is increasing with the increase of the CO_2 injection volume and tends to be gentle finally. It reaches the peak before the transition zone disappears. The higher the displacement pressure is, the more obvious the Increasing tendency is. The interfacial tension in the front edge of the transition zone is reducing with the increasing of displacement pressure. The above results show that it is easier to reach miscibility in a higher pressure and larger CO_2 injection volume.

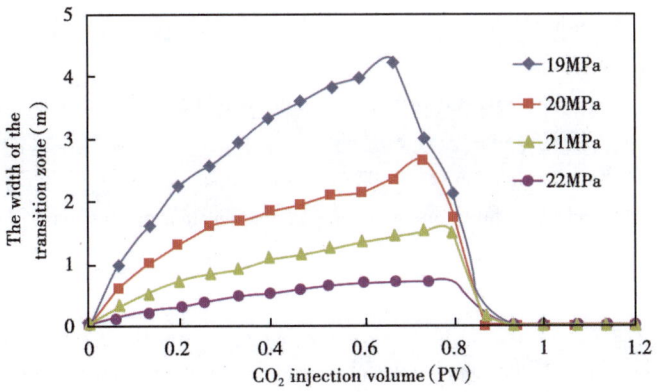

Fig. 3   The Width of the Transition Zone vs. CO_2 Injection Volume

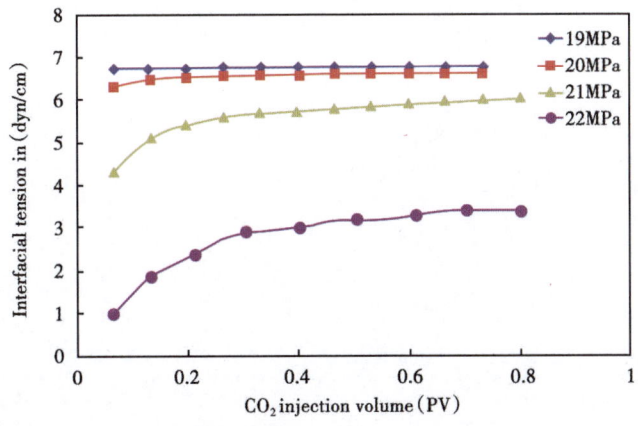

Fig. 4   The Interfacial Tension in the Front Edge of the Transition Zone vs. CO_2 Injection Volume

3   Conclusions

(1) The MMP of one of oil samples in Xinjiang oilfield obtained by slim tube experiment and numerical simulation is 19. 1 MPa and 18. 6 MPa respectively. The relative error between them is

2. 62% which indicates that the numerical model can simulate the process of slim tube experiment correctly.

(2) CO_2 and crude oil cannot reach miscible phase when they just start to contact which lead to higher value of interfacial tension. The transition zone moves forward with flooding and disappears in the outlet.

(3) The width of the transition zone is increasing with the increase of the CO_2 injection volume and is reducing with the increase of displacement pressure before the mass transfer channel is built. It indicates that it is easier to reach miscible phase in higher pressure and larger CO_2 injection volume.

References

[1] Shen P P, Liao X W. The technology of carbon dioxide stored in geological media and enhanced oil recovery [M]. Petroleum Industry Press, 2009: 47-52.

[2] Todd M, Longstaff W. The development, testing and application of a numerical simulator for predicting miscible flood performance [C]. SPE 3484, 1972: 874-882.

[3] John S A, John N D. The deterioration of miscible zones in porous media [C]. SPE 904-G, 1957.

[4] Guo P, Yang X F. Minimum miscibility pressure study of gas injection in reservoirs [M]. Beijing: Petroleum Industry Press, 2005.

[5] Tan J, Bentsen R G. A new mathematical model for 1D miscible displacement [J]. Journal of Canadian Petroleum Technology, 1999.

[6] Rao D N. A new technique of vanishing interfacial tension for miscibility determination [J]. Fluid Phase Equilibria, 1997 (139): 311-324.

[7] Hua Y, Johns R T. Simplified method for calculation of minimum miscibility pressure or enrichment [C]. SPE 77381, 2002.

CO$_2$—原油细管试验混相判定新方法

黄海东　陈百炼　章　杨　谢新春　任韶然

（中国石油大学（华东）石油工程学院）

摘　要：最小混相压力（MMP）是 CO$_2$ 混相驱设计中的一个重要参数，细管试验是其标准确定方法。通过常规细管试验及 MMP 值附近的加密试验，发现采收率在 MMP 附近呈曲线变化，使得采用线性拟合取交点判定 MMP 的方法具有较大主观性，且结果会偏小。此外，原油性质及装置设计会明显影响最终采收率，不可将采收率作为混相评价标准。以改进的三次样条插值方法为基础，提出了以插值曲线曲率最小处为 MMP 的新判定方法，并采用新方法重新处理了文献中的细管试验数据，将计算结果与文献给出值进行了对比。结果表明，传统的线性拟合取交点的判定方法使结果平均偏小 1.4MPa，而新方法可消除主观因素的影响，更适合作为标准判定方法。

关键词：CO$_2$ 驱；细管试验；最小混相压力；判定准则；三次样条插值

0　引言

CO$_2$ 驱是一种有效的提高采收率方法[1,2]，大庆、胜利、吉林等油田均已开展 CO$_2$ 驱室内实验、矿场试验[3]。在 CO$_2$ 混相驱的应用中，最小混相压力（MMP）的确定是非常重要的[4,5]。细管试验最能反映 CO$_2$ 驱替原油的过程，是确定 MMP 的标准方法[1,4,6,7]。细管试验中常用的混相判定方法有：（1）CO$_2$ 突破时采收率达到 80% 或最终采收率达到 90%~95%；（2）将试验数据点分为混相与非混相两部分，认为两部分采收率均线性变化，分别做直线拟合，两直线交点处对应的值认为是最小混相压力。在应用过程中，发现前一种方法标准不够明确，后一种方法则会遇到试验数据明显非线性、与该方法的假定前提不符的问题。针对以上问题，笔者通过细管加密试验及对三次样条插值数学方法的运用，提出了一种更加标准化的 CO$_2$—原油细管试验混相评判标准。

1　细管试验的传统 MMP 判定方法

为了对传统 MMP 判定方法进行分析，对吉林油田 3 个原油样品进行了细管试验（图 1、图 2），并对大老爷府区块原油样品在接近 MMP 的驱替压力下进行了细化加密试验。

1.1　判定方法一：以采收率为标准

该方法的判断标准为：CO$_2$ 突破时采收率达到 80% 或最终采收率达到 90%~95% 的驱替压力为 MMP。

分析发现，该方法存在以下几个缺陷：

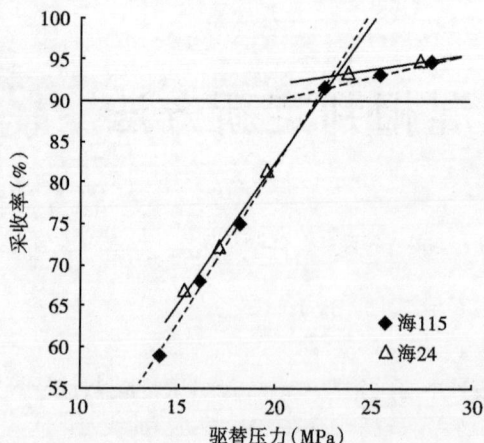

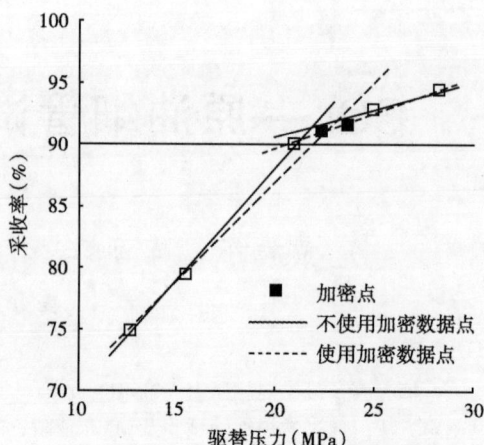

图 1　海 115 区块与海 24 区块原油细管试验结果　　　　图 2　大老爷府区块原油细管试验结果

（1）如果以 CO_2 突破时采收率达到 80% 为标准，则存在难以直接通过试验找到该点的问题（细管试验中能够直接控制的是压力），这就需要试验者根据经验进行估计，这会造成一定误差；而如果进一步加密试验点则会明显加大试验工作量。例如，对于图 1 中的海 24 区块油样试验结果，在采收率 82%~93% 范围内没有数据点，需要估计一个值作为 MMP，而该区间对应的驱替压力差高达 4.2MPa，难以准确估计 CO_2 突破时采收率为 80% 的驱替压力。

（2）如果以最终采收率达到 90%~95% 为标准，则存在该区间内驱替压力范围较大的问题。例如，图 1 中海 24、海 115 区块原油的试验结果中，采收率介于 90%~95% 的驱替压力范围分别为 22~28MPa 和 22~29MPa，而对于在该压力范围内取何具体值则没有相关的方法。

（3）采收率较大程度地受到原油性质、试验装置设计和操作条件等的影响[1,6,8-10]，单一的采收率值难以适用于不同的试验结果。如管长在 745~2980mm 之间变化时，相应采收率变动范围为 87%~94%。此外，细管直径、驱替速度和孔隙度等都会明显影响最终采收率的大小[9]。

1.2　判定方法二：线性拟合取交点

该方法的判断标准为：根据采收率上升快慢将数据分为混相与非混相两部分，分别做线性拟合，以两条拟合直线的交点对应的值为 MMP。这是目前文献中[3,7,9]最常用的方法。

该准则假定混相前后采收率的变化均为线性，在最小混相压力处发生突变。但结合试验结果对该方法进行分析，发现驱替压力接近 MMP 时会出现采收率变化的明显曲线段，即采收率的变化率是渐变而非突变，理论分析也证实了该现象[5]。这说明该方法的假设不合理，这会导致以下问题：

（1）判断结果的不确定性。由图 2 可见，在驱替压力为 22.3MPa 和 23.6MPa 处进行的细化加密试验数据点与前后的试验数据点都不能呈现出较好的线性关系，如果将其归入未混相部分或者混相部分，则相应部分的线性拟合结果都会与不使用细化加密试验数据点的拟合直线有较大差异，进而对判定结果产生较大影响。此时需要试验者根据经验进行判断，即是

46

否将其加入混相或者非混相数据点，或者将该点舍弃。由此可见，该现象使判断结果具有不确定性。

（2）取值点偏向于非混相部分。传统判断方法相当于取拟合直线交点处的垂线与采收率曲线的交点（如图3中点A所示）为临界点，取其驱替压力为MMP，但从图3中可见，该点明显偏向于非混相部分。由于实际采收率变化为曲线，该变化过程的临界点应该位于曲线段中部某位置（如图3中点B所示）。也就是说，这种方法会造成得到的MMP比实际值偏低。

由此可见，以上传统方法均存在明显缺点。因此，提出一种具有明确物理意义和规范化处理过程的新方法是很有必要的。

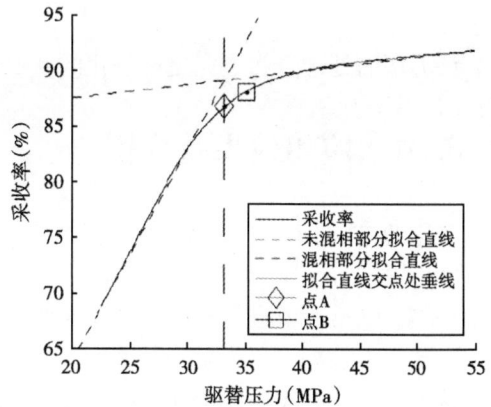

图3　取拟合直线交点处压力造成的误差

2　新方法的提出

CO_2—原油体系细管试验混相判定新方法判断过程如下：

（1）利用预插值及三次样条插值方法对试验数据点进行插值。

三次样条插值是一种分段插值方法，可以离散数据点为插值节点，得到一条符合数据点变化规律的光滑曲线。该方法在插值节点间采用三次多项式进行插值，在节点处附加二阶连续可导的条件，可以保证插值结果的稳定性。

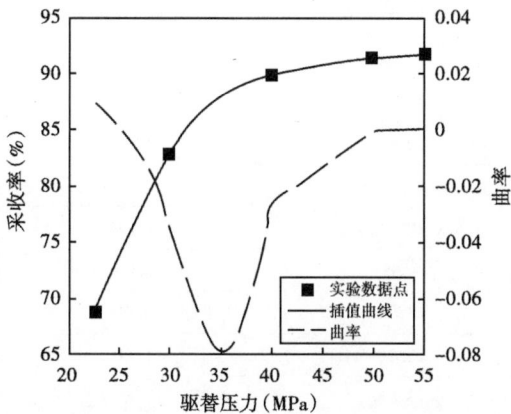

图4　改进三次样条插值法应用示意图

三次样条插值方法求解时需要附加边界条件，而细管试验并不能直接提供该条件。通过判定方法二分析可知，采收率在MMP附近呈曲线变化，但是在数据点的两端驱替压力远小于或者大于MMP时，采收率与驱替压力呈现良好的直线关系。因此，可以进行线性预插值得到两端斜率作为边界条件，即在试验数据两侧各取两个数据点作线性插值，取插值直线的斜率作为最低、最高驱替压力数据点处的斜率，以此作为边界条件再进行三次样条插值。插值结果示意图如图4所示。

（2）计算各处曲率变化。

由判定方法二分析可知，在非混相或者混相驱替中采收率变化均接近线性，也就是说，采收率插值曲线弯曲程度较小；MMP是CO_2驱替过程从非混相驱变为混相驱的临界点，在该临界点处采收率从快速上升转变为缓慢上升，则MMP处的曲线弯曲程度应该是最大的。曲线弯曲程度可以用曲率κ表示：

$$\kappa = \frac{f''}{(1 + f'^2)^{3/2}}$$

式中，f'' 为插值曲线的二阶导数；f' 为插值曲线的一阶导数。

计算结果如图 4 所示。曲率数值越小，说明插值曲线弯曲程度越大，则曲率最小处即对应于 MMP。

该方法的处理过程是完全规范化的，试验结论不会受到主观因素的影响。

3 新方法应用效果及分析

应用新方法对一些文献中的细管试验数据[3,6,9,11-19]进行了重新处理（表 1）。将得到的结果与文献给出值进行对比，如图 5 所示。

由图 5 中数据点拟合直线公式可知：直线斜率接近于 1，直线截距为 1.3825MPa，说明传统方法的判断值比新方法平均偏小约 1.4MPa，这恰好验证了判定方法二的分析，即利用传统判定方法得到的 MMP 会比真实值偏小，证明了新方法的合理性。

此外，应该注意到数据点分散于拟合直线

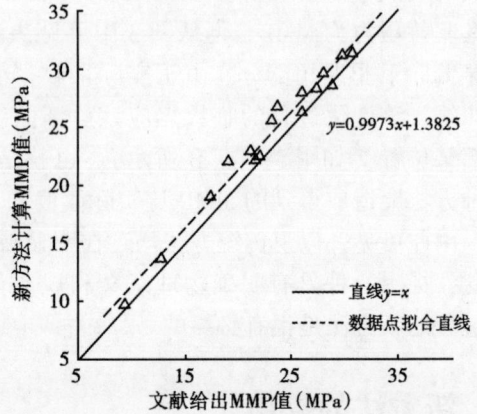

图 5 改进三次样条插值法与文献给出 MMP 值的对比

两侧，说明不同人的判断标准不同，使得传统方法中含有较大的主观因素，导致 MMP 判断结果有或高或低的波动。而新方法不会受到主观因素影响，消除了这种波动，更适合作为标准方法。

表 1 新方法与传统判定方法结果对比

来源文献	文献给出值（MPa）	新方法计算结果（MPa）	偏差（MPa）
郝永卯 等	26	26.30	0.30
徐辉 等	22.11	22.62	0.51
曾悠悠 等	28.13	29.60	1.47
郭龙	30.76	31.40	0.64
徐阳	28.9	28.80	−0.10
R. S. Metcalfe 等	9.37	9.85	0.48
M. Dong 等	12.8	13.70	0.90
M. Dong 等	17.5	19.10	1.60
M. Dong 等	21.2	23.08	1.88
郭平 等	23.82	26.82	3.00
郭平 等	23.15	25.60	2.45
郭平 等	19.12	21.98	2.86
郭平 等	30.01	31.34	1.33
郭平 等	33.5	35.14	1.64
苗国锋	21.7	22.30	0.60
巢忠堂 等	27.34	28.36	1.02
郭庆安	26.15	28.00	1.85

4 结论

（1）由于原油性质、试验装置设计、操作条件等影响，细管试验最终采收率各不相同，不能用采收率来衡量是否混相。

（2）细管试验中采收率的变化是渐变过程，对非混相、混相部分分别进行直线拟合的方法存在结果不确定性。

（3）对传统直线拟合方法的分析表明其判定结果会偏小，新方法的应用分析表明传统的直线拟合方法判定结果比新方法平均偏小 1.4MPa，两处结论相符，验证了新方法的合理性。

（4）新方法有明确物理意义和规范化处理过程，能较好描述采收率渐变的特征，可以消除主观因素带来的结果波动，适合作为细管试验中 MMP 判定的标准方法。

参 考 文 献

[1] 郝永卯，陈月明，于会利. CO_2 驱最小混相压力的测定与预测 [J]. 油气地质与采收率，2005，12（6）：64-66.

[2] 李春芹. CO_2 混相驱技术在高 89-1 块特低渗透油藏开发中的应用 [J]. 石油天然气学报，2011，33（6）：328-329.

[3] 郝永卯，薄启炜，陈月明. CO_2 驱油实验研究 [J]. 石油勘探与开发，2005，32（2）：110-112.

[4] 孙业恒，吕广忠，王延芳，等. 确定 CO_2 最小混相压力的状态方程法 [J]. 油气地质与采收率，2006，13（1）：82-84.

[5] Dzulkamain I, Awang M B, Mohamad A. Uncertainty in MMP prediction from eos fluid characterization [A]. Proceeding of SPE Enhanced Oil Recovery conference [C]. 2011.

[6] 郭平，张恩永，吴莹，等. 大港油田二氧化碳驱最小混相压力测定 [J]. 西南石油学院学报，1999，21（3）：19-21.

[7] Randall T E, Bennion D B. Laboratory factors influencing slim tube test results [J]. Journal of Canadian Petroleum Technology, 1989. 28 (4): 60-70.

[8] Rathmell J J, Stalkup F I, Hassinger R C. A laboratory investigation of miscible displacement by carbon dioxide [A]. Proceeding of 46th Annual Fall Meeting of the Society of Petroleum Engineers of AIME [C]. 1971.

[9] 郭平，杨学锋，冉新权. 油藏注气最小混相压力研究 [M]. 北京：石油工业出版社，2005.

[10] Orr Jr F M, Silva M K, Lien C L. Equilibrium phase compositions of CO_2/crude oil mixtures-part 2: comparison of continuous multiple-contact and slim-tube displacement tests [J]. Society of Petroleum Engineers Journal, 1983, 23 (2): 281-291.

[11] 徐辉，聂军，廖顾舟. 台兴油田阜三段油藏 CO_2 驱油最低混相压力确定实验研究 [J]. 石油地质与工程，2010，24（1）：113-114.

[12] 曾悠悠，欧成华，陈伟，等. 注 CO_2 混相驱最小混相压力模拟研究 [A]. 成都：2009 年油气藏地质及开发工程国家重点实验室第五次国际学术会议，2009.

[13] 郭龙. 渤南油田义 34 块特低渗透油藏二氧化碳混相驱实验 [J]. 油气地质与采收率，2011，18（1）：37-40.

[14] 徐阳. 低渗油藏 CO_2 近混相驱提高采收率机理研究 [D]. 中国石油大学（华东），2011.

[15] Metcalfe R S. Effects of impurities on minimum miscibility pressures and minimum enrichment levels for CO_2 and rich-gas displacements [J]. Society of Petroleum Engineers Journal, 1982, 22 (2): 219-225.

[16] Dong M，Huang S，Dyer S B，et al. A comparison of CO_2 minimum miscibility pressure determinations for Weybum crude oil ［J］. Journal of Petroleum Science and Engineering，2001，31（1）：13-22.

[17] 苗国锋. 龙虎泡油田高台子油层二氧化碳驱最小混相压力研究 ［J］. 石油地质与工程，2010，24（3）：70-72.

[18] 巢忠堂，陈其荣，刘爱武，等. 注 CO_2 提高采收率机理室内研究 ［J］. 江汉石油学院学报，2003.25（S2）：66-67.

[19] 郭庆安. 松南气田 CO_2 驱油试验研究 ［J］. 石油天然气学报，2010，32（4）：130-132.

CO$_2$ 驱替沥青质原油长岩心实验及数值模拟

黄 磊[1] 贾 英[2] 全一平[3] 孙 雷[4]

(1. 中国石油勘探开发研究院；2. 中国石化石油勘探开发研究院；
3. 中国海洋石油国际有限公司；4. 西南石油大学石油工程学院)

摘 要：针对 CO$_2$ 驱油过程中出现的沥青质固相沉积现象，综合考虑注气过程中沥青质在多孔介质中的沉积、吸附及堵塞效应，建立了 CO$_2$ 驱替含沥青质原油多相多组分渗流模型。以长岩心实验为基础，拟合实验数据表明，使用本文建立的模型计算的采收率与实验值更接近。同时对 CO$_2$ 气驱过程中沥青质沉积规律进行了探讨，模拟研究表明，注入 CO$_2$ 过程中，沥青质沉积首先发生在注入端；沥青质吸附也首先发生在注入端，但随着注入气量的增加，存在解吸附过程；沥青质沉积前缘与注气前缘一致，说明注入 CO$_2$ 是导致原油中沥青质沉积的主要原因。

关键词：沥青质；注 CO$_2$；沉积规律；数值模拟

0 引言

近年来，CO$_2$ 驱油技术成为提高原油采收率研究及应用领域的热点，不仅源于 CO$_2$ 降低原油黏度、降低油气两相界面张力、膨胀原油以及可能发生的混相作用等[1]有利于提高原油采收率的因素，还归因于驱替过程中 CO$_2$ 在油藏的滞留作用并由此带来的温室气体减排效应。然而，注 CO$_2$ 可能引发重质原油，特别是含沥青质原油中沥青质组分沉淀，沉淀的沥青质可能沉积、吸附在岩石内表面，造成地层堵塞，严重影响生产。因此，开展 CO$_2$ 驱替沥青质原油固相沉积规律研究对生产实践过程中应对沥青质沉积带来的影响具有一定指导意义。

本文以长岩心驱替实验为基础，采用前期研究建立的 CO$_2$ 驱替沥青质原油多相多组分渗流模型进行数值模拟[2-4]，通过拟合实验结果对模型的相关热力学参数进行修正，并以此为基础讨论了沥青质沉积吸附分布规律。

1 CO$_2$ 驱替含沥青质原油长岩心实验

CO$_2$ 驱替含沥青质原油长岩心实验模拟了油藏条件下，注气速度及压力对原油采收率的影响。

1.1 实验用油样基本性质

实验用油样为井口油气配样，油样所在地层压力为 15MPa，地层温度为 66℃，饱和压力为 11.1MPa，溶解气油比为 39.65m^3/m^3，地层油体积系数为 0.97，地层油密度为 0.79g/

cm³，地层油黏度为 6.6mPa·s。单次脱气后死油密度为 0.86g/cm³，脱气油含蜡量为 19.98%，沥青质含量为 27.18%。地层油中 CO_2 含量为 0.97%，C_1+N_2 含量为 30.88%，C_2—C_{10}组分含量为 14.16%，C_{11+} 为 53.99%。配样具有一定代表性。

1.2 实验设备及测试结果

1.2.1 实验设备及流程

长岩心驱替实验是在加拿大 Hycal 长岩心驱替装置上完成的，实验装置示意见图 1。

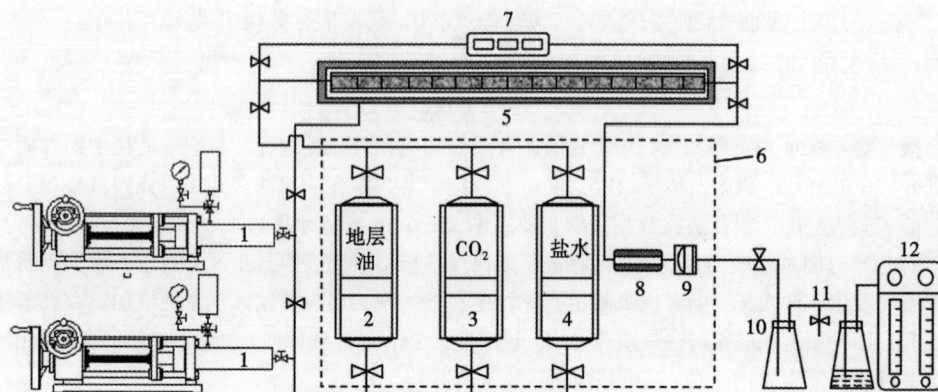

图 1 长岩心驱替实验装置示意图

1—驱替泵；2—地层油容器；3—CO_2 气容器；4—盐水容器；5—长岩心夹持器；6—恒温箱；
7—压力传感器；8—观察窗；9—回压阀；10—分离瓶；11—取样口；12—气量计

1.2.2 实验结果

长岩心实验结果见表 1。实验结果表明：CO_2 注入量相近时，体系压力的增加有利于提高原油采收率，但是二者不呈线性关系。当回压从 10MPa 升至 15MPa 时，采收率提高了 4.62%；当回压从 15MPa 升高至 20MPa 时，采收率提高了 13.69%，压力增加导致的采收率增加效果非常明显；不过压力继续升高时，采收率增加的幅度显著减小，回压从 20MPa 升高至 25MPa 时，采收率仅提高 1.17%。

表 1 长岩心实验结果

岩心编号	岩心长度 （cm）	平均渗透率 （mD）	平均孔隙度 （%）	回压 （MPa）	注入孔隙体积倍数 （HCPV）	采收率 （%）
1	43.2	8.9	17.1	10	3.3	59.8
2	42.9	8.9	16.9	15	3.1	64.4
3	45.8	8.7	17.2	20	3.1	78.1
4	43.2	8.9	17.1	25	2.9	79.2

受探测手段的限制，目前长岩心驱替实验仅能考察诸如出口端回压、采收率之类的宏观参数，难以定量地确定每块岩心中沉积的沥青质含量。为获得长岩心中沉积沥青质的分布规律，需要借助数值计算手段，模拟沥青质在岩心中的沉积动态。

2 CO₂ 驱替沥青质原油多相多组分渗流模型

目前，关于沥青质具体组分方面的研究尚无定论，不过在沥青质沉积方面，人们普遍认为沥青质堵塞储层孔隙具有两种可能性：（1）沥青质在喉道处积聚、阻塞（图2）；（2）沥青质在岩石孔隙表面上吸附沉积（图3）。

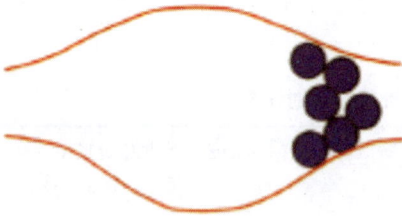

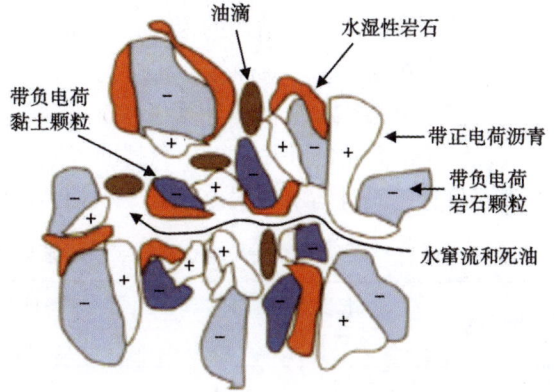

图 2　沥青质孔喉阻塞机理　　　　　图 3　沥青质沉积吸附（据 Lenontaritis 等）

Dubey[5] 和 Gonzalez[6] 等通过实验证实了沥青质在多孔介质内表面吸附应服从单分子层吸附规律（Langmuir Ⅰ型）。本文参考 L. X. Nghiem 建立的沉积模型[7]，将沉淀的沥青质固相 N_s 分为 3 部分：吸附在岩石孔隙表面的固相沥青质 N_{sa}；在孔隙喉道处被机械捕获的同相沥青质 N_{se}；以悬浮态存在于油相中的同相沥青质 N_{sf}。

$$N_s = N_{sa} + N_{se} + N_{sf} \tag{1}$$

上式各项计算请参考文献 [2，4]。

关于沥青质对多孔介质的堵塞效应的处理方法以及综合考虑沥青质在多孔介质中的沉积、吸附及堵塞效应，建立的 CO₂ 驱替含沥青质原油多相多组分渗流模型请参考文献 [8]，在此不再赘述。使用 Cernansky 和 Siroky[9] 建立的沉积量与渗透率关系式模拟堵塞过程：

$$1 - \frac{\hat{\phi}}{\phi_o} = E\left(1 - \frac{K}{K_o}\right) + \frac{1 - E}{\exp G - 1}\left\{\exp\left[G\left(1 - \frac{K}{K_o}\right)\right] - 1\right\} \tag{2}$$

$$\hat{\phi} = \phi - (N_{sa} + N_{se}) \cdot u_s \tag{3}$$

式中，$\hat{\phi}$ 为未被沉积吸附的沥青质占据的孔隙度；ϕ_o 为未被沉积吸附时的孔隙度；K 为孔隙度为 $\hat{\phi}$ 时的渗透率；K_o 为基质渗透率；E 为未被固相堵塞的孔隙空间系数，为可调参数；G 为被固相堵塞的孔隙空间系数，为可调参数。

将建立的 $2n_c + 3$ 个方程进行差分离散化处理，形成带有 $2n_c + 3$ 个未知变量方程组，求解变量为 P，N_1，N_2，…，N_{n_c}，N_{1g}，N_{2g}，…，$N_{n_{cg}}$，通过 Newton 法迭代求解。

3 CO_2 驱替沥青质原油长岩心实验结果拟合及沥青质沉积规律

应用第 2 节所述 CO_2 驱替含沥青质原油多相多组分渗流模型模拟 CO_2 驱长岩心实验过程，在对相关实验参数拟合的基础上，讨论岩心各区域沥青质吸附沉积的分布规律。

3.1 实验结果拟合

首先开展长岩心实验结果的拟合，确保模型的合理可靠性。根据实验基础数据，建立长岩心模型。在拟合 CO_2—原油体系相态的基础上，分别应用常规组分模型和 CO_2 驱替沥青质原油多相多组分渗流模型模拟计算了 CO_2 驱替含沥青质原油长岩心实验的采收率，计算结果见表 2。由表 2 可知，采用 CO_2 驱替沥青质原油多相多组分渗流模型计算得到的采收率与实验值更接近，并且与实验误差小于 5%，模型可用于下一步沥青质沉积规律探讨。

表 2 长岩心驱替实验结果与数值模拟计算结果对比表

岩心编号	岩心长度 (cm)	回压 (MPa)	实验测得采收率 (%)	本文数值模拟计算的采收率 (%)	常规组分模型计算的采收率 (%)
1	43.2	10	59.8	58.5	63.3
2	42.9	15	64.4	64.1	67.9
3	45.8	20	78.1	80.3	82.8
4	43.2	25	79.2	82.1	84.3

3.2 长岩心实验沥青质沉积规律

由于 CO_2 注入导致系统热力学平衡被打破而沉淀析出的沥青质在岩石孔隙中有 3 种存在状态：较小颗粒的沥青质固相跟随液相一起流动，由于吸附作用沉积在岩石孔隙表面以及由于机械捕获作用积聚在孔喉处。其中，吸附沉积的沥青质以及孔喉处聚集的沥青质会改变孔隙的空间结构从而影响原油的流动，降低 CO_2 驱替效率。因此，本文从机械捕获导致的沉积及吸附导致的沉积两方面入手，探讨沥青质固相在长岩心中的分布规律。

3.2.1 机械捕获作用导致沉积的沥青质固相在长岩心中的分布

提取有代表性的网格（注入端、出口端、长岩心内部第 2，3，5，10 及 15 号网格），作出上述各个网格中因机械捕获导致沉积的沥青质质量随 CO_2 注入量的变化关系图（图 4）。图 5 显示了不同 CO_2 注入量时长岩心内各网格的沥青质沉积质量。从图 4 可以看出，当注入 0.1 HCPV CO_2 时，注入端由于机械捕获作用导致沉积的固相沥青质迅速增长，随着 CO_2 注入量的增加，注入端沉积的沥青质保持稳定；当注入 0.9 HCPV CO_2 之后，注入端沉积沥青质逐渐减少。紧邻注入端的第二个网格内，随着 CO_2 注入量增加沉积的沥青质含量先增加后稳定；当注入端沥青质沉积量减少时，第二网格中的沉积沥青质质量开始增加。

上述结果表明，CO_2 的持续注入使得出口端沉积的沥青质又重新溶解到原油中并跟随原油一起流动，并在接下来的位置发生了再沉积。从图 5 可以看出，注入端附近，沉积沥青质含量最多；随着注入 CO_2 增加，中间及出口端逐渐出现沉积沥青质。

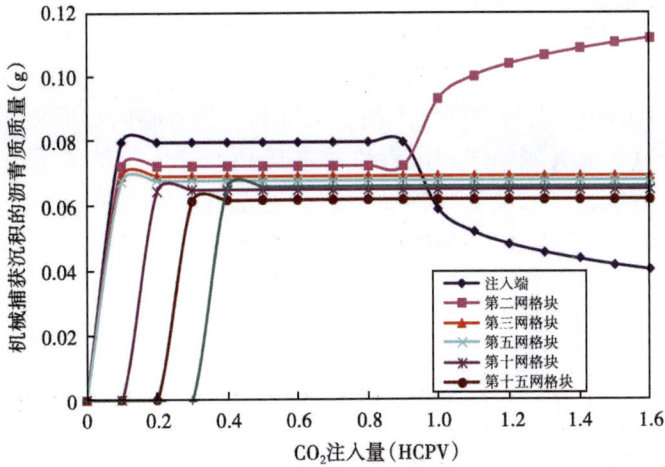

图 4 代表网格内沉积沥青质质量与 CO_2 注入量的关系

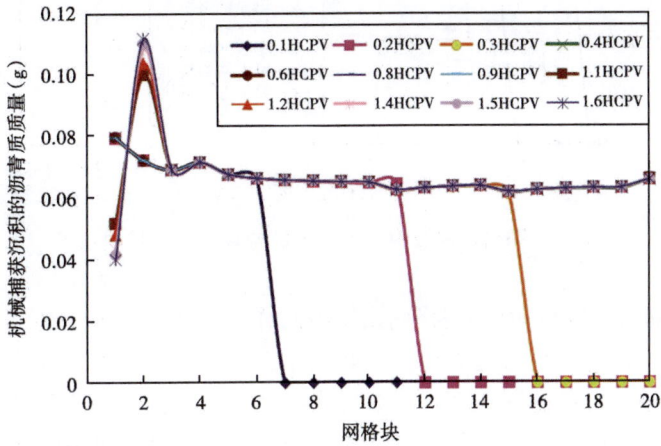

图 5 不同 CO_2 注入量时各网格沥青质沉积质量

上述计算结果显示的，沥青质沉积动态与 Satoru Takahashi 等人[8]岩心驱替实验数据及 R. K. Srivatava 等[10]CAT 岩心扫描图所反映的长岩心沉积沥青质分布规律吻合。

图 6 为不同 CO_2 注入量情况下气相饱和度分布与沥青质沉积量分布的对应关系图。由

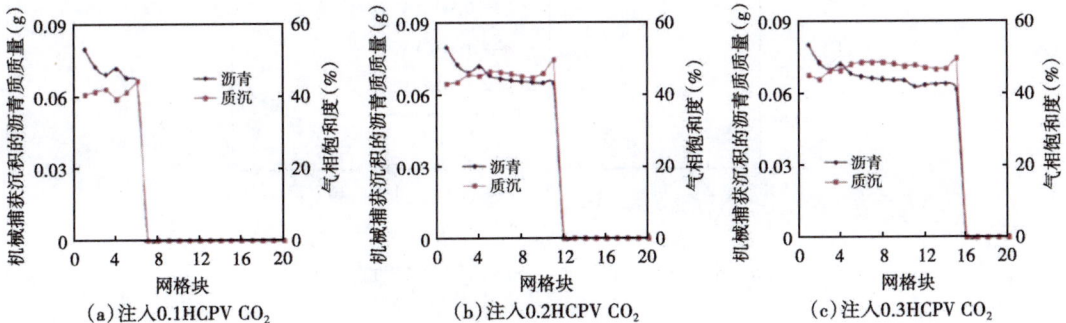

（a）注入0.1HCPV CO_2　　　　（b）注入0.2HCPV CO_2　　　　（c）注入0.3HCPV CO_2

图 6 不同网格内沥青质沉积质量及气相饱和度

图 6 可知，注入 CO_2 突破前，沥青质的沉积前缘与注气前缘相一致，CO_2 的引入是导致原油中沥青质沉积的主要原因。

3.2.2 吸附作用导致沉积的固相沥青质分布

图 7 描述了选取的代表网格内（注入端、出口端、长岩心内部第 2，3，5，10 及 15 号网格）由于吸附作用导致沉积的固相沥青质质量与 CO_2 注入量的关系，从图 7 可以看出，随着 CO_2 注入量的增加，注入端吸附的沥青质含量先迅速增加然后稳定下来再逐渐减少。中间网格块内随着注入过程的不断持续而逐渐吸附沥青质，出口端最后才出现吸附作用导致沉积的沥青质。

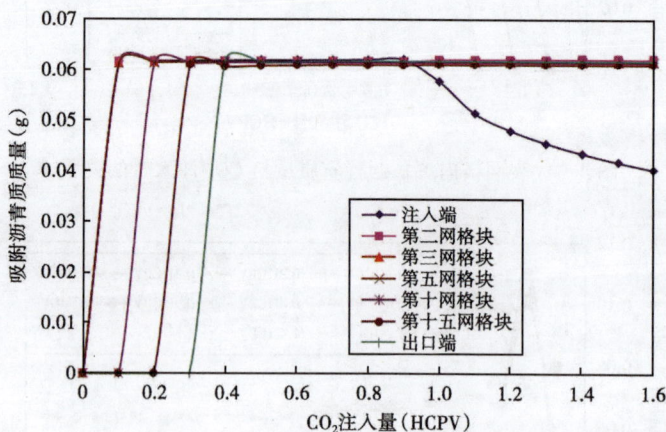

图 7　各网格块内吸附沥青质质量与 CO_2 注入量关系图

图 8 为注入不同量的 CO_2 时各网格内由于吸附作用导致沉积在孔喉表面的沥青质质量。从图 8 中可以看出，在气体突破之前（注入 0.4 HCPV 之前）吸附导致沉积的固相沥青质分布规律与机械捕获导致沉积的固相沥青分布规律类似。但是，随着注气量的增加，由于解吸附效应的存在，注入端的吸附沥青质含量会在达到最大值后呈现出不断减少的趋势。

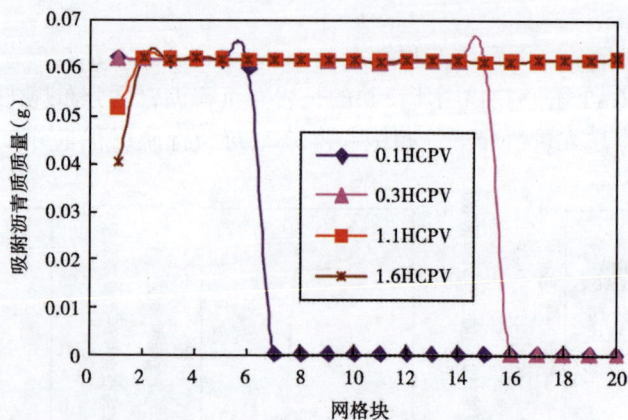

图 8　不同 CO_2 注入量下各网格内吸附沥青质质量

4 结论

（1）CO_2 注入过程中，机械捕获导致的沥青质沉积首先发生在注入端，并逐渐向驱替下游扩展；不过，注入端附近，沥青质沉积最严重。

（2）CO_2 注入过程中，沥青质吸附也首先发生在注入端，并逐渐向出口端延伸，但随着注入气量的增加，存在解吸附效应。

（3）沥青质的沉积前缘与注气前缘一致，表明注入 CO_2 是导致原油中沥青质沉积的主要原因。

（4）在注 CO_2 过程中，应重视沥青质沉积吸附对驱替的影响，特别是在注入井附近，沥青质的沉积吸附可能堵塞井筒附近的储渗空间，影响 CO_2 的驱替效率。建议在施工过程中，注入化学剂减轻沥青质沉积危害。

参 考 文 献

[1] 李士伦，张正卿，冉新权，等. 注气提高石油采收率技术 [M]. 成都：四川科学技术出版社，2001.

[2] 贾英，孙雷，孙良田，等. 注 CO_2 过程中含沥青质原油渗流规律数值模拟研究 [J]. 石油勘探与开发，2007，34（6）：734-739.

[3] Minssicux L. Removal of asphalt deposits by cosolvent squeeze：mechanisms and screening [C]. SPE 69672, 2001.

[4] 黄磊，沈平平，贾英，等. CO_2 注入过程中沥青质沉淀预测 [J]. 石油勘探与开发，2010，37（3）：349-353.

[5] Dubey S T, Waxman M H. Asphaltene adsorption and desorption from mineral surfaces [C]. SPE 18462, 1991.

[6] Geozalez G，Moreira M B C. The adsorption of asphaltenes and resins on various minerals // Asphaltenes and Asphalts I [M]. Elsevier Science, 1994.

[7] Nghiem L X，Kohse B F，Farouq S M，et a1. Asphaltene precipitation：phase behavior modeling and compositional simulation [C]. SPE 59432, 2000.

[8] Satom T，Yoshihisa H，Shunya T，et a1. Characteristics and impact of asphaltene precipitation during CO_2 injection in sandstone and carbonate cores：an investigative analysis through laboratory tests and compositional simulation [C]. SPE 84895, 2003.

[9] Cernansky A，Siroky R. Deep-bed filtration on filament layers on particle polydispersed in liquids [J]. Int. Chem. Eng., 1985（2）：365-375.

[10] Srivatava R K，Huang S S，Dong Mingzhe. Asphaltene deposition during CO_2 flooding [C]. SPE 59092, 1999.

CO_2 驱地下流体相态特征研究

姜凤光[1]   胡永乐[2]

(1. 中国石化石油勘探开发研究院；2. 中国石油勘探开发研究院)

摘　要：目前国内注入 CO_2 提高采收率开发项目由于矿场试验周期短，对开发机理和气驱油特征规律认识还不够深入，尤其对 CO_2 驱替过程中的流体相态特征研究鲜有报道。通过油藏工程方法和数值模拟方法，利用相平衡计算模型和 PVTi 流体分析软件包，分析 CO_2 含量对地下流体相态的影响。研究表明，对于任何 CO_2—原油体系都存在临界点，在临界点下方，当低于一定温度时，随着 CO_2 注入量的增加，在一定的压力范围内，CO_2 与原油体系会出现三相区；同时 CO_2 含量对 CO_2—原油体系黏度及体积膨胀系数有很大的影响，随着 CO_2 溶解量的增加，CO_2—原油体系黏度降低，膨胀系数增加。该研究可对 CO_2 驱油藏开发起到一定的指导作用。

关键词：CO_2 驱；相平衡；采收率；流体；相态特征；三相驱

0　引言

为实现油气增产和 CO_2 减排的双赢效果，将 CO_2 气体注入油藏提高原油采收率已成为 CO_2 资源化利用的重要途径之一[1-3]。CO_2 驱油过程主要包括混相驱和非混相驱，其中混相驱可以形成稳定的驱油带，微观驱替效率可达 90% 以上，而非混相驱提高采收率幅度较低。目前国内注 CO_2 提高采收率开发项目由于矿场试验周期短，对开发机理和气驱油特征规律认识不清，现场应用效果远低于预期试验结果。因此，利用油藏工程和数值模拟方法，研究不同 CO_2 含量下的 P–X 相图及相态包络线，分析 CO_2 含量对地下流体相态特征和渗流特征的影响，有利于更好地发挥 CO_2 驱提高采收率的优势[4-6]。

1　相平衡计算模型

在任一给定的油藏压力 p 和油藏温度 T 状态下，体系中气、液摩尔分数满足 $n_g + n_1 = 1$。根据流体相平衡热力学原理，建立相平衡计算热力学模型[7-10]：

$$\begin{cases} F_1(x_i, y_i, p, T) = f_{1_1} - f_{1_g} = 0 \\ F_2(x_i, y_i, p, T) = f_{2_1} - f_{2_g} = 0 \\ \cdots\cdots\cdots\cdots \\ F_i(x_i, y_i, p, T) = f_{i_1} - f_{i_g} = 0 \\ \cdots\cdots\cdots\cdots \\ F_n(x_i, y_i, p, T) = f_{n_1} - f_{n_g} = 0 \\ F_{n+1}(x_i, y_i, p, T) = \sum \dfrac{z_{0i}(k_i - 1)}{1 + (k_i - 1)n_g} = 0 \end{cases} \quad (1)$$

其中：
$$k_i = y_i / x_i$$

式中，p 为油藏压力，MPa； T 为油藏温度，℃；n_g，n_l 为气、液相摩尔分数；i 为组分数；$i=1$，2，…，n；x_i 为气相中 i 组分的物质的量；y_i 为液相中 i 组分的物质的量；z_{0i} 为油气体系原始组成；k_i 为平衡常数；F_i 为气、液相逸度相等平衡条件目标函数；F_{n+1} 为气、液相组成归一化平衡条件目标函数（$\sum(y_i - x_i) = 0$）。

2 CO_2—原油体系相态特征研究

纯 CO_2 的相态主要受压力、温度的影响[11-13]，随着温度和压力的变化，CO_2 呈现出气态、液态、固态和超临界态 4 种状态，CO_2 的临界温度为 31.06℃，临界压力为 7.39 MPa。当 CO_2 注入油藏，CO_2—原油体系相态除了受温度、压力影响外，还受 CO_2 摩尔分数（用符号 C 代替）的影响。以普鲁德霍湾 MilnePoint 油田 Kupa 储层为算例[14]，该油藏原始地层压力为 12.07MPa，地层温度为 30℃，饱和压力为 6.18MPa，原油黏度为 3mPa·s。利用 PVTi 流体分析软件包，研究不同 CO_2 摩尔分数下 CO_2—原油体系流体相图特征，分析 CO_2 摩尔分数对地下流体相态特征的影响（图 1）。

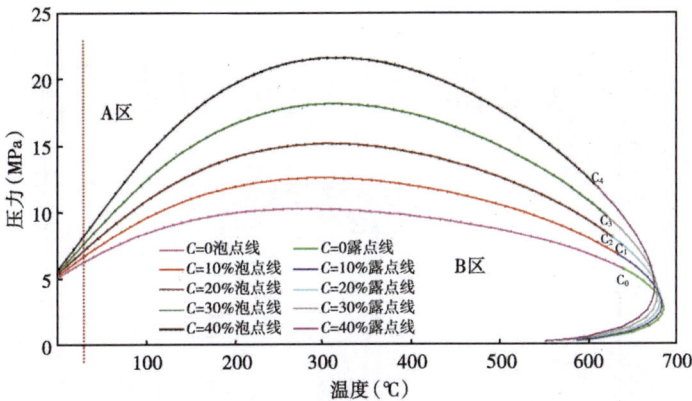

图 1 CO_2—原油体系相态包络图

由图 1 可知，在原始条件下，地下流体为液相（A 区），位于临界点 C0 的上方，随着 CO_2 注入量的增多，CO_2—原油体系压力—温度相图临界点不断向左移动（由 C_0 到 C_4），两相区（B 区）不断扩大。同时随着 CO_2 注入量的增多，CO_2—原油体系饱和压力不断增加（图 2）。表明 CO_2 对地下流体的相态特征有一定的影响，注入量越大，影响越大。

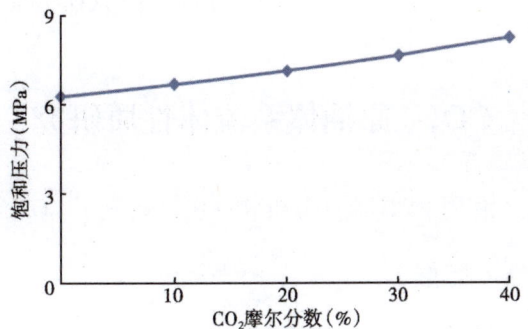

图 2 CO_2—原油体系饱和压力变化曲线

3 CO₂—原油体系渗流特征研究

当温度位于 CO_2 临界温度附近或低于 CO_2 临界温度时，随着压力的变化，CO_2—原油体系将出现三相状态。仍以普鲁德霍湾 MilnePoint 油田 Kupa 储层为算例[14]，利用相平衡计算模型计算压力—摩尔分数相图，分析 CO_2 含量对地下流体渗流特征的影响（图3）。随着体系中 CO_2 物质的量增加，CO_2 与原油体系存在一个临界点，在临界点的下方，地下流体随着压力的降低，分离出部分气态 CO_2，由单一液相 L（CO_2 与原油混相 A 区）变为气液（富含烃类的液相 L_1 和 CO_2 气相 V）两相（B 区）；在临界点的上方，地下流体随着压力的降低，有一部分液态 CO_2 转化为气态的 CO_2，由液—液（富含烃类的液相 L_1 和 CO_2 液相 L_2）两相（C 区）变为液—气—液（富含烃类的液相 L_1、CO_2 气相 V 和 CO_2 液相 L_2）三相（D 区），压力继续降低，再变成气—液两相（B 区）。注 CO_2 驱油藏开发过程中，一旦出现三相区，将会极大地降低 CO_2 注入能力，影响油田开发效果。因此注 CO_2 驱油藏开发时，应保持一定的地层压力，生产压差不能过大，尽量避开三相区。

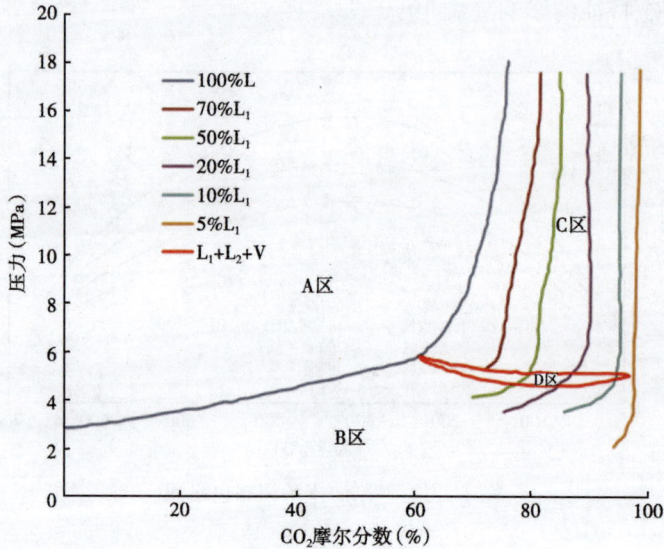

图3　不同 CO_2 含量压力—摩尔分数关系

4 CO₂—原油体系流体性质研究

利用 PVTi 流体分析软件包，研究了 CO_2 含量对 CO_2—原油体系流体性质的影响[15-19]。

4.1 黏度

随着 CO_2—原油体系中 CO_2 溶解量的增加，地层原油黏度降低，流度比得到改善。当 CO_2—原油体系中 CO_2 摩尔分数超过 60%，随着 CO_2 含量的增加，黏度降低幅度减小（图4）。研究表明，体系中 CO_2 摩尔分数为 60% 时降黏效果最好，与原始条件下黏度相比，原油黏度降低了 48%，说明油藏注 CO_2 开发时存在一个最佳注入量，在该注入量下地层原油降黏效果最好，当 CO_2 注入量超过所处压力下的溶解气量时，油气会分离成两相，影响降黏效果。因此，油藏进行注气开发方案设计时需对 CO_2 注入量进行优化。

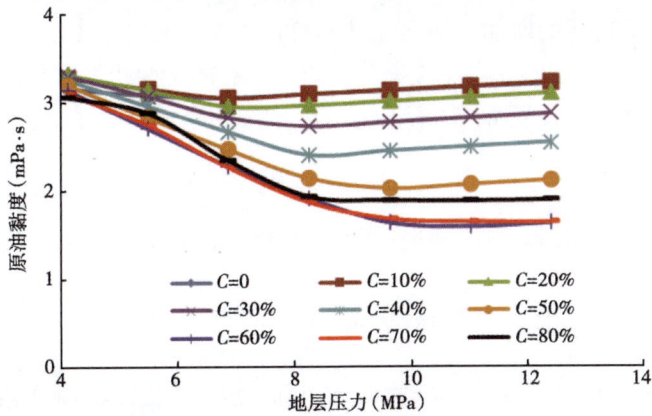

图 4　CO_2—原油体系黏度变化曲线

4.2　原油相对体积

CO_2 溶解于原油后，与原始状态的原油相比，体积系数增加。CO_2—原油体系膨胀系数随着原油平均分子质量减小（轻质组分增多）而增加，随 CO_2 在原油中的摩尔分数增加而增大。当 CO_2—原油体系中 CO_2 的摩尔分数为 40%，地层压力为 0.5MPa 时，与原始条件相比，原油的相对体积膨胀了近 11 倍，表明 CO_2 具有很强的膨胀地层原油、增加可动油的能力，可有效增加地层的弹性能量（图 5）。

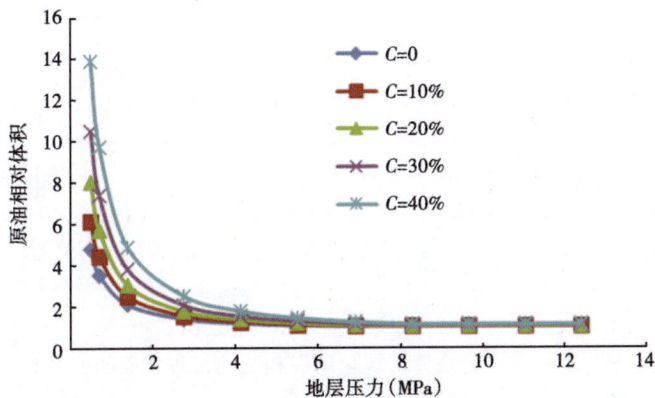

图 5　原油相对体积变化曲线

5　结论

（1）CO_2 含量对原油相态有一定影响，CO_2 含量越大，影响越大。随着 CO_2 含量的增加，CO_2—原油体系临界点不断向上移动，两相区不断扩大，饱和压力不断增加。

（2）CO₂—原油体系存在临界点，在临界点下方，随着压力的降低，地下流体由单一液相变为气—液两相；在临界点上方，当低于一定温度时，随着压力的变化会出现三相区。油藏进行注气开发时，生产压差不能过大，要注意避开三相区，以免影响油田开发效果。

（3）随着CO₂溶解量的增加，地层原油黏度降低，流度比得到改善，但是体系中CO₂的摩尔分数超过60%时，黏度降低幅度减弱。CO₂—原油体系膨胀系数随着原油平均分子质量减小（轻质组分增多）而增加，随着CO₂在原油中的摩尔分数增加而增大。

参 考 文 献

[1] 许志刚，陈代钊，曾荣树. CO₂的地质埋存与资源化利用进展 [J]. 地球科学进展，2007，22（7）：698-705.

[2] 任韶然，张莉，张亮. CO₂地质埋存：国外示范工程及其对中国的启示 [J]. 中国石油大学：自然科学版，2010，34（1）：93-95.

[3] 曾荣树，孙枢，陈代钊，等. 减少二氧化碳向大气层的排放——二氧化碳地下储存研究 [J]. 中国科学基金，2004，18（4）：196-200.

[4] 王庆，吴晓东，刘长宇，等. 高含CO₂原油井筒流动压力和温度分布综合计算 [J]. 石油钻采工艺，2010，32（1）：65-69.

[5] 吴晓东，王庆，何岩峰. 考虑相态变化的注CO₂井筒温度压力场耦合计算 [J]. 中国石油大学学报：自然科学版，2009，33（1）：73-77.

[6] 余华杰，王星，谭先红，等. 高含CO₂凝析气相态测试及分析 [J]. 石油钻探技术，2013，41（2）：104-108.

[7] Danesh A. 油藏流体的PVT与相态 [M]. 沈平平，韩冬，译. 北京：石油工业出版社，2000：151-154.

[8] Duraya Al-Anazi B, Pazuki G R, Nikookar M, et al. The prediction of the compressibility factor of sour and natural gas by an artificial neural network system [J]. Petroleum Science and Technology, 2011, 29（4）：325-336.

[9] Rushing J A, Newsham K E, Van Fraassen K C, et al. Natural gas Z factors at HP/HT reservoir conditions: comparing laboratory measurements with industry standard correlations for a dry gas [A]. SPE 114518, 2008: 356-372.

[10] Heidaryan E, Moghadasi J, Rahimi M. New correlations to predict natural gas viscosity and compressibility factor [J]. Journal of Petroleum Science and Engineering, 2010, 73（1-2）：67-72.

[11] 韩宏伟，张金功，张建锋，等. 济阳拗陷二氧化碳气藏地下相态特征研究 [J]. 西北大学学报：自然科学版，2010，40（3）：493-496.

[12] 张镜澄. 超临界流体萃取 [M]. 北京：化学工业出版社，2001：30-36.

[13] 李孟涛，单文文，刘先贵，等. 超临界二氧化碳混相驱油机理实验研究 [J]. 石油学报，2006，27（3）：80-83.

[14] Baris Guler, Peng Wang, Mojdeh Delshad, et al. Three-and four-phase flow compositional simulations of CO₂/NGL EOR [C]. SPE 71485, 2001：274-280.

[15] 吕渐江，唐海，李春芹，等. 考虑相态变化的CO₂气藏物质平衡方程 [J]. 新疆石油地质，2010，31（6）：618-620.

[16] 苏崇华. 海上低渗透油藏挖潜技术及其应用 [J]. 石油钻采工艺，2007，29（6）：73-76.

[17] 卞小强，杜志敏. 高含CO₂天然气相变及其物性参数实验测试 [J]. 新疆石油地质，2013，34（1）：63-65.

[18] 苏玉亮，吴晓东，侯艳红，等．低渗透油藏 CO_2 混相驱油机制及影响因素 [J]．中国石油大学：自然科学版，2011，35（3）：99-102.

[19] 杨胜来，杭达震，孙蓉，等．CO_2 对原油的抽提及其对原油粘度的影响 [J]，中国石油大学学报：自然科学版，2009，33（4）：85-88.

CO_2 驱过程中不同相态流态对采收率的影响

徐 阳[1] 任韶然[2] 章 杨[2] 郭 平[3]

(1. 中国石化集团国际石油勘探开发有限公司；2. 中国石油大学（华东）
石油工程学院；3. 中国石化胜利油田分公司 地质科学研究院)

摘 要：CO_2驱过程中相态、流态的变化均会对最终采收率有所影响。为研究不同相态及流态条件下采收率的变化情况，以压力判断相态的标准、雷诺数为判断流态的方法进行人造低渗透岩心 CO_2 驱替实验。分析实验结果认为：不同混相条件原油采收率不同，混相压力条件下采出效果最佳；以雷诺数作为划分流态的标准，实验条件下，采收率随驱替速率的增加呈现先增后减的变化，存在提高采收率的最佳流速。

关键词：CO_2；岩心驱替实验；相态；流态；采收率

0 引言

CO_2 驱已成为一种成熟的提高采收率方法[1-3]。大量国内外文献表明，CO_2 驱可以提高采收率 15%~25%。作为一种常规方法，CO_2 岩心驱替实验可以研究不同实验条件下 CO_2 驱提高采收率的效果，在一定程度上体现 CO_2 驱替效果随实验条件的变化趋势，并为矿场试验提供施工参数。

1 实验设备与方法

实验装置如图 1 所示，利用高压 PVT 分析仪和美国 HP 气相色谱仪对实验原油进行常规

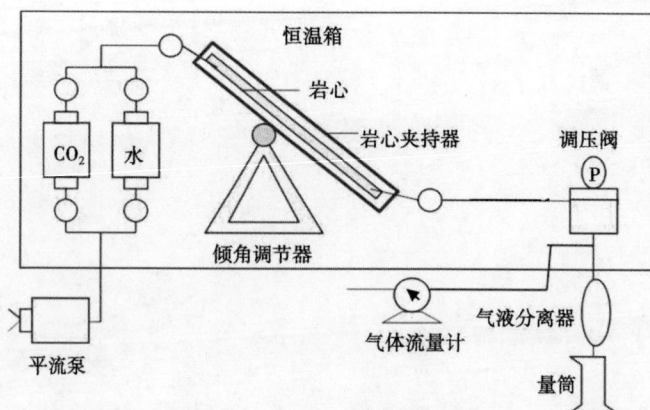

图 1 岩心驱替流程

PVT 分析，实验流体样品采用胜利油田某区块所取井口油气样品按照饱和压力配制。油藏条件为：油层深度 2700~3100m，地层压力 29MPa，油层温度 126℃，泡点压力 11.61MPa，细管实验得到最小混相压力为 28.9MPa[4]。主要实验内容包括：实验用油的配制与相关参数的分析，人造低渗透岩心 CO_2 驱替实验[5-6]。

2 实验结果与讨论

2.1 实验样品的配制

泡点压力对气油比很敏感，是气油比的单调递增函数。一般泡点压力下容易得到准确数据，因此，按拟合目的泡点压力的原则，配制本次实验所用原油，基本物性见表 1。

表 1 原油物性

饱和压力 p_b（MPa）	地层油体积系数 B_o	溶解气油比 R_s（m³/t）	地层油密度 ρ（g/cm³）	地层油黏度 μ（mPa·s）
11.290	1.144	46.695	0.792	1.980

2.2 不同相态 CO_2 岩心驱替实验

2.2.1 实验研究思路

在模拟油藏条件下进行人造低渗透岩心驱替实验，物理模拟不同压力下（10~30MPa）非混相、混相 CO_2 驱替过程，考察不同注入压力下相态与驱油效率之间关系，对比分析不同相态条件下驱油效果与驱油机理。

2.2.2 实验结果及分析

实验进行了 7 组低渗透岩心 CO_2 驱替实验，压力分别为 10MPa，15MPa，19.6MPa，22.5MPa，25MPa，27MPa 和 30MPa。实验结果见表 2。

表 2 岩心驱替实验结果

序号	驱替压力 p（MPa）	采收率（%）
1	10.00	50.59
2	15.00	56.00
3	19.60	68.89
4	22.50	77.65
5	25.00	84.80
6	27.00	90.02
7	30.00	92.25

根据不同驱替压力条件下采收率随 PV 数的变化关系，可作出其相关关系曲线，选取有代表性的 3 个驱替压力，结果如图 2 所示。

如图 2 所示：在气体突破前，随着注入孔隙体积（PV）数的增加，采收率基本呈线性增加；气体突破后，累计驱油效率增加趋势变缓，采收率增幅大幅降低；在其他注入条件相同的情况下，CO_2 的突破时间随着驱替压力的增大而延长；15MPa（非混相驱）下的最终采收率为 56.00%，30MPa（混相驱）下采收率达到 92.25%，比 15MPa 条件下最终采收率高 36.25%，提高采收率幅度大幅上升。由此可见，原油最终采收率随着 CO_2 驱替压力的增加而增加，并且混相条件下的原油采收率要远高于非混相条件下的采收率。

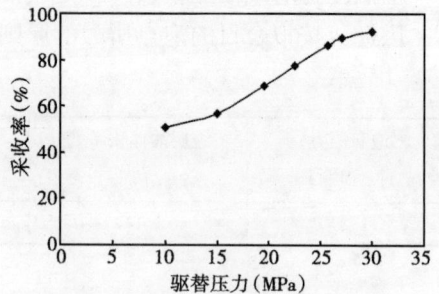

图 3 为累计采收率与驱替压力关系曲线，从曲线的变化趋势分析，此过程大致分为 3 个阶段：驱替压力较低（10~15MPa）情况下，处于非混相驱替压力范围，气窜现象严重，压力对采收率的影响难以体现，导致采收率增幅较小；驱替压力范围介于 MMP（最小混相压力）与 15MPa 之间，采收率与驱替压力几乎呈线性关系，采收率增幅较大；当驱替压力高于 MMP 时，采收率很难有所提高。

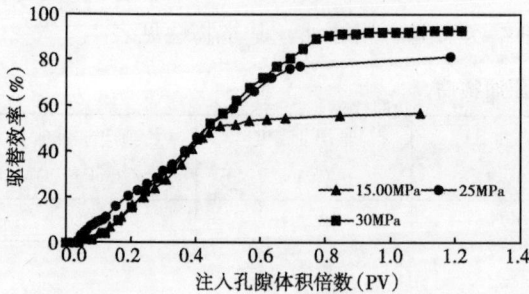

图 2　不同驱替压力下累计采收率与 PV 数关系　　　　图 3　累计采收率与驱替压力关系

2.3　不同流态 CO_2 岩心驱替实验

2.3.1　实验研究思路

非线性渗流使低渗透油藏的驱油效率明显低于中高渗透油藏，降低了原油的采收率，增加了开发难度；雷诺数是储层、流体物性条件及开发条件的综合反映，可作为流态的综合判断依据。以往研究[7-8]表明，阻力系数与雷诺数的关系可以将流动划分为非达西低速渗流区、达西渗流区和非达西高速渗流区 3 个不同的状态区。非达西低速渗流区主要为薄层流动，摩阻力与相间摩擦附加力起主要作用；达西渗流区主要为层流流动，摩阻力起主要作用；而非达西高速渗流区域主要为紊流，摩阻力和惯性力起主要作用。划分 3 个状态区的两个临界雷诺数值分别为 1×10^{-5} 和 0.02~0.03。

利用 CO_2 驱替装置，共进行了 8 组 CO_2 长岩心驱替实验，实验压力 20MPa。分别设定 CO_2 注入速度 0.02mL/min、0.03mL/min、0.17mL/min、1.00mL/min 和 2.00mL/min，最大流量与最小流量间约为 100 倍关系，目的在于研究不同流态 CO_2 驱油影响规律，探究不同雷诺数范围内不同流速、流态对采收率影响的机理。为了验证实验数据的准确性，重复实验 3 组，分别为 0.03mL/min、0.17mL/min 和 1mL/min。另外，为了更好地与现场相结合，特将注入流速换算成注入通量，依次为 0.058m/d、0.087m/d、0.499m/d、2.843m/d 和 5.870m/d。

2.3.2　实验结果及分析

1）CO_2 注入流速与采收率的关系

根据不同 CO_2 注入通量下的注入孔隙体积倍数（PV）与采收率关系数据，选用 5 组不同注入通量实验数据，二者之间关系如图 4 所示。

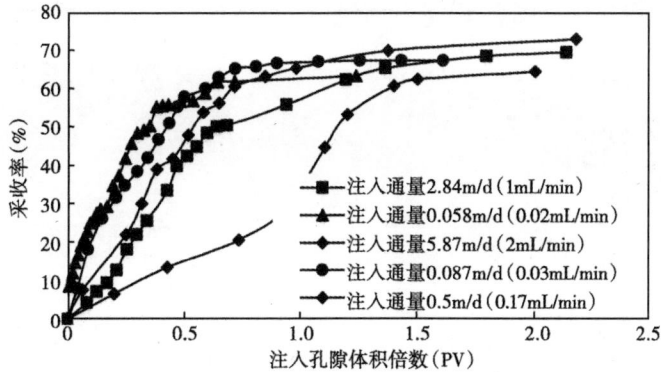

图 4 不同 CO_2 注入通量下 PV 数与采收率关系

2）CO_2 流态与采收率的关系

卡佳霍夫提出的雷诺数：

$$Re = \frac{v\sqrt{K}\rho}{17.50\mu\phi^{\frac{3}{2}}} \tag{1}$$

式中，Re 为雷诺数；K 为渗透率，D；ϕ 为孔隙度；ρ 为流体密度，g/cm^3；v 为渗流速度，cm/s；μ 为流体的黏度，$mPa \cdot s$。

由式（1）计算不同 CO_2 注入通量下的雷诺数，并作出 CO_2 雷诺数与采收率关系表和关系图，见表 3 和图 5。

实验用人造低渗透岩心渗透率为 $0.52 \sim 4.61mD$，利用数值模拟软件获得实验温度、压力条件下 CO_2 的质量密度 ρ 为 $0.294g/cm^3$，黏度 μ 为 $0.0684\ mPa \cdot s$。

表 3 和图 5 揭示了实验条件下岩心中 CO_2 流体雷诺数与采收率的关系，趋势呈 "凸" 形，实验流体雷诺数的范围为 $4.25 \times 10^{-6} \sim 3.12 \times 10^{-4}$，由图 5 可以看出，当雷诺数小于 1×10^{-5} 数量级时，采收率随着雷诺数的增大而增大，而当雷诺数大于 1×10^{-5} 数量级时，采收率则随雷诺数的增大而减小。究其原因，可以用低渗透低速条件下气体的渗流规律来解释。

表 3 CO_2 注入通量与采收率关系

注入流量 （mL/min）	0.02	0.03	0.17	1.00	2.00	重复实验		
						0.03	0.17	1.00
注入通量（m/d）	0.058	0.087	0.499	2.843	5.870	0.087	0.500	2.843
CO_2 雷诺数	0.43×10^{-5}	0.69×10^{-5}	2.27×10^{-5}	18.35×10^{-5}	31.20×10^{-5}	1.23	5.03	22.36
采收率（%）	62.39	67.06	74.61	70.00	64.38	67.20	73.98	70.29

当雷诺数小于 1×10^{-5} 数量级时，气体渗流应符合非线性规律，此时气体分子平均自由程较大，可比拟于孔隙介质的孔径，气体分子与孔隙壁的碰撞可能性增大，宏观渗流规律所依据的连续介质条件不再完全得到满足，该类碰撞对渗流规律产生影响，在宏观上表现为滑脱效应[9]。可认为驱替过程中产生了除黏滞阻力以外的附加力，而附加力体现在真实启动压力和滑脱力综合作用的形式。附加力使渗流阻力增大，在驱替压力不变的情况下不利于驱

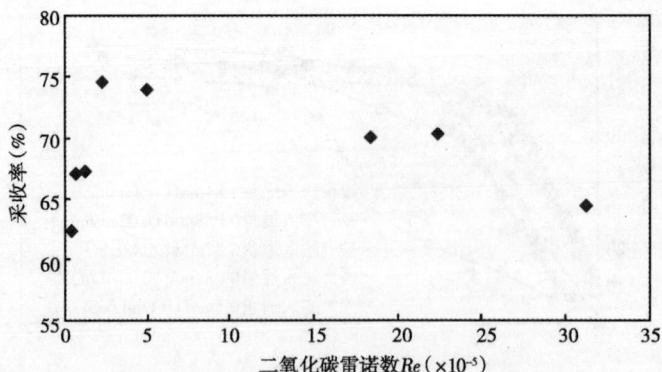

图 5　计算的 CO_2 雷诺数 Re 与采收率的关系

出微小孔隙中的油，导致采收率降低。雷诺数越小，则滑脱效应越严重，产生的附加阻力越大，采收率越低。

当雷诺数大于 1×10^{-5} 数量级时，气体渗流符合线性规律，此时气体流速较大，雷诺数较大，气体分子平均自由程小，渗流的流体可视为连续介质，气体分子与孔隙壁的碰撞机会很小，其对渗流的影响可以不考虑。气体流动时不同流速层气体分子之间因碰撞交换动量，造成定向动量迁移，表现为流体的黏滞现象，它服从牛顿内摩擦定律，由连续介质理论可得达西线性渗流规律。此时只有黏滞力起作用，渗流阻力较非线性渗流时小，在驱替压力不变的情况下有利于驱出微小孔隙中的油，能够提高采收率。但是，并非 CO_2 气体流速越大越好，随着雷诺数的增大，容易导致 CO_2 气窜和严重黏性指进，导致采收率降低。

3　结论

（1）针对不同相态 CO_2 驱替研究，驱替压力较低的非混相驱，10MPa 与 15MPa 下采收率相差不大，在气体突破前产油量基本一致；对于混相驱，采收率较非混相驱有大幅提高，随着压力的升高，采收率逐渐升高。

（2）针对不同流态 CO_2 岩心驱替实验研究，当雷诺数小于 1×10^{-5} 数量级时，气体渗流符合非线性规律，采收率较低；当雷诺数大于 1×10^{-5} 数量级时，气体渗流符合达西线性渗流规律。在一定雷诺数范围内，流速增加有利于提高采收率，但是若注入速率较高时易形成气窜，黏性指进现象严重，采收率会有所降低。

（3）针对本实验，人造低渗透岩心渗透率范围为 $0.52 \sim 4.61$ mD，实验流体雷诺数的范围为 $4.25 \times 10^{-6} \sim 3.12 \times 10^{-4}$，存在最佳流速 0.5m/d，此时采收率最高。

参 考 文 献

[1] 李士伦，郭平，戴磊，等. 发展注气提高采收率技术 [J]. 西南石油学院学报，2002，22（3）：41-46.

[2] 高慧梅，何应付，周锡生. 注二氧化碳提高原油采收率技术研究进展 [J]. 特种油气藏，2009，16（1）：6-11.

[3] 李菊花，李相方，刘斌，等. 注气近混相驱油藏开发理论进展 [J]. 天然气工业，2006，26（2）：108-110.

［4］郝永卯，陈月明，于会利，等．CO_2 驱最小混相压力的测定与预测［J］．油气地质与采收率，2005，12（6）：64-66.

［5］郝永卯，薄启炜，陈月明，等．CO_2 驱油实验研究［J］．石油勘探与开发，2005，32（2）：110-112.

［6］沈平平，黄磊．二氧化碳—原油多相多组分渗流机理研究［J］．石油学报，2009，30（2）：247-251.

［7］王道成，李闽，谭建为，等．气体低速非线性渗流研究［J］．大庆石油地质与开发，2007，26（6）：74-77.

［8］冯文光．油气渗流力学基础［M］．北京：科学出版社，2007.

［9］陈代殉，王章瑞，高家碧．多孔介质中低速气体的非线性渗流［J］．西南石油学院学报，2002，24（5）：40-42.

修正混合规则的 BWRS 型状态方程及其在 CO_2—原油体系相态计算中的应用

廉黎明[1,2]　秦积舜[1,2]　刘同敬[3]　李　实[1,2]　姬泽敏[1,2]　第五鹏翔[4]

(1. 中国石油勘探开发研究院提高石油采收率国家重点实验室；
2. 国家能源二氧化碳驱油与埋存技术研发（实验）中心；
3. 中国石油大学（北京）提高采收率研究中心；4. 中国地质大学能源学院)

摘　要：CO_2—原油体系中含有非烃和较多重组分，非理想性较强。以对非理想体系有较好计算效果的 BWRS 状态方程为基础，考虑温压条件和组分间的相互作用，修正混合规则，使其更加适用于 CO_2—原油体系的相态计算，提高方程的预测精度。应用油田现场油样实测气液相组分数据，以三参数 PR 和 PT 方程的计算均值代表常用状态方程，分别采用基于修正混合规则的 BWRS 方程和常用状态方程对油气组分和密度等参数进行计算。结果表明：修正混合规则的 BWRS 方程中各组混合参数都可还原为维里系数对组分的依赖形式，理论性更强；方程计算精度显著提高，对非烃类气体和重组分有较好的适应性，能满足工程应用要求。

关键词：CO_2—原油体系；BWRS 状态方程；混合规则；逸度计算

0　引言

气液体系的相态研究是对不同温度、压力条件下物质状态及组分变化的表征，是分析和认识体系的重要手段[1]。在此过程中，研究目标不再是处于恒定条件下的理想状态纯物质，需要引入状态方程研究温度、压力与物质组成之间的关系[2]，引入混合物质计算规则（混合规则）研究各组分、相间的相互作用[3]。现有的状态方程可分为 van del Waals（vdW）型及其改进型状态方程[4-6]、varial 型状态方程[7-8]、多参数状态方程[9-11]（BWR，BWRS 等）以及具有严格统计力学基础的状态方程（微扰硬链和转子链等）[12-13]。CO_2—原油体系是典型的非理想体系[14]，且处于高温高压状态[15]。油藏中非烃类和重组分含量较高，笔者引入 BWRS 状态方程进行相态研究，对其各个参数的混合规则进行修正，使方程能够适用于描述 CO_2—原油体系的相态变化。

1　BWRS 状态方程

BWRS 状态方程[16-17]形式如下：

$$p_i = \rho_i RT + \left(B_{0i}RT - A_{0i} - \frac{C_{0i}}{T^2} + \frac{D_{0i}}{T^3} - \frac{E_{0i}}{T^4} \right)\rho_i^2 + \left(b_i RT_i - a_i - \frac{d_i}{T_i} \right)\rho_i^3 +$$

$$\alpha_i \left(a_i + \frac{d_i}{T} \right)\rho_i^6 + \frac{c_i \rho^3}{T^2}(1 + \gamma_i \rho_i^2)\exp(-\gamma_i \rho_i^2) \tag{1}$$

式中，p_i 为组分 i 的压力，Pa；T 为体系温度，K；ρ_i 为组分的密度，g/cm³；B_{0i}，A_{0i}，C_{0i}，D_{0i}，E_{0i}，b_i，a_i，d_i，α_i，c_i 和 γ_i 为 BWRS 状态方程的相关系数。

该方程共有 11 个需要实验回归的参数，其与临界参数和偏心因子的关联如下：

$$
\begin{cases}
\rho_{ci}B_{0i} = A_1 + B_1\omega_i \\[2mm]
\dfrac{\rho_{ci}A_{0i}}{RT_{ci}} = A_2 + B_2\omega_i \\[2mm]
\dfrac{\rho_{ci}C_{0i}}{RT_{ci}^3} = A_3 + B_3\omega_i \\[2mm]
\rho_{ci}^2\gamma_i = A_4 + B_4\omega_i \\[2mm]
\dfrac{\rho_{ci}^2 D_{0i}}{RT_{ci}^4} = A_9 + B_9\omega_i \\[2mm]
\dfrac{\rho_{ci}E_{0i}}{RT_{ci}^5} = A_{11} + B_{11}\omega_i\exp(-3.8\omega_i)
\end{cases}
\qquad
\begin{cases}
\rho_{ci}^2 b_i = A_5 + B_5\omega_i \\[2mm]
\dfrac{\rho_{ci}^2 a_i}{RT_{ci}} = A_6 + B_6\omega_i \\[2mm]
\rho_{ci}^3\alpha_i = A_7 + B_7\omega_i \\[2mm]
\dfrac{\rho_{ci}^2 c_i}{RT_{ci}^3} = A_8 + B_8\omega_i \\[2mm]
\dfrac{\rho_{ci}^2 d_i}{RT_{ci}^2} = A_{10} + B_{10}\omega_i
\end{cases}
\tag{2}
$$

式中，T_{ci} 为组分 i 的临界温度；ρ_{ci} 为组分 i 的临界密度；ω_i 为组分 i 偏心因子；A_1—A_{11} 和 B_1—B_{11} 为通用常数。

式（1）、式（2）的研究对象是纯物质或者混合物中的纯组分。对于多相多组分的实际体系，由于相间和组分间的相互作用，须引入组分混合规则对所涉及的物理量进行处理。

2　混合规则及其修正

2.1　现用 BWRS 状态方程的混合规则

目前，BWRS 状态方程各参数所配套的混合规则具有如下形式：

$$
\begin{cases}
A_{0m} = \displaystyle\sum_{i=1}^{n}\sum_{j=1}^{n} x_i x_j A_{0i}^{1/2} A_{0j}^{1/2}(1-k_{ij}) \\[3mm]
B_{0m} = \displaystyle\sum_{i=1}^{n} x_i B_{0i} \\[3mm]
C_{0m} = \displaystyle\sum_{i=1}^{n}\sum_{j=1}^{n} x_i x_j C_{0i}^{1/2} C_{0j}^{1/2}(1-k_{ij})^3 \\[3mm]
\gamma_m = \Big(\displaystyle\sum_{i=1}^{n} x_i \gamma_i^{1/2}\Big)^2 \\[3mm]
D_{0m} = \displaystyle\sum_{i=1}^{n}\sum_{j=1}^{n} x_i x_j D_{0i}^{1/2} D_{0j}^{1/2}(1-k_{ij})^4 \\[3mm]
E_{0m} = \displaystyle\sum_{i=1}^{n}\sum_{j=1}^{n} x_i x_j E_{0i}^{1/2} E_{0j}^{1/2}(1-k_{ij})^5
\end{cases}
\qquad
\begin{cases}
a_m = \Big(\displaystyle\sum_{i=1}^{n} x_i a_i^{1/3}\Big)^3 \\[3mm]
b_m = \Big(\displaystyle\sum_{i=1}^{n} x_i b_i^{1/3}\Big)^3 \\[3mm]
c_m = \Big(\displaystyle\sum_{i=1}^{n} x_i c_i^{1/3}\Big)^3 \\[3mm]
d_m = \Big(\displaystyle\sum_{i=1}^{n} x_i d_i^{1/3}\Big)^3 \\[3mm]
\alpha_m = \Big(\displaystyle\sum_{i=1}^{n} x_i \alpha_i^{1/3}\Big)^3
\end{cases}
\tag{3}
$$

式中，k_{ij} 为二元交互作用参数；x_i 为 i 组分摩尔分数；x_i 为 j 组分摩尔分数；A_{0m}，B_{0m}，C_{0m}，γ_m，D_{0m}，E_{0m}，a_m，b_m，c_m，d_m 和 α_m 为 BWRS 状态方程相关参数适用于混合物质的形式。

式（3）中，二元参数项（密度平方项的系数也即体积平方倒数的系数）可以化为维里系数的依赖形式，而其他项则不能转化为此种关系，从而此混合规则在理论上不够严格；三元参数项（密度立方项的系数也即体积立方倒数的系数）没有考虑到分子间的相互作用，混合规则也不够精确。因此，需要对 BWRS 方程现用的混合规则进行修正。

2.2 修正的 BWRS 状态方程混合规则

2.2.1 状态方程混合规则的改进

CO_2—原油体系是典型的非理想性体系，各组分间存在性质差异，非烃类气体的存在扩大了这种差异性，而高温高压条件也对组分间的作用产生影响。基于此，需要将组分间的相互作用表征为状态方程的参数。

在 BWRS 状态方程中，参数 A_{0m}，B_{0m}，C_{0m}，D_{0m}，E_{0m} 和 γ_m 表征物质两个分子之间的作用，参考式（2），若参数与温度相关联，则需要引入二元交互作用参数进行修正，使其更符合实际。

$$\begin{cases} A'_{0m} = \sum_{i=1}^{n} \sum_{j=1}^{n} x_i x_j A_{0i}^{1/2} A_{0j}^{1/2} (1 - k_{ij}) \\[2mm] B'_{0m} = \sum_{i=1}^{n} x_i x_j B_{0i}^{1/2} B_{0j}^{1/2} \\[2mm] C'_{0m} = \sum_{i=1}^{n} \sum_{j=1}^{n} x_i x_j C_{0i}^{1/2} C_{0j}^{1/2} (1 - k_{ij})^3 \\[2mm] D'_{0m} = \sum_{i=1}^{n} \sum_{j=1}^{n} x_i x_j D_{0i}^{1/2} D_{0j}^{1/2} (1 - k_{ij})^4 \\[2mm] E'_{0m} = \sum_{i=1}^{n} \sum_{j=1}^{n} x_i x_j E_{0i}^{1/2} E_{0j}^{1/2} (1 - k_{ij})^5 \\[2mm] \gamma'_m = \sum_{i=1}^{n} \sum_{j=1}^{n} x_i x_j \gamma_i^{1/2} \gamma_j^{1/2} \end{cases} \tag{4}$$

式中，A'_{0m}，B'_{0m}，C'_{0m}，D'_{0m}，E'_{0m} 和 γ'_m 为改进后 BWRS 状态方程的二元系数。

通过变换回复，式（4）可化为第二维里参数的依赖形式。

参数 a_m，b_m，c_m，d_m 和 α_m 表征体系中三分子之间的相互作用，参考式（2）与式（3），若参数与温度相关联，则需要引入三元交互作用参数进行修正，以提高预测精度。但是三分子间的相互作用不好直接表征，笔者利用二元交互作用参数的叠加来表征三元交互影响。

$$\begin{cases} a'_m = \sum_{i=1}^{n} \sum_{j=1}^{n} \sum_{l=1}^{n} a_i^{1/3} a_j^{1/3} a_l^{1/3} \sqrt[3]{(1 - k_{ij})(1 - k_{jl})(1 - k_{li})} \\[2mm] b'_m = \sum_{i=1}^{n} \sum_{j=1}^{n} \sum_{l=1}^{n} b_i^{1/3} b_j^{1/3} b_l^{1/3} \\[2mm] c'_m = \sum_{i=1}^{n} \sum_{j=1}^{n} \sum_{l=1}^{n} c_i^{1/3} c_j^{1/3} c_l^{1/3} [(1 - k_{ij})(1 - k_{jl})(1 - k_{li})] \\[2mm] d'_m = \sum_{i=1}^{n} \sum_{j=1}^{n} \sum_{l=1}^{n} d_i^{1/3} d_j^{1/3} d_l^{1/3} [\sqrt[3]{(1 - k_{ij})(1 - k_{ji})(1 - k_{li})}]^2 \\[2mm] \alpha'_m = \sum_{i=1}^{n} \sum_{j=1}^{n} \sum_{l=1}^{n} \alpha_i^{1/3} \alpha_j^{1/3} \alpha_l^{1/3} \sqrt[3]{(1 - k_{ij})(1 - k_{jl})(1 - k_{li})} \end{cases} \tag{5}$$

式中，a'_m，b'_m，c'_m，d'_m 和 α'_m 为改进后 BWRS 状态方程的三元系数。

通过变换回复，式（5）可化为第三维里参数的依赖形式。

可以看出，修正后的 BWRS 状态方程混合规则完全符合维里系数对组成的依赖关系，而原规则只有部分参数符合维里系数对组成的依赖关系，因此修正的混合规则比原规则更符合理论严格性。

二元交互作用参数可根据多个含 CO_2 体系的气液平衡实验数据，获得了 BWRS 模型对相应体系的二元交互作用参数 k_{ij} 及其随蒸气压数据的变化关系式，并将其与偏心因子进行关联（图 1）。

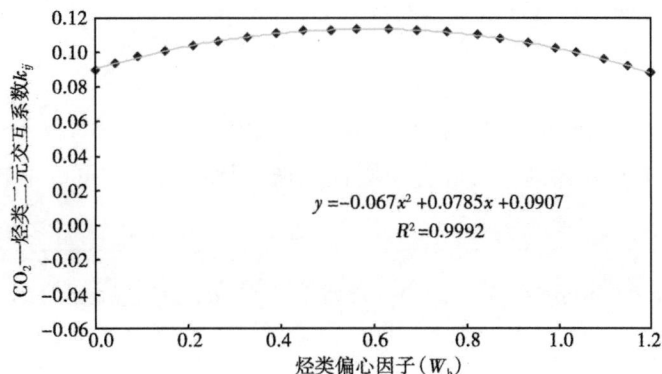

图 1　CO_2—烃类二元交互系数 k_{ij} 关系曲线

2.2.2　其他物理量的混合规则

对处于高温高压状态的混合物体系，由于各个组分的临界性质不同，还应该对温度和压力进行混合计算，得到温压条件的混合规则。

拟临界温度采用通用的定义，如下所示：

$$T_{cm} = \left\{ \frac{\left[\sum_j y_j \left(T_{cj}^{5/2} / p_{cj} \right)^{1/2} \right]^2}{\sum_j y_j \left(T_{cj} / p_{cj} \right)} \right\}^{2/3} \tag{6}$$

式中，T_{cm} 为拟临界温度，K。

拟临界压力 p_{cm}（Pa）采用通用的定义，即：

$$p_{cm} = \frac{T_{cm}}{\sum_j y_j \left(T_{cj} / p_{cj} \right)} \tag{7}$$

通过上述方程，可以将原本用于纯物质的状态方程应用于描述混合物的相态变化。

3　应用

3.1　闪蒸方程

闪蒸计算时，通过对泡点和露点条件下组分与相关系的推导，可得到 Rachford-Rice 方程[18]，即：

$$\sum_i (y_i - x_i) = \sum_i \frac{z_i(K_i - 1)}{1 + N_V(K_i - 1)} = \sum_i \frac{z_i(K_i - 1)}{K_i - N_L(K_i - 1)} = 0 \tag{8}$$

式中，z_i 为组分 i 的总摩尔分数；N_V 为气体无量纲体积；N_L 液体无量纲体积；K_i 组分 i 的气液平衡常数。

3.2 混合物的逸度及逸度因子

根据热力学相平衡原理，所研究的 CO_2—原油体系处于平衡状态时，体系的每一组分在各相中的逸度相等[6,19]，气液两相平衡条件为：

$$f_i^V = f_i^L \tag{9}$$

式中，f_i^V 为组分 i 的气相逸度；f_i^L 为组分 i 的液相逸度。

将 BWRS 状态方程及其修正的混合规则代入逸度计算方程，可以推导出如下计算组分逸度的公式：

$$
\begin{aligned}
RT \ln f_i = {} & RT \ln(\rho RT x_i) + 2\rho \sum_{i=1}^n x_i \Big[(B'^{1/2}_{0m} B^{1/2}_{0i}) RT - (A'^{1/2}_{0m} A^{1/2}_{0i})(1 - k_{ij}) - \\
& \frac{(C'^{1/2}_{0m} C^{1/2}_{0i})}{T^2}(1 - k_{ij})^3 + \frac{(D'^{1/2}_{0m} D^{1/2}_{0i})}{T^3}(1 - k_{ij})^4 - \frac{(E'^{1/2}_{0m} E^{1/2}_{0i})}{T^4}(1 - k_{ij})^5 \Big] + \\
& \frac{3\rho^2}{2} \sum_{i=1}^n x_i \Big\{ (b'^{2/3}_m b^{1/3}_i) RT - (a'^{2/3}_m a^{1/3}_i) \sqrt[3]{(1 - k_{ij})(1 - k_{jl})(1 - k_{li})} - \\
& \frac{(d'^{2/3}_m d^{1/3}_i)}{T} [\sqrt[3]{(1 - k_{ij})(1 - k_{jl})(1 - k_{li})}]^2 \Big\} + \\
& \frac{3\rho^5}{5} \sum_{i=1}^n x_i \Big\{ (a'^{2/3}_m a^{1/3}_i) \sqrt[3]{(1 - k_{ij})(1 - k_{jl})(1 - k_{li})} + \frac{(d'^{2/3}_m d^{1/3}_i)}{T} [\sqrt[3]{(1 - k_{ij})(1 - k_{jl})(1 - k_{li})}]^2 + \\
& \Big(a'_m + \frac{d'_m}{T} \Big)(\alpha'^{2/3}_m \alpha^{1/3}_i) \sqrt[3]{(1 - k_{ij})(1 - k_{jl})(1 - k_{li})} \Big\} + \\
& 3\rho^2 \sum_{i=1}^n x_i \Big\{ \frac{(c'^{2/3}_m c^{1/3}_i)[1 - k_{ij}(1 - k_{jl})(1 - k_{li})]}{T^2} \Big[\frac{1 - \exp(-\gamma'_m \rho^2)}{\gamma \rho^2} - \frac{\exp(-\gamma'_m \rho^2)}{2} \Big] \Big\} - \\
& \frac{2c_m}{\gamma_m T^2} \Big(\frac{\gamma_i}{\gamma'_m} \Big)^{1/2} \Big[1 - \Big(1 + \gamma'_m \rho^2 + \frac{1}{2}\gamma'^2_m \rho^4 \Big) \exp(-\gamma'_m \rho^2) \Big]
\end{aligned}
\tag{10}
$$

联立逸度系数、组分与气液平衡常数，得到：

$$K_i = \frac{y_i}{x_i} = \frac{\phi_i^L}{\phi_i^V} = \frac{f_i^L/x_i}{f_i^V/y_i} \tag{11}$$

式中，ϕ_i^V 为组分 i 的气相逸度系数；ϕ_i^L 为组分 i 的液相逸度系数。

将式（11）取对数并写成残差形式，得：

$$R_i = \ln K_i + \ln\phi_i^V - \ln\phi_i^L \tag{12}$$

式中，R_i 为组分 i 的计算残差。

将根据状态方程及修正的混合规则所推导的式（10）代入式（12），并对 Rachford-Rice 方程进行迭代求解，每次迭代过程中，通过式（11）求取新的气液平衡常数，带入下一步迭代过程中。

联合式（4）至式（12），通过 Newton-Raphson 迭代或者超松弛迭代[20]，可计算体系的相态参数。

4 实例计算与分析

在井流物测试条件（温度20℃，压力0.1MPa）和地层压力条件（温度地层温度，压力25MPa）下，将三参数 PR 和 PT 计算平均值作为常用状态方程计算值，并分别采用基于修正混合规则的 BWRS 状态方程以及常用状态方程对油样组分进行计算，并将计算结果进行比较。为加快迭代速度，将油样组分合并为 CO_2，C_1+N_2，C_2—C_7，C_8—C_{10}，C_{11}—C_{22} 和 C_{23+} 6 个拟组分，分别用三参数 PR 和 PT 等状态方程以及 BWRS 状态方程计算 4 个油样各组分的气、液相摩尔分数，并与实验数据对照（表1至表4，图2和图3），最终得到各组分的气、液相摩尔组成及原油密度（表5）。

表1 油样1实验值与计算值的对比

压力 （MPa）	温度 （℃）	组分	实验值（%）			常用状态方程 计算值（%）		修正 BWRS 方程 计算值（%）	
			气+液	液	气	液	气	液	气
0.1	20	CO_2	0.1720	0.0000	0.3903	0.0030	0.3210	0.0006	0.3764
		C_1+N_2	36.3360	0.1065	82.1634	0.2700	68.0720	0.1392	79.3451
		C_2—C_7	17.1050	16.8502	17.4275	0.6520	31.5830	13.6105	20.2586
		C_8—C_{10}	12.1420	21.7272	0.0187	25.9390	0.0020	22.5696	0.0154
		C_{11}—C_{22}	20.3850	36.4976	0.0000	43.5260	0.0230	37.9033	0.0046
		C_{23+}	13.8600	24.8186	0.0000	29.6110	0.0000	25.7771	0.0000
30	74.8	CO_2	0.1720	0.1940	0.0150	0.2522	0.0380	0.2056	0.0196
		C_1+N_2	36.3360	41.2740	0.9870	52.4014	0.3680	43.4995	0.8632
		C_2—C_7	17.1050	19.4940	0.0006	21.5446	0.0010	19.9041	0.0007
		C_8—C_{10}	12.1420	0.0080	98.9970	0.0064	99.5930	0.0077	99.1162
		C_{11}—C_{22}	20.3850	23.2330	0.0000	15.9097	0.0000	21.7683	0.0000
		C_{23+}	13.8600	15.7960	0.0000	9.8857	0.0000	14.6139	0.0000

表2 油样2实验值与计算值的对比

压力 （MPa）	温度 （℃）	组分	实验值（%）			常用状态方程 计算值（%）		修正 BWRS 方程 计算值（%）	
			气+液	液	气	液	气	液	气
0.1	20	CO_2	22.2029	0.0000	65.2971	0.3410	48.9780	0.0682	62.0333
		C_1+N_2	8.6630	0.0263	25.4261	0.0710	19.1860	0.0353	24.1781
		C_2—C_7	14.4779	17.1722	9.2485	0.4880	31.6120	13.8353	13.7212
		C_8—C_{10}	11.6353	17.6155	0.0283	21.1180	0.0210	18.3160	0.0268
		C_{11}—C_{22}	24.2047	36.6754	0.0000	43.8020	0.2020	38.1007	0.0404
		C_{23+}	18.8162	28.5107	0.0000	34.1790	0.0000	29.6443	0.0000

75

压力 (MPa)	温度 (℃)	组分	实验值（%）			常用状态方程 计算值（%）		修正BWRS方程 计算值（%）	
			气+液	液	气	液	气	液	气
30	112.6	CO_2	22.2029	22.2030	48.4100	20.0910	41.3448	21.7806	46.9970
		C_1+N_2	8.6630	8.6630	29.6360	2.3636	21.6724	7.4031	28.0433
		$C_2—C_7$	14.4779	14.4780	21.8550	10.3063	36.8314	13.6437	24.8503
		$C_8—C_{10}$	11.6353	11.6350	0.0050	10.9841	0.0051	11.5048	0.0050
		$C_{11}—C_{22}$	24.2047	24.2050	0.0940	32.8358	0.1128	25.9312	0.0978
		C_{23+}	18.8162	18.8160	0.0001	23.4193	0.0001	19.7367	0.0001

表3 油样3实验值与计算值的对比

压力 (MPa)	温度 (℃)	组分	实验值（%）			常用状态方程 计算值（%）		修正BWRS方程 计算值（%）	
			气+液	液	气	液	气	液	气
0.1	20	CO_2	0.1530	0.0000	0.5380	0.0030	0.3770	0.0006	0.5058
		C_1+N_2	19.0110	0.0000	66.7300	0.1790	47.1020	0.0358	62.8044
		$C_2—C_7$	21.5790	17.1840	32.6120	0.9060	52.4150	13.9284	36.5726
		$C_8—C_{10}$	13.2530	18.4830	0.1210	22.1320	0.0080	19.2128	0.0984
		$C_{11}—C_{22}$	28.6290	40.0350	0.0000	47.7560	0.0980	41.5792	0.0196
		C_{23+}	17.3750	24.2970	0.0000	29.0230	0.0000	25.2422	0.0000
30	97.3	CO_2	0.1530	0.1530	0.3180	0.1620	0.3140	0.1548	0.3172
		C_1+N_2	19.0110	19.0110	67.6930	15.1260	50.6738	18.2340	64.2892
		$C_2—C_7$	21.5790	21.5790	31.9260	15.6500	48.8838	20.3932	35.3176
		$C_8—C_{10}$	13.2530	13.2530	0.0030	10.0160	0.0369	12.6056	0.0098
		$C_{11}—C_{22}$	28.6290	28.6290	0.0610	36.5201	0.0915	30.2072	0.0671
		C_{23+}	17.3750	17.3750	0.0000	22.5259	0.0000	18.4052	0.0000

表4 油样4实验值与计算值的对比

压力 (MPa)	温度 (℃)	组分	实验值 (mol)			常用状态方程 计算值 (mol)		修正BWRS方程 计算值 (mol)	
			气+液	液	气	液	气	液	气
0.1	20	CO_2	0.0660	0.0000	0.3242	0.0020	0.2770	0.0004	0.3148
		C_1+N_2	18.7000	0.1519	90.8681	0.3310	79.7060	0.1877	88.6357
		$C_2—C_7$	4.9550	3.9777	8.7549	0.4300	19.9850	3.2681	11.0009
		$C_8—C_{10}$	12.2830	15.4256	0.0528	15.9810	0.0020	15.5367	0.0426
		$C_{11}—C_{22}$	31.2480	39.2805	0.0000	40.6480	0.0290	39.5540	0.0058
		C_{23+}	32.7480	41.1642	0.0000	42.6080	0.0000	41.4530	0.0000

压力 （MPa）	温度 （℃）	组分	实验值 （mol）			常用状态方程 计算值（mol）		修正BWRS方程 计算值（mol）	
			气+液	液	气	液	气	液	气
30	81	CO_2	0.0660	0.0660	0.1670	0.0834	0.1170	0.0695	0.1570
		C_1+N_2	18.7000	18.7000	92.9110	14.7405	80.1940	17.9081	90.3676
		$C_2—C_7$	4.9550	4.9550	6.9160	4.8170	19.6810	4.9274	9.4690
		$C_8—C_{10}$	12.2830	12.2830	0.0003	11.7790	0.0003	12.1822	0.0003
		$C_{11}—C_{22}$	31.2480	31.2480	0.0050	35.4784	0.0077	32.0941	0.0055
		C_{23+}	32.7480	32.7480	0.0000	33.1017	0.0000	32.8187	0.0000

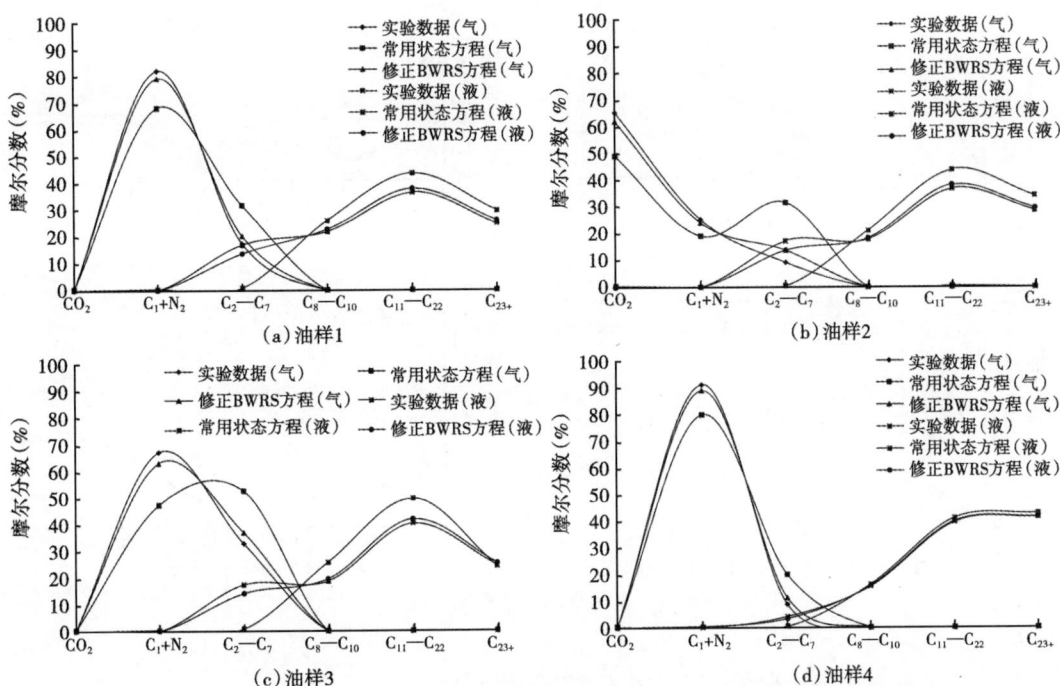

图2　4个油样在常温常压下的常用状态方程、基于修正混合规则BWRS状态
方程计算值与实验数据的对比

表5　油样密度的实验值与计算值得对比

样品 编号	压力 （MPa）	温度 （℃）	实验值 （g/cm³）	常用状态方程		修正BWRS方程	
				计算值（g/cm³）	相对误差（%）	计算值（g/cm³）	相对误差（%）
1	0.1	20	0.842	0.90319	7.27	0.851179	1.09
	30	74.8	0.7427	0.76537	3.05	0.750635	1.07
2	0.1	20	0.8542	0.85811	0.46	0.85596	0.21
	30	112.6	0.7986	0.82882	3.78	0.806155	0.95

样品编号	压力（MPa）	温度（℃）	实验值（g/cm³）	常用状态方程		修正 BWRS 方程	
				计算值（g/cm³）	相对误差（%）	计算值（g/cm³）	相对误差（%）
3	0.1	20	0.8571	0.90433	5.51	0.864185	0.83
	30	97.3	0.7757	0.84166	8.50	0.785594	1.28
4	0.1	20	0.8668	0.82845	4.42	0.857213	1.11
	30	81	0.8243	0.82256	0.21	0.823517	0.09

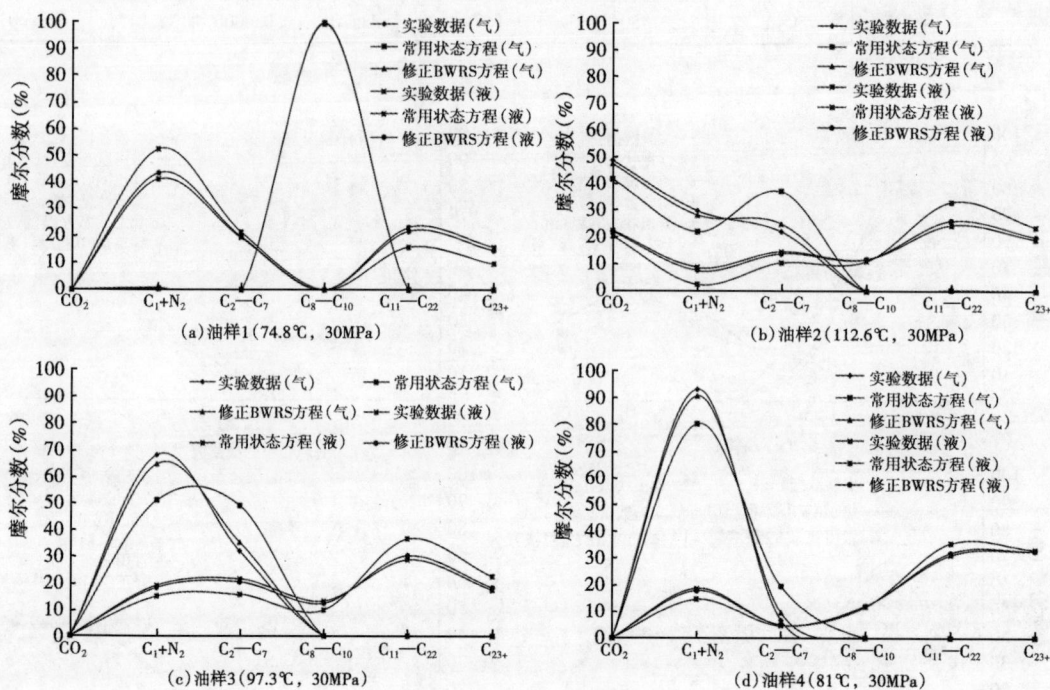

图 3　计算值与实验数据的对比

分析表 1 至表 5 中数据和图 2、图 3 的结果得到：

（1）常用状态方程计算这 4 个油样时误差较大，计算组分时的平均误差超过 30%，计算密度时的平均误差约为 7%；而采用基于修正混合规则的 BWRS 方程计算时所得结果误差大大减小，计算组分时的平均误差小于 10%，计算密度时的平均误差不到 2%，比较接近于实际数据。

（2）常用状态方程计算时出现较大偏差，原因是这 4 个油样中重组分含量都比较高，再加上非烃类气体的影响，体系非理想性大大增强；而基于修正混合规则的 BWRS 方程则有效缓解了这些影响，可得到满意的计算结果，符合工业精度。

（3）基于修正混合规则的 BWRS 方程在计算非烃组分和重组分时，精度也会有所下降，但仍然处在可以接受的范围之内，依然可满足工业精度要求。

5　结论

（1）基于修正混合规则的 BWRS 状态方程应用于 CO_2—原油体系时，对气相和液相参数都有较为满意的计算结果。

（2）应用二元叠加的方法可以将三元相互作用参数量化表征出来，可以克服三组分同时作用参数无法求取的障碍，且计算精度高，实用性强。

（3）BWRS 状态方程的混合规则在修正后，可以还原为维里系数的依赖关系，理论严格性得到提升。

（4）在非烃气体和重组分含量高时，常用状态方程会出现较大计算误差；但基于修正混合规则的 BWRS 状态方程可以取得良好的效果，可以大幅度提高计算精度，并且当体系非理想性增强时，这一优势进一步扩大。

（5）本方法的适应对象是含有非烃类气体和重组分的非理想性 CO_2—原油体系，对于某些胶质、沥青质含量较高的 CO_2—原油体系，由于胶质、沥青质吸附电荷产生的极性以及分子链增加后所产生的极性，都会对计算结果产生较大影响。

参 考 文 献

[1] 沈平平，廖新维. 二氧化碳地质埋存与提高石油采收率技术 [M]. 北京：石油工业出版社，2009.

[2] 秦积舜，李爱芬. 油层物理学 [M]. 2 版，东营：中国石油大学出版社，2006：37-61.

[3] 童景山，李敬. 流体热物理性质的计算 [M]. 北京：清华大学出版社，1982：50-52.

[4] Prausnitz J M. Molecular thermodynamics of fluid phase equilibria [M]. Englewood Cliffs：Prentice-Hall, Inc.，1969.

[5] Reid R C, Prausnitz J M, Poling B E. The properties of gases and liquids [M]. 4th ed. New York：McGraw-Hill, 1987.

[6] 郭天民. 多元气—液平衡和精馏 [M]. 北京：石油工业出版社，2002.

[7] 任闽燕，王珍，徐阳，等. 改进的 CO_2 驱相对渗透率模型及其应用 [J]. 中国石油大学学报：自然科学版，2011，35（4）：108-112.

[8] Reed T M, Gubbins K E. Applied statistical mechanics [M]. New York：McGraw-Hill, 1973.

[9] Holub R, Vonka P. The chemical equilibrium of gaseous systms [M]. Reidel, 1976.

[10] 薛卫东，朱正和，邹乐西. 超临界 CO_2 热力学性质的理论计算 [J]. 原子与分子物理学报，2004，21（2）：295-300.

[11] 汤定国. 用多参数状态方程求解汽液平衡的数值方法 [J]. 华东石油学院学报，1987，11（3）：46-50.

[12] 王利生，郭天民. 基于扰动硬链理论的活度系数模型 [J]. 中国石油大学学报：自然科学版，1995，19（1）：87-92.

[13] 郭天民. 立方型转子链状态方程及其应用（一）[J]. 化学工程，1985（1）：1-13.

[14] 杨胜来，杭达震，孙蓉. CO_2 对原油的抽提及其对原油黏度的影响 [J]. 中国石油大学学报：自然科学版，2009，33（4）：85-88.

[15] 吴晓东，王庆，何岩峰. 考虑相态变化的注 CO_2 井井筒温度压力场耦合计算模型 [J]. 中国石油大学学报：自然科学版，2009，33（1）：73-77.

[16] Starling K E, Han M S. Thermo data refined for LPG. Pt. 15. Industrial applications [J]. Hydrocarbon Process, 1972, 51（5）：129.

［17］Bennedict M，Webb G B，Rubin L C. An empirical Equation for thermodynamic properties of light hydrocarbons and their mixtures I. methane，ethane，propane and n-butane［J］. J. Chem. Phys.，1940（8）：334.

［18］施文，桓冠仁，郭尚平. 二氧化碳—烃—水系统相平衡闪蒸计算方法研究［J］. 石油勘探与开发，1992，19（3）：48-55.

［19］傅鹰. 化学热力学导论［M］. 北京：科学出版社，2010：118-129.

［20］何伟，孙雷. 牛顿—拉夫森算法在相平衡计算中的应用研究［J］. 石油勘探与开发，1999，26（4）：68-71.

The Analysis of Dynamic Displacement Characteristics and Influential Factors for CO_2 Foam Flooding in Low Permeability Reservoirs

Liu Li[1]   Pi Yanfu[1]   Wan Xue[2]   Gong Ya[2]   Liu Yingjie[2]

(1. Key Laboratory of Enhanced Oil and Gas Recovery of Ministry of Education of China, Northeast Petroleum University; 2. Key Laboratory of Enhanced Oil and Gas Recovery of Ministry of Education of China, Northeast Petroleum University)

Abstract: In view of low permeability reservoirs, this paper had conducted foam performance test and physical simulation experiments for CO_2 foam system of three kinds of different foaming agent under condition of three kinds of temperature, contrasted the effect of enhanced recovery, and analyzed how the kind of foaming agent, temperature, pressure, the stability of foam that crude oil influenced and other aspects influenced the displacement efficiency. The results show that, the foaming volume of the foaming agent called FP388 is the biggest, and FP388's half-life period is in the middle, its aggregative index of foam is highest, it means that it is easier for FP388 to form stable foam than the other two kinds of foam system under the same condition; At the same level of permeability of cores, the three kinds of CO_2 foam system could enhance the recovery ratio for more than 20% after water flooding, it turns out that the three kinds of foam system have good ability to enhance the recovery, and the effect of the foam system which is composed by FP388 is the best; And this paper had haven in-depth analysis of how the formula of foam system, temperature, pressure, crude oil and other aspects effected the displacement efficiency, and obtained the following understandings: From the view of the types of foaming agent and temperature, the foaming agent called FP388 in the former three kinds of foaming agent is the most stable; From the view of the pressure, the stability of foam increases with the increase of pressure until it reaches the stable system; From the view of the crude oil, how crude oil influences the stability of foam depends on whether the crude oil is the state of dissolved gas or the state of oil emulsion and whether the crude oil has formed the stable false emulsion membrane.

Keywords: low permeability; CO_2 foam flooding; displacement efficiency; influential factors.

0   Introduction

As a kind of gas drive, CO_2 flooding has the same displacement mechanism as the general gas drive, besides that some special displacement mechanism which is because of its physical and chemical characteristics that CO_2 is easy to soluble in oil and water[1]. CO_2 can greatly enhance the oil recovery by reducing the viscosity of crude oil, reducing the oil-water interfacial tension, reaching miscible with oil and other displacement mechanisms, and CO_2 flooding has been one of the most promising methods to enhance oil recovery factor[2]. Foam flooding is an efficient method in enhancing the oil recovery, and it can control the fluidity of gas and improve gas breakthrough and viscous fingering in heterogeneous reservoirs during the gas drive. The indoor experimental results in profile control of CO_2 foam show that, the seepage characteristic of CO_2 foam is very special, and CO_2 foam has selective plugging for layer and can greatly enhance the oil recovery[3-4]. So the scholars both at home and abroad have paid further attention on the study and application in CO_2 flooding and CO_2 foam flooding[5-7], and it needs to be studied further that the choice of foaming agent for CO_2 foam flooding, the influential factors on the performances of CO_2 foam and the aspect of improving the recovery mechanism for foam. Therefore, this paper has compared the dynamic displacement effect of the three kinds of CO_2 foam system in low permeability reservoirs through the laboratory experiment, and analyzed the related influential factors of oil displacement efficiency.

1 Experiment Section

1.1 Experimental Equipments and Materials

1.1.1 Experimental equipments

Air compressor; Gas cylinder which is equipped with high CO_2 gas and is used for air source; Piston containers whose volume is 1000mL and 400mL; Foam generator; Constant -flux pump and ISCO pump; Plastic pipeline and steel pipeline; Calorstat; Six-way valve and pressure gage; Equipment for saturating formation water; Gas metering device.

1.1.2 The main experimental materials

Foaming agents: FP388, FP246, FP275; Foam stabilizer: WP125; Viscosity stabilizing agent: Thiocarbamide;

The experimental oil: Simulated oil; The experimental water: Simulated water which is prepared according to the requirements of salinity in oil field;

The experimental cores: Artificial cores glued by quartz sand, the standard cylindrical cores (ϕ2.5cm×10cm). The permeability of the core is averagely around 10mD which is selected according to the average permeability of simulation demonstration area, the number of core is 9, they are 1-9 according to the experiment scheme numbers.

1.2 Experimental Schemes

In order to compare the displacement efficiency of CO_2 foam system made by three kinds of different foaming agent in low permeability cores, we have conducted water flooding and then CO_2 flooding under the temperature condition of 45℃, 75℃, 100℃, specific experimental schemes is that water displaces the oil until the moisture content is 98%, and then foam displaces oil until the moisture content is 98%.

1.3 Experimental Process

According to the requirement of the experimental conditions, the experimental process of foam flooding is as shown as Fig. 1.

In chart 1, there are respectively CO_2 gas and solution of foaming agent system in piston container 3 and piston container 4, the function of pump 1, 2 makes the proportion between solution of foaming agent system and CO_2 gas be 1:10, and then drive them into foaming core in core holder 5, so the system of CO_2 foam. When the stable foam is observed at Valve 13, then open the Valve 14 in order to make the stable foam enter into experimental core in 8. The main function of 6 and 7 is measuring the pressure at injection side in the foam system. Back pressure control valve controls the back pressure at delivery end of the core, according to bottom hole flowing pressure of production well in the demonstration zone, the back pressure which is 8MPa is set.

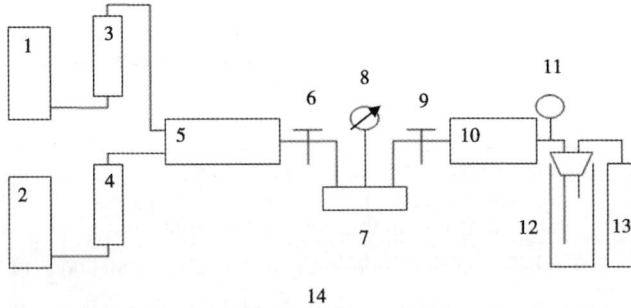

Fig. 1　Diagram of Experiment Process for Foam Flooding

1, 2—ISCO pump; 3—Piston container filled with CO_2; 4—Piston container filled with foaming agent; 5—Core holder;

7—Six-way valve; 6, 9—Valve ; 8—Pressure gage; 10—Core holder; 11—Back pressure control valve;

12—Liquid metering device; 13—Gas metering device; 14—Calorstat

1.4　Experimental Procedure

The experimental process is the same for the three kinds of CO_2 foam system, the specific experimental steps are as follows:

(1) The performance test for foam;

(2) Put the core into core holder, make the casing pressure be 3.5MPa, and vacuum for 2h;

(3) Saturate water for the core, and then cultivate it;

(4) Saturate oil for the former cultivated core, and calculate the initial oil saturation for this core, then put the core into the calorstat for 8–10h;

(5) Conduct water flooding experiment, and record the data which includes the pressure at the injection side, fluid production, oil production every 30min, and stop the water flooding experiment when the water content ratio reach at 98%;

(6) Use the foam-making apparatus (the foaming core) to make foam, connect the delivery end of the foam with the core holder when the stable foam is produced, and conduct the experiment of CO_2 foam flooding, and record the experimental data which includes the pressure at the injection side, fluid production, oil production, gas production every 30min, and stop the experiment when the water content ratio reach at 98%;

(7) Clean up and calculate the experimental data.

2 Experimental Results

The test data logging of foaminess and foaming stability for the three kinds of foaming agent under the normal temperature and pressure is as shown as Table 1.

Table 1 The Test Result of Foaminess and Foaming Stability for the Three Kinds of Foaming Agent (Normal Temperature and Pressure)

The type of foaming agent	Foaming volume V_f (mL)	Half-life period $t_{1/2}$ (s)	Aggregative index FCI (mL · s)
FP388	930	499	348052.5
FP246	910	439	299617.5
FP275	730	310	169725

Annotations: the mass concentration of foaming agent used in testing is 0.4%.

It can be seen from Table 1 that under the normal temperature and pressure, the foaming volume of t FP388 is the biggest between the three kinds of foaming agent, and FP388's half-life period is in the middle, its aggregative index of foam is highest, it turns out that FP388's overall performance is the best.

Cores of the same level permeability are selected, in view of the three kinds of CO_2 foam system, the half-life period of CO_2 foam system and the biggest foaming volume in the normal temperature and pressure are tested under the temperature condition of 45℃, 75℃, 100℃, the data logging is as shown as Table 2.

Table 2 The Experimental Data of Half-Life Period for Foaming Agent (Normal Pressure)

The type of foaming agent	Half-life period $t_{1/2}$ (s)		
	45℃	75℃	100℃
FP388	964	483	167
FP246	923	409	113
FP275	1055	566	221

It can be seen from Table 2 that under the normal pressure, CO_2 foam system's stability gets worse with the temperature increases, and its half-life period decrease sharply; half-life period of the foam downs quickly when the temperature is from 45℃ to 100℃, and its stability changes sharply; for the three kinds of foaming agent solutions whose concentrations are the same, FP275's half-life period is the biggest at the same temperature, it means that FP275's foam stability is the best; FP246's stability is fairly with FP388's only when the temperature is 45℃, but when the temperature increases, FP246' stability decreases sharply, and the stability of foam formed by FP246 is the worst between the three kinds of foaming agent; with the increase of temperature, the difference of stability between FP388 and FP275 decreases gradually.

Cores of the same level permeability are selected, in view of the three kinds of CO_2 foam system, the experiments of water flooding and then CO_2 flooding have been conducted under the tem-

84

perature condition of 45℃, 75℃, 100℃, the experimental result is as shown as Table 3.

Fig. 2 is drawn according to the data in Table 3, and then further compare the displacement effect between the three kinds of foam system, as shown as Fig. 2.

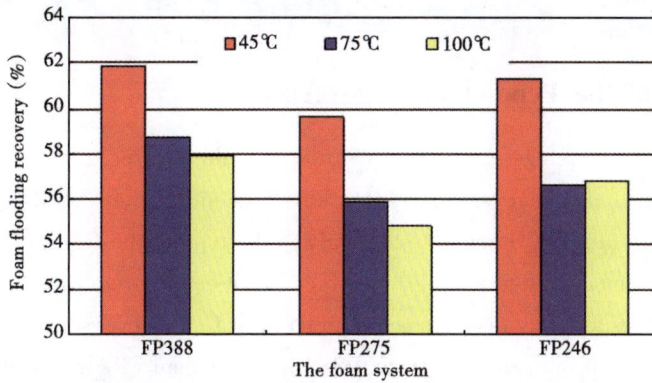

Fig. 2　Comparison Diagram of Overall Recovery for the Three Kinds of Foam System

It can be seen from the Table 3 and Fig. 2 that when the temperature of indoor displacement experiment is 45℃, 75℃ or 100℃, the laws of displacement efficiency for the three kinds of foam system are similar, it is that the effect of foam system made up by FP388 foaming agent is the best, the effect of foam system made up by FP246 foaming agent takes second place, the effect of foam system made up by FP275 foaming agent is the worse.

Table 3　The Data of Displacement Experiment for the Three Kinds of CO_2 Foam System

Experimental scheme	Permeability (mD)	Temperature (℃)	The type of foaming agent	The degree of reserve recovery		
				Water flooding	Increased	Total
1	10. 2		FP388	36. 6	25. 3	61. 9
2	10. 3	45	FP275	35. 0	24. 6	59. 6
3	10. 1		FP246	37. 2	24. 1	61. 3
4	10. 4		FP388	35. 9	22. 8	58. 7
5	10. 3	75	FP275	34. 9	21. 0	55. 9
6	10. 2		FP246	34. 8	21. 8	56. 6
7	10. 1		FP388	36. 3	21. 6	57. 9
8	10. 5	100	FP275	34. 8	20. 0	54. 8
9	10. 2		FP246	35. 9	20. 9	56. 8

3　Analysis of Influential Factors

Laboratory oil displacement experiment is an important way to evaluate the CO_2 foam flooding, in the process of oil displacement, the formula and component of CO_2 foam system, injection condition, technological conditions, reservoir characteristic and others influence the effect of the foam flooding [8]. This study has mainly conducted oil displacement experiment for three kinds of conditions and three kinds of foaming agent system. In the experiment, the degree of mineralization for formation water is certain and the levels of permeability for core are the same, therefore this part has studied displacement efficiency of CO_2 foam flooding only considering the formula of foam system, temperature, pressure and crude oil.

3. 1   The Effect of the Type of Foaming Agent

Only the type of foaming agent is different in the formula of three kinds of foam system in this paper, and other type of chemical reagent and working concentration are all the same, so the difference between the three kinds of formula system only is the type of the foaming agent. According to the displacement experiment results (Table1 and Fig. 2), it can be seen that the types of foaming agent are different and respectively displacement efficiency of CO_2 foam system are different, manifestation at three kinds of temperature all is that FP388 is better than FP246 and FP246 is better than FP275, namely the displacement efficiency of the foam system which uses the FP388 as the foaming agent is always the best between the three, the foam system which uses the FP246 as the foaming agent takes the second place, and the foam system which uses the FP275 as the foaming agent is the worst.

3. 2   The Effect of Temperature

The foam system is typical thermodynamic instability system. As the temperature increases, the solubleness of the foaming agent increases, the adsorptive capacity at the gas−liquid interface decreases, and the film strength decreases; in the heat, the viscosity of liquid between the films decreases, the rate of delivery for liquid film accelerates, and that makes foam stability decrease, and its performance is that the foam in CO_2 foam system is easy to break out during the displacement process and influence its ultimate displacement efficiency.

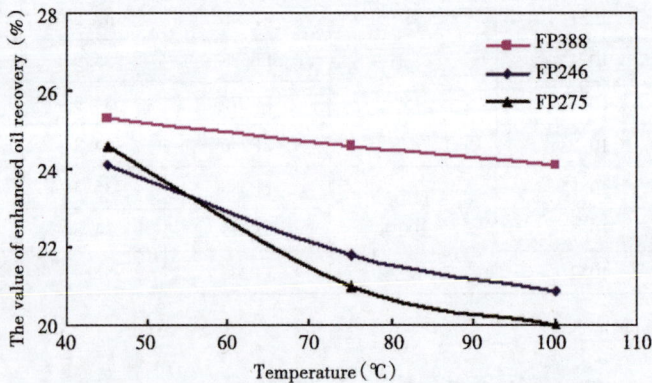

Fig. 3   The Relationship between the Value of Enhanced Oil Recovery for the
Three Kinds of Foam System and Temperature

It can be seen from Fig. 3 that when the temperature increases from 45℃ to 100℃, the values of enhanced oil recovery for the three kinds of foam system are all decreased, and the higher the

86

temperature is, the lower the value of enhanced oil recovery is, but the system made by FP388 is more stable comparing with the other kinds of foam system.

3.3   The Effect of Pressure

It is generally believed that when the pressure increases, the foam stability increases[9].

During the displacement process, and when the permeability of the core is around 10mD, it needs higher injection pressure in order to inject CO_2 foam system, during the foam flooding process, resistance factor is bigger and the pressure in the system is higher, and those are good for the stability of foam. For the actual heterogeneous reservoir, the most of foam enters into more tend to be high permeability layer, so the higher resistance factor are formed at high permeability layer, and this forces subsequent injected fluid into the low permeability layer, which expands the swept volume, and then enhances oil recovery.

Taking high temperature and high pressure test results in PVT canister as a reference [10], it also verifies that when the pressure increases, the foam stability increases. Take the foam system of CO_2+FP388 as an example, the relationship between the pressure under the three kinds of temperature and percentage of liquid volume is as shown as Fig. 4.

It can be seen from Fig. 4 that for the foam system of CO_2+FP388, when the temperature is a certain, the percentage of liquid volume in PVT canister gets less and less with the increase of pressure, it turns out that more solution of foaming agent forms stable foam with

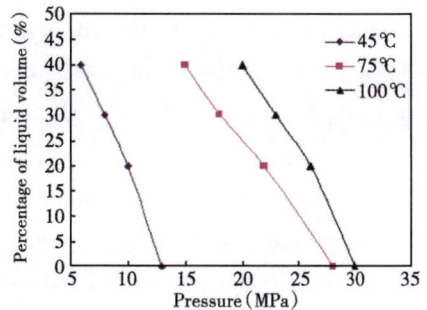

Fig. 4   The Relationship between the Pressure for the Foam System of CO_2+FP388 and Percentage of Liquid Volume

the increase of pressure; When the pressure increases to critical threshold of pressure low limit, all of solution of foaming agent forms stable foam with CO_2 gas; When the pressure continues to increase, there will be no liquid precipitine, so it can be seen that increasing pressure can actually increase the stability of foam until the foam reaches the stable system. That is because of after increasing the pressure, the density of gas molecule increases, and it also benefits that the molecules of foaming agent can arrange closely, and the interfacial tension of the formed foam system gets bigger, it is easier to form stable foam.

3.4   Crude Oil's Effect on Foam Stability[11]

The reaction between foam and crude oil first occurs between foam film and crude oil. The effect that crude oil on the stability of foam depends on that whether the crude oil is dissolution oil or emulsible oil, and whether it has been formed stable false emulsion membrane which is as shown as Fig. 5[12-13].

Only when there have formed stable false emulsion membranes between foam and crude oil, the stability of foam will be stronger, or the existence of the crude oil is to reduce the stability of foam. So when the foam meets with the crude oil, some foam will be more stable, and others will be very

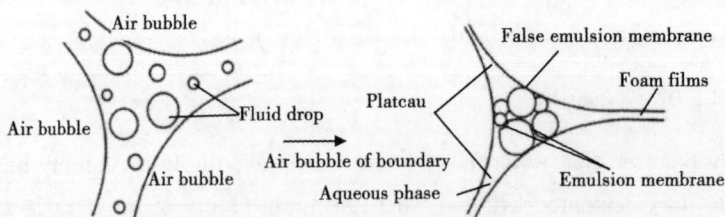

Fig. 5　False Emulsion Membrane, Emulsion Membrane, and the Structure in
Plateau after Flowing for Three-Phase Froth

unstable; when the same foam meets with different crude oil, the stability of the foam will be different.

In the process of this experiment, it has been found that when the injected CO_2 foam system meets the residual oil in the core, its stable performance will decrease and influence the displacement efficiency of foam. During the testing experiment of lower limit of permeability for foam, when the foam system has been injected into cores for a long time, there will be part of the foam discharges with liquid at the fluid end of the core, along with the continuing experiment, the foam output will increase until it is stable at a certain foam volume; But in the process of foam flooding experiment, there are only trace amounts of foam in produced liquid until the last stage of the moisture content increased. Those turns out that when it is conducted that foam flooding after water flooding, once CO_2 foam system meets the residual oil in the core, its stable performance will decrease, which can decrease the recovery efficiency of foam flooding.

3.5　The Analysis of Causes

Analyze the reasons for the above situations, we think:

(1) FP388's foaming volume is the biggest between the three kinds of foam system, and its half-life period is in the middle, its aggregative index of foam is the highest, and by the PVT experiment, it is found that its critical pressure of lower limit is the lowest to form stable foam (as shown as Fig. 5: it is 13MPa when the temperature is 45℃, and is 28MPa when the temperature is 75℃, and is 30MPa when the temperature is 100℃), it means that it is the easiest to form stable foam at the same condition, this is why its displacement efficiency is the best.

(2) FP246's foaming volume is in the middle between the three kinds of foam system, and its half-life period is the lowest, but its overall performance of aggregative index for foam is in the middle, and by the PVT experiment, it is found that its critical pressure of lower limit (it is 18MPa) to form stable foam is higher than FP388's only when the temperature is 45℃, its critical pressure of lower limit is the same as FP388's, it means that it is easier to form stable foam at the same condition, so its overall performance of displacement efficiency is better.

(3) FP275's foaming volume is the lowest between the three kinds of foam system, though its half-life period is the longest, its overall performance of aggregative index for foam is the lowest, and by the PVT experiment, it is found that its critical pressure of lower limit is the highest to form stable foam (it is 20MPa when the temperature is 45℃, and is 28MPa when the temperature is

75℃, and is 35MPa when the temperature is 100℃), it means that it is difficult to form stable foam for FP275 system at the same condition, so its overall performance of displacement efficiency is the worst between the three kinds of system; but because its molecular chain is not destroyed under the condition of low temperature 45℃, and it still has higher displacement capacity, so its displacement efficiency is a little better than FP246.

4 Conclusions

(1) Three kinds of foam system are mentioned in this paper, and the displacement efficiency for the foam system whose foaming agent is FP388 is the best, this is because FP388's foaming volume is the biggest between the three kinds of foam system, and its half-life period is in the middle, its aggregative index of foam is the highest, it means that it is the easiest to form stable foam at the same condition.

(2) In the cores of the same level permeability, all those three kinds of CO_2 foam system can enhance the oil recovery ratio above 20%, it means that they have a better ability to enhance the recovery ratio, and the effect of the foam system whose foaming agent is FP388 is the best.

(3) This paper has analyzed and testified that the type of foaming agent, temperature, pressure, the existence of the crude oil and other factors can influence the displacement efficiency of foam flooding. In the aspect of the type of foaming agent and temperature, the foaming agent called FP388 is the most stable between the three kinds of foaming agent which are mentioned in this paper; In the aspect of pressure, the foam stability will increase with the pressure increasing until the foam system reaches the stable system; In the aspect of crude oil, the effect that crude oil on the stability of foam depends on that whether the crude oil is dissolution oil or emulsible oil, and whether it has been formed stable false emulsion membrane.

References

[1] Wang Guanhua. The study of profile control and flooding technology for supercritical CO_2 foam [D]. China University of Petroleum, 2011.

[2] Zhang Yang, Zhang Liang, Chen Bailian, et al. The study of performance evaluation and experimental method for CO_2 foam in the normal temperature and pressure [J]. Journal of Chemical Engineering Colleges and Universities, 2014, 28 (3): 535-541.

[3] Li Hong. The comparison and analysis of mobility ratio between flooding and foam flooding in heterogeneous reservoirs [J]. Production Test Technology, 1998, 19 (4): 24-26.

[4] Li Chun, Yi Xiangyi, Liu Wei, et al. The selection of foaming agent for CO_2 foam flooding in Caoshe Oilfield [J]. Oilfield Chemistry, 2007, 24 (3): 255-257.

[5] Wang Haitao, Yi Xiangyi, Li Xiangfang, et al. CO_2 foam plugging experiments in the high temperature and high salinity reservoirs [J]. Xingjiang Petroleum Geology, 2009, 30 (5): 641-643.

[6] Enick R M, Olsen D K, Ammer J R, et al. Mobility and conformance control for CO_2 EOR via thickeners foams and gels—A literature review of 40 years of research and pilot tests [C] //SPE Improved Oil Recovery Symposium. Society of Petroleum Engineers, 2012.

[7] Zhang Yang, Song He, Li Dexiang, et al. The stability test of CO_2 foam in high pressure based on non-ionic

surfactant [J]. Journal of China University of petroleum: Natural Science, 2013, 37 (4): 119–123.

[8] Li Lizhong. The study of displacement mechanism for foam complex flooding and factors affecting displacement efficiency [D]. Daqing Petroleum Institute, 2003.

[9] Song He, Zhang Yang, Chen Bailian, et al. The experiment research of high temperature and high salinity CO_2 foam performance [J]. Oilfield Chemistry.

[10] Liu Li, Song Kaoping, Wang Yu, et al. The study of P–V characteristics in high temperature for CO_2 foam system [J]. Science technology and engineering, 2014, 14 (30): 135–138.

[11] Fan Xijing. How the crude oil impacts on the stability of foam [J]. Oilfield Chemistry, 1999, 14 (4): 384–388.

[12] Jia Xingang, Yan Yongli, Qu Chengtun, et al. The progress of the mechanism of the crude oil to stabilization of aqueous foam [J]. Household and personal care chemical industry, 2010, 40 (1): 54–59.

[13] Lobo L A, Nikolov A D, Dmitrov A S, et al. Contact angle of a air bubbles attached to an air–water surface in foam applications [J]. Langmuir, 1990 (6): 995–1001.

[14] Schramm L L. Foams fundamentals and applications in the petroleum industry [M]. Washington DC: ACS, 1994.

CO_2 地质埋存中扩散规律实验研究

李　实　许世京　陈兴隆

(中国石油勘探开发研究院提高石油采收率国家重点实验室)

摘　要：盐水层中二氧化碳的物理运移会影响二氧化碳的溶解和流动过程。因此，二氧化碳在地层水中的运移规律对二氧化碳地质埋存量有重要影响。本研究通过实验方法，以深部咸水层二氧化碳地质封存为背景，针对超临界压力二氧化碳和水在深部盐水层中的驱替过程和运移规律进行了实验研究，研究了距离埋存点不同位置的二氧化碳和水的分布规律。研究发现，在二氧化碳的运移中，浮力和浓度梯度均起到重要作用，二氧化碳竖直向上扩散的速度远大于竖直向下和沿水平方向扩散的速度。在二氧化碳的地质埋存中，二氧化碳竖直方向上的扩散起主要作用。

关键词：CCUS；地质埋存；深部盐水层；垂向扩散

1　实验装置

二氧化碳扩散实验采用的主要实验装置为耐压容器、驱替泵及水浴加热系统。耐压容器的最大工作压力为70MPa，水浴的最高工作温度为100℃；耐压容器可360°旋转，在一侧相同间隔处有多个取样测量点。二氧化碳扩散实验流程如图1所示。

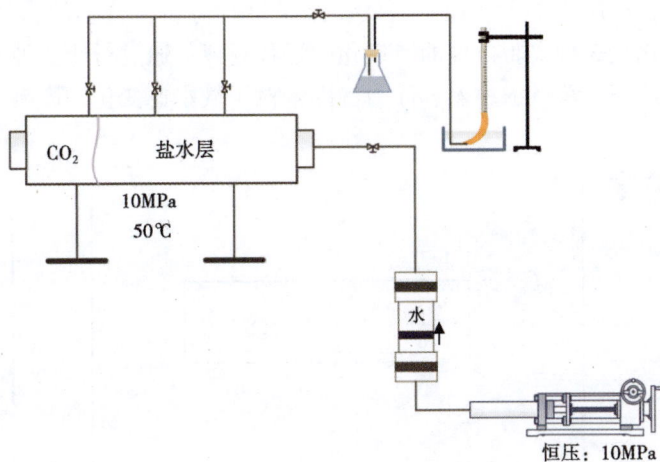

图1　二氧化碳埋存实验流程图

2 实验程序

在正常的大气条件下，二氧化碳在热力学上非常稳定，密度为 $1.872kg/m^3$。当温度和压力均大于临界值（$T_c = 31.1℃$，$p_c = 7.38$ MPa）时，二氧化碳处于超临界状态，密度明显上升，根据地层已知的压力和温度范围，二氧化碳密度在 $150\sim800kg/m^3$ 范围内变化[12]。

孔隙度测量：岩石孔隙体积的大小通常以孔隙度来表征。而所谓孔隙度，指的是填砂模型的孔隙体积与填砂模型中充填物视体积的比值，即 $\phi = V_p/V_f$。本文所使用的填砂模型的基本参数见表1。

表1　二氧化碳埋存实验填砂模型基本参数

长度 （mm）	直径 （mm）	孔隙体积 （mL）	孔隙度 （%）
1000	100	2906	37

实验方法：将耐压容器填砂并抽真空，然后饱和盐水。升温至地层温度（50℃）并恒温 4h 以上，用泵注入盐水至地层压力（10MPa）。打开出口阀门，在恒压状态注入 600mL 二氧化碳。关闭出口阀门，维持系统压力为 10MPa。间隔一定时间从取样测量点取混合液，分别测量锥形瓶中的液体体积，并用排水法测二氧化碳的体积。由此获得随着二氧化碳的扩散，地层水/气相体积的变化规律。

3 实验结果分析

首先定义一个无量纲的量 L/H，其中 L 是取样点到二氧化碳注入点的垂向距离，H 是耐压容器底面的半径。此无量纲量便于将实验中的结果推广至更大的空间尺度。距离二氧化碳注入点的距离越远，L/H 值越大。通过对实验结果进行分析，分别给出从顶部、沿水平方向、从底部注二氧化碳 3 种实验条件的二氧化碳运移规律；通过对比 3 种实验条件下同一取样点的气水比变化速率，获得浓度差、浮力等因素对于气水比变化的影响。

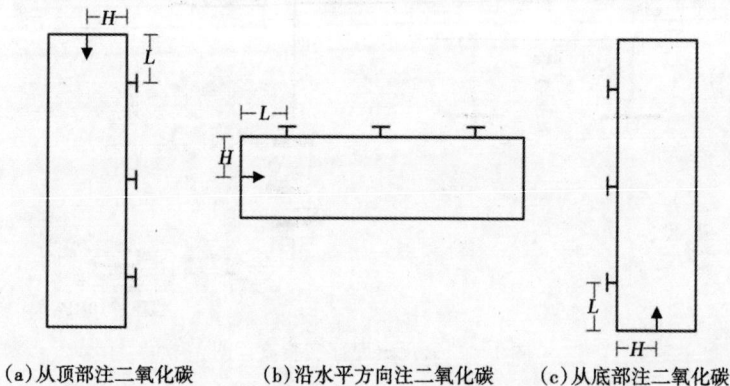

(a)从顶部注二氧化碳　　(b)沿水平方向注二氧化碳　　(c)从底部注二氧化碳

图2　沿不同方向注二氧化碳示意图

3.1 从顶部注二氧化碳的运移规律

二氧化碳在运移过程中，影响其扩散的主要因素是浮力引起的对流作用与浓度差引起的扩散作用。二氧化碳从顶部注入后，最初以游离态存在。由于重力作用，注入的二氧化碳会聚集在储层的上部。可以认为二氧化碳与盐水层之间形成了一个非常薄的水平界面。在此界面下方，二氧化碳溶解在盐水中，形成一个随厚度、时间增长的扩散边界层。

由于盐水的密度随着二氧化碳浓度的增加而增加，因此在此扩散边界层处的盐水溶解大量二氧化碳后密度比边界层下面的盐水大，上层密度大的水和下层密度小的水之间会发生对流。密度大的含二氧化碳的水向下运移，密度小的不含二氧化碳的水向上运移，与上层的二氧化碳接触。这种密度差驱动的对流可以加快二氧化碳向盐水中的传质，可以比二氧化碳单纯的溶解作用快几个数量级。

实验结果表明，在同样的时刻，距离气顶越近的取样点得到的气水比越高，气水比随着时间的增加而增大，距离埋存点较远的地方，汽水比很低而且增加缓慢；对于同一个取样点而言，气水比随着时间的增加而增加，最后趋于稳定。对于距离注入点最近的取样点（L/H =4），取样点处气水比单位时间内的视增量，$\delta = \delta_d - \delta_b = 1.63 \times 10^{-6}$（mg/l）/s，其中 δ_d 代表浓度差的影响，δ_b 代表浮力的影响。对于从顶部注二氧化碳而言，浮力与浓度差的作用相反，对二氧化碳的运移起阻碍作用，因此在计算式中 δ_b 前为负号。

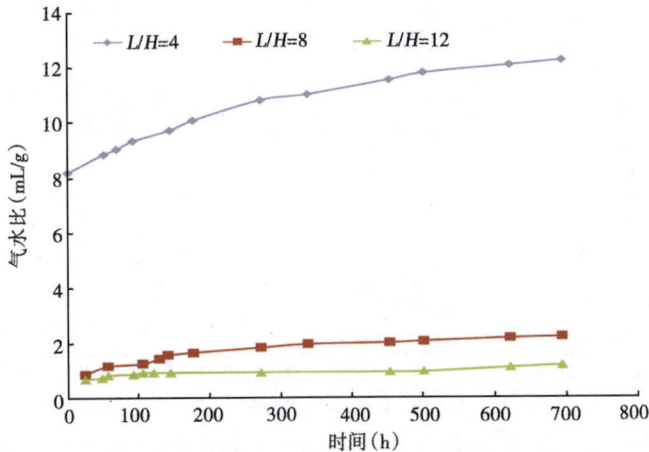

图3　不同取样点的气水比随时间的变化规律（从顶部注二氧化碳）

3.2 沿水平方向注二氧化碳的运移规律

沿水平方向注二氧化碳，取样点位于压力容器的上部。在运移过程中不需要考虑浮力的影响，仅考虑由于浓度梯度引起的扩散，取样点处气水比单位时间内的视增量 $\delta = \delta_d = 2.95 \times 10^{-6}$（mg/l）/s，其中 δ_d 代表浓度差的影响。距离埋存点较远的地方，气水比很低而且增加缓慢。

3.3 从底部注二氧化碳的运移规律

由图5可以看出，在同样的时刻，距离气顶越近的取样点得到的气水比越高；气水比随着时间的增加而增加。取样点处气水比单位时间内的视增量 $\delta = \delta_d + \delta_b = 9.7 \times 10^{-5}$（mg/l）/s，其中 δ_d 代表浓度差的影响，δ_b 代表浮力的影响。对于从底部注二氧化碳而言，浮力与浓度梯度的作用相同，对二氧化碳的垂向运移起促进作用。

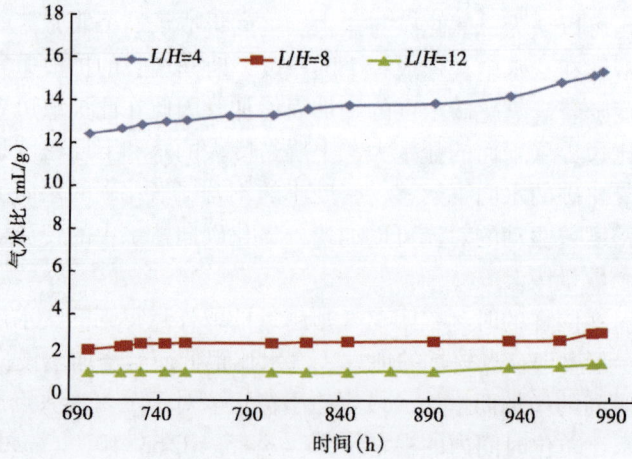

图 4　不同取样点的气水比随时间的变化规律（沿水平方向注二氧化碳）

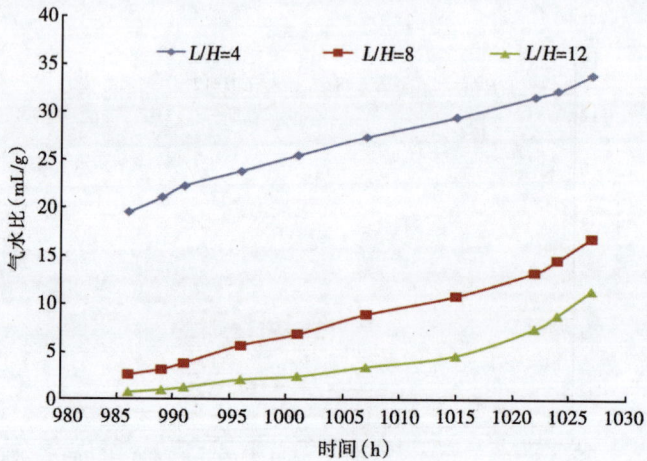

图 5　不同取样点的气水比随时间的变化规律（从底部注二氧化碳）

4　结论

从以上分析可以看出，二氧化碳沿水平方向扩散的速度大于竖直向下扩散的速度，但两者基本处于相同的数量级；二氧化碳竖直向上扩散的速度远大于竖直向下和沿水平方向扩散的速度，两者相差 1~2 个数量级。二氧化碳的地质埋存中，二氧化碳竖直向上的扩散起主要作用。因此为了保证二氧化碳不泄漏，封存地点储层之上需要一个低渗透盖层以阻止二氧化碳在浮力作用下向上运移。在 CO_2 封存初期，绝大多数 CO_2 以游离状态存在，在浮力作用下向上运移直到被盖层所阻挡。经过长时间地与水接触，部分 CO_2 溶解于水中。由于 CO_2 饱和的地下水比未饱和的地下水密度略大，在重力作用下，CO_2 饱和的地下水将向下运移。

这种对流可以促进 CO 溶解于水中，利于长期封存[6]。

参 考 文 献

[1] Metz B, Dvidson O R, Bosch P R, et al. Summary for policymakers//Climate change 2007: Mitigation. contribution of working group III to the fourth assessment report of the intergovernmental panel on climate change [R]. Cambridge University Press, Cambridge, United Kingdom and New York, NY, USA. IPCC, 2007.

[2] IPCC. Carbon Dioxide Capture and Storage [M]. UK: University Press, 2005.

[3] Taku Ide S, Kristian Jessen, Franklin M. Orr. Storage of CO_2 in saline aquifers: Effects of gravity, viscous, and capillary forces on amount and timing of trapping [J]. *International Journal of Greenhouse Gas Control*, 2007 (4): 481-491.

[4] Holt T, Jensen J I, Lindeberg E. Underground storage of CO_2 in aquifers and oil reservoirs. *Energy Convers. Mgmt.* 1994, 36 (6-9): 535-538.

[5] HORTON C W, ROGERS F T. Convection currents in a porous Medium [J]. *Journal of Applied Physics*, 1945 (16): 367.

[6] Ennis-King Jonathan, Paterson L. Role of convective mixing in the long-term storage of Carbon dioxide in deep saline formations [J]. SPE Journal, 2005, 10 (3): 349-356.

CO_2 地质埋存中浓度梯度
引起的垂向扩散规律研究

李 实　许世京　陈兴隆　张　可　俞宏伟　李 军

(中国石油勘探开发研究院提高石油采收率国家重点实验室)

摘　要：本文以二氧化碳在盐水层中的驱油与埋存为研究背景，运用数值分析方法，开展在垂向浓度梯度作用下二氧化碳在各向同性的多孔介质中的扩散对流作用规律研究。研究表明，为了使垂向浓度扩散产生，需要满足的条件是垂向浓度梯度 $\lambda > 4D\pi^2\mu/(Kg\rho_0\gamma H^2)$，其中 D 为分子扩散系数，μ 为流体黏度，K 为岩石渗透率，ρ_0 为流体初始密度，γ 是流体密度随浓度的变化值，H 是流体的垂直厚度。在二氧化碳地质埋存中，垂向的扩散能力对于二氧化碳的储存容量有很大影响，本文给出了该扩散作用发生的条件。

关键词：CCUS；地质埋存；深部盐水层；垂向扩散；浓度梯度

0　引言

二氧化碳捕集、利用与封存（Carbon Capture Utilization and Storage，CCUS）技术是一项新兴的、具有大规模二氧化碳减排潜力的技术，有望实现化石能源的低碳利用，被广泛认为是应对全球气候变化、控制温室气体排放最现实的重要技术之一[1]。全球气候变化问题日益严峻，已经成为威胁人类可持续发展的主要因素之一，削减温室气体排放以减缓气候变化成为当今国际社会关注的热点。有关研究显示，未来几十年化石能源仍将是人类最主要的能量来源，要控制全球温室气体排放，除大力提升能源效率、发展清洁能源、提高自然生态系统固碳能力外，CCUS 技术将发挥十分重要的作用。我国的能源结构以煤为主，随着国民经济的快速发展，CO_2 排放量将长期处于高位；发展煤炭行业 CO_2 捕集与石油行业 CO_2 驱油及埋存产业链的关键技术，有利于优化能源结构和 CO_2 减排，保障我国能源安全；鉴于未来可能形成的全球性低碳产业，发展 CCUS 技术将是提升我国低碳产业竞争力的重要机遇，CCS-EOR 是中长期温室气体减排最现实和最重要途径之一。

二氧化碳被封存于地下之后，有以下 4 种可能的存在方式：（1）以超临界状态存在于储层中；（2）以残余气形式束缚于空隙中；（3）溶解于地下流体中；（4）与原生矿物发生化学反应形成次生矿物[2]。深部咸水层封存 CO_2 成为地质封存技术中封存量最大、最有前景的途径，CO_2 在多尺度非均质深部咸水层中的运移与俘获特征将是碳捕捉封存技术的研究重点。

二氧化碳注入盐水层后，其传质过程受到若干个体积力的共同作用[3]。在较短的时间尺度内，黏度、重力、毛细管压力以及地层中的温度梯度起综合作用。在大尺度的各向同性盐水层，相对于黏性力而言，重力的影响对于二氧化碳的运移起决定性作用，导致二氧化碳覆盖在盐水之上。但是由于在竖直方向上二氧化碳存在浓度梯度，底部未饱和的盐水会运移至气水界面上部，使二氧化碳向低浓度方向扩散，这样就增大了二氧化碳在盐水层中的储存量。为了加强盐水层的存储能力，必须加强二氧化碳在垂向上的扩散。Holt 等人使用油藏模拟程序预测二氧化碳在盐水层中的分布，但是并没有对垂直方向上的扩散条件进行建模[4]，因此，必须专门研究二氧化碳在垂直方向上的扩散以确定油藏模拟程序是否给出可信的预测。

1　数理模型

由于二氧化碳与盐水之间存在密度差，在浮力作用下，竖直方向上易形成较大的浓度梯度。以下只考虑由于浓度差引起的扩散作用。

多孔介质内流体的运动学方程[5]：

$$u = - (K/\mu)(\partial p/\partial x) \tag{1}$$

$$v = - (K/\mu)(\partial p/\partial y) \tag{2}$$

$$w = - (K/\mu)(\partial p/\partial z + g\rho) \tag{3}$$

式中，K 代表岩石渗透率；μ 代表水的黏度。

假设流体密度随着二氧化碳的浓度变化：

$$\rho = \rho_0\left(1 + \frac{\partial p}{\partial c}c\right) \tag{4}$$

式中，c 代表 CO_2 在水中的浓度；$\frac{\partial \rho}{\partial c}$ 代表密度对浓度的偏导数。

假设在深部盐水层中的盐水是不可压缩的，因此有：

$$\frac{\partial u}{\partial x} + \frac{\partial v}{\partial y} + \frac{\partial w}{\partial z} = 0 \tag{5}$$

扩散方程为：

$$\frac{\partial \rho}{\partial t} + \frac{1}{\phi}\left(u\frac{\partial \rho}{\partial x} + v\frac{\partial \rho}{\partial y} + w\frac{\partial \rho}{\partial z}\right) = \frac{D}{\phi}\left(\frac{\partial^2 \rho}{\partial x^2} + \frac{\partial^2 \rho}{\partial y^2} + \frac{\partial^2 \rho}{\partial z^2}\right) \tag{6}$$

式中，D 代表分子扩散系数；ϕ 代表岩石的孔隙度。

2　方程求解

为了求解式（1）至式（6），将浓度 c 分成稳定的浓度 λz（由垂直方向的浓度梯度 λ 引起）和浓度的波动值 c^* 两部分。令：

$$c = \lambda z + c^* \tag{7}$$

引入势函数：

$$\Omega = p + g\rho_0 z + g\rho_0 \gamma \int \lambda z \mathrm{d}z \tag{8}$$

97

在式（8）中做出假设，认为密度随浓度的变化是常数，令 $\dfrac{\partial \rho}{\partial c} = \gamma$。

根据式（4）、式（7）和式（8），式（1）至式（3）可以写为：

$$u = -(K/\mu)(\partial \Omega/\partial x) \tag{9}$$

$$v = -(K/\mu)(\partial \Omega/\partial y) \tag{10}$$

$$w = -(K/\mu)(\partial \Omega/\partial z + g\rho_0\gamma c^*) \tag{11}$$

式（6）变为：

$$\frac{\partial c}{\partial t} + \frac{1}{\phi}\left(u\frac{\partial c}{\partial x} + v\frac{\partial c}{\partial y} + w\frac{\partial c}{\partial z}\right) = \frac{D}{\phi}\left(\frac{\partial^2 c}{\partial x^2} + \frac{\partial^2 c}{\partial y^2} + \frac{\partial^2 c}{\partial z^2}\right) \tag{12}$$

根据式（7）可得：

$$\frac{\partial c^*}{\partial t} + \frac{1}{\phi}\left(u\frac{\partial c^*}{\partial x} + v\frac{\partial c^*}{\partial y} + w\frac{\partial c^*}{\partial z} + w\lambda\right) = \frac{D}{\phi}\left(\frac{\partial^2 c^*}{\partial x^2} + \frac{\partial^2 c^*}{\partial y^2} + \frac{\partial^2 c^*}{\partial z^2}\right) \tag{13}$$

假设 u，v，w，c^* 以及它们的变化率都很小，于是式（13）变为：

$$\frac{\partial c^*}{\partial t} + \frac{1}{\phi}w\lambda = \frac{D}{\phi}\left(\frac{\partial^2 c^*}{\partial x^2} + \frac{\partial^2 c^*}{\partial y^2} + \frac{\partial^2 c^*}{\partial z^2}\right) \tag{14}$$

根据流动稳定性理论，认为 u，v，w，c^* 和 Ω 均与 $e^{ilx}e^{imy}e^{nt}$ 成正比，于是式（9）至式（11）变为：

$$u = -i(Kl/\mu)\Omega \tag{15}$$

$$v = -i(Km/\mu)\Omega \tag{16}$$

$$w = -(K/\mu)(\partial \Omega/\partial z) - Kg\rho_0\gamma c^*/\mu \tag{17}$$

因此式（5）变为：

$$ilu + imv + \partial w/\partial z = 0 \tag{18}$$

通过式（15）至式（18），可得：

$$\Omega = -(\mu/K)[1/(l^2 + m^2)](\partial w/\partial z) \tag{19}$$

由式（17）和式（19）可得：

$$w = [1/l^2 + m^2)](\partial^2 w/\partial z^2) - Kg\rho_0\gamma c^*/\mu \tag{20}$$

于是式（14）变为：

$$nc^* + \frac{1}{\phi}w\lambda = \frac{D}{\phi}\left(\frac{\partial^2}{\partial z^2} - l^2 - m^2\right)c^* \tag{21}$$

我们认为在流体层的顶部（$z = H$）和底部（$z = 0$），w 和 c^* 均为 0，假设 w 和 θ 与 $\sin sz$ 成正比，其中 $s = r\pi/H$，r 为常数，于是式（20）变为：

$$w = -[s^2/(l^2 + m^2)]w - Kg\rho_0\gamma c^*/\mu \tag{22}$$

由式（21）可得：

98

$$nc^* + \frac{1}{\phi}\omega\lambda = -\frac{D}{\phi}(l^2 + m^2 + s^2)c^* \tag{23}$$

求解式（22）：

$$w = -\frac{(l^2 + m^2)Kg\rho_0\gamma c^*}{(l^2 + m^2 + s^2)\mu} \tag{24}$$

结合式（23）：

$$n = \frac{1}{\phi(l^2 + m^2 + s^2)}\left[(l^2 + m^2)Kg\rho_0\gamma\lambda/\mu - D(l^2 + m^2 + s^2)^2\right] \tag{25}$$

如果 λ 为负值，则 n 为负值，于是由浓度梯度引起的任意扰动都会随着时间以指数形式减小。如果 λ 是正值，则 n 有可能是正值，扰动有可能发生，这意味着只有 λ 是正值，由浓度梯度产生的传质才能发展下去。

令，$l^2 + m^2 + s^2 = \sigma$，可得：

$$n = \frac{1}{\phi\sigma}\left[(\sigma - s^2)Kg\rho_0\gamma\lambda/\mu - D\sigma^2\right] \tag{26}$$

n 为正值的条件是：

$$(\sigma - s^2)Kg\rho_0\gamma\lambda/\mu - D\sigma^2 > 0 \tag{27}$$

即

$$(\sigma - s^2)\lambda > D\sigma^2\mu/(Kg\rho_0\gamma) \tag{28}$$

流动稳定的临界点是 $n = 0$，在式（26）中令 $n = 0$，则：

$$(\sigma - s^2)\lambda - D\sigma^2\mu/(Kg\rho_0\gamma) = 0 \tag{29}$$

在式（29）中对 σ 求导：

$$\lambda - 2D\sigma\mu/(Kg\rho_0\gamma) = 0 \tag{30}$$

求解上述两个方程，可得：

$$\begin{cases} \sigma = 2s^2 \\ l^2 + m^2 = s^2 \end{cases} \tag{31}$$

对于给定的 s，浓度梯度引起的传质现象发生的必要条件是：

$$\lambda > D\sigma^2\mu/(Kg\rho_0\gamma s^2) = 4Ds^2\mu/(Kg\rho_0\gamma) \tag{32}$$

当 $r = 1$ 时，$s = r\pi/H$ 有最小值 π/H；由公式（32）可知，只有当 λ 满足以下条件时，由浓度梯度引起的传质现象才能发生：

$$\lambda > 4D\pi^2\mu/(Kg\rho_0\gamma H^2) \tag{33}$$

在此，可以定义使垂向扩散发生的最小的浓度梯度 $4D\pi^2\mu/(Kg\rho_0\gamma H^2)$ 为突破浓度梯度。

3 结论

本文研究了超临界二氧化碳盐水层封存过程中，超临界二氧化碳垂向扩散时垂向浓度梯度岩石渗透率、流体黏度和流体密度的关系，分析了这些因素对于二氧化碳垂直扩散的影响。研究表明：

（1）为了使垂向浓度扩散产生，需要满足的条件是垂向浓度梯度 $\lambda > 4D\pi^2\mu/(Kg\rho_0\gamma H^2)$，其中，使垂向扩散发生的最小的浓度梯度 $\lambda > 4D\pi^2\mu/(Kg\rho_0\gamma H^2)$ 为突破浓度梯度。

（2）在岩石孔隙度相同的情况下，影响突破浓度梯度增加的因素有岩石渗透率、流体黏度和流体密度等。

参 考 文 献

［1］ Metz B, Dvidson O R, Bosch P R, et al. Summary for policymakers // Climate change 2007: Mitigation. contribution of working group III to the fourth assessment report of the intergovernmental panel on climate change ［R］. Cambridge University Press, Cambridge, United Kingdom and New York, NY, USA. IPCC, 2007.

［2］ IPCC. Carbon Dioxide Capture and Storage ［M］. UK: University Press, 2005.

［3］ Taku Ide S, Kristian Jessen, Franklin M Orr Storage of CO_2 in saline aquifers: Effects of gravity, viscous, and capillary forces on amount and timing of trapping ［J］. International Journal of Greenhouse Gas Control, 2007, 1 (4): 481-491.

［4］ Holt T, Jensen J I, Lindeberg E. Underground storage of CO_2 in aquifers and oil reservoirs ［J］. Energy Convers. Mgmt., 1994, 36 (6-9): 535-538.

［5］ Horton C W, Rogers F T. Convection currents in a porous medium ［J］. Journal of Applied Physics, 1945 (16): 367.

Measurement and Prediction Model of Carbon Dioxide Solubility in Aqueous Solutions Containing Bicarbonate Anion

Tang Yong[1]   Bian Xiaoqiang[1]   Du Zhimin[1]   Wang Changquan[2]

(1. The State Key Laboratory of Oil & Gas Reservoir Geology and Exploitation Engineering, Southwest Petroleum University; 2. Petroleum Engineer College, Yangtze University)

Abstract: The solubility of CO_2 in aqueous solutions containing bicarbonate anion was determined at 308. 15K, 328. 15K, 368. 15K, and 408. 15 K up to 40 MPa. In comparison with Cl^-, HCO_3^- has a larger size and higher molecular weight, which leads to the opposite salting-out effect. HCO_3^- can suppress CO_2 dissolution because of equilibrium reaction between CO_2, and H2O to form HCO_3^-. Cooperation of the three effects leads to slightly variable solubility of CO_2 in the solution containing HCO_3^-. It is found that the Duan model can more accurately predict the CO_2 solubility at low salt concentration, while the PR-HV model shows smaller error at high salt concentration. Moreover, the solubility of CO_2 in natural ground water was measured at different temperatures and pressures, and the results verified the presented mathematical model.

Keywords: Carbon dioxide; Solubility; Water; Anion effect; PR-HV model

0 Introduction

The concentration of carbon dioxide (CO_2) in the atmosphere has increased considerably because of fossil fuel combustion, leading to a significant climate change. Serious efforts are likely to be made during the coming decades to reduce CO_2. A particular technology for achieving these reductions in the relatively near future is Carbon Capture and Sequestration (CCS)[1]. Geological storage, which injects CO_2 into deep saline aquifers, depleted hydrocarbon reservoirs, or deep coalbeds, is currently attracting much attention because of the large capacity[2].

Effective estimation of the total amount of the stored CO_2 requires the solubility of CO_2 in aqueous solutions. Generally, CO_2 dissolution into the aqueous phase is limited by its solubility, which in turn depends upon the water temperature, pressure, and salt concentration and composition. These parameters can change significantly from one location to another. Salt composition and concentration in deep formation waters may thus vary substantially among different formations depending on their depth, thermal conditions, geomorphology, and reservoir minerals.

Currently, many systems, such as CO_2-H_2O[3-9], $CO_2-H_2O-NaCl$[10-12], CO_2-H_2O-KCl[11, 13], $CO_2-H_2O-Na_2SO_4$[14], CO_2-seawater[15] and CO_2-brine[6], have been studied at different ranges of pressure and temperature. Many models including the Duan model[16,17] and the

Peng−Robison[18] Equation of state combined with Huron−Vidal[19] mixing rules (PR−HV model) for predicting CO_2 aqueous solubility have been developed in studies over recent decades using a combination of phase−equilibrium and empirical relationships[16,17,20−26]. Despite these contributions, the limitation of the current work is the relative rarity of investigations on the effect of anion of salts. In addition to halogen ion (Cl^-, Br^-, I^-) and SO_4^{2-}, three noteworthy anionic examples are nitrate (NO_{3-}), acetate (CH_3COO^-), and phosphate, which exhibit different effects on CO_2 solubility[27,28]. To the best of our knowledge, no measurement and prediction model of CO_2 solubility for salt solutions containing bicarbonate anion (HCO_{3-}) are reported in the current literatures, although prediction of CH_4 and N_2 solubility for salt solutions containing bicarbonate anion was successfully achieved[29]. Indeed, HCO_3^- is commonly present in natural ground waters although NaCl is often the major component, and HCO_{3-} can also be formed because of reaction between CO_2 and water. Therefore, it can be anticipated that HCO_3^- should show important effects on CO_2 solubility.

In this paper, the solubility of CO_2 was determined in aqueous solution containing HCO_3^- in the range of 308.15−408.15℃ and 8−40 MPa. Moreover, available experimental data from various conditions were used to compare the performance of several models in predicting CO_2 solubility. The Duan and PR−HV models can provide accurate prediction for CO_2 solubility in aqueous solutions containing HCO_3^-. It is interesting to note that the Duan model is shown to fit better at low salt concentration, while the PR−HV model is more suitable for high salt concentration.

1   Experimental

1.1   Materials

CO_2 (purity 99.99%) was supplied by Chengdu Dongfang Electric Gas Co. Ltd. The salts of NaCl, KCl, $NaHCO_3$, $KHCO_3$ and $CaCl_2$ supplied by Sinopharm Chemical Reagent Co. Ltd. were of analytical grade with a purity of >99% (Table 1). This purity is acceptable, with the uncertainty taken to be approximately 0.5% for the NaCl and KCl and 2% for the $CaCl_2$ and $MgCl_2$. The reagents were used without further purification. The water used to prepare solutions was twice−distilled water (conductivity<5mS/cm). The solvent mixture was prepared gravimetrically.

Table 1   Source and Purity of the Reagents Used

Compounds	CAS Number	Source	Molar mass (g/mol)	Purity (mass fraction)	Analysis method
CO_2	124−38−9	Chengdu Dongfang Electric Gas Co. Ltd	44.01	0.9999	GC[①]
NaCl	7647−14−5	Sinopharm Chemical Reagent Co. Ltd	58.5	0.995	None
KCl	7447−40−7	Sinopharm Chemical Reagent Co. Ltd	74.551	0.995	None
$NaHCO_3$	144−55−8	Sinopharm Chemical Reagent Co. Ltd	84.01	0.999	None
$KHCO_3$	298−14−6	Sinopharm Chemical Reagent Co. Ltd	100.12	0.99	None
$CaCl_2$	10043−52−4	Sinopharm Chemical Reagent Co. Ltd	110.98	0.99	None

① Gas−liquid chromatography.

1.2 Preparation of Aqueous Solutions

Aqueous solution samples with different concentrations were prepared by mixing the desired a-mount of NaCl, $MgCl_2$, $CaCl_2$, and $NaHCO_3$. They were weighted on an analytic balance with sensitivity ±0.1 mg and then dissolved in distilled water.

The ground water samples collected from the Jilin Oilfield Group formation, China, was measured by ICP−AES and the titrimetric method to determine the components. ICP−AES measurement for determining metal ionic was performed on an IRIS Advantage (HR) spectrometer. The compositions and concentration of aqueous solutions are listed in Table 2.

Table 2 Molar Concentration of Ions in Aqueous Solutions at Temperature
$T=298$ K and pressure $p=0.101$MPa[1]

Samples	Ionic	c (mol/kg)
No. 1[2]	Na^+	0.0595±0.0002
	Ca^{2+}	0.0899±0.0005
	Mg^{2+}	0.1040±0.0003
	Cl^-	0.3913±0.0006
	HCO_3^-	0.0595±0.0002
	Total[3]	0.7041±0.0004
No. 2[2]	Na^+	0.1190±0.0004
	Ca^{2+}	0.1798±0.0006
	Mg^{2+}	0.2079±0.0002
	Cl^-	0.7825±0.0003
	HCO_3^-	0.1190±0.0005
	Total[3]	1.4083±0.0004
Groundwater	K^+	0.0153±0.0005
	Na^+	0.1261±0.0006
	Ca^{2+}	0.0087±0.0003
	Mg^{2+}	0.0012±0.0005
	Fe^{2+}	0.0005±0.0002
	Cl^-	0.0917±0.0004
	SO_4^{2-}	0.0051±0.0004
	HCO_3^-	0.0098±0.0003
	Total[3]	0.2586±0.0004

[1] Standard uncertainties u are u (T) = 0.1 K, u (p) = 0.01MPa.

[2] No. 1 and No. 2 aqueous solution samples were obtained by mixing the desired amount of NaCl, $MgCl_2$, $CaCl_2$, and $NaHCO_3$ in distilled water.

[3] Molar concentration of the total ions in aqueous solutions.

1.3 Apparatus and Procedures

The experiments were performed with apparatus based on the static approach. Details of the experimental arrangement were given before[30]. A schematic of the experimental set-up for CO_2 solubility measurement is shown in Fig. 1. The apparatus was composed mainly of a CO_2 gas cylinder, a RUSKA pump (a computer-controlled metering syringe pump), a JEFRI equilibrium view cell with a sapphire window, a thermostatic air bath, a sampling bomb, a pressure transducer, gas meters and an electronic balance. The temperature of the thermostatic air bath was controlled by a temperature controller.

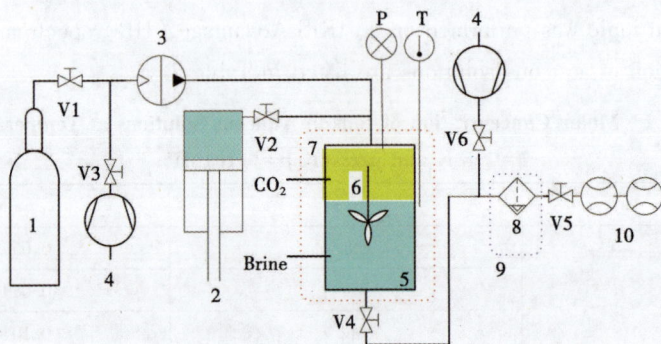

Fig. 1　Schematic Diagram of the Experimental Apparatus

1—CO_2 gas cylinder; 2—RUSKA pump; 3—pressure amplifier; 4—vacuum pump;

5—JEFRI equilibrium view cell with a sapphire window; 6—magnetic stirrer; 7—thermostatic air bath;

8—sampling cylinder; 9—ice bath; 10—gasometer; P—pressure gauge; T—temperature transducer

In a typical experiment, the equilibrium cell and lines are evacuated using a vacuum pump prior to introducing the aqueous solution and CO_2 gas. The desired amount of salt solution was introduced into the view cell, the view cell was placed in the air bath, whose temperature was controlled to the desired temperature with an immersed thermocouple by an electric furnace, then CO_2 was charged into the system by using the pressure amplifier and the cell pressure was controlled to reach the desired pressure. The aqueous phase in the view cell was stirred by the magnetic stirrer. The stirring of the aqueous phase ensured a homogeneous temperature inside the view cell. The temperature was held constant to within ±0.1K by a thermocouple and the pressure was monitored by a transducer with a precision of 0.01 MPa. The system was considered to have reached equilibrium if the pressure of the system had been unchanged for 3 hours. Then the pressure of the system was recorded. The valve of the sampling cylinder was opened slowly to collect a sample in a sampling cylinder. At the same time, the volume of the view cell was adjusted to keep the pressure unchanged during the sampling process. The sampling valve was closed after a large enough liquid sample had been collected in the sampling cylinder. The sampling cylinder was then removed for composition analysis. To determine the composition of the liquid phase, the mass of the sampling cylinder was first measured using an electronic balance with an accuracy of 0.1 mg. The mass of the sample in the sample cell was known from the mass difference of the sampling cylinder before and after sampling. After the sample was retrieved from the view cell, the total amount of CO_2 dissolved in the sample, (V_{CO_2}), consisted of three parts:

$$V_{CO_2} = V_1 + V_2 + V_3 \tag{1}$$

104

where V_1, V_2, and V_3 are the CO_2 released from the cylinder, the CO_2 in the gas phase in the sampling cylinder, and the CO_2 dissolved in the brine remaining in the cylinder at ambient temperature and atmospheric pressure, respectively. All the volume terms in Eq. (1) are at the conditions of 293.15 K and 0.101 MPa. V_1 can be obtained as follows:

$$V_1 = \frac{293.15}{T_{ambient}} \cdot \frac{p_{ambient}}{0.101} \cdot V_{gasometer} \qquad (2)$$

where $p_{ambient}$, $T_{ambient}$ and $V_{gasometer}$ are ambient pressure in MPa, ambient temperature in Kelvin, and cumulative volumes measured by the gasometer in mL, respectively.

V_2 can be expressed as follows from the volume balance:

$$V_2 = \left(V_{cylinder} - \frac{W_{brine}}{\rho_{brine}} \right) \frac{293.15}{T_{icebath}} \qquad (3)$$

where $V_{cylinder}$ is the sampling cylinder volume in ml, $T_{icebath}$ is the temperature in Kelvin, and W_{brine} and ρ_{brine} are the weight and the density of the brine saturated with CO_2 at ambient pressure and ice bath temperature in g and g/cm^3, respectively.

V_3 can be obtained from the mass balance as follows:

$$V_3 = m_{CO_2} W_{water} \times 24.056 \qquad (4)$$

where m_{CO_2} and W_{water} are the measured molality of dissolved CO_2 in mol/kg and the weight of water in g, respectively.

$$W_{walter} = \frac{W_{brine}}{1 + (44.01 m_{CO_2} + 84 m_{NaHCO_3})/1000} \qquad (5)$$

where m_{NaHCO_3} is $NaHCO_3$ molality in mol/kg.

m_{CO_2} can be estimated from Henry's law

$$m_{CO_2} = \frac{p_{atm} - p_w^s}{H_{CO_2}} \qquad (6)$$

where p_{atm} and p_w^s are the atmospheric pressure and the vapor pressure of water at room temperature, both in MPa. The Henry constant in brine, H_{CO_2}, is estimated by the following formula[30]:

$$H_{CO_2} = (192.876 + 0.024125 m_{NaHCO_3} - 0.00752 m_{NaHCO_3}^2) +$$
$$(-9624.4 + 0.000199 m_{NaHCO_3})/T + (0.01441 - 0.00211 m_{NaHCO_3}) T$$
$$+ (-28.749 + 0.1446 m_{NaHCO_3}) \ln T \qquad (7)$$

where the temperature T is in Kelvin. Once V_{CO_2} was calculated by Eq. (1), the CO_2 solubility in terms of moles of CO_2 per kg water (mol/kg) is given by $V_{CO_2}/(24.056 W_{awter})$. It should be noted that the dissolved CO_2 in brine solution remaining in the sampling cylinder, V_3, is a non-negligible contribution to the total CO_2 solubility.

2 Modeling

In this work, the Duan[16-17], PR-HV[18-19] and Chang models[23] presented in Appendix have been used to fit the experimental data and compared with each other.

3　Results and Discussion

3.1　Reliability of Experiments

Fig. 2 compared the CO_2 solubility in distilled water at 308.15 K and 328.15 K pressures up to 50 MPa from our measurements and those in the literature[4,31,28,13]. The measured solubilities in water were in well agreement with the literature data, indicating that the apparatus in this work was reliable for CO_2 solubility measurement at low temperatures.

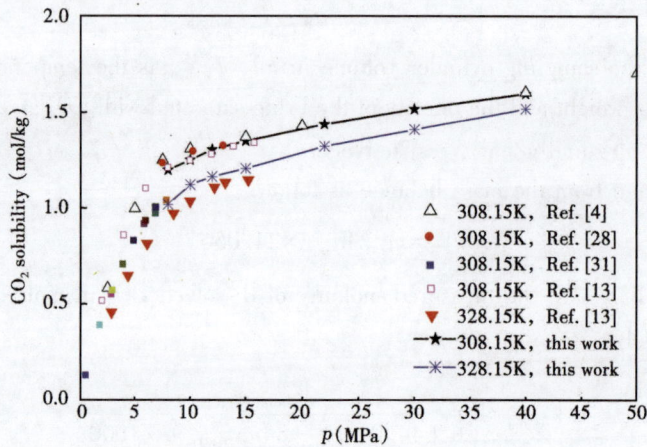

Fig. 2　Comparison of CO_2 Solubility with references. [4, 13, 28, 31] in Distilled Water at 308.15 K and 328.15 K and 0.579~50.0 MPa

In addition, since experimental data at temperatures 368.15 K and 408.15 K from the literature are scarce, and for convenience, experimental solubility of CO_2 in distilled water was measured at 368.15 K and 408.15 K with pressures up to 40 MPa and compared with correlation of the Duan model to evaluate the accuracy and reliability of the CO_2 solubility apparatus. The accuracy of the calculations was evaluated by the absolute average relative deviation (AARD) defined as follows:

$$AARD = \frac{100}{N} \sum_{i=1}^{N} \frac{\left| x_i^{calc.} - x_i^{exp.} \right|}{x_i^{exp.}} \tag{8}$$

where $x^{calc.}$ and $x^{exp.}$ are the calculated and experimental CO_2 solubility data, respectively.

As shown in Fig. 3, experimental solubility of CO_2 in distilled water at 368.15 K and 408.15 K in this work was in well agreement with that calculated by the Duan model[16-17] with $AARD = 2.37\%$, indicating that the apparatus in the present work was also reliable for CO_2 solubility measurement at high temperatures. Therefore, our method was reliable for CO_2 solubility measurement at high pressures and high temperatures. The measured CO_2 solubility can be repeated within 3% in our measurements.

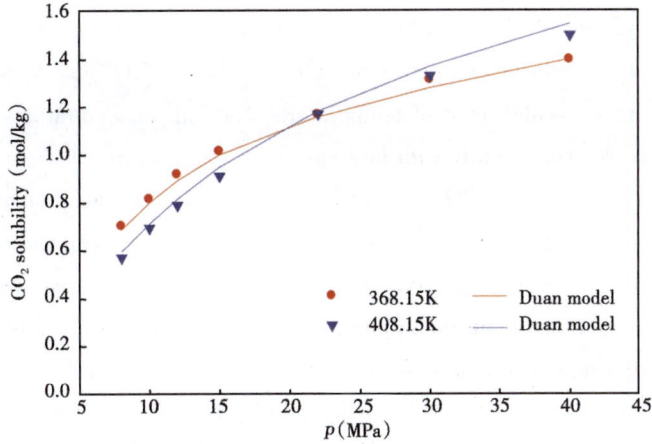

Fig. 3   Comparison of CO_2 Solubility of Experimental Measurements with the Duan Model
in Distilled Water at 368. 15 K and 408. 15 K over a Pressure Range of 8−40 MPa

3. 2   The Solubility of CO_2 in the Aqueous Solutions of Different Salts

The solubility of CO_2 in two aqueous solution samples (No. 1 and No. 2) with different salt concentrations was measured at different temperatures and pressures, and the results are listed in Table 3.

Table 3   The Solubility of CO_2 in Aqueous Solutions Containing HCO_3^- at
308. 15~408. 15 K over a Pressure Range of 8~40 bar[1]

Solubility (mol/kg)	p (MPa)	T (K)			
		308. 15	328. 15	368. 15	408. 15
No. 1[2]	8	1. 1011±0. 0002	0. 8932±0. 0002	0. 6332±0. 0001	0. 4917±0. 0001
	10	1. 1489±0. 0003	0. 9805±0. 0002	0. 7367±0. 0002	0. 6108±0. 0002
	12	1. 1893±0. 0004	1. 0451±0. 0002	0. 8182±0. 0002	0. 7040±0. 0002
	15	1. 2510±0. 0004	1. 1116±0. 0002	0. 9014±0. 0002	0. 7915±0. 0003
	22	1. 3992±0. 0003	1. 2571±0. 0003	1. 0727±0. 0003	1. 0057±0. 0002
	30	1. 5009±0. 0006	1. 3502±0. 0004	1. 2198±0. 0004	1. 2291±0. 0003
	40	1. 6155±0. 0005	1. 4340±0. 0005	1. 3129±0. 0004	1. 3036±0. 0004
No. 2[2]	8	1. 0325±0. 0001	0. 8194±0. 0001	0. 6028±0. 0002	0. 4600±0. 0001
	10	1. 0895±0. 0002	0. 9302±0. 0002	0. 6948±0. 0002	0. 5520±0. 0001
	12	1. 1308±0. 0002	1. 0023±0. 0003	0. 7806±0. 0004	0. 6406±0. 0002
	15	1. 1816±0. 0003	1. 0503±0. 0004	0. 8768±0. 0005	0. 7254±0. 0002
	22	1. 2867±0. 0004	1. 1640±0. 0003	1. 0173±0. 0003	0. 9023±0. 0003
	30	1. 3688±0. 0005	1. 2478±0. 0004	1. 1174±0. 0004	1. 0407±0. 0003
	40	1. 4340±0. 0005	1. 3316±0. 0005	1. 1733±0. 0006	1. 1942±0. 0005

① Standard uncertainties u are u (T) = 0. 1K, u (p) = 0. 01MPa.

②No. 1 and No. 2 aqueous solution samples refer to Table 2.

Experimental data on the solubility of CO_2 in $NaHCO_3$ solution are scarce. Fig. 4 and Fig. 5 show the dependence of the solubility of CO_2 in aqueous solutions on temperature, pressure, and salt concentration. As reported previous observations, the solubility of CO_2 increases consistently with an increase of pressure at the whole range of temperatures and salt concentrations[3-9]. However, the solubility of CO_2 decreases consistently with an increase of temperature at a range of low pressure (p < 30 MPa). At high pressure, the CO_2 solubility increases with increasing temperature. Similar observations have been previously reported[16]. Compared to the results with the CO_2 solubility in distilled water, decreasing values were observed by addition of salts at the same pressure and temperature. At the whole range of temperature and pressure, the solubility of CO_2 in the solutions decreases significantly with increasing concentration of the salts, which can be attributed to the so-called salting-out effect[32-33]. This can be reasonably explained by the fact that when the ions are solvated, some of the water becomes unavailable for the solute which is then salted out from the aqueous phase.

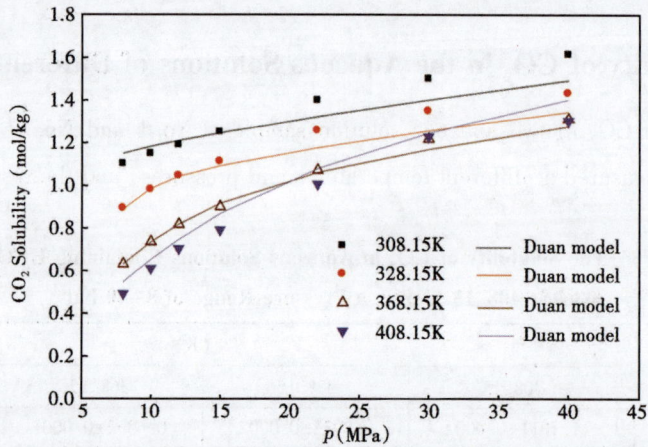

Fig. 4　Comparison of CO_2 Solubility of Experimental Measurements with the Duan Model in No. 1 Solution with Total Ions Molar Concentration 0. 7041 mol/kg at 308. 15 – 408. 15 K over a Pressure Range of 8-40 MPa

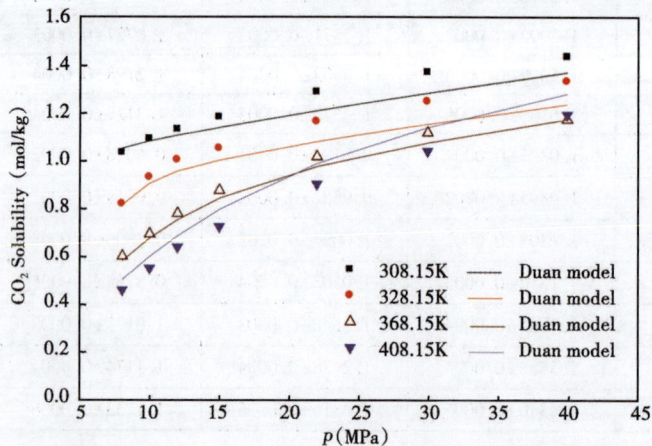

Fig. 5　Comparison of CO_2 Solubility of Experimental Measurements with the Duan Model in No. 2 Solution with Total Ions Molar Concentration 1. 4083 mol/kg at 308. 15 – 408. 15 K over a Pressure Range of 8-40MPa

The reliability of the models presented was further tested using natural ground water, whose chemical composition was determined and is listed in Table 2. For natural ground water, the three dominant anionic were Cl^-, SO_4^{2-}, and HCO_3^-. The solubility of CO_2 in natural ground water was measured at 308.15−408.15 K over a pressure range of 8−50 MPa (Fig. 6). Fig. 6 also shows that the determined solubility of CO_2 was very consistent with the mathematical models presented, which further supported that the above models were reliable.

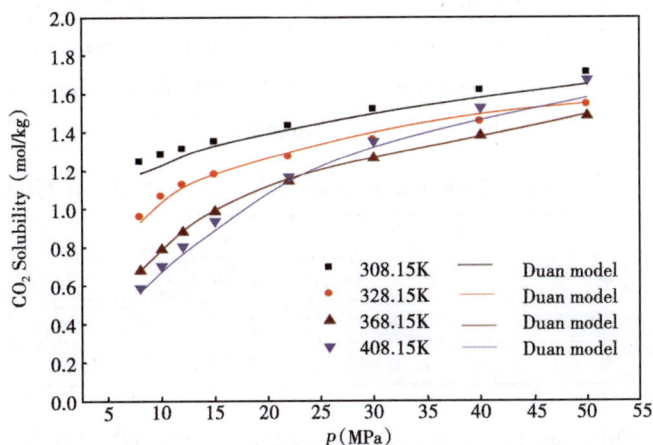

Fig. 6 Comparison of CO_2 Solubility of Experimental Measurements with the Duan Model in Natural Ground Water with Total Salinity of 0.2586 mol/kg at 308.15−408.15 K over a Pressure Range of 8−50 MPa

In the present work, three common models, the PR−HV, Chang and Duan models, were used to correlate experimental data of CO_2 solubility in aqueous solution containing HCO_3^-. A comparison was done of experimental solubility of CO_2 in aqueous solutions with different salinities and the predicted results of these three models is given in Table 4.

Table 4 The AARD Results between Three Models and Experimental Results

Samples[1]	Total salinity[2] (mol/kg)	AARD (%)[3]		
		PR−HV	Chang	Duan
Distilled water	0± 0.0001	2.95	9.61	2.60
No. 1	0.7041± 0.0004	2.75	11.01	3.63
No. 2	1.4083± 0.0004	3.18	12.98	5.04
Ground water	0.2586± 0.0004	1.89	8.74	2.32

①Four aqueous solutions include distilled water and other samples referred to Table 2.

②The total salinity of each sample.

③ $AARD = \sum_{i=1}^{N} (| x_i^{calc} - x_i^{exp} | / x_i^{exp})/N \times 100\%$, where $x^{calc.}$ and $x^{exp.}$ are the calculated and experimental CO_2 solubility data, respectively, and N is the number of data points.

As can be seen, the agreement between the experimental and calculated values was satisfactory with *AARD* less than 6% for both the PR−HV and Duan models. The results of the PR−HV model were similar to the Duan model for distilled water. In combination with Fig. 3 (in the case of distilled water), the PR–HV and Duan models were also able to predict CO_2 solubility in excellent agreement with experimental data over a wide range of pressures and temperatures. It is interesting to note that the Duan model could more accurately predict the CO_2 solubility at low salt concentration, while the PR–HV model showed smaller error at high salt concentration (see Fig. 4 and Fig. 5). On the basis of experimental observation, salt concentration of ~0. 7 mol/kg should be the critical concentration. In addition, the Chang model was the worst among three models in this work with *AARD* more than 8%. So, the correlation of the PR–HV model was more accurate than that of the Duan model for brine with high salinity over a wide range of temperatures and pressures. Critical parameters for the PR–HV and Duan models are listed in Table 5, and mixing rules and modified model coefficients for the PR–HV model are listed in Table 6.

Table 5 Critical Parameters for PR–HV and Duan Models

Component	T_c (K)	p_c (MPa)	ω	T_b (K)	M_w (g/mol)
H_2O	647. 29	22. 085	0. 344	373. 15	18. 01
NaCl	700. 00	3. 546	1. 000	1686. 15	58. 44
CO_2	304. 20	7. 377	0. 225	194. 65	44. 01

Table 6 Mixing Rules and Model Coefficients for the PR–HV Thermodynamic Model for the Water–CO_2–NaCl Systems

Mixing rules/ Coefficients	Component	H_2O	NaCl	CO_2
Mixing rules	H_2O	–	–	–
	NaCl	HV[1]	–	–
	CO_2	HV	HV	
k_{ij}[2]	H_2O	–	–	–
	NaCl	−0. 2169	–	–
	CO_2	0. 2000	2. 1000	
α for HV[3]	H_2O	–	–	–
	NaCl	−0. 7806	–	–
	CO_2	0. 0270	0. 0162	–
g/R for HV[4] (K)	H_2O	–	−19. 72	−4247. 21
	NaCl	94. 7	–	–
	CO_2	4104. 13	0. 9	2446. 82
g–T/R for HV[4]	H_2O	–	0. 00	8. 80
	NaCl	0. 00	–	0. 00
	CO_2	−6. 00	0. 00	–

[1]The Huron and Vidal mixing rule.

[2]k_{ij}: Binary interaction parameters.

[3]α is a non-randomness parameter.

[4]parameters of Huron –Vidal mixing rule.

Previous literatures have shown that ion types have an important influence on the Setschenow constant (K_s) $(\ln (R_{sw}/R_{sb}) = K_s \times C_s$, where R_{sw} and R_{sb} refer to the solubility of the solute in pure water and in a salt solution of concentration C_s, respectively), thus leading to a different salting-out effect[23]. To further demonstrate the effect of HCO_3^- anion, the solubility of CO_2 in the solutions containing NaCl, $MgCl_2$, and $CaCl_2$ was required as comparison. Duan et al. [17] previously presented a theoretical model for accurate prediction of the solubility of CO_2 in the solutions with NaCl, KCl, $MgCl_2$, and $CaCl_2$ in a wide range of temperatures, pressures, and salt concentrations. Therefore, the solubility of CO_2 in the solutions with NaCl, $MgCl_2$, and $CaCl_2$ at the same conditions of temperature, pressure, and salt concentration using Cl^- instead of HCO_3^- was calculated by the Duan model. Fig. 4 and Fig. 5 clearly showed that the calculated results agreed well with our determined data when pressure is below 15 MPa. The relative error is below 6% when pressure is above 15. It is known that the size of HCO_3^- is larger than that of Cl^- and the hydration action of HCO_3^- is smaller than that of Cl^-. Therefore, there are more free water molecules acting on CO_2 molecules in aqueous solution containing HCO_3^-, which makes the salting-out effect smaller. Besides, the molecular weight of HCO_3^- is larger than that of Cl^-, so there are fewer ions in the Cl^- solution, which also decrease salting-out effect. On the other hand, HCO_3^- can suppress CO_2 dissolution because of equilibrium reaction between CO_2 and H_2O to form HCO_3^-. Therefore, cooperation of three effects leads to nearly invariable solubility of CO_2 in solution containing HCO_3^- and Cl^-. This claim can be further confirmed by experimental data. Fig. 7 shows a comparison of CO_2 solubility in $NaHCO_3$ and NaCl solutions under the same conditions of 333. 15 K and total salinity of 0. 7186 mol/kg. The measured results of CO_2 solubility in $NaHCO_3$ at low pressure (<2.0MPa) (Fig. 7) are very consistent with the reported values[34]. Slight solubility differences of CO_2 in $NaHCO_3$ and NaCl solutions can be observed in a wide pressure range, which also supports cooperation of the three effects.

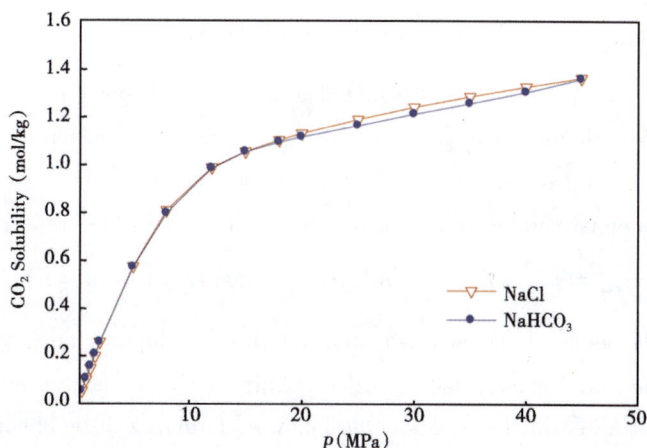

Fig. 7 CO_2 Solubility in $NaHCO_3$ and NaCl Solutions under the Same Conditions of 333. 15 K,

0. 2−45MPa and Total Salinity of 0. 7186 mol/kg

4　Conclusions

The effect of HCO_3^- anion on the solubility of CO_2 in aqueous solutions has been studied at different temperatures and pressures. In comparison with Cl^-, HCO_3^- has a large size and high molecular weight, which leads to the opposite salting-out effect. HCO_3^- can suppress CO_2 dissolution because of equilibrium reaction between CO_2, and H_2O to form HCO_3^-. Cooperation of the three effects leads to slightly variable solubility of CO_2 in solution containing HCO_3^-. The Duan model could more accurately predict the CO_2 solubility at low salt concentration (<0.7 mol/kg), while the PR-HV model showed smaller errors at high salt concentration. Our study provides the possibility to predict the solubility of CO_2 in natural ground water with reasonable accuracy over a wide range of temperatures and pressures.

Appendix A. Duan Model

A thermodynamic model was presented for the solubility of CO_2 in aqueous salt solutions in a wide temperature-pressure-ionic strength range (from 273 to 533 K, from 0 to 2000 bar, and from 0 to 4.3 molality of salts) by the Duan research group (herein, called the Duan model)[16,17]. The model is based on a specific particle interaction theory for the liquid phase and a highly accurate equation of state for the vapor phase.

At equilibrium, the chemical potential of CO_2 in the gas phase, $\mu_{CO_2}^V$, and that in the liquid phase, $\mu_{CO_2}^L$, are equal:

$$\mu_{CO_2}^V = \mu_{CO_2}^L \tag{A1}$$

The chemical potential can be written in terms of fugacity in the vapor phase:

$$\mu_{CO_2}^V = \mu_{CO_2}^{V(0)}(T) + RT\ln(y_{CO_2}p) + RT\ln\phi_{CO_2}(T, p, y_{CO_2}) \tag{A2}$$

where $\mu_{CO_2}^{V(0)}$ is the standard chemical potential of CO_2 in the vapor phase, T is the system temperature in Kelvin, R is the universal gas constant, y_{CO_2} is the mole fraction of CO_2 in vapor phase, p is the system pressure in bar, and ϕ_{CO_2} is the fugacity coefficient of CO_2.

The chemical potential can be expressed in terms of activity in the liquid phase:

$$\mu_{CO_2}^L = \mu_{CO_2}^{L(0)}(T, p) + RT\ln m_{CO_2} + RT\ln\gamma_{CO_2}(T, p, m) \tag{A3}$$

where $\mu_{CO_2}^{L(0)}$ is the standard chemical potential of CO_2 in the liquid phase, γ_{CO_2} is the activity coefficient of CO_2 in the liquid phase, and m is the molality of CO_2 or salts in the liquid phase.

Combining Eq. (A2) with Eq. (A3), the following formula can be obtained.

$$\ln\frac{y_{CO_2}p}{m_{CO_2}} = \frac{\mu_{CO_2}^{L(0)}(T, p) - \mu_{CO_2}^{V(0)}(T)}{RT} - \ln\phi_{CO_2}(T, P, y_{CO_2}) + \ln\gamma_{CO_2}(T, p, m) \tag{A4}$$

For convenience, $\mu_{CO_2}^{V(0)}(T)$ was set to zero in Eq. (A4) and $\ln\phi_{CO_2}(T, p, y_{CO_2})$ an be calculated from the EoS for pure CO_2[35] as below:

$$\ln \phi_{CO_2}(T, p) = Z - 1 - \ln Z + \frac{c_1 + c_2/T_r^2 + c_3/T_r^3}{V_r} + \frac{c_4 + c_5/T_r^2 + c_6/T_r^3}{2V_r^2} +$$

$$\frac{c_7 + c_8/T_r^2 + c_9/T_r^3}{4V_r^4} + \frac{c_{10} + c_{11}/T_r^2 + c_{12}/T_r^3}{5V_r^5} + \tag{A5}$$

$$\frac{c_{13}}{2T_r^3 c_{15}}\left[c_{14} + 1 - \left(c_{14} + 1 + \frac{c_{15}}{V_r^2}\right)\exp\left(-\frac{c_{15}}{V_2^2}\right)\right]$$

$$Z = \frac{p_r V_r}{T_r} = 1 + \frac{c_1 + c_2/T_r^2 + c_3/T_r^3}{V_r} + \frac{c_4 + c_5/T_r^2 + c_6/T_r^3}{V_r^2} + \frac{c_7 + c_8/T_r^2 + c_9/T_r^3}{V_r^4}$$

$$+ \frac{c_{10} + c_{11}/T_r^2 + c_{12}/T_r^3}{V_r^5} + \frac{c_{13}}{T_r^3 V_r^2}\left[\left(c_{14} + \frac{c_{15}}{V_r^2}\right)\exp\left(-\frac{c_{15}}{V_r^2}\right)\right] \tag{A6}$$

$$p_r = \frac{p}{p_{c,CO_2}} \tag{A7}$$

$$T_r = \frac{T}{T_{c,CO_2}} \tag{A8}$$

$$V_r = \frac{V}{RT_{c,CO_2}/p_{c,CO_2}} \tag{A9}$$

where p_{c,CO_2} and T_{c,CO_2} are the critical pressure and critical temperature for CO_2, respectively ($p_{c,CO_2} = 73.773$ bar, $T_{c,CO_2} = 304.20$ K). The parameters of Eq. (A5) and Eq. (A6) are presented in Table A1.

Table A1 Model parameters for Eq. (A5), Eq. (A6) and Eq. (A11)

Coefficient	Value	Coefficient	Value	Coefficient	Value
c_1	8.99288497×10^{-2}	c_6	9.49887563×10^{-2}	c_{11}	8.93353441×10^{-5}
c_2	-4.94783127×10^{-1}	c_7	5.20600880×10^{-4}	c_{12}	7.88998563×10^{-5}
c_3	4.77922245×10^{-2}	c_8	-2.93540971×10^{-4}	c_{13}	-1.66727022×10^{-2}
c_4	1.03808883×10^{-2}	c_9	-1.77265112×10^{-3}	c_{14}	1.39800000
c_5	-2.82516861×10^{-2}	c_{10}	-2.51101973×10^{-5}	$c15$	2.96000000×10^{-2}
d_0	-38.640844	d_1	5.8948420	d_2	59.876516
d_3	26.654627	d_4	10.637097		

On condition that the water vapor pressure of the mixtures is the same as pure water saturation pressure, y_{CO_2} can be approximately calculated by

$$y_{CO_2} = \frac{p - p_{H_2O}}{p} \tag{A10}$$

where pure water saturation pressure, p_{H_2O}, can be calculated from the following empirical equation:

$$p_{H_2O} = \frac{p_{c,\,H_2O}T}{T_{c,\,H_2O}}\left[\,1 + d_0(-t)^{1.9} + d_1t + d_2t^2 + d_3t^3 + d_4t^4\,\right] \tag{A11}$$

$$t = \frac{T - T_{c,\,H_2O}}{T_{c,\,H_2O}} \tag{A12}$$

where p_{c,H_2O} and T_{c,H_2O} are the critical pressure and critical temperature of water, respectively ($p_{c,H_2O} = 220.85\text{bar}$, $T_{c,H_2O} = 647.29\text{K}$). The model parameters of Eq. (A11) are listed in Table A1.

$\ln \gamma_{CO_2}\ (T,\ p,\ m)$ can be obtained through a virial expansion of excess Gibbs energy[36]:

$$\ln \gamma_{CO_2}(T,\ p,\ m) = \sum_c 2\lambda_{CO_2-c}m_c + \sum_a 2\lambda_{CO_2-a}m_a + \sum_c \sum_a \zeta_{CO_2-a-c}m_cm_a \tag{A13}$$

where λ and ζ are second-order and third-order interaction parameters, respectively, c and a stand by cations and anions, respectively. Substituting Eq. (A13) for Eq. (A4), the following formula can be obtained:

$$\ln \frac{y_{CO_2}p}{m_{CO_2}} = \frac{\mu_{CO_2}^{L(0)}(T,\ p)}{RT} - \ln\phi_{CO_2}(T,\ p) + \sum_c 2\lambda_{CO_2-c}m_c$$
$$+ \sum_a 2\lambda_{CO_2-a}m_a + \sum_c \sum_a \zeta_{CO_2-a-c}m_cm_a \tag{A14}$$

In Eq. (A14), the parameters λ, ζ and $\mu_{CO_2}^{L(0)}/\ (RT)$ can be calculated using the Pitzer's method[37] as follows:

$$Parameters(T,\ p) = e_0 + e_1T + e_2T^2 + e_3/T + e_4/(630 - T) + e_5p$$
$$+ e_6p\ln T + e_7p/T + e_8p/(630 - T) + e_9p^2/(630 - T)^2 + e_{10}T\ln p \tag{A15}$$

Eq. (A14) and Eq. (A15) are called the Duan model for the correlation of CO_2 in aqueous solution in this work. The coefficients in Eq. (A15) are listed in Table A2.

Table A2　Interaction parameters for Eq. (A15)

coefficient	$\mu_{CO_2}^{L(0)}/\ (RT)$	λ_{CO_2-Na}	$\zeta_{CO_2-Cl-Na}$
e_0	28.9447706	−0.411370585	3.36389723×10^{-4}
e_1	−0.0354581768	6.07632013×10^{-4}	-1.98298980×10^{-5}
e_2	1.02782768×10^{-5}	0	0
e_3	−4770.67077	97.5347708	0
e_4	33.8126098	0	0
e_5	9.04037140×10^{-3}	0	0
e_6	-1.14934031×10^{-3}	0	0
e_7	−0.307405726	−0.0237622469	2.12220830×10^{-3}
e_8	−0.090731486	0.0170656236	-5.24873303×10^{-3}
e_9	9.32713393×10^{-4}	0	0
e_{10}	0	1.41335834×10^{-5}	0

Appendix B. Chang model

Chang et al. [23] developed a set of empirical correlations for estimating CO_2 solubility in NaCl brine as functions of temperature, pressure, and salinity (herein, called the Chang model). According to the correlations, the solubility of CO_2 in water is estimated as

$$R_{sw} = 14.5\kappa p \left[1 - \beta \sin \left(\frac{\pi}{2} \cdot \frac{14.5\varepsilon p}{14.5\varepsilon p + 1} \right) \right]_{\text{if } p < p^0} \quad (B1)$$

$$R_{sw} = \kappa p^0 (1 - \beta^3) + \chi (14.5p - p^0)_{\text{if } p \geqslant p^0} \quad (B2)$$

with

$$\kappa = \sum_{i=0}^{4} (\kappa_i \times 10^{-3i} TF^i) \quad (B3)$$

$$\beta = \sum_{i=0}^{4} (\beta_i \times 10^{-3i} TF^i) \quad (B4)$$

$$\varepsilon = 10^{-3} \sum_{i=0}^{4} (\varepsilon_i \times 10^{-3i} TF^i) \quad (B5)$$

$$p^0 = 0.04389 \cdot \frac{\sin^{-1}(\beta^2)}{\varepsilon \cdot \left[1 - \frac{2}{\pi} \cdot \sin^{-1}(\beta^2) \right]} \quad (B6)$$

$$TF = \frac{9}{5}(T - 275.15) + 32 \quad (B7)$$

$$\chi = \kappa \left\{ 1 - \beta \left[\sin \left(\frac{\pi}{2} \cdot \frac{\varepsilon p^0}{\varepsilon p^0 + 1} \right) + \frac{\pi}{2} \cdot \frac{\varepsilon p^0}{(\varepsilon p^0 + 1)^2} \cos \left(\frac{\pi}{2} \cdot \frac{\varepsilon p^0}{\varepsilon p^0 + 1} \right) \right] \right\} \quad (B8)$$

where R_{sw} is the solubility of CO_2 in pure water in scf (standard cubic foot) of CO_2 per STB (standard barrel) of water, p is the system pressure in bar, and T is the system temperature in Kelvin. The correlation coefficients are shown in Table B1.

Table B1 Correlation coefficients of Chang model.

coefficient	$i=0$	$i=1$	$i=2$	$i=3$	$i=4$
κ_i	1.163	−16.630	111.073	−376.859	524.889
β_i	0.965	−0.272	0.0923	−0.1008	0.0998
ε_i	1.280	−10.757	52.696	−222.395	462.672

The calculated solubility of CO_2 in distilled water can be adjusted further for the effects of salinity to obtain the solubility of CO_2 in aqueous salt solutions as follows:

$$\lg \left(\frac{R_{sb}}{R_{sw}} \right) = -0.028S \cdot TF^{-0.12} \quad (B9)$$

115

where R_{sb} and S are CO_2 solubility in scf of CO_2 per STB of brine and the salinity of brine in weight percent of solid.

Appendix C. PR–HV model

The PR EoS was developed in 1976 by Peng and Robinson[18] and can be expressed as

$$p = \frac{RT}{v_m - b_m} - \frac{a_m}{v_m(v_m + b_m) + b_m(v_m - b_m)} \quad (C1)$$

where p is the system pressure in bar, v_m is the molar volume in cm^3/mol, T is the system temperature in kelvin, a_m is the energy parameter for a mixture, b_m is the covolume parameter, and R is the universal gas constant.

The parameters a_m and b_m are obtained from Huron–Vidal mixing rules[19]:

$$a_m = b_m \left[\sum_{i=1}^{n} \left(x_i \cdot \frac{a_i \alpha_i}{b_i} \right) - \frac{G_\infty^E}{c_0} \right] \quad (C2)$$

$$b_m = \sum_{i=1}^{n} x_i b_i \quad (C3)$$

$$c_0 = \frac{1}{2\sqrt{2}} \ln \left(\frac{2 + \sqrt{2}}{2 - \sqrt{2}} \right) = 0.6232 \quad (C4)$$

where x_i represents the mole fraction of component i and n the number of components in the mixture.

The value of the excess Gibbs energy at infinite pressure can be calculated in exactly the same way as in the NRTL equation[38].

$$G_\infty^E = \sum_{i=1}^{n} x_i \frac{\sum_{j=1}^{n} x_j G_{ji} C_{ji}}{\sum_{k=1}^{n} x_k G_{ki}} \quad (C5)$$

$$G_{ji} = b_j \exp \left(-\alpha_{ji} \cdot \frac{C_{ji}}{RT} \right) \quad (C6)$$

$$C_{ji} = g_{ji} - g_{ii} \quad (C7)$$

$$g_{ii} = -c_0 \frac{a_i}{b_i} \quad (C8)$$

$$g_{ij} = -2 \frac{(b_i b_j)^{0.5}}{b_i + b_j} \cdot (g_{ii} g_{jj})^{0.5} (1 - k_{ij}) \quad (C9)$$

The pure component parameters a_i and b_i are determined by the following equations:

$$a_i = 0.457235 \frac{R^2 T_{ci}^2}{P_{ci}} \alpha_i(T) \quad (C10)$$

116

$$b_i = 0.077796 \frac{RT_{ci}}{P_{ci}} \tag{C11}$$

$$\alpha_i(T) = \left[1 + m_i (1 - \sqrt{T/T_{ci}}) \right]^2 \tag{C12}$$

$$m_i = 0.37464 + 1.54226\omega_i - 0.26992\omega_i^2 \tag{C13}$$

In Eq. (C9), k_{ij} is the binary interaction parameter. The binary interaction parameters were optimized using the Nelder−Mead simplex method[39] by minimizing the objective function, OF, as below:

$$OF = \sum_{i=1}^{N} \left(\frac{p_i^{calc} - p_i^{exp}}{p_i^{exp}} \right)^2 \tag{C14}$$

where N is the number of experimental data points, and $p^{calc.}$ and $p^{exp.}$ are the calculated and experimental bubble−point pressures, respectively.

With the PR−HV equation, the fugacity coefficient of a component i is given by the following relation:

$$\ln \phi_i = \frac{b_i}{b_m} (Z_m - 1) - \ln \left[Z_m (1 - \frac{b_m}{v_m}) \right]$$
$$- \frac{1}{2\sqrt{2}RT} \left[\frac{a_i}{b_i} - \frac{RT \ln\gamma_i}{c_0} \right] \cdot \ln \left[\frac{v_m + (\sqrt{2} + 1)b_m}{v_m - (\sqrt{2} - 1)b_m} \right] \tag{C15}$$

where Z_m is the compressibility factor of a mixture, which can be obtained as follows:

$$Z_m^3 - (1 - B_m) Z_m^2 + (A_m - 2B_m - 3B_m^2) Z_m - (A_m B_m - B_m^2 - B_m^3) = 0 \tag{C16}$$

$$A_m = \frac{a_m p}{R^2 T^2} \tag{C17}$$

$$B_m = \frac{b_m p}{RT} \tag{C18}$$

At thermodynamic equilibrium for a mixture, the partial fugacity of CO_2 in the vapor phase and aqueous phase are equal.

$$f_{CO_2}^{V} = f_{CO_2}^{aq} \tag{C19}$$

The fugacity can be calculated through the fugacity coefficient:

$$f_{CO_2} = x_{CO_2} \phi_{CO_2} p \tag{C20}$$

The fugacity coefficient can be obtained through Eq. (C15).

Graphic Abstract

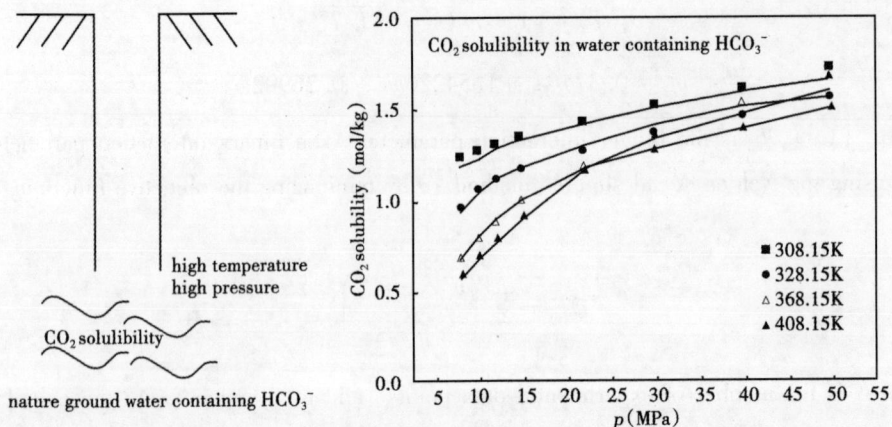

References

[1] Kopp A, Class H, Helmig R. Investigations on CO_2 storage capacity in salineaquifers. Part 1. Dimensional analysis of flow processes and reservoir characteristics [J]. Int. J. Greenhouse Gas Control, 2009 (3): 263-276.

[2] Nicolas S, Karsten P, Jonathan E. K., CO_2-H_2O mixtures in the geological sequestration of CO_2. I. Assessment and calculation of mutual solubility from 12 to 100℃ and up to 600 bar [J]. Geochim. Cosmochim. Acta, 2003 (67): 3015-3031.

[3] Wiebe R, Gaddy V L. The solubility in water of carbon dioxide at 50, 75 and 100℃, at pressures to 700 atmospheres [J]. J. Am. Chem. Soc., 1939 (61): 315-318.

[4] Wiebe R, Gaddy V L. The solubility of carbon dioxide in water at various temperatures from 12 to 40℃ and at pressures to 500 atmospheres. Critical phenomena [J]. J Am Chem Soc, 1940 (62): 815-817.

[5] Diamond L W, Akinfiev N N. Solubility of CO_2 in water from −1.5 to 100 ℃ and from 0.1 to 100 MPa: evaluation of literature data and thermodynamic modeling [J]. Fluid Phase Equilib, 2003 (208): 265-290.

[6] Li Z W, Dong M Z, Li S L, et al. Densities and solubility for binary systems of carbon dioxide + water and carbon dioxide + brine at 59℃ and pressures to 29 MPa J Chem Eng Data, 2004 (49): 1026-1031.

[7] Bamberger A, Sieder G, Maurer G. High-pressure (vapor + liquid) equilibrium in binary mixtures of (carbon dioxide + water or acetic acid) at temperatures from 313 to 353 K [J]. J Supercrit Fluid, 2000 (17): 97-110.

[8] Chapoy A, Mohammadi A H, Chareton A, et al. Measurement and modeling of gas solubility and literature review of the properties for the carbon dioxide-water system [J]. Ind Eng Chem Res, 2004 (43): 1794-1802.

[9] Dohrn R, Büenz A P, Devlieghere F, et al. Experimental measurements of phase equilibria for ternary and quaternary systems of glucose, water CO_2 and ethanol with a novel apparatus [J]. Fluid Phase Equilib, 1993 (83): 149-158.

[10] Rumpf B, Nicolaisen H, Oal C, et al. Solubility of carbon dioxide in aqueous solutions of sodium chloride: experimental results and correlation [J]. J Solution Chem, 1994 (23): 431-448.

[11] Kiepe J, Horstmann S, Fischer K. Experimental determination and prediction of gas solubility data for CO_2 +

H_2O mixtures containing NaCl or KCl at temperatures between 313 and 393 K and pressures up to 10 MPa [J]. Ind Eng Chem Res, 2002 (41): 4393-4398.

[12] Bando S, Takemura F, Nishio M. Solubility of CO_2 in aqueous solutions of NaCl at (30 to 60) ℃ and (10 to 20) MPa [J]. J Chem Eng Data, 2003 (48): 576-579.

[13] Liu Y, Hou M, Yang G, et al. Solubility of CO_2 in aqueous solutions of NaCl, KCl, $CaCl_2$ and their mixed salts at different temperatures and pressures [J] J of Supercrit Fluid, 2011 (56): 125-129.

[14] Bermejo M. D, Martin A, Florusse L J. The influence of Na_2SO_4 on the CO_2 solubility in water at high pressure [J]. Fluid Phase Equilib, 2005 (238): 220-228.

[15] Murray C N, Riley J P. The solubility of gases in distilled water and sea water. IV. Carbon dioxide [J]. Deep-Sea Res, 1971 (18): 533-541.

[16] Duan Z H, Sun R. An improved model. calculating CO_2 solubility in pure water and aqueous NaCl solutions from 273 to 533 K and from 0 to 2000 bar [J]. Chem Geol, 2003 (193): 257-271.

[17] Duan Z H, Sun R, Zhu C, et al. An improved model for the calculation of CO_2 solubility in aqueous solutions containing Na^+, K^+, Ca^{2+}, Mg^{2+}, Cl^-, and SO_4^{2-} [J]. Mar Chem, 2006 (98): 131-139.

[18] Peng D Y, Robinson D B. A new two-constant equation of state [J]. Ind Eng Chem Fundam, 1976 (15): 59-64.

[19] Huron M J, Vidal J. New mixing rules in simple equations of state for representing vapor-liquid equilibria of strongly non-ideal mixtures [J]. Fluid Phase Equilib, 1979 (3): 255-271.

[20] Harvey A H, Prausnitz J M. Thermodynamics of high-pressure aqueous systems containing gases and salts [J]. AIChE J, 1989 (35): 635-644.

[21] Enick R M, Klara S M. CO_2 solubility in water and brine under reservoir conditions [J]. Chem Eng Commun, 1990 (90): 23-33.

[22] Soreide I, Whitson C H. Peng Robinson predictions for hydrocarbons, CO_2, N_2, and H_2S with pure water and NaCl brine, Fluid Phase Equilib [J]. 1992 (77): 217-240.

[23] Chang Y B, Coats B K, Nolen J S. A compositional model for CO_2 floods including CO_2 solubility in water [J]. SPE Reservoir Eval Eng, 1998 (1): 155-160.

[24] Darwish N A, Hilal N. A simple model for the prediction of CO_2 solubility in H_2O-NaCl system at geological sequestration conditions [J]. Desalination, 2010 (260): 114-118.

[25] Akinfiev N N, Diamond L W. Thermodynamic model of aqueous CO_2-H_2O-NaCl solutions from -22 to 100℃ and from 0.1 to 100 MPa [J]. Fluid Phase Equilib, 2010 (295): 104-124.

[26] Spycher N, Pruess K. A phase-partitioning model for CO_2-brine mixtures at elevated temperatures and pressures: Application to CO_2-enhanced geothermal system [J]. Transp Porous Med, 2010 (82): 173-196.

[27] Xia J, Rumpf B, Maurer G. Solubility of carbon dioxide in aqueous solutions containing sodium acetate or ammonium acetate at temperatures from 313 to 433 K and pressures up to 10 MPa [J]. Fluid Phase Equilib, 1999 (155): 107-125.

[28] Ferrentino G, Barletta D, Balaban M O, et al. Measurement and prediction of CO_2 solubility in sodium phosphate monobasic solutions for food treatment with high pressure carbon dioxide [J]. J Supercrit Fluid, 2010 (52): 142-150.

[29] Sun R, Hu W, Duan Z. Prediction of nitrogen solubility in pure water and aqueous NaCl solutions up to high temperature, pressure, and ionic strength [J]. J Solution Chem, 2001 (30): 561-573.

[30] Yan W, Huang S, Stenby E H. Measurement and modeling of CO_2 solubility in NaCl brine and CO-saturated NaCl Brine density [J]. Int J Greenhouse Gas Con, 2011 (5): 1460-1477.

[31] Valtz A, Chapoy A, Coqueleta C. Vapour-liquid equilibria in the carbon dioxide-water system, measurement and modelling from 278.2 to 318.2K [J]. Fluid Phase Equilibria, 2004 (226): 333-344.

[32] Gorgenyi M, Dewulf J , van Langenhove H, et al. Aqueous salting-out effect of inorganic cations and anions on non-electrolytes [J]. Chemosphere, 2006 (65): 802-810.

[33] Hasseine A, Meniai A H, Korichi M. Salting-out effect of single salts NaCl and KCl on the LLE of the systems (water + toluene + acetone), (water + cyclohexane + 2-propanol) and (water + xylene +methanol) [J]. Desalination, 2009 (242): 264-276.

[34] Han Xiaoying, Yu Zhihui, Qu Jingkui, et al. Measurement and correlation of solubility data for CO_2 in NaH-CO_3 aqueous solution [J]. J Chem Eng Data, 2011 (56): 1213-1219.

[35] Duan Z H, Møller N, Weare J H. An equation of state for the $CH4-CO_2-H_2O$ system: I. Pure systems from 0 to 1000℃ and 0 to 8000 bar [J]. Geochim Cosmochim Acta, 1992 (56): 2605-2617.

[36] Pitzer K S. Thermodynamics of electrolytes. I. Theoretical basis and general equations [J]. J Phys Chem, 1973 (77): 268-277.

[37] Pitzer K S, Peiper J C, Busey R H. Thermodynamic properties of aqueous sodium chloride solutions [J]. J Phys Chem Ref Data, 1984 (13): 1-102.

[38] Renon H, Prausnitz J M. Local compositions in thermodynamic excess functions for liquid mixtures [J]. AIChE J 1968 (14): 135-144.

[39] Nelder J A, Mead R. A simplex method for function minimization [J]. Comput J, 1965 (7): 308-313.

气藏中 CO_2 封存过程气水
互溶特性实验研究

王长权[1]　杜志敏[1]　汤　勇[1]　石立红[2]　孙　扬[1]

（1. 西南石油大学油气藏地质及开发工程国家重点实验室；2. 西南石油大学）

摘　要：利用 PVT 装置开展了不同 CO_2 含量下 CO_2—烃—水体系在不同条件下气水互溶特性实验，研究气藏注 CO_2 封存过程中 CO_2—烃—水体系互溶规律。结果表明，相同温度、压力下，随 CO_2 的不断注入，气相中 CO_2 含量和水蒸气含量不断增加，液相中 CO_2 在水中的溶解度越大，CH_4 溶解度越小，地层条件下 CO_2 含量为 68%（摩尔分数）的气样比 CO_2 含量为 23%（摩尔分数）的气样的 CO_2 溶解度增加 1.116%（摩尔分数），而 CH_4 的溶解度减小 0.131%（摩尔分数）。CO_2 和 CH_4 在水中的溶解度均随着压力升高而增大，随着温度升高而减小；在 CO_2 临界点附近，CO_2 在水中的溶解度变化显著，40℃下 CO_2 含量为 23%（摩尔分数）的气样的 CO_2 溶解度在 6~9MPa 下增加了 0.138%（摩尔分数），而在 9~12MPa 下仅增加 0.092%（摩尔分数），且压力越大增加量越小。高温低压时受水蒸发作用影响，气相中 CO_2 及 CH_4 含量随温度升高而急剧降低，而随压力升高则缓慢上升，当压力高于 18MPa 后，气相中 CO_2 及 CH_4 含量基本保持不变。

关键词：注 CO_2—EHR；地质封存；气水互溶

0　引言

废弃油气藏中注 CO_2-EHR 同时进行 CO_2 地质封存是目前处理温室气体在大气中排放最有前景的方法之一。溶解封存是油气藏 CO_2 地质封存重要的封存机制[1-5]。油气藏储层中总是存在地层水，且水体体积往往比油气藏体积大很多倍[6-7]，在进行注 CO_2—EHR 和地质封存潜力评价时，气水互溶影响不容忽视。随着 CO_2 的不断注入，CO_2 与油气藏中的烃类和地层水不断接触，CO_2 的浓度将不断发生变化，导致 CO_2—烃—水互溶特性也发生改变。对于 CO_2—天然气共存体系，CO_2 在水中的溶解度要比烃类在水中的溶解度大得多，因此 CO_2—烃混合体系中的 CO_2 大量溶于水相，气水互溶，影响 CO_2 的封存潜力。目前，国内外已有很多学者对 CO_2—H_2O，CH_4—CO_2—H_2O，C_2H_6—CO_2—H_2O 等体系的相平衡及气相中水蒸气含量和液相中气体含量进行了研究。但基本都以单相 CH_4 或 C_2H_6 与 CO_2 的混合物进行研究[8-13]，而对 CO_2—天然气混合体系中 CO_2 在水中溶解度的研究甚少，尤其对饱和水蒸气下的混合气相中水蒸气、CO_2 及 CH_4 溶解的研究更少，无法确定气水互溶规律，使得在气藏中进行 CO_2—EHR 及地质封存潜力评价时存在不确定性。因此，研究在气藏中注 CO_2—EHR 及地质封存过程中气水互溶特性对 CO_2 封存潜力评价有极为重要的意义。CO_2—烃—水互溶特性是注 CO_2—EHR 及地质封存潜力评价的基础。通过开展不同 CO_2 含量混合气体系的气水比及气相中 CO_2 和 CH_4 含量变化实验，研究气藏中注 CO_2—EHR 及地质封存过程中

的 CO_2 浓度、温度及压力对混合体系中气水互溶特性的影响规律，为气藏中的 CO_2 封存潜力评价研究提供依据。下文中所有的气体含量及溶解度等百分含量均为物质的量的百分含量。

1 实验部分

1.1 实验仪器及样品

实验仪器主要包括高温高压 PVT 装置、气液分离装置、HP6890 气相色谱仪、气量计以及电子天平。气样：以某气田气样与工业用纯 CO_2 按不同比例在地层温度为 100℃、压力为 41.4MPa 下复配而成。水样：实验室配制 1g/L NaCl 的盐水，密度为 $1.004g/cm^3$。

1.2 实验条件及主要步骤

压力从地层压力 41.4MPa 降至 6.0MPa，温度从地层温度 100℃降至 20℃。

实验主要步骤：（1）将 CO_2 与现场取得的天然气样品按不同比例进行复配，获得不同 CO_2 含量的天然气样品；（2）将气样及过量水样转入 PVT 筒中，温度、压力恒定在实验设定温度、压力，至少搅拌 1h 后进行单次脱气实验及气相色谱分析，测定并计算 CO_2 及 CH_4 在水相中的溶解量及气水比；（3）翻转 PVT 筒 180°，稳定 10min 后进行单次闪蒸实验及气相色谱分析，测定并计算气相中水的含量、CO_2 含量及 CH4 含量；（4）改变温度、压力条件后再重复步骤（2）和（3）。

2 结果与讨论

2.1 地层条件下气样中水蒸气含量变化及组分组成

随 CO_2 的注入，CO_2 的浓度不断增加，引起地层条件下饱和水蒸气含量不断发生变化，通过对不同 CO_2 含量的混合气样过饱和地层水，同时测定气相中水蒸气含量及气样的组分组成，获得了饱和水蒸气的 3 组不同 CO_2 含量混合气体的组分组成，结果见表 1。

表 1 饱和气态水的富含 CO_2 天然气样组分

样品	H_2O(%)	CO_2(%)	N_2(%)	C_1(%)	C_2(%)	C_3(%)	iC_4(%)	nC_4(%)	iC_5(%)	nC_5(%)	C_6(%)
气样 1	1.5526	22.9629	4.3268	69.9311	1.0583	0.0098	0.0089	0.1378	0.0030	0.0020	0.0069
气样 2	2.0090	39.1964	3.3701	54.4691	0.8243	0.0077	0.0069	0.1074	0.0023	0.0015	0.0054
气样 3	2.4612	68.2772	1.6773	27.1089	0.4103	0.0038	0.0034	0.0534	0.0011	0.0008	0.0027

从表 1 中可以看出，随着 CO_2 的不断注入，CO_2 含量不断增加，混合气体系中的饱和水蒸气的含量也不断上升，说明 CO_2 的注入会引起气体体系中水蒸气的含量不断上升形成水气互溶。

2.2 液相中的气水比测试

随着 CO_2 的不断注入，CO_2 浓度不断增加导致 CO_2 及 CH_4 等气相在地层水中的溶解量不断发生变化，从而改变了 CO_2 在水中的溶解封存潜力。通过对混合体系液相中气水比测试，分析不同 CO_2 浓度、温度及压力对混合体系中 CO_2 及 CH_4 在水中溶解度的变化规律，结果见图1和图2。

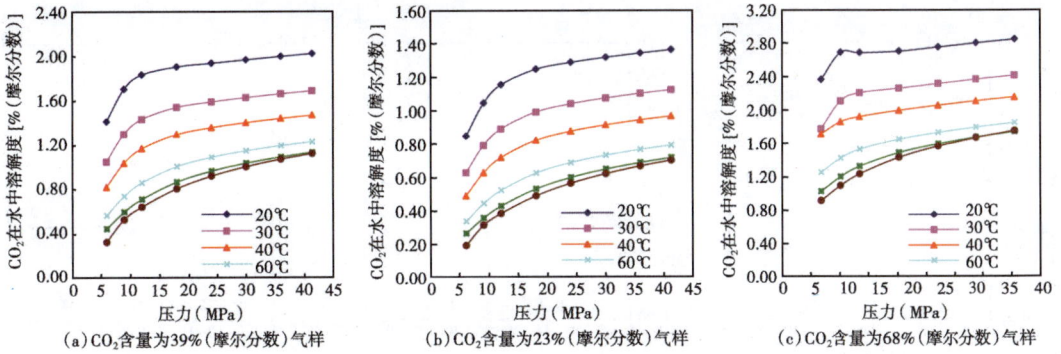

图1　不同 CO_2 含量下 CO_2 在水中的溶解度随压力和温度变化曲线

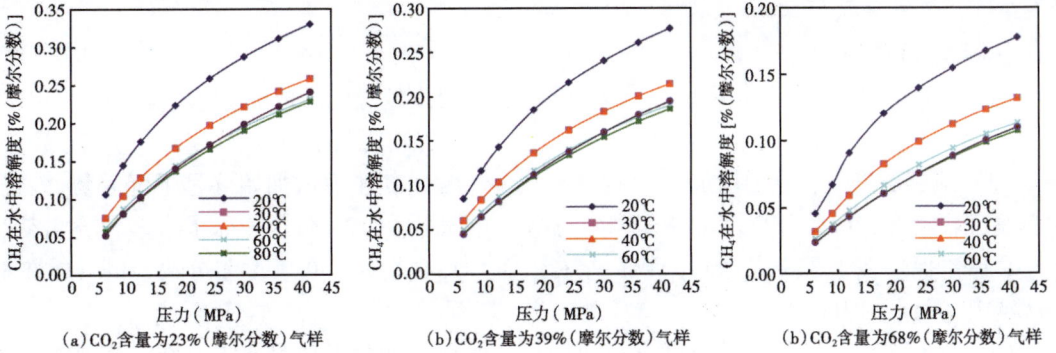

图2　不同 CO_2 含量下 CH_4 在水中的溶解度随压力和温度变化曲线

从图1和图2中可以看出：

（1）相同温度压力条件下，CO_2 含量越高，CO_2 在水中的溶解度越大，而 CH_4 的溶解度越小；地层条件下，CO_2 浓度从23%增加到68%时，CO_2 在水中溶解量可增加1.05%。

（2）相同压力及 CO_2 含量下，CO_2 及 CH_4 在水中的溶解度均随温度升高逐渐减小，低温下 CO_2 在水中的溶解量随温度的升高降低幅度大，而高温下其降低幅度逐渐趋于平缓；对比不同温度下 CO_2 在水中的溶解量变化图，温度越高，溶解量随压力升高而增加得越快。

（3）相同温度及 CO_2 含量下，CO_2 及 CH_4 在水中的溶解度均随压力升高而上升，且低压下 CO_2 在水中的溶解量随压力的升高增幅明显，而高压下曲线增幅不大。

（4）在 CO_2 的临界点值（7.38MPa，31.1℃）附近，受 CO_2 超临界特性的影响，CO_2 在水中的溶解度随温度、压力变化显著，变化幅度较大；当压力和温度处于 CO_2 过临界点时，其气态特征增强，CO_2 在水中的溶解量变化趋势逐渐变缓；由于 CH_4 的临界点值（4.6MPa，−82℃）低，通常情况下，温度和压力都高于 CH_4 的临界值，处于超临界态下的 CH_4 在水中的溶解量在低压下随温度升高降幅较小，而高压下随温度的升高降幅增加，但总体上变化趋势不会出现明显的突变。

2.3 气相中水蒸气含量测试

CO_2 含量的增加会引起气相中水蒸气含量变化，对注 CO_2—EHR 有一定的影响，因此实验测定了混合体系气相中水蒸气含量变化，结果见图 3。

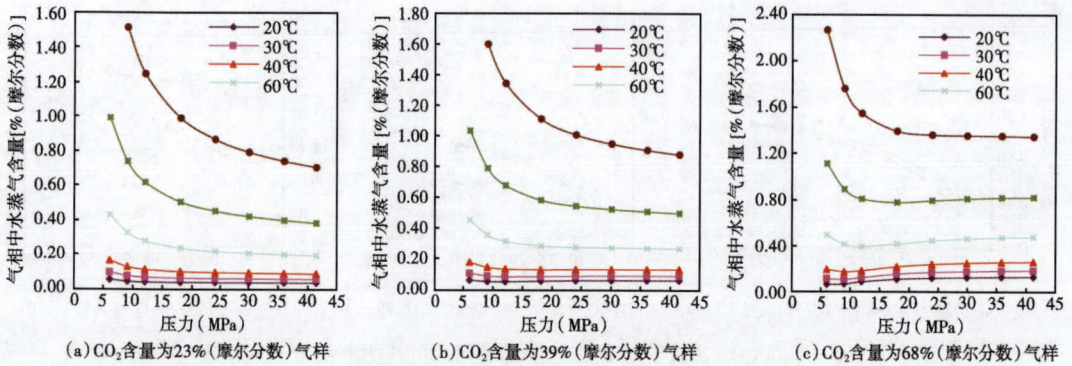

图 3　不同 CO_2 含量下气相中的水蒸气含量随压力和温度变化曲线

从图 3 中可以看出：

（1） CO_2 含量越高，气相中水蒸气含量越大，且温度越低变化越小，地层温度压力下 CO_2 含量为 23%（摩尔分数）的气样中水蒸气含量比 68%（摩尔分数）的气样低 0.64%（摩尔分数）。

（2）地层温度下，压力增加，气相中的水蒸气含量随压力增加始终呈下降趋势，且压力越高降幅越缓，CO_2 含量为 68%（摩尔分数）的气样在 18~41.4MPa 下水蒸气含量变化极小，基本保持在 1.35%（摩尔分数）左右，而 CO_2 含量为 23%（摩尔分数）的气样在 18~41.4MPa 变化约为 0.28%。

（3）气相中水蒸气含量随温度升高而增加，低温下变化近似呈水平直线，而高温下下降幅度较大。

2.4 气相中 CO_2 及 CH_4 含量

CO_2 浓度的增加导致 CH_4 含量降低，但相同 CO_2 含量下由于 CO_2 及 CH_4 在水中溶解，导致气相中 CO_2 及 CH_4 含量不断变化，从而影响了 CO_2 的封存能力。通过对不同气样进行气相中 CO_2 及 CH_4 含量测试，得到不同 CO_2 含量混合体系气相中 CO_2 及 CH_4 含量随压力和温度变化，结果见表 2。

表 2　不同 CO_2 含量下气相随压力和温度变化结果

压力 （MPa）	CO₂ 含量［%（摩尔分数）］											
	23% CO_2				39% CO_2				68% CO_2			
	20℃	40℃	60℃	100℃	20℃	40℃	60℃	100℃	20℃	40℃	60℃	100℃
41.4	23.30	23.29	23.27	23.15	39.95	39.93	39.88	39.64	69.89	69.80	69.65	69.05
30.0	23.30	23.29	23.27	23.14	39.96	39.93	39.88	39.61	69.89	69.81	69.67	69.05
18.0	23.30	23.29	23.26	23.09	39.96	39.93	39.88	39.55	69.90	69.83	69.70	69.02
9.0	23.30	23.28	23.24	22.97	39.96	39.93	39.85	39.36	69.93	69.86	69.70	68.76
6.0	23.30	23.28	23.22	22.91	39.96	39.92	39.81	39.20	69.93	69.85	69.65	68.41

124

表 3 不同 CO_2 含量下气相中 CH_4 含量随压力和温度变化结果

压力 （MPa）	CH_4 含量［%（摩尔分数）］											
	23% CO_2				39% CO_2				68% CO_2			
	20℃	40℃	60℃	100℃	20℃	40℃	60℃	100℃	20℃	40℃	60℃	100℃
41.4	71.02	70.98	70.90	70.54	55.57	55.53	55.45	55.10	27.77	27.73	27.67	27.42
30.0	71.02	70.98	70.89	70.47	55.57	55.53	55.45	55.06	27.77	27.73	27.67	27.42
18.0	71.02	70.97	70.87	70.33	55.58	55.53	55.44	54.97	27.78	27.74	27.68	27.41
9.0	71.01	70.95	70.80	69.96	55.58	55.52	55.39	54.70	27.79	27.75	27.68	27.30
6.0	71.00	70.92	70.73	69.82	55.57	55.50	55.34	54.47	27.79	27.75	27.66	27.16

从表 2 和表 3 中可以看出：对于不同 CO_2 含量混合体系，气相中 CO_2 及 CH_4 含量随温度升高而急剧降低，且温度越高降低得越快；气相中 CO_2 及 CH_4 含量随压力升高则缓慢上升，且压力越高增幅越小，当压力高于 18MPa 后，气相中 CO_2 及 CH_4 含量基本保持不变。分析原因主要为：由于 CO_2 的存在，在高温低压下，地层水蒸发作用显著，气相中水蒸气的含量大大增加，导致气相中的 CO_2 及 CH_4 含量相对变低。

2.5 溶解气水比测试

由于 CO_2 比 CH_4 在水中的溶解度高很多，因此随 CO_2 的注入，溶解气水比会发生明显变化，为了确定不同条件下液相中溶解气情况，进行了不同 CO_2 含量气样在不同温度、压力条件下的溶解气水比测试，结果见表 4。

表 4 不同 CO_2 含量气样溶解气水比变化结果

压力 （MPa）	不同温度下 23% CO_2（m³/m³）				不同温度下 39% CO_2（m³/m³）				不同温度下 68% CO_2（m³/m³）			
	20℃	40℃	60℃	100℃	20℃	40℃	60℃	100℃	20℃	40℃	60℃	100℃
41.4	22.91	16.52	13.77	12.69	31.12	22.68	19.05	17.68	31.12	22.68	19.05	17.68
30.0	22.36	15.98	13.19	11.95	30.54	22.08	18.39	16.77	30.54	22.08	18.39	16.77
18.0	21.67	15.27	12.44	11.00	29.84	21.32	17.53	15.60	29.84	21.32	17.53	15.59
9.0	20.85	14.41	11.52	9.860	29.05	20.41	16.47	14.15	29.05	20.41	16.47	14.15
6.0	19.78	13.24	10.29	8.418	28.16	19.19	15.01	12.26	28.16	19.19	15.01	12.26

由表 4 可以看出：随 CO_2 含量的不断增加，溶解气水比不断上升，且不同 CO_2 含量混合体系溶解气水比随温度升高而降低，随压力增加而增大。主要原因为：CO_2 与水的互溶能力比 CH_4 与水的互溶能力强，随 CO_2 含量的增加，CO_2 在水中的溶解度越大，使得气水比越高，这将有利于 CO_2 地质封存。

3 结论

（1）随 CO_2 的不断注入，CO_2 含量不断增加，导致体系气相中饱和水蒸气含量不断上升，液相中 CO_2 溶解量不断增加，而 CH_4 溶解量不断降低，说明随 CO_2 的不断注入，溶解封存机制作用越明显，封存潜力随 CO_2 的不断注入而不断上升。

（2）相同 CO_2 含量条件下，CO_2 及 CH_4 在水中的溶解量受温度、压力的影响，且随温度增加而降低，随压力的增加而增加；在 CO_2 临界点附近，CO_2 在水中的溶解量随温度压力变化显著；当压力和温度高于 CO_2 的临界点值时，CO_2 在水中的溶解量变化趋势逐渐平缓，变化幅度较小。

（3）在注 CO_2—EHR 及地质封存过程中，由于地层水和 CO_2 的同时存在，一方面，在高温低压下，地层水蒸发作用显著；另一方面，CO_2 和地层水互溶能力强，随 CO_2 含量不断变化，CO_2 在地层水中的溶解度也发生变化，从而影响 CO_2 地质封存能力。

参 考 文 献

［1］ Metz B, Davidson O, de Coninck H, et al. IPCC special report on carbon dioxide capture and storage ［M］. New York: Cambridge University Press, 2005: 1-431.

［2］ 张炜，李义连. 二氧化碳储存技术的研究现状和展望 ［J］. 环境污染与防治，2006，28（12）：950-953.

［3］ Orr F M Jr. Storage of carbon dioxide in geologic formations ［C］. SPE 88842, 2004: 90-97.

［4］ Al-Hashami A, Ren S R, Tohidi B. CO_2 injection for enhanced gas recovery and geo-storage: reservoir simulation and economics ［C］. SPE 94129, 2005: 1-7.

［5］ 刘洪林，王红岩，李景明. 利用碳封存技术开发我国深层煤层气资源的思考 ［J］. 特种油气藏，2006，13（4）：6-9.

［6］ 汤勇，杜志敏，张哨楠，等. 高温气藏近井带地层水蒸发和盐析研究 ［J］. 西南石油大学学报，2007，29（2）：96-99.

［7］ 李琴，李治平，胡云鹏，等. 深部盐水层 CO_2 埋藏量计算方法研究与评价 ［J］. 特种油气藏，2011，18（5）：6-10，32.

［8］ Duan Z, Moller N, Weare J H. An equation of state for the CH4-CO_2-H2O system: I. pure systems from 0 to 1000℃ and 0 to 8000bar ［J］. Geochim Cosmochim Acta, 1992 (56): 2605-2617.

［9］ Dhima A, de Hemptinne J C, Jose J. Solubility of hydrocarbons and CO_2 mixtures in water under high pressure ［J］. Ind Eng Chem Res, 1999 (38): 3144-3161.

［10］ Bamberger A, Sieder G, Maurer G. High-pressure (vapor plus liquid) equlibrium in binary mixtures of (carbon dioxide puls water or acetic acid) at temperatures from 313 to 353K ［J］. J Supercrit Fluids, 2000 (17): 97-100.

［11］ D'Souza R, Patrick J R, Teja A S. High pressure phase equilibria in the carbon dioxide-n-hexadecane and carbon dioxide-water systems ［J］. Can J Chem Eng, 1988 (66): 319-323.

［12］ Duan Z, Sun R. An improved model calculating CO_2 solubility in pure water and aqueous NaCl solutions from 273 to 533K and from 0 to 2000bar ［J］. Chemical Geology, 2003 (193): 257-271.

［13］ Duan Z, Sun R, Zhu C, et al. An improved model for the calculation of CO_2 solubility in aqueous solutions containing Na^+, K^+, Ca^{2+}, Mg^{2+}, Cl^-, and SO_4^{2-} ［J］. Marine Chemistry, 2006 (98) 131-139.

An Experimental Study of CO_2-Brine-Rock Interaction at in Situ Pressure-Temperature Reservoir Conditions

Yu Zhichao[1]   Liu Li [1]   Yang Siyu[2]   Li Shi [2]   Yang Yongzhi [2]

(1. College of Earth Sciences, Jilin University; 2. State Key Laboratory of EOR,
Research Institute of Petroleum Exploration & Development)

Abstract: A detailed investigation of CO_2-brine-rock interactions, through a core flooding laboratory experiment, was carried out under simulated reservoir conditions (100°C and 24 MPa). Changes in the ionic chemistry of the outlet solution, combined with core scanning electron microscopy (SEM) and bulk-rock X-ray diffraction (XRD) analysis of the core pre- and post-experiment reveal new insights into CO_2-brine-rock interactions. Minerals such as potassium (K) feldspar, albite, calcite, and ankerite are variably dissolved after the experiments. Calcite is the mineral most affected by dissolution, followed by ankerite, whereas dissolution of feldspar minerals is minimal. Small amounts of kaolinite and solid phases were generated as a result of the experiment. The solid phases are mainly comprised of C, O, Na, Cl, Al, and Si, and are presumed to be the transitional products in the formation of carbonate minerals. The very low fluid penetration rate in the experiment resulted in significantly reduced contact area between the acidic CO_2 fluid and minerals in the cores, resulting in low rates of feldspar dissolution. Core permeability decreased substantially throughout the experiment, although core porosity remained unchanged. The permeability reduction is the result of precipitation of new mineral phases (e.g., kaolinite and solid phases), and potentially also the presence of clay particles released by the dissolution of carbonate cement, which have then been transported in the fluid flow path and accumulated at pore throats. The results provide new insights into CO_2 trapping mechanisms in depleted oil and gas reservoirs, and into the potential formation damage that may result from massive injections of CO_2 into reservoirs during enhanced oil recovery programs.

Keywords: CO_2-flooding; reservoir sandstone; solid phases; clay particles; permeability

0   Introduction

Atmospheric carbon dioxide (CO_2) is perhaps the most important factor in driving recent anthropogenic global warming[1], and further increases in atmospheric CO_2 are also projected to adversely affect future life on Earth[2]. Geological sequestration or underground storage of CO_2 in depleted oil and gas reservoirs results in improved oil recovery and reduced net carbon emissions into the atmosphere[3-5]. CO_2-flooding is a process in which CO_2 gas and brine are injected into reservoirs through wells, and has been an important technique for enhanced oil recovery (EOR) since the 1980s[6]. Presently, CO_2 capture in reservoirs during EOR is one of the most important techniques for CO_2 sequestration, and is second only in importance to the injection of CO_2 into deep saline aquifers[7]. However, CO_2 is a special gas that has significantly more influence upon the host rocks and pore waters than petroleum fluids[8]. After injection of CO_2 into depleted oil and gas reservoirs, the initial physico-chemical equilibrium between the saline formation fluid and reservoir rock may be disturbed and triggered chemical reactions among the injected CO_2, saline formation fluid, and reservoir rock[4]. Such interactions might eventually lead to changes in the physical and chemical properties of the reservoir system[3-4]. Reservoir permeability might be particularly susceptible to these interactions, and a decrease in permeability would have a serious impact on the long-term CO_2 storage capacity[9-10], safety, and stability of the reservoir[4].

Precise knowledge of the CO_2-induced interactions between injected CO_2, saline formation fluid and reservoir rock sand under reservoir conditions, and of the resulting changes in the chemical and physical properties of the reservoir system is therefore a prerequisite for any secure operation of a storage site[4]. Unfortunately, few experimental studies have investigated CO_2-brine-rock interactions under reservoir conditions. As such existing information on CO_2-induced interactions at real storage sites is indirect and primarily based on gas or fluid samples recovered from observation wells[8,11-12]. Moreover, these experimental investigations of long-term (geological scale) reactions during CCS (carbon capture and sequestration) have only been investigated by numerical simulations[13]. The reliability of such numerical simulations depends mainly on the availability of large amounts of data, the assumed model parameters, and details of the geological model. It is thus important that the results of numerical simulation models should be verified by laboratory experiments. Therefore, experimental studies of CO_2-brine-rock interactions at simulated reservoir pressure (p) and temperature (T) conditions are an important and elegant way to study the problem outlined above.

This contribution investigates the geochemical and hydrodynamic changes during the CO_2-formation water-feldspar/carbonate interaction induced by massive injection of CO_2 into feldspar-rich sandstone of the Qingshankou Formation (Qing 1 reservoir) from the southern Songliao Basin in China during EOR operations. The Songliao Basin is a high geothermal gradient basin, with an average geothermal gradient of 4.2℃/100m[14]. Drill hole measurements of temperature and pressure conditions of the Qing 1 reservoir are 100℃ and 24 MPa at the sampling horizon (2427.54 m depth). The southern Songliao Basin contains numerous CO_2-rich gas accumulations that are characterized by inorganic CO_2 ($\delta^{13}C_{CO_2} = -11.36‰$ to $-2.20‰$) [15], and associated mantle helium with R/Ra of 1.19 to 4.96[16]. However, the Qing 1 reservoir contains < 1 mol% CO_2, and is a site of a CO_2 injection EOR project, which is similar to the EnCana Weyburn project[17].

To explore the short-term CO_2 EOR-related fluid-rock interaction processes occurring in feldspar-rich geological reservoirs during the injection of CO_2, we have designed and carried out an experimental study of CO_2-brine-rock interactions at simulated reservoir conditions (100℃ and 24 MPa) through a core flooding laboratory experiment. The focus of our study was to document the mineralogical and chemical changes that result from CO_2-brine-rock interactions and, in particular, how these interactions impact on lithological porosity and permeability in the geological sequence and

128

the potential for CO_2 storage. The results of our study provide basic geological information for CO_2 trapping mechanisms in depleted oil and gas reservoirs, as well as the potential formation damage resulting from massive injections of CO_2 into reservoirs during EOR programs.

1 Geological Setting and Reservoir Petrography

The Songliao Basin is the largest Mesozoic−Cenozoic sedimentary basin in the Cathaysian system and the largest oil−producing province in China[18]. The Changling fault depression is regionally oriented in a north−northeast direction (area: $1.3 \times 10^4 km^2$) and is the largest system in the basin with the most abundant occurrences of natural CO_2 gas[19]. The Qing 1 reservoir is located in the middle of the Changling fault depression, and adjacent to the Qian´an oil field to the northeast and the Huazijing terrace to the east (Fig. 1).

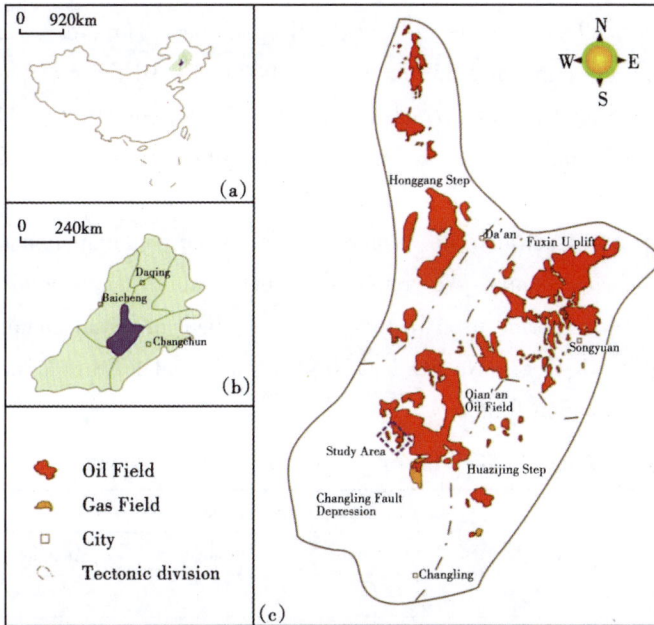

Fig. 1　Map of the S Corner of the Songliao Basin Showing the Study Area in Changling Fault Depression
(a) location of Songliao Basin (green marked); (b) location of southern part of Songliao Basin (blue marked);
(c) tectonic division of southern part of Songliao Basin and Study area

Sandstones of the Qing 1 reservoir are moderately well or well sorted, being very fine to fine−grained feldspathic sandstones. The main detrital constituents are monocrystalline quartz grains (average 38.7 vol %), albitic plagioclase (average 29.7 vol %), and K−feldspars (average 4.5 vol %). Authigenic minerals in the Qing 1 reservoir sandstones include quartz, feldspar, illite, calcite, and ankerite. Ankerite and calcite are the main authigenic minerals. Quartz and feldspar cements typically represent much less than 1% of the rock volume, and occur as thin fringes on the edge of quartz or feldspar grains. In addition, quartz cements may occur as microcrystalline pore−filling cements.

Porosity and permeability measurements of the Qing 1 reservoir have been made by the Research Institute of Petroleum Exploration & Development (RIPED, Beijing, China) and are available from well completion reports. The porosity of the Qing 1 reservoir ranges from 8% to 10% (average 10.4%). Porosity is mostly primary intergranular macroporosity, although this has been variably reduced by compaction and cementation. The permeability of the Qing 1 reservoir ranges from 0 to 30 mD (average 4.3 mD).

2 Experimental Design and Techniques

2.1 Sample Descriptions

The rock samples used in this experiment were provided by the RIPED and were taken from the Qing 1 Formation reservoir (2427.54 m depth) of the Jilin Oil Field in China. It should be noted that these samples were collected before the CO_2 EOR operations. Three core samples were used in the experiments, which are numbered S3-2, 6-2, and S11-1. All the core samples were 2.5 cm in diameter (D) and the total length (L) of the core was 17.82 cm. Bulk-rock X-ray diffraction (XRD) analysis showed that the average mineral composition of the core samples was 57.8% quartz, 4.0% potassium (K) -feldspar, 32.6% plagioclase (albite), 1.9% calcite, 4.3% clay minerals, and trace quantities of ankerite (Table 1). To avoid contamination with organic matter, crude oil present in the core samples was removed by petroleum ether prior to starting the experiment. The synthetic reservoir brine used in the experiment had an initial composition of 0.09 g/L $CaCl_2$, 0.06 g/L $MgCl_2$, 0.40 g/L Na_2SO_4, 8.20 g/L NaCl, and 1.88 g/L $NaHCO_3$, and is comparable to the Qing 1 Formation fluid.

Table 1   The Mineral Composition of Core Samples

No.	Mineral types and content (wt%)					
	Quartz	K-feldspar	Albite	Calcite	Ankerite	Clay minerals
S_{3-2}	57.5	2.9	35.2	—	—	4.4
6-2	56.3	4.9	31.1	1.3	—	6.4
S_{11-1}	59.7	4.2	31.6	2.4	—	2.1
average	57.8	4.0	32.6	1.9	—	4.3

Note: Values performed from Petro China Research Institute of Petroleum Exploration; wt%—Weight percent; "–" —Trace.

2.2 Experimental Equipment and Procedure

Fig. 2 shows a schematic diagram of the experimental set-up. The experiment was conducted in the State Key Laboratory for EOR of RIPED. The main parts of the experimental apparatus were a cylindrical container (1030mL), two syringe pumps (one for injection and one for confining pressure control), a Hassler core holder, a high-temperature oven, a pressure sensor, a back-pressure regulator, and a gas flow meter. The injection syringe pump was equipped with a "Smart Key" controller, which allows accurate control of the flow rate to within ± 0.5% of the set-point.

The core assemblage consisted of three cores wrapped with a Teflon sheet, which were placed into the Hassler core holder and saturated with synthetic reservoir brine. Synthetic reservoir brine and industrial-grade pure CO_2 were then injected into the cylindrical container (1030 mL) and together with the core holder were placed in the oven. Finally, all the experimental components were connected with stainless steel tubing ($D = 1$mm).

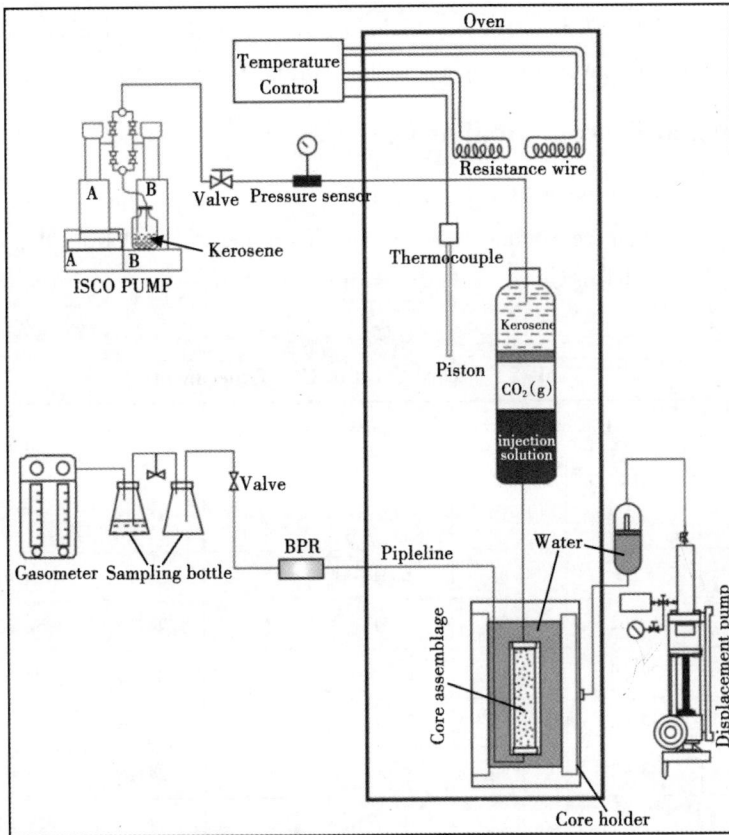

Fig. 2   Schematic Diagram of CO_2 Flooding Water-Rock Experiment

2.3 Experimental Conditions

The syringe pump was used to inject kerosene into the upper part of the container, continuously displacing the piston and injection solution at a constant flow rate through the cores during the experiment. The injection flow rate was set to $Q = 0.05$mL/min in order to avoid damage by illite and kaolinite, which are sensitivity to velocity. The temperature of the system was set to 100℃ and the total pressure (p_{tol}) of the inlet brine was set to 24 MPa by adjusting the position of the pistons. These p-T conditions are comparable with those of the Qing 1 Formation reservoir. The entire duration of the experiment (131.18 h) was determined by the volume and flow rate of the injection solution. The total injection volume in the core assemblage (352.77mL) was similar to the original volume (518.74mL) injected at a constant flow rate of 0.05 mL/min.

The injection solution was supersaturated with CO_2 under the experimental conditions ($T =$

131

100℃ and p_{tol} = 24MPa）. This can be calculated given that the CO_2 partial pressure (p_{CO_2}) is 16 MPa (i. e. , 1.1037 mol/kg of CO_2) in the formation water, and the salinity of the formation water is 10, 000 mg/L at 100℃ and 24 MPa[20]. Under these experimental conditions, the volume of synthetic reservoir brine injected (in the cylindrical container) was 518.74 mL, along with 246.31 mL of CO_2. The CO_2 dissolved in the synthetic reservoir brine was 0.57 mol, or ca. 39.77 mL in volume. Based on the above, it is possible to calculate the volume of CO_2 in the gas cap of the container, which was ca. 206.54 mL. These calculations show that the CO_2 content of the injection solution is sufficient for it to have been supersaturated during the entire CO_2-flooding experiment.

2.4 Experimental Process Monitoring

During the entire experiment, the reaction time, the injected volume (including the injected volume and corresponding pore volume), sampling measurement (including the amount of liquid and gas volume), pH, inlet pressure, outlet pressure, and displacement pressure were continuously measured (Table 2).

Table 2　Record Sheet of CO_2 Experiment

Sample	Reaction time (h)	Injected volume (mL)		Sampling measurement		Inlet pressure (MPa)	Outlet pressure (MPa)	Displacement pressure (MPa)
		Injected volume (mL)	Pore volume (PV)	Liquid (mL)	Gas (mL)			
1	3.00	8.35	0.61	6.51	5	24.14	24.01	0.13
2	7.12	18.02	1.32	8.51	5	24.16	24.00	0.16
3	11.62	29.39	2.16	10.26	7	24.15	24.03	0.12
4	15.62	38.60	2.83	8.21	6	24.16	24.01	0.15
5	19.95	48.90	3.59	9.60	7	24.16	24.01	0.15
6	24.62	59.55	4.37	10.27	25	24.17	24.01	0.16
7	35.62	87.57	6.42	27.22	10	24.17	24.01	0.16
8	59.24	148.22	10.87	55.66	26	24.19	24.03	0.16
9	83.07	216.25	15.87	65.75	6	24.22	24.00	0.22
10	107.06	285.05	20.91	61.93	5	24.15	23.92	0.23
11	131.18	352.77	25.88	63.24	5	24.13	23.95	0.18

2.5 Analytical Methods

The outlet solutions were collected into conical flasks. The volume of CO_2 degassed out of sample solutions was measured with a gas flow meter. The water samples were acidified with a 6 mol/L HCl in order to keep all ions in solution for later chemical analyses. K, Na, Ca, Mg, Fe, Mn, Al, Sr, and Ba contents were analyzed by inductively coupled plasma-atomic emission spectrophotometry (ICP-AES). Unfortunately, due to the lack of an appropriate standard, Si contents were not

132

analyzed during this experiment. The pH of the outlet solutions was measured daily using a pH meter (Orion 4-STAR, Thermo Scientific, Cambridge, USA) . The pH measurements were performed at <6h from the time of solution collection, after CO_2 degassing from the outlet solution.

Two samples (A and B) were prepared from the core assemblage (middle stream core) for mineral analysis (Fig. 3) . Sample A was taken pre-experiment, while sample B was taken post-experiment. Both of these samples were 2. 5 cm in diameter (D) and 2 mm in length (L) (Fig. 3). The length of each cutting core samples was 2 mm, and sufficient to avoid sample inhomogeneity.

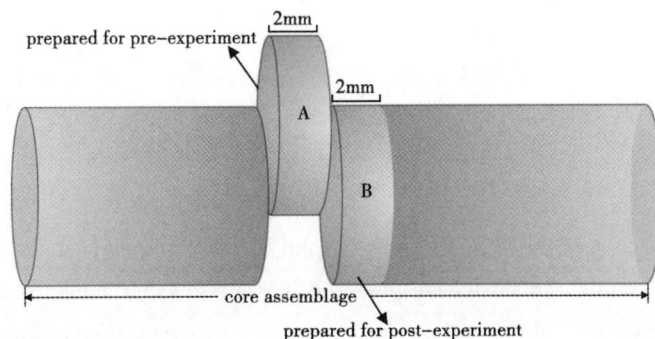

Fig. 3 Schematic Diagram of pre- and post-Experimental Sampling Strategy

The bulk-rock and clay fraction (< 2μm) mineralogy of the pre- and post-experimental core samples (A and B) were characterized in detail by quantitative XRD analysis (D/max-2500, Rigaku Corporation, Tokyo, Japan) . Core samples were crushed and sieved to<1 mm grain size. The coarsely crushed material was then ground in the micronizer mill to<40 μm in size for bulk-rock XRD analysis. The<2 μm separates (i. e. , clay fraction) from the micronized samples were then prepared for XRD analysis using centrifugation, and pipetted from an aqueous suspension onto a glass slide and air dried. The dried material was then mixed with corundum ($\alpha-Al_2O_3$, purity > 99. 5%, particle size<40 μm) in a weight ratio of 1:1. The corundum was used as an internal standard to enable the weight fraction of unknown mineral phases to be determined on an absolute basis. Core samples were characterized before and after the experiment to document mineralogical changes due to fluid-rock interactions using a scanning electron microscope (SEM; JSM6700F, JE-OL Corporation, Tokyo, Japan) . The SEM was operated at an accelerating voltage of 15 kV and a beam current of 10 nA, equipped with an energy dispersive X-ray system (EDS).

The porosity and air permeability of the cores were measured using a helium porosimeter (PHI -220, Coretest Systems, CA, USA) and a gas permeameter (KA210, Coretest Systems), respectively. The permeability of the core assemblage was recorded continuously by measuring the pressure drop between the inlet and outlet using Darcy's Law for laminar flow, as follows:

$$K=\mu LQ/S\Delta p$$

Where K is the permeability (m^2) , μ is the dynamic viscosity of the brine (Pa · s) , L is the length of the sample (m) , Q is the volumetric flow rate (m^3/s) , S is the cross-sectional area of the sample (m^2) , and Δp is the displacement pressure (Pa) .

2.6 Simulation of the Chemistry of the Solution

The geochemical code PHREEQC was used to calculate the speciation of each outlet solution. In these calculations, the amount of CO_2 degassed from the solution after sampling was used to recalculate in situ pH values, the total dissolved inorganic carbon (TIC), and the alkalinity. Using the chemistry of outlet solution, the saturation index (S. I.) of minerals was calculated using PHREEQC. The S. I. should be zero at equilibrium, whereas positive and negative S. I. values indicate supersaturation and undersaturation, respectively.

3 Results

3.1 Changes in Water Chemistry

The analytical results of the collected outlet solutions are summarized in Table 3. Significant changes in the outlet solution chemistry are observed during the experiment (Fig. 4). After injection, the pH (in situ values) of the outlet solution abruptly increases from an initial value of 3. 18 to 4. 40 after 7. 12 h [Fig. 4 (a)]. However, as the reaction continued, the pH drops to 4. 13 at 11. 62 h and then suddenly increases to the maximum value of 4. 45 at 15. 62 h [Fig. 4 (a)]. Following this, the pH gradually decreases to relatively constant values (values in the range 4. 03 and 4. 32) within 40h [Fig. 4 (a)]. Alkalinity as calculated from the measured solution pH, and the CO_2 recovered from the degassed solutions using PHREEQC, exhibited similar changes as pH. Both pH and alkalinity peaked at the same time (15. 62 h: 4. 45 for pH, $5. 19 \times 10^{-4}$ eq/L for alkalinity), and then decreased with reaction time to relatively constant values (alkalinity = $0. 72 \times 10^{-4}$ to $1. 33 \times 10^{-4}$ eq/L) within 40 h [Fig. 4 (b)].

Na, K, and Al recorded similar changes in outlet water chemistry during the experiment. Na, K, and Al concentrations peaked at the same time (4725mg/L, 33. 57mg/L, and 0. 32 mg/L, respectively, at 7. 12 h) (Fig. 4 (c) to Fig. 4 (e)], and then decreased with reaction time to relatively constant values (4033mg/L, 15mg/L, and 0. 08 mg/L, respectively) within 40 h [Fig. 4 (c) to Fig. 4 (e)]. A similar pattern is evident for Ca, Sr, and Fe concentrations, which peaked at 15. 62 h (757. 50mg/L, 3. 79mg/L, and 29. 64 mg/L, respectively) [Fig. 4 (f) to Fig. 4 (h)]. Compared with these elements (Ca, Sr, and Fe), Mg concentrations peak at 11. 62 h [Fig. 4 (h)]. Ca, Sr, Fe and Mg concentrations then decrease with reaction time to relatively constant values (375. 40mg/L, 2. 62mg/L, 8. 89mg/L, and 18. 35mg/L, respectively) within 40h [Fig. 4 (f) to Fig. 4 (h)].

Total inorganic carbon (TIC) as calculated from the measured solution pH, and the CO_2 recovered from the degassed solutions using PHREEQC, was much less than that of the original injection solution (Table 3). This most likely reflects loss of CO_2 by diffusion through the Teflon sheet and rubber tubing when the solution passed through the core assemblage[6].

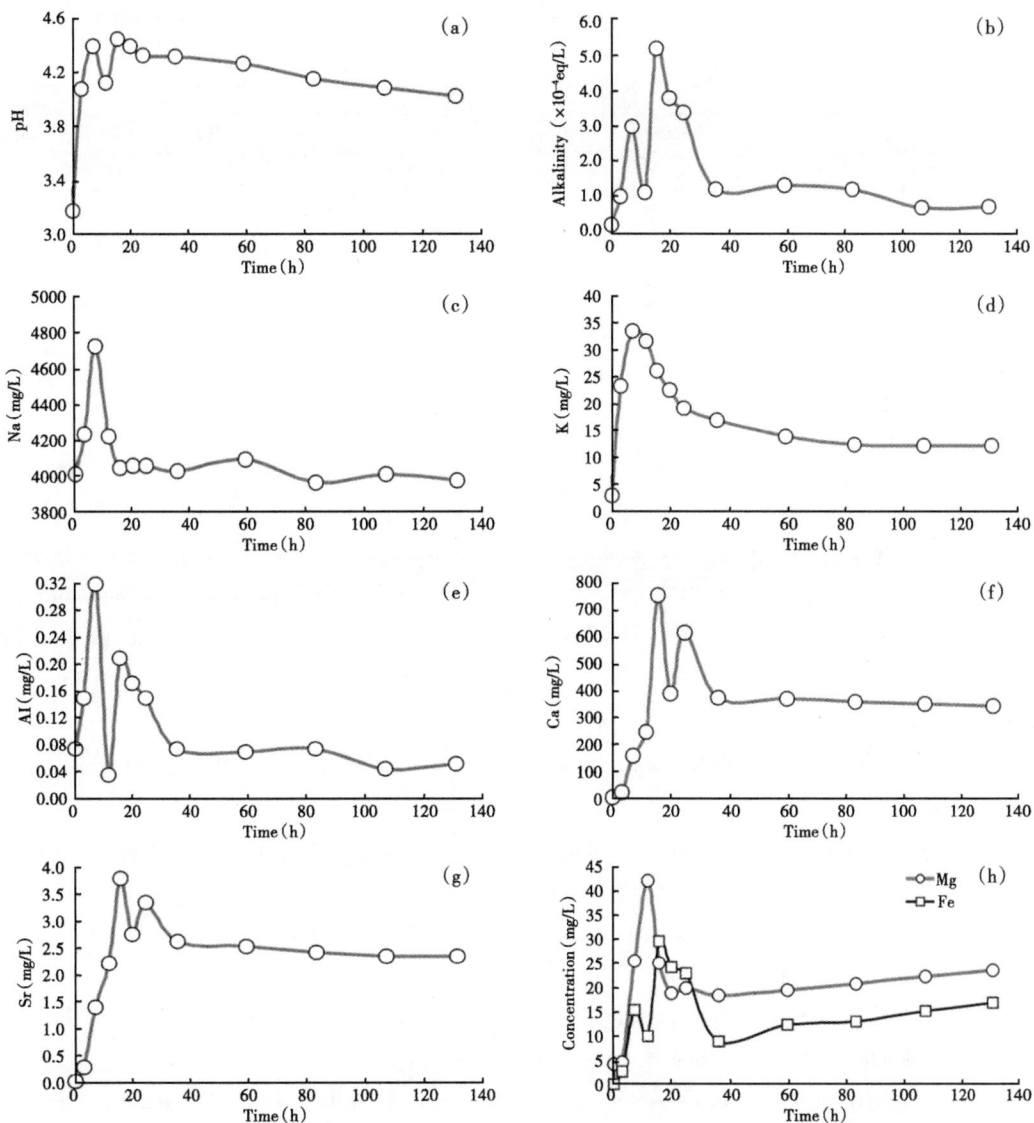

Fig. 4 pH, Alkalinity and Concentrations of Aqueous Chemical Components along the Different React Time

Table 3 Chemical Composition of Injection and Sample Solutions

Sample	Reaction time (h)	pH[①]	K (mg/L)	Na (mg/L)	Ca (mg/L)	Mg (mg/L)	Fe (mg/L)	Al (mg/L)	Sr (mg/L)	Alkalinity[②] (10^{-4} eq/L)	TIC[③] (mol/L)
Init. Sol	—	3. 18	2. 78	4016. 00	2. 56	4. 09	0. 07	0. 07	0. 01	0. 20	1. 10
1	3. 00	4. 08	23. 32	4236. 00	22. 74	4. 58	2. 68	0. 15	0. 28	1. 01	0. 03
2	7. 12	4. 4	33. 57	4725. 00	157. 50	25. 62	15. 40	0. 32	1. 40	3. 00	0. 02
3	11. 62	4. 13	31. 73	4228. 00	248. 50	42. 13	9. 90	0. 04	2. 21	1. 11	0. 03
4	15. 62	4. 45	26. 20	4052. 00	757. 50	25. 12	29. 64	0. 21	3. 79	5. 19	0. 03
5	19. 95	4. 4	22. 53	4065. 00	392. 00	18. 85	24. 20	0. 17	2. 76	3. 78	0. 03
6	24. 62	4. 39	19. 07	4061. 00	619. 10	19. 96	23. 03	0. 15	3. 34	3. 40	0. 10

Sample	Reaction time (h)	pH[1]	K (mg/L)	Na (mg/L)	Ca (mg/L)	Mg (mg/L)	Fe (mg/L)	Al (mg/L)	Sr (mg/L)	Alkalinity[2] (10^{-4} eq/L)	TIC[3] (mol/L)
7	35.62	4.32	16.90	4033.00	375.40	18.35	8.89	0.08	2.62	1.21	0.01
8	59.24	4.27	13.92	4099.00	369.70	19.55	12.41	0.07	2.53	1.33	0.02
9	83.07	4.16	12.39	3969.00	358.70	20.82	12.98	0.07	2.42	1.21	0.03
10	107.06	4.09	12.08	4018.00	350.50	22.22	15.25	0.05	2.34	0.68	0.02
11	131.18	4.03	12.04	3979.00	344.20	23.54	16.94	0.05	2.35	0.72	0.03

[1]Calculated in situ pH; [2] [3]Calculated by using the PHREEQC.

3.2 Mineralogical Changes during the Experiment

Mineral surfaces of detrital albite grains were generally clean without obvious signs of alteration prior to the experiment [Fig. 5 (a)], but after the experiment albite surfaces showed a serrated structure [Fig. 5 (b)], principally along cleavage planes. K-feldspar showed signs of incipient alteration prior to the experiment [Fig. 5 (c)], which were more pronounced after the experiment [Fig. 5 (d)]. Before the experiment, ankerite grains had a rhombohedral structure and only showed possible weak corrosion on the edges of grains [Fig. 5 (e)]. After the experiment, many tiny corrosion pits were evident on the surface of the ankerite grains [Fig. 5 (f)]. In contrast, authigenic albite and micro-quartz grains showed no signs of alteration after the experiment [Fig. 5 (g) and Fig. 5 (h)].

XRD observations identified that a higher amount of quartz was present after the experiment than before (Table 4). This is in agreement with the micrographs [Fig. 5 (g) and Fig. 5 (h)], which showed that micro-quartz grains remain unaltered. Due to the lack of Si concentration data, it is not possible to calculate the equivalent SiO_2 (aq) concentration. However, bulk-rock XRD analysis results and the qualitative imaging suggest it is reasonable to infer that the outlet fluid was over-saturated with respect to quartz during the experiment.

Bulk-rock XRD analysis after the percolation experiment identified a large decrease of albite and K-feldspar contents (Table 4). This is also in accordance with the micrographs [Fig. 5 (b) and Fig. 5 (d)], which showed that feldspar crystals were altered after the CO_2-rich brine percolation.

Table4   Quantitative XRD Analysis of pre-and post-Middle Stream Cores

Minerals	Mineral content (wt%)		Clay mineralogy of the<2 μm fraction	Mineral content (wt%)	
	Before experiment	After experiment		Before experiment	After experiment
Quartz	56.0	59.2	Illite	2.9	4.1
Albite	33.5	30.9	Kaolinite	0.0	1.0
K-feldspar	4.8	3.1	Chlorite	1.5	1.7
Calcite	1.3	0.0			
Ankerite	—	—			
Clay minerals	4.4	6.8			

Note: Values performed from Petro China Research Institute of Petroleum Exploration; wt%—Weight percent; "–" —Trace.

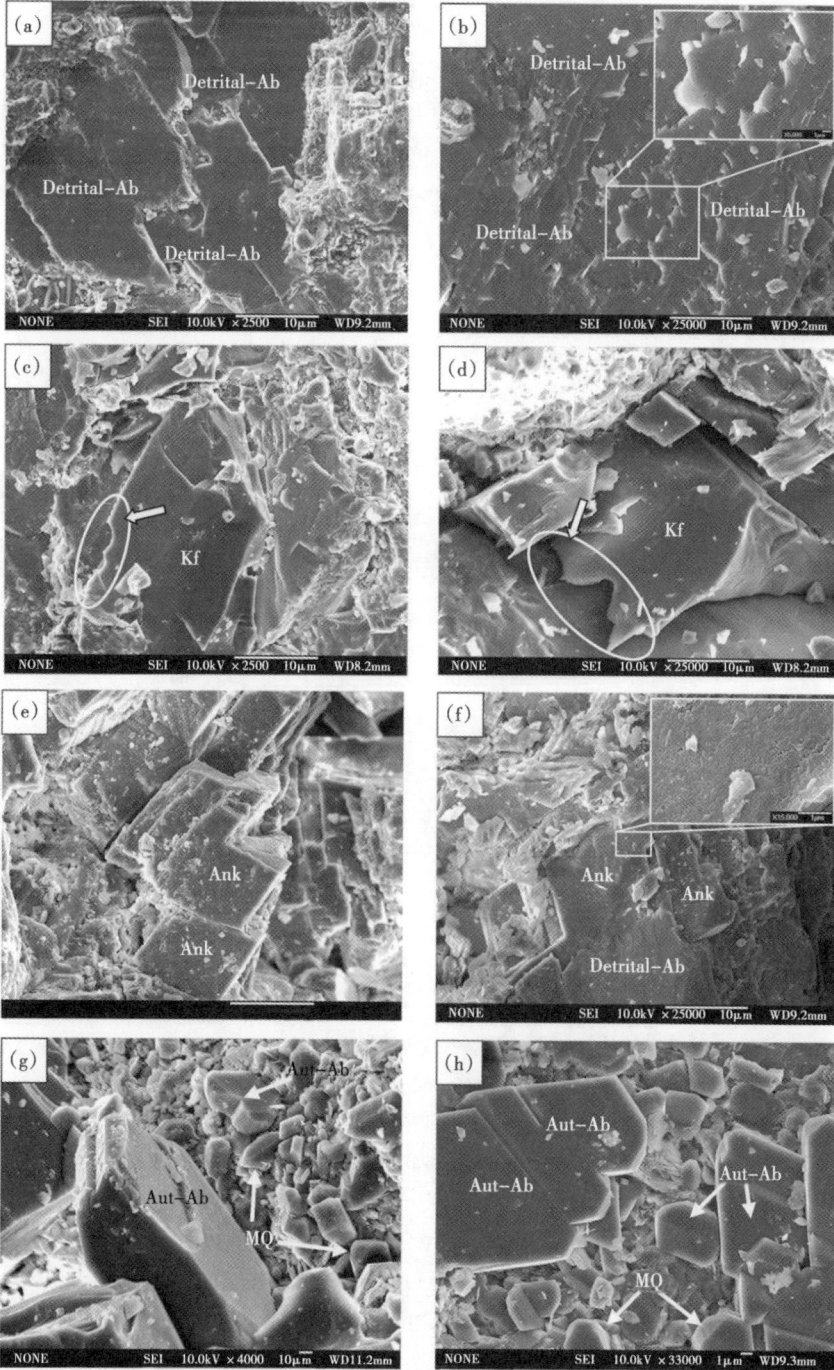

Fig. 5 Scanning Electron Photomicrographs of pre- and post-Experimental Cores

(a) detrital albite before the experiment; (b) detrital albite after the experiment; (c) K-feldspar beforethe experiment;

(d) K-feldspar after the experiment; (e) ankerite before the experiment; (f) ankerite after the experiment;

(g) authigenic albite and micro quartz before the experiment; (h) authigenic albite and micro quartz after the experiment

Ank—akerite; MQ—microcrystalline quartz; Detrital-Ab—detrital albite; Kf—K-feldspar; Aut-Ab—authigenic albite

Compared with pre-experiment XRD results, the calcite content dropped to zero after the experiment (Table 4), and XRD analysis failed to identify any X-ray peaks of calcite after the experiment (Fig. 6). Moreover, calcite was not observed in the post-CO_2 experimental SEM images. This implies that calcite was completely dissolved in the core material after the experiment. XRD analysis revealed smaller X-ray peaks diagnostic of ankerite after the experiment as compared with prior to the experiment (Fig. 6). This is in agreement with the qualitative imaging [Fig. 5 (f)], which identified ankerite alteration.

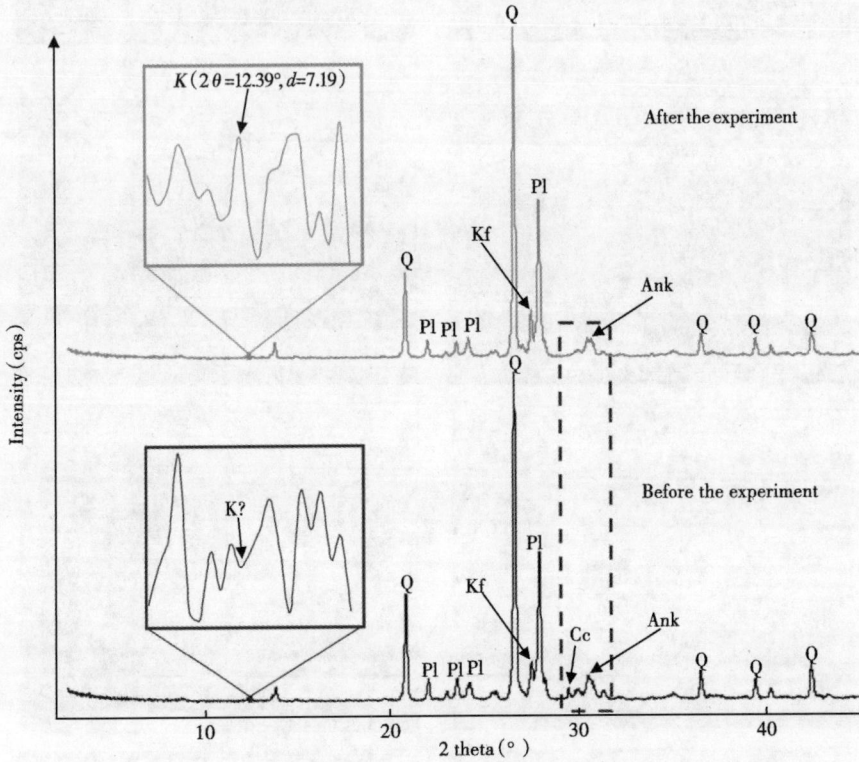

Fig. 6   Spectra Line of XRD of pre- and post-Experimental Cores

Q—quartz; Pl—plagioclase (albite); Kf—K-feldspar; Ank—ankerite; Cc—calcite; K— Kaolinite; K? —No Kaolinite

XRD analyses also revealed a more substantial fraction of clay minerals after the experiment (Table 4), although this was not observed in the post-CO_2 experimental SEM images. Given the fact that calcite was completely dissolved in the core material after the experiment, these additional clay minerals may have been released by the dissolution of carbonate cement and precipitation of new clay minerals. Although not observed in the pre- and post-CO_2 experimental SEM images, XRD analyses indicated a more significant fraction of kaolinite after the experiment (Table 4) as very weak X-ray peaks diagnostic of kaolinite at $2\theta = 12.39°$ were observed after the experiment; but were not present prior to the experiment (Fig. 6). All of these observations suggest that small amounts of kaolinite were precipitated during the experiment. In addition, quantitative XRD analysis of clay minerals in the core samples after the percolation experiment identified a large increase of illite contents, while a small increase of chlorite contents (Table 4).

138

3.3 Precipitation of New Minerals

As mentioned in the previous section, a small amount of kaolinite was precipitated during the experiment (Table 4 and Fig. 6). Similar results during CO_2-exposure experiments have been reported by Murakami et al.[21], Shiraki and Dun[6], Carroll and Knauss[22], and Bertier et al.[23]. SEM and EDS analyses reveal that micron-sized platy crystals had precipitated after the present experiment [Fig. 7 (a) and Fig. 7 (b)]. Chemical analysis by EDS did not yield any recogniza-

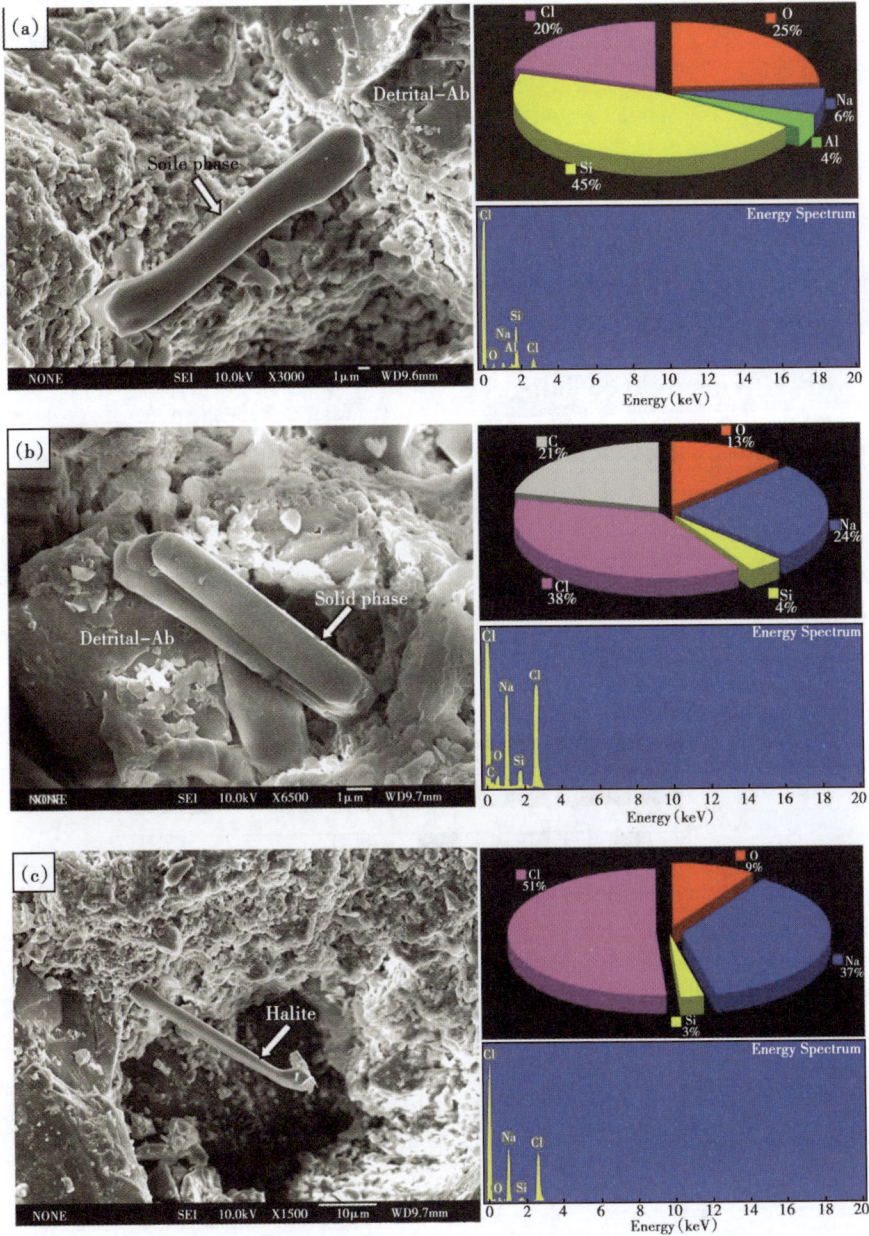

Fig. 7 SEM Image and Energy Spectrum Analysis of the Solid Phases

On the right: Chemical analysis associated with the SEM image, Detrital-Ab—detrital albite

ble mineral composition, so we consider these new phases to be "solid phases" that can be divided into two types: the first does not contain C and is mainly composed of Si, Cl, O, and small amounts of Na and Al [Fig. 7 (a)]; the second contains C and is mainly composed of C, O, Cl, and small amounts of Na and Si [Fig. 7 (b)]. Precipitation of solid phases during CO_2-exposure experiments has previously been reported by Hangx and Spiers[24] and Luquot et al. [25], and the latter study identified the phases as amorphous carbon. However, these solid phases could be related to the drying procedure after the experiment, and be a mixture of water-soluble salts, dominated by sodium chloride. However, according to the EDS analysis results, the first type of solid phases only contains 6% Na and 20% Cl [Fig. 7 (a)], while the second type of solid phases contains 24% Na and 38% Cl [Fig. 7 (b)]. Moreover, Si and C represent > 20% of these solid phases. Therefore, the solid phases cannot be mixtures of water-soluble salts that precipitated during the drying of the sample, and thus must be related to interaction of the CO_2-rich brine with the core.

Halite has also been noted as a localized precipitate in pore throats of post-experiment cores reacted with synthetic water [Fig. 7 (c)]. The texture of the halite suggests that it developed as an efflorescence crystal product during the reaction. Brine drawn up through the block by capillary reaction evaporates at the block surface in the supercritical CO_2 atmosphere and crystallizes halite[26].

3.4　Porosity/Permeability Changes

Changes in porosity and permeability are shown in Fig. 8 and Table 5. The upstream and middle cores used in the experiment show decreases in pore volume of 11% and 14.32%, respectively, which probably explains the decreases in porosity of 0.87% and 1.8%, respectively. Although the downstream core shows no significant change in porosity, the pore volume decreases by 10.43%. The permeability of the upstream and middle cores decreases by 10% and 17.84%, respectively. The largest decrease in permeability (20%) was observed in the downstream core.

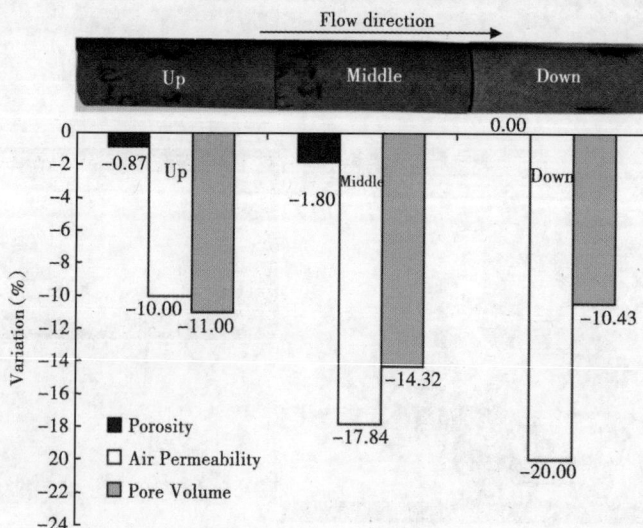

Fig. 8　Physical Properties of pre- and post-CO_2 Experimental Cores

Up—upstreamcore; Middle—middle core; Down—downstream core

Table 5 Physical Properties of pre- and post- CO$_2$ Experimental Cores

	Porosity (%)		Variation (%)	Air Permeability ($10^{-15} m^2$)		Variation (%)	Pore Volume (cm^3)		Variation (%)
	b	a	$(a-b)/b$	b	a	$(a-b)/b$	b	a	$(a-b)/b$
Upstream	11.5	11.4	−0.87	2.5	2.25	−10.00	3.91	3.48	−11.00
Middle	11.1	10.9	−1.80	6.11	5.02	−17.84	3.84	3.29	−14.32
Downstream	15.8	15.8	0.00	29	23.2	−20.00	4.89	4.38	−10.43

Note: a—after the CO$_2$ experiment; b—before the CO$_2$ experiment.

3.5 Mineral Saturation

Significant changes in the S. I. of the outlet solution were observed during the experiment (Fig. 9). The outlet solutions were supersaturated with respect to gibbsite within the starting 60 h of the experiment. However, as the reaction continued, the S. I. of gibbsite dropped to below zero after 80 h (Fig. 9), which implies that the outlet solutions were undersaturated with respect to gibbsite during the latter part of the experiment. All types of carbonate, with the exception of siderite, recorded similar S. I. changes during the experiment (Fig. 9). S. I. values peaked at the same time (at 15.62h) (Fig. 9), and then decreased with reaction time to relatively constant values within 40 h. However, the S. I. of all the carbonates were negative during the experiment (Fig. 9), which implies that the outlet solutions were undersaturated with respect to the carbonate minerals throughout experiment.

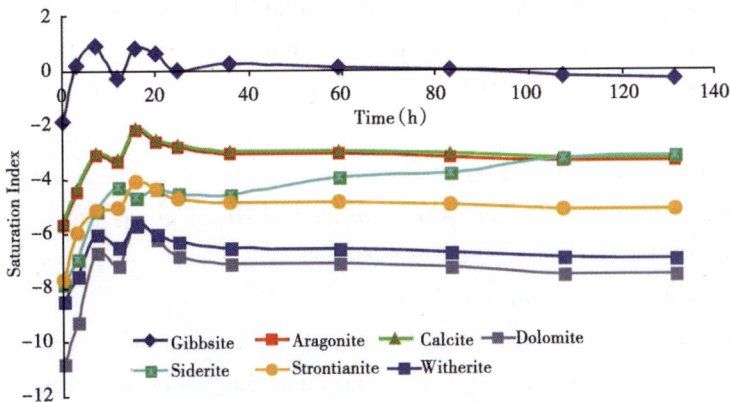

Fig. 9 Saturation Indices of Carbonate Minerals vs. Reaction Time

4 Discussion

4.1 Feldspar

Dissolution of feldspar is commonly reported in CO$_2$ experiments[3-4,6,23,27-28], numerical modeling[29-31], field tests[11-12], and observations of natural analoguesI[26,32]. The observation of changes in the composition of the outlet solution [Fig. 4 (c) to Fig. 4 (e)], combined with SEM analysis

141

[Fig. 5 (b) and Fig. 5 (d)] confirmed that feldspar crystals were altered after the percolation of CO_2-rich brine. According to the chemical composition of the rock-forming minerals, the release of Na, Al, and K can be related to the dissolution of albite and K-feldspar. In the present experiment, Na, K, and Al show similar trends with ongoing reaction time, attaining their respective peaks and equilibrium at the same time [Fig. 4 (c) to Fig. 4 (e)]. In addition, these ions show strong linear correlations among each other (based on principal components analysis), yielding correlation coefficients greater than 0.5 (Table 6). The similar patterns of change among these ions indicate that their contents were controlled by a common reaction mechanism. According to Gaus[33] and Shiraki and Dun[6], the dissolution of albite and K-feldspar is controlled by the following reactions:

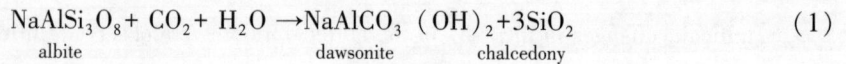

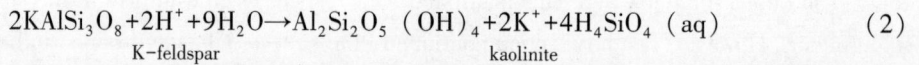

$$NaAlSi_3O_8 + CO_2 + H_2O \rightarrow NaAlCO_3 (OH)_2 + 3SiO_2 \qquad (1)$$
albite dawsonite chalcedony

$$2KAlSi_3O_8 + 2H^+ + 9H_2O \rightarrow Al_2Si_2O_5 (OH)_4 + 2K^+ + 4H_4SiO_4 (aq) \qquad (2)$$
K-feldspar kaolinite

Experiments and modeling results presented in previous studies[34] have suggested that these silicate minerals gradually undergo dissolution and over the long-term lead to the precipitation of various new mineral phases. As expressed in Eq. (1) and Eq. (2), feldspar dissolution results in clay and quartz precipitation (Table 4) in the presence of high CO_2 concentrations at underground storage p-T conditions during laboratory experiments. This is in agreement with our SEM observations and XRD analyses (Sections 4.2). As described in Section 4.3, the mineral phases (called "solid phases" in our study) are observed close to detrital albite [Fig. 7 (a) and Fig. 7 (b)], which indicates a genetic relationship between these solid phases and detrital albite. These solid phases may be alteration products of detrital albite, resulting from the interaction between detrital albite and CO_2-rich brine [e.g., Eq. (1)]. However, these solid phases have not been converted into the corresponding carbonate. Given the fact that the cation concentrations of the collected fluids are too low to have reached supersaturation of carbonate under experimental conditions (Fig. 9), these solid phases may be a product with a transitional chemical constituent between albite and carbonate. It is likely that the presence of pH buffering assemblages, such as albite, imposes tight constraints on carbonate (e.g., dawsonite) precipitation[35].

Authigenic albite showed no signs of alteration during our experiment, which is consistent with the results reported by Weibel et al. [36]. This observation may be explained by two factors. First, authigenic albite is a late-stage diagenetic mineral that is difficult to dissolve again. Feldspar in the deep strata (depth > 2000 m) of the Songliao Basin is almost pure sodium end-member albite[37-38]. The rock samples used for our experiment have been buried as deep as 2427 m. Prior to the experiment, bulk-rock XRD analysis of the cores showed that the albite content reaches up to 32.6%, while the K-feldspar content is only 4.0% (Table 1). SEM and EDS observations show that euhedral albite crystals are almost the pure sodium end-member, which indicates that albitization is well developed. Based on the above observations, the authigenic albite in the experimental cores is a late-stage diagenetic mineral and thus it is not susceptible to re-dissolution. Second, the authigenic albite in the Qing 1 Formation reservoir does not show any signs of corrosion and it is possible that only the corroded and previously dissolved parts of the feldspar grains are readily dissolved during experimental investigations[36]. Detrital albite dissolution can be spontaneous and is an exothermic reaction with a Gibbs energy of reaction of −132 kJ/mol ($\Delta G_T{}^0 < 0$) [Eq. (3)][37]:

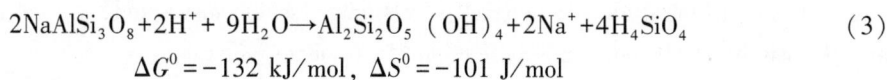

$$2NaAlSi_3O_8 + 2H^+ + 9H_2O \rightarrow Al_2Si_2O_5 (OH)_4 + 2Na^+ + 4H_4SiO_4 \qquad (3)$$
$$\Delta G^0 = -132 \text{ kJ/mol}, \ \Delta S^0 = -101 \text{ J/mol}$$

The albite dissolution process is described by the following formula:

$$\Delta G_T{}^0 = \Delta H - T\Delta S$$

where $\Delta G_T{}^0$ is the Gibbs energy change, ΔH is the enthalpy change, ΔS is the entropy change, and T is temperature. This formula shows that the dissolution of albite decreases with increasing temperature. Therefore, authigenic albite is relatively stable under high temperatures and acidic conditions, and does not readily react or dissolve. Conversely, long−duration batch experiments (ca. 2 yr) performed by Wandrey et al[28] have shown that euhedral albite precipitation occurs in sandstone systems under simulated reservoir (lower) $p-T$ conditions of 5. 5 MPa and 40℃.

4. 2 Carbonate

The most prominent mineralogical change during the CO_2−flooding experiment was carbonate dissolution[36]. Bulk quantitative analysis of the fluid samples, and XRD and SEM analysis of the core samples, identified a large decrease in the calcite and ankerite contents of the cores after the experiment. In particular, calcite was completely dissolved after the experiment (Table 4 and Fig. 6). This observation has been noted by previous studies on calcite[12,26,39] and ankerite dissolution[6].

The release of Ca, Fe, and Mg can be attributed to the dissolution of calcite and ankerite. In our experiment, Ca, Sr, and Fe behaved similarly with reaction time, reaching peak and equilibrium solution concentrations at the same time [Fig. 4 (f) to Fig. 4 (h)]. These ions also showed a strong linear correlation (using principal component analysis), with correlation coefficients of > 0. 5 (Table 6). This finding indicates that the release of these ions into solution was controlled by the same reaction mechanism. Although the solution Mg concentration does not peak at the same time as Ca, Sr, and Fe concentrations, solution Mg concentrations reach equilibrium at the same

Table 6　Correlation Coefficient Matrix of the Outlet Solution Ions

		Correlation Matrix						
		K	Na	Ca	Mg	Fe	Al	Sr
Correlation	K	1. 000						
	Na	0. 714	1. 000					
	Ca	0. 142	−0. 360	1. 000				
	Mg	0. 565	0. 208	0. 399	1. 000			
	Fe	0. 343	−0. 079	0. 873	0. 433	1. 000		
	Al	0. 540	0. 341	0. 440	−0. 035	0. 601	1. 000	
	Sr	0. 210	−0. 313	0. 958	0. 566	0. 854	0. 286	1. 000

time as Ca, Sr, and Fe [Fig. 4 (f) to Fig. 4 (h)]. A principal component analysis also showed a strong linear correlation between Mg and Ca, Fe, and in particular Sr (Table 6). Sr is geochemically similar to K and Ca, and can substitute into Ca- and K-bearing minerals such as carbonate and feldspar. In this study, Sr behaved in a similar fashion to Ca and, as such, the release of Sr and Mg may be attributed to the dissolution of carbonate minerals.

The main factor triggering these geochemical reactions is the dissolution of CO_2 in water, forming H_2CO_3 acid that dissociates in the brine. This causes a drop in the pH of the brine and acid attack of the minerals. Interactions with carbonate minerals that are generally present, to some extent, in sedimentary rocks would rapidly buffer the pH and make the brine less acidic[33] according to the following reactions:

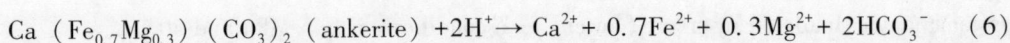

$$CO_2 + H_2O \rightarrow H^+ + HCO_3^- \qquad (4)$$

$$CaCO_3 \ (calcite) + H^+ \rightarrow Ca^{2+} + HCO_3^- \qquad (5)$$

$$Ca \ (Fe_{0.7}Mg_{0.3}) \ (CO_3)_2 \ (ankerite) + 2H^+ \rightarrow Ca^{2+} + 0.7Fe^{2+} + 0.3Mg^{2+} + 2HCO_3^- \qquad (6)$$

Calcite, and to a lesser extent ankerite, were rapidly dissolved in the CO_2-supersaturated, synthetic brine at the beginning of the experiment. Subsequently, as silicate minerals (mainly detrital albite and K-feldspar) were dissolved [Eq. (1) and Eq. (2)], the liberated ions reacted with brine to form clay minerals (e. g., kaolinite). The above reactions may be the main reasons for the abrupt increase in pH at the beginning of the experiment [Fig. 4 (a)]. However, as the reaction proceeded to steady state, the pH gradually decreased to relatively constant values [Fig. 4 (a)]. The same observations have been noted in field test results[11]. In our experiment, no carbonates, and only a small amount of kaolinite and platy solid phases, were observed. Precipitation of carbonates may have been inhibited if the solution did not attain the critical saturation state required for nucleation[40]. This is in accordance with PHREEQC geochemical modeling results (Section 4.5), which showed that the outlet solutions were undersaturated with respect to carbonate minerals throughout the experiment. However, secondary calcite precipitation was observed in batch experiments carried out on the Triassic Sherwood Sandstone (UK), which was reacted with CO_2-saturated, seawater-like fluids over a period of eight months[26].

4.3 Mineral Dissolution Rates

Given that the chemical composition of the minerals in the present core samples is known, the progress of the reaction can be determined from mass-balance calculations using the inlet and the outlet solution chemistries. Firstly, we assume that changes in the K concentration of the fluid were due solely to K-feldspar dissolution and that the Mg concentration of the fluid was controlled solely by ankerite dissolution. However, chemical contributions from albite could not be evaluated during the experiment because the high Na concentration of the inlet brine prevented the measurement of small changes in this element[25]. Secondly, the volume of minerals dissolved was calculated by multiplying the molar amount of each mineral's constituent elements by the molar volume of each mineral (64.3 cm^3/mol for ankerite and 109 cm^3/mol for K-feldspar)[6,41-42]. Using these assump-

144

tions, we calculated the reaction rate of mineral dissolution using the K and Mg solution concentration curves (Fig. 4).

The calculated dissolution rates for ankerite and K-feldspar show a peak at the same time ($3.87×10^{-6}$ cm^3/min and $3.08×10^{-6}$ cm^3/min, respectively, at 11.62 h), and then decrease with reaction time to relatively constant values [Fig. 10 (a) and Fig. 10 (b)]. This is in agreement with the dissolution volume [Fig. 10 (c) and Fig. 10 (d)], which shows that ankerite and K-feldspar crystals were clearly altered after percolation of the CO_2-rich brine [Fig. 10 (c) and Fig. 10 (d)]. The dissolution rate of ankerite during the experiment was markedly higher than that of K-feldspar [Fig. 10 (a) and Fig. 10 (b)], meaning that carbonate was more easily dissolved than feldspar. However, the time for ankerite to reach equilibrium was longer than that for K-feldspar (83.07 h for ankerite and 59.24 h for K-feldspar) [Fig. 10 (a) and Fig. 10 (b)]. This result suggests that the carbonate (ankerite in the case of our experiment) dissolution may occur over a long time. Our result is different from that of Weibel et al.[36], who concluded that lower reaction rates of silicate minerals, as compared with carbonate minerals, means that carbonate dissolution will occur in the short term, whereas carbonate precipitation is expected to take place as silicate alteration intensifies and the reactions approach equilibrium[36].

As described above, the maximum K-feldspar dissolution rate was $3.08×10^{-6}$ cm^3/min. However, the CO_2-water-rock interaction experiment performed by Shiraki and Dun[6] showed that the average K-feldspar dissolution rate was $4.12×10^{-6}$ cm^3/min, which is higher than our results. This

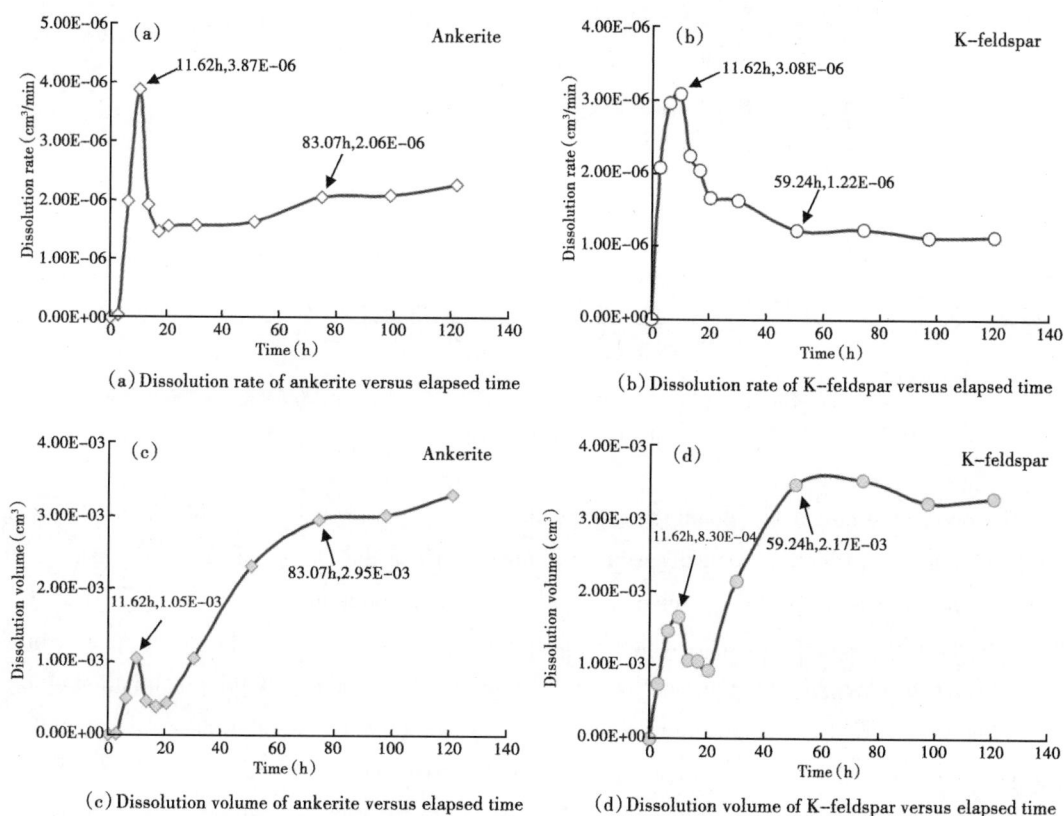

(a) Dissolution rate of ankerite versus elapsed time

(b) Dissolution rate of K-feldspar versus elapsed time

(c) Dissolution volume of ankerite versus elapsed time

(d) Dissolution volume of K-feldspar versus elapsed time

Fig. 10　Dissolved Minerals Volumes and Rate versus Elapsed Time

145

discrepancy may reflect the large differences in permeability between our experiment and that of Shiraki and Dun[6]. The average core permeability was 68 mD in the Shiraki and Dun[6] experiment, and was considerably lower in our experiment (17mD). The very low fluid penetration rate in our experiment may have significantly reduced the contact area of the acidic CO_2 fluid and minerals in the cores, resulting in the low feldspar−dissolution rate in our study.

4.4　Factors Controlling the Reduction in Core Permeability

Reduction in permeability is a common phenomenon during CO_2−flooding experiments[6,25,43−44]. As shown in Fig. 11, the core permeability in the present study decreased from 1.88×10^{-15} m^2 to 1.04×10^{-15} m^2. This is again consistent with data on the permeability of the pre− and post−experimental cores (See Table 5 and Fig. 8). The results of similar previous experiments are also consistent with this conclusion. For example, Ross et al. [43] carried out core CO_2−flooding experiments under reservoir conditions using North Sea calcareous sandstones. These authors observed notable decreases in permeability after their CO_2−flooding experiment. Sayegh et al. [44] also observed that the permeability of cores decreased rapidly at an early stage during their experimental runs (decreases to 10%−60% of the original values), and then gradually increased, although the permeability did not return to the original values. Other authors have also reported permeability decreases coupled with weak mineral (feldspar and carbonate) dissolution under simulated reservoir conditions during laboratory experiments[6,25].

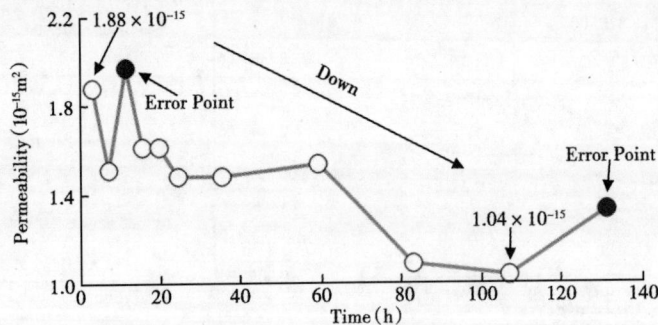

Fig. 11　Measured Permeability (K in m^2) versus Reaction Time (Error Point is Caused by the Pressure Sensor Fault)

The porosity of post−CO_2−flooding experimental cores in the present study only decreased by up to 1.8% (Fig. 8), which is basically within the uncertainty of the measurements. This result indicates that the core porosity did not change significantly throughout the experiment. The lack of a change in porosity, accompanied by a permeability decrease, is most likely due to the permeability being affected by clay particles released by the dissolution of the cement, by the precipitation of kaolinite (Fig. 6 and Table 4), and, to a lesser extent, by the formation of solid phases during the experiment [Fig. 7 (a) and Fig. 7 (b)], called microporous material by Luquot et al. [25]. Shiraki and Dun[6] suggested that kaolinite crystal growth in pore throats was responsible for reducing permeability. In our experiments, we observed the dissolution of carbonate cement (Section 4.2) and the

146

precipitation of solid phases (Section 4. 3) and possibly very minor amounts of kaolinite (Section 4. 2). However, only minor secondary precipitates were identified after our CO_2-flooding experiment. Consequently, it is unlikely that precipitation of these minerals would account for all the reduction in permeability. In contrast, substantial cement dissolution was observed throughout the experiment, with calcite having been completely dissolved after the experiment (Fig. 6 and Table 4). These carbonate materials are cementing agents that form during diagenesis and that bind sand and clay particles in sandstone reservoirs. As such, cement dissolution allows some of these particles to be released (Fig. 12), to move in the fluid flow path (Fig. 12), and to accumulate at pore throats and thereby reduce permeability[44]. However, given that the average amount of calcite initially present in the cores was only 1. 9%, this mechanism alone cannot have caused the substantial changes in permeability. In addition, XRD analysis results of the pre- and post-core samples showed that clay minerals were present in the core samples. Thus, the changes in permeability could also be the result of clay swelling. However, the main constituent of these clay minerals is illite (Table 4), which is a non-swelling mineral, and will not lead to water sensitive damage. Furthermore, the flow rate of the injection solution was not high enough to damage the velocity sensitivity controlled by illite. Thus, the clay minerals in the core cannot have been responsible for the substantial changes in permeability observed during the experiment.

Therefore, we conclude that precipitation of new minerals (kaolinite and solid phases), combined with the clay particles released by dissolution of the carbonate cement and that move in the fluid flow path and accumulate at pore throats, were responsible for the permeability reduction during our experiment.

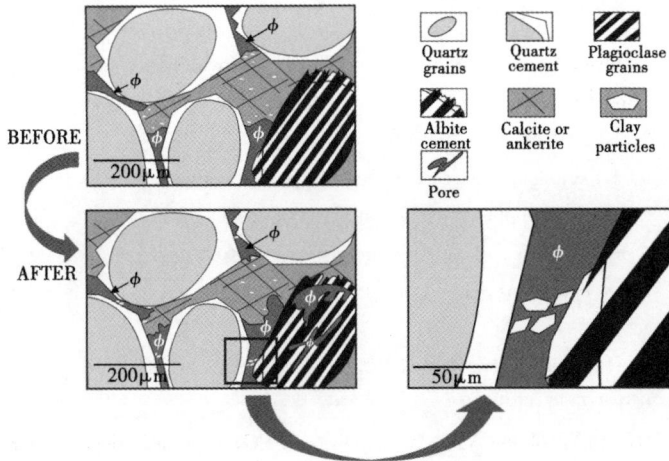

Fig. 12   Sketch Illustration of Clay Particles Released by the Dissolution of the Carbonate Cement Dissolution Accumulates at Pore Throats and Thereby Reduces Permeability

5   Conclusions

A core CO_2-flooding laboratory experiment, carried out under simulated reservoir conditions (100°C and 24 MPa) to study CO_2-brine-rock interactions, yielded the following results:

(1) Minerals such as K-feldspar, albite, calcite, and ankerite were dissolved during the CO_2 -flooding experiment. Calcite dissolution was most pronounced, followed by ankerite, with feldspar minerals showing the least signs of dissolution.

(2) Small amounts of kaolinite and solid phases were precipitated during the CO_2-flooding experiment.

(3) The very low penetration of fluid into the core resulted in a marked reduction in the contact area of the acidic CO_2 fluid and minerals in the cores, resulting in low rates of feldspar dissolution.

(4) Core permeability showed a marked decrease during the experiment, although porosity did not change. The reduction in permeability is attributed to precipitation of new minerals (kaolinite and solid phases), and to clay particles released by the dissolution of the carbonate cement, which move in the fluid flow path and accumulate at pore throats.

References

[1] Intergovernmental Panel on Climate Change (IPCC). Climate change 2007: Mitigation Contribution of Working group III to the Fourth Assessment Report of the Intergovernmental Panel on Climate Change [M]. Cambridge University Press, Cambridge, United Kingdom and New York, NY, USA.

[2] Intergovernmental Panel on Climate Change (IPCC). Contribution of Working Group I to the Third Assessment Report of the Intergovernmental Panel on Climate Change Climate change [R] // Synthesis Report. Summary for policy makers, Cambridge University Press, Cambridge, New York, NY, USA.

[3] Wigand M, Carey J W, Schütt H, et al. Geochemical effects of CO_2 sequestration in sandstones under simulated in situ conditions of deep saline aquifers [J]. Applied Geochemistry, 2008 (28): 2735-2745.

[4] Fischer S, Liebscher A, Wandrey M, et al. CO_2-brine-rockinteraction — First results of long-term exposure experiments at in situ P-T conditions of the Ketzin CO_2 reservoir [J]. Chemie der Erde; 2010 (70): 155-164.

[5] Bacci G, Korre A, Durucan S. An experimental and numerical investigation into the impact of dissolution/precipitation mechanisms on CO_2 injectivity in the wellbore and far field regions [J]. International Journal of Greenhouse Gas Control, 2010 (5): 579-588.

[6] Shiraki R, Dunn T L. Experimental study on water-rock interactions during CO_2 flooding in the Tensleep Formation, Wyoming, USA [J]. Applied Geochemistry, 2000 (15): 265-279.

[7] Michael K, Golab A, Shulakova V, et al. Geological storage of CO_2 in saline aquifers—A review of the experience from existing storage operations [J]. International Journal of Greenhouse Gas Control. 2010 (4): 659-667.

[8] Baines S J, Worden R H. The long term fate of CO_2 in the subsurface: natural analogues for CO_2 storage [J]. Geological Society 2004 (233): 59-85.

[9] Law H S, Bachu S. Hydrological and Numerical Analysis of CO_2 disposal in deep aquifer systems in the Alberta sedimentary basin [J]. Ener. Convers, 1966 (37): 1167-1174.

[10] Van Der Meer L G H. The Conditions Limiting CO_2 Storage in Aquifers [J]. Ener. Convers, 1993 (37): 959-966.

[11] Bowker K A, Shuler P J. Carbon dioxide injection and resultant alteration of the Weber sandstone, Rangely Field, Colorado [J]. The American Association of Petroleum Geologists Bulletin, 1991 (75): 1489-1499.

[12] Assayag N, Matter J, Ader M, et al. Water-rock interactions during a CO_2 injection field-test: implications on host rock dissolution and alteration effects [J]. Chemical Geology, 2009 (265) 227-235.

[13] Xu T, Pruess K. Modeling multiphase non-isothermal fluid flow and reactive geochemical transport in variably

saturated fractured rocks: 1. Methodology [J]. American Journal of Science, 2001 (301): 16-33.

[14] Guo W, Fang S, Liu Z J. Research on Thermal Evolutionary History During the Period of Quantou-Nenjiang Formation in the South of Songliao Basin [J]. Journal of oil and gas technology, 2009, 31 (3): 1-6 (in Chinese.).

[15] Dai J X, Yang C, Hu A P, et al. Distribution characteristics of natural gas in Eastern China [J]. Natural Gas Geoscience, 2009, 20 (4), 471-487 (in Chinese, with English Abstr.).

[16] Qu X Y, Liu L, Gao Y Q, et al. Geology record of mantle derived magmatogenetic CO_2 gas in the northeastern China [J]. Acta Petrol Ei Sinica, 1996, 31 (1): 61-67 (in Chinese, with English Abstr.).

[17] Cantucci B, Montegrossi G, Vaselli O, et al. Geochemical modeling of CO_2 storage in deep reservoirs: The Weyburn Project (Canada) case study [J]. Chemical Geology, 2009 (265): 181-197.

[18] Liu N, Liu L, Qu X Y, et al. Genesis of authigene carbonate minerals in the upper cretaceous reservoir, Honggang Anticline, Songliao Basin: A natural analogue for mineral trapping of natural CO_2 storage [J]. Sedimentary Geology, 2011 (237): 166-178.

[19] Yang G, Zhao Z, Shao M L. Formation of carbon dioxide and hydrocarbon gas reservoirs in the Changling fault depression, Songliao Basin [J]. Petroleum exploration and development, 2011 (38): 52-58.

[20] Duan Z H, Sun R. An improved model calculating CO_2 solubility in pure water and aqueous NaCl solutions from 273 to 533 K and from 0 to 2000 bar [J]. Chemical Geology, 2003 (193): 257-271.

[21] Murakami T, Kogure T, Kadohara H, et al. Formation of secondary minerals and its effect on anorthite dissolution [J]. Am. Mineral, 1998 (83) 1209-1219.

[22] Carroll S A, Knauss K G. Dependence of labradorite dissolution kinetics on CO_2 (aq), Al (aq) and temperature [J]. Chemical Geology, 2005 (217): 213-225.

[23] Bertier P, Swennen R, Laenen B, et al. Experimental identification of CO_2-water-rock interactions caused by sequestration of CO_2 in Westphalia and Buntsandstein sandstones of the Campine Basin (NE-Belgium) [J]. Journal of Geochemical Exploration, 2006 (89): 10-14.

[24] Hangx J T, Christopher J Spiers. Reaction of plagioclase feldspars with CO_2 under hydrothermal conditions [J]. Chemical Geology, 2009 (265): 88-98.

[25] Luquot L, Andreani M, Gouze P, et al. CO_2 percolation experiment through chlorite/zeolite-rich sandstone (Pretty Hill Formation-Otway Basin-Australia) [J]. Chemical Geology, 2012 (294-295): 75-88.

[26] Pearce J M, Holloway S, Wacker H, et al. Natural occurrences as analogues for the geological disposal of carbon dioxide [J]. Energy Convers, 1996 (37): 1123-1128.

[27] Zemke K, Liebscher A, Wandrey M, et al. Petrophysical analysis to investigate the effects of carbon dioxide storage in a subsurface saline aquifer at Ketzin, Germany (CO_2SINK) [J]. Int J Greenhouse Gas Control, doi: 10. 1016/j. ijggc. 2010. 04. 008.

[28] Wandrey M, Fischera S, Zemkea K, et al. Monitoring petrophysical, mineralogical, geochemical and microbiological effects of CO_2 exposure-Results of long-term experiments under in situ conditions [J]. Energy Procedia, 2011 (4): 3644-3650.

[29] Xu T, Apps J A, Pruess K. Numerical simulation of CO_2 disposal by mineral trapping in deep aquifers [J]. Applied Geochemistry, 2004 (19): 917-936.

[30] Xu T, Apps J A, Pruess K. Mineral sequestration of carbon dioxide in a sandstone-shale system [J]. Chemical Geology, 2005 (217): 295-318.

[31] Zerai B, Saylor B Z, Matisoff G. Computer simulation of CO_2 trapped through mineral precipitation in the Rose Run Sandstone [J]. Ohio Applied Geochemistry, 2006 (21): 223-240.

[32] Wilkinson M, Haszeldine R S, Fallick A E, et al. CO_2-mineral reaction in a natural analogue for CO_2 storage -implications for modeling [J]. Journal of Sedimentary Research, 2009 (79): 486-494.

[33] Gaus I. Role and impact of CO_2-rock interactions during CO_2 storage in sedimentary rocks [J]. International Journal of Greenhouse Gas Control, 2010 (4): 73-89.

[34] Kjoller C, Weibel R, Bateman K, et al. Geochemical impacts of CO_2 storage in saline aquifers with various mineralogy-results from laboratory experiments and reactive geochemical modeling [J]. Energy Procedia, 2010 (4): 4724-4731.

[35] Bénézeth P, Palmer D A, Anovitz L M, et al. Dawsonite synthesis and reevaluation of its thermodynamic properties from solubility measurements: implications for mineral trapping of CO_2 [J]. Geochim. Cosmochim. Acta, 2007, 71 (18): 4438-4455.

[36] Weibel R, Kjoller C, Bateman K, et al. Mineral changes in CO_2 experiments-Examples from Danish onshore saline aquifers [J]. Energy Procedia, 2011 (4): 4495-4502.

[37] Zhuo S G, Zhou S X, Jiang Y J. Diagenesis of Cretaceous Sandstone in Songliao Basin [J]. Scientia Geologica Sinica, 1992 (Supplement): 216-224 (in Chinese).

[38] Yang G F, Zhuo S G, Niu B. Albitization of Detrital Feldspar in Cretaceous Sandstone from the Songliao Basin. Geological Review, 2003, 49 (2): 155-161 (in Chinese).

[39] Flukiger F, Bernard D. A new numerical model for pore scale dissolution of calcite due to CO_2 saturated water flow in 3D realistic geometry: Principles and first results [J]. Chemical Geology, 2009 (265): 171-180.

[40] Ketzer J M, Iglesias R, Einloft S, et al. Water-rock-CO_2 interactions in saline aquifers aimed for carbon dioxide storage: Experimental and numerical modeling studies of the Rio Bonito Formation (Permian), southern Brazil [J]. Applied Geochemistry, 2009 (24): 760-767.

[41] Robie R A, Hemingway B S, Fisher J R. Thermodynamic properties of minerals and related substances at 298. 15 K and 1 bar (105 Pascals) pressure and at higher temperatures [J]. U S Geol Surv Bull, 1978 (1452): 456.

[42] Helgeson H C, Kirkham D H, Flowers G C. Theoretical prediction of the thermodynamic behavior of aqueous electrolytes at high pressures and temperatures. VI. Calculation of activity coefficients, osmotic coefficients, and apparent molal and standard [R]. and relative partial molal properties to 600°C and 5 kb [J]. Amer J Sci, 1981 (281): 1249-1516.

[43] Ross G D, Todd A C, Tweedie J A, et al. The dissolution effects of CO_2-brine systems on the permeability of U. K. and North Sea calcareous sandstones [R]. Proc. 3rd Joint SPE/DOE Symp. Enhanced Oil Recovery, 1982: 149-162. (SPE/DOE 10685).

[44] Sayegh S G, Krause F F, Girard M, et al. Rock/fluid interactions of carbonated brines in a sandstone reservoir: Pembina Cardium, Alberta, Canada [J]. SPE Formation Eval, 1990 (5):399-405.

饱和 CO_2 地层水驱过程中的水—岩相互作用实验研究

于志超[1]　杨思玉[2]　刘　立[1]　李　实[2]　杨永智[2]

(1. 吉林大学地球科学学院；

2. 中国石油勘探开发研究院提高石油采收率国家重点实验室)

摘　要：为了研究 CO_2 注入后储层岩性和物性的变化情况，利用室内岩心驱替装置，模拟了地层条件下（100℃，24MPa）饱和 CO_2 地层水驱过程中的水—岩相互作用，并对 CO_2 注入后，组成储层岩石的矿物溶蚀、溶解和沉淀情况以及渗透率变化的原因进行了研究。通过对实验前后反应液离子成分变化、岩心扫描电镜和全岩 X 射线衍射（XRD）分析表明：实验后砂岩岩心中的碳酸盐矿物出现明显的溶解现象，且方解石溶解程度最高，片钠铝石次之，铁白云石最低；反应液中 K+质量浓度的变化主要是由碎屑钾长石颗粒溶蚀造成的。实验后有少量的高岭石和中间产物生成，其中间产物的成分主要为 C、O、Na、Cl、Al 和 Si，并有向碳酸盐矿物转变的趋势。新生成的高岭石、中间产物和由碳酸盐胶结物溶解释放出的黏土颗粒一起运移至孔喉，从而堵塞孔隙，降低了岩心渗透率。通过以上实验再现了 CO_2 注入后，短时期内储层岩石中长石和碳酸盐类矿物的溶蚀和溶解过程以及新矿物沉淀情况，并且揭示了储层渗透率变化的原因，从而为 CO_2 的地下捕获机制提供地球化学依据。

关键词：CO_2 埋存；饱和 CO_2 地层水驱；储层岩石；渗透率；溶蚀速率

0　引言

CO_2 是全球气候变暖最主要的影响因素[1,2]。在即将枯竭的油气储层中将 CO_2 进行地质埋存，不仅是短时期内控制 CO_2 排放量的一种最为有效的方法[3]，而且还能够大幅度地提高原油采收率[4]。因此，这项技术已经成为仅次于深部咸水层埋存 CO_2 方法的主要工程手段[5]。CO_2 驱是 20 世纪 80 年代发展起来的一项重要技术[6]，已经受到世界各国的广泛重视[7]。但是，与油、气等地质流体相比，CO_2 是一种活性气体[8]，当其注入地下时极易与周围储层中的地层水和岩石发生反应[9]，从而打破地层水与围岩之间的物理—化学平衡，进而引发大规模的 CO_2—地层水—岩石相互作用，最终改变储层的物理和化学性质[10]。尤其是储层渗透率的变化将会严重地影响 CO_2 的地下封存能力[11]，因此就需要对 CO_2—地层水—岩石的相互作用进一步加以研究。

CO_2（超临界流体）—地层水—岩石相互作用实验是研究 CO_2 注入储层后的物性变化和评价 CO_2 地下埋存安全性的最有效手段[12]。由于以前的实验多数是在非地层条件下进行的，在地层条件下的研究则非常少见[10]，而 CO_2 注入后的储层变化情况则多数是基于观测井的液体或者气体样品的分析结果来确定的[13]。因此，在储层温度和压力条件下进行 CO_2—水—岩石相互作用实验的研究就更为必要[10]。针对该问题，笔者设计了一组在模拟地层条件

下饱和 CO_2 水驱过程中的 CO_2—水—岩石相互作用实验，研究了 CO_2 注入后，组成储层岩石的矿物溶蚀和溶解作用，新矿物沉淀现象以及地层水溶液离子的变化情况，并重点探讨了在饱和 CO_2 水驱过程中，导致储层渗透率变化的主要原因，从而为 CO_2 地下埋存提供一定的技术支持。

1 实验

1.1 实验样品

实验样品取自松辽盆地南部，样品的埋深为 1829.2~1829.4m。实验前将岩心加工成直径为 2.5cm 的柱状样，并将 3 个柱状样组合成总长度为 16.1cm 的"组合试件"。根据骨架碎屑统计，试件的岩性主要为长石砂岩。全岩 X 射线衍射（XRD）分析表明，试件主要由石英（45.2%）和斜长石（48.4%）组成，其次为钾长石（2.9%）、黏土矿物（2.4%）以及碳酸盐矿物（1.1%）。为避免有机质污染，利用石油醚对试件进行了原油清洗，清洗的温度、压力和时间为 80℃，5MPa 和 5d。组合试件中的每段岩心样品在实验前后均进行了孔隙度、孔隙体积和渗透率的测定，并进行了扫描电镜观察。其中，孔隙度、孔隙体积和渗透率的测试采用 PHI-220 氦气孔隙度仪和 KA210 气体渗透率仪；扫描电镜观察采用 Quanta450 型环境扫描电镜。

1.2 实验装置

实验装置主要由高温高压耐腐蚀圆柱形容器（容积为 1030mL）、哈氏岩心夹持器（长度为 30cm）、ISCO-100DX 型高压计量泵、BLUE M 高温烘箱、压力传感器、回压阀、DJB-80A 型手动/电动计量泵和气体流量计组成（图 1）。

在实验时，首先将组合试件按渗透率由低到高的顺序排列，并置于岩心夹持器中，同时饱和地层水；然后，将配制的地层水溶液和 CO_2 气体注入高温高压耐腐蚀圆柱形容器中，并连同岩心夹持器一起置于高温烘箱内；最后，利用不锈钢管线（直径为 1mm）将所有实验仪器连接起来。

1.3 实验条件

实验条件为：高温高压耐腐蚀圆柱形容器内的 CO_2 分压为 24.1MPa，回压为 24MPa（参照地层压力），围压为 30MPa；烘箱温度为 100℃（参照地层温度）；驱替速度为 1.67 mL/h（恒速）；实验时间为 150.68h。

在当前实验条件下，在矿化度为 10000mg/L 的地层水中，CO_2 溶解度为 1.1037mol/kg[14]。实验时，注入地层水 508mL（矿化度为 10568mg/L），注入 CO_2 体积为 487mL，溶解在地层水中的 CO_2 为 0.56mol，约 38.64mL，由此可计算出容器内 CO_2 气顶的体积，约为 448.36 mL。因此，在整个饱和 CO_2 水驱实验过程中，注入液中的 CO_2 一直处于过饱和状态。

1.4 实验过程监测与样品采集分析

在整个实验过程中，对累计反应时间、累计注入量（注入体积和相应的孔隙体积倍数）、液体量、气体量、液体样 pH 值、注入端压力、出口端压力和驱替压差等参数进行了实时监测（表 1）。

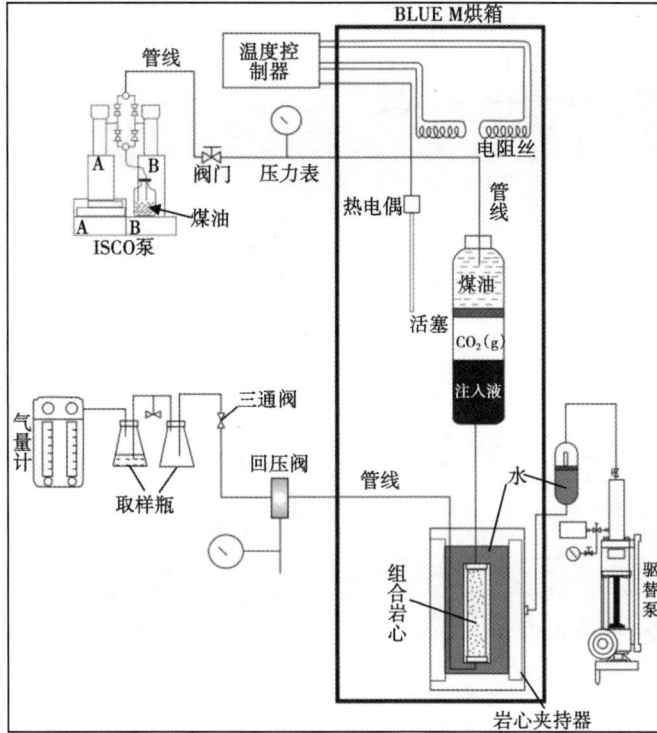

图 1　饱和 CO_2 水驱水—岩相互作用实验装置

表 1　实验参数实时监测结果

样品编号	累计反应时间（h）	累计注入量		取样计量		液体样品 pH 值	注入端压力（MPa）	出口端压力（MPa）	驱替压差（MPa）
		注入体积（mL）	孔隙体积倍数	液体量（mL）	气体量（mL）				
1	1.17	1.30	0.08	5.00	374.00	6.91	22.28	21.97	0.31
2	3.58	5.48	0.25	11.00	153.00	7.14	22.30	21.80	0.50
3	6.92	9.56	0.50	7.50	90.00	7.08	22.21	21.96	0.25
4	10.45	13.64	0.75	9.00	90.00	7.06	22.06	21.46	0.60
5	14.00	17.76	1.08	9.00	6.00	6.97	21.71	21.19	0.52
6	30.75	45.79	2.78	46.00	364.00	6.48	21.00	20.37	0.63
7	52.87	81.10	4.93	53.00	5.00	6.33	20.42	20.48	-0.06
8	78.02	121.06	7.36	50.00	335.00	6.29	20.10	20.18	-0.08
9	99.82	156.68	9.52	33.00	270.00	—	22.01	22.06	-0.05
10	131.08	262.65	15.96	36.00	7.00	6.54	22.12	22.03	0.09
11	150.68	409.31	24.87	25.00	300.00	6.35	22.19	22.03	0.16

注：初始地层水溶液的 pH 值为 6.42，9 号样品太少，pH 值未检测出。

从实验开始到注入量为 1PV 之间密集取样，水样用三角瓶收集，排出来的 CO_2 气体用气体流量计计量，然后大约每隔 30h 收集一次样品。实验产出液用 Thermo 公司生产的 Orion4 STAR 型 pH 电子酸度计测量 pH 值，每个样品的测试时间距其收集时间不超过 6h。每次离子分析前需用浓度为 1:1 的 HCl 滴定，再对其进行离子分析。水溶液离子分析所用的仪器为 Dionex500 型高效离子色谱仪，分析精度在 $10^{-3} \sim 10^{-9}$ g 之间，分析结果见图 2。

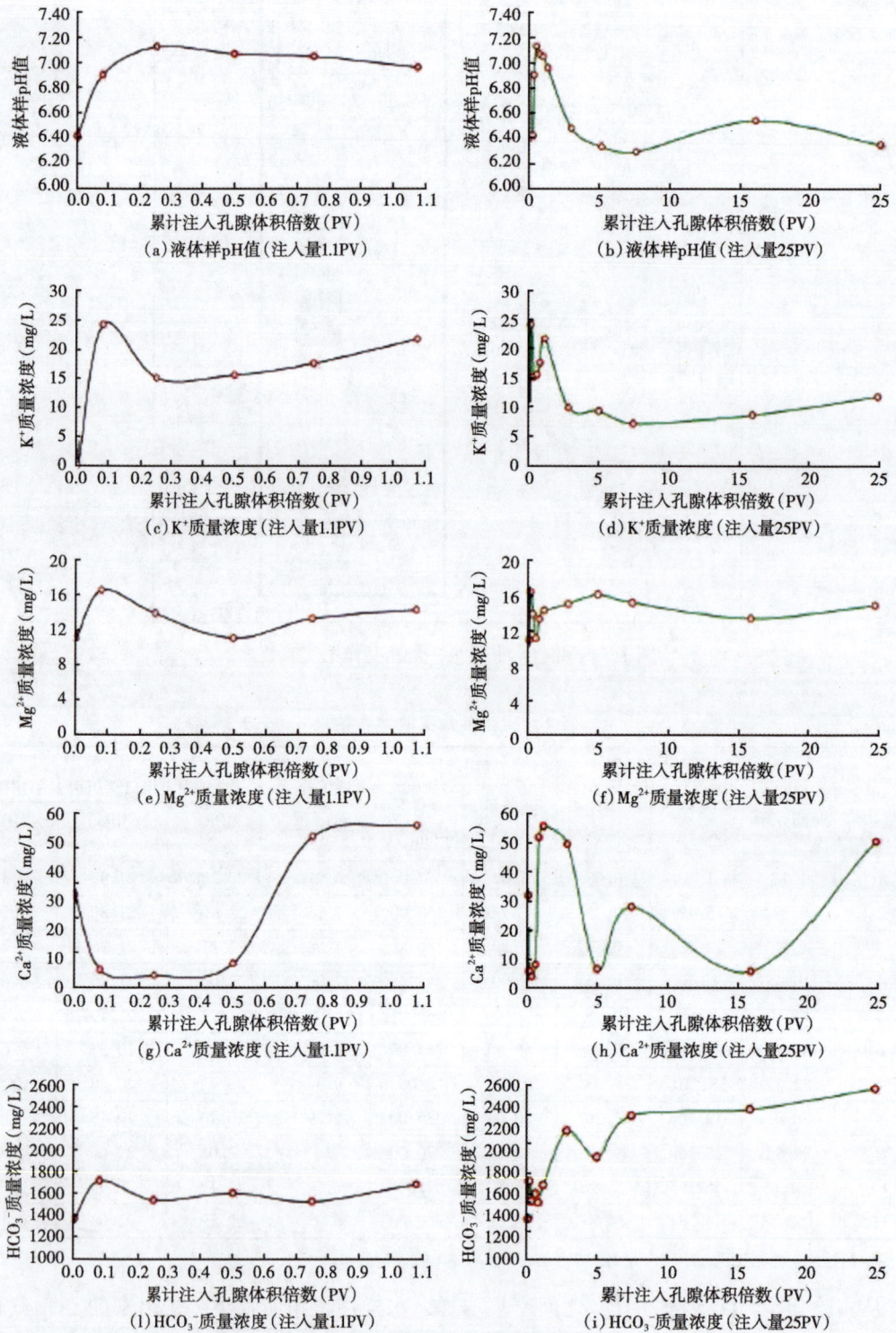

图 2　反应液 pH 值和离子质量浓度的变化

154

2 实验结果

2.1 产出液 pH 值和化学成分的变化

由于 CO_2—水—岩反应在实验刚开始到注入量为 1PV 范围内最为剧烈,因此分别绘制了注入量在 1PV 范围内和注入量大于 1PV 的产出液离子质量浓度变化图(图2)。

溶有 CO_2 的地层水溶液初始 pH 值为 6.42,为弱酸性。随着反应的进行,pH 值在反应开始后的 0.1PV 内迅速升高至 6.91,在注入量为 0.25PV 时,达到最高值 7.14,然后缓慢下降[图 2(a)]。当注入量大于 1PV 时,pH 值随反应的进行快速降低,最低可至 6.29,最后缓慢上升,最终维持在 6.4 左右,接近注入液的初始 pH 值[图 2(b)]。

注入量在 1PV 范围内,K^+ 质量浓度表现为先快速升高,然后下降,再缓慢上升的变化趋势[图 2(c)]。注入量为 0.1PV 时,K^+ 质量浓度达到最高值 24.2mg/L,之后随着反应的进行,K^+ 质量浓度迅速降低,注入量超过 0.5PV 后,K^+ 质量浓度又开始缓慢上升。当注入量大于 1PV 时,K^+ 质量浓度快速降低,在 7.36PV 时,达到最低值 7.04mg/L,之后缓慢上升,最高可达 11.7mg/L[图 2(d)]。

与 K^+ 质量浓度变化情况相似,注入量在 1PV 范围内,Mg^{2+} 质量浓度也表现为先升高后降低,再缓慢升高的变化趋势[图 2(e)]。Mg^{2+} 初始质量浓度为 11mg/L,在注入量为 0.1PV 时,其质量浓度增加到最高值 16.6mg/L,然后持续下降,当注入量大于 0.5PV 时,Mg^{2+} 质量浓度又开始缓慢升高[图 2(e)]。当注入量大于 1PV 时,Mg^{2+} 质量浓度快速增加,在 4.93PV 时达到 16.2 mg/L,接近整个反应的最高值,之后缓慢降低,最终稳定在 15 mg/L 左右[图 2(f)]。

与 K^+ 和 Mg^{2+} 相比,Ca^{2+} 和 HCO_3^- 质量浓度表现出更为独特的变化趋势。注入量在 0.5PV 范围内,Ca^{2+} 质量浓度急速下降,由初始的 32.05mg/L 降至 5~8mg/L[图 2(g)]。当注入量大于 0.5PV 时,Ca^{2+} 质量浓度快速增加,在 1.1PV 时,达到最大值 56mg/L[图 2(g)]。当注入量大于 1PV 时,Ca^{2+} 质量浓度表现为特殊的变化情况[图 2(h)],原因详见 3.3 小节。HCO_3^- 初始质量含量为 1363.5mg/L,当注入量小于 1PV 时,HCO_3^- 质量浓度表现为先逐渐增大后降低,最后缓慢升高的变化趋势。当注入量为 0.1PV 时,HCO_3^- 质量浓度达到最高值 1720mg/L,而后逐渐降低,当注入量大于 0.75PV 时,HCO_3^- 质量浓度又开始缓慢升高[图 2(i)];当注入量大于 1PV 时,随着反应的进行,HCO_3^- 质量浓度快速升高,最高可达 2574 mg/L[图 2(j)]。

2.2 矿物的溶蚀和溶解

饱和 CO_2 水驱后的试件通过扫描电镜观察到发生溶解的矿物为铁白云石和片钠铝石。其中,实验前的铁白云石菱形晶体普遍完整[图 3(a)];实验后的铁白云石菱形晶体边缘形成了凹坑(图 3(b)白色圆圈)。实验前在孔隙中观察到了较多的以束状集合体为特征的片钠铝石[图 3(c)],而在实验后的试件中仅观察到少量的单个片钠铝石晶体[图 3(d)],并且其边缘也往往表现出港湾状溶解的特征[图 3(e)]。通过实验前后试件的 X 射线衍射谱图对比也证实了铁白云石和片钠铝石的溶解。反应前铁白云石的衍射强度较大,其对应的 d 值(晶体面网间距)为 2.91,接近于标准 d 值 2.90。反应后铁白云石的衍射强度明显降低，其对应的 d 值为 2.93。实验前后片钠铝石的衍射强度也具有相似的情况（图 4）。

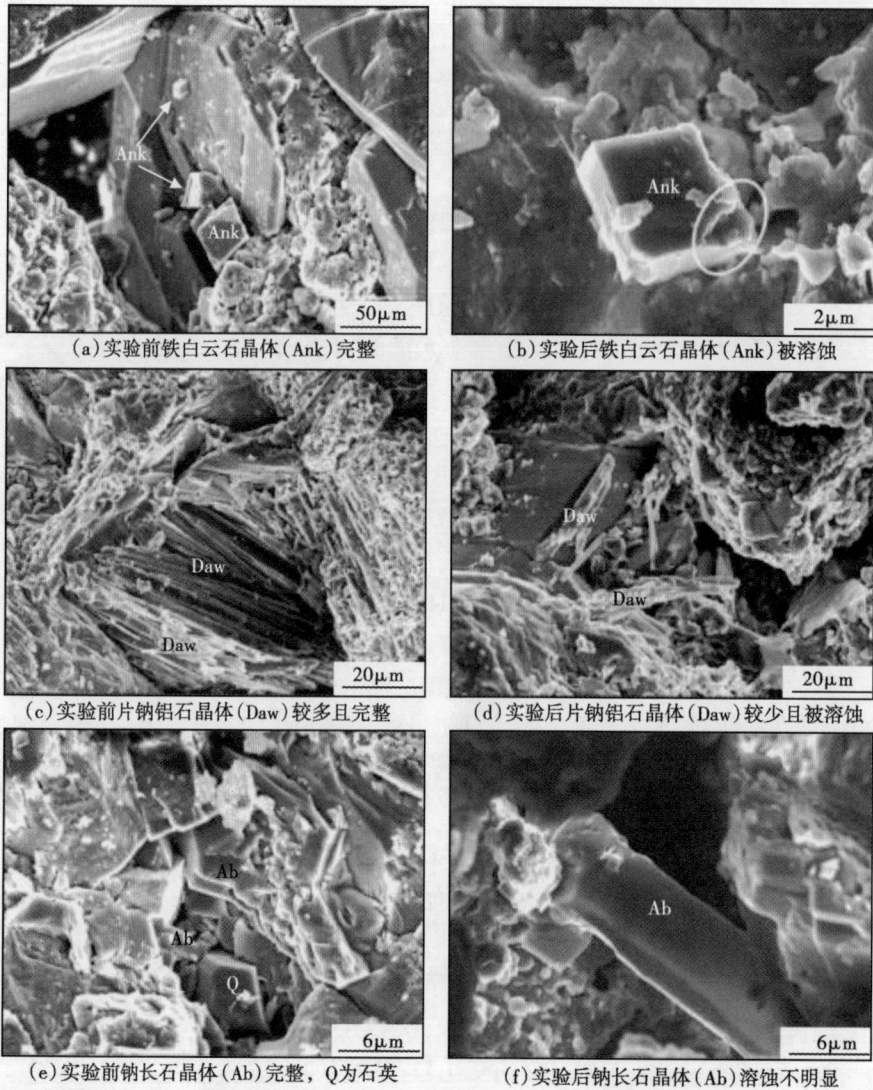

（a）实验前铁白云石晶体（Ank）完整   （b）实验后铁白云石晶体（Ank）被溶蚀

（c）实验前片钠铝石晶体（Daw）较多且完整   （d）实验后片钠铝石晶体（Daw）较少且被溶蚀

（e）实验前钠长石晶体（Ab）完整，Q 为石英   （f）实验后钠长石晶体（Ab）溶蚀不明显

图 3　饱和 CO_2 水驱实验前后的岩心形貌扫描电镜照片

在扫描电镜下未观察到方解石的溶解迹象，在实验后试件的 X 射线衍射谱图上也已检测不到方解石峰（图 4），说明该矿物已溶解殆尽。

组成岩心的主要矿物钠长石，在实验前后没有观察到明显的溶蚀现象 [图 3（e），图 3（f）]。由于试件中钾长石的含量很低，在扫描电镜和 X 射线衍射谱图上均未观察和检测到其溶蚀的迹象。值得注意的是，在产出液中检测到 K^+ 质量浓度随反应时间和注入量的增加而略微增加。但是实验所用的溶液中不含有 K^+，并且在实验的试件中也未观察和检测到其他含 K^+ 的矿物，因此，产出液中 K^+ 质量浓度的略微增加应该是钾长石溶蚀造成的。

综上所述，在本次实验中观察或检测到铁白云石、片钠铝石、方解石和钾长石均发生了不同程度的溶蚀或溶解，并且碳酸盐矿物的溶解程度明显高于钾长石，其中方解石溶解程度最高，片钠铝石次之，铁白云石最低。

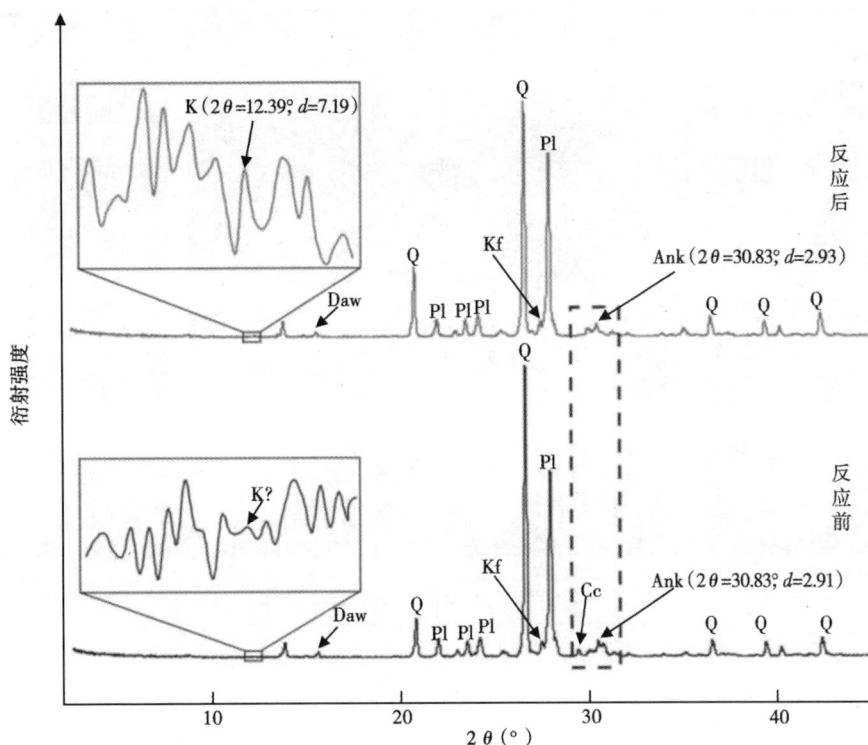

图 4　饱和 CO_2 水驱实验前后的岩心 XRD 特征谱图

Q—石英；Pl—斜长石；Kf—钾长石；K—高岭石；Ce—方解石；Ank—铁白云石；Daw—片钠铝石；K? —无高岭石

2.3　新矿物的沉淀

实验前后的扫描电镜下都没有观察到高岭石的存在，但是在实验后的 X 射线衍射谱图上显示出微弱的高岭石峰（图 4），说明实验后有少量的高岭石生成。

通过扫描电镜对比实验前后的砂岩岩心，发现实验后有呈长柱状和菱形四面体的中间产物生成（图 5）。应用能谱仪对中间产物的化学成分做了进一步的分析和研究。长柱状矿物成分分析显示：中部（点③）的化学成分为 NaCl；靠近中部的对称两点（点②和点④）的化学成分则较为复杂，主要为 O，Na 和 Cl，其中点④还含有一定量的 Ca 元素；点①和点⑤的化学成分则以 C，O，Na 和 Cl 为主，其中 C 元素占 50% 以上。对菱形矿物的成分分析显示：点⑦和点⑩的化学成分以 C，O 和 Ca 为主；点⑧和点⑨化学成分主要为 C，O，Na，Al 和 Si，而含碳量却有所降低；点⑥化学成分与点③接近，主要为 NaCl。

上述研究表明：长柱状矿物由中心向两边其含 C 量逐渐增加，且有 Ca 元素出现，有向碳酸盐矿物转变的趋势，而且这种转变趋势的程度由中部向边部逐渐增加。菱形矿物也表现为相似的碳酸盐矿物转变趋势，所不同的是靠近长方形溶蚀坑部位向方解石转变的程度较高，而没有溶蚀坑的部位则表现为含 C，O，Na，Al 和 Si 的混合物形式，且转变程度较低。

2.4　孔隙度、渗透率和孔隙体积的变化

表 2 为实验前后"组合试件"不同部位孔隙度、渗透率和孔隙体积的变化情况，可以看出：渗透率在组合试件的注入端和出口端变化幅度最大。在实验前，注入端和出口端的渗

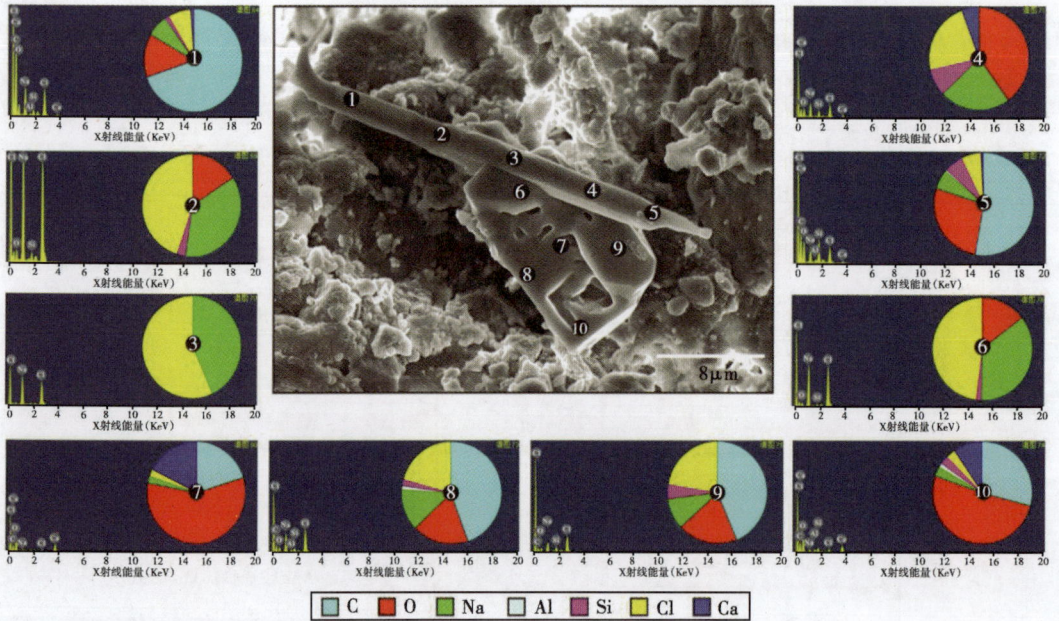

图 5　中间矿物扫描电镜照片及能谱分析

透率值分别为 16.5mD 和 17mD，实验后分别降至 15.7mD 和 16.3mD，与实验前相比降幅达 4.85% 和 4.12%。中间部分的渗透率值变化不大，与实验前相比，略有升高，增幅为 0.59%。而孔隙度和孔隙体积在组合试件的中间部位变化幅度最大，实验前的孔隙度和孔隙体积分别为 20.2% 和 4.39cm^3，而实验后分别降至 19.7% 和 4.24cm^3，与实验前相比，其孔隙度和孔隙体积分别下降了 2.48% 和 3.42%。注入端和出口端孔隙度和孔隙体积的降幅较小，其中，注入端孔隙度和孔隙体积分别下降了 0.94% 和 1.39%，出口端下降了 0.47% 和 1.07%。

表 2　饱和 CO_2 水驱实验前后的岩心物性变化

位置	孔隙度（%）		孔隙度变化率（%）	气测渗透率（mD）		渗透率变化率（%）	孔隙体积（cm^3）		孔隙体积变化率（%）
	实验前	实验后		实验前	实验后		实验前	实验后	
注入端	21.3	21.1	-0.94	16.5	15.7	-4.85	5.76	5.68	-1.39
中间	20.2	19.7	-2.48	16.9	17	0.59	4.39	4.24	-3.42
出口端	21.3	21.2	-0.47	17	16.3	-4.12	6.54	6.47	-1.07

2.5　钾长石的溶蚀速率

由于本次实验过程中主要是碎屑钾长石发生溶蚀作用，因此可以用溶液中钾离子的溶蚀速率来大体估算整个反应过程中钾长石的溶蚀程度。依据每次取样液中钾离子的物质的量和相应的钾长石摩尔体积（109cm^3/mol）[15—16]，计算得到钾长石的溶蚀速率与注入体积之间的关系（图6）。

当注入液体积倍数达到 0.08PV 时，钾长石的溶蚀速率达到最高值为 $4.83×10^{-6}cm^3$/

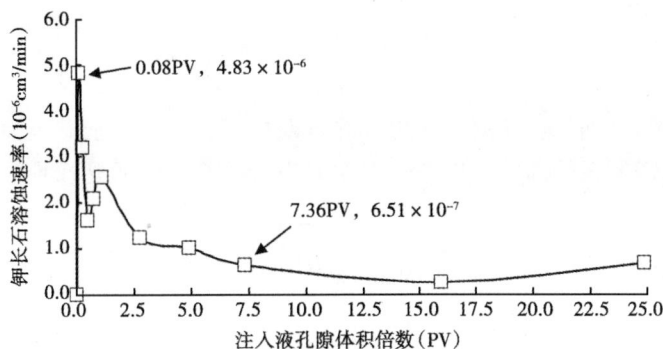

图 6 钾长石的溶蚀速率与注入体积关系

min。之后随着反应的进行，钾长石的溶蚀速率逐渐降低，到注入液体积倍数达到 7.36PV 时，钾长石的溶蚀速率趋于平衡，为 $6.51×10^{-7}cm^3/min$。上述钾长石的溶蚀速率最大值为 $4.83×10^{-6}cm^3/min$，而 Shiraki 等[6]在 CO_2—水—岩石相互作用实验中所得到的钾长石的溶蚀速率则为 $17.5×10^{-6}cm^3/min$，明显高于本文的研究成果。这是由于 Shiraki 实验用的岩心渗透率平均值为 68mD，而本实验所用的岩心渗透率平均值仅为 17mD，岩心的低渗透率大幅度地减小了 CO_2 酸性流体在岩心中与矿物的接触面积，从而导致钾长石的溶蚀速率降低。

3 讨论

3.1 钾长石溶蚀的原因

造成实验中碎屑钾长石溶蚀而自生钠长石未溶蚀的主要原因有以下三点：

（1）实验岩心中的自生钠长石是晚期成岩作用的产物，其物理化学反应已基本达到平衡，很难再次发生溶蚀作用。卓胜广[17]和杨桂芳[18]等的研究表明，随着埋藏深度的增加，松辽盆地砂岩中斜长石含量逐渐增加，而钾长石含量却逐渐降低。当埋深为 1000~1500m 时，钾长石含量明显降低，与之相比斜长石含量却明显升高。当埋深大于 2000m 时，砂岩中已基本不含钾长石。与此同时，斜长石中的钠长石组分也逐渐增加，最终形成纯钠端元的钠长石[17-18]。本次实验的岩心埋深达 1829.2m，实验前全岩 X 衍射分析显示岩心中的斜长石含量高达 48.4%，而钾长石仅为 2.9%。扫描电镜配合能谱分析研究也发现，实验岩心中自形程度较好的长石几乎都为纯钠端元的钠长石，说明岩心中长石的钠长石化非常发育，为晚期成岩产物，因此很难再次发生溶蚀作用。

（2）高温条件下，已经发生过部分溶蚀作用的碎屑长石更容易与酸性流体发生反应，从而滞后了自生长石溶蚀作用的发生[19]。钾长石（$KAlSi_3O_8$）和钠长石（$NaAlSi_3O_8$）与酸性流体作用，发生溶蚀作用的方程式[6,17]为：

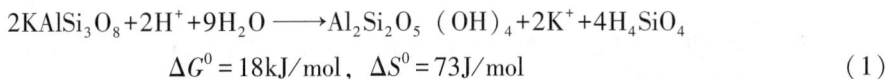

$$\Delta G^0 = 18kJ/mol，\Delta S^0 = 73J/mol \tag{1}$$

$$\Delta G^0 = -132 \text{kJ/mol}, \quad \Delta S^0 = -101 \text{J/mol} \tag{2}$$

式中，ΔG^0 是同一温度、标准压强下矿物的吉布斯自由能变；ΔS^0 是同一温度下标准压强下矿物的熵变。

上述方程式表明，钠长石的溶蚀作用能够自发进行，是一个放热反应（$\Delta G^0 < 0$），而钾长石的溶蚀作用则需要在较高的温度条件下才能进行，是一个吸热反应（$\Delta G^0 > 0$）。应用对温度敏感的自由能公式：

$$\Delta G_T^{\ 0} = \Delta H - T\Delta S \tag{3}$$

式中，$\Delta G_T^{\ 0}$ 为吉布斯自由能变，kJ/mol；ΔH 是各矿物的标准摩尔反应焓，kJ/mol；T 为温度，K；ΔS 是各矿物的标准熵变，kJ/（mol·K）。

据此分析长石的溶蚀过程可以发现，随着温度的升高钾长石溶蚀程度有增强的趋势（$\Delta G_T^{\ 0} \rightarrow 0$），而钠长石溶蚀程度则有减弱的趋势（$\Delta G_T^{\ 0} \rightarrow 0$）[17]。即高温条件下，钾长石更容易与酸性流体发生反应，从而发生溶蚀作用，而钠长石在高温酸性条件下则变得相对稳定，不易发生反应。Wandrey 等[20]模拟德国 Ketzin 的 CO_2 注入工程的地层条件，选取与实验区岩性相符的长石砂岩岩心，对其饱和 Ketzin 地层水，随后置入高压釜中与 CO_2 酸性流体发生反应，在 24 个月后发现有钠长石生成。这一结果说明，钠长石在高温、酸性条件下更易沉淀，不易发生溶蚀作用。

（3）注入溶液中 Na^+（碱性离子都以 Na^+ 代替）比 K^+ 高出约两个数量级，对反应（2）起到了一定的缓冲作用，从而抑制了钠长石的溶蚀作用。

3.2 碳酸盐矿物的溶解和沉淀

在注 CO_2 提高石油采收率过程中，方解石和白云石等碳酸盐矿物的溶解和沉淀是溶液中 Ca^{2+} 含量变化的主要原因[21]。实验刚开始时，溶液中 Ca^{2+} 质量浓度一直在减少，说明在实验的初始阶段碳酸盐矿物没有发生溶解，反而出现了沉淀。这是因为在实验刚开始时，CO_2 酸性流体与砂岩的反应比较剧烈，硅酸盐矿物溶蚀释放出的二价阳离子与碳酸根（或者是碳酸氢根）反应生成碳酸盐 [式（4）]，从而导致溶液的 pH 值快速升高[19]，溶液由弱酸性向中性转变 [图 2（a）、图 2（b）]。同时上述反应促进了碳酸的二级电离 [式（5）和式（6）]，导致 HCO_3^- 持续电离分解，产生出较多的 CO_3^{2-}，进而与溶液中 Ca^{2+} 结合生成碳酸盐，导致溶液中 Ca^{2+} 的含量不断减少。

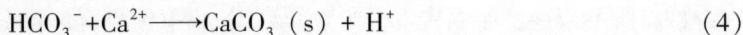

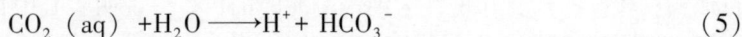

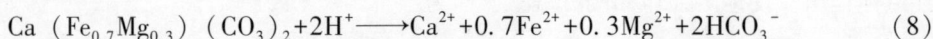

$$HCO_3^- + Ca^{2+} \longrightarrow CaCO_3 \text{ (s)} + H^+ \tag{4}$$

$$CO_2 \text{ (aq)} + H_2O \longrightarrow H^+ + HCO_3^- \tag{5}$$

$$HCO_3^- \longrightarrow H^+ + CO_3^{2-} \tag{6}$$

$$CaCO_3 + H^+ \longrightarrow Ca^{2+} + HCO_3^- \tag{7}$$

$$Ca \text{ (} Fe_{0.7}Mg_{0.3} \text{) } (CO_3)_2 + 2H^+ \longrightarrow Ca^{2+} + 0.7Fe^{2+} + 0.3Mg^{2+} + 2HCO_3^- \tag{8}$$

随着实验的进行，Ca^{2+} 和 Mg^{2+} 的质量浓度都在增加，且 Ca^{2+} 质量浓度增加得更快，说明此时组合试件中的铁白云石和方解石等碳酸盐矿物都发生了溶解现象，且方解石的溶解程度要高于铁白云石，具体反应详见式（7）和式（8）（当溶液 pH 值为 6.01~9.62 时，碳酸盐

类矿物与 CO_2 地层水溶液发生的主要反应）。已有的实验和地质实例证实[22-23]：在纯水中，天然白云石的溶解度为 320mg/L；方解石则仅为 14mg/L。当在 25℃，1atm（101325Pa）的条件下，在碳酸水中，方解石的溶解度为 900mg/L；白云石仅为 599mg/L，在有 Mg^{2+} 存在的前提下，还会增加方解石的溶解度[22-23]。因此当 CO_2 注入砂岩储层中，方解石比铁白云石更易溶解。

3.3 Ca^{2+} 质量浓度变化的原因

实验中，Ca^{2+} 质量浓度表现出非常特殊的变化情况 [图 2（g）、图 2（h）]。为了研究这一现象，当注入液体积倍数大于 1PV 时，反应液中释出的 CO_2 气体含量与 Ca^{2+} 质量浓度的变化规律进行对比，发现反应液中 CO_2 脱气量与 Ca^{2+} 质量浓度的变化有着很好的相关性。当反应液中 CO_2 的脱气量增高时，对应的反应液中 Ca^{2+} 质量浓度也增高。而当反应液中 CO_2 的脱气量降低时，对应的反应液中 Ca^{2+} 质量浓度也降低（图 7）。

反应液中 Ca^{2+} 质量浓度的变化主要与方解石的溶解和沉淀有关。在沉积岩的成岩环境中，方解石的溶解和沉淀与地下水化学特征、温度、CO_2 分压以及 pH 值有着密切的关系[24]。方解石溶解和沉淀有关的化学反应式为：

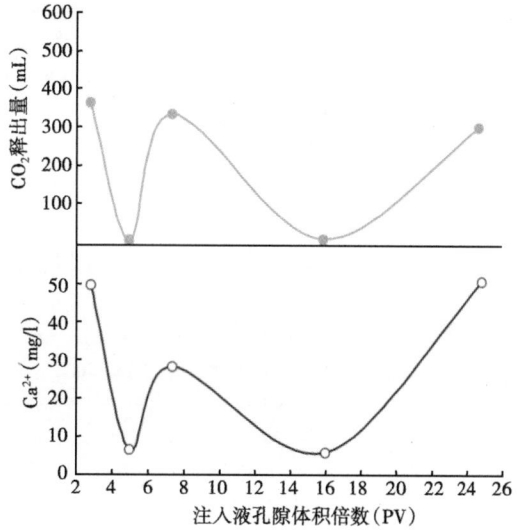

图 7　Ca^{2+} 浓度变化与 CO_2 脱气量对比

$$CO_2（aq）+H_2O = H^+ + HCO_3^-, \quad K_1 = \frac{\alpha_{H^+} + \alpha_{HCO_3^-}}{\alpha_{CO_2}\alpha_{H_2O}} \tag{9}$$

$$HCO_3^- = H^+ + CO_3^{2-}, \quad K_2 = \frac{\alpha_{HCO_3^-}}{\alpha_{CO_3^{2-}}\alpha_{H^+}} \tag{10}$$

$$CaCO_3 + 2H^+ = Ca^{2+} + H_2O + CO_{2(aq)} \tag{11}$$

$$CaCO_3 + H^+ = Ca^{2+} + HCO_3^- \tag{12}$$

$$CaCO_3 = Ca^{2+} + CO_3^{2-} \tag{13}$$

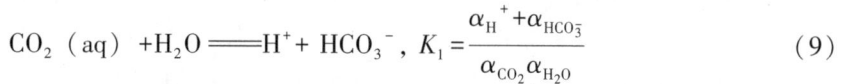

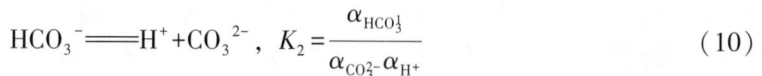

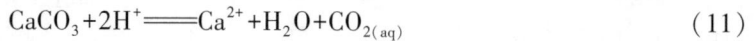

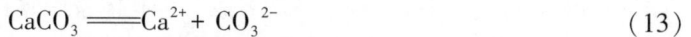

式中，K_1 和 K_2 为电离平衡常数；a_i 为离子活度。按照 Herry 定律，在稀溶液中离子活度近似于浓度，K_1 约为 K_2 的 1000 倍[24]。

假定溶液中的 Ca^{2+} 和总碳酸（ΣCO_2）的变化满足质量守恒定律，方解石的溶解—沉淀规律可用图 8 表示，其中实线代表了初始 $[\Sigma CO_2]_0 - [Ca^{2+}]_0$ 时方解石的溶解度，虚线则代表了初始 $[\Sigma CO_2]_0 - [Ca^{2+}]_0$ 时各碳酸组成；$[Ca^{2+}]_0$ 代表初始溶液中 Ca^{2+} 的总量，mol。$\Sigma CO_2 = CO_2（aq）+HCO_3^- + H_2CO_3 + CO_3^{2-}$。$(K_c)^{1/2}$ 为同等温度和压力条件下，蒸馏水中方解石的溶解度。pK 为电离平衡常数的负对数，p$K_1 = 6.01$，p$K_2 = 9.62$。

在碳酸总量相同的前提下，酸性溶液溶解方解石，碱性溶液沉淀方解石（图 8）。随着

pH 值的降低，方解石的溶解度呈指数增加。在 pH<pK_1 的强酸性溶液中方解石的溶解反应见反应方程式（11），溶解物除了 Ca^{2+} 外，还增加了溶液中 CO_2 的含量。在弱酸至中性溶液（pK_1≤pH≤pK_2）中，方解石的溶解反应为化学方程式（12），溶液中 HCO_3^- 的含量明显增加，并且方解石的溶解度随着 pH 值的增加大幅度降低。当溶液为碱性（pH≥pK_2）时，方解石的溶解度极低，基本为一恒定值（图 8）。本次实验的温压条件为 $t=100℃$，$p=24MPa$，

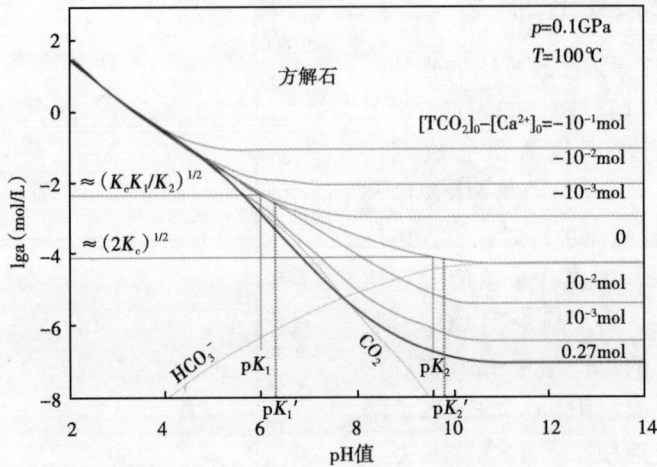

图 8　方解石溶解度与 pH 值相关图（据文献［24］修改）

而图 8 中的温压条件为 $t=100℃$；$p=100MPa$，因此在本次实验中，碳酸的电离平衡常数应该小于图 8 中的电离平衡常数。其所对应的 pH 值应该略大于图 8 中的 pK_1 和 pK_2，位于图 8 中虚线所属区域内（图 8 中 pK_1'和 pK_2'所示区间内）。图 8 中粗蓝色线是本次实验初始［∑ CO_2］$_0$-［Ca^{2+}］$_0$ 值的示意曲线（根据图 8 方解石溶解度规律推断）。可以看出，在［∑ CO_2］$_0$-［Ca^{2+}］$_0$ 值一定的条件下，随着 pH 值的增加，方解石的溶解度快速降低。由于反应液中的 CO_2 含量始终处于过饱和状态，从而导致注入岩心中的 CO_2 可能存在两种形式：一种是溶解在地层水溶液中的碳酸组分（∑ CO_2）；另一种是以超临界状态存在于岩心中的 CO_2 游离气。CO_2 的脱气量越多，说明岩心中 CO_2 游离气的含量也越多，而这部分 CO_2 在有水存在的前提下，也会与岩心中的矿物反应。随着以超临界游离气存在的 CO_2 的增多，间接地增加了反应液的酸度，降低了 pH 值，增加了方解石的溶解度，进而也增加了反应液中 Ca^{2+} 的质量浓度。同时，上述反应也遵循质量守恒定律：在 CO_2—H_2O—$CaCO_3$ 体系中，方解石的溶解可使其中的 CO_3^{2-} 进入溶液，引起体系中碳酸总量［∑ CO_2］增加，相反方解石的沉淀可使溶液中的 CO_3^{2-} 进入固相方解石中，进而引起溶液中碳酸总量降低，因此成岩流体中的碳酸总量随着方解石的溶解—沉淀平衡而变化[24]。由于在整个固—液体系中物质的量是守恒的，即溶液中 Ca^{2+} 变化的数量应与碳酸总量的变化相当[24]，因此其质量守恒方程为：

$$［∑CO_2］-［Ca^{2+}］=［∑CO_2］_0-［Ca^{2+}］_0 \qquad (14)$$

式中，［∑ CO_2］为溶液中碳酸总量，mol；［Ca^{2+}］为溶液中总 Ca^{2+} 总量，mol。

当游离态的 CO_2 加入上述反应，相当于增加了反应液中［∑ CO_2］的含量，而［∑ CO_2］$_0$-［Ca^{2+}］$_0$ 的值是固定的，为了维持上述方程式的平衡，溶液中［Ca^{2+}］总量也就会

相应增加。综合以上两种原因，最终导致了反应中 Ca^{2+} 浓度变化规律与 CO_2 脱气量变化规律一致。

3.4 新矿物的转变趋势

本次实验有含 Ca 和 C 的中间产物生成，这些中间产物的核部主要为 NaCl，边部含有一定量的 Ca 和 C 等元素。NaCl 的出现说明含 CO_2 酸性地层水在与长石砂岩反应时出现了类似盐霜的反应[27]。Pearce 等[27]研究发现：当含有 NaCl 酸性地层水在毛细管压力的作用下被抽提到矿物表面后，NaCl 溶液会充分暴露在超临界 CO_2"大气"下，发生盐霜反应，析出 NaCl 晶体。但是在 Pearce 的实验中没有出现含 Ca，C，O，Na，Cl，Al 和 Si 的中间产物，说明本次实验不仅在矿物表面析出 NaCl 晶体，同时在超临界 CO_2 流体的作用下，与之反应的矿物还有向碳酸盐转变的趋势。国外同类型的实验需要 8 个月以上的反应时间才能生成碳酸盐矿物[25]，而此次实验只进行了不到 7d（150.68h），因此不能够完全转变为碳酸盐矿物，只能生成含有 Ca，C，O，Na，Cl，Al 和 Si 等元素的固相物质。

3.5 渗透率变化的原因

本次实验中岩心渗透率发生变化的原因有以下两个方面：

（1）仪器的测量误差。与实验前相比，实验后岩心渗透率的下降幅度为 5%，基本在测量误差之内，因此实验后岩心渗透率的降低有可能是测量误差造成的。

（2）新生成的高岭石、中间产物和由碳酸盐胶结物溶解释放出的黏土颗粒一起运移至孔喉，从而堵塞孔隙，导致实验岩心渗透率降低。实验结果显示，在实验后，岩心中的碳酸盐胶结物发生了明显的溶解作用，并且有高岭石和中间产物的生成。尽管新矿物的含量较低，但其和由碳酸盐胶结物溶解所释放的黏土颗粒一起运移至孔喉时，必定会堵塞孔隙，从而导致渗透率降低。国外同类型的实验研究也证实了这一点。Ross 等[25]通过对英国北海油田的钙质砂岩进行 CO_2—水—岩实验研究后发现，实验后岩心的渗透率值明显低于实验前。Sayegh 等[26]对加拿大艾伯塔省 Pembina 油田的储层砂岩进行了类似的 CO_2—水—岩实验研究发现，实验后岩心的孔隙度、孔隙体积和渗透率值都明显降低。其中，渗透率的变化最明显，在实验开始的早期阶段，渗透率值表现为快速降低，只相当于初始值的 10%~60%，此后略有回升。Shiraki 等[6]对美国怀俄明州 Tensleep 地层砂岩进行的 CO_2—水—岩相互作用实验显示，实验后岩心的孔隙度和孔隙体积基本没有发生变化，但是渗透率值表现出明显的降低趋势。

综合以上研究成果并结合本次实验的具体情况可以看到，本次实验中岩心渗透率的降低是由于新矿物（包含中间产物）和碳酸盐胶结物溶解释放出的黏土颗粒一起堵塞孔隙所形成的综合作用的结果。

4 结论

（1）地层条件下的饱和 CO_2 水驱实验与高压釜的间歇式反应相比是一个流动的连续性反应，能够反映出 CO_2 注入后，超临界 CO_2 流体、储层岩石和地层水之间反应的真实情况。

（2）饱和 CO_2 水驱实验后，砂岩岩心中的长石和碳酸盐类矿物发生了不同程度的溶蚀、溶解作用，而自生钠长石并未发生明显的溶蚀作用。碳酸盐类矿物比长石等硅酸盐类矿物的反应程度高，其中，方解石溶解程度最高，片钠铝石次之，铁白云石最低。

（3）饱和 CO_2 水驱实验后有少量的高岭石和中间产物生成，中间产物核部为 NaCl，边部含有 Ca 和 C 等元素，并有向碳酸盐矿物转变的趋势。

（4）饱和 CO_2 水驱实验后，岩心渗透率、孔隙体积和孔隙度分别下降了 4%，3% 和 2.5%。新矿物（高岭石和中间产物）和碳酸盐矿物溶解所释放出来的黏土颗粒堵塞孔喉是导致饱和 CO_2 水驱实验后岩心渗透率降低的主要原因。

（5）饱和 CO_2 水驱实验结果表明，组成储层岩石的碎屑钾长石的溶蚀作用是非常微弱的，而方解石和铁白云石等碳酸盐矿物则发生了明显的溶解作用。上述实验结果可为 CO_2 地下埋存过程中的矿物捕获机制提供基础的地球化学信息。

参 考 文 献

[1] 叶建平，冯三利，范志强，等. 沁水盆地南部注二氧化碳提高煤层气采收率微型先导性试验研究[J]. 石油学报，2007，28（4）：77-80.

[2] 云箭，覃国军，徐凤银，等. 低碳视角下中国非常规天然气的开发利用前景 [J]. 石油学报，2012，33（3）：526-532.

[3] Stuart M V Gilfillan, Chris J Ballentine, Greg Holland, et al. The noble gas geochemistry of natural CO_2 gas reservoirs from the Colorado Plateau and Rocky Mountain provinces, USA [J]. Geochimica et Cosmochimica Acta, 2008, 72（4）: 1174-1198.

[4] 汤勇，杜志敏，孙雷，等. CO_2 在地层水中溶解对驱油过程的影响 [J]. 石油学报，2011，32（2）：311-314.

[5] Michael K, Golab A, Shulakova V, et al. Geological storage of CO_2 in saline aquifers—A review of the experience from existing storage operations [J]. International Journal of Greenhouse Gas Control, 2010, 4（4）: 659-667.

[6] Ryoji Shiraki, Thomas L Dunn. Experimental study on water-rock interactions during CO_2 flooding in the Tensleep Formation, Wyoming, USA [J]. Applied Geochemistry, 2000, 15（3）: 265-279.

[7] 沈平平，黄磊. 二氧化碳—原油多相多组分渗流机理研究 [J]. 石油学报，2009，30（2）：247-251.

[8] 曲希玉，刘立，高玉巧，等. 中国东北地区幔源—岩浆 CO_2 赋存的地质记录 [J]. 石油学报，2010，31（1）：61-67.

[9] Shelagh J Baines, Richard H Worden. The long-term fate of CO_2 in the subsurface: natural analogues for CO_2 storage [J]. Geological Society, 2004（233）: 59-85.

[10] Sebastian Fischer, Axel Liebscher, Maren Wandrey. CO_2-brine-rock interaction—First results of long-term exposure experiments at in situ P-T conditions of the Ketzin CO_2 reservoir [J]. Chemie der Erde-Geochemistry, 2010, 70（Supplement 3）: 155-164.

[11] 朱子涵，李明远，林梅钦，等. 储层中 CO_2—水—岩石相互作用研究进展 [J]. 矿物岩石地球化学通报，2011，30（1）：104-112.

[12] Wigand M, Carey J W, Schütt H, et al. Geochemical effects of CO_2 sequestration in sandstones under simulated in situ conditions of deep saline aquifers [J]. Applied Geochemistry, 2008, 23（9）: 2735-2745.

[13] Assayag N, Matter J, Ader M, et al. Water-rock interactions during a CO_2 injection field-test: Implications on host rock dissolution and alteration effects [J]. Chemical Geology, 2009, 265（1-2）: 227-235.

[14] Zhenhao Duan, Rui Sun. An improved model calculating CO_2 solubility in pure water and aqueous NaCl solutions from 273 to 533 K and from 0 to 2000 bar [J]. Chemical Geology, 2003, 193（3-4）: 257-271.

[15] Richard A. Robie, Bruce S Hemingway, James R Fisher. Thermodynamic properties of minerals and related substances at 298. 15 K and 1 Bar (105 Pascals) pressure and at higher temperatures: U. S. Geological Sur-

vey Bulletin 1452 [M]. Washington, D. C.: United States Government Printing Office, 1978: 452–456.

[16] Harold C Helgeson, David H Kirkham, George C Flowers. Theoretical prediction of the thermodynamic behavior of aqueous electrolytes by high pressures and temperatures; VI, Calculation of activity coefficients, osmotic coefficients, and apparent molal and standard and relative partial molal properties to 600 degrees C and 5kb [J]. American Journal of Science, 1981, 281 (10): 1249–1516.

[17] 卓胜广, 周书欣, 姜耀俭. 松辽盆地白垩系砂岩成岩作用 [J]. 地质科学, 1992, 27 (增刊): 216–225.

[18] 杨桂芳, 卓胜广, 牛奔, 等. 松辽盆地白垩系砂岩长石碎屑的钠长石化作用 [J]. 地质论评, 2003, 49 (2): 155–161.

[19] Weibel R, Kjøller C, Bateman K, et al. Mineral changes in CO_2 experiments—Examples from Danish onshore saline aquifers [C] // John Gale, Chris Hendriks, Wim Turkenberg. Energy Procedia: 10th International Conference on Greenhouse Gas Control Technologies, GHGT-10, Amsterdam, Netherlands, September 19–23, 2010. Elsevier Science Ltd. , 2011, (4): 4495–4502.

[20] Maren Wandrey, Sebastian Fischer, Kornelia Zemke, et al. Monitoring petrophysical, mineralogical, geochemical and microbiological effects of CO_2 exposure—Results of long-term experiments under in situ conditions [C] // John Gale, Chris Hendriks, Wim Turkenberg. Energy Procedia: 10th International Conference on Greenhouse Gas Control Technologies, GHGT-10, Amsterdam, Netherlands, September 19–23, 2010. Elsevier Science Ltd. , 2011, (4): 3644–3650.

[21] Kent A Bowker, Shuler P J. Carbon dioxide injection and resultant alteration of the Weber Sandstone, Rangely Field, Colorado [J]. AAPG Bulletin, 1991, 75 (9): 1489–1499.

[22] 闫志为, 刘辉利, 张志卫. 温度及 CO_2 对方解石、白云石溶解度影响特征分析 [J]. 中国岩溶, 2009, 28 (1): 7–10.

[23] 蔡杰兴. 方解石、白云石、菱铁矿分别与含 CO_2 水溶液反应平衡时的温度和压力 [J]. 矿物岩石, 1993, 13 (2): 37–41.

[24] 于炳松, 赖兴运. 成岩作用中的地下水碳酸体系与方解石溶解度 [J]. 沉积学报, 2006, 24 (5): 627–634.

[25] Ross, Graham D, Todd, et al. The dissolution effects of CO_2-brine systems on the permeability of U. K. and North Sea Calcareous Sandstone: SPE Enhanced Oil Recovery Symposium, Tulsa, Oklahoma, USA, April 4–7, 1982 [C]. Society of Petroleum Engineers, 1982.

[26] Sayegh S G, Krause F F, Marcel Girard, et al. Rock/Fluid interactions of carbonated brines in a sandstone reservoir: Pembina Cardium, Alberta, Canada [J]. SPE Formation Evaluation, 1990, 5 (4): 399–405.

[27] Pearce J M, Holloway S, Wacker H, et al. Natural occurrences as analogues for the geological disposal of carbon dioxide [C] // Pierce W F Riemer, Andrea Y Smith. Energy Conversion and Management: Proceedings of the International Energy Agency Greenhouse Gases: Mitigation Options Conference, London, U. K. , August 22–25, 1995. Elsevier Science Ltd. , 1996, 37 (6–8): 1123–1128.

盐水层 CO_2 封存主控因素数值模拟研究

俞宏伟　李　实　陈兴隆

（中国石油勘探开发研究院提高石油采收率国家重点实验室）

摘　要：将二氧化碳注入地下盐水层中，通过一系列物理化学反应，最终将二氧化碳封存于地下，与生物圈隔离数百数千年，甚至更长时间，这种碳的处理方式将对全球碳循环产生深远影响。由于 CO_2 在盐水层中受到孔隙半径、盐水类型等不同作用的影响，其封存形式主要分为构造封存、水动力封存、溶解封存和矿物捕获封存。本文通过对具有 $3.65×10^6 m^3$ 水体规模的氯化钠型盐水层模型进行的 8 种典型条件下 CO_2 封存数值模拟研究，定量化地评价了 1000 年时间尺度下的毛细管压力、岩石压缩系数和盐水矿化度 3 种因素对各封存形式和封存分量的影响，同时得到了 CO_2 向盖层中上逸的速度，为将来盐水层 CO_2 埋存的大规模实施提供理论依据。

关键词：盐水层；CO_2 埋存；主控因素；定量化分析

0　引言

随着全球气候问题的日益严峻，各国政府和科学家已经越来越重视对 CO_2 储层技术的研究，尤其是 CO_2 地下封存技术，并已开展了大量的可行性研究。

CO_2 封存技术是一项具有大规模应用前景的 CO_2 处置技术，有望实现化石能源使用后 CO_2 的零排放。其主要技术包括：将 CO_2 注入地下深部盐水层、废弃油气藏或正在开发的油气藏中提高原油采收率[1]，以及不能开采的煤层中提高煤层气的采收率和海洋中等 4 个方面，其中深部盐水层 CO_2 埋存的潜力最大[2]。

把二氧化碳地质封存于盐水层对全球碳循环具有深远影响。利用这种方法可把大量二氧化碳注入地下，通过一系列物理化学反应，最终将 CO_2 永久封存于盐水层中，并把二氧化碳与生物圈隔离数百数千年，甚至更长时间[3]。国际能源署（IEA）对全球范围内 CO_2 地质埋存总量的评估值为 $(1250～10850)×10^{12} t$，其中深部盐水层埋存量约占总量的 92%。图 1 为不同埋存方式埋存量占总埋存量的比例示意图。

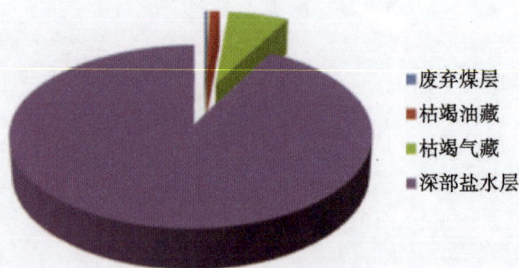

图 1　不同埋存形式埋存量所占比例

废弃煤层
枯竭油藏
枯竭气藏
深部盐水层

在 CO_2 地质埋存的选址中，人们往往把含水的砂岩层确定为埋存的目的层，适于封存二氧化碳的最小地层深度为 800m。由于 CO_2 在盐水层中受到不同作用的影响，其封存形式主要分为构造封存、水动力封存、溶解封存和矿物捕获封存。二氧化碳的这 4 种地下封存过程同时进行，但由于受到不同因素的影响，它们之间会随时间发生相互转换。

1 盐水层 CO_2 埋存主控因素

1.1 构造封存

决定 CO_2 构造封存分量的主要控制因素为水气毛细管压力的大小。部分注入的 CO_2 在毛细管压力的作用下被"固定"在岩石的微小孔隙中，这部分 CO_2 虽然以游离态存在，但不能自由移动，只有当水气饱和度发生相对变化而引起毛细管压力波动时，部分被"固定"的 CO_2 才有可能被释放以真正的游离态存在，转换为水动力封存。

1.2 水动力封存

在一定的压力上限条件下，盐水中 CO_2 含量达到过饱和，CO_2 以低密度的气态游离存在；当达到 CO_2 的临界温度和临界压力时，CO_2 以高密度的超临界状态游离存在，这样可以增加 CO_2 在盐水层中的埋存总量，但会使地层平均压力有一定的上升。

1.3 溶解封存

在盐水层埋存过程中，CO_2 被注入目标层段之后首先与地层盐水接触，随即部分 CO_2 将溶解于盐水之中，而溶解总量会受到水体规模、温度、压力和地层水矿化度等因素的影响。图 2 所示为不同温压及盐水矿化度条件下的 CO_2 溶解度。

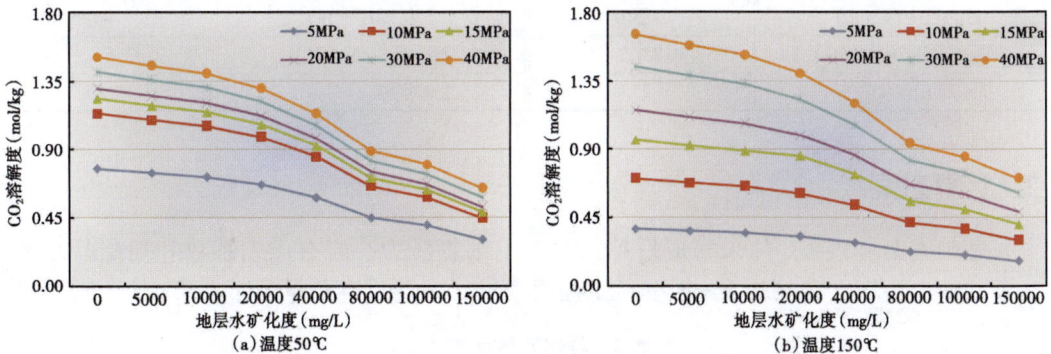

图 2　不同温度、压力及矿化度条件下的 CO_2 溶解度

由以上 CO_2 的溶解度模板得到了 CO_2 在 NaCl 盐水中的溶解度变化规律：（1）CO_2 的溶解度随着盐水矿化度的增加而减小，随着压力的增加而增加；（2）NaCl 溶液矿化度小于 80000mg/L 时，温度对 CO_2 溶解度的影响较小。

1.4 矿物捕获封存

矿物捕获是 CO_2 溶于地下水后通过水—岩相互作用，最终将"碳"以碳酸盐矿物的形式"固结"起来的 CO_2 捕获机制[4]。

注入的 CO_2 与地层水反应生成碳酸，致使地层水呈弱酸性，碳酸与原生矿物反应生成黏土和碳酸盐矿物沉淀，从而可以埋存部分注入的 CO_2，并使砂岩的孔隙度和渗透率发生改变[5]。因此，含有金属阳离子造岩矿物越多的岩石以矿物形式捕获 CO_2 的潜力就越大。

CO_2 矿物封存进行的速度虽然缓慢，但埋存量会随着时间的推移而增加。反应速率主要受到盐水层温度、压力、矿化度和地层岩石类型及其矿物组成的影响。

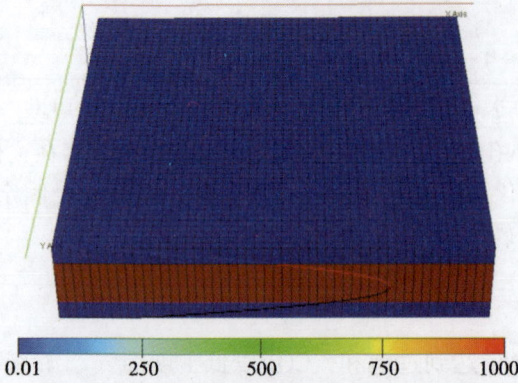

图3 模型渗透率场

2 CO_2 埋存定量化分析

2.1 地质模型的建立

为得到 CO_2 封存过程中各因素对其埋存分量的定量化影响，建立了 CO_2 盐水层埋存数值模拟计算模型（表1）。由于缺乏室内实验支持，该模型暂未考虑 CO_2 矿物捕获封存。图3所示为模型渗透率场图。

表1 模型基础参数表

参数	数值	参数	数值
网格数量（个×个×个）	50×50×50	盖层孔隙度（%）	1
网格尺寸（m×m×m）	10×10×2	盐水层渗透率（mD）	1000
流体组分	水、NaCl、CO_2	盖层渗透率（mD）	0.01
地层水矿化度（mg/L）	5000	盐水层厚度（m）	60
目的层中深（m）	1550	盖层厚度（m）	40
地层平均压力（MPa）	15.5	盐水层温度（℃）	60
盐水层孔隙度（%）	25	水体规模（$10^6 m^3$）	3.65

2.2 计算方案

在基础方案（表2）注采完成后单独改变毛细管压力、岩石压缩系数和地层盐水矿化度，添加7个计算方案（表3）以分析主控因素对 CO_2 各埋存分量的影响。

表2 基础方案参数选取

注气井数 （口）	生产井数 （口）	注气速度 （t/d）	产水速度 [m^3/（d·口）]	注采年限 （a）	CO_2 埋存预测年限 （a）
1	4	320	80	4	1000

表3 各计算方案参数选取

方案编号	方案变量	毛细管压力倍数	岩石压缩系数倍数	盐水矿化度（mg/L）
1	基础	1	1	5000
2		0.1	1	5000
3	毛细管压力	0.5	1	5000
4		2	1	5000

168

方案编号	方案变量	毛细管压力倍数	岩石压缩系数倍数	盐水矿化度（mg/L）
5	岩石压缩系数	1	0.1	5000
6		1	10	5000
7	地层盐水矿化度	1	1	10000
8		1	1	100000

2.3 计算结果分析

基础方案截至注采完成后，CO_2 埋存总量为 34.7×10^4 t，在重力的作用下 CO_2 逐步向盐水层上方移动。图 4 所示为模型 CO_2 饱和度分布随时间变化情况。表 4 为主要计算结果。

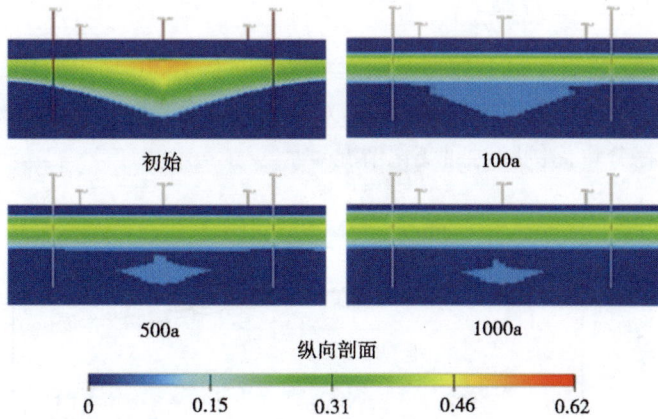

图 4 基础方案 CO_2 饱和度分布随时间变化情况

表 4 各方案计算结果对比

方案编号	方案	水中溶解 CO_2 量（t）	不可动 CO_2 量（t）	游离 CO_2 量（t） 盐水层	盖层	CO_2 平均上逸速度（t/a）
1	基 础	134185	67166	142377	3446	3.45
2	毛细管压力	121402	39150	184233	2389	2.39
3		129710	55595	158747	3122	3.12
4		136338	84979	122444	3413	3.41
5	0.1 倍岩石压缩系数	133492	66508	143666	3508	3.51
6		137889	70959	135313	3012	3.01
7	地层盐水矿化度	134084	67174	142470	3446	3.45
8		99567	78365	165796	3446	3.45

2.3.1 CO_2 构造封存分量

CO_2 埋存过程中，构造封存 CO_2 分量的主要控制因素为毛细管压力，埋存量随毛细管压力的增大而增大，1000 年时间尺度下毛细管压力为基础计算方案 2 倍时不可动 CO_2 的埋存量占总埋存量的比例达到了 0.25。图 5 所示为 CO_2 构造封存分量比例随时间变化情况。

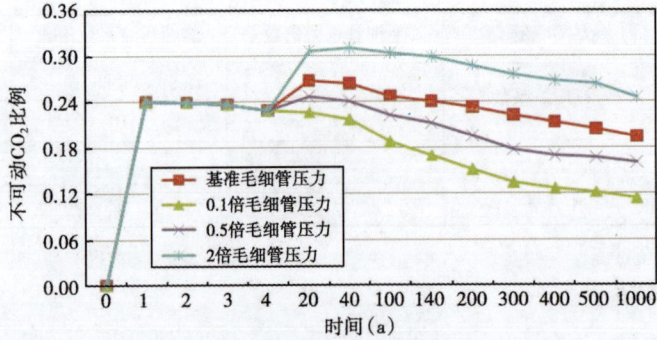

图5　CO_2 构造封存分量比例随时间变化曲线

2.3.2　CO_2 水动力封存分量

CO_2 埋存过程中，岩石压缩系数在基础方案取值的 0.1 ~ 10 倍内变化，它对各埋存分量的影响程度较小。针对这 3 个计算方案，CO_2 水动力封存分量的比例都在 42% 左右。图6所示为 CO_2 水动力封存分量比例随时间变化情况。

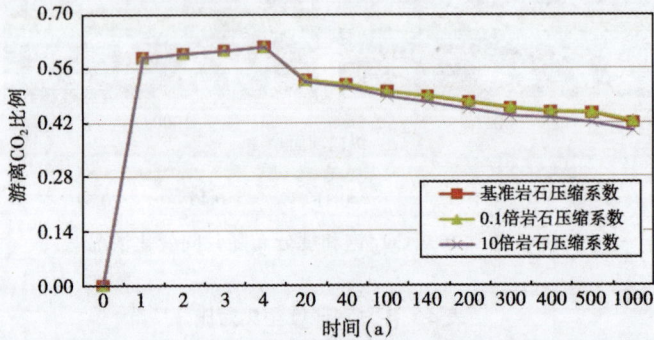

图6　CO_2 水动力封存分量比例随时间变化曲线

2.3.3　CO_2 溶解封存分量

CO_2 埋存过程中，溶解封存分量的主要控制因素为盐水矿化度，埋存量随盐水矿化度的增大而减小。图7所示为 CO_2 溶解封存分量比例随时间变化情况。

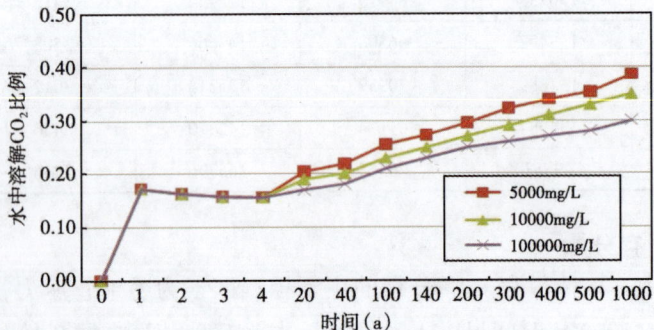

图7　CO_2 溶解封存分量比例随时间变化曲线

2.3.4 二氧化碳封存安全性

通过计算得到 8 种方案 CO_2 的最快上逸速度为 3.51t/a，仅占 CO_2 埋存总量的十万分之一，并且盐水层上部仍有近千米的地层作为天然保护，CO_2 不会上窜至地面对自然环境造成破坏。因此，从封存安全性的角度看，将 CO_2 封存于盐水层中是安全可行的。

3 结论

（1）盐水层 CO_2 埋存分为构造封存、水动力封存、溶解封存和矿物捕获封存 4 种形式。

（2）盐水层 CO_2 构造封存分量的主要控制因素为毛细管压力，埋存量随毛细管压力的增大而增大。

（3）在一定的温度、压力下，CO_2 溶解封存分量的主要控制因素为盐水矿化度，埋存量随盐水矿化度的增大而减小。

参 考 文 献

[1] 沈平平，杨永智. 温室气体在石油开采中资源化利用的科学问题 [J]. 中国基础科学，2006，8（3）：23-31.

[2] 江怀友，沈平平，罗金玲，等. 世界二氧化碳埋存技术现状与展望 [J]. 中国能源，2010，32（6）：28-32.

[3] 王涛. 盐水层 CO_2 埋存潜力及影响因素分析 [J]. 岩性油气藏，2010，22（12）：85-88.

[4] Bachu S, Gunter W D, Perkins E H. Aquifer disposal of CO_2: Hydrodynamic and mineral trapping [J]. Energy Converse Management, 1994, 35（4）: 269-279.

[5] 高玉巧，刘立，曲希玉. CO_2 与砂岩相互作用机理与形成的自生矿物组合 [J]. 新疆石油地质，2007，28（5）：579-584.

第二篇
二氧化碳驱油藏工程方法及优化设计技术

Influencing Factor Study of CO_2 -Assisted Gravity Drainage in Extra-low Permeability Reservoir

Wang Huan   Liao Xinwei   Li Mengmeng
Lu Ning   Lv Yuli   Liao Changlin

(China University of Petroleum)

Abstract: CO_2 flooding is an important method of enhancing oil recovery (EOR) in extra-low permeability reservoir, which can resolve the problems of water flooding effectively. The EOR theory and influencing factors of CO_2 -assisted gravity drainage were studied in this paper. The influencing factors, such as reservoir dip, injection volume, injecting position, well placement and reservoir effective thickness were analyzed and optimized based on the theoretical numerical model which had been established according to Tuha Niuquanhu inclined reservoir parameters. This research indicates that CO_2 -assisted gravity drainage can effectively EOR if appropriate parameters were adopted. The results of this paper can offer suggestions to similar reservoir development.

Keywords: Extra-low permeability reservoir; CO_2 -assisted gravity drainage; Influencing factor; Numerical simulation

1 Introduction

The CO_2 -assisted gravity drainage technology is studied earlier in abroad. In 1988, Chatzis and Tiffin had studied the mechanism of CO_2 -assisted gravity drainage by long core physics experiments[1,2]. Ren and Bentsen (2003) presented the criteria of gravity assisted drainage through the method of numerical simulation[3]. Rao et al proposed a gravity assisted drainage model, which can be used indifferent reservoirs[4]. Experiments show that CO_2 -assisted gravity drainage effect is better than WAG flooding. Mahmoud and Rao (2007) had studied the mechanism of CO_2 -assisted gravity drainage by using visual physical model, the results of their study show that the volumetric sweep efficiency of gravity assisted CO_2 immiscible flooding can reach 83%, while the volumetric sweep efficiency of gravity assisted CO_2 miscible flooding is nearly 100%[5].

At present, the CO_2 -assisted gravity drainage technology is studied rarely in China. Four wells CO_2 Huff and Puff filed test of Mabei block were applied in Tuha Santanghu oilfield in June 2010 and have good development effect. Taking into account that the reservoir has higher inclination and the technical problem of turning into CO_2 flooding after CO_2 Huff and Puff, it is necessary to study the technology of CO_2 -assisted gravity drainage. In this paper, a 2D reservoir numerical model was built based on the basic geological and fluid parameters of an extra-low permeability inclined reservoir in Tuha Niuquanhu oilfield and studied the sensitive parameters of CO_2 -assisted gravity drainage.

175

2　The EOR Mechanism of CO_2 Flooding

CO_2 EOR mechanism is mainly reflected in the following aspects: (1) viscosity reduction, the viscosity of crude oil dropped significantly when CO_2 dissolved in it. Generally, the higher the original viscosity of crude oil, the higher the percentage of viscosity reduction. The viscosity of crude oil can be reduced to the 1/100 to 1/10 of original one. (2) improve oil-water mobility ratio, a large amount of CO_2 dissolved in the crude oil and water, the viscosity of crude oil reduced and water viscosity increased by 20% after carbonation, so that the mobility ratio of water and oil is improved and the swept volume can be expanded. (3) oil expansion, when CO_2 comes into contact with crude oil a process of dissolution occurs thereby causing swelling, it usually can expanded 10% to 100% of original volume. (4) solution gas drive, with the reduction of reservoir pressure, the dissolved CO_2 gas escapes from the crude oil to generate a driving force and to improve the oil displacement effect[6]. (5) interfacial tension reduction, CO_2 evaporation and dissolution effects will generate ultra-low interfacial tension, the ultra-low interfacial tension is an important factor for EOR[7]. (6) CO_2 - assistedgravity drainage, in addition to the above flooding mechanism, another important mechanism is using gravity differentiation caused by the density difference between oil and gas to improve swept volume of the inclined reservoir, so as to improve the ultimate oil recovery.

3　Establish The Model of CO_2-Assisted Gravity Drainage

3.1　Formation and Fluid Properties Description

The structure characteristics of Niuquanhu reservoir of Tuha oilfield is nearly an east-west trending anticline, the north and south sides is gripped by thrust faults. The tectonic of northeast wing is moderate (formation dip 3° to 5°), while the tectonic of southwest wing is steep (formation dip 10° to 15°). The reservoir structure is simple with small-scale faults and natural fractures are not developed (see Fig. 1).

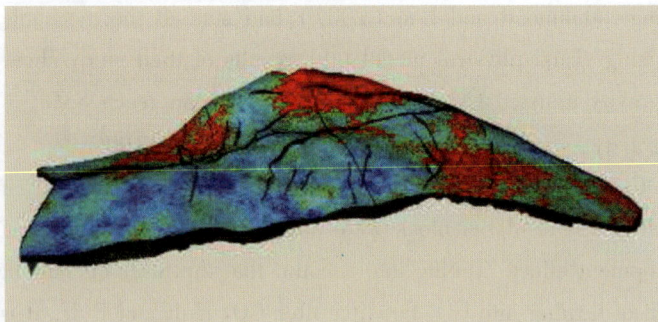

Fig. 1　Niuquanhu Structure Model

3.2 Numerical Simulation Model

According to the reservoir characteristics of Niuquanhu reservoir, numerical model of CO_2-assistedgravity drainage is established by application of Eclipse numerical software. The dimension of x, y and z is $44 \times 1 \times 4$, and the corresponding grid spacing is $10m \times 30m \times 2.54m$, but when analysis of Pay thickness, the grid spacing of z direction is variation. The plane permeability is $K_x = K_y = 3.4mD$, vertical permeability K_z is 2.0md and porosity is 14%. The original formation pressure is 15MPa, initial gas-oil ratio is 16.3 m^3/m^3, original oil saturation is 0.65 and gas saturation is 0. There is an injection well on the top of the model and a production well at the bottom, as shown in Fig. 2.

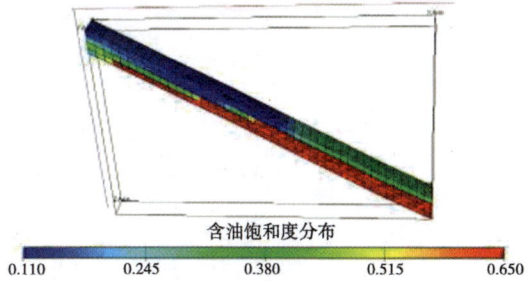

含油饱和度分布

0.110   0.245   0.380   0.515   0.650

Fig. 2 CO_2-assisted Gravity Drainage Model

The PVT experiments of Niuquanhu reservoir crude oil are conducted. 10 pseudo-components of crude oil and their mole fractions are shown in Table 1. PR3 equation of state is applied in PVT regression via PVTi module of Eclipse 2010 to match the experimental data of single flash vaporization test, multi-degas test and constant composition expansion test. The parameters are well matched to meet the accuracy required for simulation. The oil-water relative permeability curve is obtained through core displacement experiments and oil-gas relative permeability curve is obtained through Corey model calculation. A 2D compositional numerical model is established based on the above geological data, PVT data and relative permeability data.

Table 1   Pseudo-components of Crude Oil Sample

Pseudo-component	N_2	CO_2	C_1	C_2	C_3	C_4	C_5	C_6	C_{7+}	C_{28+}
Mole fraction (%)	2.55	0.03	16.68	1.21	0.96	0.75	0.5	1.5	51.16	24.67

4   Influencing Factors of CO_2-Assisted Gravity Drainage

Application of the established CO_2-assisted gravity drainage reservoir model to study the impact of the reservoir inclination, gas injection volume, gas injection position, the location of the injection wells and the reservoir thickness to CO_2-assisted gravity drainage, and the relevant parameters are optimized. To prevent the crude oil from degassing at the bottom of production well, the producer bottom-hole pressure of all the scenarios is set 8 MPa, which is above the saturation pressure. As the minimum miscible pressure (MMP) of the crude oil is high up to 29.8MPa, it is unfavorable to implement CO_2 miscible flooding, so the following scenarios are all CO_2 immiscible flooding.

4.1 Reservoir Dip

Reservoir dip is an important sensibility parameter in the process of CO_2-assisted gravity drain-

177

age. To study its influence on oil recovery, the reservoir models with different reservoir dip (10°, 30°, 60°) are established, the result is shown in Fig. 3.

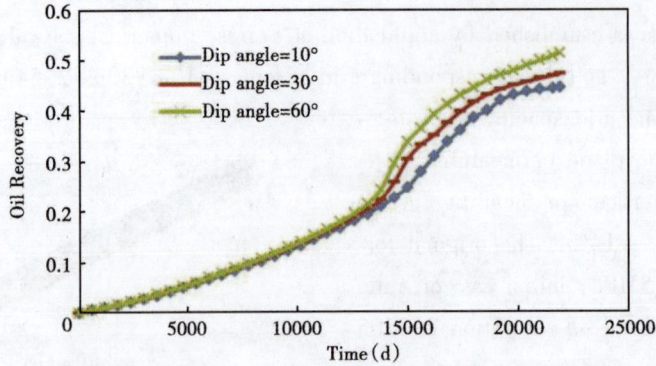

Fig. 3　The Curve of Oil Recovery with Different Reservoir Dip

It can be seen from Fig. 3 that with the increase of reservoir dip, the oil recovery is on the rise. When the reservoir dip is 60°, the maximum oil recovery is 50%. The primary cause is that when the CO_2 flooding is implemented in the inclined reservoir, the CO_2 injected on the top of the reservoir is lighter than the crude oil, in the effect of gravitational differentiation, CO_2 could displace crude oil from the top to the bottom, and it restrains the gas fingering and enhances the CO_2 sweep efficiency finally. In this sense, a bigger dip is better to develop the advantage of CO_2-assisted gravity drainage.

4.2　Gas Injection Volume

With the increase of the gas injection volume, more CO_2 would dissolve into oil which could decrease oil viscosity and interfacial tension efficiently; meanwhile, it could expand the crude oil volume, increase the driving energy near the wells and improve the oil recovery finally. However, the cost of CO_2 flooding field test depends on the CO_2 injection volume to a large extent. To choose the reasonable CO_2 injection volume, 0.2 HCPV, 0.3 HCPV, 0.4 HCPV, 0.5 HCPV and 0.6HCPV was injected respectively to conduct the optimization research. The result is shown in Fig. 4.

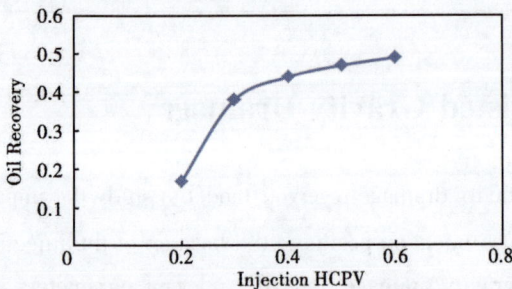

Fig. 4　The Curve of Oil Recovery with
Different Injection Volume

It can be seen from Fig. 4 that the inflection point appears in 0.4 HCPV. When the injection volume is less than 0.4 HCPV, the oil recovery increases rapidly with the increase of injection volume. But when the injection volume is more than 0.4 HCPV, the oil recovery increases slowly. Therefore, the optimal injection volume should be 0.3 HCPV to 0.4 HCPV. The main reason of the slow rise in the later period is the gas breakthrough in the oil wells when the CO_2 injection volume has reached to a certain value, thus the CO_2 cannot expand swept volume efficiently.

4.3 Gas Injection Position

CO_2-assisted gravity drainage can increase the swept volume in the reservoir with big dip. However, for the reservoir with smaller dip, CO_2 could penetrate along the top easily in the most circumstances which could cause a low swept volume and oil recovery. It is necessary to conduct a sensitivity analysis about gas injection position. The oil layer was divided into four layers from top to bottom to study the influence of different perforation layers of injection wells for the oil recovery.

It can be seen from Fig. 5 that the oil recovery of perforation the two layers on the bottom is larger than that of perforation the two layers on the top apparently. The closer to the bottom the perforation site, the higher the oil recovery. This is because when the perforation site is on the top of the reservoir, CO_2 could permeate to the oil well along the top easily, which could make gas breakthrough in advance. Therefore, for the reservoir with a small dip, the perforation position should on the bottom of the reservoir when proceeding CO_2-assisted gravity drainage.

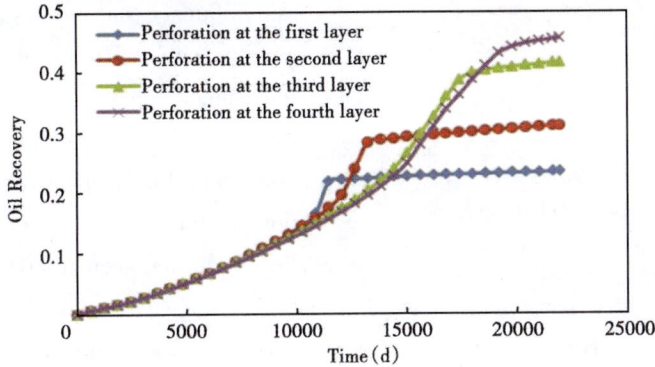

Fig. 5   The Curve of Oil Recovery with Different Injection Position

4.4 The Location of the Injection Well

When the injection well is on the roof of the reservoir, the technology of the CO_2-assisted gravity drainage can utilize the gravitational differentiation formed by oil and gas density difference to gather the top crude oil into new front oil enrichment zone and flood it to the lower reservoir evenly and produce. When the injection well is on the foot of the reservoir, the technology of the CO_2-assistedgravity drainage can put the injection gas into the higher position of the structure through the gravitational differentiation to form gas cap which could flood the oil remained in the top to the bottom and recover it. Therefore, to arrange the injection well on the roof, middle and foot respectively to contrast and analyze, see Fig. 6.

The Fig. 6 indicates that the recovery is highest when the injection well is arranged on the roof of the reservoir and the middle location takes the second place, while the foot location is worst. Consequently, it is more favorable to arrange well on the roof of the reservoir than that on the middle and foot.

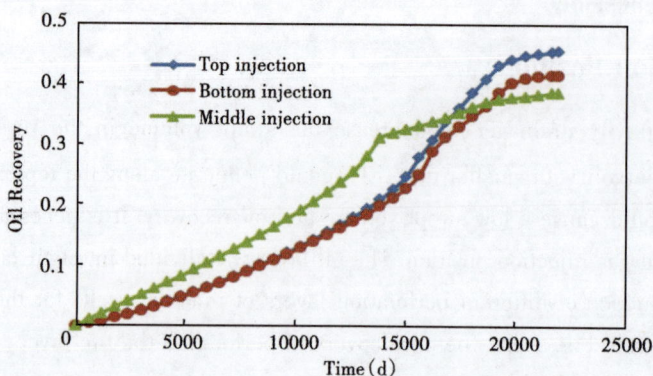

Fig. 6   The Curve of Oil Recovery with Different Location of Injection Well

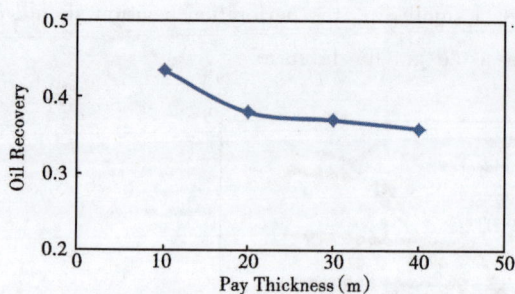

Fig. 7   The Curve of Oil Recovery with
Different Pay Thickness

4.5   Pay Thickness

The pay thickness of reservoir has a certain influence on the result of CO_2 – assisted gravity drainage. It can emerge gravity override phenomenon easily with a larger reservoir thickness. Set the pay thickness to 10m, 20m, 30m, and 40m respectively to conduct simulation calculation, the result is shown in Fig. 7.

It can be seen from Fig. 7 that the oil recovery is highest when the pay thickness is 10m, the other three pay thickness have a lower oil recovery, as when the reservoir thickness is bigger the injection gas could flood along the top of the reservoir easily, thus the bottom of the reservoir could not be swept, which could lead to a low oil recovery. The reservoir implements the CO_2–assisted gravity drainage would have a better effect when its thickness is smaller.

5   Conclusions

(1) The reservoir dip has a big influence on the effect of CO_2 –assisted gravity drainage. When the dip is small, the injected CO_2 could finger easily, bypass the crude oil and breakthrough in the oil well. When the dip is big, CO_2 could utilize the effect of gravity and enlarge swept volume, then the displacement effect would be distinct. For the reservoir with a small dip, it should be considered that to perforate the bottom of the reservoir which can enhance oil recovery mostly when implementing CO_2–assisted gravity drainage.

(2) The swept volume of CO_2 flooding depends on the volume of injection gas primarily. A large cumulative gas injection volume can obtain a big swept volume. However, the cost of CO_2 flooding field test depends on the CO_2 injection volume to a large extent, it is necessary to choose a reasonable injection volume.

180

（3）The recovery of arranging the injection well on the roof of the reservoir is better than that on the middle and foot.

（4）Generally, the thicker pay thickness would have a better gravitational differentiation impact and the displacement effect would be better. However, when the pay thickness is bigger, the injection gas could flood along the top of the reservoir easily, thus the bottom of the reservoir could not be swept, which could lead to a low oil recovery. The reservoir pay thickness has an optimal range. A 10m thickness can obtain a best displacement effect through optimization.

References

[1] Chatzis I, Kantzas A, Dullien F A L. On the Investigation of Gravity-assisted Inert Gas Injection using Micro-models, Long Berea Sandstone Cores, and Computer- assisted Tomography [C]. SPE 18284. 1988: 223-234.

[2] Tiffin, Kremesec. Mechanistic Study of the Gas-Assisted CO_2 Flooding [J]. SPE Reservoir Engineering, 1988: 524-532.

[3] Ren, Bentsen. Numerical Simulation and Screening of Oil Reservoirs for Gravity-Assisted Tertiary Gas-Injection Process [C]. SPE 81006, 2003: 1-8.

[4] Rao D N, Ayirala S C. Development of Gas Assisted Gravity Drainage (GAGD) Process for Improved Light Oil Recovery [C]. SPE 89357, 2004: 1-12.

[5] Mahmoud, Rao. Mechanisms and Performance Demonstration of the Gas-Assisted Gravity-Drainage Process Using Visual Models [C]. SPE 130593, 2007: 1-14.

[6] Menzie D E, Nielsen R F. A Study of the Vaporization of Crude Oil by Carbon Dioxide Repressuring [J]. Revue, IFP, 1968 (5): 95-102.

[7] Rosman A, Zana E, Experimental Studies of Low IFT Displacement by CO_2 Injection [J]. SPE Journal, 1977: 1-4.

Study on Enhanced Oil Recovery Technology in Low Permeability Heterogeneous Reservoir by Water–Alternate–Gas of CO$_2$ Flooding

Liao Changlin   Liao Xinwei   Zhao Xiaoliang   Lu Ning

Ding Hongna   Wang Huan   Liu Yongge

(MOE Key Laboratory of Petroleum Engineering, China University of Petroleum-Beijing)

Abstract: The common problems that water injection come across in low permeability heterogeneous oil reservoir are poor injectability, low production, high water cut and low oil recovery. The previous studies showed that Water–Alternate–Gas (WAG) of CO$_2$ flooding could enhance oil recovery in low permeability oil reservoir effectively. However, the way to use effective technology to enhance oil recovery and net present value (NPV) in low permeability heterogeneous oil reservoir became a difficult point in WAG flooding. This paper introduces allocation of injection rates (AOIR), tapered Water – Alternate – Gas (TWAG) and AOIR–TWAG to study the feasibility of the technologies in enhancing oil recovery with numerical reservoir simulation base on the previous studies. The development effects of constant WAG in the reservoirs with different variation coefficients were contrasted. And the feasibility of the three technologies to enhance oil recovery was studied. The results show that the development effect of constant WAG flooding is better than that of continuous gas injection with the increasing of the variation coefficient. And the oil recoveries of the three technologies are better than that of the constant WAG flooding. However, only the NPV of the AOIR and AOIR–TWAG are better than that of constant WAG flooding. An actual oil reservoir model of Xinjiang oil field in China was used to check the feasibility of the two technologies. The results show that the oil recoveries and NPVs of the two technologies are better than that of constant WAG flooding. This result proves that the conclution of the reservoir conceptual model is reliable, and the technology could be suitable for other similar low permeability heterogeneous oil reservoirs.

Keywords: Low Permeability Heterogeneous Reservoir; EOR; Water – Alternate – Gas of CO$_2$ Flooding; Numerical Simulation

1 Introduction

With the development of exploit patterns in oilfield, CO$_2$ flooding has become an important injection pattern in the tertiary recovery technique. WAG is a common technique that has been used for a long time for mobility control of CO$_2$ because the flow of water with the gas lowers the total mobility of the displacing fluid and thus improves the sweep efficiency (Christensen et al. , 1998). The water saturation affects the gas mobility which would delay gas breakthrough and expand the swept volume by CO$_2$. At the same time, the gas saturation affects the water mobility which would make it

182

more effective for reducing the residual oil saturation in three phases flow zones (Christensen et al. , 2000). Moreover, WAG is also better for keeping formation pressure, improving CO_2 utilization efficiency and reducing the cost of injection gas than that of continue CO_2 flooding. The advantages of WAG could make up for the deficiency of the continuous gas injection.

The summary of the previous studies (Sohrabi et al. , 2000; Zhang et al. , 2004; Taheri and Sajjadian, 2006; Zhou et al. , 2012) indicates that WAG technique is one of the key points in studying CO_2 flooding in recent years. And the way to improve oil production and economic benefit of the low permeability heterogeneous oil reservoir has become one of the difficulties. The previous researchers (Odi and Gupta, 2010; Reza and Zhang, 2012) investigated some techniques and obtained some achievements. However, most of them just study on the single technique with little combination of multiple techniques deeply. Whether the combination of multiple techniques can exert the comprehensive effect of single ones is the problem that will be investigated in this study. Basing on the previous studies (Sharma and Clements, 1996; Jahangiri et al. , 2012), the development effects of constant WAG in the reservoirs of different variation coefficients were compared. And the AOIR, TWAG and AOIR−TWAG were introduced to study the feasibility of the technologies in enhancing oil recovery with numerical reservoir simulation. And an actual oil reservoir in Xinjiang oilfield was adopted to confirm the feasibility. Finally, the sensitivity analysis of economic factors were used to study the effects of oil price and CO_2 price on the NPV.

2　Methodology

The three techniques including AOIR, TWAG and AOIR−TWAG were introduced into this study. The main purposes of these techniques are to control CO_2 mobility, delay CO_2 breakthrough, expand CO_2 swept volume, reduce recycle of gas injection and improve oil recovery and economic benefit.

The TWAG technique is used to adopt different WAG ratios in different production stages. Comparing with constant WAG, the TWAG could change the water gas ratio based on the actual production stage flexibly and regulate oilfield production work system and delay the CO_2 breakthrough. At the same time, production response speed could be accelerated or delayed artificially based on bearing capability of the field apparatuses and pipelines.

The AOIR technique is used to adopt different injection rates in different wells based on the different injectabilities. The gas injection rates of the wells located in the high permeability zones reduced, and that of the wells located in the low permeability zones is increased. The water injection rate is opposite to the gas injection rate.

The AOIR−TWAG is the technique combination of the AOIR and TWAG. The parameters of the case are all the same except the technique. The results are compared with the cases of water injection and constant WAG.

The NPV is used as the evaluation parameter for the development effect of WAG flooding. The oil revenue, the cost of CO_2 and water production and ground facilities are considered in the NPV. The NPV is calculated in the following equation (Jahangiri et al. , 2012),

$$NPV = \sum_{t=1}^{T} \frac{C}{(1+r)^t} - C_0 \qquad (1)$$

The cash inflow for the time stept is given by,

$$C = (FOPT \times \$ / bbl) - (FGIT \times \$ / ton) - (FWIT \times \$ / wat) - (FWPT \times \$ / dwat) \qquad (2)$$

3 Reservoir Description

In this study, detailed compositional simulations were performed by a commercial simulator. A reservoir conceptual model and a reservoir actual model were built to investigate the feasibility of the WAG technique based on the actual stages of the BQ oil reservoir in Xinjiang oilfield in China.

The BQ oil reservoir is located in the northwest of Junggar basin in China. The main characteristics of the reservoir are complex pore structure, low porosity and permeability and serious heterogeneity in the plane and profile. The reservoir does not have a uniform oil-water surface, and its aquifer is inactive. It belongs to light oil reservoir with low crude oil density and viscosity. Water flooding is hard to adopt in this reservoir because of low porosity and permeability. The reservoir began to produce in 1982, went through high and stable production stage, declined production stage and comprehensive treatment stage, and turned into high water cut stage in 2004. From then on, the fluid production, oil production and water injection rate were decreasing steadily.

The grid size of the reservoir conceptual model [Fig. 1 (a)] is 29×29×10, and its top depth is 1900 m. This model belongs to low permeability oil reservoir. More detailed model parameters are shown in Table 1. There are four injection wells and nine production wells in this model. And all the wells are completed and perforated in every layer. Two phase relative permeability relationships of oil-water and oil-gas are shown in Fig. 2.

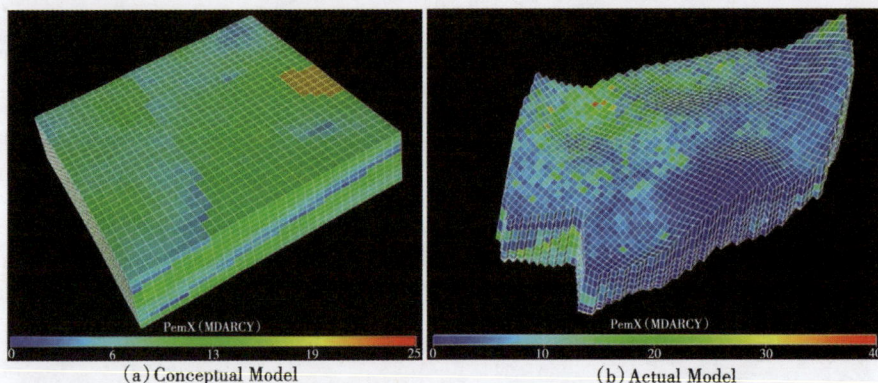

(a) Conceptual Model   (b) Actual Model

Fig. 1   3D View for Numerical Reservoir Models

The 3D view for the reservoir actual model is shown in Fig. 1 (b). The variation coefficient of permeability for this model is 0.41 which is calculated by the Lorenz Curve method (Kan et al., 2002). There are six injection wells and twenty-two production wells in this model. The fluid parameters and relative permeability of the model are the same as the conceptual model. The production

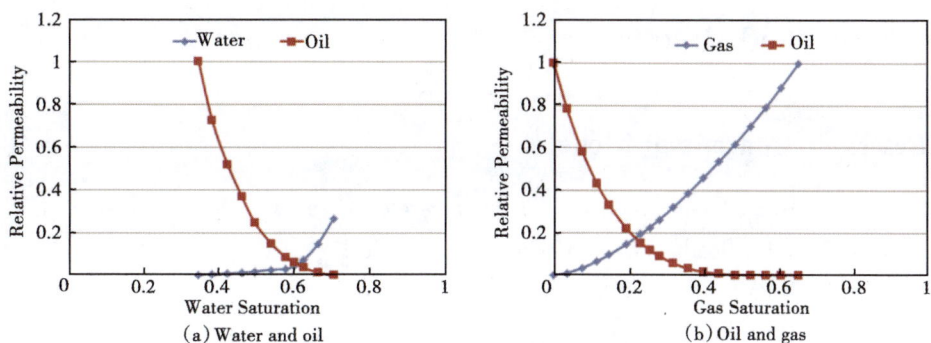

Fig. 2   Two Phase Relative Permeability Relationship

history of the actual model has been matched. The matched parameters included reserve, formation pressure, oil and water production, water cut and gas oil ratio et al.

Table 1   Reservoir Parameters

Parameter	Conceptual model	Actual model
Grid size	29×29×10	49×59×8
Top depth (m)	1900	1864
Initial pressure (Pa)	25	24. 7
Saturation pressure (MPa)	8. 4	8. 4
MMP (MPa)	21. 2	21. 2
Temperature (℃)	62. 15	62. 15
Porosity	0. 16	0. 14
Average permeability (D)	9. 1	8. 5
Initial oil saturation	0. 66	0. 64

The oil components were regrouped into eight components based on the experiment data of the oil component analysis and PVT test. The three−parameter Peng−Robinson equation of state was adopted in the simulation. The component parameters are shown in Table 2.

Table 2   Oil Pseudo Components of the Reservoir

Component	Mole fraction (%)	Molecular mass (g/mol)	Critical pressure (MPa)	Critical temperature (K)
CO_2	0. 1	44. 01	7. 39	304. 7
N_2C_1	30. 35	16. 18	1. 26	94. 4
C_2	2. 8	33. 88	3. 09	280. 51
C_{5-6}	45. 8	71. 99	1. 2	468. 98
C_{7-10}	14. 49	114. 29	2. 73	587. 89
C_{11-17}	21. 67	187. 66	1. 9	700. 02
C_{18-27}	13. 98	298. 75	1. 38	730. 43
C_{28+}	12. 01	518. 63	0. 28	787. 55

4　Scenarios and Results

4.1　Reservoir Conceptual Model

The study of reservoir conceptual model includes eleven cases (case 1 to case 11) which are shown in Table 3. Water flooding was adopted in each case before CO_2 flooding. The injected water slug size was 1.0 HCPV.

Table 3　Reservoir Simulation Cases

Case	Model	Variation coefficient	Technique pattern
1	Conceptual model	0	Constant WAG
2	Conceptual model	0.18	Constant WAG
3	Conceptual model	0.32	Constant WAG
4	Conceptual model	0.39	Constant WAG
5	Conceptual model	0.47	Constant WAG
6	Conceptual model	0.57	Constant WAG
7	Conceptual model	0.39	AOIR
8	Conceptual model	0.39	TWAG
9	Conceptual model	0.39	AOIR-TWAG
10	Conceptual model	0.39	Continuous gas injection
11	Conceptual model	0.39	Water
12	Actual model	0.41	Constant WAG
13	Actual model	0.41	AOIR
14	Actual model	0.41	AOIR-TWAG
15	Actual model	0.41	Water

4.1.1　WAG Flooding in Heterogeneous Oil Reservoir

Six cases (case 1 to case 6) were designed with different variation coefficients. The values of the variation coefficients (V_k) of the six cases are as follows: 0, 0.18, 0.32, 0.39, 0.47 and 0.57. The other parameters of the cases are all the same. Through the simulation of the cases with different Variation coefficients, the enhanced oil recoveries could be obtained (Fig. 3). The enhanced oil recovery is defined as the D-value between the oil recovery of constant WAG and that of the continuous gas injection. The result shows that the D-value is 1.4% in the homogeneous reservoir ($V_k = 0$). And it increases with the increase of variation coefficient of permeability. The advantage of the WAG in heterogeneous reservoir is obvious with increase of the variation coefficient of permeability. The D-value is 7.7% when the value of the variation coefficient of permeability is 0.57.

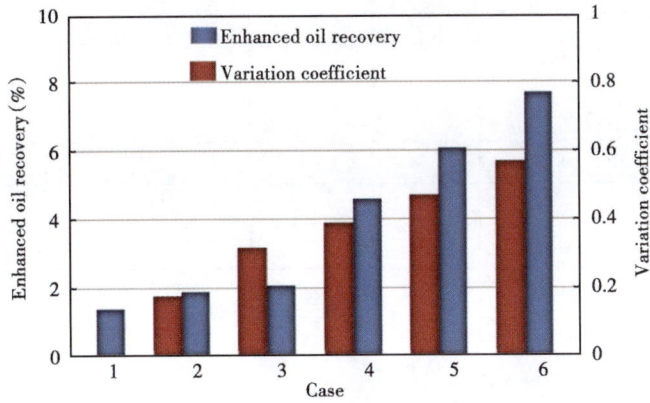

Fig. 3   Enhanced Oil Recoveries and Variation Coefficients of Different Cases

4.1.2   Techniques of WAG Flooding

The case 4 ($V_k = 0.39$) was used as the base case in this section. Three cases (case 7 to case 9, shown in Table 3) were built to simulate WAG techniques including AOIR、TWAG and AOIR-TWAG. They were designed in the maximum production capacity that may be reached. CO_2 flooding was adopted when the slug size of injection water was 1.0HCPV, with a 93.4% water cut and a 45.68% oil recovery. In this section, the base case has been optimized, and the optimized parameters are shown in Table 4.

Table 4   Optimized Parameters of the Base Case

Model	Gas injection rate (m^3/d)	Water injection rate (m^3/d)	Upper limit Gas oil ratio (m^3/m^3)	WAG ratio	WAG cycle (month)
Conceptual	20000	40	2000	1:1	12
Actual model	20000	30	1500	1:1	6

The average value of permeability of the zones where the injection wells located obtained. And the injection rate of the injection wells is calculated where the reservoir average value of permeability and the injection rate were selected in the base case. The data are shown in Table 5. The water injection rate is proportional to the average permeability, and the gas injection rate is inversely proportional.

Table 5   Injection Rate of the Injectors

Model	Well name	Average permeability (mD)	Water injection rate (m^3/d)	Gas injection rate (m^3/d)
Conceptual model	I1	19.64	40.4	19816.7
	I2	24.71	50.8	15750.7
	I3	26.16	53.8	14877.74
	I4	7.33	15.1	53096.9

187

Model	Well name	Average permeability (mD)	Water injection rate (m³/d)	Gas injection rate (m³/d)
Actual model	H42-9	11.3	31.7	15781.7
	H42-11	11.6	32.5	15373.6
	H43-8	12.1	33.9	14738.3
	H44-9	5.2	14.6	34294.9
	H44-11	8.4	23.6	21230.2
	H45-10	4.9	13.7	36394.6

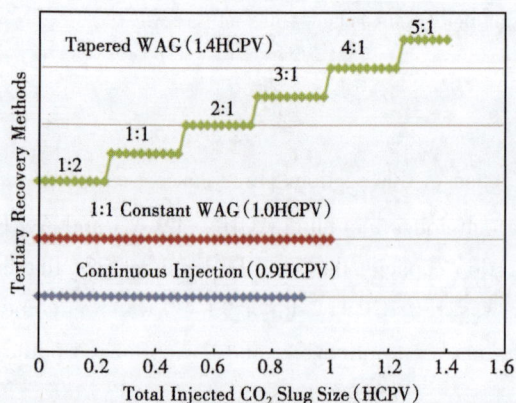

Fig. 4   Different WAG Ratios in Different Production Stages

The injected CO_2 slug size is 0.25 HCPV when the WAG ratio is 1:2, 1:1, 2:1, 3:1 and 4:1 respectively. And the injected CO_2 slug size is 0.15 HCPV when the WAG ratio is 5:1. The injected CO_2 slug size is shown in Fig. 4.

The AOIR-TWAG is the technique combinations of the AOIR and TWAG. The parameters of the case are all the same except the techniques. The results are compared with the cases of water injection and constant WAG.

4.1.3   Simulation Results

The techniques of AOIR, TWAG and AOIR-TWAG were simulated separately. The results of water flooding, continuous gas injection and constant WAG were compared. The oil recoveries and production gas oil ratios of the cases are shown in Fig. 5 and Fig. 6. The oil recovery is enhanced significantly in a short time by continuous gas injection, and the ultimate oil recovery is 52.7% when the production gas oil ratio reaches the upper limit. Its production time is 12.7 years. The ultimate oil recovery of constant WAG is 57.35% which is higher than that of continuous gas injection obviously. And its production time is 30.4 years. The ultimate oil recoveries of TWAG, AOIR and AOIR-TWAG are 59.25%, 58.51% and 58.49% respectively. They are higher than that of constant WAG in different ranges. The production times of AOIR and AOIR-TWAG are 26.4 years and 27 years which are less than that of constant WAG. The production time of TWAG is 55.1 years which is longer than that of constant WAG. The CO_2 slug size injected in the AOIR, AOIR-TWAG and constant WAG is 1.0 HCPV. The injected CO_2 slug size of TWAG and continuous CO_2 injection is 0.9 HCPV. The well was shut up when the production gas oil ratio reached the limit. The production of the oil reservoir was stopped when all the wells were reached the upper limit of the production gas oil ratio (Fig. 6).

4.1.4   Economic Benefit Analysis

The discount rate is 5% in this study when the NPV is calculated. The cash inflow is calculated

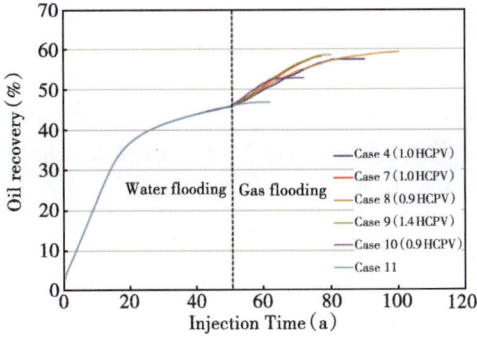

Fig. 5 Oil recoveries of the techniques

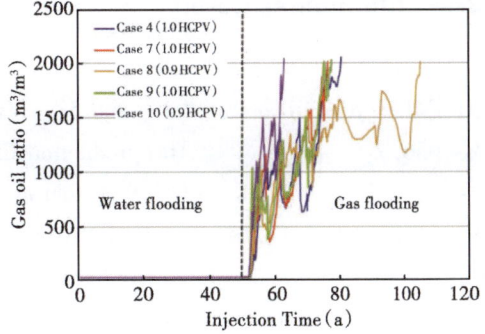

Fig. 6 Production Gas Oil Ratios of the Techniques

based on oil revenue, the cost of CO_2, the cost of water production and the cost of water process and ground facilities. The price of oil is considered to be $80 per barrel for the entire production period while the cost of water disposal is $1. 5 per barrel of produced water. The cost of water injection is $0. 25 per barrel. And the cost of CO_2 injection is assumed to be $80 per ton of CO_2. The total cost of ground facilities is $2. 5×10^6.

The NPV and ultimate enhanced oil recoveries of the cases were transformed into dimensionless values to evaluate the development effect of CO_2 WAG based on the base case. The value of each case was calculated in the largest production capacity. The dimensionless NPVs and enhanced oil recoveries of the cases are shown in Fig. 7. The results show that the ultimate oil recoveries of all the cases are higher than that of the base case. However, only the dimensionless NPVs (D-NPVs) of case 7 and case 9 are higher than that of the base case. The values are 1. 41 and 1. 44 respectively.

4. 2 Reservoir Actual Model

The oil actual reservoir in Xinjiang oilfield was simulated to confirm the feasibility of the two techniques which were optimized in the conceptual model. The cases of constant WAG, AOIR and AOIR-TWAG, and water flooding were run (case 12 to case 15). The injection rates of the AOIR were calculated and shown in Table 5. Different WAG ratios in different production stages of AOIR-TWAG are shown in Fig. 8.

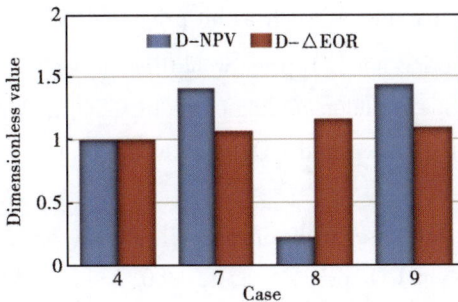

Fig. 7 The Dimensionless NPVs and Enhanced Oil Recoveries of the Cases

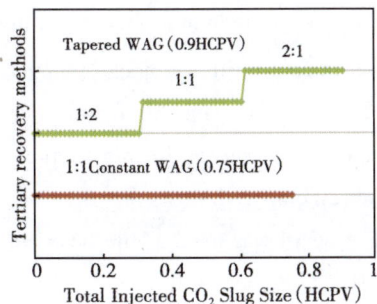

Fig. 8 Different WAG Ratios in Different Production Stages

4. 2. 1　Simulation Results

The results of the simulations are shown in Fig. 9 and Fig. 10. The ultimate oil recoveries of AO-IR-TWAG and AOIR are 59. 02% and 58. 73% respectively that are all higher than that of constant WAG (56. 97%) (Fig. 9). The production times of AOIR-TWAG and AOIR are 30. 2 and 29. 9 years respectively that all are less than that of constant WAG (36. 7 years). The injected CO_2 slug sizes of Constant WAG、AOIR and AOIR-TWAG are 0. 75 HCPV, 0. 9 HCPV and 0. 89 HCPV respectively. The NPVs and ultimate enhanced oil recoveries of the cases were transformed into dimensionless values to evaluate the development effect of WAG based on the base case. The value of each case was calculated in the largest production capacity. The total cost of ground facilities is $ 4. 3× 10^6. The D-NPVs and enhanced oil recoveries of the cases are shown in Fig. 10. The dimensionless enhanced oil recoveries of AOIR and AOIR-TWAG are 1. 07 and 1. 09 respectively. And their D-NPVs are 1. 46 and 1. 59 respectively. The results show that the dimensionless enhanced oil recoveries of AOIR and AOIR-TWAG are a little higher than that of constant WAG, but their D-NPVs are much higher than that of constant WAG.

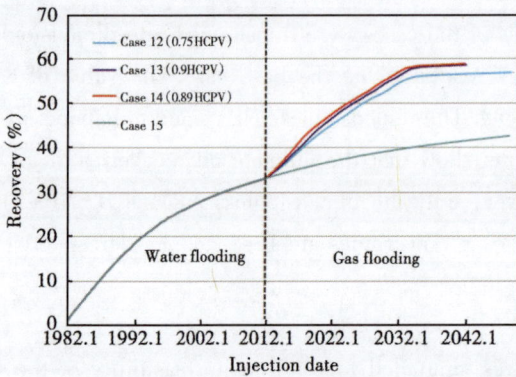

Fig. 9　Oil Recoveries of the Cases

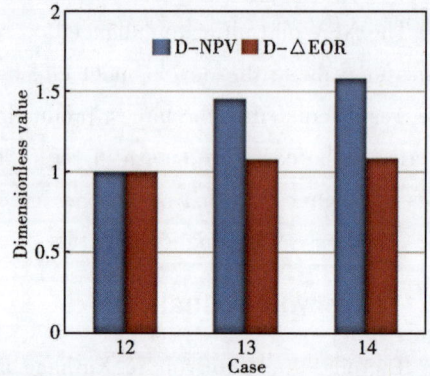

Fig. 10　The Dimensionless NPVs and Enhanced Oil Recoveries of the Cases

4. 2. 2　Effect of Oil Price and CO_2 Price

In addition to technical factors, different economic parameters such as oil price and CO_2 price have great impact on the project economics. Neither oil price nor CO_2 price would remain static over several decades. The economic factors were studied to achieve understanding of the financial performance of the EOR projects. In this section, the effects of oil price and CO_2 price to the project economics were compared.

The case 14 was used as the base case. There were four cases with different oil prices ($50, $80, $110, $140 per battle) and four cases with different CO_2 prices ($50, $80, $110, $140 per battle) run respectively. The D-NPVs of the cases for different oil prices are shown in Fig. 11. In this section, the D-NPV is the NPV of the technique per that of the contant WAG at the same perorid. The variation tendency of the D-NPV of the four cases is similar when the value reaches the

190

peak and then declines. The variation has little change in later production stage. Higher oil price will make more benefits for the project. The ultimate D−NPVs are close to each other when the oil price is in a high degree. The ultimate D−NPV is more sensitively when the oil price is in a low degree. Using the technique of AOIR−TWAG, the CO_2 injection rate is faster, and the increase of oil production is obviously in the early production period, which reaches the peak of the D−NPV much faster.

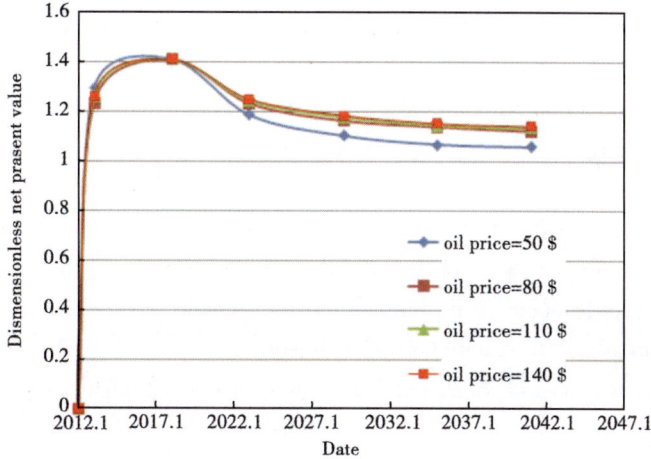

Fig. 11 Comparison of the D−NPVs for the Cases of Different Oil Prices

The D−NPVs of the cases with different CO_2 prices are shown in Fig. 12. Higher CO_2 price will make less benefit for the project. The D−NPV is more sensitive when CO_2 price is in a high degree. The peak value of D−NPV decreases with the increase of CO_2 price. And the peak is more cuspidal in low CO_2 price. It indicates that there are more benefits in low CO_2 price. The reason of cuspidal peak is that the CO_2 injection rate is faster than that of contant WAG. And the cost of CO_2 accounts for large proportion in the production cost. The D−NPVs of different CO_2 prices are almost same before reaching the peak, and change little in the later production period.

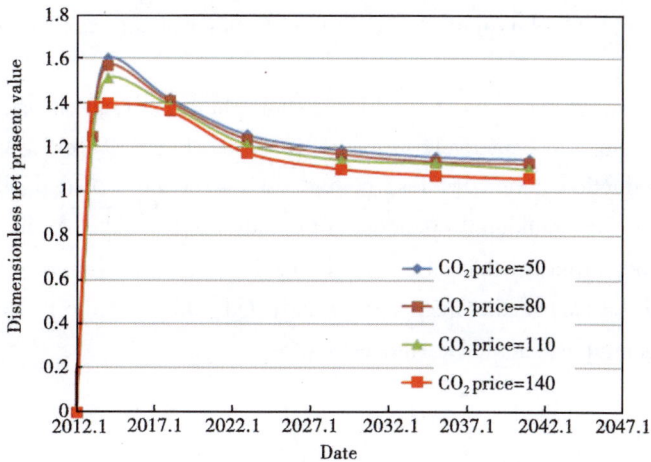

Fig. 12 Comparison of the D−NPVs for the Cases of Different CO_2 Prices

191

5 Discussion

In low permeability heterogeneous oil reservoir, water flooding is hard to enhance oil recovery further while CO_2 flooding could enhance oil recovery after water flooding which has been confirmed in this study. The Lorenz Curve method was used to calculate variation coefficient of permeability which was adopted to describe the heterogeneity of permeability distribution in reservoir scientifically. The comparison of the development effect of constant WAG and continuous gas injection in different variation coefficients demonstrates that the ultimate oil recovery of constant WAG is higher than that of continuous gas injection with increase of variation coefficients. The result indicates that the advantage of WAG flooding is obviously in the development of low permeability heterogeneous oil reservoir. The water slug of WAG can reduce the mobility of CO_2 in order to delay gas breakthrough. It will help to expand the CO_2 swept volume too. The gas slug of WAG can reduce the residual oil saturation in the three phase flow area which is lower than that in two phase flow area (oil and gas or oil and water). These two advantages contribute to the oil recovery greatly. In general, WAG flooding will make up the shortages of the continuous CO_2 injection.

The economics advantages of the D-NPV appear obviously through comparing the cases of the three techniques to the base case. There are some differences with the previous studies where single technique was used mostly, while the combination of the techniques is used in this study. The results show that the ultimate oil recoveries of the three techniques in their largest production capacities are higher than that of the base case, but only the D-NPVs of AOIR and AOIR-TWAG are higher than that of the base case with the consideration of economics benefits. It indicates that high ultimate oil recovery do not always cause high NPV always. Because longer production time will increase not only the higher ultimate oil recovery but also the higher production cost such as the cost of injection CO_2, injection water, water procession and so on.

The two techniques optimized in the conceptual model were adopted in the reservoir actual model. The results show that the ultimate oil recoveries and NPVs of the both techniques are higher than that of constant WAG. This result confirms that the conclusion of the reservoir conceptual model is reliable, and the optimized techniques can be applied to other similar low permeability heterogeneous oil reservoirs.

The results of sensitivity analysis of economic factors show that high oil price and low CO_2 price are good for the NPV. But it is sensitively in low oil price and high CO_2 price. And the NPV reaches the peak in early production period because of high CO_2 injection rate and oil production. In this study, numerical reservoir simulation was adopted to study the feasibility of the techniques. And the parameters and methods are ideal that may limit the model in practical application. The feasibility of the techniques should be confirmed in experiment and field pilot. And the problems in the practical applications of the techniques should be summarized.

6 Conclusions

(1) CO_2 flooding plays an important role after water flooding. The advantages of WAG flooding

in low permeability heterogeneous oil reservoir show up for more oil production than that of continuous CO_2 flooding. And the advantages are obvious with increase of variation coefficient of permeability.

(2) The ultimate oil recoveries of the three techniques are higher than that of constant WAG. However, only the NPVs of AOIR and AOIR-TWAG are higher than that of the base case. It indicates that the two techniques can reap more oil recoveries and economic benefits than that of constant WAG.

(3) Both the ultimate oil recovery and NPV of the two techniques used in Xinjiang oilfield are higher than that of constant WAG. It indicates that the results of conceptual model are reliable and the two techniques can be applied to other similar low permeability heterogeneous oil reservoirs.

(4) High oil price and low CO_2 price are good for the NPV. But it is sensitively in low oil price and high CO_2 price. And the NPV reaches the peak in early production stage because of high CO_2 injection rate and oil production.

Nomenclature

t = time step;

C = cash inflow for the time step, t;

r = annual or periodic discount rate;

T = cumulative investment or production period;

C_0 = initial investment;

C = cash inflow over the time step, $\$$;

$\$/bbl$ = price of oil per bbl, $\$$;

$\$/ton$ = cost of CO_2 injection per ton, $\$$;

$\$/wat$ = cost of water injection per bbl, $\$$;

$\$/dwat$ = cost of water disposal per bbl, $\$$;

$FOPT$ = cumulative oil production over time step, bbl;

$FWPT$ = cumulative water production over time step, bbl;

$FGIT$ = cumulative CO_2 injection over time step, bbl;

$FWIT$ = cumulative water injection over time step, bbl.

References

[1] Christensen J R, Stenby E H, Skauge A. Review of WAG Field Experience [C]. SPE 39883, 1998.

[2] Christensen J R, Larse M, Nicolaisen H. Compositional Simulation of Water-alternating-gas Processes [C]. SPE 62999, 2000.

[3] Jahangiri R H, Zhang D X. Ensemble based Co-optimization of Carbon Dioxide Sequestration and Enhanced Oil Recovery [J]. International Journal of Greenhouse Gas Control, 2012 (8): 22-33.

[4] Kang X, Liu D, Jiang M, Liu Z. The Application of Lorentz Curve in Reservoir Engineering [J]. Xinjiang Petroleum Geology, 2002, 23 (1): 65-67.

[5] Odi U, Gupta A. Optimization and Design of Carbon Dioxide Flooding [C]. SPE 138684, 2010.

[6] Sharma A K, Clements L E. From Simulator to Field Management: Optimum WAG Application in a Vlkst Texas CO_2 flood-a Case History [C]. SPE 36711, 1996.

［7］ Sohrabi M, Henderson G D, Tehrani D H, Danesh A. Visualization of Oil Recovery by Water Alternating Gas (WAG) Injection using High Pressure Micromodels Water Wet System ［C］. SPE 63000-MS, 2000.

［8］ Taheri A, Sajjadian V A. WAG Performance in a Low Porosity and Low Permeability Reservoir, Sirri-A Field, Iran ［C］. SPE 100212, 2006.

［9］ Zhang J, Zhou Z, Wang W, Tao L. EOR Field test of Gas-water Alternative Injection in Pubei Oilfield ［J］. Petroleum Exploration and Development, 2004, 31 (6): 85-87.

［10］ Zhou D, Yan M, Calvin W M. Optimization of a Mature CO_2 Flood—from Continuous Injection to WAG ［C］. SPE 154181, 2012.

The Optimization of CO_2 Huff and Puff Well Location in Ultra-low Permeability Reservoir

Zhao Dongfeng[1]   Liao Xinwei[1]   Wang Shaoping[2]
Yin Dandan[3]   Shang Baobing[1]

(1. MOE Key laboratory of Petroleum Engineering, China University of Petroleum (Beijing);
2. No. 2 Oil Production Plant, Changqing Oilfields, China National Petroleum Corporation;
3. EOR Research Institute, China University of Petroleum (Beijing))

Abstract: The oil recovery of CO_2 huff and puff well in ultra-low permeability reservoir is low. It is necessary to take measures to increase the development benefit. So the relevant development examples and theoretical research are taken into account in this paper, and according to liquid property and reservoir characteristics, the screening and evaluating system and grading evaluation standard of CO_2 huff and puff well location is established. Based on fuzzy comprehensive evaluation, screening steps of CO_2 huff and puff well location in ultra-low permeability reservoir is also established. The effect of the method has been proved by field practice. The result provides theoretical and technological support. And it provides reference for similar reservoirs.

Keywords: Ultra-low permeability reservoir; CO_2 huff and puff; Well location; Fuzzy comprehensive evaluation

1   Introduction

As the oil price rises, many countries speed up the development of ultra-low permeability reservoir. For the special property of CO_2, CO_2 huff and puff is a promising method for development of ultra-low permeability reservoir[1,2]. As the CO_2 mechanism of action in different reservoir differs, the effect is different[3]. So it is necessary to study screening suitable CO_2 huff and puff well block.

The article establishes a quantitative evaluation method of CO_2 huff and puff well. Firstly, each factor's evaluation standard and weight are confirmed. Secondly, CO_2 huff and puff suitability degree of wells were calculated by considering multiple factors that influenced the well based on fuzzy comprehensive judgment. Lastly, choose suitable well according to the suitability of CO_2 huff and puff.

2   The Establishment of Evaluation Indexes

Establish the evaluation index system on the basis of CO_2 stimulation mechanism, reservoir and

195

crude characteristics.

2. 1　Reservoir Characteristics

Depth: the deeper the reservoir is, the more suitable the reservoir is for CO_2 huff and puff, because the deep reservoir can stand high injection pressure. Besides the CO_2 solubility in crude oil rises when pressure rises[4].

Pressure: the pressure of CO_2 huff and puff need to be above the minimum miscibility pressure. So when the reservoir temperature is above critical temperature, the pressure should be high enough to make sure that CO_2 is miscible with oil.

Temperature: it is indicated that CO_2 can not be miscible with oil when the temperature is below 36℃. However the temperature rises, the minimum miscibility pressure rises. High temperature is bad for getting miscible with oil.

The thickness of target layer: if the thickness of stratified reservoir is more than 15m, the success rate of CO_2 huff and puff will be high, because the injected CO_2 swells the crude volume and overlaps crude and water. So there forms elastic gas drive in farther zone. While the thickness increases, the overlap will happen near wellbore area. In such a situation more remaining oil can be carried out.

2. 2　Liquid Properties

Remaining oil saturation: Thomas and Monger indicated that the recovery should have high oil saturation before CO_2 huff and puff based on core experiment and numerical simulation[5,6]. In general, if the current oil saturation is high, the oil and gas production is high and the water cut is low, which is good for CO_2 huff and puff.

Crude viscosity: the low viscosity is good for CO_2 huff and puff within the proper range.

Crude density: the viscosity rises with the density rising, and the solubility of CO_2 in oil reduces. And the viscous fingering forms easily which results in the displacement efficiency reduces[7].

The water cut of single well: too high (water cut is more than 95%) or too low water cut (water cut is less than 10%) is bad for CO_2 huff and puff according to development examples. When the water cut is near 50%, the effect is good[8,9]. Because CO_2 dissolves in oil can increase the mobility of oil and decreases the mobility of water. So the better mobility ratio can increase oil production.

The influencing factors and evaluation standard are shown in Table 1.

Table 1　The Judgment Standard of Well Location for CO_2 Huff and Puff

Category	Evaluation indexes	Evaluation grades——the influence on the well location for CO_2 huff and puff					Weight coefficient	
		Better ($Z=0.90$)	Good ($Z=0.75$)	Middle ($Z=0.60$)	Bad ($Z=0.45$)	Worse ($Z=0.30$)	First-grade	Second-grade
Reservoir characteristics	Depth (m)	>2500	2000~2500	1500~2000	1000~1500	<1000	0.45	0.25
	Pressure (MPa)	>20	15~20	10~15	5~10	<5		0.40
	Temperature (℃)	<40	40~60	60~80	80~100	>100		0.25
	Thickness (m)	>20	15~20	10~15	5~10	<5		0.10

category	Evaluation indexes	Evaluation grades——the influence on the well location for CO_2 huff and puff					Weight coefficient	
		Better ($Z=0.90$)	Good ($Z=0.75$)	Middle ($Z=0.60$)	Bad ($Z=0.45$)	Worse ($Z=0.30$)	First-grade	Second-grade
Liquid property	Remaining oil saturation(%)	>50	40~50	30~40	20~30	<20	0.55	0.35
	Crude viscosity (mPa·s)	<5	5~10	10~100	100~1000	>1000		0.30
	Crude density (g/cm³)	<0.8	0.8~0.83	0.83~0.86	0.86~0.89	>0.89		0.10
	Water cut (%)	>93	93~79	79~53	53~37	<37		0.25
Comprehensive evaluation	Evaluation grading range	0.84~1.0	0.72~0.84	0.48~0.72	0.34~0.48	<0.34	1.0	1.0
	The evaluation results of suitability	Better	Good	Middle	Bad	Worse		

3 The Weight of Evaluation Indexes and The Calculation of Comprehensive Index

3.1 Weight Assignment

The issue is divided into two grades by the analytic hierarchy process. The reservoir characteristics and liquid property are assigned as first-grade weight of 0.45 and 0.55 according to expert scoring method. The indexes of reservoir characteristics and liquid property are defined as second-grade weight. It is shown in Table 1.

3.2 Dynamic Method of Weighting

When some data of reservoir or liquid indexes is absent, dynamic method of weighting should be adopted for avoiding errors. Firstly, the weight of the absent data is defined as 0. Secondly, Summer up all the indexes weight. Lastly, each weight should be divided by the cumulative weight, which is actual weight.

3.3 The Calculation of Multiple Indexes

Establish the evaluate index R. Where u is index set; c is sample set; x_{ij} represents the ist sample and jst index.

$$
R = \begin{bmatrix}
 & u_1 & u_2 & \cdots & u_n \\
c_1 & x_{11} & x_{12} & \cdots & x_{1n} \\
c_2 & x_{21} & x_{22} & \cdots & x_{2n} \\
\vdots & \vdots & \vdots & \vdots & \vdots \\
c_m & x_{m1} & x_{m2} & \cdots & x_{mn}
\end{bmatrix} \tag{1}
$$

The reservoir multiple index is that the evaluate index $R_{\text{reservoir}}$ are multiplied by the second-grade weight $\lambda_{\text{reservoir}}$. $Z_{\text{reservoir}} = R_{\text{reservoir}} \times \lambda_{\text{reservoir}}$. Similarly, the liquid multiple index can be calculated. $Z_{\text{liquid}} = R_{\text{liquid}} \times \lambda_{\text{liquid}}$. Then multiply $Z_{\text{reservoir}}$ and Z_{liquid} with their first-grade weight. The multiple index (MI) is calculated by summing $Z_{\text{reservoir}}$ and Z_{liquid}.

$$Z_i = \sum_{j=1}^{n} (R_{ij} \times \lambda_j) \tag{2}$$

$$MI = 0.45 \times Z_{\text{reservoir}} + 0.55 \times Z_{\text{liquid}} \tag{3}$$

As the multiple index increases, CO_2 huff and puff has a better effect, and the risk is smaller, while the multiple index decreases, CO_2 huff and puff is not suitable.

4 The Development Practice of CO_2 Huff and Puff in Ultra-low Permeability Reservoir

The authors have chosen several wells for CO_2 huff and puff in ultra-low permeability reservoir of Changqing oil field, and evaluated these wells with the above method. The evaluation result is shown in Table 2.

Table 2   The Comprehensive Judgment of CO_2 Huff and Puff Wells

Well number	Y298-48	Y298-49	Y298-50	Y298-51	Y298-52	Y298-53	Y298-54	Y298-55
MI	0.68	0.78	0.65	0.79	0.43	0.66	0.71	0.46
Comment	Middle	Good	Middle	Good	Bad	Middle	Middle	Bad

If the comment is "good", the CO_2 huff and puff has economic benefit and low risk. If the comment is "middle", it still has benefit but some risk. If the comment is "bad", it may have little benefit and high risk. Based on the above analysis, choose the wells with the comment of "good" for CO_2 huff and puff.

5 Conclusions

(1) Analysis the laboratory experiment and development examples and put forward the evaluation standard for screening wells for CO_2 huff and puff in ultra-low permeability reservoir.

(2) Based on fuzzy comprehensive judgment, establish grading evaluation standard. Ascertain different factor's every grade weight. Put forward a method to calculate the suitability of CO_2 huff and puff well.

(3) According to fuzzy comprehensive judgment, calculate the suitability of wells in ultra-low permeability reservoir. The method has guiding significance for CO_2 huff and puff in field.

References

[1] Asghari K, Torabi F. Laboratory Experimental Results of Huff 'n' Puff CO_2 Flooding in a Fractured Core System [C]. SPE 110577, 2007: 1-8.

[2] Tadao Uchiyama, Yusuke Fujita, Yoshiaki Ueda, et al. Evaluation of a Vietnam Offshore CO_2 Huff 'n' Puff Test [C]. SPE 154128, 2012: 1-12.

[3] Alshmakhy A, ConocoPhillips, Maini B. A Follow-Up Recovery Method After Cold Heavy Oil Production Cyclic CO_2 Injection [C]. SPE 157823, 2012: 1-19.

[4] Huerta M, Alvarez J M, Jossy E, et al. Use of Acid Gas (CO_2/H_2S) for the Cyclic Solvent Injection (CSI) Process for Heavy Oil Reservoirs [C]. SPE 157825, 2012: 1-10.

[5] Qazvini Firouz A, Torabi F. Feasibility Study of Solvent-Based Huff-n-Puff Method (Cyclic Solvent Injection) To Enhance Heavy Oil Recovery [C]. SPE 157853, 2012: 1-18.

[6] Mohammed-Singh L, Petrotrin, Singhal A K, et al. Screening Criteria for Carbon Dioxide Huff 'n' Puff Operations [C]. SPE 100044, 2006: 1-10.

[7] ZHANG Y P, SAYEGH S G, HUANG S. Laboratory Investigation of Enhanced Light-Oil Recovery By CO_2/Flue Gas Huff-n-Puff Process [J]. JCPT, January, 2006 (19): 24-32.

[8] Wang Chunhong, Li Gaoming. Huff 'n' Puff Recovery Technique on Waterout Horizontal Wells [C]. SPE 104491, 2006: 1-7.

[9] Emanuel A S, Tang R W, Fang W S, et al. Analytic and Numerical Model Studies of Cyclic CO_2 Injection Projects [C]. SPE 22934, 1991: 489-500.

The Study of CO_2 Flooding of Horizontal Well with SRV in Tight Oil Reservoir

Wang H[1]   Liao X[1]   Zhao X[2]   Ye H[2], Dou X[2]
Zhang Q[3], Zhao D[2], Lu N[2]

(1. SPE, China University of Petroleum-Beijing; 2. China University
of Petroleum-Beijing; 3. Jilin Oilfield, PetroChina)

Abstract: As one kind of unconventional reservoirs, tight oil reservoir has become one of the main forces of oil reserves and production growth. The characteristics of tight oil reservoir are low porosity and ultra-low permeability, thus stimulated reservoir volume (SRV) should be conducted whether applying the mode of vertical wells or horizontal wells production. Tight oil reservoir is mostly developed by natural depletion or water flooding recently, but the problems are existed, including low recovery factor with natural depletion and the difficulty of water injection. To further improve the development effect of tight oil reservoir, CO_2 flooding is proposed.

Based on Chang-8 tight oil reservoir in Ordos Basin, an oil sample of typical block is selected. The PVT experiments are conducted. The compositional numerical model of five-spot pattern is established with a horizontal well in the middle and 4 vertical wells on the edge. Based on the model, several CO_2 flooding scenarios of horizontal well with different completion measures are studied. Furthermore, parameters such as the formation pressure, production rate, shut-in gas-oil ratio and total gas injection volume are optimized.

The results of this study show that the recovery factor of horizontal well with SRV is higher than those of horizontal well and conventional fractured horizontal well. The minimum miscible pressure (MMP) and the total gas injection volume are two key factors of CO_2 flooding effect. CO_2 flooding of volume fractured horizontal well in tight oil reservoir can not only improve oil recovery, but also realize CO_2 geological sequestration. It plays dual benefits of economy and environment.

The study gives new ideas of CO_2 flooding with volume fractured horizontal well for the Ordos Basin tight oil reservoir. It can be helpful for rapid and effective development of tight oil reservoirs in Ordos Basin.

Keywords: Tight Oil Reservoir; EOR; CO_2 Flooding; Simulated Reservoir Volume

1 Introduction

In recent years, it is become a hot spot to develop unconventional reservoirs, such as tight oil, tight gas and shale gas (Brohi et al., 2011; Zhou et al., 2013). Tight oil is a typical unconventional resource, which has the characteristic of good fluid properties and poor reservoir properties. The permeability and porosity of tight oil reservoir are general less than 1mD and 10%, respectively (Zou et al., 2012). Currently, the unconventional reservoirs are usually developed by horizontal wells, especially the segmented multi-cluster fractured horizontal wells, which have been widely used worldwide (Yu et al., 2013). The simulated reservoir volume (SRV) can be formed around the horizontal well after segmented multi-cluster fracturing (Du et al., 2011; Zhao et al., 2012). The technology of SRV is to achieve the important goal of increasing the contact area between matrix and fractures or fracture network as far as possible during the development of unconventional reservoirs (Saldungaray and Palisch, 2012; Wu et al., 2012). The study and application of developing tight oil and shale oil reservoirs mostly focus on the natural depletion (Suarez et al., 2013). But some studies have shown that CO_2 flooding is an effective approach of enhanced oil recovery (EOR) in tight reservoirs (Ghaderi et al., 2012; Zhao, 2012; Song and Yang, 2012). Based on the real case of Chang-8 tight oil reservoir in Ordos Basin, both PVT experiments and minimum miscible pressure (MMP) experiments were conducted for the crude oil sample of typical block. The compositional numerical model of typical five-spot well group is built, which is used to study the water flooding, CO_2 flooding and Water-Alternate-Gas (WAG) of CO_2 flooding. Different completion measures of horizontal well are analyzed. Furthermore, parameters such as the formation pressure, production rate, shut-in gas-oil ratio and total gas injection volume are analyzed.

2 Reservoir Characteristics and Numerical Simulation Model

2. 1 Reservoir Description

A tight oil reservoir of Chang-8 formation is in Ordos Basin, which belongs to the delta frontal subfacies deposition, mainly consisted of fine sandstone and local developed middle-fine sandstone. The average depth of the reservoir is 2540 m with the formation pressure of 19. 2MPa. The average matrix permeability of the reservoir is 0. 5mD and average porosity is 8. 8%. The average thickness of the reservoir is 10m. Formation oil density is 733 kg/m³ and formation oil viscosity is 1. 4mPa · s. The gas-oil ratio is 68. 17m³/m³. The formation saturation pressure is 9. 25MPa and crude oil volume factor is 1. 297.

2. 2 PVT and MMP Experiments

A crude oil sample is obtained in the tight oil reservoir. The PVT experiments are also conducted, and PR3 equation of state (EOS) is applied in PVT regression via PVTi module of Eclipse 2010 to match the experimental data of single flash vaporization test, differential liberation (DL) experiment and constant composition expansion (CCE) experiment. 9 pseudo-components of crude oil are grouped and their mole fractions are shown in Table 1. The parameters are well matched to meet the accuracy required for simulation (Table 2). It illustrates that the critical parameters of fluid can reflect the characters of real reservoir fluid.

Table 1 Pseudo-components of Crude Oil Sample

Pseudo-component	CO_2	N_2+C_1	C_2	C_3—nC_4	C_5—C_6	C_7—C_{10}	C_{11}—C_{17}	C_{18}—C_{27}	C_{28+}
Mole fraction (%)	0. 05	24. 64	7. 83	16. 99	5. 69	10. 96	15. 59	9. 62	8. 62

Table 2　Fluid Parameters after Regression

Components	MW (g/mol)	p_c (bar)	T_c (K)	V_c [m³/ (kg·mol)]	Ω_a	Ω_b	AF
CO_2	44.010	73.866	304.700	0.094	0.457	0.078	0.225
N_2+C_1	16.563	76.650	203.786	0.098	0.457	0.078	0.014
C_2	30.070	48.839	305.430	0.148	0.457	0.078	0.099
C_{3+}	49.369	71.279	339.297	0.221	0.830	0.062	0.169
C_{5+}	75.871	32.621	479.106	0.323	0.457	0.078	0.260
C_{7+}	120.118	74.482	1408.507	0.482	0.274	0.132	0.044
C_{11+}	185.069	73.427	1636.818	0.712	0.457	0.132	0.064
C_{18+}	303.675	67.394	1902.062	1.065	0.457	0.171	0.093
C_{28+}	572.433	39.176	2356.645	2.187	0.171	0.087	0.183

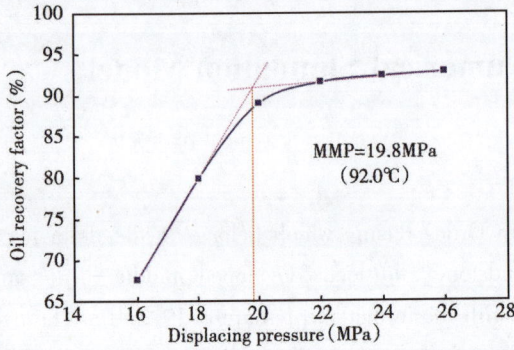

Fig. 1　Experimental Result of MMP for CO_2 Flooding

Slim tube experiment of the crude oil sample is conducted. The result shows that the CO_2 minimum miscible pressure (MMP) of the crude oil and CO_2 is 19.8MPa (Fig. 1).

2.3　Numerical Simulation Model

Based on the tight oil reservoir, three numerical models of five-spot well groups with SRV are built. Model 1, a perforated horizontal well without fracturing located in the middle of the well group and the length of the horizontal section is 540m (Fig. 2, Model 1). Model 2, there are a hydraulic fracturing horizontal well with three bi-wing transverse fractures in the toe, middle and heel of the well, respectively (Fig. 2, Model 2), and the location and length of model 2 are the same as Model 1. Model 3, a 540 m length horizontal well located in the middle of the well group. The horizontal well has been segmented multi-cluster fractured and formed four100m × 180m × 10m SRVs around the wellbore (Fig. 2, Model 3). Four vertical wells are located around them for each model. In this paper, the geometry of frac-

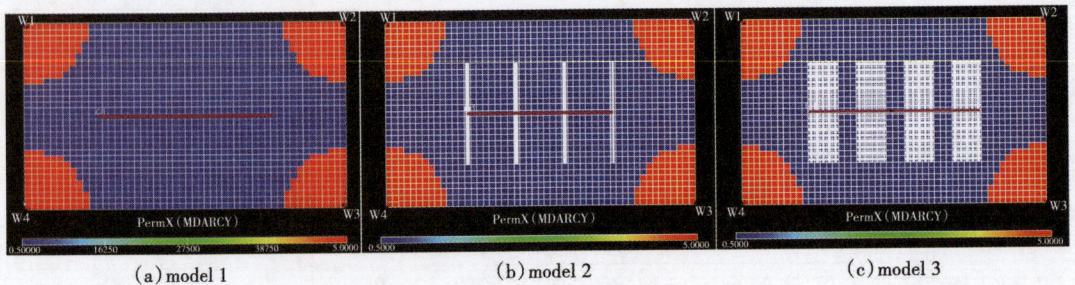

(a) model 1	(b) model 2	(c) model 3

Fig. 2　Numerical Simulation Models

ture extension is assumed to be wire-mesh networks, which forms a rule rectangular fracture network after fracturing. In order to guarantee the convergence in simulation and achieve the rectangular fracture propagation, the grids with width of 0. 1m after local grid refinement (LGR) are set to fractures. Vertical wells are also volume fractured to increase their injectivity, which is simulated simply by changing the reservoir permeability around them. Detailed parameters of these numerical models are shown in Table 3.

Table 3 Parameters of numerical simulation model

Parameters	Value	Parameters	Value
Grid spacing (m)	20×20×2	Irreducible water saturation	0. 43
Grid dimension	51×31×5	Initial reservoir pressure (MPa)	19. 2
Geologic reserve (10^4t)	26. 14	Saturation pressure (MPa)	9. 25
Temperature (℃)	80	Initial gas/oil ratio (m^3/m^3)	68. 17
Reservoir depth (m)	2540	Oil volume factor	1. 297
Effective thickness (m)	10	Density of surface oil (kg/m^3)	851. 0
Porosity (%)	8. 8	Density of formation oil (kg/m^3)	733. 0
Permeability (mD)	0. 5	Viscosity of formation oil (mPa · s)	1. 4
K_v/K_h	0. 1	Conductivity of bi-wing fracture (mD · m)	400
Initial oil saturation	0. 6	Conductivity of fracture network (mD · m)	30

3 Development Scenarios Optimization

Different completion measures, including perforated completion, conventional fracturing completion and segmented multi-cluster fracturing completion, are analyzed firstly. Then the optimized well group with superior completion are used to optimize different development modes, such as water flooding, succession CO_2 flooding and WAG of CO_2 flooding. Finally, the parameters, including the formation pressure, production rate, shut-in gas-oil ratio and total gas injection volume, are optimized. Evaluation index is mainly the recovery factor of ten years and the CO_2-oil draining efficiency. The CO_2-oil draining efficiency is calculated by the following equation.

$$E_{CO_2} = \frac{T_{oil}}{T_{CO_2}} \quad (1)$$

Where, E_{CO_2} is CO_2-oil draining efficiency, which is the reciprocal of CO_2 utilization factor, T_{oil} is total oil production amount (t), T_{CO_2} is CO_2 total injection amount (t).

3. 1 Well Completion Measures

In order to study the different horizontal well completions impact on development effect, three scenarios are designed. The first scenario is that the horizontal well is completed by perforating. The second scenario is that the horizontal well is completed by conventional fracturing. The third scenario is that the horizontal well is completed by segmented multi-cluster fracturing. The vertical wells are water injection wells, the horizontal wells are production wells, and the production control conditions of the three scenarios are the same.

As it can be seen from the Fig. 3, the scenario of the horizontal well with SRV has the highest oil recovery, the following is the scenario of the horizontal well with bi-wing fracture, and the scenario of the horizontal well with perforation completion has the lowest recovery. Therefore, the horizontal well with SRV is selected as one of the best well completion measures to develop tight oil reservoir. In the following scenarios, the wells are all volume fractured as wells in Model 3.

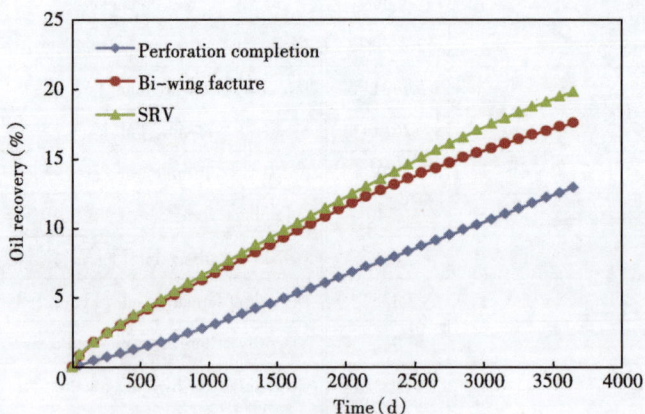

Fig. 3   Oil Recoveries of Different Horizontal Well Completions

3.2   Development Mode

The depletion development mode is usually used to develop tight oil reservoir, which is mainly due to the difficulty in injecting an oil-displacing agent to such tight reservoirs. The vertical and horizontal wells after volume fracturing can obtain SRV around them. The permeability and flow capacity of the reservoir have been greatly improved, which makes it possible to inject an oil-displacing agent to develop tight oil reservoir. Therefore, the water flooding, succession CO_2 flooding and WAG of CO_2 flooding are designed to study their effect on the development tight oil reservoirs. For succession CO_2 flooding and WAG of CO_2 flooding, the same amounts of CO_2 are controlled to inject in these two scenarios.

In Fig. 4, the scenario of WAG of CO_2 flooding has the highest oil recovery, while the scenario of water flooding has the lowest recovery. The oil recovery of WAG of CO_2 flooding is higher than that of the succession CO_2 flooding. One reason is that the same CO_2 amounts are injected in these two scenarios, and both of CO_2 and water can contribute to oil recover in WAG of CO_2 flooding, but there only CO_2 contribute to oil recover in succession CO_2 flooding. The other reason is that after gas breakthrough, the WAG of CO_2 flooding can control the produced gas-oil ratio better than that of succession CO_2 flooding. Considering obtaining the same recovery, it requires fewer CO_2 amount of the WAG of CO_2 flooding than that of succession CO_2 flooding, which reduces the cost of gas flooding. Furthermore, the WAG of CO_2 flooding is better to maintain formation pressure and to reduce the produced gas-oil ratio, so it is selected as the best development mode.

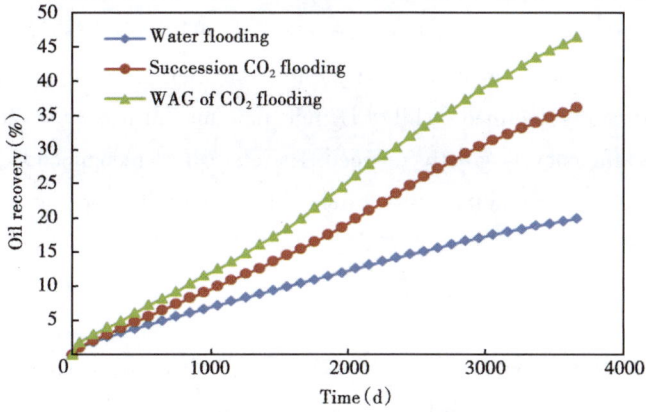

Fig. 4　Oil Recoveries of Different Development Modes

3. 3　Formation Pressure

The MMP of CO_2 for crude oil is a key factor for the WAG of CO_2 flooding. Because the value determines whether miscible flooding could be achieved. In this paper, the MMP of CO_2 for crude oil is 19. 8 MPa (Fig. 1). Three scenarios with average formation pressure of 17MPa, 20MPa and 22MPa, respectively, are shown in Fig. 5.

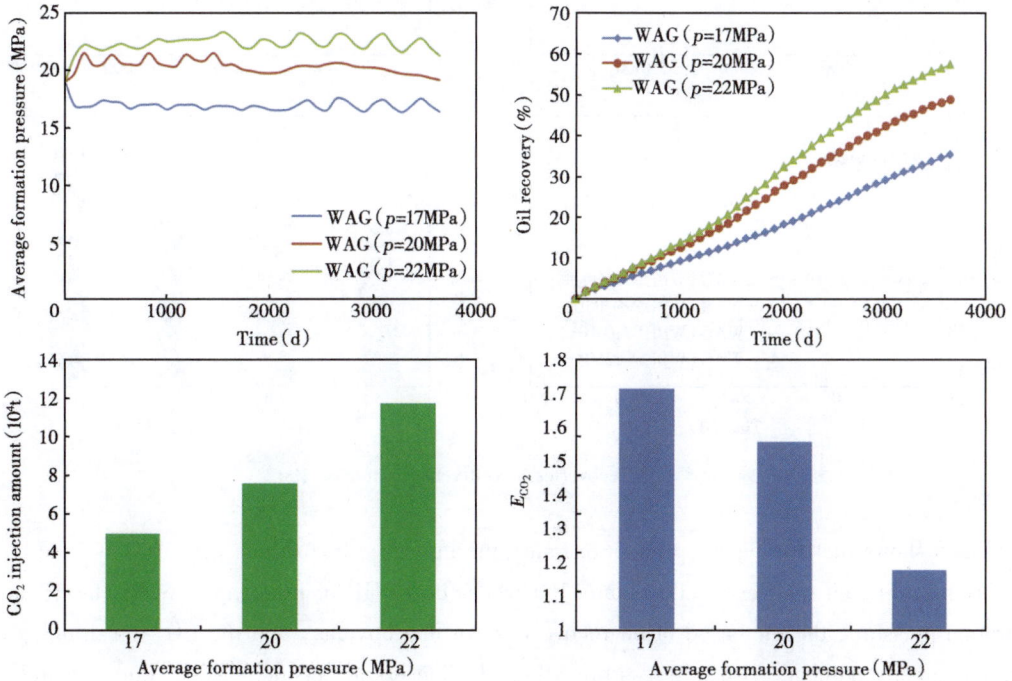

Fig. 5　Oil Recoveries of Different Average Formation Pressure

Fig. 5 shows that the value of CO_2-oil draining efficiency, between 1 and 2, is relatively large in these scenarios. The reason is that the WAG of CO_2 flooding is conducted at the initial development stage, and T_{oil} is the total oil production amount, which is contributed by water slugs and CO_2 gas slugs. The higher the average formation pressure is, the bigger the oil recovery could obtain. When the formation pressure is above MMP, if the formation pressure continues to increase, the improved oil recovery rate is declined, while CO_2 injection amount is greatly increased, as a result, the CO_2-oil draining efficiency is greatly reduced. Therefore it is recommended that when WAG of CO_2 flooding is applied, the formation pressure should be controlled on the above of MMP 1 to 2 MPa.

3. 4　Production Rate

As the production of horizontal well with SRV is controlled by flowing bottom-hole pressure (FBHP), different production rate can be obtained by adjusting the FBHP. In order to study the impact, 4 scenarios with different FBHP are designed, FBHP of which are 8 MPa, 9. 3 MPa, 10 MPa and 12 MPa, respectively.

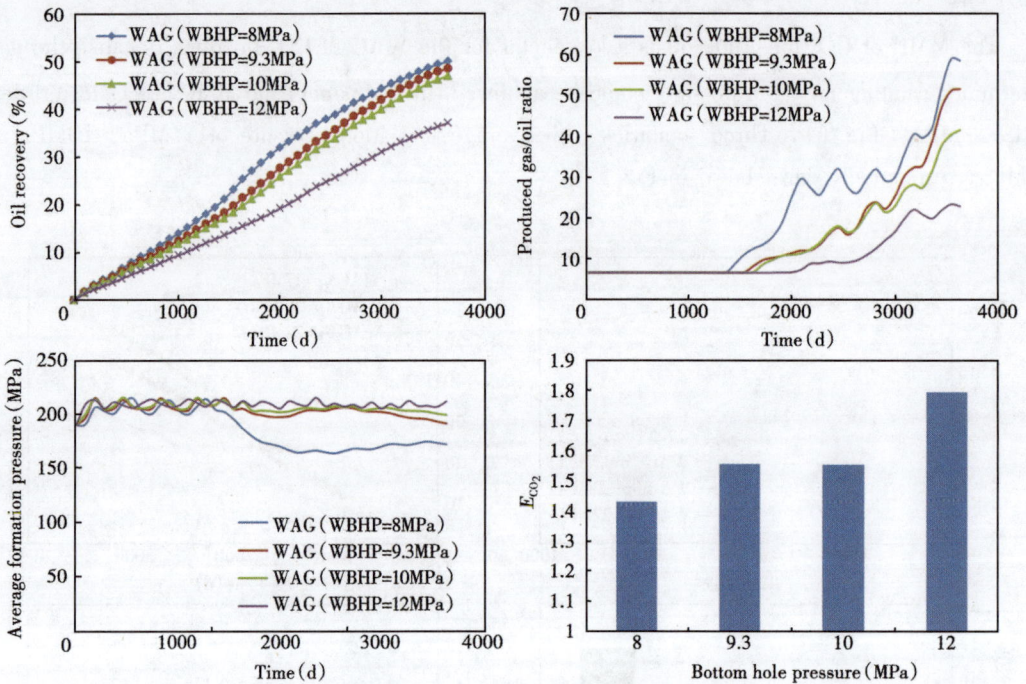

Fig. 6　Oil Recoveries of Different Production Rate

Fig. 6 shows that the higher production rate (the lower the bottomhole pressure) is, the higher the corresponding oil recovery could obtain. But when the FBHP of production well is less than the saturation pressure, the increased of oil recovery is not that obvious, and the CO_2-oil draining efficiency is greatly reduced. Also, the gas breakthrough time would become earlier. Furthermore, it is difficult to maintain the formation pressure around the MMP in the later stage. Therefore the FBHP should not be lower than the saturation, in other words, the production rate should not be too large. Through the comprehensive analysis of the above, it is preferable to control bottomhole pressure at 9. 3MPa, which is little above saturation pressure (9. 25MPa). When the displacement front reaches the SRV of the horizontal well, the CO_2 starts to breakthrough, then it should be appropriate to increase the water slug to decrease produced gas-oil ratio and maintain formation pressure.

3.5 Shut-in Gas-oil Ratio

Shut-in gas-oil ratio is also an important factor in the gas flooding. In order to evaluate its effect on the WAG of CO_2 flooding, 4 scenarios with different shut-in gas-oil ratio are designed, whose values are $200m^3/m^3$, $300m^3/m^3$, $400m^3/m^3$ and $500m^3/m^3$, respectively.

Fig. 7 shows that the oil recovery smoothly increases with the increasing of shut-in gas-oil ratio, while the CO_2-oil draining efficiency is slightly reduced. Considering these two factors, the reasonable production gas-oil ratio should be at 400 to 500 m^3/m^3 in these scenario. The shut-in gas-oil ratio, however, is not as strong sensitivity to WAG of CO_2 flooding as to succession CO_2 flooding. So in order to pursue higher oil recover, the shut-in gas-oil ratio can be properly increased.

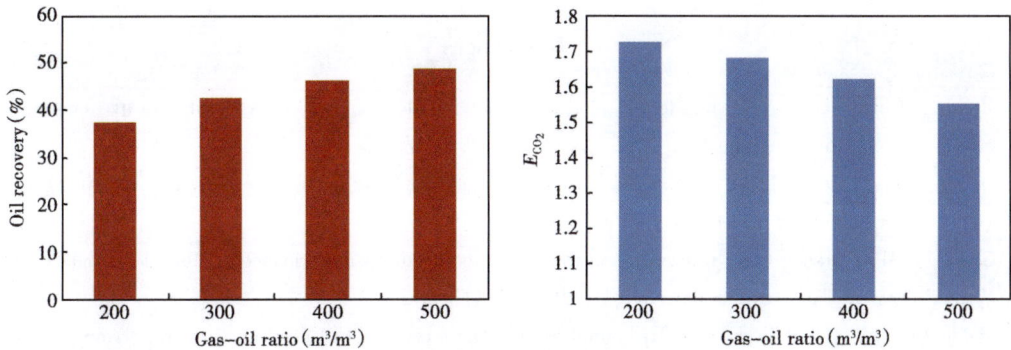

Fig. 7   Oil Recoveries of Different Shut-in Gas-oil Ratio

3.6 CO_2 Injection Amount

Based on these following 10 WAG of CO_2 flooding scenarios, the relationship between oil recovery of ten years and CO_2 injection amount is obtained. Fig. 8 shows that CO_2 injection volume is proportional to the oil recovery, the larger the CO_2 injection amount is, the higher the oil recovery could obtain.

Ensuring adequate CO_2 gas injection amount is the key to improve oil recovery. Fig. 8 shows that oil recovery increases with the increasing of CO_2 injection amount. But when the CO_2 injection amount is larger than 4×10^4t, the increase of oil recovery is slow. Therefore, the reasonable total injection amount of $(4 \sim 4.5)$ $\times10^4$t CO_2 is recommended for this five-spot pattern well group.

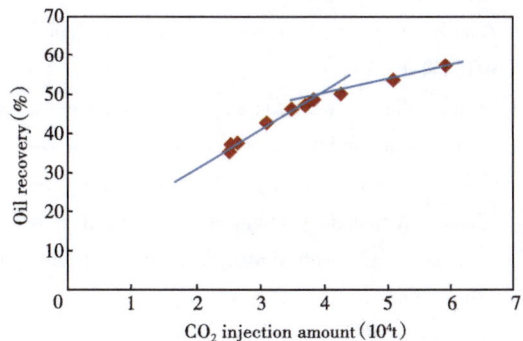

Fig. 8 CO_2 Injection Amount vs. Oil Recovery Factor

4　Conclusions

The permeability and flow capacity of the reservoir have been greatly improved after larger scale SRV measures, which makes it possible to inject an oil-displacing agent to develop of tight oil reservoir. The WAG of CO_2 flooding has better development effect than water flooding or succession CO_2 flooding. The WAG of CO_2 flooding can offset the shortcomings of low oil displacement efficiency of water flooding, and it can also improve the low sweep efficiency of succession CO_2 flooding. In order to slow down the speed of the CO_2 breakthrough and maintain the formation pressure, it should be appropriate to increase the water slug when the displacement front reaches the SRV of the horizontal wells. The MMP of CO_2 and crude oil is a key factor for the WAG of CO_2 flooding, it is better to maintain the formation pressure 1 to 2MPa high than MMP. CO_2 injection volume is proportional to the oil recovery, but there is an optimal value.

References

［1］ Brohi I, Pooladi-Darvish M, Aguilera R. Modeling Fractured Horizontal Wells As Dual Porosity Composite Reservoirs - Application To Tight Gas, Shale Gas And Tight Oil Cases ［C］. SPE 144057, 2011.

［2］ Du C, Zhang X, Zhan L, et al. Modeling Hydraulic Fracturing Induced Fracture Networks in Shale Gas Reservoirs as a Dual Porosity System ［C］. SPE 132180, 2010.

［3］ Ghaderi S M, Clarkson C R, Chen S, et al. Evaluation of Recovery Performance of Miscible Displacement and WAG Process in Tight Oil Formations ［C］. SPE 152084, 2012.

［4］ Saldungaray P M, Palisch T T. HydraulicFracture Optimization in Unconventional Reservoirs ［C］. SPE 151128, 2012.

［5］ Suarez M, Vega Velasquez L, Monti L J, et al. Modeling Vertical Multifractured Wells in Vaca Muerta Shale Oil Play, Argentina ［C］. SPE 164537, 2013.

［6］ Song C, Yang D T. Optimization of CO_2 Flooding Schemes for Unlocking Resources from Tight Oil Formations ［C］. SPE 162549, 2012.

［7］ Wu Q, Xu Y, Wang X, et al. Volume Fracturing Technology of Unconventional Reservoirs: Connotation, Design Optimization and Implementation ［J］. Petroleum Exploration and Development, 2012, 39 (3): 377-384.

［8］ Yu W, Sepehrnoori K. Optimization of Multiple Hydraulically Fractured Horizontal Wells in Unconventional Gas Reservoirs ［C］. SPE 164509, 2013.

［9］ Zhou W, Gupta S, Banerjee R, et al. Production Forecasting and Analysis for Unconventional Resources ［C］. IPTC17176, 2013.

［10］ Zou C, Zhu R, Wu S, et al. Types, Characteristics, Genesis and Prospects of Conventional and Unconventional Hydrocarbon Accumulations: Taking Tight Oil and Tight Gas in China as An Instance ［J］. ActaPetroleiSinica, 2012, 33 (2): 173-187.

［11］ Zhao G. A Simplified Engineering Model Integrated Stimulated Reservoir Volume (SRV) and Tight Formation Characterization with Multistage Fractured Horizontal Wells ［C］. SPE 162806, 2012: 1-18.

［12］ Zhao G. A Simplified Engineering Model Integrated Stimulated Reservoir Volume (SRV) and Tight Formation Characterization With Multistage Fractured Horizontal Wells ［C］. SPE 132180, 2012.

Coupled CO_2 Enhanced Oil Recovery and Sequestration in China's Demonstration Project: Case Study and Parameter Optimization

Su Kun[1]　Liao Xinwei[1]　Zhao Xiaoliang[1]　Zhang Hui[2]

(1. Petroleum Engineering Department, China University of Petroleum;

2. Exploration and Development Research Institute of Jilin Oilfield, PetroChina)

Abstract: Reservoir stimulation by carbon dioxide (CO_2) flooding can bring extra oil production because of its unique advantages of higher displacing efficiency and lower injection pressure compared to water flooding. Also, greenhouse gas control has been the focus of worldwide attention in recent decades, and CO_2 injection into reservoirs has been regarded as a favorable method of achieving sequestration. Therefore, coupled enhanced oil recovery (EOR) and sequestration becomes a cost-effective and environmentally safe method, which is feasible for developing countries, especially China. In this paper, China's first field-scale reservoir demonstration project is introduced to evaluate its performance with respect to both oil recovery and carbon sequestration. Also, given that injected CO_2 tends to immaturely break through toward production wells in reservoirs, parameters of recycled gas injection scenarios were screened and then optimized by experimentations as well as by a new optimizing method. Both of these approaches could contribute to the research and methodology required for scaling up programs in China.

Keywords: CCS; EOR; CO_2 Flooding; Numerical Simulation; Case Study

1　Introduction

Carbon sequestration in oil reservoirs is a proven safe and effective approach because of its natural advantages. [1] Given that formations can seal hydrocarbons for millions of years, the distribution of a rock cap above is also believed to prevent carbon leakage in the long-term future, which is critical for an ideal CO_2 sequestration site. Although aquifer formation is expected to be a larger potential reservoir for carbon sequestration, its less benefit than CO_2 enhanced oil recovery (CO_2 EOR) activities have impeded its development in present-day China. Therefore, in the oilfield sector, CO_2 EOR and sequestration will draw more attention for its economic advantages.

China is well-qualified in many respects to undertake CO_2 EOR and storage activities. (1) With respect to carbon sources, it is widely believed that carbon emissions in China will continue to rise, even though fossil fuel consumption is predicted to be reduced in the near future[2] (see Fig. 1). Also, many CO_2-rich gas reservoirs can provide a source of gas [see Fig. 2 (a)], which can significantly reduce the cost of gas and its transport. (2) With respect to project sites, China is among several countries in which oil reservoirs are distributed on a large scale, and thorough geological information has been obtained during the long periods of their exploitation. (3) With respect to switching opportunity, given that reservoirs by water flooding have seen high water cut,[3,4] CO_2 flooding technology has been regarded as a suitable method for sustainable development, just as it has been in the U. S. A. and many European countries.[5,6] All of these reasons contributed to the launching of a pilot program in China.

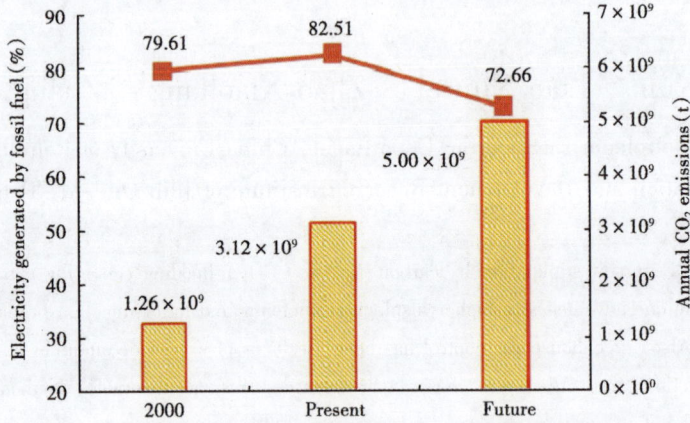

Fig. 1 Carbon Emission Trends in China

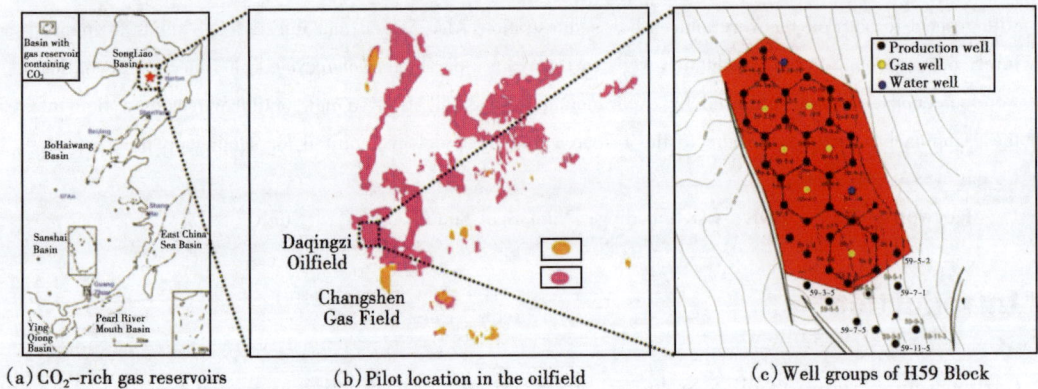

(a) CO_2-rich gas reservoirs   (b) Pilot location in the oilfield   (c) Well groups of H59 Block

Fig. 2 Location of the H59 Block and Well Groups

Before the wider application of switching to CO_2 flooding in China, the first field-scale pilot program, H59 Block, was launched. This was funded by PetroChina and supported by the government. This pilot program includes flooding/sequestration mechanism studies, surface engineering, and laboratory research. Developments recognized during this program will facilitate the scaling-up of deployment of similar projects in other target oilfields. In this paper, the production history of this block is first overviewed to provide field experience for similar projects. Then, parameters in the recycled gas scenario, including the gas/oil ratio (GOR) constraint and gas components for re-injection are chosen and optimized to maximize oil recovery and CO_2 sequestration (see Fig. 3).

210

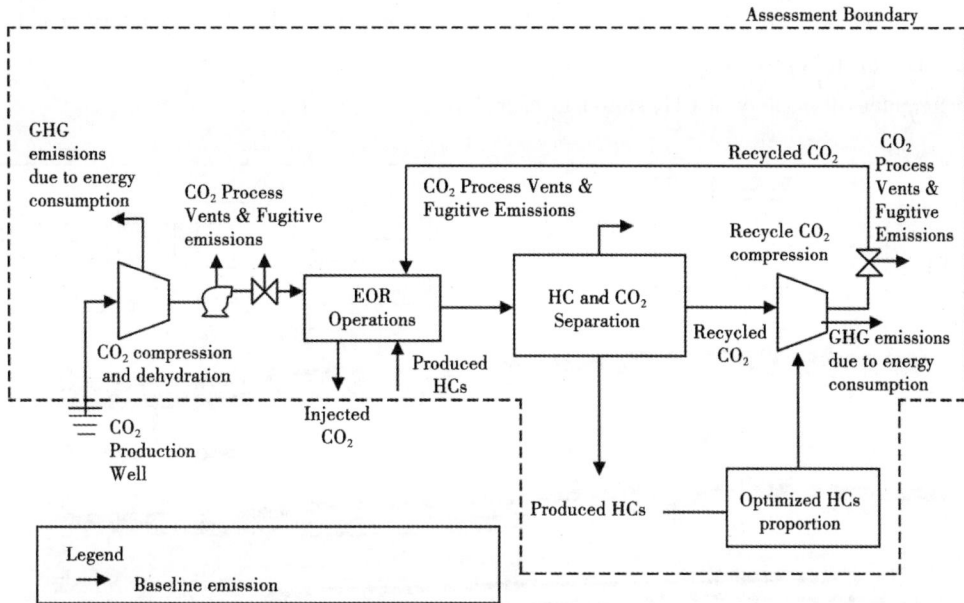

Fig. 3   Illustration of CO_2 EOR and Sequestration Operation[7]

2   Project Overview

2. 1   Geological Configuration

The pilot field is located in the southern part of the Song Liao Basin, which is isolated by centripetal faults in its western and eastern parts, respectively [see Fig. 2 (b)]. As a stable delta deposit, the sandstone thickness of this area is stable, with little variation. The testing block is stratified into 40 layers of porous and dense zones that are not continuous. Among them, four main layers are chosen for oil production. Around these four layers, which on average are 2 m thick, mudstones are distributed above and below, guaranteeing the safety of simultaneous CO_2 sequestration. Inside of them, isolated interbeds are developed so that the reservoir properties are poor in porosity by 13% and permeability by 3 mD. These characteristics lead to much higher water injection pressures being required with associated poor benefits, which will be discussed later.

The H59 Block [see Fig. 3 (a)] covers 3. 1 km2, and the oil–bearing layers are 2267 to 2490 m deep and 2 to 4 m thick. The reservoir oil is 47° in American Petroleum Institute (API) gravity and 3. 97 cP in viscosity under reservoir conditions. The dissolved GOR is 36 m^3/m^3 (under standard conditions), and the bubbling pressure of the crude oil is 7. 31 MPa.

Most of the 40 layers coexist as oil and water; thus, their initial oil saturation is lower than 50% (45% on average), which is classified as relatively low. Poor layers with relatively lower oil saturation in the mare frequently present. To improve the efficiency of injected solvents, these layers are beyond the perforation range, especially for CO_2 injection wells. Among all layers vertically, the four main layers (indexed as 7, 12, 14, and 15) chosen in the target developing are a contribute to 78% of the original oil in place (OOIP) in this area [see Fig. 4 (b)]. As interpreted by well logging, the cap rocks above are stable-distributed mudstone averaging 1.5 m thick (see Fig. 5), which guarantees the safety of CO_2 stored in each layer.

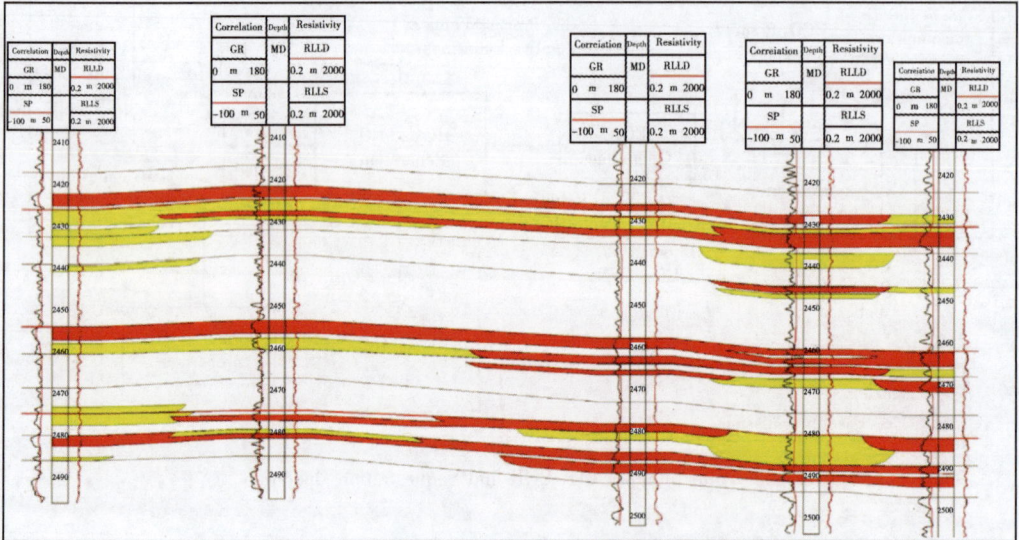

(a) Illustration of a vertical well section

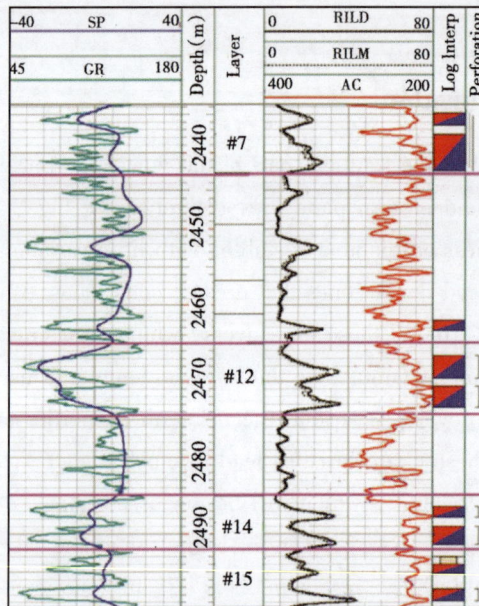

(b) Well log information of the target layers

Fig. 4　Strata Information of the H59 Block

The interbeds inside each layer commonly exist as discontinuous fine sands with a thickness of less than 1 m. Also, areal coefficients of variation (V_E) for permeability are defined as

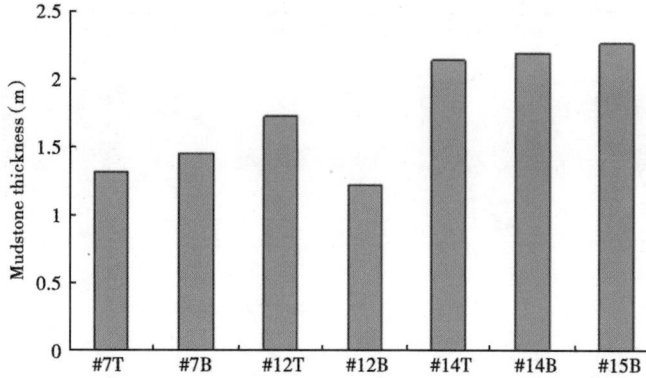

Fig. 5   Thickness of Mudstone Distributed near the Main Layers (T, top; B, bottom)

$$
\begin{cases}
V_E = S / \overline{K} \\[2mm]
s = \sqrt{\dfrac{\sum_{i=1}^{n}(K_i - \overline{K})^2}{(n-1)}} \\[4mm]
\overline{K} = \dfrac{\sum_{i=1}^{n}(h_i K_i)}{\sum_{i=1}^{n} h_i}
\end{cases}
\tag{1}
$$

where S is the permeability standard deviation of the core sample pool , K is the average value of core permeability, K_i is the permeability of the core sample, and hi is the net thickness. For the H59 Block, the V_E values for the four key layers are 0. 49 (7), 0. 55 (12), 0. 43 (14), and 0. 56 (15), which represent medium uniformity for each layer.

2. 2   Development Situation During Pure CO_2 Flooding

From the end of 2003 until October 2007, 44 wells, including 34 for production and 10 for water injection, were drilled. Among them, 37 wells (29 for production and 8 for injection) were drilled in the pattern of an inverted seven point (well spacing, 440 m; well array, 140m), which was believed to retard water channeling along the developed natural micro fractures. During the first scenario of water flooding, 40 m^3 was allocated as the daily water injection rate for each well and the monthly injection/production ratio in cumulative volume reached 0. 6. As with most reservoirs in China, water injection in this area has seen lower than expected oil increments. It brought 5. 8 tons of oil (of 9. 5 tons of liquid) production on average per well initially, which subsequently reduced to an unacceptable 2 tons. During that period, total oil production of the block was 3. 87% of OOIP, which was lower than expected (see Fig. 7). Another unfavorable characteristic was the exceptionally high water cut at 40% from the beginning. In fact, the above reasons made it necessary for development mode updating.

Because of the limitations of water flooding in this block and the advantages of an accessible CO_2 source from a nearby gas reservoir, as witch to CO_2 flooding was made in this block as an updated scenario in 6 well groups in the northern part of the block from May 2008. The CO_2 resource is

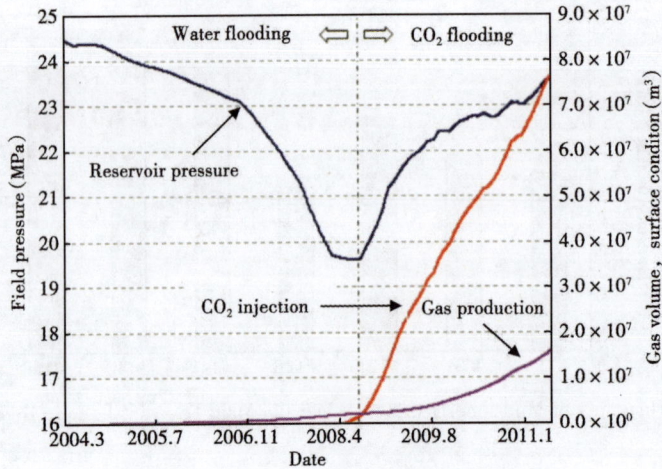

Fig. 6   Dynamic Character of the H59 Block

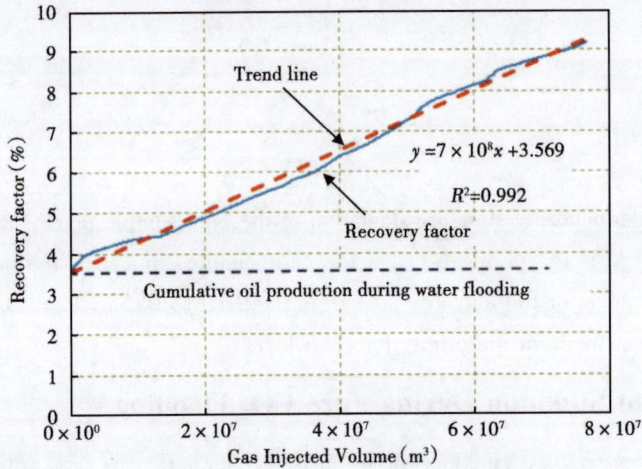

Fig. 7   Oil Production verses Gas Injection

transported by pipeline to the target area from a nearby CO_2-rich gas reservoir. This unique advantage has made the H59 Block an ideal target for pilot CO_2 flooding.

As shown in Fig. 3 (c), an inverted seven-spot pattern was used for CO_2 flooding to minimize early gas breakthrough. The testing are a contains 6 complete gas-injection well groups as well as 2 in complete water-injection groups near the boundary. The gas injection rate was set at an average of 35 tons per day for each well. As shown in soon to the initial reservoir pressure, which indicates that the injection of CO_2 was much easier than for water in this inferior reservoir formation, and miscibility was readily achieved in that the minimum miscible pressure (MMP) is 22.8 MPa as observed by experimentation. Until 2010, daily oil production rose from 2 to 5 tons and the average water cut dropped from 45% to 25%, which suggests a greater feasibility of CO_2 as a solvent than water, and cumulative oil production reached 9.94% of OOIP cumulatively after a total of 0.19 hydrocarbon component pore volume (HCPV) of gross CO_2 was injected.

214

Until June 2011, nearly 80% of the injected CO_2 has been provisionally sequestrated in nearly 3 years compared to 60% in CO_2 flooding fields in the U. S. A. [8] This phenomenon suggests that, before early breakthrough of gas, both enhanced oil recovery and carbon storage are fulfilled by most of the injected solvent being trapped in the pore space of the reservoir. [9,10] Apparently, total oil production is linearly related to the injected CO_2 volume (see Fig. 6 and Fig. 7), and cumulatively, 123. 1 thousand tons of CO_2 has been sequestrated. Analyzed from a production history point of view, ideal equivalence between EOR and environmental protection has now been achieved.

Therefore, as the first nationally funded CO_2 EOR and sequestration reservoir in China, the H59 Block in the Jilin Oil field has performed well in the past 4 years. Confidence has been gained by the company, and on the basis of developing experience, larger projects will be scheduled near this region. Until 2015, annual CO_2 sequestration can be expected to be 1200 thousand tons, with 1 million tons of incremental oil in the neighboring area, which is encouraging from both CO_2 EOR and sequestration perspectives.

3　Parameter Optimization During Recycled Gas Flooding

Two different basic considerations for a CO_2 EOR field are the desire to improve oil recovery[11,12] or to merely achieve as much carbon retention as possible in the geological structures. There is equilibrium between these two aspects for coupled CO_2 sequestration and EOR, which has recently been focused on by researchers. Ghomian et al. [13] investigated various parameters based on the construction of mathematical response model sand concluded that produced GOR constraints, well spacing, production and injection well types, and injection plan as well as key reservoir characters were among the principal parameters requiring optimization. On the basis of the previous discussion on well types and reservoir characters for the H59 Block, hydrocarbon proportions in the recycled gas and GOR constraints were screened for designing future scenarios.

3. 1　Recycled Gas Components

For recycled gas injection planned in this block, the composition of injected gas should be optimized in that it affects oil production by changing the MMP between crude oil and CO_2.

According to the mechanism of CO_2 flooding, miscible conditions can guarantee more oil production than immiscible conditions in the target reservoir. [14,15] Thus, the nearer the conditions to complete miscibility, the more oil production can be expected. Given that CO_2 is known to be mainly stored in the reservoir by occupying the place of hydrocarbons, higher oil recovery will increase the carbon sequestration amount. In this way, when reservoir conditions (including pressure and temperature) are stable, miscibility between oil and CO_2 will facilitate both oil recovery and carbon sequestration. [16]

Meanwhile, MMP greatly depends upon components of the oil and injected gas. Therefore, during the recycled gas flooding scenario, gas for re-injection should be purified to achieve miscibility in the reservoir. The MMP can be obtained mainly from empirical correlation equations[17] and exper-

215

imentations. Given the character differences between oil samples[18] and poor accuracy results using empirical equations, the MMP resulting from different composition percentages of the produced gas with crude oil was observed by slim tube experimentations (see Fig. 8).

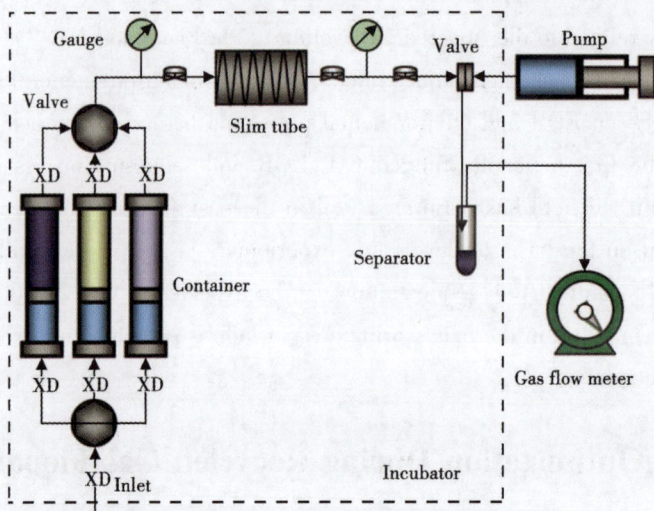

Fig. 8   Diagram of the Slim Tube Apparatus

For the hydrocarbon mixture in the diluted gas used in the experimentations, the proportions of the hydrocarbon group components are in accordance with the analysis results of production gas samples (see Table 1).

Table 1   Normalized Components in Produced Gas after Breaking through

Components	Methane	Ethane	Propane	n-Butane	n-Pentane	Nitrogen
Proportion (%)	31.695	12.654	9.828	3.317	0.86	39.435

Eight different proportions of hydrocarbon components, namely, 0, 5%, 10%, 15%, 20%, 30%, 40%, and 50% were investigated in the experimentations. Using the 5% proportion as one example, before the experimentation began, crude oil from the H59 Block was stored in the first container. In the second container, injected gas was prepared containing 95% CO_2 and 5% hydrocarbon components. First, crude oil was pumped into the slim tube until saturated conditions were reached, and 1.2 pore volumes (PV) of injected gas were injected under pressures of 13, 17, 21, 24, and 29 MPa. During the process, the amount of oil production was measured after every 0.1 PV gas was injected [see Fig. 9 (a)]. Then, the maximum recovery factor under different displacing pressures can be obtained, and MMP was determined as 23 MPa from the knee point of the curve [see Fig. 9 (b)].

Similarly, MMP experimentations using other hydrocarbon proportions were investigated using the above routine. As shown in Fig. 10, the MMP increases with an increase in the mole fraction of hydrocarbon (MPH) in the injected gas. In detail, when MPH is greater than 10%, the MMP between the injected gas and crude oil increases dramatically. This suggests that miscibility will be

216

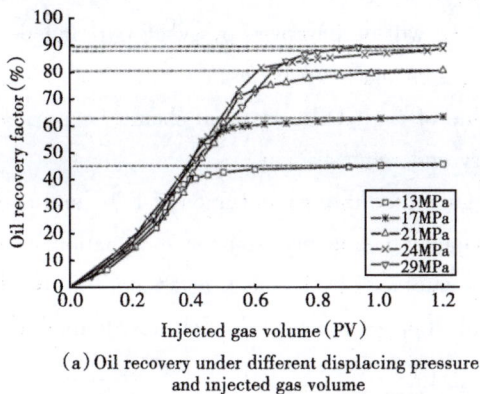

(a) Oil recovery under different displacing pressure
and injected gas volume

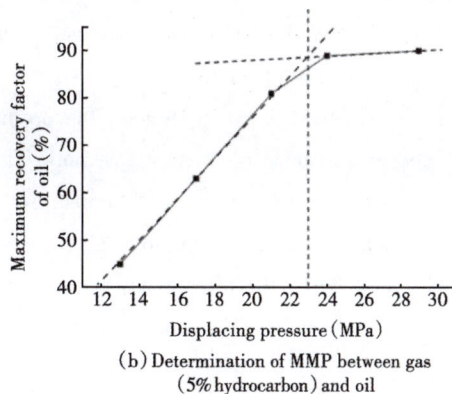

(b) Determination of MMP between gas
(5% hydrocarbon) and oil

Fig. 9   Oil Production under Different Displacing Pressures and MMP Determination

more difficult to achieve. Especially, when the hydrocarbon component fraction is greater than 30%, miscible conditions cannot be obtained because the MMP needed has already exceeded the permitted pressure (rock fracture pressure) for sequestration safety. If estimated by achieving miscibility under the current pressure, the optimum value of MPH is 7%. A higher percentage of hydrocarbons will impede the development of miscible flooding, and a lower percentage may result in more expenditure on purification, making it less economically competitive.

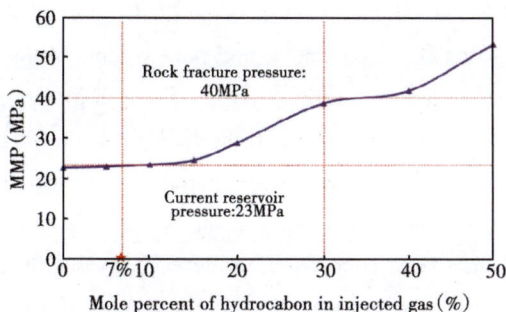

Fig. 10   Relationship between MMP
and MPH in the recycled gas

3. 2   Produced GOR Constraints

The GOR is among the principle parameters in reservoir engineering, especially for CO_2 flooding areas. For unfavorable mobility ratios between oil and CO_2, gas would definitely move faster toward production wells and limit the overall invaded area, thus reducing the flooding efficiency of the injected solvent. For continental sediments in which interbeds are developed, heterogeneities could lead to faster gas breaking toward production wells. In this block until June 2011, the average block GOR was up to 250 (m^3/m^3) and CO_2 was produced in high percentages (see Fig. 11). If a GOR constraint is absent in the scenario design, the gas production rate will increase monotonically, bringing extra pollution and burdening the

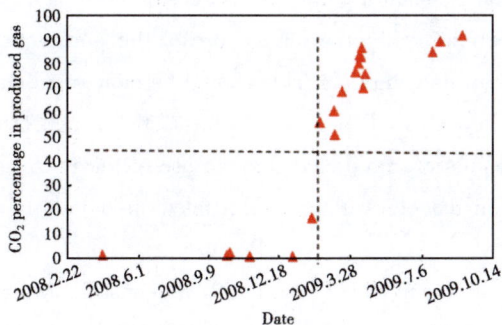

Fig. 11   Percentages of CO_2 in Gas Samples

field gas disposal facilities. Otherwise, if production wells are set to shut when exceeding a reasonable upper limit of the GOR, not only use of injected CO_2 will be improved by avoiding production along high permeability channels but also the invading area may expand as a result of the changed flow regime in the reservoir. Hence, the optimum value of the GOR constraint should be previously evaluated to maximize the benefits of both oil recovery and sequestration.

Screening of the optimal GOR constraint should guarantee the maximization of EOR and carbon sequestration amounts. Because of the absence of financial parameters, economic valuation is unavailable, and here, the alternative evaluation initially employed by Kovscek and Cakici[19] is adapted for GOR screening. Their paper introduced an evaluation parameter f, which was defined as

$$f = \omega_1 \frac{N_p^*}{OIP} + \omega_2 \frac{V_{CO_2}^R}{V^R} \qquad (2)$$

where ω_1 ($0 \leqslant \omega_1 \leqslant 1$) and $\omega_2 = 1 - \omega_1$ are weights, N_p^* is the cumulative oil production, OIP is the volume of oil in place at the start of the solvent injection scheme, and and are the volumes of CO_2 sequestration and pore volume of the reservoir, respectively. The value of the weighting terms ω_1 and ω_2 is previously set according to preferences between EOR and carbon sequestration, which together contribute to the ranking of the scenarios. Therefore, the value of f indicates the combined benefits of both oil production and carbon sequestration. If a larger weight is placed on ω_1, oil production will be more crucial during scenarios screening through the value of f and vice versa. Under each condition, an increasing value of f indicates more benefits weighted by oil production and carbon sequestration, which is meaningful for scenario screening. Here, equal emphasis ($\omega_1 = \omega_2 = 0.5$) is placed on both aspects.

Considering the pore volume variance as a pressure change during exploitation, and in Eq. 2 should be updated rather than being treated as a constant. Because the volumes of CO_2 injected and produced are observed from surface conditions, should be treated as its converse under reservoir conditions. Therefore, is updated as below

$$V^R = V_0^R e^{-C_p(p-p_0)} \qquad (3)$$

where is the pore volume under the initial reservoir pressure, C_p is the rock compressibility coefficient, and p is the updated reservoir pressure at each step. With adjustment by the combination of Eqs 2 and 3, such a parameter is considered to allow for more accurate evaluation of the use of the pore volume rather than its potentials. This should be more significant during the CO_2 EOR process. In this section, the exploitation results are predicted through use of the numerical simulation model, which is illustrated in detail in the Appendix.

As shown in Fig. 12, lower GOR constraints result in higher values of f. The value of f corresponding to a constraint of 800 GOR slowly ascends until the reservoir

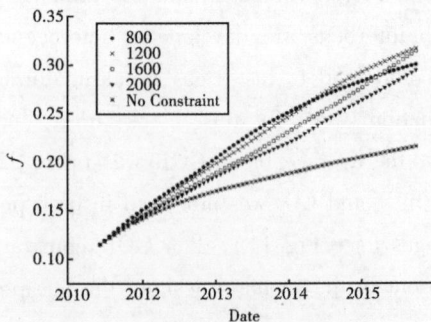

Fig. 12　Value of f under Different GOR Constraints

218

pressure reaches an upper limit. Hence, although the oil increment differs among cases, carbon sequestration places greater weight on integral evaluation. As shown in Fig. 12, from 2015, the 1200 GOR constraint is preferred for its highest value of f compared to other conditions.

In conclusion, in the recycled gas injection scenario, 6 wells will be scheduled for gas re-injection (MPH = 7%) at a rate of 35 tons per day, while the oil wells are set to produce under a constant bottom hole pressure of 8 MPa to avoid gas dissolution in the reservoir. The GOR constraint used will be 1200 (m^3/m^3). Then, the developing performance can be predicted through use of the simulation model, and comparisons can be made between CO_2 EOR and storage.

4   Evaluation of Results at Different Stages

When the volume of injected solvent is less than 0.3 HCPV (the middle section in Fig. 13), the growth rate of the recovery factor is obviously higher than for subsequent scenarios. As shown in Fig. 14, during the first 7 years of pure gas flooding, 58% of the total carbon dioxide sequestration is achieved during the pure CO_2 flooding stage, which is 75% of the injected CO_2.

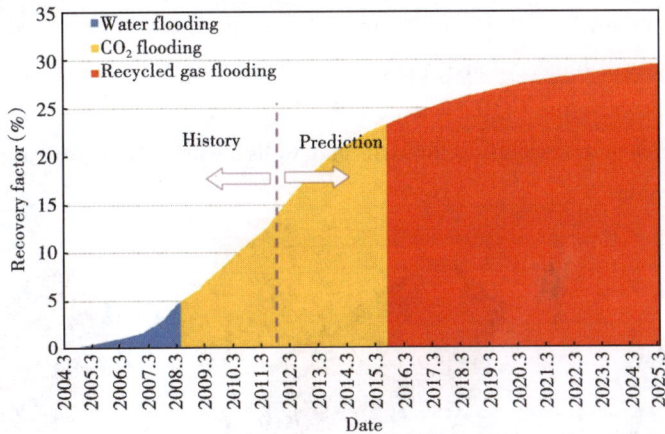

Fig. 13   Recovery Factor during Each Stage

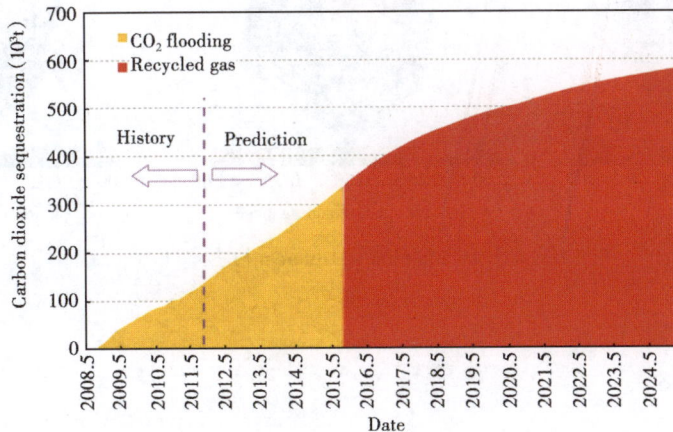

Fig. 14   Carbon Sequestration during Each Stage

219

During recycled gas flooding, the oil production rate becomes slowest among all periods (see Table 2). That is mainly because, after a long period of gas flooding, oil wells tend to shut because of overwhelming gas production, which sharply reduces oil production. In contrast, subsequent scenarios are expected to be inferior to the previous stage on the CO_2 EOR aspect. Therefore, in recycled gas development, it functions much better for carbon sequestration than for oil recovery.

Table 2 Performance Comparisons among Different Stages

Item	Recovery factor (%)	Gross CO_2 requirement per increment of oil ($10^3 ft^3$ bbl)	Annual oil production (%OOIP)	CO_2 sequestration (10^3t)
Water flooding	5	—	1. 23	—
CO_2 flooding	18. 4	6. 2	2. 63	340
Recycled gas flooding	6. 5	7. 8	0. 63	240

Fig. 15 shows the carbon dioxide distribution in one production layer at three different stages. From Fig. 15 (a), the conclusion can be made that, during the production history, the displacing area is finite and the invaded space is confined by well groups. After 5 years of continuous CO_2 flooding [see Fig. 15 (b)], the invading area of CO_2 is predicted to be obviously enlarged, even though it is still constrained by the group pattern. When it comes to the third stage [see Fig. 15 (c)], injected CO_2 will break through toward the oil wells and the invading area will expand toward the boundary. Because the GOR constraint is active, extra regions excluded by the well groups see carbon distribution by transcending through shut wells, which brings extra carbon sequestration as well as oil production.

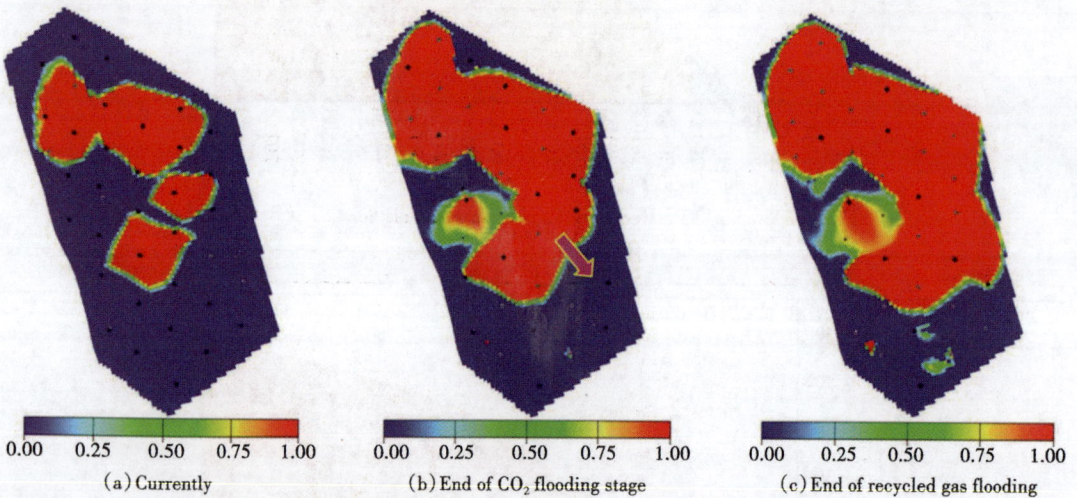

0.00 0.25   0.50 0.75   1.00   0.00 0.25   0.50 0.75   1.00   0.00 0.25 0.50   0.75   1.00

(a) Currently   (b) End of CO_2 flooding stage   (c) End of recycled gas flooding

Fig. 15 CO_2 Distribution at Different Stages (Layer 7)

5 Conclusions

(1) Carbon sequestration during CO_2 flooding is a practical method and of great potential in

China. For water-flooding oilfields characterized as high water cut and low permeability, the scenario switching to CO_2 flooding can be expected to be a sustainable method for oil production and simultaneous carbon sequestration, both of which would currently be beneficial for China.

(2) CO_2 EOR and sequestration saw encouraging benefits in the H59 Block during 3 years of demonstration activity. With respect to oil recovery, the daily production rate of oil increased to 5 tons with a decreasing water cut to 25% compared to 2 tons of oil production and 45% water cut during water flooding. In addition, the carbon sequestration percentage was nearly 80% during the 3 years. Scaling-up deployment could be launched in target reservoirs based on the pilot scheme to abate anthropogenic carbon emissions.

(3) During recycled gas injection, the gas component for re-injection and GOR constraints should be optimized. Through slim tube experimentations on different MPH, a trend of MMP values was obtained and optimized as 7%. An updated screening method for the GOR constraint showed a preferred value of 1200 (m^3/m^3). Under such parameters, recycled gas flooding is predicted to lower the oil production rate for well shutting. However, with respect to carbon sequestration, it still functions well and contributes to 42% of the total carbon dioxide sequestration.

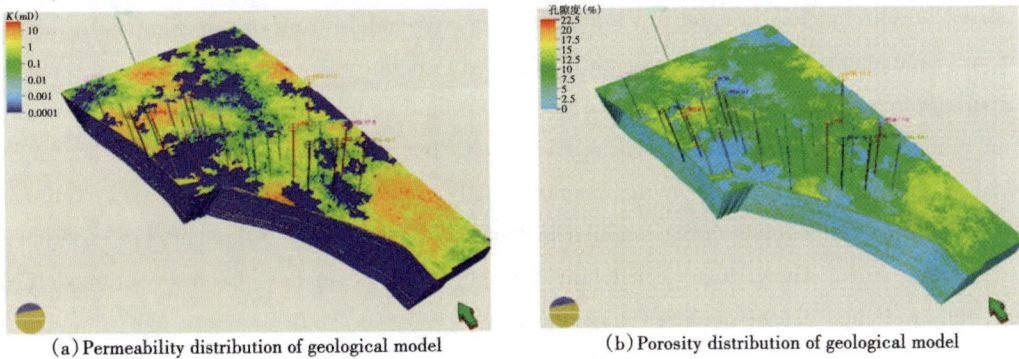

(a) Permeability distribution of geological model (b) Porosity distribution of geological model

Fig. 16   Properties of the H59 Block Model

Table 3   Fluid Parameters after Regression

Components	MW (g/mol)	Ω_a	Ω_b	T_c (K)	p_c (bar)	V_c [m^3/ (kg $\cdot$ mol)]	Z_c	AF
CO_2	44.01	0.46	0.08	304.7	73.87	0.09	0.274	0.257
CH_4+N_2	17.82	0.73	0.13	190.45	44.25	0.1	0.271	0.019
C_2H_6	30.07	0.56	0.05	404.57	48.84	0.15	0.215	0.113
C_{3+}	44.1	0.63	0.05	555.38	42.46	0.2	0.184	0.174
C_{4+}	58.12	0.5	0.08	419.04	40.14	0.26	0.297	0.223
C_{5+}	77.01	0.5	0.08	483.74	34.59	0.33	0.281	0.305
C_{7+}	114.15	0.76	0.1	526.1	24.52	0.46	0.258	0.472
C_{11+}	237.14	0.65	0.09	610.11	21.35	0.87	0.365	0.707
C_{29+}	532.76	0.66	0.09	745.85	8.98	1.93	0.28	0.72

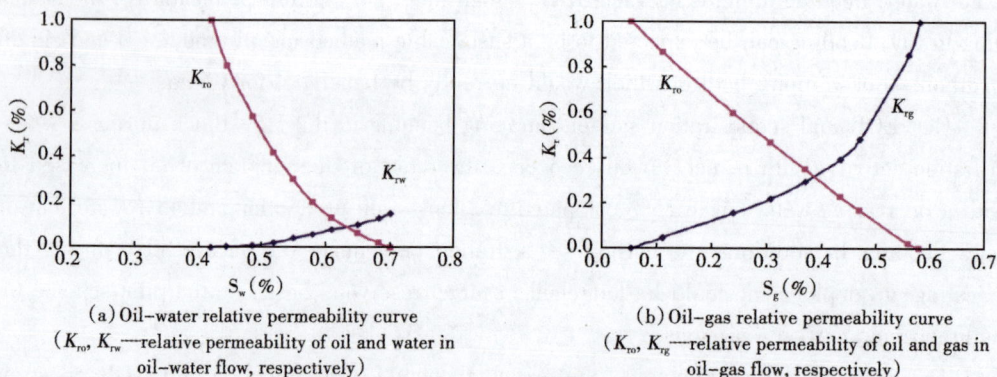

(a) Oil-water relative permeability curve
(K_{ro}, K_{rw}—relative permeability of oil and water in oil-water flow, respectively)

(b) Oil-gas relative permeability curve
(K_{ro}, K_{rg}—relative permeability of oil and gas in oil-gas flow, respectively)

Fig. 17　Relative Permeability Curves by Core Experimentations

Appendix

The reservoir model is built based on integrated information during exploration, including well log, seismic analysis, laboratory experimentations as core analysis, etc. The up scaling model is dimensioned as $57 \times 113 \times 14$, and two geological property models are shown in Fig. 16. In the model, 4 to 8 components (pseudo-components) are used to define the reservoir fluid. Each component has a string of parameters that must be specified in the fluid characterization process. These parameters for pure components are usually considered fixed, but for pseudo-components, they are obtained by regression with experimentation results. Finally, regression results are shown in Table 3. In addition, relative permeability curves applied in the model were obtained through core experimentations (see Fig. 17). Then, history matching was performed to update the reservoir model, and hence, the prediction can be achieved via software.

Nomenclature

MW = molecular weight;

Ω_a and Ω_b = non-dimensional constants in the equation of state;

T_c = critical temperature, K;

p_c = critical pressure, bar;

V_c = critical volume, $m^3/(kg \cdot mol)$;

Z_c = critical Z factor AF = a centric factor;

$HCPV$ = hydrocarbon component pore volume, rm^3;

$OOIP$ = original oil in place, rm^3.

References

[1] Grigg R B. Long-Term CO_2 Storage Using Petroleum Industry Experience; New Mexico Petroleum Recovery Research Center: Socorro, NM, 2002.

[2] Carbon Monitoring for Action (CARMA). 5 Highest CO_2 Emitting Power Sectors by Country; http: //carma. org/region (accessed Oct 19, 2012).

［3］ Hu W R. Necessity and feasibility of PetroChina Mature Field Redevelopment ［J］. Pet. Explor. Dev. , 2008, 35 (1): 1-5.

［4］ Hook M, Xu T. Development Journey and Outlook of Chinese Giant Oilfields ［J］. Pet. Explor. Dev. , 2010, 37 (2): 237-249.

［5］ Babadagli T. Mature Field Developed—A Review ［C］. Proceedings of the SPE Europec /EAGE Annual Conference, 2005: 13-16.

［6］ Davis D W. Project Design of a CO_2 Miscible Llood in a Water Flooded Sandstone Reservoir ［C］. Proceedings of the SPE/DOE Improved Oil Recovery Symposium, 1994: 17-20.

［7］ International Petroleum Industry Environmental Conservation Association and American Petroleum Institute. Part II: Carbon Capture and Geological Storage Emission Reduction Family; June 2007; http: //www. ipieca. org/ system/files/publications/CCS-FINAL_ merged. pdf (accessed Oct 19, 2012).

［8］ Hadlow R E. Update of Industry Experience with CO_2 Injection ［C］. Proceedings of the SPE Annual Technical Conference and Exhibition, 1992.

［9］ Claridge E L. Prediction of Recovery in Unstable Miscible Flooding ［J］. SPE J, 1972, 12 (2): 143-155.

［10］ Gorecki C D, Hamling J A. Integrating CO_2 EOR and CO_2 Storage in the Bell Creek Oil Field ［C］. Proceedings of the Carbon Management Technology Conference, 2012: 7-9.

［11］ Smith R L. SACROC Initiates Landmark CO_2 Injection Project ［J］. Pet. Eng, 1971, 43 (13): 43-47.

［12］ Graue D J, Blevins T R. SACROC Tertiary CO_2 Pilot Project ［C］. Proceedings of the SPE Symposium on Improved Methods of Oil Recovery, 1978.

［13］ Ghomian Y, Sepehrnoori K, Pope G A. Efficient Investigation of Uncertainties in Flood Design Parameters for Coupled CO_2 Sequestration and Enhanced Oil Recovery ［C］. Proceedings of the SPE International Conference on CO_2 Capture, Storage, and Utilization, 2010.

［14］ Stalkup F I, Miscible Displacement. Society of Petroleum Engineers (SPE) , 1984.

［15］ Jarrell P M, Practical Aspect of CO_2 Flooding ［J］. Society of Petroleum Engineers (SPE) , 2002.

［16］ Malik Q M, Islam M R. CO_2 Injection in the Weyburn Field of Canada: Optimization of Enhanced Oil Recovery and Greenhouse Gas Storage with Horizontal Wells ［C］. Proceedings of the SPE/DOE Improved Oil Recovery Symposium, 2000.

［17］ Robl F W, Emanuel A S, Van Meter O E Jr. The 1984 National Petroleum Council Estimate of Potential EOR for Miscible Processes ［J］. J. Pet. Technol, 1986, 38 (8): 875-882.

［18］ Ping G, Miao L. A Study on the Miscible Conditions of CO_2 Injection in Low-permeability Sandstone Reservoirs ［J］. Oil Gas Geol. , 2007, 28 (5): 687-692.

［19］ Kovscek A R, Cakici M D. Geologic Storage of Carbon Dioxide and Enhanced Oil Recovery. II . Co-optimization of Storage and Recovery ［J］. Energy Convers. Manage, 2005, 46 (11-12): 1941-1956.

基于多指标正交试验设计的 CO_2 非混相驱注气参数优化

李蒙蒙　廖新维　王万福　陈昌照　王　欢

（中国石油大学（北京）石油工程学院）

摘　要： 根据吐哈油田牛圈湖油藏储层和流体特征，应用油藏数值模拟和多指标正交试验相结合的方法对影响生产动态指标的主控因素进行了分析。结果表明，关井气油比是 CO_2 连续注气非混相驱替的主要影响因素，水气交替段塞比、注气周期、生产井井底流压是 CO_2 水气交替（WAG）非混相驱替的主要影响因素。应用优化的注气参数进行了水驱、CO_2 驱与 CO_2 WAG 驱开发方案设计，通过采出程度、换油率、气油比、CO_2 埋存系数、CO_2 滞留率和平均地层压力等指标的综合对比，确定 CO_2 WAG 非混相驱为最优开发方案，并进行了开发方案指标预测。研究结果对吐哈油田提高采收率技术具有一定指导意义。

关键词： CO_2非混相驱；影响因素；多指标正交试验；数值模拟

0　引言

CO_2 驱油作为一项日趋成熟的采油技术已受到世界各国的广泛关注。CO_2 驱从机理上可以分为混相驱和非混相驱。我国油藏中原油的突出特点是黏度高、蜡和胶质含量高、凝固点高。这些特点决定了我国多数油藏中的原油与 CO_2 的最小混合压力过高，我国多数油田不适合于 CO_2 混相驱。Nezhad 等[1] 在 2006 年对 CO_2 WAG 非混相驱进行了室内实验研究，研究发现 CO_2 WAG 非混相驱在二次采油阶段的驱油效果比在三次采油阶段的应用效果要好。Chen 等[2] 在 2009 年建立了考虑净现值以及其不确定性的目标函数，应用改进的遗传算法对 CO_2 驱注气井的注入量和生产井的井底流压进行了优化，并以此作为油藏数值模拟的控制条件，进行 CO_2 驱优化研究。尚庆华等[3] 在 2010 年基于油藏数值模拟和正交试验设计方法，考虑不同油藏和流体物性对油井产能的影响，建立了 CO_2 驱油井产能方程。马超群等[4] 在 2012 年建立了油藏机理模型，应用正交实验设计分析方法，确定了最优气水交替注气方案。参考前人的研究成果，本文采用油藏数值模拟与多指标正交实验设计相结合的方法，以吐哈油田牛圈湖油藏为例，建立了牛圈湖油藏组分数值模型，对 CO_2 非混相驱可行性进行了相关研究。

1　试验区块油藏地质特征

吐哈油田牛圈湖油藏构造特征整体为近东西向展布的宽缓背斜，南北两侧受背冲逆断裂夹持。储层构造简单，断层规模小，天然裂缝不发育。试验区块位于牛圈湖东区（图1），为 4 个菱形反九点注采井网，排距 150m，井距 450m，井排方向与压裂裂缝方向平行。试验区块共有油井 22 口，水井 6 口，其中 H43-8 和 H45-10 为油井转注水井。油藏埋深 1500 ~ 1850m，地质储量 193×10^4t。油层有效厚度为 11.3m，渗透率为 3.4mD，孔隙度为 13.7%。地面原油密度为 0.87g/cm³，黏度为 22.3mPa·s。油藏原始地层压力 17.8MPa，泡点压力为 6.9MPa，油藏温度为 45℃。依据细管实验数据，得到牛圈湖区块地层原油 CO_2 驱的最小混相压力为 29.8MPa，因此可以判定在地层压力 17.8MPa 下注 CO_2 为非混相驱。

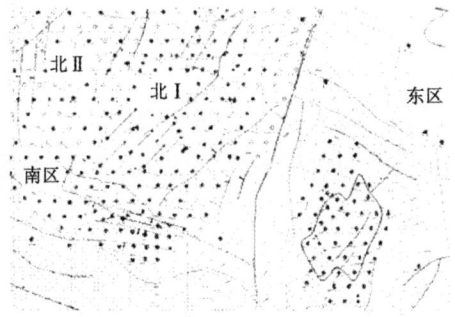

图 1　牛圈湖试验区块构造井位图

　　根据牛圈湖油藏储层及流体特征，收集油藏地质模型数据、相渗数据、原油高压物性数据和油藏生产动态数据，应用 Eclipse 数模软件，建立实验区块实际油藏的组分数值模型[5]。试验区块的基本网格步长 30m×30m，纵向上根据开发层系粗化为 8 个小层，模型网格采用角点网格系统。在平面 I 方向划分为 49 个单元，J 方向划分为 59 个单元，Z 方向划分为 8 个单元，总共网格数为 49×59×8＝23128 个。

2　牛圈湖试验区块 CO_2 非混相驱注气参数优化

　　由于单因素分析不能得到各因素对试验指标的影响程度以及最优的 CO_2 非混相驱注气参数，而正交试验设计方法，可以使试验数据点均匀分布，用较少的试验次数获得最优的试验结果，因此本文应用正交实验设计对注气参数进行了优化，首先进行 CO_2 水气交替（WAG）非混相驱注气参数优化，CO_2 WAG 非混相驱的注气影响因素较多[6]，本文选取注气时间、注气速度、关井气油比、井底流压、水气交替段塞比和注气周期 6 个因素进行正交试验方案设计[7]，对每个因素选取 4 个水平（表 1），按 L32（4^9）进行正交试验方案设计，注气试验方案及试验结果见表 2。

表 1　CO_2 WAG 非混相驱注气试验因素和水平

因素单位	注气时间（a）	注气速度（m³/d）	关井气油比	井底流压（MPa）	段塞比	注气周期水平
1	10	20000	500	13	1:1	2
2	15	15000	1000	11	1:2	4
3	20	10000	1500	9	2:1	6
4	25	5000	2000	7	1:3	8

表 2　CO_2 WAG 非混相驱注气试验方案及结果

试验号	因素						评价指标			
	注气时间（a）	注气速度（m³/d）	关井气油比	井底流压（MPa）	注气周期	段塞比	地层压力（MPa）	采出程度（%）	CO_2 滞留率（t/t）	换油率（t/t）
1	10	20000	500	13	2	1:1	32.9	8.3	0.77	38.8
2	10	15000	1000	11	4	1:2	26.7	14.3	0.51	53.4
3	10	10000	1500	9	6	2:1	21.7	10.6	0.52	128.5

试验号	因素						评价指标			
	注气时间 （a）	注气速度 （m³/d）	关井气 油比	井底流压 （MPa）	注气周期	段塞比	地层压力 （MPa）	采出程度 （%）	CO$_2$滞留率 （t/t）	换油率 （t/t）
4	10	5000	2000	7	8	1:3	12.1	12.8	0.21	98.8
5	15	20000	500	11	6	1:2	36.6	7.2	0.91	25.0
6	15	15000	1000	13	8	1:1	31.2	12.4	0.57	52.7
7	15	10000	1500	7	2	1:3	14.0	20.7	0.26	73.4
8	15	5000	2000	9	4	2:1	16.9	12.8	0.38	183.0
9	20	20000	1000	9	2	1:3	37.9	23.9	0.58	41.0
10	20	15000	500	7	4	2:1	28.9	24.1	0.59	104.6
11	20	10000	2000	13	6	1:2	23.9	14.4	0.27	40.6
12	20	5000	1500	11	8	1:1	19.7	13.0	0.31	94.0
13	25	20000	1000	7	6	2:1	20.2	32.6	0.31	90.4
14	25	15000	500	8	8	1:3	38.8	8.1	0.90	22.4
15	25	10000	2000	11	2	1:1	21.4	18.3	0.25	57.3
16	25	5000	1500	13	4	1:2	23.2	13.3	0.32	57.8
17	10	20000	2000	13	4	1:3	30.7	12.2	0.49	32.2
18	10	15000	1500	11	2	2:1	21.7	10.4	0.40	77.7
19	10	10000	1000	9	8	1:2	22.0	13.5	0.51	74.3
20	10	5000	500	7	6	1:1	19.3	13.0	0.66	188.5
21	15	20000	2000	11	8	2:1	27.8	12.4	0.46	70.2
22	15	15000	1500	13	6	1:3	28.8	16.4	0.41	30.9
23	15	10000	1000	7	4	1:1	15.1	21.3	0.33	115.8
24	15	5000	500	9	2	1:2	21.4	15.4	0.55	115.2
25	20	20000	1500	9	4	1:1	20.2	28.3	0.25	59.6
26	20	15000	2000	7	2	1:2	14.9	27.0	0.21	55.5
27	20	10000	500	13	8	2:1	35.0	6.0	0.90	43.6
28	20	5000	1000	11	6	1:3	28.8	9.9	0.64	34.4
29	25	20000	1500	7	8	1:2	34.8	32.5	0.43	42.9
30	25	15000	2000	9	6	1:1	18.3	28.1	0.22	62.6
31	25	10000	500	11	4	1:3	38.3	8.8	0.85	28.7
32	25	5000	1000	13	2	2:1	23.5	12.1	0.37	102.2

　　由于单一的试验指标不能得到较为理想的试验结果，因此采用了多指标分析。在多指标正交试验中，各指标的最优试验方案之间可能存在一定的矛盾，所以分析试验结果时需要兼顾各项指标，找出使每个指标都尽可能好的试验方案。本次试验采用综合评分的方法，确定相应指标的组合系数或权，然后对每号试验进行综合评分[8]，评分公式如下：

$$试验得分 = \sum_i (\omega_i \times 第\ i\ 个指标) \tag{1}$$

本次试验共选用平均地层压力、采出程度、CO_2滞留率和换油率4个试验指标。在试验指标中，采出程度加权系数为3，CO_2滞留率和换油率加权系数为1.5，地层压力加权系数为1。

$$综合评分=3×采出程度+1.5×CO_2滞留率+1.5×换油率+地层压力 \quad （2）$$

本次实验的评分标准见表3，试验结果的极差分析见表4。通过极差分析可以看出，各因素对综合试验指标的影响顺序依次是D>E>F>A>C>B，因此井底流压、段塞比和注气周期为影响CO_2 WAG驱的主要因素，其他3个因素对试验结果影响较小。通过对各因素水平的均值比较分析，可以得出最优水平组为$A_4B_1C_3D_4E_3F_1$。通过以上分析得出，吐哈油田牛圈湖典型区块CO_2 WAG驱的最优参数为注气年限为25年，注气速度为20000m³/d，总注气量为72.1×10⁴t，关井气油比为1500m³/m³，井底流压为7MPa，水气交替段塞比为2:1，注气周期为6个月。

表3 CO_2 WAG 非混相驱注气试验评分标准

地层压力（MPa）	评分	采出程度（%）	评分	CO_2滞留率（t/t）	得分	换油率（t/t）	评分
10~15	2	5~10	4	0.2~0.34	2	20~55	2
15~20	4	10~15	8	0.34~0.48	4	55~90	4
20~25	8	15~20	12	0.48~0.62	6	90~125	6
25~30	4	20~25	16	0.62~0.76	8	125~160	8
>30	2	>25	20	>0.76	10	>160	10

表4 CO_2 WAG 非混相驱注气试验结果分析

试验号	因素									实验结果
	A 注气时间	B 注气速度	C 关井GOR	D 井底流压	E 段塞比	F 注气周期	空列	空列	空列	
1	1	1	1	1	1	1	1	1	1	18
2	1	2	2	2	2	2	2	2	2	20
3	1	3	3	3	3	3	3	3	3	26
4	1	4	4	4	4	4	4	4	4	18
5	2	1	1	2	2	3	3	4	4	18
6	2	2	2	1	1	4	4	3	3	18
7	2	3	3	4	4	1	1	2	2	24
8	2	4	4	3	3	2	2	1	1	26
9	3	1	2	3	4	1	2	3	4	26
10	3	2	1	4	3	2	1	4	3	32
11	3	3	4	1	2	3	4	1	2	20
12	3	4	3	2	1	4	3	2	1	20
13	4	1	2	4	3	4	2	4	1	36
14	4	2	1	3	4	4	3	1	2	18
15	4	3	4	2	1	1	4	3	3	26

试验号	因素									实验结果
	A注气时间	B注气速度	C关井GOR	D井底流压	E段塞比	F注气周期	空列	空列	空列	
16	4	4	3	1	2	2	1	3	4	22
17	1	1	4	1	4	2	3	2	3	18
18	1	2	3	2	3	1	4	1	4	24
19	1	3	2	3	2	4	1	4	1	26
20	1	4	1	4	1	3	2	3	2	30
21	2	1	4	2	1	4	3	3	2	20
22	2	2	3	1	4	3	2	4	1	22
23	2	3	2	4	1	2	3	1	4	28
24	2	4	1	3	2	1	4	2	3	32
25	3	1	3	3	1	2	4	4	2	34
26	3	2	4	4	2	1	3	3	1	28
27	3	3	1	1	3	4	2	2	4	18
28	3	4	2	2	4	3	1	1	3	18
29	4	1	3	4	2	4	2	1	3	28
30	4	2	4	3	2	1	3	1	2	4	30
31	4	3	1	2	4	2	4	3	1	18
32	4	4	2	1	3	1	3	4	2	26
均值1	22.5	24.75	23	20.25	25.5	25.5	23.75	22.5	24.25	
均值2	23.5	24	24.75	20.5	24.25	24.75	24.5	24.75	24	
均值3	24.5	23.25	25	27.25	26	25	22.75	23.5	24.75	
均值4	25.5	24	23.25	28	20.25	20.75	25	25.25	23	
极差	3.00	1.5	2	7.75	5.75	4.75	2.25	2.75	1.75	

在对牛圈湖试验区块 CO_2 WAG 非混相驱进行注气参数优化后，运用同样方法进行 CO_2 连续注气非混相驱注气参数优化设计。选取注气时间、注气速度、关井气油比和井底流压 4 个影响因素[9]，并对每个因素选取 4 个水平，按 L16 (4^5) 进行正交试验方案设计。本次试验依然采用综合评分的方法，评分公式和试验评价指标与 CO_2 WAG 非混相驱相同。根据实验所得综合评分，得到牛圈湖典型区块 CO_2 连续注气非混相驱的最优参数为注气年限为 20 年，注气速度为 15000 m^3/d，总注气量为 131.4×10^4t，关井气油比为 2000 m^3/m^3，井底流压为 7MPa。

3 牛圈湖试验区块提高采收率潜力评价

应用试验优化的注气参数进行试验区块提高采收率方案设计，应用油藏数值模拟方法，对方案动态指标进行预测，对该区块提高采收率潜力进行评价。牛圈湖试验区块开发方式为早期注水开发，因此在此基础上提出 3 种提高采收率的技术方案。

方案一：注水开发。油井以原来的工作制度定产量生产，注水井以原来的工作制度定注入量注水，使地层压力保持在一定水平，注水年限为 25 年。

方案二：水驱后 CO_2 连续注气开发. 生产井以定压生产，井底流压保持在饱和压力以上为7MPa，注气速度为 $15000m^3/d$，总注入量为 131.4×10^4t，关井气油比为 $2000m^3/m^3$，生产年限为20年。

方案三：水驱后 CO_2 水气交替注气开发。生产井以定压生产，井底流压保持在饱和压力以上为7MPa，注气速度为 $20000m^3/d$，注水速度为 $20m^3/d$。总注气量为 72.1×10^4t，总注水量为 73×10^4t。关井气油比为 $1500m^3/m^3$，水气交替段塞比为2:1，注气周期为6个月，生产年限为25年。

以上3种技术方案分别代表了3种不同的开发方式在最优注入工艺情况下的开发方案，不同开发方案的预测结果见图2~图4和表5。

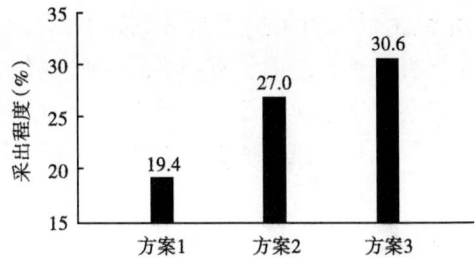

图2　不同开发方案采出程度变化柱状图

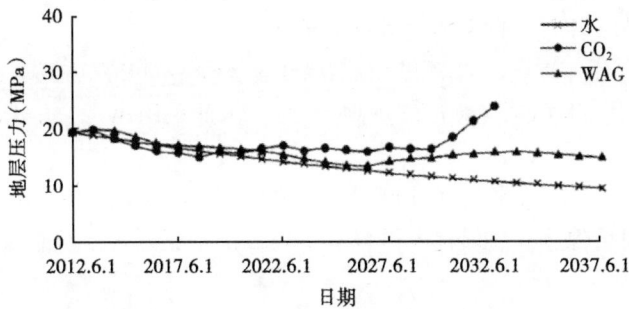

图3　不同开发方案地层压力变化曲线图

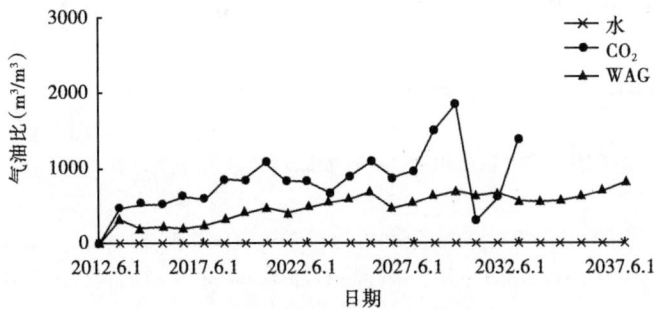

图4　不同开发方案气油比变化曲线图

表5　不同开发方案指标预测数据表

注气方式	累计增液量 (10^4m^3)	累计增油量 (10^4m^3)	累计增气量 (10^8m^3)	综合含水 (%)	累计注水量 (10^4t)	累计注气量 (10^4t)	埋存量 (10^4t)	地层压力 (MPa)	气油比 (%)	采出程度 (%)	换油率 (t/t)	CO_2滞留率 (t/t)	埋存系数 (t/t)
水驱	183.6	37.2	0.08	91	152.3	—	—	10	26	19.4	—	—	—
CO_2驱	60.5	54.3	4.4	10.6	—	129.4	45.7	24	1447	27	36.6	0.4	0.23
WAG驱	114.8	62.6	2.9	53.2	72.3	70.3	15.6	15	841	30.6	77.5	0.2	0.08

由以上 3 种开发方案的预测结果可以看出，CO_2 驱和 CO_2 WAG 驱替方式的采出程度以及地层压力保持水平均比水驱效果要好。CO_2 WAG 驱的采出程度、换油率和气油比等指标均比 CO_2 驱效果好，但 CO_2 滞留率以及埋存系数较 CO_2 驱偏低，这主要是因为 CO_2 驱中 CO_2 注入量比 CO_2 WAG 驱的总注入量要大很多，因此埋存量也比较大。综合考虑以上指标，采用 WAG 方式注 CO_2 开发效果好。CO_2 水气交替开发 25 年后，综合含水为 53.3%。此时由于生产井气油比较大，有一半的生产井关井，累计增油量增加缓慢，可停止注气，转注水开发。进行二次水驱[10]开发，可以有效地驱替残余油，并且能够将岩石孔隙中的气体驱替出来，为原油提供流动通道，提高原油采收率。

4 结论

（1）根据吐哈油田油藏地质开发特征，建立试验区块实际油藏的组分数值模型。采用多指标正交试验设计方法，进行 CO_2 非混相驱注气参数优化，得到 CO_2 连续注气和 WAG 驱的注气优化参数。关井气油比为 CO_2 连续注气驱主要影响因素，而 CO_2 WAG 驱的主要影响因素为水气交替段塞比、注气周期和生产井井底流压。

（2）对比水驱，CO_2 连续注气驱和 CO_2 WAG 驱 3 种不同开发方式，CO_2 WAG 驱采出程度比水驱要高 11.2%，比 CO_2 连续注气驱高 3.6%。因此采用 WAG 方式注 CO_2 开发，可以达到很好的驱油效果。

（3）进行 CO_2 WAG 驱方案指标预测发现，开发 25 年后生产井由于气油比上升而关井数量大幅增加。此时可停止注气转注水开发，进行二次水驱开发，可有效驱替原油，提高原油采收率。

参 考 文 献

[1] Nezhad S A T. Experimental Study on Applicability of Water—Alternating—CO_2 Injection in the Secondary and Tertiary Recovery [J]. SPE 103988, 2006：1-4.

[2] Chen Shennan, Li Heng. Production Optimization and Uncertainty Assessment in a CO_2 Flooding Reservior [J]. SPE 120642. 2009：1-13.

[3] 尚庆华，吴晓东，韩国庆，等. CO_2 驱油井产能及影响因素敏感性分析 [J]. 石油钻探技术，2010，39 (1)：83-87.

[4] 马超群，黄 磊. 气水交替 CO_2 注入界限研究 [J]. 油气地球物理，2012，10 (2)：83-87.

[5] Wang J, Mcvay D A. Compositional Simulation and Optimization of Secondary and Tertiary Recovery Strategies in Monument Butte Field, Utah [J]. SPE 117775. 2008：1-19.

[6] Ghaderi S M, Clarkson C R. Optimization of WAG Processfor Coupled CO_2 EOR—Storage in Tight Oil Formations：An Experimental Design Approach [J]. SPE 161884, 2012：1-17.

[7] 王万中. 实验的设计与分析 [M]. 北京：高等教育出版社，2004，95-103.

[8] 苑玉凤. 多指标正交试验分析 [J]. 湖北汽车工业学院学报，2005，19 (4)：53-56.

[9] 廖海婴. 腰英台 DB34 井区 CO_2 驱替油藏数值模拟研究 [J]. 西安石油大学学报，2010，25 (5)：50-53.

[10] Ren W, Cunha L B. Numerical Simulation and Sensitivity Analysis of Gravity—Assisted Tertiary Gas—Injection Process [J]. SPE Reservoir Evaluation & Engineering, 2004：184-192.

低渗透非均质性油藏扩大波及体积技术

廖长霖　廖新维　赵晓亮　卢　宁　赵东锋

（中国石油大学（北京）石油工程学院）

摘　要：为了有效利用水气交替技术提高低渗透、非均质性油藏的采收率，引入注入速度差异控制（AOIR）、段塞比变化的水气交替驱（TWAG）和 AOIR—TWAG 技术研究了提高采收率的可行性。结果表明，渗透率变异系数越大，水气交替驱相对于连续注气驱的效果越好，3 种技术的采收率均高于常规的水气交替驱。但是，如果考虑经济效益，则只有 AOIR 和 AOIR—TWAG 技术的净现值高于常规的水气交替驱。通过新疆油田某油藏的模拟验证了概念模型研究的可行性，对提高低渗透、非均质性油藏气驱波及体积具有指导意义。

关键词：低渗透；非均质性；水气交替驱；CO_2；波及体积

0　引言

研究表明，CO_2 水气交替驱能够提高低渗透油藏的采收率[1—5]。采用水气交替注入方式能够控制油气流度差异（即水的饱和度影响着气的流度），能够防止 CO_2 沿高渗通道渗流指进，有利于扩大 CO_2 波及体积。水气交替驱还有利于保持地层压力，提高 CO_2 的利用效率，节约注入气及产出气的处理成本，弥补连续注气的不足。但是，如何采用有效的水气交替驱技术提高低渗透、非均质性油藏产量及经济效益是水气交替的难点。许多学者的研究虽取得了一定效果[6—12]，但多是对单一技术的研究，多种技术组合时的综合效果有待于研究。因此，采用油藏数值模拟技术首先研究不同渗透率变异系数的低渗透非均质性油藏的 CO_2 水气交替驱开发效果，然后分别引入注入速度差异控制、段塞比变化的水气交替驱、AOIR—TWAG 技术，并把经济净现值作为 CO_2 的水气交替驱开发效果的评价指标研究其开发效果。通过对新疆油田某油藏的模拟验证了这些技术在低渗透非均质性油藏中对于控制 CO_2 的指进、延迟气窜、扩大 CO_2 波及体积的可行性。

1　3 种技术的评价指标

1.1　注入速度

注入速度差异控制是对高渗区加大注水速度、降低注气速度，而对低渗区降低注水速度、加大注气速度[13]。通过统计注入井控制区域的平均渗透率，以油藏平均渗透率和常规水气交替驱注入速度为基准计算各注入井的注水、注气速度公式为：

$$v_{iw} = (\overline{K_1}/\overline{K}) \cdot v_w \tag{1}$$

$$v_{ig} = (\overline{K}/\overline{K}_1) \cdot v_g \qquad (2)$$

式中，v_{iw} 为第 i 口井的注水速度；v_{ig} 为第 i 口井的注气速度；i 为注入井编号；$\overline{K}_i$ 为第 i 口井注入井范围内的地层平均渗透率；$\overline{K}$ 为油藏的平均渗透率；v_w 为常规水气交替驱注水速度；v_g 为常规水气交替驱注气速度。

在不同的生产时期，根据实际生产情况采用了不同段塞比的水气交替技术[14,15]，因此能够及时调整油田生产措施，防止在 CO_2 流速较高的区域过早发生气窜，降低指进的发生，同时能够根据现场设备及管线的承受能力加快或延迟驱替响应速度。AOIR—TWAG 技术就是同时采用了注入速度差异控制和段塞比变化的水气交替驱技术。

1.2 经济净现值

经济净现值作为油藏开发评价指标，既考虑了原油、CO_2、注入水的价格，又考虑到生产时间、产出水的处理费用以及地面设施、防腐设备等费用，其计算公式为：

$$N = \sum_{i=1}^{T} \frac{C}{(1+r)i} - C_0 \qquad (3)$$

$$C = (V_{OPT} \times a) - (V_{GIT} \times b) - (V_{WPT} \times c) - (V_{WIT} \times d) \qquad (4)$$

式中，r 为贴现率；T 为生产年限，a；t 为生产时间，a；C_0 为首次投资费用，美元；C 为年现金流入量，美元；a 为油价，美元/m^3；b 为 CO_2 价格，美元/t；c 为注入水价格，美元/m^3；d 为产出水处理费，美元/m^3；V_{OPT} 为累计产油量，m^3；V_{WPT} 为累计产水量，m^3；V_{GIT} 为累计注入气量，t；V_{WIT} 为累计注水量，m^3。

2 油藏描述

基于新疆油田某油藏的油藏特征分别建立概念模型和实际模型，模拟了水气交替技术的可行性（为了简化，这 2 个模型都忽略了毛细管压力的作用）。概念模型采用 29m×29m×10m 的油藏网格模拟低渗透油藏，顶部深度为 1900m，其网格参数见表 1。该模型共有 4 口注入井和 9 口生产井，所有井在每一层都采取了射孔措施。

表 1 油藏基本参数

模型类别	顶部深度 （m）	原始地层压力 （MPa）	饱和压力 （MPa）	最小混相压力 （MPa）	油藏温度 （℃）	孔隙度 （%）	平均渗透率 （mD）	初始含油 饱和度（%）
概念模型	1900	25.0	8.4	21.2	62.15	16	9.1	66
实际模型	1864	24.7	8.4	21.2	62.15	14	8.5	64

采用 49m×59m×8m 的油藏网格建立了新疆油田某油藏局部井组模型，根据洛伦兹曲线法[16]求得实际模型的渗透率变异系数为 0.41。该模型共有 6 口注入井和 22 口生产井，采用的流体组分和相渗参数与概念模型相同。对实验区域进行了储量、油藏压力、产油量、产水量、含水率等水驱历史拟合。单井拟合精度超过 90%，能够正确反映油藏的情况，可以作为 CO_2 驱油模拟的基础模型。

依据完整的原油组分分析和高压物性实验数据，采用流体模拟软件将全油组分重组为 8

个拟组分，通过回归拟合得到状态方程的参数见表 2。因为 CO_2 在原油中的混相能力将对原油采收率产生重大影响，所以以通过细管实验确定了原油的最小混相压力为 21.6 MPa，同时通过数值模型模拟细管实验验证其可靠性，得到的最小混相压力为 21.2 MPa，相对误差为 1.85%。说明该原油拟组分能够代表实际原油，可以作为油藏数值模拟的原油组分。

表 2 油藏流体拟组分状态方程参数

原油组分	摩尔分数（%）	摩尔质量（g/mol）	临界压力（MPa）	临界温度（K）	状态方程系数 A	状态方程系数 B	偏心因子	临界体积 [m³/（kg·mol）]	临界压缩因子
CO_2	0.0010	44.01	7.39	304.70	0.457	0.077	0.04	0.094	0.27
N_2C_1	0.3035	16.18	1.26	94.40	0.514	0.019	0.14	0.097	0.15
C_{2-4}	0.0280	33.88	3.09	280.51	0.546	0.036	0.09	0.162	0.21
C_{5-6}	0.0458	71.99	1.20	468.98	0.457	0.077	0.03	0.308	0.09
C_{7-10}	0.1449	114.29	2.73	587.89	0.457	0.077	0.05	0.459	0.25
C_{11-17}	0.2167	187.66	1.90	700.02	0.457	0.077	0.12	0.720	0.23
C_{18-27}	0.1398	298.75	1.38	730.43	0.457	0.077	0.06	1.051	0.23
C_{28+}	0.1203	518.63	0.28	787.55	0.910	0.109	0.79	1.988	0.08

3 模拟方案与结果

3.1 油藏概念模型

油藏概念模型研究共有 11 个设计方案（见表 3 的方案 1—方案 11）。每个方案均先进行水驱模拟，当基础方案的连续注水量为 1.0 倍含烃孔隙体积时，含水率达到 93.4%，采收率达到 45.68% 时进行注气驱模拟。

表 3 油藏模拟方案

方案	变异系数	技术类型	方案	变异系数	技术类型
1	0	常规水气交替驱	9	0.39	AIOR-TWAG
2	0.18	常规水气交替驱	10	0.39	连续注气驱
3	0.32	常规水气交替驱	11	0.39	连续注水驱
4	0.39	常规水气交替驱	12	0.41	常规水气交替驱
5	0.47	常规水气交替驱	13	0.41	注入速度差异控制
6	0.57	常规水气交替驱	14	0.41	AOIR-TWAG
7	0.39	注入速度差异控制	15	0.41	连续注水驱
8	0.39	渐变的水气交替驱			

3.1.1 非均质性储层的水气交替驱

由表 3 可知，方案 1—方案 6 的变异系数分别为 0，0.18，0.32，0.39，0.47 和 0.57，6 个方案的其他参数都相同，通过模拟不同渗透率的变异系数方案得到的提高采收率变化情况如图 1 所示，图中的提高采收率差值表示同一个模型下常规水气交替驱与连续注气驱采收率之间的差值。由图 1 可知，渗透率的变异系数为 0 时的常规水气交替驱与连续注气驱采收率的差值最小（为 1.4%）。随着渗透率的变异系数的增大，储层的非均质性增强，常规水气交替驱与连续注气驱采收率的差值逐渐增大，显示出水气交替在低渗透率、非均质性油藏开发中的优势。当渗透率的变异系数为 0.57 时，常规水气交替驱相对连续注气驱的采收率提高值达到最大（为 7.7%）。

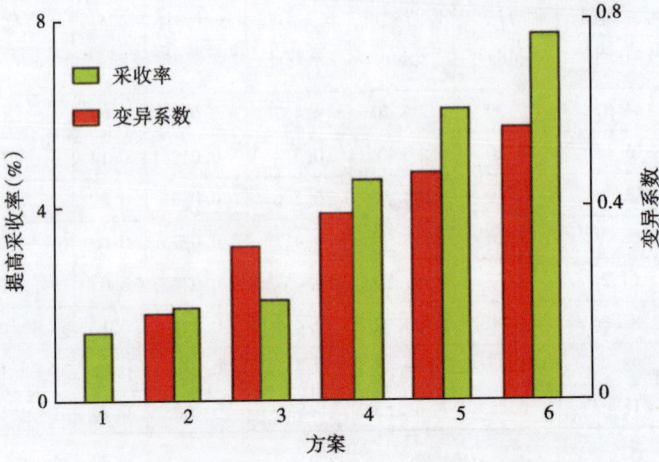

图 1　提高采收率值与渗透率变异系数变化

3.1.2　3 种水气交替技术对比

以方案 4 作为基础方案，分别模拟 AOIR、TWAG 及 AOIR—TWAG 建立 3 个方案（方案 7—方案 9），并与连续注水驱、连续注气驱和常规水气交替方案作对比。采用 TWAG 技术时，分别以水气段塞比为 1:2、1:1、2:1、3:1 和 4:1 注入 0.25 倍含烃孔隙体积，最后以 5:1 注入 0.15 倍含烃孔隙体积。统计基础方案的 4 口注入井控制区域内的平均渗透率，然后根据全区的平均渗透率和平均注入速度计算出每口注入井的注水速度和注气速度，进而得到注入井的注气、注水数据（表 4）。

表 4　注入井的注入速度

模型类别	井名	平均渗透率（mD）	注水速度（m³/d）	注气速度（m³/d）
概念模型	I1	19.64	40.4	19816.7
	I2	24.71	50.8	15750.7
	I3	26.16	53.8	14877.8
	I4	7.33	15.1	53096.9
实际模型	H42—9	11.30	31.7	15781.7
	H42—11	11.60	32.5	15373.6
	H43—8	12.10	33.9	14738.3
	H44—9	5.20	14.6	34294.9
	H44—11	8.40	23.6	21230.2
	H45—10	4.90	13.7	36394.6

通过模拟得到了不同方案下采收率的变化曲线（图 2）。由于研究中均采用各种技术的最大生产能力进行模拟，所以连续注气能够在短时间内较大幅度地提高采收率。当达到设定的生产气油比极限时，连续注气驱的采收率为 52.7%，生产时间为 12.7a；常规水气交替驱的采收率为 57.35%，生产时间为 30.4a。由此可见，常规水气交替驱的采收率明显高于连续注气驱，而 AOIR、TWAG、AOIR—TWAG 技术的采收率又在不同程度上高于常规水气交替驱。TWAG 技术的采收率为 59.25%，AOIR 技术的采收率为 58.51%，AOIR-TWAG 技术的采收率为 58.49%。AOIR 技术的生产时间少于常规水气交替驱（为 26.4 a），TWAG 技术的生产时间则大于常规水气交替驱（为 55.1 a），AOIR—TWAG 技术的生产时间少于常规水气交替驱（为 27.4 a）。

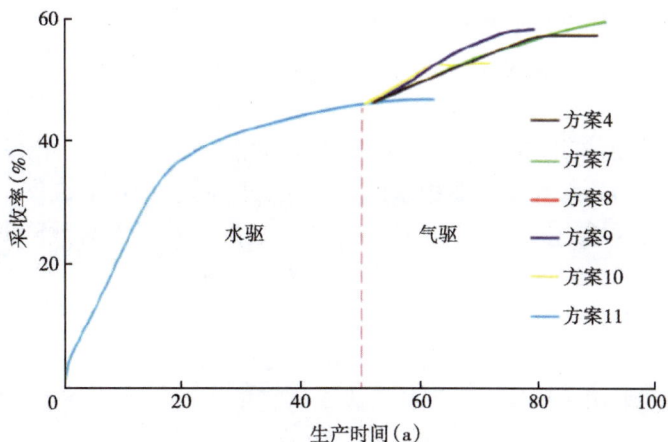

图 2 采收率变化对比曲线

3.1.3 经济效益分析

在计算净现值时，贴现率的取值为 5%。现金流入量的计算基于原油、CO_2、注入水的价格以及产出水、地面设施的处理费用等。整个生产时期，原油的价格定为 503.2 美元/m^3，CO_2 的价格为 80 美元/t，注入水的价格为 1.57 美元/m^3，产出水的处理费用为 9.44 美元/m^3。4 个井组的初始投入费用为 $2.5×10^6$ 美元，研究中设定的相关费用可随实际价格变动，因此能够对经济净现值作出准确的预测。

为了评价 CO_2 水气交替驱的开发效果，基于基础方案（方案 4）将采用各个方案提高的采收率和净现值无量纲化。对比各个方案的无量纲净现值及无量纲采收率可知，方案 7—方案 9 的无量纲采收率都高于基础方案，但只有方案 7 和方案 9 的无量纲净现值高于基础方案（分别为 1.41 和 1.44）。方案 8 的无量纲采收率达到最大值，但因生产时间过长导致其无量纲净现值为最小值。

3.2 油藏实际模型

通过模拟实际模型得到了不同方案（表 3 的方案 12—方案 14）的采收率变化情况。

AOIR—TWAG 技术的采收率最大值为 59.02%，AOIR 技术的采收率最大值为 58.73%，均高于常规水气交替驱（为 56.97%）。AOIR 和 AOIR—TWAG 技术的生产时间为 30.2 年和 29.9 年，少于常规水气交替驱（为 36.7 年）。常规水气交替驱、AOIR 和 AOIR—TWAG 技

术的 CO_2 注入量分别为 0.75 倍、0.9 倍和 0.89 倍含烃孔隙体积。为了评价水气交替驱的开发效果，基于基础方案（方案 12）的净现值和采收率将方案 13 和方案 14 的净现值和采收率无量纲化。每个方案都是以最大生产能力计算的，区块的初始投入费用为 4.3×10^6 美元。对比各个方案的无量纲净现值及无量纲采收率可知，AOIR 和 AOIR—TWAG 技术的无量纲采收率为 1.07 和 1.09，无量纲净现值为 1.46 和 1.59。虽然两种技术的无量纲采收率略高于常规水气交替驱，但其无量纲净现值明显高于常规水气交替驱。

4 讨论

通过对比不同渗透率变异系数的常规水气交替驱与连续注气驱的开发效果可知，随着变异系数的增大，常规水气交替驱相对于连续注气驱提高的采收率逐渐增大，表明在非均质性油藏中更能体现出水气交替驱的优势。因为非均质性油藏渗透率分布的不均匀性，采用连续注气驱时，其生产前期产油量大、石油增产速度快，但容易导致注入气体顺着优势通道渗流至井底而使 CO_2 的波及体积较小（在非均质性严重的油藏更严重）。而水气交替驱能够弥补连续注气驱的不足，因为水气交替驱中水的存在大大降低了 CO_2 的流度，使 CO_2 波及到连续注气驱未能波及的区域，扩大了 CO_2 波及体积；气的存在又能有效降低三相流动区域的剩余油饱和度，而且三相的剩余油饱和度均低于两相（油—气、油—水）的剩余油饱和度。因此，两者对采收率都有很大贡献。

引入常规水气交替驱、AOIR 和 AOIR—TWAG 技术，对比了单个技术及其技术组合的可行性。结果表明，采用 3 种技术得到的采收率均高于基础方案。但是，在考虑了经济效益的无量纲净现值分析中，AOIR 和 AOIR—TWAG 的无量纲净现值高于常规水气交替驱，所以具有相对较高的采收率并不意味着能够带来较高的经济效益，因为较长的生产时间在带来较高采收率的同时也增加了生产成本。

根据油藏概念模型优化得到的 2 个较好的技术在实际油藏的模拟结果表明，这两种技术的采收率均比常规水气交替驱高，而且 AOIR—TWAG 因注入见效快、注入时间相对较短具有优势，表明该技术适用于其他类似的低渗透非均质性油藏。但是，采用油藏数值模拟进行相关技术研究比较理想化，可能在实际应用中具有局限性。相关技术的操作及可行性还需实验验证和现场总结。

5 结论

（1）CO_2 驱能够在水驱后的低渗透油藏中发挥重要作用，且在低渗透非均质性油藏中采用水气交替驱的注入方式更能体现其优势。

（2）注入速度差异控制、段塞比变化的水气交替驱和 AOIR—TWAG 技术都能够有效降低指进现象的发生率，防止 CO_2 过早发生气窜，扩大 CO_2 在油藏中的波及体积，有效提高原油采收率。3 种技术的采收率均高于常规水气交替驱。

（3）注入速度差异控制和 AOIR—TWAG 的净现值大于常规水气交替驱，表明这 2 种技术在有效提高采收率的同时提高了经济效益。

（4）利用概念模型优化的 AOIR 和 AOIR—TWAG 技术在实际油藏的采收率和经济净现值都高于常规水气交替驱，验证了概念模型研究的可靠性。该技术适用于其他类似的低渗透非均质性油藏。

参 考 文 献

［1］ Christensen J R, Stenby E H, Skauge A. Review of WAG Field Experience ［R］. SPE 39883, 1998.

［2］ Christensen J R, Larsen M, Nicolaisen H. Compositional Simulation of Water. Alternating－Gas Processes ［R］. SPE 62999, 2000.

［3］ Skauge A, Laisen J A. There－phase Relative Permeabilities and Trapped Measurements Related to WAG Processes ［R］. Int. Symp. of the Society of the Core Analysts, 1994.

［4］ Jiang H F, Nuryaningsih L, Adidharma H. The Study of Timing of Cyclic Injections in Miscible CO_2 WAG ［R］. SPE 153792, 2012.

［5］ Odi U, Gupta A. Optimization and Design of Carbon Dioxide Flooding ［R］. SPE 138684, 2010.

［6］ Genrich J F. A Simplified Model to Predict Heterogeneity Efects on WAG Flooding Performance ［C］. The 61 st Annual Technical Conference and Exhibition of the society of Petroleum Engineers, 1986.

［7］ Surguchev L M, Korbol R, Haugen S, et a1. Screening of WAG Injection Strategies for Heterogeneous Reservoirs ［R］. SPE 25075, 1992.

［8］ Taheri A, Sajjadian V A. WAG Performance in a Low Porosity and Low Permeability Reservoir, Sirri－A Field, Iran ［R］. SPE 100212, 2006.

［9］ 计秉玉, 王凤兰, 何应付. 对 CO_2 驱油过程中油气混相特征的再认识 ［J］. 大庆石油地质与开发, 2009, 28 (3): 103-109.

［10］ 龚蔚, 蒲万芬, 彭陶钧, 等. 就地生成二氧化碳技术提高采收率研究 ［J］. 大庆石油地质与开发, 2008, 27 (6): 104-107.

［11］ 张国强, 孙雷, 姚为有, 等. 小断块油藏 CO_2 吞吐过程压力与含油饱和度分布 ［J］. 大庆石油地质与开发, 2008, 27 (2): 110-112.

［12］ 张硕, 单文文, 张红丽, 等. 特低渗透油藏 CO_2 近混相驱油 ［J］. 大庆石油地质与开发, 2009, 28 (1): 114-117.

［13］ Jahangiri H R, Zhang D X. Ensemble based co－optimization of Carbon Dioxide Sequestration and Enhanced Oil Recovery ［J］. International Journal of Greenhouse Gas Control, 2012 (8): 22-33.

［14］ Sharma A K, Lucille E C. From Simulator to Field Management: Optimum WAG Application in a Vlkst Texas CO_2 Flood—A Case History ［R］. SPE 36711, 1996.

［15］ Zhou D, Yan M, Calvin W M. Optimization of a Mature C02 Flood—from Continuous Injection to WAG ［R］. SPE 154181, 2012.

［16］ 康晓东, 刘德华, 蒋明煊, 等. 洛伦茨曲线在油藏工程中的应用 ［J］. 新疆石油地质, 2002, 23 (1): 65-68.

自生 CO_2 结合表面活性剂复合
吞吐数值模拟

汤 勇[1]　汪 勇[2]　邓建华[3]　杨付林[3]　薛 芸[3]

（1. 西南石油大学油"气藏地质及开发工程"国家重点实验室；2. 西南石油大学
石油工程学院；3. 中国石化江苏油田分公司石油工程技术研究院）

摘　要：层内自生 CO_2 吞吐是低渗小断块油田提高采收率的新型技术，既发挥了 CO_2 混相或非混相驱油机理，又解决了 CO_2 气源和注入过程井筒腐蚀等问题。针对目前自生 CO_2 以及结合表面活性剂复合吞吐的参数优化研究很少等情况，探索了应用化学驱、热采模型对层内自生 CO_2 以及注入表面活性剂的操作工艺参数进行优化的方法。基于某实际低渗透油藏单井径向模型和室内自生 CO_2 及表面活性剂实验测试结果，在对流体相态、自生 CO_2 过程、表面活性剂驱相关参数及生产动态拟合的基础上，开展生气剂注入量、生气剂浓度、注入速度、关井时间、采液强度对复合吞吐效果影响的模拟研究。结果显示，生气剂注入量、注入浓度和采液强度对复合吞吐效果影响较大，并且存在最佳的取值范围，其中最佳生气剂量为 $200 \sim 250t$、摩尔分数为 $3.0\% \sim 5.0\%$、采液强度为 $7m^3/d$；适当增加注入速度和延长关井时间有助于提高吞吐效果。该研究对于自生 CO_2 复合吞吐的优化以及该技术在小断块低渗油藏提高采收率领域的推广应用具有重要意义。

关键词：自生 CO_2；表面活性剂；复合吞吐；操作工艺参数；数值模拟

0 引言

层内自生 CO_2 提高原油采收率技术是向地层中注入反应溶液，使其在油藏条件下充分反应而释放出 CO_2 气体。CO_2 气体能溶解于原油之中，降低原油黏度，膨胀原油体积，从而达到提高原油采收率的目的[1]。该技术具有无需天然 CO_2 资源、产气量可控、工艺简单、注入性好的优点。CO_2 吞吐可在不同的油藏条件下成功地采出原油，并且经济上也是可行的，它较之于其他提高采收率措施，具有投资少、见效快、返本期短及技术要求不高的优点，因此风险性很小[2]。表面活性剂提高采收率的主要原理是利用驱替流体与被驱替原油体系之间具有低界面张力 IFT 的特性，从技术上讲，表面活性剂驱最适合三次采油，基本上不受含水率的限制，可以获得很高的水驱残余油采收率[3]。自生 CO_2 复合表面活性吞吐技术综合了自生、吞吐、表面活性剂 3 方面工艺的优点，其适应性更强，是一种极具潜力的技术，应用推广前景较大。

自生 CO_2 分为单液法和双液法，利用生气体系评价实验、原油膨胀实验、原油降黏实验[4,5]、结垢实验、长岩心物理实验等对生气体系筛选、评价，确定生气方式、生气剂种类、浓度、生气温度以及反应动力学等相关参数。吞吐中表面活性剂的加入起到了提高洗油效率、解除近井带伤害的作用。目前国内针对自生 CO_2 复合表面活性剂吞吐数值模拟研究较少，本文针对某一低渗透油藏，在生气剂筛选、评价室内实验的基础上，进行了单井自生 CO_2 复合表面活性剂吞吐的数值模拟研究，对主要注采参数进行敏感因素分析。

1 模型基本参数

1.1 地质模型

根据实际油藏，利用 CMG 软件的热采模型 STSRS，建立单井平面径向模型，网格数为 135（15×1×9）个，x 方向上网格步长为 4.0~50.0m，z 方向上网格步长根据实际油藏厚度建立，步长范围 0.5~3.5m，总厚度 9.9m，有效厚度 6.4m，油藏埋藏深度为 2730m，泄油半径为 300m，原始含油饱和度为 46%，地质储量为 $8.54×10^4$t。岩石骨架体积热容量为 $2.35×10^6$J/（$m^3 \cdot ℃$），岩石热传导系数为 $6.6×10^5$J/（$m \cdot d \cdot ℃$），油相热传导系数为 8035J/（$m \cdot d \cdot ℃$），水相热传导系数为 $5.35×10^4$J/（$m \cdot d \cdot ℃$），平均孔隙度为 12%，平均渗透率为 4mD。初始平均地层压力为 28.00MPa，目前平均地层压力为 7.00MPa，平均地层温度为 85℃，含油饱和度为 44%，地面原油密度为 0.8417g/cm^3，地层温度下原油黏度为 9.5mPa·s，地层温度下油相饱和压力为 3.68MPa，原始气油比为 15.2cm^3/cm^3。

1.2 地层流体相态

利用 Winprop 相态模拟软件，对油藏流体等组成膨胀、注气膨胀、饱和压力和单次闪蒸等实验数据进行计算拟合，得到能够代表储层流体特性的状态方程参数[6]（表 1，1 atm = 101325 Pa）。模拟时将油相归并成 5 个组分，分别为：CO_2（0.7%），N_2—C_1（11.3%），C_2—C_6（7.2%），C_7—C_{10}（28.3%），C_{11}—C_{28}（52.5%）。

表 1　油藏流体拟组分主要特征参数

组分	摩尔质量（g/mol）	临界压力（atm）	临界温度（K）	临界体积（m^3）	偏心因子	方程系数 a	方程系数 b
CO_2	44.010000	72.800000	304.200000	0.094000	0.225000	0.457236	0.077796
N_2—C_1	16.148929	45.299833	190.002430	0.098915	0.008283	0.457236	0.077796
C_2—C_6	65.504920	31.858350	517.546390	0.286349	0.266378	0.457236	0.077796
C_7—C_{11}	91.431800	30.484690	644.708790	0.446760	0.303370	0.457236	0.077796
C_{11}—C_{28}	234.133530	16.804922	806.343200	0.857081	0.512470	0.457236	0.077796

2 自生 CO_2 反应参数计算

层内自生 CO_2 反应参数主要有生气速度、反应活化能、反应焓，分解温度等，反应参数的计算将直接影响数值模型的准确性和可靠性。其中生气速度为生气反应在地层温度压力条件下的反应速度，主要通过拟合室内实验而得到，反应活化能、反应焓等热力学参数可以通过建立反应动力学模型和热力学计算得到。本次自生 CO_2 采用单液法，反应如式（1）所示，生气剂 A 分解反应生成 B，CO_2 和 H_2O，反应为吸热反应。根据室内生气剂评价实验，建立数值模型拟合生气反应生气速度，通过拟合生气体系压力变化曲线（图1），得到反应频率因子为 0.06。室内评价实验中，由于生成的 NH_3 溶解于水中，体系压力下降，因此压力曲线后期拟合度较差。通过计算得到热力学参数，计算出反应活化能为 116.89 J/mol，反应焓为 -172690 J/mol。

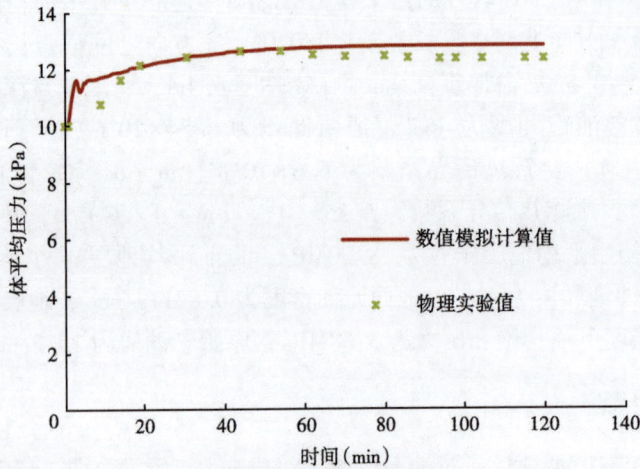

$$A = B + CO_2 + H_2O \tag{1}$$

图 1　体系压力拟合曲线

3　表面活性剂驱参数计算

表面活性剂驱的控制参数主要有油水界面张力及插值参数、低界面张力时相渗曲线及插值参数、高界面张力时相渗曲线及插值参数等[7-9]，通过拟合自生 CO_2、表面活性剂复合吞吐长岩芯物理模拟实验[10-12]，得到相关参数。

根据长岩心物理实验建立长岩心数值模型，首先水驱、然后注入含表面活性剂的生气剂溶液，关井一段时间后，开井放喷，共进行 3 次吞吐。表 2 为数值模型与长岩心物理实验采收率拟合结果，图 2 为采收率拟合曲线。通过拟合得到非润湿相高、低界面张力时相渗内插值参数分别为 -2.92 和 -7.22。

表 2　长岩心数值模型与物理实验拟合结果数据表

方案	水驱采收率（%）	一次吞吐采收率（%）	二次吞吐采收率（%）	三次吞吐采收率（%）
物理实验	54.44	62.44	67.33	68.67
数模计算	53.70	62.21	66.50	68.02

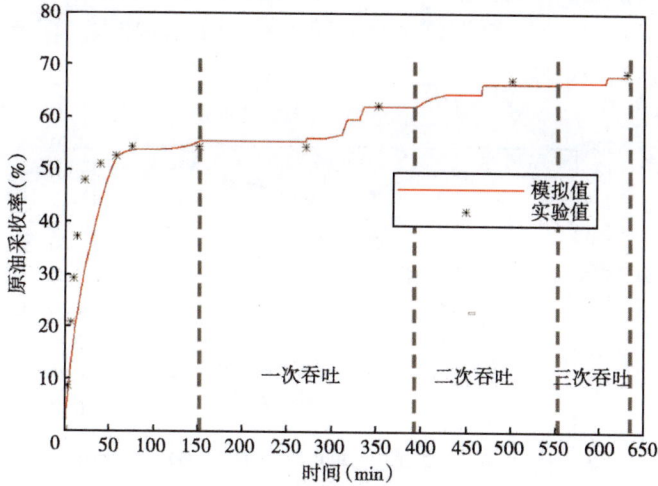

图2　长岩心数值模型与物理实验采收率拟合曲线

4　注入参数敏感性研究

利用所建立的模型，在完成历史拟合的基础上，结合生产实际情况，对影响自生 CO_2 吞吐效果的操作工艺参数进行优化研究，包括注入生气剂量、生气剂浓度、注入速度、关井时间和采液强度。评价指标主要有累计产油量、增产油量、生气剂换油率等，预测时间为 1 年。

4.1　生气剂注入量

生气剂的注入量直接影响到 CO_2 生成量的多少，而 CO_2 生成量是影响自生 CO_2 吞吐效果的主要参数之一。取注入生气剂溶液摩尔浓度为 5.0%，计算比较 7 种不同生气剂注入量（100t，150t，200t，250t，300t，400t 和 500t）下该井累计产油量、增产油量、生气剂换油率。

图3 为不同注入量方案下累计产油量对比曲线，可见累计产油量随生气剂注入量的增加而增加。

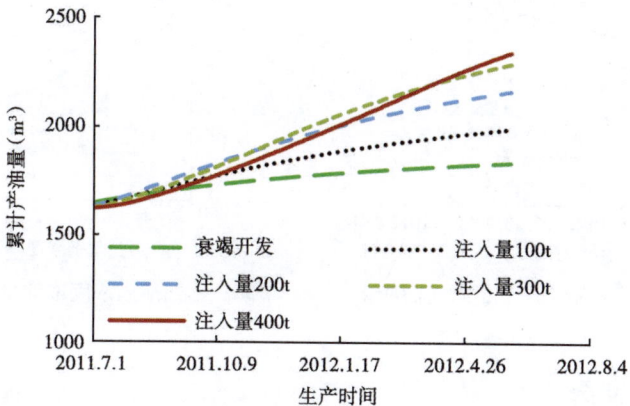

图3　不同注入量累计产油量对比曲线

图 4 为不同注入量增油量与换油率变化曲线。随注入量增加，增产油量和换油率均先增加后减小，由于模拟时间限制，实际增油量应随注入量增大而增大。综合考虑增油量、换油率和实际操作可行性和经济性，最佳注入量为 200~250t。

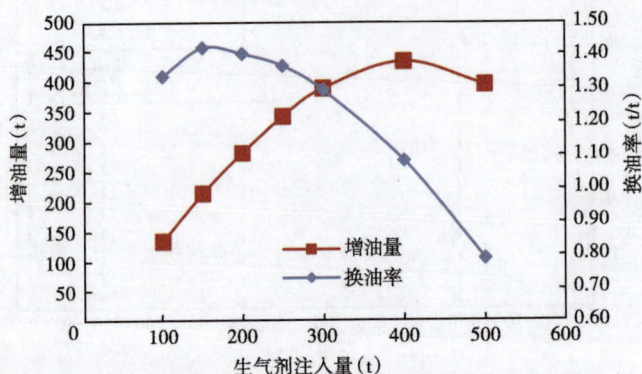

图 4　不同注入量增油量与换油率变化曲线

4.2　生气剂浓度

在生气剂注入量确定的情况下，生气浓度直接影响注入量的大小。若浓度太小则注入溶液总量很大，对现场施工要求较高；若浓度太大，则注入总液量太少，注入生气剂都分布在近井地带，可能导致吞吐半径减小，影响吞吐效果。利用已经优选的注入量（250t），计算比较了摩尔分数为 2.0%、2.5%、3.0%、3.5%、4.0%、5.0%、10.0%、15.0% 和 20.0% 时累计产油量、增产油量、生气剂换油率。

随着生气剂浓度的增加，累计产油量先增加后减小（图 5）。

图 6 为不同浓度下增油量与换油率变化曲线。随着生气剂浓度的增加，增产油量与换油率的变化一致，摩尔分数在 3.0% 左右出现拐点，随后换油率与增产油逐渐降低，综合考虑增油量和换油率，最佳的摩尔分数为 3.0%~5.0%。

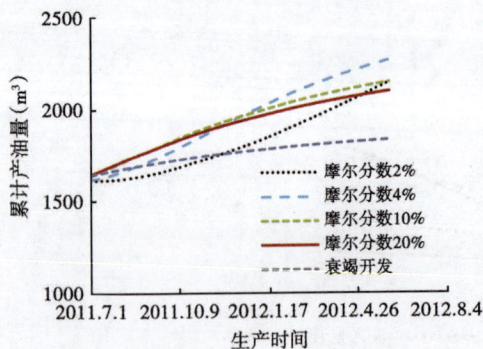

图 5　不同生气剂浓度累计产油量对比曲线

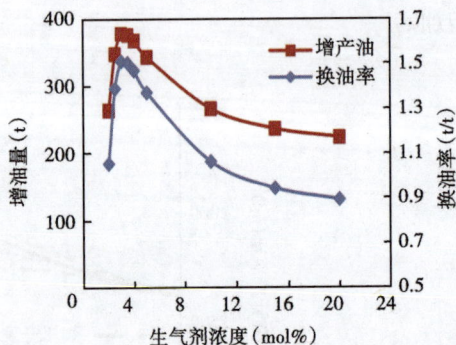

图 6　不同注入浓度增油量与换油率变化曲线

4.3　生气剂注入速度

采用已经优选的参数（生气剂注入量 250t，摩尔分数 4.0%），模拟计算生气剂注入速

242

度为 $200m^3/d$，$300m^3/d$，$400m^3/d$，$500m^3/d$、$600m^3/d$ 和 $700m^3/d$ 下的增产油量和换油率。图 7 为不同注入速度下累计产油量对比曲线，图 8 为不同注入速度下增油量与换油率变化曲线。

由图 7 可见，各个方案累计产油量曲线变化一致且相差较小，说明生气剂注入速度对吞吐效果影响不大。由图 8 可见，随着注入速度的增加，换油率和增油量均增大，但变化幅度不大，因此注入速度可以取现场最大注入量 $480m^3/d$。

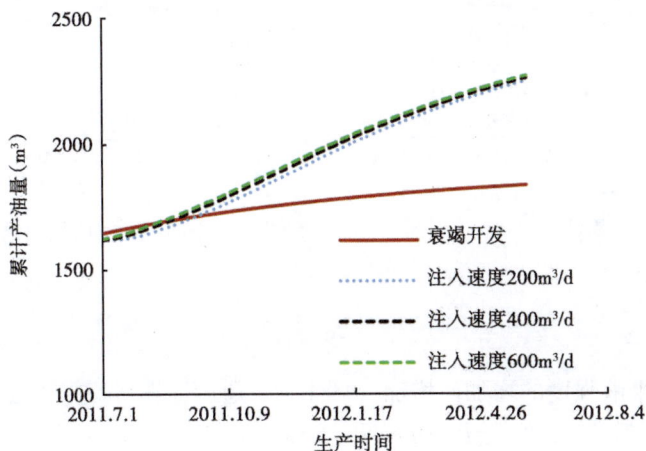

图 7　不同注入速度累计产油量对比曲线

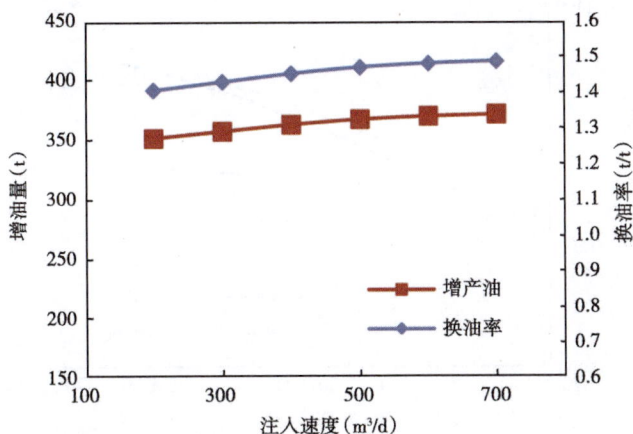

图 8　不同注入速度增油量与换油率变化曲线

4.4　关井时间

由于 CO_2 产生后需要一定的时间的分子扩散传质作用，才能溶于原油起到膨胀、降黏、抽提在蒸发作用，因此需要关井一段时间以保证 CO_2 的吞吐效果。然而关井时间太长，则可能由于浸泡期长而使注入的 CO_2 扩散到油层深部和边界，降低了井周地层 CO_2 弹性驱动能量和近混相条件，影响到油井的产量。焖井 2 天，7 天和 15 天的累计产油量曲线如图 9 所示。累计产油量随关井时间增加而增加，关井时间越久 CO_2 与原油接触越充分，溶解降黏作用越明显，吞吐效果越好。

模拟计算 7 种关井时间（2 天、4 天、7 天、10 天、15 天、20 天和 30 天）的增油量与换油率曲线如图 10 所示。随着焖井时间增加，增油量与换油率均出现上升。考虑到实际生产情况，焖井时间取 7~10 天。

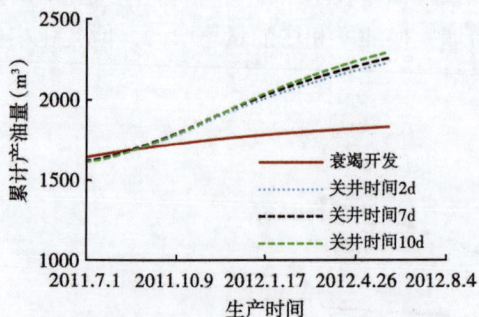

图 9　不同关井时间累计产油量对比曲线　　　图 10　不同关井时间增油量与换油率变化曲线

4.5　采液强度

累计产油量随采液强度的增加而增加（图 11）。通过计算对比采液强度为 $2m^3/d$，$3m^3/d$，$5m^3/d$，$6m^3/d$，$7m^3/d$，$8m^3/d$，$10m^3/d$，$15m^3/d$ 和 $20\ m^3/d$ 时增产油量与换油率，进行采液强度优选。

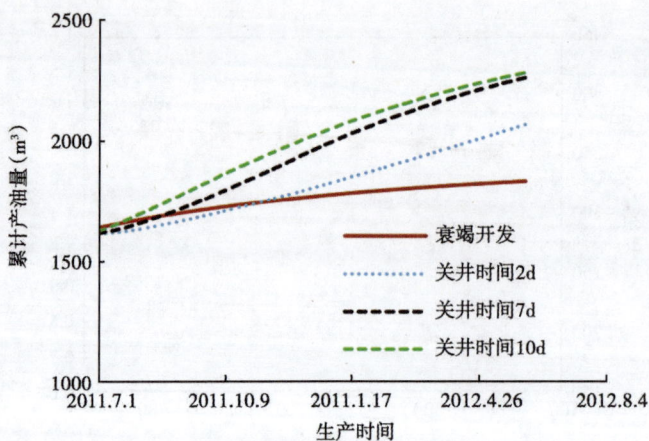

图 11　不同采液强度累计产油量对比曲线

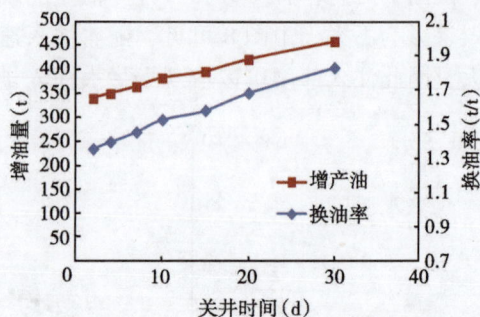

图 12 为不同采液强度下增油量与换油率变化曲线。随采液强度增加，增产油和换油率迅速增加，然后出现拐点，当采液强度超过 $7m^3/d$ 后，增油量与换油率变化很小（$1×10^4m^3$），考虑到实际生产时，若产液量太大可能导致压力降低过快，因此取最佳采液强度为 $7m^3/d$。

4.6　CO_2 吞吐增油效果

根据敏感因素分析结果，采用生气剂注入量 250t，注入摩尔分数 4%，注入速度 $480m^3/d$，关井时间 7 天，采液强度 $7m^3/d$ 进行模拟，采用 CO_2 累计产油量和衰竭开发的累计产油

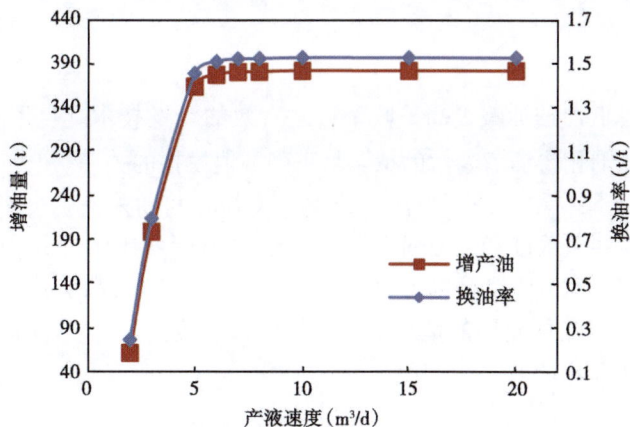

图 12 不同采液强度增油量与换油率变化曲线

量和日产油量如图 13 和图 14 所示。在 1 年内产油量达 673.4m³，增产油 447.3m³，增产近 3 倍；日产油量最高增加 3 倍，换油率达 1.52t/t，因此该区块实施自生 CO_{2+} 表面活性剂复合吞吐的前景较好。

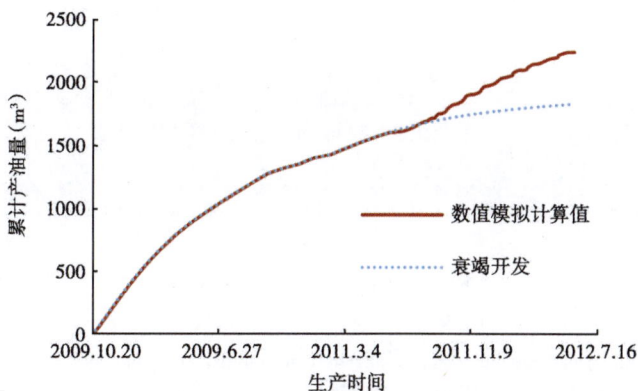

图 13 累计产油量对比曲线

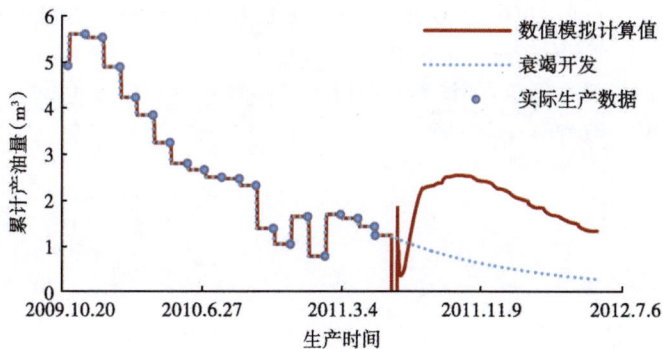

图 14 日产油量对比曲线

5 结论

（1）CO_2吞吐是低渗透油藏提高采收率行之有效的开发技术。自生CO_2体系评价实验生气速度与表面活性剂相渗插值参数的拟合是正确数值模拟的关键。

（2）生气剂注入量、注入浓度、注入速度、关井时间和采液强度影响复合吞吐效果，其中生气剂注入量、注入浓度和采液强度影响较大，并且存在最佳的取值范围。最佳生气剂量、生气剂摩尔分数、采液强度分别为$200\sim250t$，$3.0\%\sim5.0\%$和$7m^3/d$。

（3）在设备和地层条件允许的情况下，增加注入速度有利于提高油井产量；适当延长关井时间有助于提高吞吐效果。

参 考 文 献

[1] 林波，蒲万芬，赵金洲，等. 利用就地CO_2技术提高原油采收率 [J]. 石油学报，2007，28（2）：98-101.

[2] 李士伦，张正卿，冉新权. 注气提高采收率技术 [M]. 成都：四川科学技术出版社，2001，227-228.

[3] 叶仲斌. 提高采收率原理 [M]. 北京：石油工业出版社，2007（3-4）：93-103.

[4] 宋丹，蒲万芬，徐晓峰，等. 自生CO_2驱油技术体系及驱油效果研究 [J]. 石油钻采工艺，2007，29（1）：82-85.

[5] 沈德煌，张义堂，张霞，等. 稠油油藏蒸汽吞吐后转注CO_2吞吐开采研究 [J]. 石油学报，2005，26（1）：83-86.

[6] 廖海婴，吕成远，赵淑霞，等. 用数值模拟研究低渗透油藏CO_2驱影响因素 [J]. 新疆石油地质，2011，32（5）：520-522.

[7] Van Q N, Labrid J. A numerical Study of Chemical Flooding Comparison with Experiments [R]. SPE 10202, 1983：461-474.

[8] Amaefule Jude, Lyman L Handy. The Effect of Interfacial Tensions on Relative Oil/Water Permeabilities of Consolidated Porous Media [C]. SPE 9783, 1982.

[9] Handy L L, Amaefule J O, Ziegler V M, et al. Thermal Stability of Surfactants for Reservoir Application [C]. SPE 7867, 1982.

[10] 李淑霞，姜汉桥，叶惠民，等. 真12块表面活性剂驱数值模拟研究 [J]. 石油大学学报：自然科学版，2003，27（5）：44-46.

[11] 徐怀颖，张宝生. 川东北气田伴生CO_2注入川中油田 EOR 可行性分析——以普光气田伴生CO_2注入桂花油田为例 [J]. 天然气工业，2012，32（1）：101-103.

[12] 宋兆杰，李治平，赖枫鹏，等. 水驱油藏转注CO_2驱油参数优化与效果评价 [J]. 西安石油大学学报：自然科学版，2012，27（6）：42-47.

数值弥散对 CO_2 混相驱过程的影响

汪 勇[1] 汤 勇[1] 王长权[1] 徐 建[2] 姚陆峰[2]

(1. 西南石油大学油气藏地质及开发工程国家重点实验室;

2. 中国石化华东石油局钻井工程公司)

摘 要: 数值弥散效应会影响油藏数值模拟结果的精度。基于实际油藏流体和 CO_2—原油体系室内实验,建立一维均质模型,研究数值弥散对 CO_2 多级接触混相驱替过程的影响,比较数值弥散与物理弥散对 CO_2 混相驱替的影响。结果表明,数值弥散会影响驱替过程中相间传质,数值弥散越大, CO_2 会越早突破,导致采收率越低,驱替前缘变得模糊化。数值弥散与物理弥散作用相似,具有相同的数量级,二者存在细微的差别。

关键词: CO_2;混相驱;数值弥散;物理弥散

0 引言

在数值计算过程中,通常利用有限差分法对质量守恒方程进行求解。因此,不同差分格式所带来的截断误差是不可避免的,截断误差也称为"数值弥散"[1,2]。在模拟混相驱替、组分现象和其他高精度的模拟研究中,数值弥散特别重要[3]。在 CO_2 多级接触混相驱替[4-6]过程中, CO_2 与原油之间形成驱替过渡带,在过渡带中, CO_2 与原油重复接触,靠组分的传质作用而最达到混相,在驱替前缘存在明显的界面,可以利用饱和度、密度、界面张力、组分含量等来表征混相驱替过程[7-9]。在 CO_2 驱数值模拟过程中,数值弥散可能会影响模拟结果,甚至出现较大的偏差。

1 对流—扩散差分方程的数值弥散效应

在混相驱替、热采和非混相驱替过程中,对流—扩散方程的差分方程的截断误差是不可忽略的,可以通过选择合适的网格大小和时间步长来控制其达到最小[1]。Peaceman 推导出了在对流—扩散两相流过程中,数值弥散系数[10]表达式为:

$$D_{num} = vf' \Delta x \left[\left(W - \frac{1}{2} \right) + vf' (\Delta t / \Delta x) \left(\theta - \frac{1}{2} \right) \right] \tag{1}$$

式中, D_{num} 为数值弥散系数, m^2/s; v 为流体渗流速度, m/s; W 为差分方程的距离修正参数; θ 为差分方程的时间修正参数; Δx 为空间步长, m; Δt 为时间步长, s。

f 是关于变量 u 的函数, u 为通用函数, $u = u(x, y, z, t)$, $f' = df/du$。

由式(1)可以看出,通过减小 Δx 可以降低数值弥散效应,但是,若 Δx 降低到合理值后,可能模型的网格数有会增加,在实际油藏数值模型中,往往难以达到。另外,可以通过

调整距离修正系数 W 来降低数值弥散效应，但是其前提是要保证差分方程的稳定性，以免引起数值震荡和失真。

考虑物理弥散的一维对流—扩散方程[7]为：

$$D_{phys} \frac{\partial^2 u}{\partial x^2} - vf' \frac{\partial u}{\partial x} = \frac{\partial u}{\partial t} \tag{2}$$

式中，D_{phys} 为物理弥散系数，m^2/s。

若考虑数值弥散，则一维对流—扩散方程可简写为：

$$(D_{phys} + D_{num}) \frac{\partial^2 u}{\partial x^2} - vf' \frac{\partial u}{\partial x} = \frac{\partial u}{\partial t} \tag{3}$$

当物理弥散和数值弥散均较小时，驱替前缘的界面将会变得明显；若数值弥散非常严重，甚至超过了物理弥散，即 $D_{num} > D_{phys}$，则驱替前缘将会变得模糊，前缘界面将不再明显。

2 模型参数

通过建立一维水平均质模型来研究数值弥散对 CO_2 多级接触混相过程的影响，利用改变驱替方向上网格数来表征数值弥散效应的强弱[11,12]。模型长 100m，设置网格数分别为 40，50，100，200 和 400 共 5 个模型。模型流体采用真实油藏流体，共划分 10 个拟组分（表 1），流体的性质用 Peng-Robinson 状态方程[13]模拟。油气相对渗透率值采用线性的相对渗透率曲线，忽略毛细管力，模型中只有油、气两相，孔隙度为 15%，渗透率 30mD。

表 1　模拟原油拟组分及组成

组分	物质的量百分比（%）	组分	物质的量百分比（%）
N_2—C_1	11.09	C_{14}—C_{15}	7.62
C_2—C_6	9.03	C_{16}—C_{19}	14.12
C_7	4.51	C_{20}—C_{23}	8.72
C_8—C_9	16.44	C_{24}—C_{29}	7.96
C_{10}—C_{13}	17.7	C_{30}—C_{34}	2.81

模型初始压力为 35MPa，温度为 96℃，原油饱和压力为 3.84MPa，原油黏度为 5.89mPa·s，CO_2 与原油体系的最小混相压力为 26.65 MPa。CO_2 从模型的左端注入，右端为采出端，驱替过程中，注入量为 0.05m³/d，通过模型自动控制产油量来保持模型压力在 34MPa 左右。

模型解的结构类似于隐式压力/显式饱和度方程（IMPES）解的结构，求解采用的是一阶有限差分方法。

3 数值弥散对 CO_2 混相驱过程的影响

首先，不考虑物理弥散，只考虑数值弥散，图 1~图 3 分别为注入 0.5 倍孔隙体积时模型中饱和度、界面张力、密度分布曲线。由模拟结果可以看出，CO_2 混相属于蒸发—凝析式多级接触混相[4]，过渡带前部以凝析式为主，过渡带后部以蒸发式为主；驱替前缘的油气界面张力和油相密度明显降低。

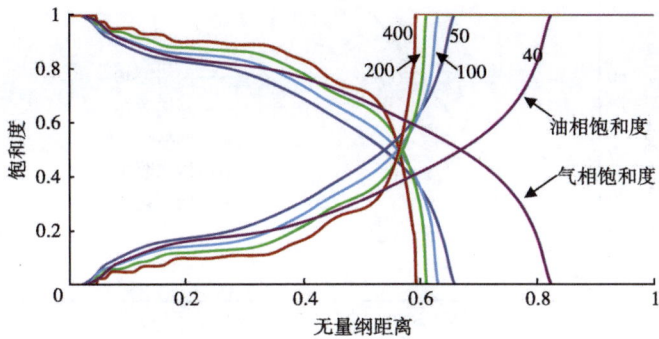

图 1　模拟中油相和气相饱和度分布

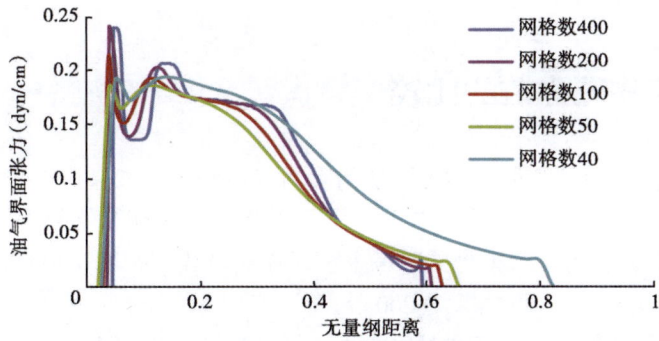

图 2　模拟中界面张力变化

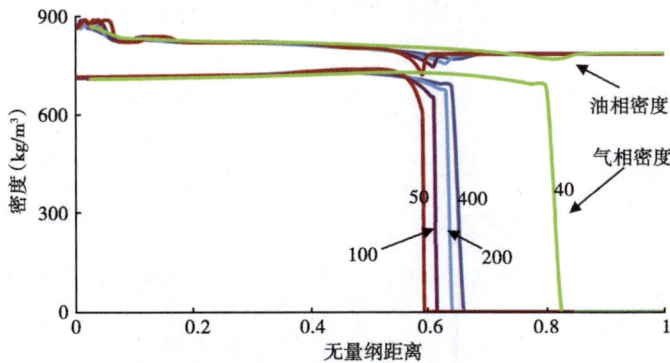

图 3　模拟中油相和气相密度分布

　　网格越大，则驱替前缘位置越靠前，驱替效率降低，CO_2 会越早突破，导致采收率降低；当网格数为 40 时，其驱替前缘与其他几个方案有着明显的差异，说明数值弥散较大，模拟结果往往会出现较大偏差（图 4），因此，该模型的网格存在一个最大值，以保证模拟结果达到一定的精度。另外，网格越大，最小的界面张力和油相密度均增加，驱替前缘变得模糊化，说明数值弥散对 CO_2—原油相间传质有较大的影响。

　　因此，在模拟过程中，利用组分模拟器对网格大小进行敏感性研究是很有必要的。

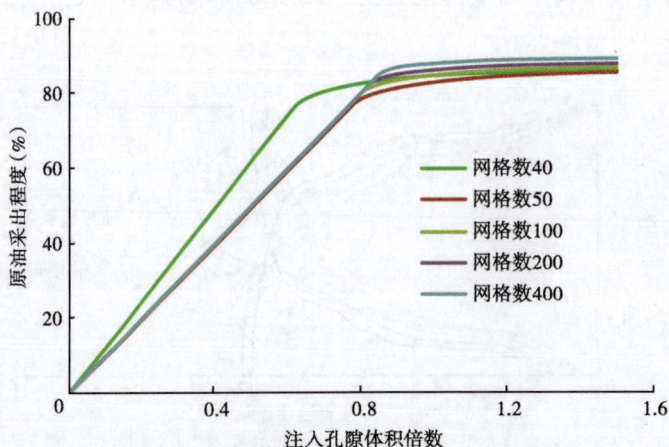

图 4　不同网格数下原油采出程度曲线

4　数值弥散与物理弥散的比较

数值模拟过程中，数值弥散与物理弥散的影响有相似之处，但并非一样[7,14]，为了比较二者的在 CO_2 混相驱过程中的异同，本文比较了两个模型的计算结果，一个模型只考虑数值弥散，网格数为100；另一个模型同时考虑物理弥散与数值弥散，模型中的物理弥散为纵向弥散[15]，弥散值为 0.4m，网格数为 200。

图 5 为模拟的采出程度与日产油对比曲线，可以看出，2 方案日产油量与采出程度变化很接近，数值弥散与物理弥散作用相似，数值弥散可以近似的补偿物理弥散效应，二者具有相同的数量级。图 6 和图 7 为 2 方案模拟的气相饱和度和密度分布，可以看出，物理弥散与数值弥散均能够使饱和度、密度分布模糊化，但二者存在在细微的差别，其对饱和度和密度分布及驱替前缘的模糊化有区别，数值弥散的效应对驱替前缘的模糊化更明显。

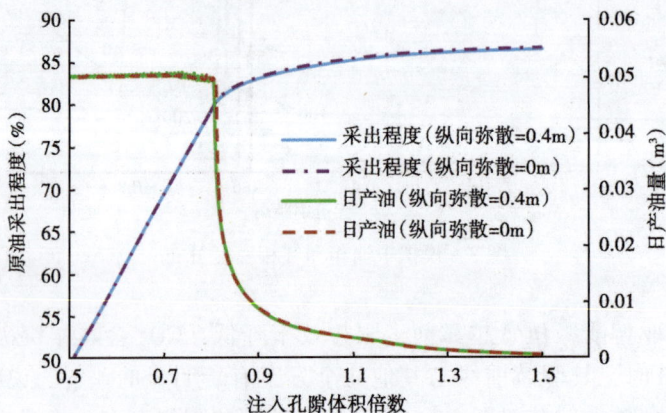

图 5　生产指标变化曲线（数值弥散与物理弥散对比）

实际油藏中的物理弥散是不确定的、不断变化的，其与分子扩散、孔隙结构、饱和度、渗流速度等有关系，本文模拟只是近似模拟。在考虑物理弥散的 CO_2 混相驱模拟过程中，采用密网格可以使模拟值更准确。

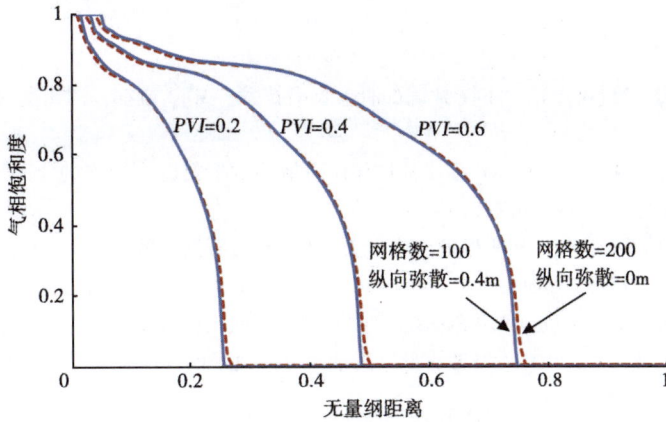

图 6　模拟中气相饱和度分布（PVI 指孔隙体积倍数）

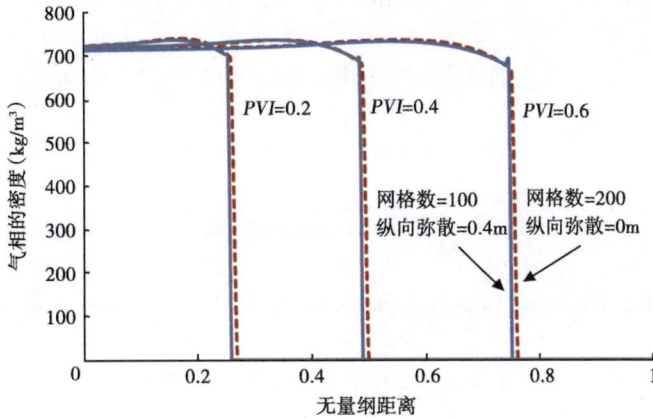

图 7　模拟中气相密度分布

5　结论

（1）在 CO_2 多级接触混相驱过程中，数值弥散越大，驱替前缘位置越靠前，驱替效率降低，CO_2 会越早突破，导致采收率降低；在具体的模型中，存在一个最大的数值弥散值。另外，数值弥散越大，最小的界面张力和油相密度均增加，驱替前缘变得模糊化，说明数值弥散对 CO_2-原油相间传质有较大的影响。

（2）数值弥散与物理弥散作用相似，二者具有相同的数量级，数值弥散可以近似地补偿物理弥散效应，二者均能够使驱替前缘模糊化，但存在细微的差别。因此，为保证模拟精度，有必要对模型的数值弥散进行敏感性研究。

参 考 文 献

［1］ Lantz R B. Quantitative Evaluation of Numerical Dispersion（Truncation Error）［J］. SPE 2811, 1971：315-320.

［2］ Orr F M J. The Theory of Gas Injection Processes ［M］. Copenhagen, Denmark：Tie-Line Publications, 2007：213-230.

［3］ 哈立德 阿齐兹，安东尼 塞特瑞. 油藏数值模拟 ［M］. 袁士义，王家禄，译. 北京：石油工业出版社，2004，267-268.

［4］ 李士伦，张正卿，冉新权，等. 注气提高石油采收率技术 ［M］. 成都：四川科学技术出版社，2001，52-54.

［5］ 李向良，王庆奎，李振泉，等. CO_2 多次抽提作用对地层油析蜡温度影响的实验研究 ［J］. 大庆石油地质与开发，2007，26 (3)：107-110.

［6］ 汤勇，孙雷，周涌等. 注气混相驱机理评价方法 ［J］. 新疆石油地质. 2004，25 (4)：414-416.

［7］ Haajizadeh M, Fayers F J, Cockin A P, et al. On the Importance of Dispersion and Heterogeneity in the compositional Simulation of Miscible Gas Processes ［J］. SPE 57264, 1999.

［8］ 吴文有，张丽华，陈文彬. CO_2 吞吐改善低渗透油田开发效果可行性研究 ［J］. 大庆石油地质与开发，2001，20 (6)：51-53.

［9］ 徐永成，王庆，韩军，等. 应用 CO_2 吞吐技术改善低渗透油田开发效果的几点认识 ［J］. 大庆石油地质与开发，2005，24 (4)：69-71.

［10］ Peaceman D W. Fundamentals of Numerical Reservoir Simulation ［M］. Elsevier Scientific Publishing. 1977：74-81.

［11］ 李菊花，李相方，姜涛，等. 油藏注烃气混相驱非稳态驱替特征机理研究 ［J］. 大庆石油地质与开发，2010，27 (4)：95-99.

［12］ 李福恺，贾文瑞. 注气非混相驱数值模拟机理研究 ［J］. 大庆石油地质与开发，1994，13 (3)：46-49.

［13］ Peng D Y, Robinson D B. A New Two-Constant Equation of State ［J］. Industrial Engineering Chemistry Fundamental, 1976, 15 (1)：59-64.

［14］ Ghanbarnezhad M R. Numerical Dispersion Impact on Local Mixing in Heterogeneous Reservoirs ［J］. SPE 149420. 2011.

［15］ 李菊花，杨红梅，刘滨，等. 油藏注气混相驱考虑扩散作用的数值模拟研究 ［J］. 油气地质与采收率，2010，17 (6)：54-56.

改进的 CO_2—原油最小混相压力计算模型

陈百炼[1] 黄海东[1] 章 杨[1] 王 蕾[1]

任韶然[1] 黄安源[2] 孙平平[3]

(1. 中国石油大学（华东）；2. 中石化胜利油田分公司孤岛采油厂；

3. 渤海钻探第三钻井公司)

摘 要：利用试验测定的多个油田 CO_2—原油最小混相压力（MMP）数据，建立了改进的 MMP 预测模型，并将该模型与其他模型进行了对比。结果表明，改进模型相比于其他模型具有更高的计算精度和稳定性，尤其在 20~30MPa 的混相压力范围内比其他模型计算精度更高，平均误差仅为 2.32%。对国内外 35 个典型的陆相和海相原油组分组成及相应 MMP 的统计分析发现，由于陆相原油组分中 C_2—C_6 的平均含量要明显低于海相的，重组分 C_{16+} 的含量又明显高于海相的，导致 CO_2 与陆相原油的 MMP 明显高于海相原油的。利用该模型对吉林油田和胜利油田不同区块油样的 CO_2 驱最小混相压力进行了计算，其计算值与试验值的误差均在平均误差之内，验证了改进的模型对我国陆相油田原油的适用性。

关键词：二氧化碳驱油；最小混相压力；预测模型；陆相原油

0 引言

最小混相压力（简称 MMP，量符号为 $p_{m,min}$）是评价和设计 CO_2 混相驱中的一个重要参数[1]。由于常规试验方法确定 MMP 需要消耗大量时间，且花费较大，探索一种快速而准确的模型来确定 CO_2 与原油的 MMP 是非常必要的。从国内外 CO_2—原油 MMP 预测模型的研究现状来看，从最早的仅仅与温度相关联[2,3]，到后来通过某些数学方法（如 ACE 法）建立起与油藏温度、原油组分及 CO_2 流体组成等参数相关联的模型[4-6]，其预测精度在一定程度上得到提高，但计算过程比较繁琐，或对原油组分测试的要求较高，且在预测较高的 MMP 时偏差较大，不太适用于我国陆相油田原油。该研究依据 CO_2 与原油传质扩散混相机制，在 Alston 模型及 GA 模型的基础之上，选取油藏温度 t、原油中的挥发组分（C_1+N_2）摩尔分数 x（C_1+N_2）、中间组分（C_2—C_6）摩尔分数 x（C_2—C_6）及 C_{7+} 的相对分子质量 M_r（C_7）为主要参数，运用回归理论建立了精度高、计算简单、适用压力范围广的改进 MMP 预测新模型。

1 改进模型的建立

1.1 模型的引出

影响 CO_2 和原油 MMP 的主要因素有油藏温度、原油组分和注入 CO_2 流体的组成[4,7—10]。Lee，Yelling 等[2,3]在前期的 MMP 模型研究中，主要考虑了温度这一参数。Metcalfe 等[11]则认为在进行 MMP 预测中，除了温度以外，还必须考虑原油中的轻馏分和中间组分。Alston 等[8]通过细管试验证明 CO_2 与原油的 MMP 随着原油中挥发组分与中间组分摩尔分数之比的增加而增加，计算模型中采用 C_{5+} 的相对分子质量 M_r（C_{5+}）比使用原油 API 重度更好。Emera 等[4]在 MMP 模型研究中，将油藏温度、C_{5+} 相对分子质量以及挥发组分和中间组分摩尔分数之比作为重要参数，采用遗传算法，在 Alston 模型基础之上发展了精度更高的 GA 模型。而 Alston 模型和 GA 模型都是以挥发组分（C_1+N_2）和中间轻质组分（C_2—C_4）的摩尔分数之比作为主要参数之一，认为原油中挥发组分和中间轻质组分的敏感程度一样，MMP 与二者的比值呈幂指数关系。Rathmell 等[12]研究发现，在原油组分中，不仅中间轻质组分 C_2—C_4 对 MMP 有降低作用，C_5 和 C_6 对 MMP 也有降低作用。另外，根据秦积舜等[13]对 CO_2 与原油传质混相过程的研究，在 CO_2—原油混合流体不断推进过程中，原油中不断地有新的轻质组分溶解到 CO_2 与原油混相液中，而轻质组分主要为 C_2—C_6，具有强传质作用，极易混相。可以看出，降低 CO_2—原油 MMP 的不仅仅是 C_2—C_4，C_5 和 C_6 也具有较强的传质作用，能显著降低 CO_2—原油 MMP。为此，在这些理论和试验观察分析基础之上，对 Alston 模型及在该模型基础上发展起来的 GA 模型进行改进，考虑到国内外原油组分通常表征到 C_{7+}，故在改进的模型中，将 t，$x_{(C_1+N_2)}$，$x_{(C_2—C_6)}$ 及 M_r（C_{7+}）作为主要参数建立更为精确和简单实用的 CO_2—原油 MMP 预测模型。

1.2 模型的建立

在改进模型中，认为挥发组分（C_1+N_2）和中间轻质组分（C_2—C_6）对 MMP 影响的敏感性不一样，需要单独考虑二者对 MMP 的影响，即赋予不同的指数函数值，而不是将二者的摩尔分数之比作为一个参数。试验样本取自国内外文献中 39 组 CO_2—原油 MMP 的试验数据[8,11,12,14—24]，具体见表 1（其中，试验的最小混相压力量符号定义为 $p_{m,min}^{exp}$）。

表 1　文献中 CO_2—原油 MMP 试验数据

文献来源	t（℃）	x（%）			M_r		$p_{m,min}^{exp}$（MPa）
		C_1+N_2	C_2—C_4	C_2—C_6	C_{5+}	C_{7+}	
8	67.8	31	22.9	23	203.81	210	16.90
8	112.2	32.7	28.1	28	213.5	220	24.15
8	110.0	32.51	35.64	36	180.6	185	20.21
8	71.1	41.27	6.99	7	221	227	23.45
8	102.2	51.28	9.84	10	205	210	28.17
8	80.0	53.36	8.6	9	240.7	245	26.76
11	32.2	10.5	14.28	24.44	187.77	206	6.90
11	40.6	10.5	14.28	24.44	187.77	206	8.28
11	57.2	10.5	14.28	24.44	187.77	206	11.86
11	49.0	34.34	22.82	26.39	187.25	200	11.04
11	65.6	34.34	22.82	26.39	187.25	200	13.45
12	42.8	17.07	20.95	28.83	204.1	222	10.35

文献来源	t（℃）	x（%）			M_r		$p_{m,min}^{exp}$（MPa）
		C_1+N_2	$C_2—C_4$	$C_2—C_6$	C_{5+}	C_{7+}	
14	42.8	17	—	28.8	204	222	10.69
14	32.2	11	—	25	188	206	7.58
14	110.0	33	—	36	181	185	20.20
14	71.1	41	—	7	221	227	23.44
14	102.2	51	—	10	205	210	28.17
14	67.8	31	—	23	204	210	16.89
14	112.2	33	—	28	214	220	24.15
14	80.0	53	—	9	241	245	26.75
14	31.1	12	—	24.2	205	240	8.10
14	73.3	49	—	8.84	210	218	24.13
14	40.0	24	—	30.7	202	221	9.07
14	54.4	30	—	37.3	169	190	11.78
14	48.9	16	—	31	214	227	10.58
14	40.0	8.3	—	31	191	205	8.78
15	59.0	5.45	11.35	18.93	205	220	12.80
16	34.4	16.78	10.76	17.7	212.56	227	10.00
17	48.9	12.5	22.62	31	205.1	227	10.59
18	71.1	4.4	13.9	24.68	207.9	227.94	15.52
19	54.4	29.48	31.82	40.05	171.2	197.4	11.00
20	42.8	19.35	26.8	33.12	196.1	221	10.62
21	118.3	34.2	28.6	28.58	171.1	192	23.45
22	81.1	9.82	15.59	26.78	—	220	15.96
23	59.0	5.45	11.47	21.67	—	230	11.70
23	61.0	6.13	10.3	26.34	—	203	12.00
23	63.0	9.56	18.62	28.28	—	215	14.20
24	98.9	18.71	11.61	17.51	222	242	22.30
24	97.3	19.01	11.82	17.74	229	249	22.10

注：“—”表示相关文献中没有给出该数据。

选择 t，$M_{r(C_{7+})}$，$x_{(C_1+N_2)}$ 以及 $x_{(C_2—C_6)}$ 作为新模型的回归变量，模型的主要形式为：

$$p_{m,\,min} = at^b\big[M_{r(C_{7+})}\big]^c\big[x_{(C_1+N_2)}\big]^d\big[x_{(C_2—C_6)}\big]^{-e} \qquad (1)$$

经过回归计算，得到：

$$p_{m,\,min} = 3.9673\times10^{-2}t^{0.8293}\big[M_{r(C_{7+})}\big]^{0.5382}\big[x_{(C_1+N_2)}\big]^{0.1018}\big[x_{(C_2—C_6)}\big]^{-0.2316} \qquad (2)$$

式中，$p_{m,\,min}$ 代表 CO_2—原油最小混相压力，MPa；t 代表油藏温度，℃；$M_{r(C_{7+})}$ 代表 C_{7+} 的相对分子质量，1；$x_{(C_1+N_2)}$ 代表挥发组分 C_1 和 N_2 的摩尔分数，%；$x_{(C_2—C_6)}$ 代表中间轻质组

分（C_2—C_6）的摩尔分数,%；a，b，c，d，e 代表回归系数。

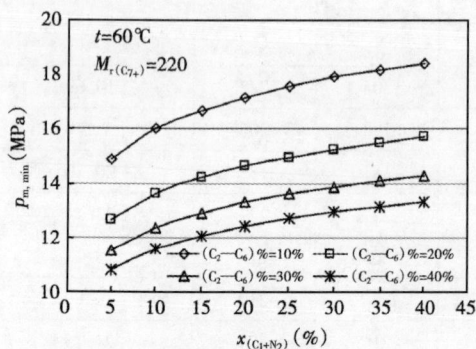

图 1　CO_2—原油 MMP 随原油中挥发组分及中间轻质组分摩尔分数的变化关系

越小，则 $p_{m,min}$ 也就越小。

整个拟合的相关系数为 0.9866，精度较高，拟合公式能反映样本数值规律。

分析式（2）可以看出，t 对 $p_{m,min}$ 的影响程度最大，$M_{r(C_{7+})}$ 的影响程度次之。另外，中间轻质组分（C_2—C_6）的敏感性比挥发组分（C_1+N_2）的要大，这正好验证了模型建立之前的假设。图 1 给出了采用该模型计算得到的 CO_2—原油 MMP 随挥发组分及中间轻质组分变化的关系，从图中可以看出，原油组分中的中间轻质组分（C_2—C_6）对 MMP 具有降低作用，而挥发组分（C_1+N_2）则使 MMP 值升高。在 t 和 $M_{r(C_{7+})}$ 一定的情况下，$x_{(C_2—C_6)}$ 越大，$x_{(C_1+N_2)}$

2　改进模型的应用

2.1　改进模型与其他模型的对比

改进模型 MMP 计算值和试验值对比如图 2 所示，可以看出计算值和试验值所对应的点很好地分布在直线 $y=x$ 附近。为验证模型计算的准确性，选取常用的其他模型进行对比，如 Cronquist 模型[26]、Yelling 模型、Alston 模型、Glaso 模型、GA 模型、Yuan 模型和 CCMMP

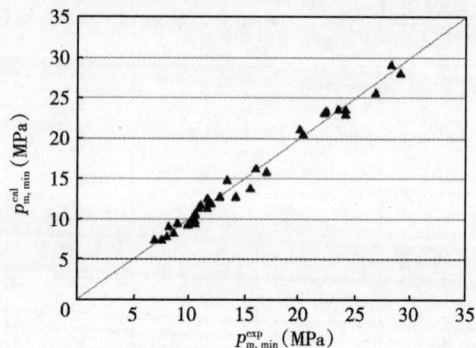

图 2　改进模型 MMP 计算值与试验值关系

模型。从对比结果可以看出，采用新模型计算产生的平均误差为 4.12%，标准偏差为 5.24%；而采用其他 7 种模型产生的平均误差分别为 15.46%、14.57%、9.63%、7.49%、6.18%、11.16% 和 9.85%，标准偏差分别为 14.75%、14.35%、11.02%、8.48%、8.13%、13.61%、11.52%，这表明改进的模型具有更高的计算精度和更好的稳定性。

图 3 给出了采用改进模型和其他 7 种模型的 CO_2—原油 MMP 计算值（$P_{m,min}^{cal}$）与试验值的对比。分析图 3 可以看出，改进模型除外，GA 模型和 Glaso 模型的计算准确度相比于其他几个模型要高一些，Alston 和 CCMMP 模型次之。对于 Alston 和 GA 这两个模型在计算 MMP 值小于 20MPa 时的准确度相对较高，但当 MMP 值高于 20MPa 后，计算精度明显降低；而 Yuan 和 CCMMP 模型则刚好相反，计算 MMP 值高于 20MPa 的准确度要高于 MMP 值小于 20MPa 的准确度；Yelling 模型的计算值比试验值整体偏低，高于 20MPa 之后偏离程度更大。

通过上述分析可以看出，改进模型相比其他模型来说，不仅在计算低于 20MPa 的 MMP 值时有较高的精度，在计算高于 20MPa 的 MMP 时也具有更高的精度。为进一步证明改进模型在计算高于 20MPa 的 MMP 时的优越性，对表 1 中高于 20MPa 和低于 20MPa 的 MMP 分别进行误差分析，结果表明改进模型在计算 MMP 超过 20MPa 时，其平均误差仅为 2.32%，而其他几个模型平均误差分别为 16.45%、26.29%、15.34%、8.19%、9.77%、7.16% 和

6.76%，均明显高于改进模型的平均误差。

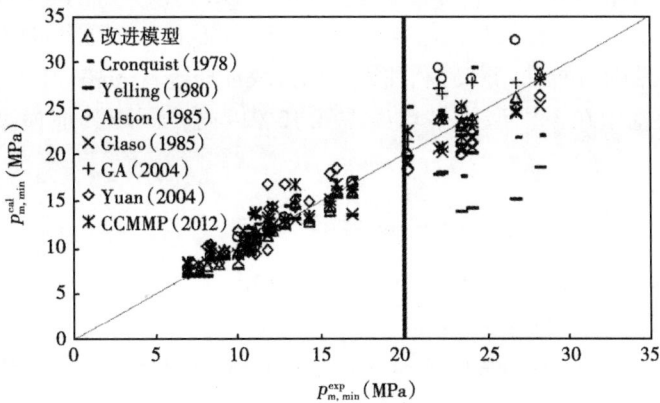

图 3 不同模型 MMP 计算值与试验值对比

2.2 改进模型的适用性

通过对国内外 35 个典型的陆相和海相原油组分组成及相应 MMP 的统计分析发现，CO_2 与陆相原油的 MMP 明显高于海相原油的，这主要是由于陆相原油组分中 C_2—C_6 的平均含量要明显低于海相的，重组分 C_{16+} 的含量又明显高于海相的（表 2），根据上文中的分析，原油组分中 C_2—C_6 对 MMP 有降低的作用，重组分 C_{16+} 对 MMP 有增加的作用，因而导致陆相原油的 MMP 会明显高于海相原油的。由于我国油田多为陆相油田，MMP 普遍偏高，采用该模型进行预测时，能较准确地计算出 MMP 值的大小。

表 2 陆相和海相原油组分组成分布对比

原油类型	x（%）							
	C_1+N_2		C_2—C_6		C_7—C_{15}		C_{16+}	
	主要分布	平均	主要分布	平均	主要分布	平均	主要分布	平均
陆相	15~30	22.28	10~20	13.15	25~35	30.52	25~40	32.51
海相	20~30	26.21	25~30	27.18	30~35	34.78	5~15	10.41

为进一步验证该模型的适用性，选用吉林油田 H115，H24 和 D-LYF 三个区块和胜利油田 G89 区块原油，采用细管试验对 CO_2—原油的 MMP 进行测定。结果表明，采用该方法计算的结果与试验结果非常接近，相对误差均在该模型的平均误差 4.12% 以内，说明该方法更适用于我国陆相油田原油的 MMP 预测。

3 结论

（1）根据我国油田常用原油组分的表征参数，建立了改进的 MMP 计算模型。与现有的计算模型相比，改进的模型在计算 CO_2—原油 MMP 时能产生更小的平均误差和标准偏差，分别为 4.12% 和 5.24%，具有更高的精度和稳定性，简单实用。

（2）改进的模型相比其他模型在计算高于 20MPa 的 MMP 时，优越性明显，平均误差仅为 2.32%。

（3）由于中间轻质组分（C_2—C_6）含量较低和重质组分含量较高，我国陆相原油 MMP 普遍偏高，采用该模型进行预测时，能较准确地计算出陆相油田原油 MMP 的大小。

（4）采用该方法对吉林油田和胜利油田不同区块油样进行 MMP 计算，其结果与试验值非常接近，相对误差均在该模型的平均误差 4.12% 以内，而用其他模型计算的误差相对较大。

参 考 文 献

［1］李春芹 . CO_2 混相驱技术在高 89-1 块特低渗透油藏开发中的应用 ［J］. 石油天然气学报（江汉石油学院学报），2011，33（6）：328-329.

［2］Lee J. Effectiveness of Carbon Dioxide Displacement under Miscible and Immiscible Conditions ［R］. Report RR -40, Petroleum Recovery Inst, Calgary, 1973.

［3］Yelling W F, Metcalfe R S. Determination and Prediction of CO_2 Minimum Miscibility Pressures ［J］. J. Pet. Technol., 1980, 32（1）：160-168.

［4］Emera M K, Sarma H K. Use of Genetic Algorithm to Estimate CO_2-oil Minimum Miscibility Pressure-A Key Parameter in Design of CO_2 Miscible Flood ［J］. J. Pet. Sci. Eng., 2005, 46（1）：37-52.

［5］Shokir EM. CO_2-oil Minimum Miscibility Pressure Model for Impure and Pure CO_2 Streams ［J］. J. Pet. Sci. Eng., 2007, 48（1）：173-185.

［6］鞠斌山，秦积舜，李治平，等 . 二氧化碳—原油体系最小混相压力预测模型 ［J］. 石油学报，2012，33（2）：274-277.

［7］Johnson J P, Pollin J S. Measurement and Correlation of CO_2 Miscibility Pressures ［J］. SPE 9790, 1981.

［8］Alston R B, Kokolis G P, James C F. CO_2 Minimum Miscibility Pressure：a Correlation for Impure CO_2 Streams and Live Oil Systems ［J］. Soc Pet Eng J, 1985, 25（2）：268-274.

［9］Sebastian H M, Wenger R S, Renner T A. Correlation of Minimum Miscibility Pressure for Impure CO_2 Streams ［J］. J Pet Technol, 1985, 37（11）：2076-2082.

［10］Nasrifar K, MoShfeghian M. Application of an Improved Equation of State to Reservoir Fluids：Computation of Minimum Miscibility Pressure ［J］. J. Pet. Sci. Eng., 2004, 42（1）：223-234.

［11］Metcalfe R S, Yarborough L. Discussion ［J］. J. Pet. Technol., 1974, 26（12）：1436-1437.

［12］Rathmell J J, Stalkup F I, Hassinger R C. A Laboratory Investigation of Miscible Displacement by Carbon dioxide ［J］. SPE 3483, 1971.

［13］秦积舜，张可，陈兴隆 . 高含水后 CO_2 驱油机理的探讨 ［J］. 石油学报，2010，31（5）：797-800.

［14］Yuan H, Johns R T, Egwuenu A M, et al. Improved MMP Correlations for CO_2 Floods using Analytical gas Flooding Theory ［J］. SPE 89359, 2004.

［15］Dong M, Huang S, Dyer S B, et al. A comparison of CO_2 Minimum Miscibility Pressure Determinations for Weyburn Crude Oil ［J］. J. Pet. Sci. Eng., 2001, 31（1）：13-22.

［16］Shelton J L, Yarborough L. Multiple Phase Behavior in Porous Media during CO_2 or Rich-gas Flooding ［J］. J. Pet. Technol., 1977, 29（9）：1171-1178.

［17］Henry R L, Metcalfe R S. Multiple-phase Generation During Carbon Dioxide Flooding ［J］. Soc. Pet. Eng. J., 1983, 23（4）：595-601.

［18］Graue D J, Zana E T. Study of a Possible CO_2 Flood in Rangely Field ［J］. J. Pet. Technol., 1981, 33（7）：1312-1318.

［19］Dicharry R M, Perryman T L, RonquiIle J D. Evaluation and Design of CO_2 Miscible Flood Project-SACROC unit Kelly-Snyder Field ［J］. J. Pet. Technol., 1973, 25（11）：1309-1318.

［20］Gardner J W, 0rr F M, Pdtel P D. The Effect of Phase Behavior on CO_2-flood Displacement Efficiency ［J］. J. Pet. Technol. , 1981, 33（11）: 2067-2081.

［21］Thakur G C, Lin C J, Patel Y R. CO_2 Mini-test, Littles Knife Field, ND: a Case History ［J］. SPE 12704, 1984.

［22］尤启东, 吕广忠, 栾志安. 求解最小混相压力方法的改进 ［J］. 西安石油学院学报: 自然科学版, 2003, 18（2）: 32-35.

［23］郝永卯, 陈月明, 于会利. CO_2 驱最小混相压力的测定与预测 ［J］. 油气地质与采收率, 2005, 12（6）: 64-66.

［24］张可. CO_2 与地层油体系界面特征及应用研究 ［D］. 北京: 中国科学院渗流流体力学研究所, 2011.

［25］Holm L W, Josendal V A. Effect of Oil Composition on Miscible-type Displacement by Carbon Dioxide ［J］. J. Pet. Technol. , 1982, 22（1）: 87-98.

［26］Cronquist C. Carbon Dioxide Dynamic Displacement with Light Reservoir Oils ［A］. Proceeding of the 1978 US DOE Annual Symposium ［C］. Tulsa: 1978.

气驱开发油藏井网密度数学模型

王高峰　马德胜　宋新民　杨思玉　秦积舜　胡永乐

(中国石油勘探开发研究院提高采收率国家重点实验室)

摘　要：井网密度对气驱开发效果有决定性影响。首次提出了考虑当前采出程度和混相程度的气驱采收率计算公式。以气驱采收率计算为基础，从技术经济学观点考察了注气项目在评价期内的投入产出情况，建立了气驱井网密度与净现值之间联系。利用驻点法求极值方法得到了气驱开发油藏的经济最优井网密度和经济极限井网密度的数学模型。研究成果对于注气潜力评价和油藏注气开发实践有重要指导意义。

关键词：气驱采收率；气驱油藏井网密度；技术经济；数学模型

0　引言

注气提高石油采收率实践正在国内蓬勃开展。实践表明，井网密度对气驱开发效果有决定性影响。由于国内规模性气驱实践经验少加之气驱本身的复杂性，气驱采收率计算方法和气驱井网密度的计算方法还未见报道。储层地质情况、转驱时剩余储量和气驱混相程度决定了注气的可行性，也决定了气驱井网密度。本文首先提出气驱采收率计算方法，再以之为基础，应用技术经济学分析方法研究了注气开发油藏井网密度问题。首次给出了注气开发油藏的经济最优和经济极限井网密度的数学模型，很好地指导了国内的注气实践。

1　理论推导

1.1　气驱采收率计算

由苏联学者提出的形如 $E_{Rw} = E_{Dw} \cdot e^{-\frac{\alpha}{s}}$ 的水驱采收率计算式已为国内外众多学者所完善和发展[1-5]。这一类型的采收率计算式之所以有重要实用价值是由于其在采收率和井网密度之间建立了联系。气驱井网密度的油藏工程研究也离不开注气采收率的标定。气驱采收率受控于储层地质条件、转驱时剩余储量和混相程度以及气驱体积波及系数决定的。考虑到注气与注水在波及体积上的不同，引入波及系数修正因子 η 来体现这一差异，其经验取值按表1。而储层地质条件已在水驱采收率计算公式中得以反映。那么，只要再计入转驱时剩余储量和气驱混相程度的影响，就可以通过修正水驱采收率公式后得到气驱采收率。据此提出水驱转气驱后的采收率计算式为：

$$E_{Rg} = \eta \frac{E_{Dg}}{E_{Dw}} \cdot \frac{S_o}{S_{oi}} E_{Rw} \tag{1}$$

式（1）在气驱与水驱采收率之间建立了联系，使前人在水驱采收率计算方面的大量研究成果可直接用于气驱采收率工程计算，也为评价气驱潜力提供了重要依据。

表 1　波及系数修正因子 η 经验取值

油层	薄油层	高倾角油层	正韵律厚油层	复合韵律厚油层
η	1.0	1.0	1.0~1.1	1.0

1.2　气驱投入产出分析

应用技术经济学分析方法对注气项目在评价期内的投入产出进行分析，以在总利润现值和井网密度之间建立联系。用采油速度 R_{vg} 反映注气开发年产量的变化情况，并按 $1000m^3$ 烃气折成 1t 油近似考虑伴生气回收。评价期内油气销售总收入为原油加上回收烃气的总值为

$\sum_{j=1}^{n} [P_o \alpha_{og} N R_{vg}]_j (1+i)^{-j}$，$\alpha_{og} = \alpha_o + 0.001 \alpha_g C_H R_t$。总经营成本为 $\sum_{j=1}^{n} [P_m \alpha_{og} N R_{vg}]_j (1+i)^{-j}$。

若平均单井固定投资（含钻井、地面工程建设与非安装设备投资）为 P_w，建设期为 T_c，固定投资贷款偿还期 T，则固定投资及建设期利息为 $\sum_{j=1}^{T} \frac{sA}{T} P_w (1+i_0)^{T_c} (1+i)^{-j}$。固定资产残值率按 3% 计算[6]，则回收固定资产余值为 $0.03 sA P_w (1+i)^{-n}$。考虑到开发前期花费的流动资金要在开发后期回收，投入产出分析时可不计流动资金。原油销售税金包括增值税、营业税、城市维护建设税和教育费附加[6]。油田开发建设项目一般不计营业税，若将基于含税价格的增值税、教育费附加和城市维护建设税"三税"税率之和记为 r_{st}，则原油销售税金 $r_{st} \sum_{j=1}^{n} [P_o \alpha_{og} N R_{vg}]_j (1+i)^{-j}$。资源税和石油特别收益金总额为 $\sum_{j=1}^{n} [Q \alpha_{og} N R_{vg}]_j (1+i)^{-j}$。

注气项目在评价期内的总收入为原油及伴生气总销售收入与回收固定资产余值之和。总支出则包括总经营成本、固定投资及利息、总销售税金、资源税和石油特别收益金。总利润现值 NPV 为总收入减去总支出：

$$NPV = \sum_{j=1}^{n} \{ [(1-r_{st})P_o - P_m - Q] \alpha_{og} N P_{vg} \}_j (1+i)^{-j} - $$
$$ sA P_w \left[\frac{(1+i_0)^{T_c}}{T} \sum_{j=1}^{T} (1+i)^{-j} - 0.03(1+i)^{-n} \right] $$

1.3　气驱井网密度数学模型

总利润最大时的井网密度就是经济最优井网密度。通过求解 $\frac{\partial NPV}{\partial s} = 0$，可得经济最优井网密度 s_r，式（2）即为经济最优气驱井网密度数学模型，可用 Newton 法迭代求解，而采油速度可由气驱采收率和递减规律得到。

$$\begin{cases} s_r^2 = \dfrac{\alpha N \sum\limits_{j=1}^{n} \{ [(1-r_{st})P_o - P_m - Q] \alpha_{og} R_{vg} \}_j (1+i)^{-j}}{A P_w \left[\dfrac{(1+i_0)^{T_c}}{T} \sum\limits_{j=1}^{T} (1+i)^{-j} - 0.03(1+i)^{-n} \right]} \\ E_{Rg} = \eta \dfrac{E_{Dg}}{E_D} \cdot \dfrac{S_o}{S_{oi}} E_{Rw}, \quad E_{Rg} = \sum\limits_{j=1}^{n} R_{vgj} \end{cases} \quad (2)$$

总利润为零时的井网密度为经济极限井网密度。通过求解 $NPV=0$ 得经济极限井网密度 s_c。式（3）是经济极限气驱井网密度数学模型，可用 Newton 法迭代求解：

$$\begin{cases} s_c = \dfrac{N \sum\limits_{j=1}^{n} \left\{ \left[(1-r_{st})P_o - P_m - Q \right] \alpha_{og} R_{vg} \right\}_j (1+i)^{-j}}{AP_w \left[\dfrac{(1+i_0)^{T_c}}{T} \sum\limits_{j=1}^{T} (1+i)^{-j} - 0.03(1+i)^{-n} \right]} \\[4mm] E_{Rg} = \eta \dfrac{E_{Dg}}{E_{Dw}} \cdot \dfrac{S_o}{S_{oi}} E_{Rw} , \quad E_{Rg} = \sum\limits_{j=1}^{n} R_{vgj} \end{cases} \qquad (3)$$

2 应用实例

将本文模型应用于东部某 CO_2 驱油藏。计算得该特低渗油藏的经济最优气驱井网密度为 11 口/km^2，经济极限气驱井网密度 19 口/km^2。当油价 45 美元/桶时，经济最优与经济极限井网密度相等，表明油价过低时，该注气项目不能盈利。矿场注气试验按此结果部署井网，取得了良好效果。

表 1　CO_2 混相驱技术经济参数与取值

含油面积（km^2）	2	资源税和特别收益金（元/t）	169
地质储量（10^4t）	90	固定投资贷款偿还期（a）	5
渗透率（mD）	3	建设期年限（a）	2
原油商品率	0.98	单井固定投资（万元）	350
气油比（m^3/t）	36	贴现率	0.12
经营成本（元/t）	1200	原油价格（元/t）	2523

3 结论

（1）所提出的气驱采收率计算公式在气驱采收率与水驱采收率之间建立了联系，并考虑了气驱混相程度和转驱时采出程度的影响，使得气驱采收率的计算有据可依。

（2）首次给出了气驱开发油藏的经济最优与经济极限井网密度数学模型，实际应用表明本文气驱井网密度数学模型可用于指导注气实践。

符号注释

α——常数；

s——井网密度，口/km^2；

E_{Rg}，E_{Rw}——气驱、水驱采收率；

E_{Dg}，E_{Dw}——气驱、水驱驱油效率；

n——评价期，a；

R_t——气油比，m^3/t；

C_H——产出气中烃类体积分数；

262

α_{o}——原油商品率；

α_{g}——烃气商品率；

S_{oi}——原始含油饱和度；

S_{o}——当前含油饱和度；

N——地质储量，10^4t；

A——含油面积，km^2；

P_{o}——油价，元/t；

s_{r}，s_{c}——经济最优和经济极限井网密度，口/km^2；

P_{w}——平均单井固定投资，万元/井；

i_{0}——固定投资贷款利率；

i——贴现率；

Q——吨油上缴资源税和特别收益金，元；

j——年份。

参 考 文 献

[1] 俞启泰. 计算水驱砂岩油藏合理井网密度与极限井网密度的一种方法 [J]. 石油勘探与开发，1986，13（4）：49-54.

[2] 钟萍萍，彭彩珍. 油藏井网密度计算方法综述 [J]. 石油地质与工程，2009，23（2）：60-63.

[3] 李士伦，郭平. 再论我国发展注气提高采收率技术 [J]. 天然气工业，2006；26（12）：30-34.

[4] 张盛宗. 合理井网密度与最终采收率的定量关系 [J]. 石油学报，1987，8（1）：45-50.

[5] 李周荣. 油田开发后期合理井网密度确定方法 [J]. 油气地质与采收率，2002，9（3）：78-80.

[6] 中国石油天然气总公司计划局规划院. 石油工业建设项目经济评价方法与参数 [M]. 北京：石油工业出版社，2004.

低渗透油藏气驱产量预测新方法

王高峰[1,2]　胡永乐[1,2]　宋新民[1,2]

秦积舜[1,2]　杨思玉[1,2]　马德胜[1,2]

（1. 中国石油勘探开发研究院提高石油采收率国家重点实验室；

2. 国家能源 CO_2 驱油与埋存研发（实验）中心）

摘　要：注气开发油藏产量预测油藏工程理论方法报道极少。为增加注气方案可靠性，提高注气效益，从气驱采收率计算公式入手，利用采出程度、采油速度和递减率之间相互关系，推导出气驱产量变化规律。结合岩芯驱替实验成果和油田开发实际经验，提出了气驱增产倍数严格定义及其工程计算近似方法。发现低渗油藏气驱增产倍数由气和水的初始驱油效率之比，以及转驱时广义可采储量采出程度决定。统计得到国内外 18 个注气项目气驱增产倍数理论值和实际值平均相对误差 6.90%。绘制了气驱增产倍数查询图板，以便于从事注气开发人员使用。研究成果为气驱产量预测提供了油藏工程理论依据，从而对控制和优化低渗油藏注气项目投资有重要意义。

关键词：低渗透油藏；气驱产量；战略规划；采收率；驱油效率；可采储量采出程度；气驱增产倍数

0　引言

气驱过程复杂性使人们对其生产动态的认识一直处于经验阶段[1-5]，Janelle Nagrampa 等关联的预测 Weyburn 油田短期平均气驱产量经验关系，不能描述产量随时间的变化，且该经验式仅适于同 Weyburn 油田开发历程和性质都接近的油藏[1]，具有明确物理意义的气驱产量预测油藏工程理论方法尚未见报道。目前气驱开发相关研究主要靠室内实验和数值模拟手段[6-13]，工作中还发现数值模拟预测生产动态与实际往往有较大出入，在低渗透油藏尤为突出。作为油田多种计划和方案的龙头，油藏开发方案得失对油田经济影响很大。为增加注气方案可靠性，提高注气项目收益，从油藏工程基本原理出发，推导出气驱产量变化规律；提出气驱增产倍数及其工程计算方法，通过国内外多个注气实例验证理论的合理性和有效性。在吉林等油田多个注气区块的应用表明，本文成果对注气项目投资的优化和控制有重要意义，对低渗透油藏气驱配产实践也有指导意义。

1　理论分析

1.1　气驱产量预测理论依据

将转驱时油藏视为新油藏。将气驱波及体积与水驱波及体积之比称为气驱波及体积修正因子，根据"采收率等于驱油效率和体积波及系数的乘积"这一油藏工程基本原理，可得气驱阶段采收率计算式[7]：

$$E_{Rg} = \eta \frac{S_o}{S_{oi}} \frac{E_{Dg}}{E_{Dw}} E_{Rwn} \tag{1}$$

式中，E_{Rg} 为基于原始地质储量的气驱采收率；E_{Rwn} 为基于转驱时剩余地质储量的水驱采收率；S_{oi}，S_o 分别为原始与转驱时平均含油饱和度；E_{Dg}，E_{Dw} 分别为转驱时气和水的驱油效率（基于原始含油饱和度），%；η 为气驱波及体积修正因子。

式（1）推导略去。

考察评价期内采出程度变化情况。根据岩芯驱替实验成果，转驱时气和水的驱油效率显然可视为定值，气驱波及体积修正因子 η 亦视作常数。因采收率是由采出程度增长而来，将采收率指标视为变量，式（1）对时间求导数有：

$$\frac{dE_{Rg}}{dt} = \eta \frac{S_o}{S_{oi}} \frac{E_{Dg}}{E_{Dw}} \frac{dE_{Rwn}}{dt} \tag{2}$$

任意 t 时刻气驱采出程度的增量显然可写作：

$$dR_g = R_{vg}dt = dE_{Rg} \tag{3}$$

任意 t 时刻水驱采出程度的增量可写作：

$$dR_{wn} = R_{vwn}dt = dE_{Rwn} \tag{4}$$

式中，R_g 为基于原始地质储量气驱采出程度；R_{wn} 为基于转驱时剩余地质储量的水驱采出程度；R_{vg} 为基于原始地质储量气驱采油速度；R_{vwn} 为基于转驱时剩余地质储量的水驱采油速度。

联立式（2）~式（4），并整理得：

$$R_{vg} = \eta \frac{S_o}{S_{oi}} \frac{E_{Dg}}{E_{Dw}} R_{vwn} \tag{5}$$

式（5）两端同乘以原始地质储量 $V_P S_{oi}$ 有：

$$R_{vg} V_P S_{oi} = \eta \frac{E_{Dg}}{E_{Dw}} R_{vwn} V_P S_o \tag{6}$$

根据前述采油速度的涵义，由式（6）可得到：

$$Q_{og} = \eta \frac{E_{Dg}}{E_{Dw}} Q_{ow} \tag{7}$$

式中，V_P 为油藏孔隙体积，m^3；Q_{og} 为 t 时刻气驱产量水平，m^3/d；Q_{ow} 为同期的水驱产量水平，m^3/d。

须指出，"同期的水驱产量"为假设油藏不注气而继续注水时油藏整体产量，可由水驱递减规律预测得到。

在此，提出气驱增产倍数 F_{gw}，并将其定义为气驱产量水平与同期水驱产量水平的比值，其定义由式（8）给出：

$$F_{gw} = \frac{Q_{og}}{Q_{ow}} = \eta \frac{E_{Dg}}{E_{Dw}} \tag{8}$$

式（8）对时间取导数可得气驱产量绝对递减率：

$$\frac{dQ_{og}}{dt} = F_{gw} \cdot \frac{dQ_{ow}}{dt} \tag{9}$$

式（8）和式（9）实质为评价期内气驱产量和水驱产量的一一对应关系。低渗透油藏气驱产量与同期的水驱产量之比为恒定值，且此值为气驱增产倍数，联合气驱增产倍数和水驱递减规律可在理论上把握气驱产量。水驱产量是长期摸索确定的合理值，气驱增产倍数为固定值，气驱产量可被唯一确定，水驱开发经验为气驱所借鉴。由式（9）知，当气驱增产倍数大于 1.0 时，比如混相驱产量绝对递减率将高于水驱情形，这解释了为什么绝大多数混相驱产量曲线比水驱产量曲线陡峭[2]。

1.2 气驱增产倍数确定方法

1.2.1 气驱波及体积修正因子取值

由于驱油效率室内可测，若获知气驱波及体积修正因子，便可按照式（8）求算气驱增产倍数。气驱波及体积修正因子受重力分异、黏性指进和扩散作用影响，在此简要评述 3 个因素在油藏注气开发过程中所能起的作用。

（1）浮力与毛细管力的对比：压汞曲线上"阈压"的存在表明多孔介质中非湿相驱替润湿相必须克服一定的启动压力，阈压用毛细管压力计算[14]。作为非湿相上浮也是驱替行为，也须克服阈压。以油气接触弯月面为底面选择厚度为 dh 油相微元为研究对象。此微元原处于静水平衡态，当存在游离气时，微元在垂向上所受合力为下部气柱上浮形成的推力与毛细管力之差。国内低渗透油层内单砂体有效厚度通常在 1.0m 左右，则微元下部单位长度气柱受到向上合力为浮力与自身重力之差，即有效浮力。作用于油相微元的垂向合力则为有效浮力与毛细管力之差。油气共存时，储层岩石为油湿，接触角常小于 75°，现取 60°。以非混相 CO_2 驱为例，地下油气密度差取 210kg/m³；油气界面张力通常小于 15.0mN/m，CO_2 非混相驱界面张力取值 6.0mN/m。浮力应用阿基米德原理计算，毛细管力应用 Laplace 公式计算，发现只有当孔喉半径超过 3.0μm，即储层为中高渗透时[15]，有效浮力才大于毛细管力（表 1），气体方能克服阈压推动油气界面上移。故在低渗透介质中气顶无法仅靠浮力自然形成，这应是少见带气顶低渗透油藏的一个原因。

（2）浮力与生产压差的对比：将进入油相的气泡分为若干高为 dh 的立方体微元。每个微元分担的有效浮力 $\Delta F_v = (\rho_o - \rho_g) g (dh)^3$，$\rho_o$ 和 ρ_g 分别为油和气的密度（kg/m³）；g 为重力加速度，m/s²；气泡在水平方向随油相一起运动，微元所受水平合力 $\Delta F_h = \text{grad}pdh$（dh）²，$\text{grad}p$ 为注采压差梯度（MPa/m）；则纵横力比 $\Delta F_v / \Delta F_h = (\rho_o - \rho_g) g / \text{grad}p$。结合油田开发实际情况计算知注采压差梯度通常大于 0.02MPa/m，有效浮力梯度不足注采压差梯度 6%（表 2）。因此，重力分异无法形成对生产有现实意义的驱替。开发地质专家薛培华统计分析喇嘛甸油田 11 口检查井资料提出"交互韵律式"剩余油分布模式[16]，即未水洗层与水洗层多呈间互状分布，且水洗剖面韵律性与物性剖面韵律性一致，这也证明重力分异在油田开发中作用很小[17]，更不存在依靠重力作用开发的低渗透油藏。

（3）垂向与水平渗透率之比：一般地，碎屑岩油藏物性越差，垂向与水平渗透率之比越小。对于低渗储层，此比值通常为 0.01~0.30；在同样压力梯度下，垂向流速不足水平速度的 1/3。

（4）小层内夹层的作用：渗透性极差的物性夹层或泥质夹层在低渗储层内是普遍存在的，构成流体上浮或下沉的天然地质遮挡，进一步限制重力分异对纵向波及系数的改变。

（5）气体蒸发萃取作用：油藏条件下，注入气不断蒸发萃取原油组分使自身被富化。实测和模拟计算知道气驱前缘附近气相黏度在 $0.1mPa \cdot s$ 附近，与地层水黏度为同一数量级（水黏度是前缘气相黏度的 3~6 倍），削弱了气体黏性指进对于波及体积修正因子的影响。

（6）水气交替注入的作用：水气交替注入是改善气驱效果的主体技术。水气交替可抑制黏性指进和控制气窜，扩大波及体积；气驱实践中多轮次的水气交替注入（交替周期一般为 2~4 个月）将使气驱波及体积与水驱趋同，气驱波及体积修正因子趋于 1.0。

（7）气相扩散作用：在漫长的成藏过程中，时间累积效应大，很多学者都认识到扩散作用是地下天然气运移的一个普遍过程[18]；但在油田开发这几十年内，扩散作用甚小。在油藏条件下测量 CO_2 在原油中的扩散系数及数值模拟研究均认为扩散作用对于孔隙型油藏注气开发的影响微不足道[8,13]。

表 1 非混相 CO_2 驱有效浮力与毛细管力的对比

界面张力（mN/m）	6.0	6.0	6.0	6.0	6.0
孔喉半径（μm）	0.1	1.0	3.0	10.0	100.0
毛细管力（MPa）	0.06	0.006	0.002	0.0006	0.00006
有效浮力（MPa）	0.002	0.002	0.002	0.002	0.002

（8）气驱油藏物性下限：通常认为低渗透油藏实施注气能改善驱替剖面，矿场确有吸气剖面为证。但气体在微孔喉差油层中能运移多远并无结论。近年来，在吉林红岗和大庆宋芳屯两个超低渗透区块（渗透率小于 1.0mD）的注气工作表明，物性过差油藏靠注气实现经济有效开发仍有极大困难。此外，在低渗透油藏开发地质研究中，水驱的渗透率下限常取 0.1mD，此下限之下的储量占总地质储量比例甚小。即便这些极差储量有所动用，对采收率的贡献也非常小。

总之，重力分异和扩散作用不会对低渗透油层注气产生有现实意义的影响，注入气体黏性指进则被相态变化和水动力学调控等因素削弱，多轮次水气交替注入使气驱波及体积趋于水驱情形，这便是低渗透油藏气驱波及体积修正因子接近 1.0 的原因。另须指出，上述论证仅为气驱增产倍数的工程计算提供依据，而不是为了证明气驱和水驱波及体积完全相等。

表 2 CO_2 驱过程中浮力与注采压差的对比

油田	注采压差梯度（MPa/m）	CO_2 密度（kg/m³）	油气密度差（kg/m³）	有效浮力梯度（MPa/m）	纵横力比（%）
长垣	0.027	650	150	0.0015	5.4
情字井	0.037	610	150	0.0015	4.0
柳赞北	0.061	580	170	0.0017	2.7

1.2.2 气驱增产倍数计算公式

为便于应用，转驱时的驱油效率须与初始驱油效率（指油藏未动用时）和转驱时水驱采出程度相关联。

驱油效率属微观层面上的概念，其近似值由岩心驱替实验给出（严格讲，岩心驱替中仍有波及体积概念）；多轮次水气交替注入又会消除注气时机对于残余油饱和度的影响，并且实验发现气驱残余油饱和度与交替注入的水气段塞比无关[4]，故气驱残余油饱和度可视为定值。依据驱油效率定义得：

$$E_{Dg} = \frac{S_{oi}E_{Dgi} - S_{oi}R_{ews}}{S_{oi}} \tag{10}$$

根据水驱油效率的定义有：

$$E_{Dw} = \frac{S_{oi}E_{Dwi} - S_{oi}R_{ews}}{S_{oi}} \tag{11}$$

将式（10）、式（11）带入式（8），并将气驱波及体积修正因子 η 取值为 1.0 可得：

$$F_{gw} = \frac{E_{Dgi} - R_{ews}}{E_{Dwi} - R_{ews}} \tag{12}$$

式中，E_{Dgi}，E_{Dw} 为气和水的初始驱油效率；R_{ews} 为转驱时基于原始地质储量的波及区水驱采出程度。

由于采出原油仅来自注水波及区域，故波及区采出程度高于油层整体采出程度。关于波及区域的确定存在两种观点：一种认为波及系数为采收率与驱油效率之比，即实际波及系数严格等于理论波及系数；另一种则认为波及系数接近 1.0，此观点来自对油藏实际加密效果的分析。加密井含水率往往低于老井，却远高于油藏初始含水（表3），并且具有初始产状的加密井比例极低，很难准确预测和钻遇。这表明剩余油分布并没有呈现大面积未动用或高度富集状态，波及区面积接近整个油层。

表3 水驱加密井与老井产状对比

油藏名称	投产初期		老井平均（加密时）		加密井平均	
	含水率（%）	日产油（t）	含水率（%）	日产油（t）	含水率（%）	日产油（t）
英东萨尔图	53.2	6.70	81.3	1.75	85.1	1.47
喇嘛甸萨3组	3.7	19.60	95.1	3.02	85.7	4.01
双河	1.3	35.80	80.3	9.50	62.1	14.10
安塞王窑	12.4	3.63	57.1	1.85	41.7	2.10

可见，上述观点都有理论和实验或油田开发实践方面的证据。综合这两种观点，应认为实际波及区域高于理论波及区域，并且波及区域不同位置动用程度存在差别。据此，可将实际波及系数表示为理论波及系数和剩余波及系数的加权平均：

$$E_V = E_{V0} + \omega(1 - E_{V0}) \tag{13}$$

式中，E_V 为实际波及系数；E_{V0} 为理论波及系数；ω 为权值，$0<\omega<1.0$。

权值 ω 反映了理论波及区域之外的储量动用程度，主要由注采参数变化对地下流场的水动力学调整引起（即液流方向改变），故其受控于储层物性级别和非均质性、井网砂体匹配程度以及油田开发时间等因素。开发时间越长，井网与砂体越匹配，注采参数变化的时间累积作用越大，ω 越大；另外，权值 ω 也反映了剩余油分布的均匀性，剩余油分布越均匀，ω 越大；对于采出程度很低的油藏和高采出程度的成熟油藏[19]，剩余油分布总体上是均匀的，推荐 $\omega=1.0$。

相应于式（13）中实际波及系数的波及区采出程度与油藏整体采出程度的关系可根据物质平衡得到：

$$R_{ews} = R_{e0}/E_V \qquad (14)$$

式中，R_{e0} 为转驱时基于原始地质储量的油层整体采出程度，%。

理论波及系数等于基于原始地质储量的水驱采收率与初始水驱油效率之比：

$$E_{V0} = \frac{E_{Rw}}{E_{Dwi}} \qquad (15)$$

式中，E_{Rw} 为转驱时基于原始地质储量的水驱采收率。

联立式（13）~式（15），并整理得：

$$R_{ews} = \frac{R_{e0}E_{Dwi}}{E_{Rw} + \omega(E_{Dwi} - E_{Rw})} \qquad (16)$$

将式（16）代入式（12）得到：

$$F_{gw} = \frac{E_{Dgi} - \dfrac{R_{e0}E_{Dwi}}{E_{Rw} + \omega(E_{Dwi} - E_{Rw})}}{E_{Dwi} - \dfrac{R_{e0}E_{Dwi}}{E_{Rw} + \omega(E_{Dwi} - E_{Rw})}} \qquad (17)$$

式（17）即为低渗透油藏气驱增产倍数计算式。将该式右端分子和分母同除以初始水驱油效率得到：

$$F_{gw} = \frac{R_1 - R_2}{1 - R_2} \qquad (18)$$

式（18）中，$R_1 = E_{Dgi}/E_{Dwi}$，即气和水初始驱油效率之比。$R_2 = R_{e0}/\left[E_{Rw}+\omega(E_{Dwi}-E_{Rw})\right]$，可称之为广义可采储量采出程度。所以，气驱增产倍数由这两个比值唯一确定。根据式（18）绘制了气驱增产倍数实用查询图板（图1）。图板横坐标为转驱时的广义可采储量采出程度 R_2；同一曲线上的数据点具有相同的初始驱油效率之比 R_1。可见，随着采出程度增加，气驱增产倍数呈快速增长趋势，这与实际气驱动态一致。国内大多数水驱开发油藏 R_2 值都低于 0.9，R_1 值通常在 1.5 附近，由式（17）知，不应期待注气后油藏整体产量会超过注气前水驱产量的 6.0 倍。

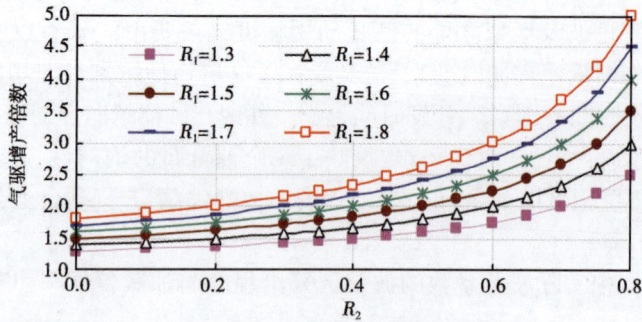

图 1　气驱增产倍数查询图板

2　应用示例

2.1　气驱油藏早期配产

配产是油田开发设计的中心问题[20]。结合多年来注气方案设计和跟踪注气动态的经验，统计了国内外 18 个成功的二氧化碳驱与烃气驱项目的产量变化情况[1-3,5-12]，并应用气驱增产倍数计算式（17）和定义式（8），研究了这些注气项目的早期配产问题。将定义式（8）中气驱产量定义为注气后产量峰值附近一年内平均日产量，水驱产量则取为注气前一年内的平均日产量。注气实践表明，从开始注气到出现气驱产量峰值所需时间通常不超过两年（表 4），而大多数连续稳定注气项目所用时间为 6~8 个月[2]，故此处忽略水驱产量递减因素，而直接用注气前水驱产量代替高峰期气驱产量对应的水驱产量（即同期的水驱产量），以简化计算。

表 4　6 个油藏达到气驱产量峰值所用时间统计

油藏名称	黑 59	黑 79	Devonial	L. S. Tensleep	Sacroc	Weyburn
所需时间（d）	330	430	110	290	145	390

首先应用式（17）计算出气驱增产倍数理论值（权值 ω 均取为 1.0）；再根据式（8）将气驱增产倍数理论值乘以注气前水驱产量得到气驱见效早期产量预测值，发现气驱产量预测值和实际值平均相对误差 6.90%（表 5）；然后将实际气驱产量除以水驱产量得到气驱增产倍数实际值，与其理论值对比得到平均相对误差为 6.90%。这表明本文方法及其理论前提的有效性与合理性，也表明将低渗透油藏气驱波及体积修正因子作常数处理且取值为 1.0 的可行性。

表 5 中韦本（Weyburn）油田位于加拿大，拥有世界上最著名的 CO_2 驱油与埋存项目。实测初始 CO_2 驱油效率为 85.0%，初始水驱油效率为 60.0%，开始注气时水驱采出程度为 26.0%。气驱增产倍数理论值为 1.73，实际值为 1.67，相对误差 3.59%。黑 79 南南区块实施近混相驱，CO_2 驱油效率为 78.0%，初始水驱油效率为 57.0%，开始注气时水驱采出程度为 11.0%，计算出气驱增产倍数理论值为 1.47，实际值为 1.43，相对误差 2.8%。图 2 为气驱增产倍数理论值与实际值对比。

270

根据注气时采出程度的不同，表5中注气项目大致分4种类型：（1）Lost Soldier tensleep 油藏和濮城1-1区块注气时采出程度高于40.0%，属高度成熟油藏注气类型，气驱增产倍数超过3.0；（2）Weyburn油田转驱时采出程度在20%以上，属于近成熟油藏注气类型；（3）黑79南南区块注气时水驱采出程度12.0%，属于水驱动用到一定程度油藏注气类型；（4）黑59区块和葡北油藏注气时采出程度低于5.0%，属弱未动用油藏注气类型，由图1知，该类型的气驱增产倍数不会很高。

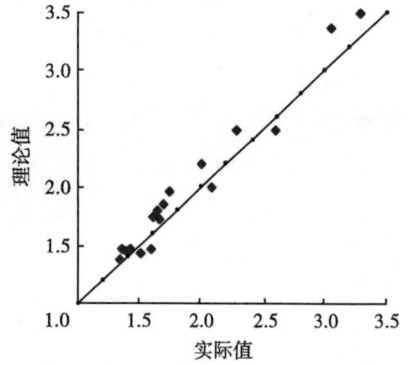

图 2　气驱增产倍数理论值与实际值对比

表 5　气驱见效早期合理配产计算结果

序号	油藏名称	气驱增产倍数		气驱产量（m³/d）		
		理论值	实际值	实际值	预测值	ADD（%）
1	Weyburn	1.73	1.67	4646	4813	3.59
2	Dollaride Devonial	1.8	1.64	190	209	9.76
3	Wertz tensleep	2.21	2.01	1598	1757	9.95
4	Lost Soldier tensleep	3.5	3.30	1574	1669	6.06
5	Means San andres	1.46	1.40	2226	2321	4.29
6	North Cross	1.85	1.70	378	412	8.82
7	Lick Creek Meakin	2.49	2.29	328	357	8.73
8	Slaughter Estate	2	2.10	70	67	4.76
9	North Ward Estes	1.74	1.62	1028	1104	7.41
10	Dever Unit-6P（MA）	1.96	1.75	111	125	12
11	Sacroc-4P	2.5	2.60	413	397	3.85
12	黑59	1.48	1.59	80	74	6.92
13	黑79南南	1.47	1.43	92	95	2.8
14	树101	1.43	1.51	49	47	5.3
15	萨南	1.47	1.36	35	38	8.09
16	濮城1-1	3.38	3.07	15	16	10.1
17	草舍	1.74	1.61	77	83	8.07
18	葡北	1.39	1.34	566	587	3.73
	平均值	1.98	1.89	749	787	6.9

注：ADD＝Abs（预测值/实际值-1）×100%。

2.2　气驱产量中长期预测

特低渗 H 油藏位于吉林油田，评价认为实施 CO_2 驱可使地层压力恢复到最小混相压力22.5MPa 以上，实现混相驱开发。应用室内长岩心驱替实验测得初始 CO_2 混相驱油效率为

271

80.0%，初始水驱油效率为55.0%；该油藏开始注气时的水驱采出程度为3.0%，水驱采油速度为1.7%，标定水驱采收率为20.3%。应用本文方法预测中长期气驱产量的步骤为：（1）首先借鉴同类型油藏水驱开发经验得到水驱产量变化情况；（2）再将初始驱油效率和水驱采出程度代入式（17）求出气驱增产倍数为1.481（权值$\omega = 1.0$）；（3）最后根据式（11）将步骤（1）中水驱产量乘以气驱增产倍数即得气驱产量剖面（图3）。本文方法预测该区块气驱采油速度能达到2.5%，实际气驱采油速度达到2.7%，气驱投产后实际产量情况和预测结果符合度较高（图3）。数值模拟法预测结果过乐观（年采油速度达到4.3%），不再给出。目前，对气驱过程中复杂相态变化、三相以上渗流和微观驱油机理不能完整而准确地进行数学描述，加上低渗透储层地质认识不确定性更大，三维地质模型难以真实反映储层非均质性等因素是造成低渗透油藏注气数值模拟结果不可靠的主要原因。根据本文油藏工程方法成功优化和控制了对该注气项目的投资额度，比数值模拟法节省投资××千万元。

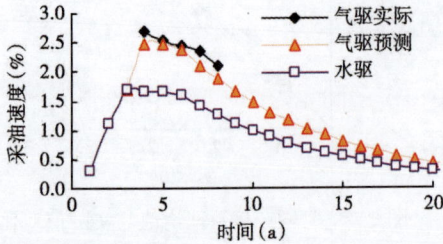

图3　H油藏气驱产量变化情况

3　结论

（1）本文定义的气驱增产倍数是确定低渗透油藏（1~50mD）气驱整体产量的油藏工程理论依据。低渗透油藏气驱增产倍数由气和水的初始驱油效率之比以及转驱时广义可采储量采出程度决定。尽可能实施混相驱是提高气驱增产倍数的内在要求，低渗透油藏气驱波及体积修正因子取值1.0是可行的。

（2）气驱产量水平仅取决于气驱增产倍数和注气前水驱产量水平以及水驱递减特征。低渗透油藏气驱增产倍数定义式和计算式的提出使得在理论上把握气驱产量成为可能，为进行气驱提高采收率潜力评价、注气方案编制和注气开发战略研究提供了全新工具和依据。

参 考 文 献

[1] James Brodie, BGarat Jhaveri, TIM Moulds. Review of Gas Injection Projects in BP [J]. SPE 154008, 2012：904-920.

[2] Fred I Stalkup, Michael H Stein. CO_2 flooding [M]. Texas of USA：SPE, 1998：1-294.

[3] Janelle Nagrampa, Koorosh Asghari. Development of a Correlation between Performance of CO_2 Flooding and the Past Performance of Water Flooding in Weyburn Oil Field [J]. SPE 99789, 2007：807-813.

[4] Robert Ehrlich, Tracht J H Laboratory and Field Study of the Effect of Mobile Water on CO_2-flood Residual Oil Saturation [J]. SPE 11957, 1984：1797-1808.

[5] 李士伦，孙雷，郭平. 再论我国发展注气提高采收率技术 [J]. 天然气工业，2006，26（12）：30-34.

[6] 魏浩光，岳湘安，赵永攀，等. 特低渗透油藏天然气非混相驱实验 [J]. 石油学报，2011，32（2）：307-310.

[7] 王高峰，马德胜，宋新民. 气驱开发油藏井网密度数学模型 [J]. 科学技术与工程，2012，12（11）：50-54.

[8] 俞宏伟. 黑59二氧化碳驱开发规律实验及数值模拟研究 [D]. 北京：中国石油勘探开发研究院，2011.

［9］郭平，杜志敏，张茂林．葡北油田气水交替注烃混相驱开发研究［J］．西南石油学院学报，2007，26（4）：25-27.

［10］张奉东，王震亮．苏北盆地草舍油田 CO_2 混相驱替试验与效果分析［J］．实验石油地质，2010，32（3）：296-299.

［11］孙锐艳，王宪中，马晓红．黑 59 区块二氧化碳驱地面工程技术［J］．油气田地面工程，2012，31（2）：37-38.

［12］李锋，杨永超．濮城油田沙一段油藏濮 1-1 井组二氧化碳驱研究及先导试验效果分析［J］．海洋石油，2009，29（4）：56-60.

［13］郭平，汪周华，沈平平．高温高压气体—原油分子扩散系数研究［J］．西南石油大学学报，2010，32（1）：73-79.

［14］何更生．油层物理［M］．北京：石油工业出版社，1994.

［15］李传亮．孔喉比对地层渗透率的影响［J］．油气地质与采收率，2007，14（5）：78-80.

［16］薛培华．油层驱替状况分析的检查井方法［R］．北京：中国石油勘探开发研究院，2006.

［17］张景存，赵永胜．关于重力对于水淹厚度的影响问题［J］．石油勘探与开发，1984，11（2）：73-79.

［18］柳广弟，李剑，李景明．天然气成藏过程有效性的主控因素与评价方法［J］．天然气地球科学，2005，16（1）：1-6.

［19］韩大匡．深度开发高含水油田提高采收率问题的探讨［J］．石油勘探与开发，1995，22（5）：47-55.

［20］王高峰，胡永乐，李治平．油气藏能量方程与一种新的配产理论初探［J］．西南石油大学学报，2007，29（5）：57-59.

CO_2 混相驱拟含气率与
采出程度图版的建立

孙 雷[1] 冯 乔[1] 陈国利[2] 张 华[2] 靳美玉[3]

(1. 西南石油大学；2. 中国石油吉林油田公司；
3. 库尔勒巴音郭楞职业技术学院)

摘 要：针对 Y 区块注 CO_2 混相驱动态开发评价方法不够全面不够系统、利用水驱童氏图版评价气驱开发效果不够准确等特点，首先定义了 A 型气驱特征曲线，在此基础上推导了适合 Y 区块的 CO_2 混相驱拟含气率和采出程度的关系式，绘制了 CO_2 混相驱特征理论图版，用改进后的图版预测了 Y 区块注 CO_2 后所提高的最大采收率，并将预测的采收率与现场预测的采收率进行比较，其结果比较吻合，说明该图版能很好地预测 CO_2 混相驱动态开发效果，同时对该区块下一步的开发及评价具有指导意义。

关键词：CO_2 混相驱；拟含气率；采出程度；提高的最大采收率；图版

0 引言

在注水开发油藏的过程中，应用油田实际生产资料以统计性的方法研究油藏的开发动态是国内外较常用的方法之一，目前已经建立了许多快速预测开发指标的水驱特征曲线关系式和图版。但关于 CO_2 混相驱油藏的动态还没有成熟的油藏工程方法来对其进行评价[1]，常用的方法是通过数值模拟或实验方法来评价油藏开发效果[2-5]，而且 CO_2 混相驱的研究重点都放在其驱替机理以及混相压力上[6,7]，利用驱替特征曲线和图版法进行开发动态的评价还比较少见。杨国绪等[8]对国内外十几个气顶油藏进行了统计分析，归纳出气驱特征曲线在气顶油藏开发中的应用。李菊花等[9]借鉴甲型水驱特征曲线关系式的理论推导基础，建立了水气交替驱特征曲线关系式。但这些都不适合评价 CO_2 混相驱油动态效果，本文以 Y 区块为例，根据该区块的生产实际探索并建立一个辅助预测油藏开发动态的 CO_2 混相驱规律图版，目的是为低渗油田注 CO_2 混相驱开发效果评价提供技术支持。

1 童氏含水率和采出程度的关系式

目前，水驱开发规律相对气驱开发规律来说要成熟得多，童宪章[10]研究了甲型水驱曲线，有：

$$\lg W_p = a + bN_p \tag{1}$$

并在甲型水驱曲线的基础上推导出生产水油比与累计产油量的关系式：

$$\lg F_{wo} = bN_p + E \tag{2}$$

其中

$$E = a + \lg \ (2.303b)$$

如果油田的原始地质储量为 N，则当累积产油量为 N_p 时，当时的采出程度可以用 $R_o = N_p/N$ 表示，把 $R_o = N_p/N$ 代入式（2）可得：

$$\lg F_{wo} = MR_o + E \tag{3}$$

式（3）即为乙型水驱特征曲线，式中的常数 $M = bN$，它是甲型水驱曲线的斜率 b 和水驱地质储量的乘积，童宪章通过统计国内外 25 个油藏单元（国外 8 个，国内 17 个）的资料发现 M 值近似为 7.5。

当取经济极限综合含水率 $f_w = 98\%$ 时，水油比 $F_{wo} = 49$，此时的采出程度 R_o 可看作是最终采收率 R_M，即 $R_o = R_M$，代入式（3）得：

$$1.69 = 7.5R_M + E \tag{4}$$

式（3）减去式（4）得含水率与采出程度的关系式：

$$\lg \frac{f_w}{1-f_w} = 7.5 \ (R_o - R_M) \ + 1.69 \tag{5}$$

式（5）即童氏含水与采出程度关系式。

童氏水驱特征图版被广泛用来描述水驱油田动态开发效果，随着科学研究的不断发展，不少学者对童氏水驱图版进行了更深层次的研究，发现式（5）中的 7.5 只适合描述油田开发中后期以及高含水期的水驱规律，而对于低渗透油田以及开发初期已经不适用了，因此对童氏图版进行了一系列的改进，那么无论是改进前还是改进后的水驱规律图版都是否仍可用来描述气驱，特别是 CO_2 混相驱的开发效果可通过以下论证来说明。

X 油田 Y 区块是典型的低渗透储层，也是目前国内较先实施 CO_2 混相驱开发的先导性试验区块，该区块从投产至 2008 年 4 月底一直是水驱开发，从 2008 年 5 月到目前采用 CO_2—水交替驱开发。

王柏力[11]经过研究发现，用 13.0 代替 7.5 能更好地反映 X 油田的水驱规律，其含水率与采出程度关系式为：

$$\lg \frac{f_w}{1-f_w} = 13.0(R_o - R_M) \ + 1.69 \tag{6}$$

赫恩杰等[12]对童氏水驱图版进行了改进，拓宽了童氏图版的应用范围，改进后适用于 X 油田的童氏采出程度与含水率的关系式为：

$$\lg \left(\frac{f_w}{1-f_w} + m \right) = 1.69 + 13.0(R_o - R_M) \ + n \tag{7}$$

其中

$$m = 10^{1.69 + 13.0(R_i - R_M) - \lg[1 - 10^{13.0(R_i - R_M)}]}$$

$$n = -\lg \ [1 - 10^{13.0(R_i - R_M)}]$$

X 油田基本没有无水采油期，因此利用式（7）绘制含水率与采出程度图版如图 1 所示。

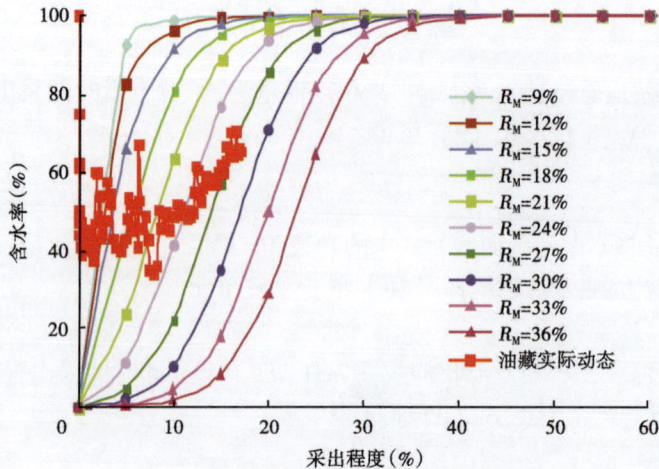

图1　改进后的 X 油田水驱特征理论图版

从图1可看出，改进后水驱特征图版标定的最终采收率为 27.0%。Y 区块实际标定的水驱采收率为 19.1%，利用水驱特征图版法计算 Y 区块 CO_2 混相驱开发比水驱开发所能提高的最大采收率为 7.9%。

借鉴水驱特征图版的研究方法推导建立 CO_2 混相驱的特征图版，进一步探索能够快速预测注 CO_2 后所提高的采收率的分析方法，从而进一步评价 CO_2 混相驱动态开发效果，为该区块下一步 CO_2 驱开发效果评价提供借鉴。

2　CO_2 混相驱拟含气率与采出程度关系式推导

将国内成功实施 CO_2 混相驱开发的油田的实际生产数据进行统计整理，发现在 CO_2 混相带未突破前，不存在统计规律，但混相带突破后累计产气量（抽提气+CO_2）和累计产油量在半对数坐标中常会出现一条近似的直线段，且暂定义其为 A 型气驱特征曲线，如图2和图3所示。

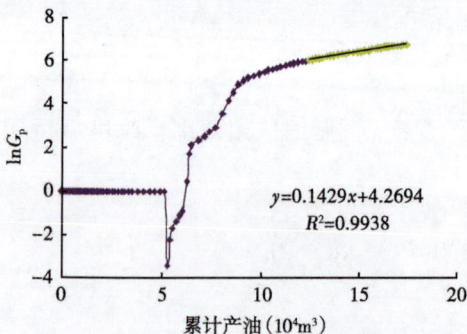

图2　X 油田 A 型气驱曲线图

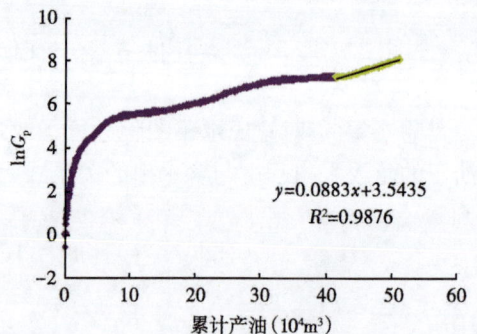

图3　Z 油田 A 型气驱曲线图

图2和图3所表示的就是典型的 CO_2 混相驱特征曲线，暂称为 A 型气驱特征曲线，其中的直线段可以表示为：

$$\ln G_p = A + BN_p \tag{8}$$

将式（8）两边同时对时间 t 求导，得：

$$\frac{1}{G_p} \cdot \frac{dG_p}{dt} = B \frac{dN_p}{dt} \tag{9}$$

又知

$$\frac{dG_p}{dt} = Q_g$$

$$\frac{dN_p}{dt} = Q_o$$

因此式（9）可变为：

$$\frac{Q_g}{G_p} = BQ_o \tag{10}$$

定义拟含气率：

$$f_{rg} = \frac{Q_g}{Q_o + Q_g} \tag{11}$$

将式（11）代入式（10），整理变形得：

$$G_p = \frac{f_{rg}}{1 - f_{rg}} \cdot \frac{1}{B} \tag{12}$$

对式（12）两边取对数，得：

$$\ln G_p = \ln \frac{f_{rg}}{1 - f_{rg}} + \ln \frac{1}{B} \tag{13}$$

对比式（8）和式（13），则有：

$$\ln \frac{f_{rg}}{1 - f_{rg}} + \ln \frac{1}{B} = A + BN_p \tag{14}$$

将式（14）变形整理，得：

$$\ln \frac{f_{rg}}{1 - f_{rg}} = A + BN_p + \ln B \tag{15}$$

如果油藏的地质储量为 N，则当累计产油量为 N_p 时，其采出程度可表示为 $R_o = N_p/N$，将 $N_p = R_o N$ 代入式（15），变形可得拟含气率和采出程度的关系式：

$$\ln \frac{f_{rg}}{1 - f_{rg}} = A + CR_o + \ln B \tag{16}$$

式（16）中，$C = BN$，它是 A 型气驱特征曲线的斜率 B 与地质储量的乘积，通过 A 型气驱特征曲线可求得 B，若还已知目标油藏的地质储量，那么就可计算出目标油藏的 C 值。

如果令 $D = A + \ln B$，将其代入式（16）可变为：

$$\ln \frac{f_{rg}}{1-f_{rg}} = CR_o + D \tag{17}$$

从式（17）可以看出 $\ln\left(\frac{f_{rg}}{1-f_{rg}}\right)$ 与采出程度 R_o 在半对数坐标中呈直线关系。如果作出 $\ln\left(\frac{f_{rg}}{1-f_{rg}}\right)$ 与采出程度 R_o 的关系曲线，也可计算出 C 值（表1）。

表1　计算 C 值的方法

名称	求法
方法一	$C = NB$
方法二	作 R_o 与 $\ln\left(\frac{f_{rg}}{1-f_{rg}}\right)$ 曲线回归求 C

取经济极限拟含气率 $f_{rg} = 98\%$，则有 $R_o = R_{ing}$，代入式（16），得：

$$3.892 = A + CR_{ing} + \ln B \tag{18}$$

式（16）减去式（18），即可得 CO_2 混相驱拟含气率和采出程度关系式：

$$\ln \frac{f_{rg}}{1-f_{rg}} = C\ (R_o - R_{ing})\ + 3.892 \tag{19}$$

将确定的 C 值代入式（19），就可得到目标油藏的 CO_2 混相驱拟含气率和采出程度图版（称为 B 型气驱特征曲线）。同样为了进一步拓宽该图版的应用范围，对该图版进行改进，改进后的 CO_2 混相驱拟含气率和采出程度关系式为：

$$\ln\left(\frac{f_{rg}}{1-f_{rg}} + P\right) = 3.892 + C\ (R_o - R_{ing})\ + Q \tag{20}$$

其中

$$P = e^{3.892 + C(R_i - R_{ing}) - \ln\left[1 - e^{C(R_j - R_{ing})}\right]}$$

$$Q = -\ln\left[1 - e^{C(R_j - R_{ing})}\right]$$

3　CO_2 混相驱拟含气率和采出程度标定图版的应用

X 油田 Y 区块具有优越的油气生成、运移、聚集、成藏的条件，但储层物性差，为低孔、低渗透储层，试验区内共有 6 个试验井组，开发井 31 口，其中 25 口生产井、6 口注入井。从投产至 2008 年 4 月注水进行开发，从 2008 年 4 月底开始注 CO_2，随后又转为 CO_2—水交替注入，Y 区块的无气采收率为 0.5%。虽然 CO_2 比一般烃类气体易溶于水，但是它在原油中的溶解度大于其在水中的溶解度，而且 CO_2 可以从水溶液中转溶于原油中，因此我们将溶于水的 CO_2 忽略不计[13]。

3.1　用方法一计算 C 值

该区块原始地质储量为 $119.4 \times 10^4 m^3$，从图 2 可看出 $B = 0.1429$，则可计算 $C = NB = 17$，

将计算的 C 值代入式（20），得出 Y 区块改进后的 CO_2 混相驱拟含气率和采出程度关系式为：

$$\ln\left(\frac{f_{rg}}{1-f_{rg}}+P\right)=3.892+17(R_o-R_{ing})+Q \tag{21}$$

其中

$$P=e^{3.892+17(R_j-R_{ing})-\ln[1-e^{17(R_j-R_{ing})}]}$$

$$Q=-\ln[1-e^{17(R_j-R_{ing})}]$$

利用式（21）绘制 Y 区块改进后的 CO_2 混相驱拟含气率和采出程度图版如图 4 和图 5 所示。

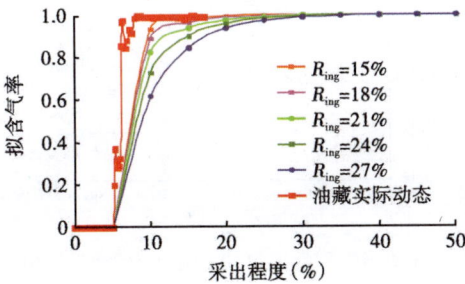

图 4　Y 区块改进的 CO_2 混相驱特征图版（据方法一）

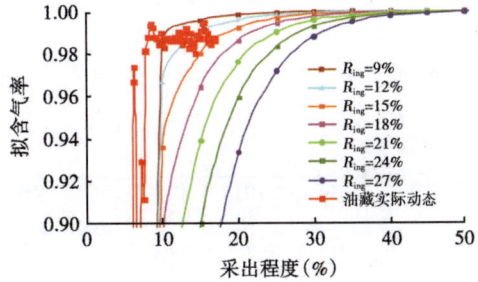

图 5　局部放大后 Y 区块改进的 CO_2 混相驱特征图版（据方法一）

由图 5 可知，通过方法一确定的改进后 CO_2 混相驱特征图版标定的采收率为 15%，即 CO_2 混相驱比水驱提高的最大采收率为 15%。

3.2　用方法二计算 C 值

利用实际生产数据通过式（17）绘制出 Y 区块 $\ln\left(\dfrac{f_{rg}}{1-f_{rg}}\right)$ 与采出程度 R_o 的关系曲线（图 6）。

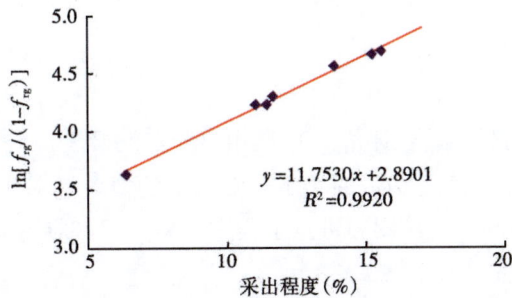

图 6　X 油田采出程度 R_o 与 $\ln\left(\dfrac{f_{rg}}{1-f_{rg}}\right)$ 的关系曲线图

从图 6 可看出 $C=11.8$，将该值代入式（20）得出 Y 区块改进后 CO_2 混相驱拟含气率和采出程度关系式为：

$$\ln\left(\frac{f_{rg}}{1-f_{rg}}+P\right)=3.892+11.8\left(R_o-R_{ing}\right)+Q \tag{22}$$

其中

$$P=e^{3.892+11.8(R_j-R_{ing})-\ln[1-e^{11.8(R_j-R_{ing})}]}$$

$$Q=-\ln\left[1-e^{11.8(R_j-R_{ing})}\right]$$

利用式（22）绘制 Y 区块改进后 CO_2 混相驱拟含气率和采出程度标定的图版如图 7 和图 8 所示。

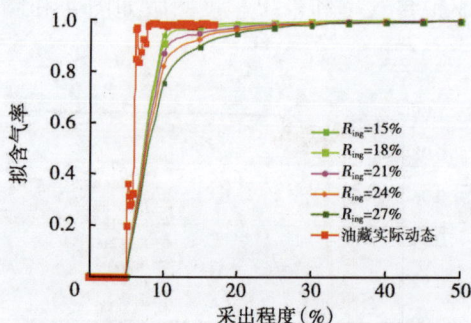

图 7　Y 区块 CO_2 混相驱特征图版（据方法二）　　图 8　局部放大后 Y 区块 CO_2 混相驱特征图版
（据方法二）

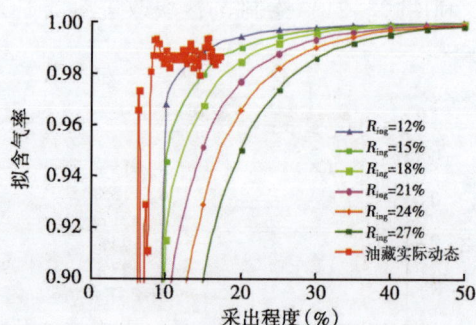

由图 8 可知，利用方法二确定的改进后 CO_2 混相驱特征图版标定的采收率为 15%，即 CO_2 混相驱比水驱提高的最大采收率为 15%。

结合其他采收率预测方法，Y 区块 CO_2 混相驱开采可比水驱采收率提高 10% 以上，通过以上分析可看出，水驱特征图版法标定的采收率偏低，初步认为水驱特征图版法已经不适合用来评价 CO_2 混相驱开发效果。通过文献调研可知，中国在各油田和集团公司勘探开发研究院进行了三次采油潜力二次评价，平均可提高采收率 16.38%[14]，与上述结果比较吻合，进一步说明本文所给出改进方法的可行性。

综上所述，利用改进后的 CO_2 混相驱拟采收率和采出程度图版在评价 CO_2 混相驱动态开发效果的应用中是可行的，可以为油田的下一步 CO_2 混相驱开发效果评价提供技术支持。

4　结论

（1）改进后的水驱图版不适合评价 CO_2 混相驱开发效果。

（2）推导了适合评价 Y 区块 CO_2 混相驱开发效果的拟含气率和采出程度图版，即 B 型气驱特征曲线，并在此基础上对该图版进行了改进，拓宽了其应用范围，使得其更好地预测 CO_2 混相驱比水驱提高的最大采收率。

（3）提出两种求取 CO_2 混相驱图版中 C 值的方法，两种方法求取的提高采收率与实际预测提高的采收率较接近，因此认为利用改进后 CO_2 混相驱拟含气率和采出程度图版来评价 CO_2 混相驱动态开发效果是行之有效的，对油田的进一步开发具有重要意义。

符号说明

W_p，G_p，N_p——分别为累计产水量、累计产气量、累计产油量，$10^4 \mathrm{m}^3$；

a，b——甲型水驱特征曲线的截距、斜率；

F_{wo}——水油比，无量纲，$F_{wo} = \dfrac{f_w}{1-f_w}$；

M——储量系数，无量纲；

E——采收率系数，无量纲；

A，B——A 型气驱特征曲线直线段的截距、斜率；

m，n，P，Q，C——修正常数；

R_i——注水开发油田无水采收率，其大小决定于地层油水黏度比，一般为 $0 \sim 8\%$；

R_j——注水后注气开发油田无气采收率；

R_o——采出程度，%；

R_M——最终采收率，无量纲；

R_{ing}——气驱比水驱提高的最大采收率，无量纲；

Q_g——日产气量，$10^4 \mathrm{m}^3$；

Q_o——日产油量，$10^4 \mathrm{m}^3$；

Q_w——日产水量，$10^4 \mathrm{m}^3$；

f_{rg}——拟含气率，%；

f_w——含水率，%。

参 考 文 献

[1] 郭平，李士伦，杜志敏，等. 低渗透油藏注气提高采收率评价 [J]. 西南石油大学学报：自然科学版，2002，24（5）：46-50.

[2] Warner H R. An Evaluation of Miscible CO_2 Flooding in Water Flooded Sandstone Reservoirs [C]. SPE 6117, 1977.

[3] Cui Maolei, Ding Yunhong, Yang Zhengming, et al. Numerical Simulation Study on CO_2 Flooding in Ultra-low Permeability Reservoirs [J]. Procedia Environmental Sciences, 2011：1469-1472.

[4] 杜建芬，陈静，李秋，等. CO_2 微观驱油实验研究 [J]. 西南石油大学学报：自然科学版，2012，34（6）：131-135.

[5] 郝永卯，薄启炜，陈月明. CO_2 驱油实验研究 [J]. 石油勘探与开发，2005，32（2）：110-112.

[6] 克林斯 M A. 二氧化碳驱油机理及工程设计 [M]. 程绍进，译. 北京：石油工业出版社，1989.

[7] 苏畅，孙雷，等. 李士伦. CO_2 混相驱多级接触过程机理研究 [J]. 西南石油大学学报：自然科学版，2001，23（2）：33-36.

[8] 杨国绪，甄鹏，赵爱婷. 气驱特征曲线在油田开发中的应用 [J]. 石油勘探与开发，1994，21（1）：71-74.

[9] 李菊花，康凯锋，高文君，等. 水气交替驱特征曲线关系式的理论推导及应用 [J]. 石油天然气学报，2010，32（5）：139-142.

[10] 童宪章. 油井产状和油藏动态分析 [M]. 北京：石油工业出版社，1981.

[11] 王柏力. 童氏含水与采出程度关系图版的改进与应用 [J]. 大庆石油地质与开发，2006，25（4）：62-64.

[12] 赫恩杰，蒋明，熊铁，等. 童氏图版的改进及应用 [J]. 新疆石油地质，2003，24（3）：232-233.

[13] 高慧梅，何应付，周锡生. 注二氧化碳提高原油采收率技术研究进展 [J]. 2009，16（1）：6-12.

[14] 李士伦，郭平，戴磊，等. 发展注气提高采收率技术 [J]. 西南石油大学学报，2000，22（3）：41-45.

低渗透油藏 CO_2 驱单层吸气能力影响因素分析

俞宏伟　李　实　陈兴隆

（中国石油勘探开发研究院国家提高采收率重点实验室）

摘　要： 超临界态的 CO_2 可与地层原油实现混相，消除常规水驱、气驱中存在的两相界面，室内实验驱油效率接近100%，具有大幅提高采收率的潜力。但在矿场实际应用中，由于多种因素的存在导致各层的吸气量与其物性关系随注入时间的增大出现严重偏差，物性较好的层吸气比例会变得越来越大，而物性较差的层吸气比例会变得越来越小，单层的吸气能力呈现出强烈的不均衡性，从而影响 CO_2 驱的波及体积，导致其提高采收率的效果变差。在室内并联岩心实验、现场产、吸剖面与数值模拟计算结果对比的基础上，分析得到层间非均质性、流度比、重力、CO_2 分子扩散及毛细管力为单层吸气能力的主要影响因素，其中非均质性的影响最大。并针对这些因素提出吉林油田 CO_2 混相驱矿场试验的调整意见，为扩大 CO_2 驱波及体积提供较为可靠的方法。

关键词： 低渗透；CO_2 驱；吸气能力；影响因素

0　引言

对于中国数十亿吨的低渗透油藏储量来说，CO_2 驱比水驱具有更明显的技术优势。2008年，中国石油在吉林油田开展了 CO_2 混相驱先导试验，目前已经取得了初步成果。但在矿场实践中，由于多种因素的存在导致单层吸气量出现严重的不均衡性，造成 CO_2 驱提高采收率的效果变差。层间非均质性、流度比、重力、CO_2 分子扩散及毛细管力为单层吸气能力的主要影响因素，本文通过现场吸气剖面分析、室内并联岩心实验及油藏数值模拟计算研究得到各因素的影响程度，并提出吉林油田 CO_2 混相驱矿场试验的调整意见，为扩大 CO_2 驱波及体积提供较为可靠的方法。

1　试验区各层吸气能力分析

先导试验区油藏属低孔/特低渗透油藏，平均孔隙度为12.77%，平均渗透率小于3mD，注水开发困难。注气目的层分为7号、12号、14号和15号4个小层，由于超临界的 CO_2 具有气体的性质，黏度低，流度大，所以 CO_2 驱较水驱具有明显的优势，注入初期5口注气井的 CO_2 平均注入指数为水驱的6.1倍，见图1。

由现场对吸水、吸气剖面的监测看：CO_2 注入初期各层吸气剖面比吸水剖面均匀，并且在注入一年后整体变化不大，但由于层间非均质性的客观存在造成纵向上各层的吸气能力差别仍旧较大，从实施 CO_2 驱后6个月的监测剖面看，物性较好的7号和14号小层的吸气量占接近总注入量的70%，而另外两个小层的吸气量只占总量的30%，见图2。由于这种情况

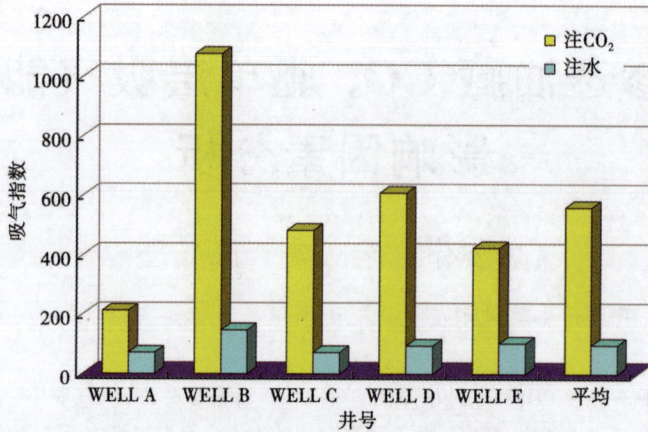

图 1 现场 CO_2 和水注入性对比曲线

的存在导致 CO_2 在各层的波及体积出现严重的不均衡性，最终导致高渗层采出程度高，低渗层采出程度低，并且在高渗层中形成 CO_2 的窜流通道，严重影响试验区 CO_2 驱提高采收率能力的发挥。

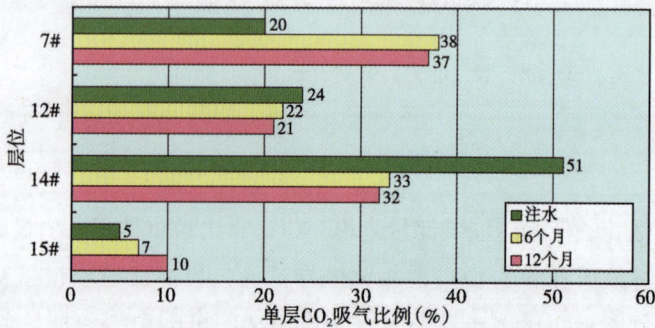

图 2 不同生产时期 B 井吸水和吸气剖面

数值模拟历史拟合计算结果与现场实际监测剖面趋势较为一致，可以利用该方法对吸气剖面进行预测，见图 3。

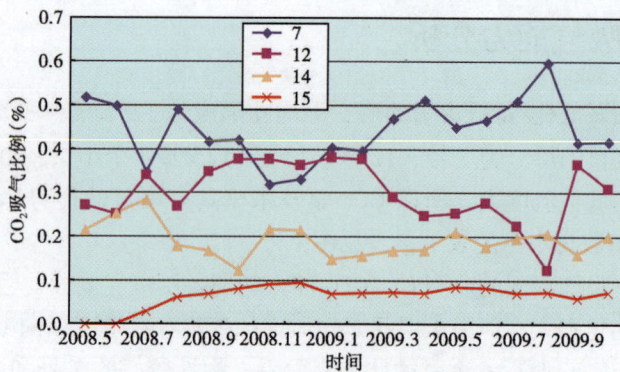

图 3 数值模拟方法计算不同生产时期 B 井吸气剖面

284

2 单层吸气能力物理模拟实验

本实验针对先导试验区油藏低渗透/特低渗透及非均质特点，采用低渗透/特低渗透并联双管长岩心物理模型，开展了 CO_2 混相/非混相驱替及水气交替注入实验，探索试验区吸气能力分布规律及通过控制流度比扩大 CO_2 驱波及体积的可行性。模型由 1 只低渗长岩心（1#岩心），和一只特低渗长岩心（2#岩心）组成，两个长岩心的平均渗透率分别为 12.2mD 和 1.67mD。

在 101.6℃、岩心出口端压力恒定在 21.2MPa 和 30MPa 条件下先进行的低渗透/特低渗透并联双管长岩心 CO_2 驱替，待 1#岩心基本不产油后再进行 CO_2/水交替注入，驱替过程中的累计采出程度变化曲线见图 4 和图 5。

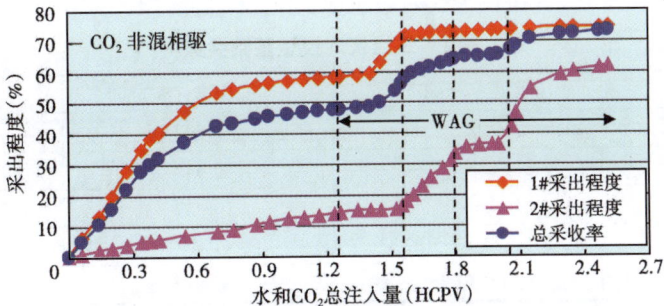

图 4 CO_2 非混相+水气交替注入采收率曲线

CO_2 非混相驱替过程中，注气量为 0.38HCPV 时在渗透率相对较高的 1#岩心发生 CO_2 气突破，突破很早，说明注入的 CO_2 在 1#岩心发生气窜，而在整个连续 CO_2 驱阶段，2#特低渗透岩心未发生 CO_2 突破，而且原油采出程度很低，表明注入的 CO_2 在 1#岩心气窜后导致波及体积减少，总体采收率很低。

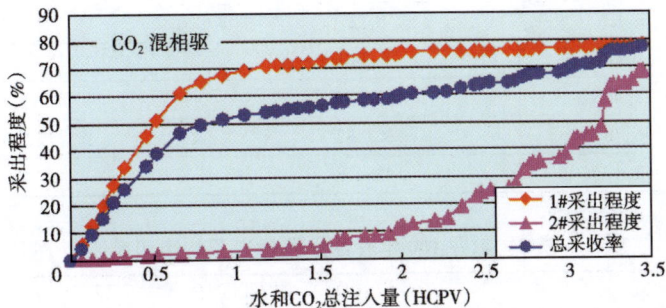

图 5 CO_2 混相+水气交替注入采收率曲线

CO_2 混相驱替过程中，注气量为 0.52HCPV 时在 1#岩心发生 CO_2 突破，相比 21.2MPa 下的 CO_2 非混相驱，30MPa 非混相驱的 CO_2 突破明显较晚，说明非混相驱比混相驱更易发生气窜。而在整个连续 CO_2 驱阶段，2#特低渗透岩心同样未发生 CO_2 突破，波及体积小，采收率低。

通过以上两组实验证明：（1）无论是混相驱替还是非混相驱替，油藏非均质性对单层吸气能力的影响都很大。连续气驱阶段，CO_2沿渗透率较高的1#岩心发生气窜，并且1#岩心的采出程度远高于特低渗2#岩心的采出程度。（2）通过水气交替注入方式控制驱替剂流度比能够有效地降低油藏非均质性对单层吸气能力的影响，扩大低渗层的波及体积，从而达到提高采收率的目的。

3 单层吸气能力影响因素数值模拟研究

以先导试验区实际储层沉积及物性条件为基础建立 60×1×20 的二维机理模型，将每个渗透率级别的储层分为一个区，研究在储层非均质性不变的条件下重力、CO_2分子扩散（实验测得 CO_2 在试验区含油岩心中的分子扩散系数为 $1.527×10^{-9} m^2/s$）及毛细管力对单层吸气能力的影响。模型初始含气饱和度为 0，主要参数见表 1。

表 1   二维机理模型基础参数

分区	渗透率（mD）	孔隙度（%）	含油饱和度（%）
1	0.5	0.112	0.48
2	1	0.115	0.5
3	3	0.117	0.51
4	6	0.12	0.53
5	10	0.122	0.55

计算模型采用一注一采的模式，在相同的注入采出强度条件下生产 3 年后关井并继续预测 100 年，分别研究重力、重力+CO_2分子扩散、重力+ CO_2 分子扩散+毛细管力对先导试验区单层吸气能力的影响。计算结果见表 2 和表 3。

表 2   注入 3 年后各分区 CO_2 平均饱和度

分区	CO_2 平均饱和度（%）		
	重力	重力+分子扩散	重力+分子扩散+毛细管力
1	1.725	1.727	1.549
2	3.335	3.340	2.774
3	13.463	13.492	9.552
4	36.783	36.582	37.548
5	42.477	42.418	45.947

表 3   封井 100 年后各分区 CO_2 平均饱和度

分区	CO_2 平均饱和度（%）		
	重力	重力+分子扩散	重力+分子扩散+毛细管力
1	3.184	3.189	2.772
2	7.234	7.249	6.940
3	20.706	20.763	22.586
4	32.419	32.296	36.909
5	34.827	34.653	29.400

针对 CO_2 先导试验区储层地质情况，在 10^3 年的时间尺度条件下，以模型中第 5 分区的 CO_2 平均饱和度变化为对比基础，从计算结果可以看出，对试验区单层吸气能力的影响从大到小的排序为：油藏非均质性 29.4%，重力 7.65%，毛细管力 5.253%，CO_2 分子扩散 0.174%，并且除油藏非均质性外其他 3 个因素都能够减缓单层吸气不均衡的趋势，但分子扩散作用很小，几乎可以忽略不计。

4 结论

（1）先导试验区储层物性差，油藏非均质性较严重，CO_2 注入能力远高于注水能力，混相驱开发较水驱开发具有明显的优势；

（2）油藏非均质性是造成单层吸气能力不均衡的主要因素，通过水气交替注入方式控制驱替剂流度比能够有效地扩大低渗层的波及体积，从而达到提高采收率的目的。先导试验区可采用该注入方式进一步提高原油采收率；

（3）重力、毛细管力、CO_2 分子扩散都能够减缓单层吸气不均衡的趋势，但分子扩散作用很小，几乎可以忽略不计。

参 考 文 献

[1] Cao Xue liang, Guo Ping, et al. An Analysis of Prospect of EOR by Gas Injection in Low-Permeability Oil Reservoir [J]. Natural Gas Industry, China, 2006, 26 (3): 100-102.

[2] Wu Zhi bai. High Development Theory and Application for Low Permeability Reservoir [M]. Beijing Petroleum. Industry: Press, 2009.

[3] Li Dao pin, et al. The Development of Low Permeability Sandstone Reservoir [M]. Beijing: Petroleum Industry Press, 1997.

[4] Chang Yih Bor, Lim M T, Pope G A, Sepehrnoori K. CO_2 Flow Patterns Under Multiphase Flow: Heterogeneous Field-Scale Conditions [J]. SPE Reservoir Engineering, 1994, 9 (3): 208-216.

[5] Tchelepi H A, Orr F M Jr. The Interaction of Viscous Fingering, Permeability Heterogeneity, and Gravity Segregation in 3D [C]. SPE 25235, 1993.

低渗透油藏 CO_2 驱过程中含水变化规律的新认识

俞宏伟　杨思玉　李　实　杨永智

（中国石油勘探开发研究院国家提高采收率重点实验室）

摘　要： 通过国内外多个油田水驱后实施 CO_2 驱的矿场实例的研究，发现注入 CO_2 后很多油井存在含水率下降的现象，这种现象发生在 CO_2 驱的整个开发阶段，包括注气初期阶段 CO_2 未到达生产井、生产井见气阶段以及高气油比阶段。本文重点讨论生产井见气前的开发规律，在该阶段油藏内以油水两相渗流为主，不受三相渗流作用的影响。通过现场动态反应并结合室内细管和长岩心实验，揭示了 CO_2 与油、水两种介质的相互作用，从而分析得出 CO_2 驱过程中油水的变化规律。

关键词： 低渗透油藏；CO_2 驱；含水变化规律

0　引言

二氧化碳混相驱是一种能大幅度提高原油采收率的三次采油方法。自 20 世纪 60 年代以来，我国在大庆、中原、华北、江苏、华东等油田相继开展了二氧化碳驱研究和小型矿场试验，取得了一些成果和经验[1]。

国内外多个油田的开发实践证明，高含水油田实施 CO_2 驱后，油田含水率将显著下降[2,3]。目前在松辽盆地一些低渗透油藏试验区实施 CO_2 驱后，短时间内油井产量和含水率均呈现出"惊人"的变化，且不同开发阶段的生产井含水率下降规律不同，为了诠释这些动态反应，本文从 CO_2 驱油机理入手[4-7]，通过细管实验和长岩心驱替实验分析不同渗透率级别油藏在不同开发阶段实施 CO_2 驱过程中的驱替规律[8]，从而总结出低渗透油藏 CO_2 驱的开发特征，指导现场动态调整。图 1 所示为美国 Slaughter Estate 单元 CO_2 驱生产动态曲线。

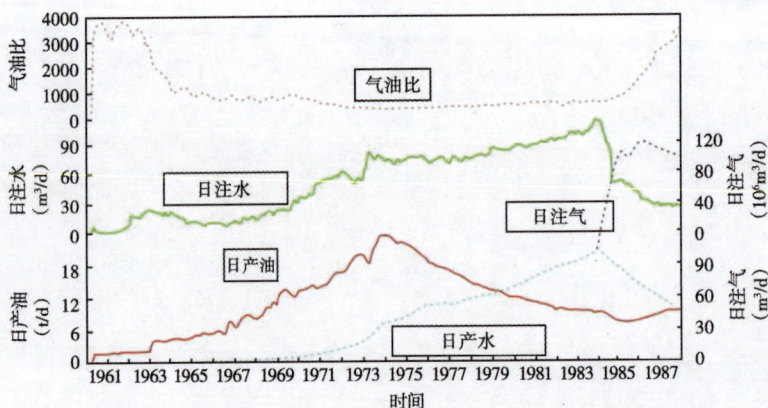

图 1　美国 Slaughter Estate 单元 CO_2 驱生产动态曲线

1 中低含水阶段油藏进行 CO_2 驱的变化规律

开发一段时间后处于中低含水阶段的油田，包括初期投产即含水的低含油饱和度油田，此时油藏中的可动油饱和度大于可动水的饱和度[9-11]，在这个阶段开始实施 CO_2 驱，在注入少量 CO_2 后，部分生产井就会出现含水率下降的现象，其中个别井下降幅度较大，产量上升快，见图2。根据动态资料和监测资料显示，在注入 CO_2 初期，驱替前缘处于注入井近井地带而尚未到达生产井时，生产井附近的原油中尚未溶解 CO_2，此时仍旧以油水两相渗流为主，产出的原油和水的性质也未发生改变。为了研究 CO_2 与油、水两种介质的反应，设计了细管实验研究流体间的相互作用，具体做法是：在油藏温度和混相压力条件下将细管中饱和地层原油，先进行0.3倍孔隙体积（Pore Volume，以下简称PV）的水驱，而后连续注入1.2PV CO_2 进行气驱，见图3。

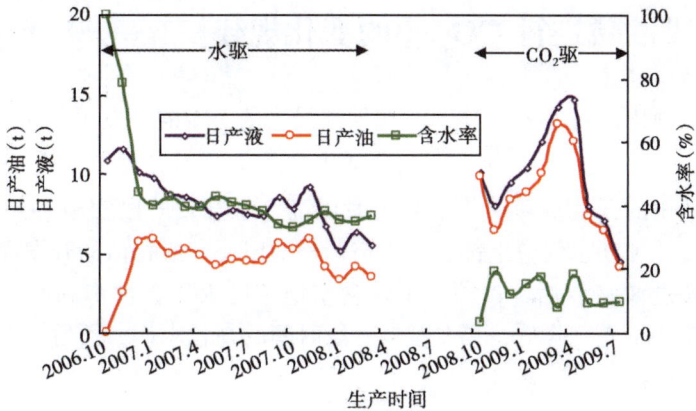

图2　A油田典型井产量、含水变化曲线

图3　细管实验驱替示意图

从细管实验驱替过程分析，注入水在注入0.98PV（水和 CO_2 的注入孔隙体积之和）时突破，但突破后产水量很少，并随 CO_2 注入量的增加而逐步减少；随后在注入总量为1.18PV时 CO_2 突破，突破后立即产出大量气，气油比急剧升高，瞬时达到500以上；直到注入总量为1.37PV时，水才大量产出。虽然水先于 CO_2 注入，但大多数的注入水滞后于 CO_2 产出，见图4。注入端流体注入的顺序是：油—水—CO_2，出口端产出物的产出顺序是：油—极少量水—大量 CO_2-大量水，通过细管实验可以看出在注入少量水后再进行 CO_2 驱，相当于油水共存、中低含水阶段的油藏，注入的 CO_2 能够穿越水段塞直接接触地层原油，与水相比 CO_2 与原油的反应更为敏感。

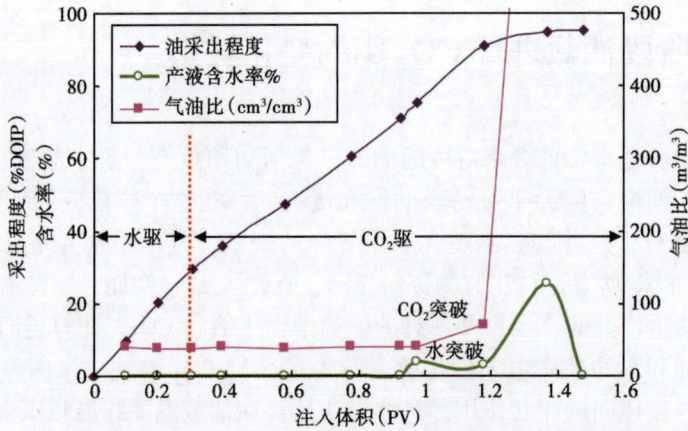

图4 0.3PV 水+1.2PV CO$_2$ 细管实验驱替曲线

2 高含水阶段油藏进行 CO$_2$ 驱的变化规律

国外绝大多数油田已将 CO$_2$ 驱作为三次采油最为的有效的开发手段而加以利用。95%以上的 CO$_2$ 驱项目是在油藏进入高含水末期、油田即将废弃时开展的，在采出了更多剩余油的同时也延长了油田的开发寿命[12,13]。目前我国高含水开发阶段实施 CO$_2$ 驱的油田数量有限，但是从驱替特征和动态反应上都表现出与国外油田相同或相似的开发规律，在高含水阶段特别是含水率超过 80% 以后，油藏内平均含水饱和度大于含油饱和度平均值，实施 CO$_2$ 驱后，油井含水率并不立刻降低，仍要维持一段时间的高含水期，而后会出现含水率迅速下降，产油量上升的现象[1]，见图5。

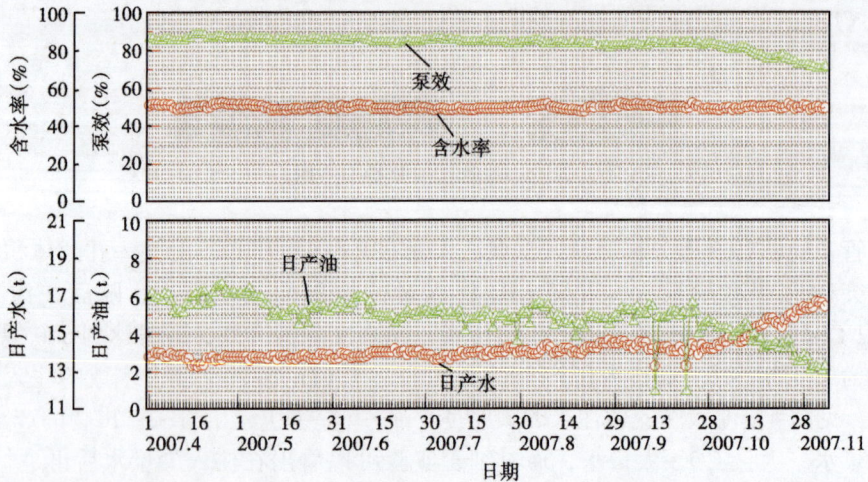

图5 B 油田 CO$_2$ 驱突破前产量、含水变化曲线

当油藏内含有少量可动水时，即地层中平均含油饱和度大于平均含水饱和度时，可以发生 CO$_2$ 直接穿越水而快速驱动地层油的现象。注气井通常都是注水井后期改造的，在经过

长期注水后，注入井周围区域为高含水饱和度带，在最初的阶段 CO_2 作为驱替力还是主要推动注入井附近孔隙内的水流向生产井，而生产井附近孔道内的油、水比例基本没有变化，就开发指标而言，在这一阶段油井的含水率和产量与注气前相比变化不大。随着 CO_2 驱替前缘逐渐向生产井推近，在 CO_2 没有突破前，CO_2 完全溶于水和油中，不论混相还是非混相驱，CO_2 在油水两相中处于动态平衡，油藏中仍旧以两相形式存在：富含 CO_2 水相和富含 CO_2 油相，没有游离的 CO_2 气相存在，这时油井含水率降低和产油量增加也证实了上述中低含水期 CO_2 直接穿越水而驱动地层油的机理。

根据不同油藏 CO_2 混相驱现场的动态反应，气体突破前后的产量和含水率变化幅度差别很大。采集了国内两个进行 CO_2 驱油田的岩心进行了的长岩心水驱后 CO_2 驱的对比实验，研究驱替规律，A 油田为特低渗透油藏，岩心平均渗透率 1.47mD，B 油田为一般低渗透油藏，岩心平均渗透率 14.6mD。实验首先进行水驱，至含水率达到 98% 以后，在保持混相驱的条件下，注入 CO_2 气体，直至产油量不再增加为止。

在本文所关注的水驱后进行 CO_2 驱至 CO_2 突破前这个阶段，两种渗透率的岩心表现出了一个共同的趋势，就是都存在一个含水率基本不降，产量很低的阶段，这一阶段主要是 CO_2 的溶解过程，出口端附近的油水组成不变，产量和含水率也没有太大变化，和注气前一致。在这个阶段原油的采出程度很低，特低渗岩心只有 0.7%，而一般低渗透岩心几乎没有产量，全部产水。当驱替逐渐稳定后，CO_2 的作用显现出来，含水率开始下降。在这个阶段不同渗透率岩心驱替过程的差异性非常明显，对于特低渗透长岩心在含水率开始下降至 CO_2 突破前，采出程度很高达到 21.93%，而气体突破后采出程度提高的幅度有限，只有 8%；对于一般低渗透长岩心而言，在含水率开始下降至 CO_2 突破前，采出程度只有 6%，气体突破后提高的幅度很大，接近 30%。如图 6、图 7 和表 1 所示。

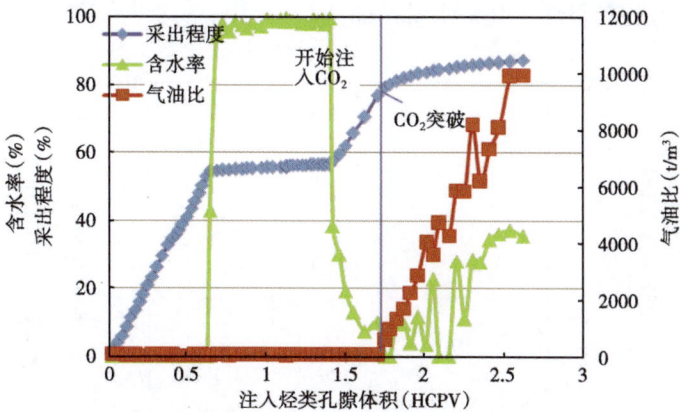

图 6　特低渗透长岩心驱替曲线（K=1.47mD）

表 1　长岩心驱替开发指标对比

岩心编号	渗透率（mD）	水突破注入量（PV）	保持高含水状态 CO_2 注入量（PV）	保持高含水状态提高采收率幅度（%）	含水下降到气体突破前提高采收率幅度（%）
1#	1.47	0.64	0.27	0.69	21.93
2#	14.6	0.49	0.6	0	6

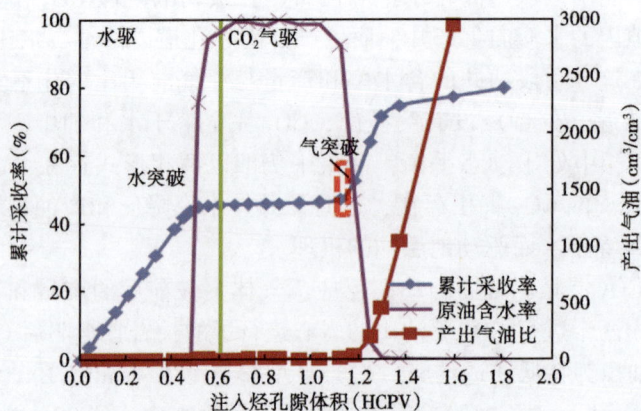

图 7　一般低渗透长岩心驱替曲线（$K=14.6mD$）

根据我国低渗透油藏的开采经验，在可以通过水驱建立起驱替压力系统、并可实现能量持续补充的低渗透储层，这类储层的开发特征与中高渗透油藏开发规律基本一致[14]。在水驱动态反应上，上述研究的 B 油田就是呈现中高渗透油田的开发规律，长岩心实验结果与国外中高渗透率油田的混相驱生产动态规律基本一致，即注入初期，CO_2 的主要作用是驱油提供能量，在生产井见 CO_2 气后，CO_2 与原油的混合过渡带前缘到达生产井，通常混相驱的过渡带长且相对稳定，因此随着气油比的逐渐增加，仍旧能够采出大量原油，在这个阶段如果能够采取适合的开发技术政策，如合理控制生产压差、注采比、保持地层压力等，可取得一段相对较长的高产稳产时间。待 CO_2—原油过渡带逐渐被采出后，气油比快速升高，产出物以 CO_2 为主，CO_2 驱替的作用减弱，少量原油通过蒸发作用被裹带出来。图 8 为美国 Slaughter Estate 单元 CO_2 驱油、气、水量产曲线。

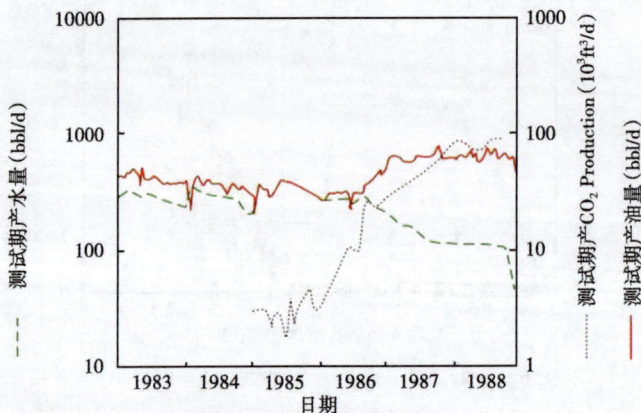

图 8　美国 Slaughter Estate 单元 CO_2 驱油、气、水量产曲线

3　结论

我国未动用储量中 65% 是低渗透储量，利用现有技术开发难度大，因此在今后一段时间内低渗透油藏是我国实施 CO_2 驱的主力战场，但由于目前进行的 CO_2 驱项目以小规模、

292

先导性试验为主，进行时间短，整体开发规律尚无法描述与总结。因此利用室内实验的研究结果与现场动态进行对比分析、把现场中暴露的开发问题与实验相结合，以实验过程表征 CO_2 驱油规律，解释动态现象，找出解决问题的办法，是在当前阶段认识 CO_2 驱开发规律的技术方法和有效手段。

通过上述室内实验与现场结合的研究方法，揭示了低渗透油田在不同含水阶段实施 CO_2 混相后的变化规律。(1) CO_2 驱始于低含水阶段、油藏中的可动油饱和度大于可动水的饱和度时，CO_2 与原油的接触更为敏感，可以穿过地层水驱替地层油，部分油井会出现产油量增加、含水率大幅度下降而产液量变化不大的规律。(2) CO_2 驱始于高含水阶段的驱替规律与低含水阶段的有所差别，存在含水率基本不降，产量低、与注气前变化不大的低产阶段，待驱替逐渐稳定后，呈现含水率开始快速下降、产量上升的规律，在这个阶段不同渗透性油藏 CO_2 驱的差异性非常明显：对于特低渗透长岩心而言，在含水率开始下降至 CO_2 突破前的采出程度占 CO_2 驱阶段总采出程度的88%，而气体突破后的采出程度占 CO_2 驱阶段总采出程度的12%；对于一般低渗透长岩心而言，在含水率开始下降至 CO_2 突破前的采出程度占 CO_2 驱阶段总采出程度的16.7%，气体突破后产量增加幅度很大，占 CO_2 驱阶段总采出程度的83.3%。

参 考 文 献

[1] 郭永辉. 草舍油田 CO_2 驱油先导试验区工艺技术研究及效果 [J]. 内蒙古石油化工, 2008 (10): 14-15.

[2] Stein M H, Frey D D, Walker R D, et al. Slaughter Estate Unit CO_2 Flood: Comparison between Pilot and Field-Scale Performance [C]. SPE19375, 1999.

[3] 谢尚贤, 韩培慧, 等. 大庆油田萨南东部过渡带注 CO_2 驱油先导性矿场试验研究 [J]. 油气采收率技术, 2007, 4 (3): 13-19.

[4] 杨承志, 岳清山, 沈平平. 混相驱提高石油采收率 (上册) [M]. 北京: 石油工业出版社, 1991, 51-75.

[5] 杨承志, 岳清山, 沈平平. 混相驱提高石油采收率 (下册) [M]. 北京: 石油工业出版社, 1991, 166-187.

[6] 克林斯 M A. 二氧化碳驱油机理及工程设计 [M]. 程绍进, 译. 北京: 石油工业出版社, 1989, 28-66.

[7] 李孟涛, 单文文, 刘先贵, 等. 超临界二氧化碳混相驱油机理实验研究 [J]. 石油学报, 2006, 27 (3): 80-83.

[8] 王进安, 袁广均, 等. 长岩心注二氧化碳驱油物理模拟实验研究 [J]. 特种油气田, 2001, 8 (2): 75-78.

[9] 石国新, 孙来喜, 李成. 石南 J_2t_2 低渗透砂岩油藏含水变化特征及水淹模式分析 [J]. 物探化探计算技术, 2009, 31 (4): 393-399.

[10] 俞启泰, 赵明, 林志芳. 水驱砂岩油田含水变化规律与采收率多因素分析 [J]. 石油勘探与开发, 1992, 19 (3): 63-68.

[11] 郭龙. 特低渗透砂岩油藏初期含水变化机理实验研究 [J]. 西南石油大学学报: 自然科学版, 2008, 30 (4): 127-129.

[12] Abel Salazar, Sharon Haggard, Vishnu Simlote, et al. CO_2 Flooding Revitalizes Old Giant Field. OGJ, 2006, 104 (15): 42-44.

[13] Enhancing Oil Recovery through Thirty Years of CO_2 Flooding Success [R]. Shell International Exploration and Production Company, 2007.

[14] 唐仁选, 张勇, 许国辰, 等. 草舍油田泰州组油藏 CO_2 驱动态监测及分析 [J]. 试采技术, 2007, 28 (增刊): 7-9.

CO$_2$ 驱后储层及剩余油物性变化探讨

陈 亮[1] 孙 雷[1] 王 英[2] 百宗虎[1] 叶雪松[3]

(1. 西南石油大学研究生院；2. 成都理工大学能源学院；
3. 中国石油长庆固井公司)

摘 要： 目前对注 CO$_2$ 提高原油采收率相关问题研究已经比较深入，而对 CO$_2$ 驱后储层及剩余油物性变化的研究还不多。据统计，成功进行 CO$_2$ 驱后剩余油饱和度约 30%，为进一步提高原油采收率，研究 CO$_2$ 驱后储层及剩余油物性的变化是非常必要的，主要从以下 3 个方面进行分析：CO$_2$ 驱后储层物性变化、CO$_2$ 驱后剩余油物性的变化及剩余油与储层之间的关系变化。

关键词： CO$_2$ 驱；储层；剩余油；物性变化

1 CO$_2$ 驱后储层物性变化

目前，CO$_2$ 驱的主要方式有 CO$_2$ 气水交替驱、CO$_2$ 与碱性表面活性聚合物驱、CO$_2$ 泡沫复合驱，它们对提高采收率起到了相当大的作用，但同时，也对储层造成较大的伤害：

（1）地层中水敏矿物遇水膨胀及流动，特别是蒙皂石的吸水膨胀、高岭石的运移，造成地层孔渗降低，储层伤害。

（2）泡沫对储层孔隙堵塞，使储层渗透率降低。

（3）由于 H$_2$CO$_3$ 是弱酸，同时，储层中 H$_2$CO$_3$ 可溶物分布具有非均值性、溶蚀具有选择性，使得溶蚀表面凹凸不平，增大了岩石的比表面。

（4）CO$_2$ 溶于地层水，使 pH 值下降，地层水呈弱酸性，使矿物快速溶解，在一定程度上增加了储层孔隙体积，同时，CO$_3^{2-}$ 与 Mg^{2+}、Ca^{2+} 结合生成沉淀，最终改变储层物性。

根据谷丽冰 CO$_2$ 岩心驱替实验分析，CO$_2$ 驱使储层孔隙度小幅度增加，而渗透率却大幅度下降，最大降幅达到 33.61%[1]。

2 CO$_2$ 驱后剩余油物性的变化

剩余油的物性主要包括剩余油组成变化、极性变化及黏度变化。其中，剩余油的组成是黏度变化、极性变化的基础。研究剩余油物性变化是进一步提高原油采收率的关键。

2.1 剩余油组成变化

在 CO$_2$ 驱中，当压力达到一定数值，CO$_2$ 能够萃取并且汽化原油中不同组分的轻烃，从而降低界面张力，达到提高采收率的目的，这是 CO$_2$ 驱提高原油采收率最主要的机理。CO$_2$ 是非极性物质，不是极性物质的溶剂，因此只能萃取原油中的非极性物质，而原油中的重质组分［沥青质（包含含 NSO 的非烃类物质）、胶质、石蜡及高分子烃］大部分都是极性较强的物质，特别是沥青质和胶质。实验表明，在 CO_2 驱过程中，其萃取的主要碳分布是 C_2—C_{30} 的非极性物质，即饱和烷烃，从而，使剩余原油重质组分增加。

目前对 CO_2 驱后剩余油组分的研究非常少，但这并不表示不重要。研究剩余油的组分组成对进一步提高采收率、有效评价及优选"四采"方案起到决定性作用。

在资料不足的情况下研究剩余油组分，目前主要有以下两种方法：

（1）类比法。目前研究最为广泛的是水驱采油，而水驱采油与 CO_2 驱存在一定的相似性，即轻质部分优先采出，重烃较难采出。从不同采出程度下原油样品中正构烷烃分布特征图不难发现（图 1），随采出程度的增加，采出油中的大于 C_{25} 组分逐渐增加，而小于 C_{15} 组分逐渐减少[2]，因此，留在地下的原油主要是大于 C_{25} 的重组分；同时，由于 CO_2 非极性，原油中重烃为极性并且岩石表面也为极性，因此在 CO_2 驱过程中，原油中重质组分不断与岩石表面结合，从而阻碍了轻烃对重烃的黏附作用，加大了重烃随轻烃采出的难度，经过 CO_2 的长期抽提，形成了以重质组分为主的剩余油。

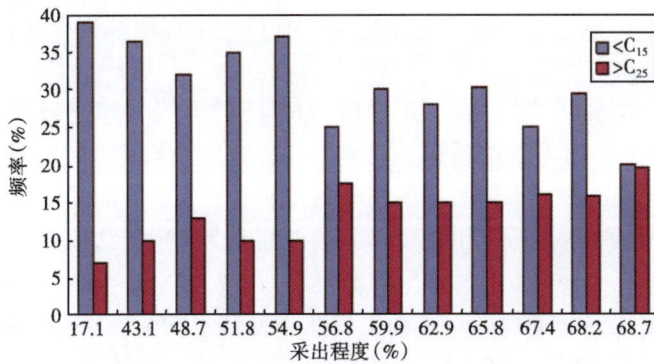

图 1　不同采出程度下原油样品中正构烷烃分布特征

（2）间接法。通过 CO_2 驱采出油成分分析及物性变化研究地层剩余油组成。F. S. Palmer 对美国路易斯安那州采用 CO_2 混相驱的 SU 油藏研究发现，CO_2 采出原油相对密度下降程度达到 10%，从原油中沥青质的含量也能间接反映出剩余油的组成变化，据 Hwang 对得克萨斯州 McElory 油田注 CO_2 试验结果研究，采出油沥青质含量降幅达到 50%，这都表明 CO_2 驱会导致剩余油中重质组分增加（图 2）；由 CO_2 气洗作用形成的苏丹 ME 盆地古近系某稠油油藏也有力证明 CO_2 会使剩余油重质组分大大增加，以至改变油藏的性质[3]。

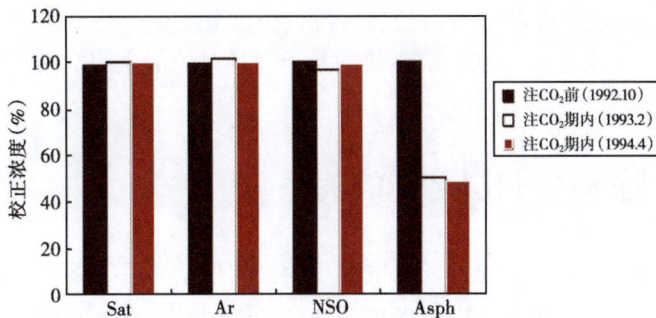

图 2　McElory 油田注 CO_2 前后原油组成变化（Asph—沥青质）

2.2 剩余油极性及黏度变化

石油是一种比较稳定的胶体分散体系。其中，分散相是沥青质及胶质的重组分，分散介质是饱和烃、芳香烃及胶质中的轻质组分，石油胶体体系的稳定性取决于分散介质和分散相之间的动态平衡，当体系动态平衡被破坏，原油性质也会发生很大的变化。CO_2 对分散介质的抽提及溶解，改变了体系的物理化学性质，打破了体系的平衡，使得重组分外部的溶剂化分子层厚度减小，甚至遭到破坏，增大了重组分之间相互接触的几率。根据 Hildebrand 提出的非电解质溶液一般理论，即：

$$\Delta H_{mix} = V_{mix}(\sigma_A - \sigma_B)^2 \varphi_A \varphi_B \tag{1}$$

式中，ΔH_{mix} 为混合热，J/mol；V_{mix} 为混合物的摩尔体积，mol；σ_A 为溶剂的溶解度参数，Hildebrand；σ_B 为溶质的溶解度参数，Hildebrand；φ_A 为溶剂的体积分数；φ_B 为溶质的体积分数。

只有当 AB 两种物质溶解度参数相等或者相近时，才会形成理想溶液，一般要求 $|\sigma_A - \sigma_B|_{max} < (1.7 \sim 2)$，若大于 2，则溶液沉淀。由于 CO_2 的强烈抽提作用，使得溶剂溶解度参数增大，溶质的溶解度参数大幅减小，体系 $|\sigma_A - \sigma_B|$ 远大于 2，导致体系重组分沉淀。体系重组分的增加，加剧分子之间的吸附作用，同时使重组分中氢键酸与氢键碱结合，形成超分子化合物；原油中饱和烃、芳香烃及胶质中的轻质组分均是非极性或极性很弱的物质，而重组分则是极性很强的物质，体系平衡的破坏及非极性物质的抽提，必然导致原油体系极性的改变，同时，原油中重组分的沉淀更是原油极性改变体现的结果。

剩余油组分及极性的变化，必然会导致其黏度改变，特别是重组分的出现，甚至会改变原油性质，当重组分达到一定程度，原油呈现非牛顿体系的流变特征。Burg 等指出，原油黏度实质上取决于偶极—偶极和偶极—诱发偶极相互作用，这些分子相互作用是偶极分子（即原油中的杂原子）和可极化分子产生的[4]。沥青质与胶质中偶极—偶极相互作用最强，是影响原油黏度的主要内在因素。重组分分子之间主要是靠氢键结合，分子间作用力大，分子结构越复杂，接触面积也大幅度提高，分子层间内摩擦力显著增强，剩余油黏度明显增加，同时，由于重组分的出现，使得 dV/dL 不再是常数，从而使得剩余油体系中出现较强的黏度微观非均质性。同时，根据潘志清等[5]提出的油质系数（TPI）与黏度（μ）关系式为：

$$\mu = 0.31 \times (TPI)^{-11.96} \tag{2}$$

式中，TPI 为轻质组分含量/（轻质组分含量+重组分含量）。

该式表明，原油中重组分与黏度成指数关系，并且对原油重组分敏感性极强。即使 TPI 只增加 0.05，黏度也会增加 2 倍以上。

3 CO_2 驱后剩余油与储层之间的关系

CO_2 驱使储层物性及剩余油性质发生很大改变，其中，储层最显著的变化就是润湿性的改变，剩余油最显著的变化则是组分、黏度及极性的改变。剩余油与储层之间吸附作用变化是刻画二者变化关系的最好纽带。

CO_2 驱后储层岩石对原油的吸附性更强，主要有以下原因：

（1）随着轻质组分被抽提的程度加大，重组分在剩余油中存在状态变为伸展状，增大了重组分与储层岩石表面接触的机会，改变了重组分在岩石表面的吸附层结构。

（2）重组分一个显著特点是极性很强，特别是含硫、氮、氧及其他非碳氢元素化合物，其偶极距更大，极性更强；岩石颗粒也是由极性物质构成，根据"相似相吸"原理，二者之间的吸附作用更强、亲油性更强。

（3）在原油胶体平衡体系未打破之前，岩石与原油之间为极性-非极性体系，主要以物理吸附为主，但是在体系平衡打破之后，由于重组分中杂原子及微量元素的参与、氢键的出现、甚至有离子的参与，使得二者之间的吸附类型可能会趋于化学吸附。种种因素表明，剩余油与储层岩石的分离难度增加。

4　认识及建议

（1）CO_2 驱会改变储层物性，特别是孔隙度及渗透率，对常规砂岩储层来讲，其渗透率降低，物性变差。

（2）CO_2 驱使剩余油重组分增加，极性增强，黏度增大。

（3）CO_2 驱后剩余油与储层岩石吸附作用改变，随着重组分增加，甚至会改变吸附类型形成化学吸附，吸附能力大大增强。

（4）目前大多数黏度模型没有考虑分子结构及极性对黏度的影响，深入探究黏度与分子结构及分子极性之间的关系，建立更加适合复杂流体的黏度模型。

（5）积极开展注 CO_2 前期评价，探究剩余油组分变化及极性变化的深层次原因，探索高分子胶体吸附作用的微观机理，指导生产实践。

参 考 文 献

[1] 谷丽冰，李治平，侯秀林，等. 二氧化碳驱引起储层物性改变的实验室研究 [J]. 石油天然气学报，2007，29（3）：258-260.

[2] 刘晓艳，李宜强，冯子辉，等. 不同采出程度下石油组分变化特征 [J]. 沉积学报，2000，18（2）：235-236.

[3] 高金玉，史卜庆，王林，等. CO_2 脱沥青作用形成重油油藏的一个实例 [J]. 现代地质，2009，23（5）：923-927.

[4] 罗平亚，敬加强，朱毅飞，等. 原油组成对其粘度影响的灰色关联分析 [J]. 油气田地面工程，2000，19（6）：12-13.

[5] 潘志清，苏秀方，梅博文，等. 含油气层油质系数与原油密度和黏度的关系 [J]. 江汉石油学报，1995，17（4）：19-21.

第三篇
二氧化碳驱油井筒与地面工艺设计、高效防腐及安全性评价技术

含 CO_2 混合气超临界注入技术研究及认识

孙锐艳　　王世刚　　马晓红　　李艳杰

(中国石油吉林油田公司勘察设计院)

摘　要： 相比 CO_2 液相注入工艺而言，采用超临界注入工艺更节约投资和运行成本。目前，虽然国外在 CO_2 驱油和埋存领域已经进行过深入研究和工业化推广，但国内已实施的 CO_2 驱油项目主要采用液相注入工艺。本文从理论上提出了 CO_2 气层气、CO_2 驱产出伴生气超临界注入相态控制研究方法，通过绘制注入介质的相包络曲线和含水量关系图，进行多级压缩过程中的相平衡分析、水合物控制和 CO_2 防腐工艺研究，同时提出了多级压缩时压缩机最佳入口温度。

关键词： CO_2 驱；超临界；注入；相平衡；压缩

0　引言

CO_2 超临界注入技术是指把 CO_2 从气态压缩至超临界态后注入地下的处理工艺。目前，CO_2 超临界注入技术在国外已经成为一种切实可行的酸气回注技术，主要用于 CO_2 驱三次采油和 CO_2 封存项目，在美国、加拿大、挪威等国家应用比较普遍，并且大多数 CO_2 驱项目都使用储层中产出的高纯度 CO_2 和工业性 CO_2 作为气源。比较典型的案例为加拿大 Weyburn 油田 CO_2 驱油和地质埋存示范工程，注入来自美国北达科他合成燃料厂净化装置的 CO_2，该项目采用管道长距离输送（输送压力大于 15.2MPa，管道长度达 320km），注入的 CO_2 纯度为 96%，含有少量的 H_2S、N_2 和烃类气体，注入状态为超临界状态，注入量 5000t/d。

国内已用于现场 CO_2 驱项目主要采用液相注入工艺。但种种研究表明，采用超临界管道输送和注入联合方式是一种最为经济的 CO_2 回注技术之一。

1　含 CO_2 混合气回注系统工艺设计要点

在 CO_2 驱三次采油和 CO_2 地质埋存项目中，含 CO_2 混合气回注系统采用的气源通常为储层中产出的高含 CO_2 混合气和工业回收的烟道气。含 CO_2 混合气回注系统设计的关键问题是相平衡计算分析，一方面气源组分复杂、机械杂质含量高，另一方面饱和含水的低压含 CO_2 混合气含水量高，对压缩工艺设计和设备安全运行具有较大影响，设计时需要考虑相平衡分析和预处理等关键因素，通常从以下几方面进行分析。

1.1　绘制相包络曲线

在进行含 CO_2 混合气回注系统工艺设计时，首先是根据气体组分绘制相包络曲线（p—

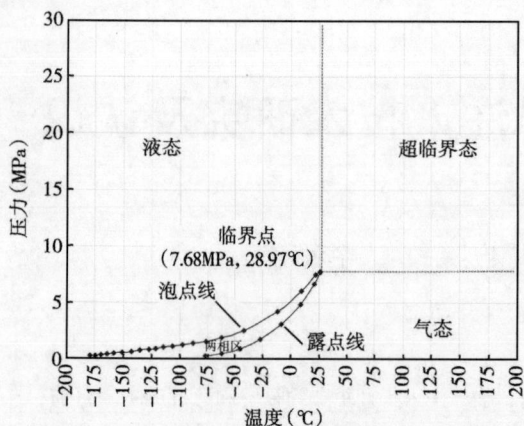

图 1　长深 4 井含 CO_2 混合气相包络曲线图

T 关系图）。在图 1 中可明显看出混合气体的相态临界点位于气液两相同时存在时的最高压力点和最高温度点。只有当压力和温度处于相包络线内时，含 CO_2 气体混合物才以两相状态存在。研究认为，对于混合气体压缩机设计，绘制真实介质的相包络曲线是正确计算和修正多级压缩级间工艺参数、合理控制相态的重要前提。

1.2　控制含水量

如果含 CO_2 混合气中有水存在，会带来两方面问题：一是含有游离水的酸气腐蚀性很强；二是含 CO_2 混合气可能生成像冰一样的水化物，易堵塞管道和冷却器，破坏设备。因此，在进行含 CO_2 混合气压缩机设计时需要脱水，采用压缩脱水或附加脱水系统使酸气的露点达到可控要求。

1.3　相平衡分析与控制

主要以含 CO_2 混合气体的相包络曲线为相态参数控制依据，反复调整压比，计算和修正压缩机各级进出口参数，工艺上主要采用控制温度的办法来控制压缩机各级入口相态参数，级间设冷却器和分离器，分离掉压缩和冷却产生的液体，确保多级增压时压缩机各级入口参数处于非两相区和非液相区。

1.4　预处理

由于储层产出的含 CO_2 混合气或烟道回收气在未经特殊处理时，一般含有较多的大直径机械杂质，如固体颗粒和液滴等，会对压缩机正常运转造成极大影响，甚至损坏叶轮，破坏压缩机结构。因此，需要根据压缩机进气条件要求，对含 CO_2 混合气进行预处理，除去大直径的液体和固体颗粒。

2　CO_2 气层气超临界注入相态研究

以吉林油田黑 59 试验区开展的 CO_2 驱试验为例，如果气源采用的是长深 2 井、长深 4 井产出的储层气，对试验区采用 CO_2 超临界注入工艺进行如下研究。

2.1　绘制相包络曲线

长深 4 井含 CO_2 混合气：CO_2 含量 97.183%，甲烷 0.877%，氮气 1.94%（体积分数）。采用 AQUAlibrium 3 软件计算含 CO_2 混合气性质，从绘制的相包络线（图 1）可以看出，介质的临界点压力 7.68MPa、临界点温度 28.87℃，两相区位于泡点线下方和露点线上方区域。

2.2　含水量分析

从长深 4 井产出含 CO_2 混合气（饱和含水时）含水量跟温度的关系图来看（图 2）：温度升高，介质含水量增加；且压力为 0.25MPa 时含水量及变化率明显高于 2.5MPa 时的含水量及变化率。从长深4井含 CO_2 混合气（饱和含水时）含水量跟压力的关系来看：在临界压力以下，含水量随压力升高呈下降趋势；临界压力以上，含水量随压力升高有上升趋势；且50℃含水量高于40℃（图2）。

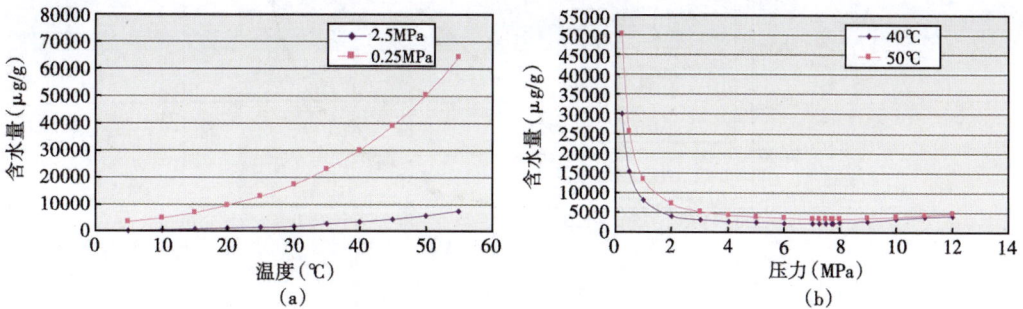

图2 长深4井产出介质含水量跟温度的关系（a）和含水量跟压力的关系（b）

2.3 相平衡分析与控制

控制相态的关键是控制压缩机入口温度。由于温度是液相含 CO_2 混合气相态和含水量的关键因素，必须从相态控制和含水量控制两方面着手研究，以确定合理的压缩机入口温度范围。

从不同组分含 CO_2 混合气的相包络曲线可以看出：随着介质 CO_2 含量减小、烃类气和氮气含量增加时，介质的临界温度呈下降趋势；当介质温度高于纯 CO_2 临界点温度31.25℃时，介质就不能处于液相区或两相区，当介质温度处于40℃及以上时，可保证介质始终处于气相或超临界状态（表1、图3）。

表1 模拟含 CO_2 混合气组分与临界参数表及相包络曲线

编号	气体组分摩尔分数			临界压力（MPa）	临界温度（℃）
	CO_2	CH_4	N_2		
1	0.5961	0.3399	0.064	9.756	-8.06
2	0.67688	0.27192	0.0512	9.448	1.77
3	0.75766	0.20394	0.0384	9.008	10.4
4	0.83844	0.13596	0.0256	8.493	18.05
5	0.91922	0.06798	0.0128	7.939	24.85
6	1	0	0	7.38	31.25

因此，结合含水量分析结论，研究认为将压缩机入口温度控制在40℃附近左右有利于含 CO_2 混合气相态控制和含水量控制。

以吉林油田黑59区 CO_2 驱试验区地层压力需求为例，压缩机最高排气压力需要达到25MPa，注入量 $2×10^4m^3/d$，由于压缩机入口介质为长深4井至黑59区 CO_2 管道末端来气，压力在2.5~3.0MPa之间，若将温度控制为40℃、压力节流2.5MPa进入压缩机，经计算，

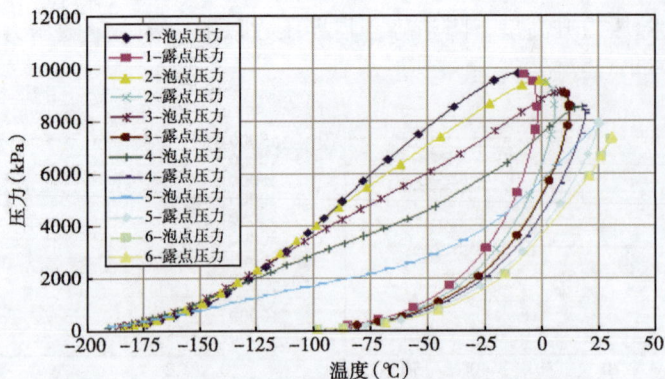

图3 临界参数表及相包络线

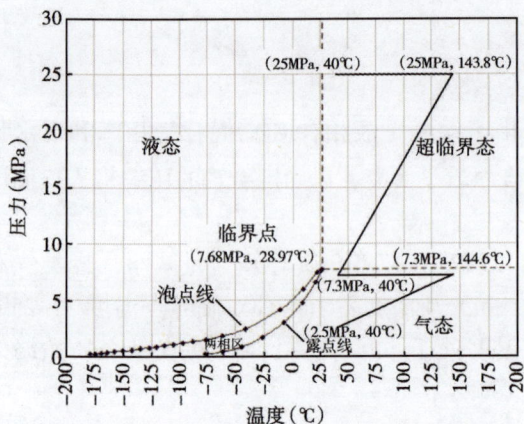

图4 长深4井含CO_2混合气超临界注入相态
控制示意图

采用两级压缩，排气压力即可达到25MPa。为了保证级间温度、压力参数不处于两相区和液相区，压缩机入口、级间须设冷却器和分离器，分离掉由于增压产生的液体（图4）。

2.4 主要工艺流程简介

该系统流程主要包括预处理系统和增压系统两部分。主要工艺流程如下：界外来气在预处理系统经除尘、除液和除雾后，进入增压系统；在增压系统经换热至40℃进入压缩机，经一级压缩至7.3MPa后冷却至40℃，再经气液分离后进入二级压缩机，压缩到25MPa后，再经冷却，然后去站外注入干线管网。

3 CO_2驱伴生气超临界注入工艺相态研究

以吉林油田在黑59区CO_2驱试验区开展的国内首个CO_2超临界注入试验项目为例进行研究，一方面实现黑59区CO_2驱伴生气循环注入和CO_2零排放，另一方面验证国产高压CO_2压缩机的使用性能，为大规模开展CO_2驱油与埋存扩大试验和工业化推广提供技术依据。

在吉林油田黑59区CO_2驱试验区，油井产出伴生气中CO_2含量变化很不稳定，单井伴生气CO_2含量波动范围在5%~90%之间。而CO_2驱油藏研究表明：为了易于实现混相，注入介质中CO_2含量不宜低于一个数量指标。研究认为，在开展CO_2驱伴生气超临界注入试验时，宜将压缩机入口介质CO_2含量提高到94%以上，预留一定波动范围，从而避免由于地面系统运行不平稳而影响驱替效果。因此，为了达到这一指标，需要将油井产出伴生气与长深2井、长深4井管道来的纯CO_2气充分混合，才能使压缩机入口介质CO_2含量提高到94%以上。显然，现场具备开展超临界试验的条件。本文对工艺上实现的可能性做了以下研究。

3.1 绘制相包络曲线

原料气组分：CO_2 含量 94%（体积分数），其余是烃和 N_2。采用 AQUAlibrium 3 软件计算含 CO_2 混合气性质，从绘制的相包络线可以看出，介质的临界点压力 7.74MPa、临界点温度 26.45℃，两相区位于泡点线下方和露点线上方区域（图 5）。

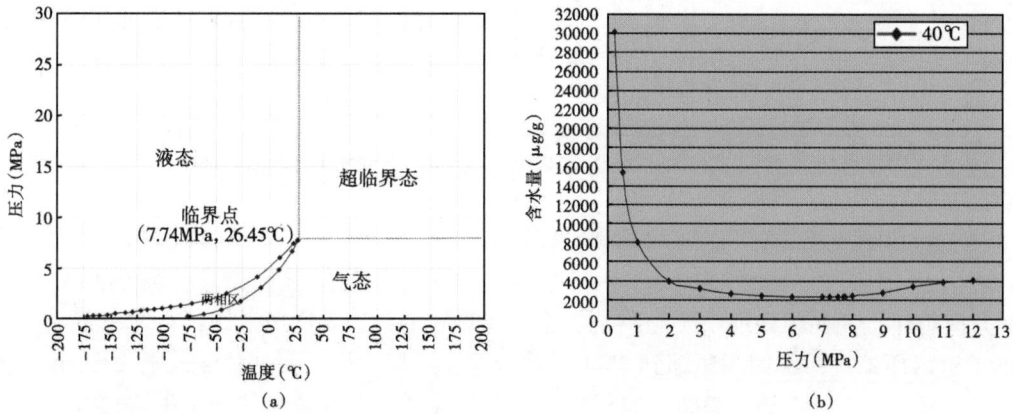

图 5　混合后 CO_2 驱伴生气相包络曲线（a）及 40℃含水量跟压力的关系（b）

3.2 含水量分析

原料气压力范围为 0.2~0.3MPa；原料气温度不大于 40℃；原料气含有一定量的游离水。可见，进入压缩机的原料气一定是饱和含水状态。经计算，需要四级压缩才能将伴生气从 0.2~0.3MPa 压缩到 25MPa 以上，其中第三级出口压力为 7~9MPa。由 40℃条件下该组分的饱和含水状态下的含水量关系图（图 5b）可知，6~8MPa 时介质含水量最低（在 2200~2800μg/g 之间），显然，在该组分介质压缩过程中，三级压缩冷却后，可将含水量降到最低，四级压缩冷却后，不会再有游离水出现。研究认为，针对该组分，采用压缩脱水工艺，即可保证压缩机出口无游离水存在。

3.3 相平衡分析及控制

由于该试验区注入介质要求最高压力达到 25MPa、温度不大于 40℃，故压缩机出口压力以 25MPa 计算，当注入量为 $5 \times 10^4 m^3/d$ 时，需要采用四级压缩才能将伴生气从 0.2~0.3MPa 压缩到 25MPa 以上。以伴生气入口压力为 0.3MPa 计算，得到如下结果（表 2）。

表 2　CO_2 驱伴生气超临界注入时压缩机各级参数表

参数	1 级		2 级		3 级		4 级	
	入口	出口	入口	出口	入口	出口	入口	出口
压力（MPa）	0.3	1.001	1.001	2.901	2.901	8.601	8.601	25.1
温度（℃）	40	138.8	40	143.5	40	147.2	40	107.2

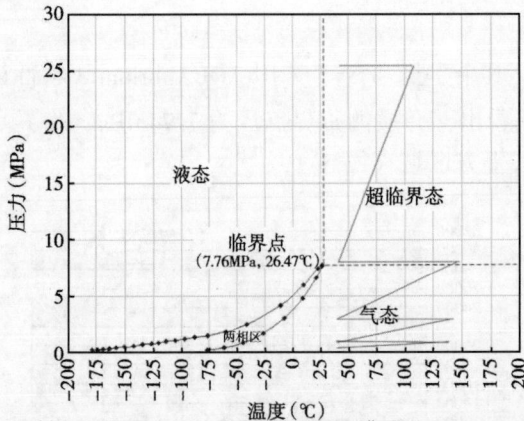

图 6　CO_2 驱伴生气超临界注入相态控制示意图

根据相态控制示意图（图 6）可看出，由于压缩机各级入口温度为 40℃，各级压缩冷却前后的含 CO_2 伴生气始终处于非液相区和非两相区，其中三级出口介质处于低压超临界区，该区域的含 CO_2 混合介质性质同样是符合压缩条件的。

因此，研究认为当 CO_2 驱伴生气 CO_2 含量不低于 94% 时，可采用压缩机超临界注入，排气压力能够达到 25MPa 以上。

3.4　主要工艺流程简介

CO_2 驱伴生气超临界注入工艺流程同样包括预处理系统和增压系统两部分。界外来油井产出伴生气在预处理系统经除尘、除液和除雾后，进入增压系统；在增压系统经换热至 40℃进入压缩机，经四级压缩到 25MPa 后，再经冷却，然后去站外注入干线管网。其中，分别在一级、二级、三级压缩机出口设冷却器和气液分离器，分离掉由于压缩和冷却形成的液体，用于级间相态、水化物和腐蚀控制。

4　结论及建议

以本文列举实例来看，采用超临界注入工艺回注 CO_2 气层气和 CO_2 驱伴生气用于三次采油是可实现的。在工艺设计，需注意以下几点：

（1）研究含 CO_2 混合气超临界注入工艺时，宜先绘制真实气体介质的相包络曲线，然后计算多级压缩级间参数，用真实气体介质的相包络曲线来修正个别参数，重新计算，以获得合理的级间工艺参数，达到相态控制目的。

（2）多级压缩时，压缩机各级入口参数须控制在相包络曲线的非两相区和非液相区域，工艺上可采用控制温度的办法控制介质相态，且压缩机各级入口温度控制在 40℃左右为宜。

（3）可采用压缩脱水工艺来控制饱和含水状态下含 CO_2 混合气的含水量，从而防止压缩机级间形成水合物和 CO_2 腐蚀，必要时，可在压缩机入口或级间附加深度脱水装置满足特殊工艺要求。

（4）由于 CO_2 气层气或 CO_2 驱伴生气可能携带大直径固体颗粒或液滴，因此建议在介质进入压缩系统前进行除尘、除液和除雾等预处理，保证系统安全平稳运行和设备使用寿命。

参 考 文 献

[1] John J. Carrol, Shouxi Wang, 汤林. 酸气回注——酸气处理的另一种途径 [J]. 天然气工业. 2009, 29 (10)：96-100.

[2] John J. Carrol, Wu Ying. Acid Gas Injection in Engineering Design Aspect Semina. Dec [C]. 2009.

[3] 康万利, 任韶然, 李玉星, 等. CO_2 气田气集输处理工艺及 CO_2 驱地面配套技术研究报告 [R]. 2009

[4] 杨俊兰, 马一太, 曾宪阳, 等. 超临界压力下 CO_2 流体的性质研究 [J]. 流体机械. 2008, 36 (01), 53-57.

[5] 梁国斌. CO_2 压缩机缸体腐蚀原因分析及处理 [J]. 化肥设计. 2002, 40 (6), 39-41.

CO_2 管道输送技术研究及应用

王世刚[1]　孙锐艳[1]　张浩男[2]　牛秉辉[3]

(1. 中国石油吉林油田勘察设计院；2. 中国石油大庆油田采油三厂；
3. 中国石油大庆油田工程建设公司)

摘　要：CO_2 有气态输送、液态输送和超临界输送三种管道输送方式，宜根据整个工程系统的实际情况综合研究确定 CO_2 管输方式。在列举的 5 套 CO_2 管输及注入方案中，随管输距离增加，采用低压超临界输送、高压超临界注入方式经济性最优，其次是气态输送、超临界注入方式。从大情字井油田 CO_2 管输试验来看，采取深埋保温方式是 CO_2 管输时控制相变的有效措施之一。

关键词：CO_2；管输；相态；超临界；注入费用

0　引言

二氧化碳（CO_2）管道输送系统类似于天然气和石油制品输送系统，包括管道、中间加压站及其辅助设备。由于 CO_2 临界参数较低，其输送可通过气态、液态和超临界态三种相态实现。气态 CO_2 在管道内的最佳流态处于阻力平方区，液态 CO_2 与超临界 CO_2 则在水力光滑区。由于 CO_2 在高压下密度大、黏度小，当管输压力在 8 MPa 以上时，可保持较高的 CO_2 输送效率，从而降低管径，提高输量，对杂质要求不高，管线不需保温，因此超临界输送从经济和技术两方面都明显优于气态输送和液态输送。另外，超临界输送可保持管道末端高压，使 CO_2 直接注入地层，无需增设注入压缩机。但是，采用何种输送方式最经济，需根据工程项目的 CO_2 气源、注入或封存场所实际情况优化而定。

1　CO_2 相态特征

纯 CO_2 $p - T$ 相图如图 1 所示。

从 CO_2 相态特征上可将 CO_2 分为气态、液态（低温低压）、高压密相（或高压液态）、超临界状态、固态五种状态。其中气态、液态和高压密相状态是 CO_2 输送过程最常见的气体相态。CO_2 的饱和线始于三相点，终于临界点。在三相点上，气相、液相、固相三相呈平衡状态。在临界点处，气相和液相的性质非常接近，两相之间界面消失，成为一相。当温度和压力高于临界温度和临界压力时，CO_2 进入超临界态。超临界 CO_2 是一种可压缩的高密度流体，它的物理性质处于气体和液体之间，既有与气体相当的高扩散性和低黏度，又有与液体相近的密度和对物质优良的溶解能力。当改变温度、压力，使得 CO_2 从气态绕过临界点进入超临界状态，然后依次进入高压密相和液态时，整个过程中 CO_2 的密度、黏度是一个逐渐变化的过程；但当 CO_2 从气态穿过饱和线进入液态，或从液态穿过饱和线进入气态时，CO_2 相态发

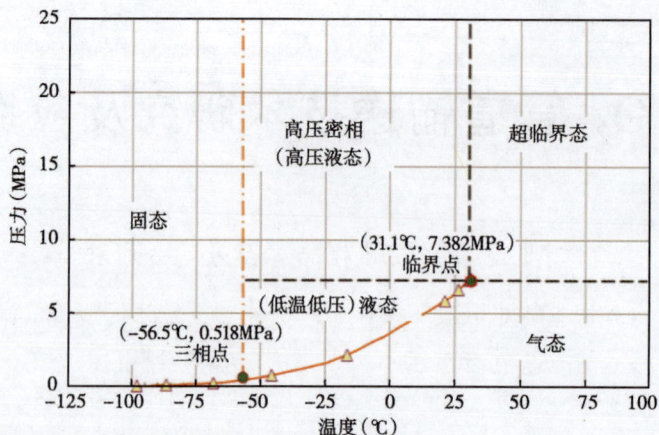

图 1　纯 CO_2 的 p-T 相图

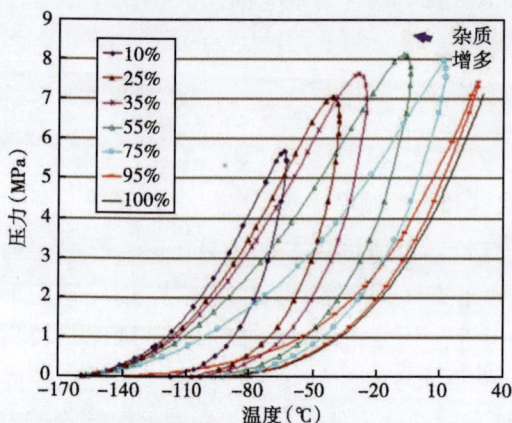

图 2　含 CO_2 气体相包络线

生跃变，CO_2 密度和黏度也发生突变，这种现象在 CO_2 输送和压缩时都应该避免。

当 CO_2 气体中含有其他气体时，CO_2 的相态特征就会发生变化：出现气液两相区，即在某一条件下，CO_2 出现气相、液相共存的现象，如图 2 所示。当杂质气体的含量由 0 逐渐增加到 90% 时，CO_2 气体临界温度降低至 95~100℃，并出现两相区，且两相区的范围随着杂质气体含量的增加而增大。两相区的出现使得 CO_2 气体在单相状态下可操作的温度和压力范围缩小，对 CO_2 输送工艺和操作要求进一步提高。因此，杂质气体含量越小，越有利于 CO_2 输送。

2　CO_2 管输方案优化

针对吉林油田 CO_2 捕集与驱油项目中长岭气田净化厂副产品 CO_2 循环利用，以大情字井油田为目标试验区，进行 CO_2 管输方案优化研究。

2.1　管输方式

气源采用长岭气田天然气净化厂 CO_2 回收气，温度 40℃，压力 0.2MPa，水露点温度 -10℃，CO_2 纯度 99%；试验区 CO_2 注入量 $10×10^4 m^3/d$，注入压力 25 MPa。根据 CO_2 相态特征，设计 5 种输送注入方案：（1）气态输送，液态增压注入；（2）气态输送，气态增压超临界注入；（3）液态输送，液态增压注入；（4）高压超临界输送，直接注入；（5）低压超临界输送，高压超临界注入。

2.2 设计结果分析

CO_2 输送注入流程的模拟计算结果见表 1。

表 1　CO_2 输送注入方案设计结果表

方案	参数	集气站			管线输送		注入站	
		进口	出口	设备	出口	设备	出口	设备
1	压力（MPa）	0.2	6	三级压缩机	5.4～5.7	X65 钢 $\phi152\times4.5$	25	液化装置增压泵
	温度（℃）	40	40		18.63～-3.09		-29.22～-29.42	
2	压力（MPa）	0.2	6	三级压缩机	5.4～5.7	X65 钢 $\phi152\times4.5$	25	二级压缩机
	温度（℃）	40	40		18.63～-3.09		40	
3	压力（MPa）	0.2	3	二级压缩机液化装置	2.7	X65 钢 $\phi127\times4$	25	增压泵
	温度（℃）	40	-40		-19.46～-27.71		-2.99～-13.08	
4	压力（MPa）	0.2	26.9	四级压缩机	25.1	X65 钢 $\phi108\times12$	—	—
	温度（℃）	40	40		19.96～-4.95		—	
5	压力（MPa）	0.2	9	三级压缩机	8.11～8.28	X65 $\phi114\times5$	25	增压泵
	温度（℃）	40	40		20.84～-4.60		45.89～9.57	

注：管线长度 100 km，埋地管线，仅方案 4 采取保温措施。

2.2.1 相态分析

CO_2 采用方案 2 气态输送时，夏季运行（冻土层温度 20 ℃）时，CO_2 在管线中可以始终保持气态，不会液化；但冬季运行（冻土层温度-5℃）时，由于环境温度低，CO_2 在管线 10km 处就开始出现气液两相共存状态，气液两相区从 10km 处一直延伸至 40km 处，之后管线中的 CO_2 全部以液态形式存在。为避免 CO_2 发生相变，可采取的措施有：（1）降低管线入口压力；（2）提高管线入口温度；（3）在管线中途增加加热站；（4）安装保温层；（5）深埋冻土层以下。CO_2 采用方案 3 液态输送时，不论夏季和冬季，都需要采取保温措施，防止 CO_2 在输送过程中出现相变。采用方案 4 高压超临界输送时或方案 5 低压超临界输送时，输送过程中 CO_2 由超临界状态逐渐进入高压密相状态，虽然 CO_2 的相态发生改变，但是在这一过程中，CO_2 的密度、黏度、压力等性质是个缓慢变化的过程。研究认为，超临界到高压密相的相态转变不会对 CO_2 的输送造成影响，因此管线不需采取保温措施。

2.2.2 成本分析

从单位 CO_2 输送注入费用来看，方案 5 输送成本最低，采用低压超临界输送，需要钢材量较少，基建费用较低，并随输送距离增加经济性更加明显；方案 4 采用大于 25MPa 高压输送，虽然输送效率很高，且流程简单，在较短输送距离内具有很高经济性，但要求较大的管线壁厚，随着距离增加，其管线铺设费用迅速增大，致使单位 CO_2 输送注入费用迅速上升。方案 2 采用低压输送，由于需要在管线末端进一步增压以达到注入要求，虽然在较短输送距离内经济性并不突出，但随着管线长度增加，管线铺设费用上优势明显增加，其单位 CO_2 输送注入费用明显低于其他方案。方案 3 采用液态输送和液态注入方式，虽然管线的输送效率较高，但是基建费用（需液化装置和保温措施）和操作费用（液化费用）都很高，使得这一方案经济性最差。

2.3 大情字井油田 CO_2 管输试验

大情字井油田建成两条 CO_2 输气管线：长深 2 井—长深 4 井—黑 59 井管线和长岭净化站—黑 46 井—长深 4 井—黑 47 井管线。

(1) 长深 2 井—长深 4 井—黑 59 井管线。长深 2 井产出 CO_2 气经节流后湿气输送至长深 4 井，与该井来气汇合后经分子筛脱水，干气输送至黑 59 试验区。其中长深 4 井—黑 59 井管线长 8 km，材质 L245，2008 年建成投产；长深 2 井—长深 4 井管线长 8km，湿气输送，材质 316L，2009 年建成投产。该管道设计压力 6.4MPa，埋深-2 m，目前输量约 $6×10^4 m^3/d$，输送相态稳定，管线运行良好。

(2) 长岭净化站—黑 46 井—长深 4 井—黑 47 井管线。长岭净化站回收的 CO_2 气经增压脱水后，采用管线输送至黑 46 井和黑 47 井的 CO_2 驱循环注入站。该管线全长 26km，设计压力 4.0MPa，埋深-2m，设计输量 $60×10^4 m^3/d$，管线材质采用 L245。由于黑 46 井和黑 47 井的 CO_2 驱循环注入站未投产，该管线暂时建成未投入运行。

3 结论

(1) 进行 CO_2 管输设计时，应避免 CO_2 相态突变。

(2) CO_2 有气态输送、液态输送和超临界输送三种管输方式，宜根据整个工程系统的实际情况综合研究确定 CO_2 管输方式。

(3) CO_2 由超临界态逐渐进入高压密相状态时，CO_2 的密度、黏度具有渐变特征。

(4) 在列举的五套 CO_2 管输及注入方案中，随管输距离增加，采用低压超临界输送、高压超临界注入方式经济性最优，其次是气态输送、超临界注入方式。

(5) 从大情字井油田 CO_2 管输试验来看，采取深埋保温方式是 CO_2 管输时控制相变的有效措施之一。

吉林油田模拟 CO_2 驱采出水性质研究

祁雪洪[1]　康万利[1]　董朝霞[1]　王宪中[2]　孙锐艳[2]　王世刚[2]

(1. 中国石油大学（北京）EOR 中心，2. 中国石油吉林油田勘察设计研究院)

摘　要：利用 PVT 装置，对吉林油田 CO_2 驱试验驱采出水采用 CO_2 与水作用来模拟采出水性质变化，探讨了 20 ℃温度条件下不同压力采出水性质及离子组成的变化规律。结果表明，通入 CO_2 后采出水的 pH 值降低值为 1，随着压力的增加，pH 值有先减小后增大的趋势；电导率变化趋势与 pH 值类似。通入 CO_2 产生的固体颗粒是 $Fe(OH)_3$，并伴有其他的杂质；压力的突变可使得钙、镁等离子沉淀，结垢现象严重。随着 CO_2 压力的增加，产生的固体颗粒粒径先由小变大，出现两极分化，后均匀化；分散度先减小后增大。固体颗粒的含量及大小直接影响油水的乳化程度及乳状液的稳定性。

关键词：CO_2 驱；采出水；离子分析；固体颗粒；结垢；污水处理

0　引言

将 CO_2 注入油田提高原油采收率（EOR）是石油工业的一项成熟技术，同时也可以实现 CO_2 的地下埋存。CO_2 驱采出液性质与水驱采出液具有很多不同之处，有关 CO_2 对采出油性质影响国内外研究较多，如 CO_2 能够有效地降低油水界面张力[1]和原油的黏度，抽提作用能使胶质、沥青质沉淀[2-3]等，而对采出水性质研究相对较少。因此，本文针对吉林油田 CO_2 驱试验驱采出水，采用 CO_2 与水作用来模拟采出水的性质变化，为 CO_2 驱采出污水处理提供理论依据。

1　室内实验

1.1　材料

实验所用材料包括：CO_2 气体（纯度为 99.95%），北京氦普北分气体工业有限公司生产；盐酸，分析纯，北京现代东方精细化学品有限公司生产；污水，乾安黑 59 缓冲罐出口水。

1.2　仪器

实验所用仪器：超高压全视 PVT 分析仪，法国 SANCHEZ TECHNOLO GIES 公司制造；酸度计为 pHS—3C 型，上海雷磁仪器厂制造；DDSJ-308A 电导率仪，上海精密科学仪器有限责任公司制造；Zetasizer Nano ZS 纳米粒度及 Zeta 电位分析仪，英国马尔文仪器有限公司制造。

1.3 方法

（1）CO_2 与水的作用。

将一定量的油田采出污水置于 PVT 分析仪高压釜中，在 20℃温度条件下，用高压泵通入 CO_2，使压力分别达到设定值（如 8MPa、10MPa、12MPa、15MPa、20MPa、25MPa）。观察高压釜体积随时间的变化，当体积基本不变时，脱气，将水放出，备用，用于水性质分析。

（2）采出水性质分析。

用酸度计测定采出水模拟液的 pH 值，用电导率仪测其电导率，用离子色谱进行各离子含量分析。

（3）粒径测定。

用 Zetasizer Nano ZS 纳米粒度及 Zeta 电位分析仪进行水中沉淀颗粒粒径的测定。测定不同压力下，CO_2 驱的采出水模拟液中固体颗粒粒径的大小，并进行对比分析。

（4）固体含量测定。

将 CO_2 驱的采出水模拟液用 $0.45\mu m$ 的微孔滤膜过滤，把微孔滤膜放入烘箱内烘干，取微孔滤膜过滤前后质量差，得各压力下所产固体颗粒的含量。

2 结果与讨论

2.1 采出水性质

从不同压力下采出水的性质变化可以看出，通 CO_2 后采出水的 pH 值由 7 下降为 6，降低了 1 个单位；采出水的 pH 值随压力变化不大，但有先减小后增大、最后基本恒定的趋势。其变化受 CO_3^{2-}、HCO_3^- 的含量影响。CO_2 溶于水显弱酸性，使得采出水的 pH 值下降。少量的 CO_2 会形成碳酸，使得 pH 值下降。CO_2 在水中的溶解是有限的，随着压力的增大，过量的 CO_2 会发生以下反应：

$$CO_3^{2-} + CO_2 + H_2O \rightleftharpoons 2HCO_3^- \tag{1}$$

此反应为可逆反应。平衡向右移动使得 HCO_3^- 量增加，HCO_3^- 水解显碱性，使得 pH 值略有大。

从处理后水的 pH 值、离子含量及电导率随压力变化曲线可以看出，Ca^{2+}、Mg^{2+}、Fe^{3+} 及碳 CO_3^{2-}、HCO_3^- 的变化较明显。Ca^{2+} 含量表现为先下降，又缓慢上升，最后趋于不变；Mg^{2+} 含量略有减少；Fe^{3+} 含量几乎为零。通入 CO_2 后水中 CO_3^{2-} 含量下降，而 HCO_3^- 上升；CO_3^{2-}、HCO_3^- 的总量增加。这是因为反应（1）为可逆反应，由于 CO_2 压力增大使得 CO_3^{2-}、HCO_3^- 总量增加。

从电导率变化可以看出，通 CO_2 前后污水的电导率变化很大。随着压力的增大，与 CO_2 作用后污水的电导率也有先减小后增大的趋势。电导率是溶液中离子含量的反应。由于 Ca^{2+}、Mg^{2+}、Fe^{3+} 等离子的减少及 HCO_3^- 的增加，使得采出水的电导率发生相应的变化。

2.2 沉淀固体颗粒粒径

从不同压力下沉淀固体颗粒粒径的大小分布可以看出，随着 CO_2 压力的增加，产生的固体颗粒粒径先由小变大并出现两极分化，而后均匀化；分散度先减小后增大。压力为 8MPa 时，粒径大小有 1.0~350μm，只有一个峰值，粒径最高峰只占总数的 2%，粒径分散度大；10MPa 时出现两个峰，低粒径峰范围 1.2~13μm，高粒径峰范围 25~330μm，峰值分别 2.7%、3.1%，粒径分散度略有减少；12MPa 时粒径相对 10MPa 峰值向两侧移动，低粒径峰范围 1.0~2.8μm，高粒径峰范围 25~94μm，峰值 4.5%、10.3%，粒径大小出现两极分化；15MPa 时，粒径进一步分化，出现三个峰，低峰范围 0.35~1.1μm，次低峰范围 2.0~3.7μm，高峰范围 45~150μm，峰值 3.6%、1.4%、11.5%，此时分散度达到最小，分化最严重；20MPa 时粒径两峰向中间靠拢，分散度加大；25MPa 时两峰并为一峰，峰范围 0.25~40μm，峰值 2.4%，分散度达最大，不过小于 8MPa 时的分散度。低压力条件下，固体颗粒产生的晶核较少，随着沉淀的析出发生聚集，大颗粒会越聚越大，出现两极分化的现象；在高压力条件下，产生较多的晶核，固体颗粒均匀增大。

2.3 沉淀物

不同压力下形成的沉淀呈现棕色絮状，将沉淀过滤后取出加入盐酸，沉淀迅速溶解，溶液的颜色为黄色，加入 NH_4SCN 溶液显红色，可推断为 $Fe(OH)_3$ 沉淀。由水中离子分析结果可以看出，Fe^{3+} 含量由原来的 5.56μg/mL 减小到 1μg/mL 以下，变化明显。用湿润的 pH 试纸，置于容器口附近，观察颜色无变化，说明无酸性气体的产生，证明无碳酸盐沉淀形成。

图 1 为 pH 值与水中 HCO_3^-、CO_3^{2-}、H_2CO_3、CO_2 的平衡关系。从图 1 可以看出，在 pH 值为 6 时水中 CO_3^{2-} 的含量很低。25℃温度条件下，$CaCO_3$ 的 k_{sp} 为 $8.7×10^{-9}$，$MgCO_3$ 的 k_{sp} 为 $2.6×10^{-5}$。以图 1 中 Ca^{2+}、CO_3^{2-}、Mg^{2+} 的含量可以计算出 Ca^{2+}、CO_3^{2-} 的 k_{sp} 远远小于 $8.7×10^{-9}$，Mg^{2+}、CO_3^{2-} 的 k_{sp} 远远小于 $2.6×10^{-5}$。可见，不可能出现碳酸盐沉淀。

尽管 Ca^{2+}、Mg^{2+} 的量有所减少，其实 Ca^{2+}、Mg^{2+} 的存在以及 CO_2 增加是结垢的内因，但不一定造成结垢，还必须具备某些外部条件，沉淀反应如下

图 1 pH 值与水中 HCO_3^-、CO_3^{2-}、H_2CO_3、CO_2 的平衡关系图

$$Ca(HCO_3)_2 \rightleftharpoons H_2O+CO_2+CaCO_3 \qquad (2)$$

$$Mg(HCO_3)_2 \rightleftharpoons H_2O+CO_2+MgCO_3 \qquad (3)$$

由平衡式看出，伴随水中 CO_2 的减少，就会有 $CaCO_3$ 产生。可见，造成结垢的外部原因是压力突然下降或温度突然上升使水中溶解 CO_2 突然减小，从而导致了可溶性 $Ca(HCO_3)_2$ 生成 $CaCO_3$ 沉淀。在实验过程中，脱气使压力骤降，部分沉淀可能残留于 PVT 容器管壁上，使得 Ca^{2+}、Mg^{2+} 含量减少。此外，大庆油田小井距注水驱后注 CO_2 油田矿场试验中，当油井含水率达 80% 以上时，地面管线、油气分离器、输油泵等处均有类似水垢的坚硬沉淀物产生，经分析，沉淀物大部分为碳酸盐，$CaCO_3$ 占 80%~90%，$MgCO_3$ 占 5%~10%。所以 CO_2 驱采出液从地下到地上的过程中，随着压力的突然降低，CO_2 溢出，将会伴随碳酸盐沉

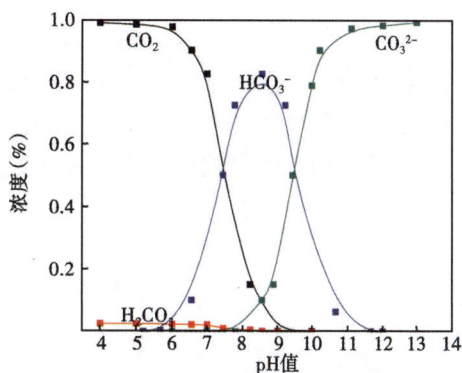

淀出现严重的结垢现象。

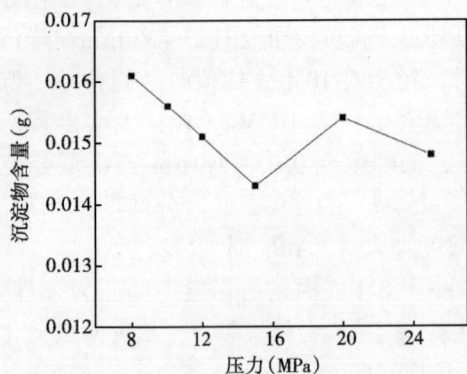

图 2　通入 CO_2 后所产生沉淀颗粒含量
随压力变化曲线图

生成固体颗粒的含量随压力变化如图 2 所示。沉淀量呈先下降、后上升、再下降的趋势。根据 Fe^{3+} 的最初浓度，并按 Fe^{3+} 全部沉淀计算，25mL 的采出液应得 0.015g 沉淀。由图 2 可看出，各压力产生的固体颗粒含量与其相当，说明产生的固体中大部分为 $Fe(OH)_3$ 沉淀，沉淀中还有其他微量杂质。

应用颗粒变化也可对图 2 结果进行解释。压力在 8~15MPa 时，固体含量逐渐减少，除与 pH 值的变化有关，也与固体颗粒的粒径大小有关。由于在 12~15MPa 范围内，固体颗粒大小产生了两极分化，出现了小于微孔滤膜孔径的颗粒，小的颗粒未被截留在滤膜上，故所得的固体含量有所减少。压力为 20MPa、25MPa 时固体颗粒出现均衡，较小的颗粒减少，故固体含量呈现上升趋势。

3　结论

（1）通入 CO_2 后采出水的 pH 值降低大约 1 个单位，随着压力的增加，pH 值有先减小后增大的趋势；电导率变化趋势与 pH 值相似；Ca^{2+}、Mg^{2+}、Fe^{3+} 及 CO_3^{2-}、HCO_3^- 的变化明显。作用后采出水酸性增大，可加快金属腐蚀，腐蚀产物呈絮状沉淀，增加污水处理难度。

（2）通入 CO_2 产生的固体颗粒是 $Fe(OH)_3$，并伴有其他的杂质；压力的突变可使得 Ca^{2+}、Mg^{2+} 等离子沉淀，结垢现象严重。

（3）随着 CO_2 压力的增加，产生的固体颗粒粒径先由小变大并出现两极分化，而后均匀化；分散度先减小后增大。固体颗粒的含量及大小直接影响油水的乳化程度及乳状液的稳定性。

参考文献

［1］孙长宇，王文强，陈光进，等. 注 CO_2 油气藏流体体系油/水和油/气界面张力实验研究［J］. 中国石油大学学报：自然科学版，2006，30（5）：109-112.

［2］Huang Lei, Shen Pingping1, Jia Ying. Prediction of asphalteneprecipitation during CO_2 injection［J］. Petrol Explor Develop, 2010, 37(3): 349-353.

［3］谢尚贤，颜五和，韩培慧. 关于用二氧化碳三次采油的腐蚀与结垢问题［J］. 大庆石油地质与开发，1989，8（4）：59-66.

Ni，Mo 和 Cu 添加对 13Cr 不锈钢组织和抗 CO_2 腐蚀性能的影响

张旭昀[1]   高明浩[1]   徐子怡[2]   王   勇[1]   毕凤琴[1]

(1. 东北石油大学；2. 中国石油大学（北京）)

摘   要：采用正交实验法，通过等离子电弧炉制备添加不同含量 Ni、Mo 和 Cu 元素的 13Cr 不锈钢，研究合金的微观组织、结构特征以及饱和 CO_2 采出液中的腐蚀电化学和高温高压浸泡腐蚀行为。结果表明：合金组织主要为马氏体和铁素体，随 Ni、Mo 和 Cu 含量不同而变化；$Ni_4Mo_{1.2}Cu_{1.4}$ 合金中马氏体含量较高，硬度达到 $296.48HV_{1.0}$；所有合金均呈现出明显钝化特征，$Ni_4Mo_{1.2}Cu_{1.4}$ 合金具有最低的维钝电流密度 $2.99×10^{-6}$ A/cm 和最高的点蚀电位 0.35V（SCE），钝化稳定性最高；制备合金在高温高压下浸泡腐蚀速率为 $0.041\sim0.053$mm/a，低于 0.076 mm/a；Ni、Mo 和 Cu 元素加入提高了合金的自腐蚀电位，降低了腐蚀倾向，其中 Cu 对于改善合金耐蚀性能作用最为突出。

关键词：合金化；13Cr；组织；CO_2 腐蚀

0   引言

随着深层含 CO_2 气藏的开发，注 CO_2 强化采油工艺在吉林和大庆油田得到了一定的推广和应用，采用 CO_2 驱油在提高采收率的同时还可有效减排。但大量的 CO_2 注入地层，使得油田集输管线的 CO_2 腐蚀问题变得日益突出。13Cr 不锈钢作为一种新型开发的石油专用管材[1-2]，具有良好的抗 CO_2 腐蚀性，2004 年被宝钢国产化后，已被塔里木、胜利、文昌和东方等油田所采用[3-4]。普通 13Cr 主要依靠表面形成一层非晶态的钝化膜提高耐蚀性，但高温时的均匀腐蚀，中温时的点蚀和低温时的硫化物应力开裂（Sulfide Stress Cracking, SSC）仍限制其广泛应用[5]。

超级马氏体不锈钢是通过对普通 13Cr 添加 Ni、Mo 等合金元素发展而来，不同温度下具有改进的耐蚀性能[6]。A. Ikeda 等[7]认为钢材中加入 Cr，Mo 对 CO_2 腐蚀有一定抵抗作用。Ni 加入会促进 CO_2 腐蚀，但在含 13%~20%Cr 钢中，Ni 和 Cu 同时加入会极大提高其抗 CO_2 腐蚀性能[8-10]。尽管 Cr 是不锈钢主要合金元素，但含量过高冲击韧性降低。Ni 提高了钢在还原剂存在环境中的耐腐蚀性，改善抗冲击和延展性[11]。Mo 扩散速率慢可使 $M_{23}C_6$ 延迟析出，具有强化和修复钝态膜的功能，使耐蚀性提高[12,13]。赵国仙等[14,15]等研究表明一定条件下 H_2S、Cl^- 的存在会降低超级 13Cr 不锈钢的点蚀电位，材料表面的点蚀坑作为裂纹源，显著增加了应力腐蚀开裂的倾向，但对于不同元素、不同成分含量对合金耐蚀性及组织结构的影响尚不明确。

本工作采用正交实验法，通过等离子电弧熔炼炉制备添加不同含量 Ni，Mo 和 Cu 元素的美国石油协会（American Petroleum Institute）标准 API-13Cr 不锈钢，测定合金的成分、

微观组织和结构特征，以及饱和 CO_2 采出液中的电化学腐蚀行为和高温高压浸泡腐蚀行为，来研究 Ni、Mo、Cu 元素含量对合金组织和耐蚀性的影响规律，为超级 13Cr 不锈钢的进一步开发与利用提供参考。

1 实验

1.1 合金制备

采用非自耗电弧熔炼炉制备合金铸锭，原料为 99.99%（质量分数）纯金属，成分配比采用 L_4（2^3）正交实验表（表1）。为确保熔炼均匀，至少熔炼 3 次。熔炼过程中控制碳含量最低，Cr 含量 12.7 %，余量为 Fe。

表1 L_4（2^3）正交实验设计表（质量分数,%）

No	Ni	Mo	Cu
$Ni_2Mo_{1.2}Cu_{0.8}$	2.0	1.2	0.8
$Ni_2Mo_2Cu_{1.4}$	2.0	2.0	1.4
$Ni_4Mo_{1.2}Cu_{1.4}$	4.0	1.2	1.4
$Ni_4Mo_2Cu_{0.8}$	4.0	2.0	0.8

1.2 组织观察

采用 Spectro m10 光谱仪分析熔炼合金铸锭的成分；用 ZEISS Axiovert 25 CA 金相显微镜观察合金组织，腐蚀剂为 1 号卡林腐蚀剂（$CuCl_2$ 1.5g+盐酸 33mL+H_2O 33mL+乙醇 33mL）；采用 S-3400II 型扫描电镜（SEM）和 Oxford INCA 350 能谱仪（EDS）观察分析合金微观组织和微区成分，组织在草酸 10g+H_2O 100mL 中经电解腐蚀（电压 5.4V，时间 5min）；合金物相 X 射线衍射（XRD）在 RINT2000 衍射仪上进行；硬度在 WILSON 401MVA 显微维氏硬度计上进行，载荷为 9.81N，加载 10s。

1.3 电化学测试

合金电化学腐蚀性能在 Gamry PCI4/7500 电化学工作站上进行。采用标准三电极系统（辅助电极为铂片，参比电极为饱和甘汞电极，工作电极为有效面积为 1cm² 的合金）。为更好对比，本测试选择 0Cr13 不锈钢作为对比材料。测试介质选取大庆采油四厂四区二队 X4-2-25 油井采出液，测试在密封电解池中进行，保证 CO_2 处于饱和状态。测试前将工作电极打磨露出光亮均匀金属面，丙酮除油无水乙醇脱脂后，用去离子水冲洗，滤纸吸干。测试时主要测试合金的电化学阻抗谱（EIS）和动电位极化曲线，先测试试样的 E_{corr}—t 曲线，当 100s 内 E_{corr} 值的变化小于 1mV 时，可认为 E_{corr} 稳定。EIS 扫描频率范围为 10mHz~1kHz。动电位极化曲线扫描范围为 -0.5~1.2V，扫描速率为 0.33mV/s，所有测试结果均至少重复3次。

1.4 高温高压浸泡腐蚀实验

普通 13Cr 不锈钢及制备合金加工成挂片试样。实验前，挂片表面打磨光亮，用去离子水清洗，丙酮除油，酒精脱水，冷风吹干置于干燥箱中，24h 后取出称重和测量。

316

高温高压腐蚀实验在 GSH 型强磁力回转搅拌反应釜中进行，腐蚀介质同为上述采出液，实验前通入高纯氮 10h 除氧，实验温度 80℃，CO_2 分压为 1.5MPa，介质流速 1m/s，腐蚀时间为 240h。

实验后，取出试样清洗除油，用 S3400 型扫描电镜观察腐蚀产物。最后，腐蚀试片常温下在 H_2O 250mL+浓盐酸 250mL +六次甲基四胺 10g 中清洗腐蚀产物，烘干后称重，用失重法计算腐蚀速率。为减小误差，平行试样为 3 个。

2　结果与讨论

2.1　合金成分

表 2 为光谱仪测得 4 种合金化学成分及换算得到的铬和镍当量。由舍夫勒相图[16] 可知，4 种成分合金均处于马氏体和铁素体的双相区域。

表 2　4 种合金化学成分及换算得到的铬、镍当量表（质量分数, %）

Steel	C	Cr	Ni	Mo	Cu	Mn	Fe	[Cr]	[Ni]
$Ni_2Mo_{1.2}Cu_{0.8}$	0.006	12.85	2.286	1.264	1.0844	0.136	82.1584	14.170	2.534
$Ni_2Mo_2Cu_{1.4}$	0.008	12.18	3.149	1.892	1.7397	0.148	80.6627	14.124	3.463
$Ni_4Mo_{1.2}Cu_{1.4}$	0.008	13.17	4.108	1.256	1.2553	0.138	79.8471	14.481	4.417
$Ni_4Mo_2Cu_{0.8}$	0.007	13.08	3.933	2.015	0.8502	0.144	79.7548	15.150	4.215

2.2　合金微观组织

图 1 为不同成分合金的金相组织。经 1 号卡林腐蚀后，马氏体呈暗灰色，铁素体被染色，奥氏体不腐蚀。薄板状组织为马氏体的光学显微特征，其亚结构为高密度层错，极易在 Fe—Mn—C 合金或 Fe—Cr—Ni 合金中出现。$Ni_2Mo_{1.2}Cu_{0.8}$ 组织与其他 3 种不同，主要为铁素体并含有少量薄板状马氏体，组织较均匀，有利于耐蚀性的提高，但大量的铁素体会降低钢的力学性能。$Ni_2Mo_2Cu_{1.4}$ 组织中铁素体含量少于 $Ni_2Mo_{1.2}Cu_{0.8}$，马氏体含量显著增加，且呈现出鲜明的树枝状组织，但 $Ni_2Mo_2Cu_{1.4}$ 中铁素体区域含有大量位相不同的层片状亚结构，易形成腐蚀微电池。

在草酸电解腐蚀条件下，马氏体相腐蚀严重，奥氏体被适度腐蚀，铁素体不受腐蚀。图 2 为 $Ni_4Mo_2Cu_{0.8}$ 合金的扫描电子显微镜照片，由图可知，合金晶界分明，形成明显的层片状马氏体，呈树枝状结晶，沿轴向呈一定角度生长，与上述金相组织一致。微区 EDS 成分测试表明，Mo 和 C 相伴而生，主要由于 Mo 加入后使 C 富集，避免与 Fe、Cr 的结合生成碳化物。考虑到热加工性，含有铁素体的 13Cr 钢在热轧过程中极易出现断裂，因此石油专用管通常采用马氏体单相不锈钢[8]，$Ni_4Mo_{1.2}Cu_{1.4}$ 和 $Ni_4Mo_2Cu_{0.8}$ 组织形貌类似，都含有相对均匀的板条状马氏体组织。

2.3　合金显微硬度

显微硬度测试结果表明，与 0Cr13 硬度（252.76 $HV_{1.0}$）相比，只有 $Ni_2Mo_{1.2}Cu_{0.8}$ 的硬度（194.04 $HV_{1.0}$）降低，$Ni_2Mo_2Cu^{1.4}$、$Ni_4Mo_{1.2}Cu_{1.4}$ 和 $Ni_4Mo_2Cu_{0.8}$ 硬度均提高，分别为

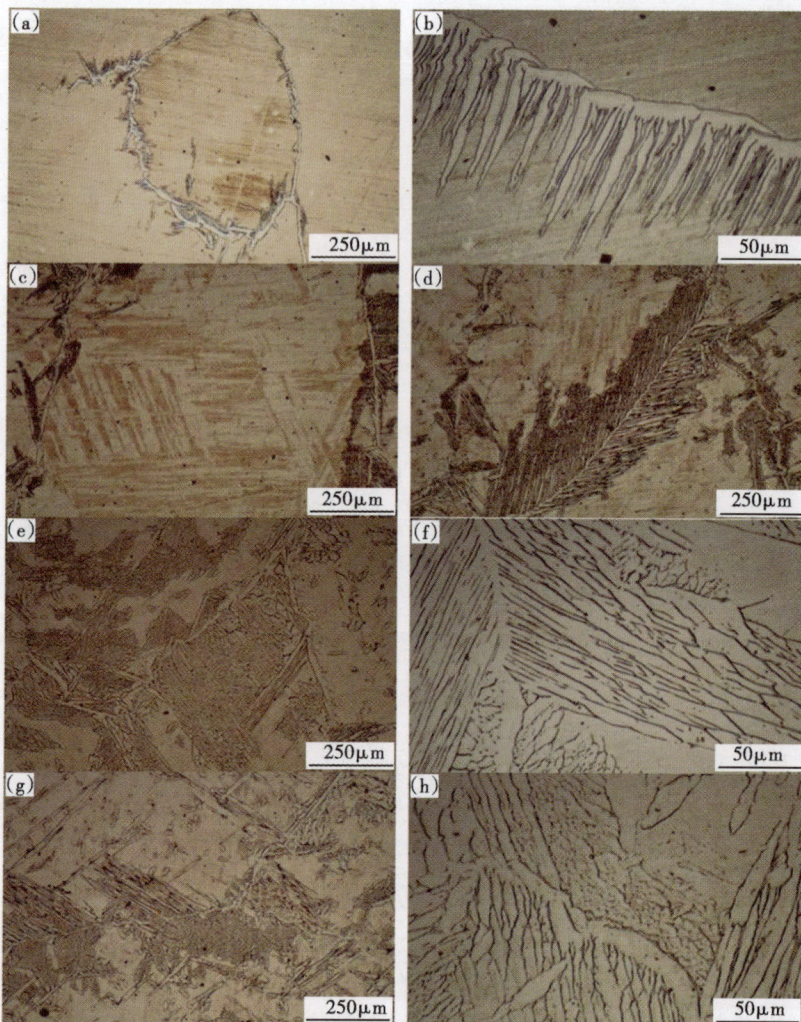

（a）、（b）为$Ni_2Mo_{1.2}Cu_{0.8}$；（c）、（d）为$Ni_2Mo_2Cu_{1.4}$；（e）、（f）为$Ni_4Mo_{1.2}Cu_{1.4}$；（g）、（h）为$Ni_4Mo_2Cu_{0.8}$

图 1　合金的光学组织显微照片

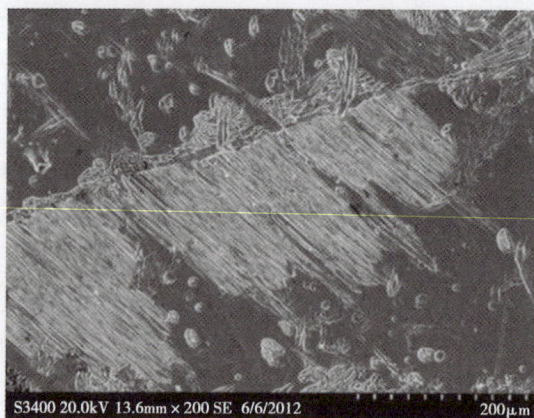

图 2　$Ni_4Mo_2Cu_{0.8}$ 合金的扫描电子显微镜形貌

295.52，296.48HV$_{1.0}$和 268.68HV$_{1.0}$。这是由于 Ni$_2$Mo$_{1.2}$Cu$_{0.8}$中含有大量铁素体，而其他成分含有较高马氏体改善了合金的硬度。

2.4 合金的物相分析

图 3 为合金物相 XRD 分析图。结果显示，4 种成分的合金物相单一均匀，只有两个明显且尖锐的衍射峰，相组成主要为 Ni—Fe—Cr 和 Cr，晶格类型为体心立方，此外 Ni$_4$Mo$_2$Cu$_{0.8}$中还含有明显的马氏体相（C$_{0.055}$Fe$_{1.945}$）。

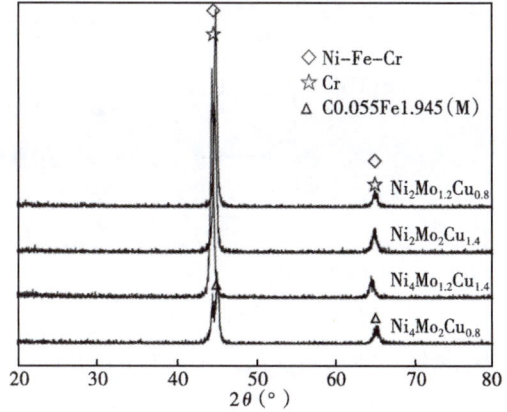

图 3 合金物相的 XRD 分析图

2.5 电化学腐蚀性能分析

图 4 为 0Cr13 不锈钢及 4 种合金在饱和 CO$_2$ 采出液中测得的动电位极化曲线图。

由图 4 可知，0Cr13 不锈钢和 4 种合金在饱和 CO$_2$ 采出液中均呈现出明显钝化特征，合金的自腐蚀电位（E_{corr}）均高于 0Cr13 不锈钢，在发生钝化以前的活化阶段腐蚀电流密度均减小，说明合金的腐蚀倾向明显降低。经过 CView 软件拟合后，4 种成分的合金致钝电流密度比 0Cr13 不锈钢（1.45×10^{-5}A/cm^2）低 1 个数量级，说明 4 种合金更易产生钝化。当合金表面形成稳定的钝化膜后，4 种合金的维钝电流密度均小于 0Cr13 不锈钢（6.78×10^{-6} A/cm^2），Ni$_4$Mo$_{1.2}$Cu$_{1.4}$合金呈现出最小的维钝电流密度。另外，4 种合金的点蚀电位均明显高于 0Cr13 不锈钢（0.15V），较小的维钝电流密度和较高的点蚀电位反映出 4 种合金具有稳定的钝化特征，且发生点蚀倾向降低。因此，添加 Ni、Mo 和 Cu 后，提高了合金的自腐蚀电位和钝化稳定性，降低了腐蚀和点蚀发生的倾向。Ni$_4$Mo$_{1.2}$Cu$_{1.4}$合金维钝电流密度最小（2.99×10^{-6}A/cm^2），击穿电位最高（0.35V），钝化区间最宽（0.90V），钝化稳定性最高。

相应的电化学阻抗谱测试结果如图 5 所示。可知，所有材料在高频区均呈现出单一容抗弧特征。高的容抗弧半径反映出材料具有高的反应阻力，4 种合金的容抗弧半径均明显高于 0Cr13 钢，其中 Ni$_4$Mo$_{1.2}$Cu$_{1.4}$最大，说明其耐蚀性最好。

图 5 的电化学阻抗谱可用图 6 的等效电路借助 ZView 软件进行拟合，拟合结果见表 3。

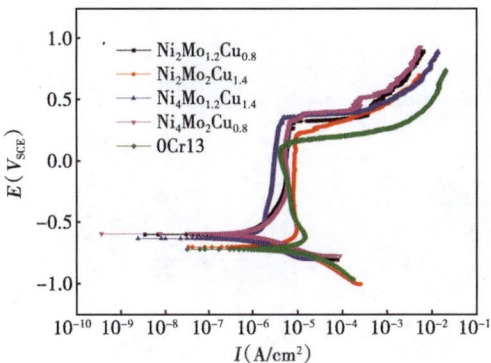

图 4 饱和 CO$_2$ 采出液中合金的动电位极化曲线图

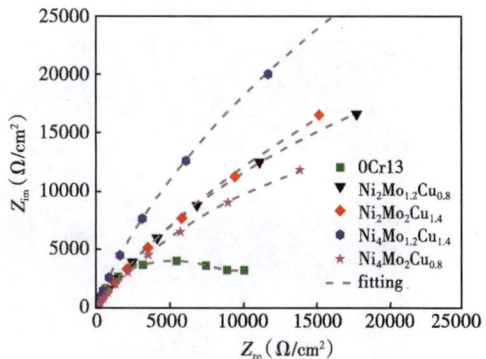

图 5 饱和 CO$_2$ 采出液中的合金电化学阻抗谱图

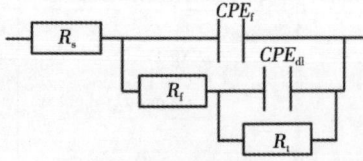

图 6 腐蚀界面等效电路图

图 6 中，R_s 表示溶液电阻，R_t 表示电荷传递电阻，CPE_{dl} 表示与 R_t 并联的双电层电容，R_f 表示钝化膜层的电阻，CPE_f 表示与 R_f 并联的双电层电容，其中 CPE 代表频响特性相对于纯电容有偏移的等效元件。

几种材料的电容值差异不大，在此不再讨论。$Ni_4Mo_{1.2}Cu_{1.4}$ 所对应的 R_t 及 R_f 显著高于 0Cr13 钢，且均为最大值；$Ni_2Mo_{1.2}Cu_{0.8}$ 和 $Ni_4Mo_2Cu_{0.8}$ 的 R_t 较高，R_f 相对于 0Cr13 钢无显著提高。因此，$Ni_4Mo_{1.2}Cu_{1.4}$ 的耐蚀性最优，而 $Ni_2Mo_2Cu_{1.4}$ 的耐蚀性最差，这与 $Ni_4Mo_{1.2}Cu_{1.4}$ 形成均匀的马氏体组织是分不开的，组织越均匀金属表面各区域间的电位差越低，减小腐蚀微电池形成的几率，有利于均匀致密钝化膜的形成，提高膜层的稳定性即合金的耐蚀性。

表 3　电化学阻抗谱拟合数据表

Specimen	R_s ($\Omega \cdot cm^2$)	R_t ($k\Omega \cdot cm^2$)	R_f ($k\Omega \cdot cm^2$)
0Cr13	30.89	12.754	9.268
$Ni_2Mo_{1.2}Cu_{0.8}$	26.37	82.396	0.058
$Ni_2Mo_2Cu_{1.4}$	27.25	10.256	0.073
$Ni_4Mo_{1.2}Cu_{1.4}$	33.11	97.186	35.82
$Ni_4Mo_2Cu_{0.8}$	51.3	52.131	2.684

2.6　高温高压腐蚀分析

图 7 是 0Cr13 与 $Ni_4Mo_{1.2}Cu_{1.4}$ 挂片表面的高温高压腐蚀形貌。本实验条件下，普通 13Cr 不锈钢和制备合金以均匀腐蚀为主，未发现明显点蚀坑。0Cr13 腐蚀产物明显比制备合金疏松，且存在分层现象，说明腐蚀产物与基体结合力不足，产生脱落，导致裸露基体再次发生腐蚀；$Ni_4Mo_{1.2}Cu_{1.4}$ 表面腐蚀产物附着牢固，平整且致密，晶粒大小均匀，能起到良好的保护作用，其他 3 种制备合金的腐蚀形貌与之类似。

失重法计算 0Cr13 不锈钢和 4 种制备合金的平均浸泡腐蚀速率分别为 0.0713, 0.0527, 0.0519, 0.0413mm/a 及 0.0499 mm/a，测试结果均低于国内石油天然气行业标准 SY/T 5329—1994 中所规定的 0.076mm/a。结果表明，添加 Ni、Mo 和 Cu 元素的 13Cr 不锈钢具有

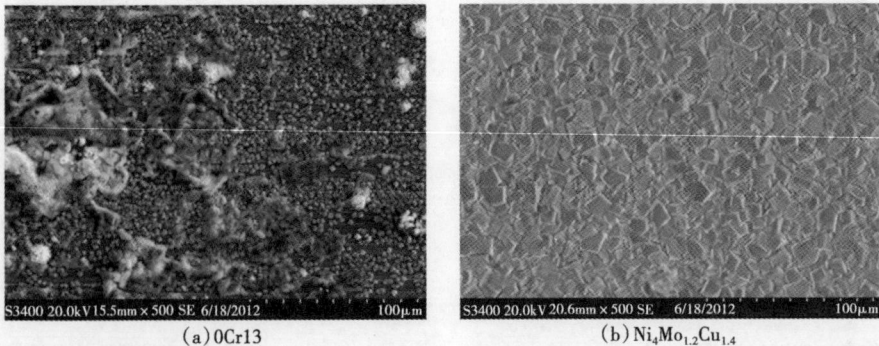

（a）0Cr13　　　　　　　　　　　（b）$Ni_4Mo_{1.2}Cu_{1.4}$

图 7　腐蚀产物微观形貌

更加优异的抗 CO_2 腐蚀性能。

2.7 元素含量对耐蚀性的影响规律

采用正交实验的直观分析法，分别计算 3 个因素所对应的性能评价参数的平均值，计算出平均极差，分析 Ni、Mo 和 Cu 元素对合金耐蚀性的影响趋势及程度。表 4 为 Ni、Mo 和 Cu 各参数平均极差值。平均极差为极差除以因素的变化量，其意义在于某元素增加单位百分含量各参数所对应的增量，正值代表正比，负值代表反比。

表 4 Ni、Mo 和 Cu 各参数平均极差的计算结果

元素	R_p ($\Omega \cdot cm^2$)	E_p (V)	ΔE (V)	R_f ($k\Omega \cdot cm^2$)	I_p (A)	E_{op} (V)	v_{corr} (mm/a)
Ni	11552.45	0.020775	0.029618	9.59	-1.39×10^{-6}	0.05	-0.019
Mo	-39216.5	-0.02394	-0.00829	-20.70	3.11×10^{-6}	-0.03	0.044
Cu	42849.97	-0.25042	0.141108	27.63	6.40×10^{-8}	-0.11	-0.013

合金的腐蚀速率可以利用式（1）通过维钝电流密度计算得到，其表征的是合金在表面形成钝化膜后的均匀腐蚀速率：

$$v_{corr} = \frac{I_p}{\rho_{alloy} n_{alloy} F} \tag{1}$$

式中，v_{corr} 为腐蚀速率；ρ_{alloy} 为合金的密度，g/cm^3；n_{alloy} 为合金的电荷克当量；F 为法拉第常数。

n_{alloy} 的值可以通过式（2）计算得到：

$$n_{alloy} = \sum_j \left(\frac{f_j n_j}{a_j} \right) \tag{2}$$

式中，f_j 为材料中第 j 种合金元素的质量分数；n_j 为阳极溶解过程反应的电荷数，假设为定值；a_j 为第 j 种元素的原子质量。

由表 4 的计算结果可知，Cu 元素质量分数提高 1% 对于提高极化电阻 R_p 的作用比 Ni 明显，而 Mo 含量的升高反而会降低极化电阻。同时，提高 Cu 含量还可以降低致钝电位 E_p，拓宽钝化区间 ΔE，使钝化膜层电阻 R_f 强烈升高，降低腐蚀介质与金属基体间的物质传输速度，加强钝化膜稳定性，缓解腐蚀。Ni 含量升高会加剧钝化膜形成的难度，Mo 则会缩小钝化区间，降低钝化稳定性。尽管随 Cu 含量升高维钝电流密度 I_p 变大，击穿电位 E_{op} 降低，但影响均非常微弱，不会削弱合金的耐蚀性，计算结果可知，Ni 和 Cu 含量升高均会降低腐蚀速率 v_{corr}。综上所述，Cu 对于进一步缓解腐蚀作用最明显，具有良好的经济性。

3 结论

（1）合金组织主要为马氏体和铁素体，随 Ni、Mo 和 Cu 含量不同而变化，$Ni_2Mo_{1.2}Cu_{0.8}$ 以铁素体为主，组织结构均匀；$Ni_4Mo_{1.2}Cu_{1.4}$ 和 $Ni_4Mo_2Cu_{0.8}$ 中马氏体含量较高，热加工性更好，$Ni_4Mo_{1.2}Cu_{1.4}$ 硬度最高。

（2）Ni、Mo 和 Cu 元素的加入使合金自腐蚀电位升高，腐蚀倾向降低；所有合金均呈

现出明显的钝化特征，维钝电流密度减小，点蚀电位升高，$Ni_4Mo_{1.2}Cu_{1.4}$合金钝化稳定性最高。

（3）Ni、Mo 和 Cu 元素的添加可以有效改善合金的耐 CO_2 腐蚀性，温度 80℃，CO_2 分压 1.5MPa 条件下，制备合金在高温高压下浸泡腐蚀速率为 0.041～0.053mm/a，低于我国石油行业标准规定的 0.076mm/a。

（4）本实验条件下，耐蚀性最好的合金配比为 4%Ni、1.2%Mo、1.4%Cu，其中 Cu 对于改善合金耐蚀性能作用最为突出，考虑到经济性，可继续提高 Cu 含量，适当降低 Ni、Mo 含量。

参 考 文 献

［1］ DENPO K, OGAWA H. Fluid flow effects on CO_2 corrosion resistance of oil well materials ［J］. Corrosion, 1993, 49（6）: 442-449.

［2］ MASAMURA K, HASHIZUME S, INOHARA Y, et al. Estimation models of corrosion rates of 13Cr alloys in CO_2 environments ［A］. Corrosion 99 ［C］. San Antonio, Tx: NACE International, 1999. 583.

［3］ 林冠发，相建民，常泽亮，等. 3 种 13Cr110 钢高温高压 CO_2 腐蚀行为对比研究 ［J］. 装备环境工程，2008, 5（5）: 1-4.

［4］ 周波，崔润炯，刘建中. 增强型 13Cr 钢抗 CO_2 腐蚀套管的研制 ［J］. 钢管，2006, 36（6）: 22-26.

［5］ 刘艳朝，赵国仙，薛艳，等. 超级 13Cr 钢在高温高压下的抗 CO_2 腐蚀性能 ［J］. 腐蚀研究，2011, 25（11）: 29-34.

［6］ ZHANG H, ZHAO Y L, JIANG Z D. Effects of temperature on the corrosion behavior of 13Cr martensitic stainless steel during exposure to CO_2 and Cl^- environment ［J］. Materials Letters, 2005, 59（27）: 3370-3374.

［7］ IKEDA A, UEDA M. Effect of microstructure and Cr content in steel on CO_2 corrosion ［A］. Corrosion 96 ［C］. Denver, CO: NACE International, 1996.

［8］ HARA T, ASAHI H, KAWAKAMI A, et al. Effects of alloying elements on carbon dioxide corrosion in 13% to 20% chromium containing steels ［J］. Corrosion, 2000, 56（4）: 419-428.

［9］ HARA T, ASAHI H, SUEHIRO Y, et al. Effect of flow velocity on carbon dioxide corrosion behavior in oil and gas environments ［J］. Corrosion, 1994, 56（8）: 860-866.

［10］ ASAHI H, HARA T, SUGIYAMA M. Corrosion performance of modified 13% Cr OCTG ［A］. Corrosion 96 ［C］. Denver, CO: NACE International, 1996.

［11］ 田世昌. 奥氏体耐蚀合金的应用 ［J］. 石油化工设备技术，1994, 15（5）: 43-45.

［12］ 徐金璋. 合金元素和组织对马氏体不锈钢的耐蚀性和硬度的影响 ［J］. 上海钢研，2001,（2）: 40-44.

［13］ PARK J Y, PARK Y S. The Effects of heat-treatment parameters on corrosion resistance and phase transformations of 14Cr-3Mo martensitic stainless steel ［J］. Materials Science and Engineering, 2007, 449-451: 1131-1134.

［14］ 刘亚娟，吕祥鸿，赵国仙，等. 超级 13Cr 马氏体不锈钢在入井流体与产出流体环境中的腐蚀行为研究 ［J］. 材料工程，2012,（10）: 17-21, 47.

［15］ 吕祥鸿，赵国仙，王宇，等. 超级 13Cr 马氏体不锈钢抗 SSC 性能研究 ［J］. 材料工程，2011,（2）: 17-21, 25.

［16］ BERES L. Proposed modification to schaeffler diagram for chrome equivalents and carbon for more accurate prediction of martensite content ［J］. Welding Journal, 1998, 77（7）: 273-276.

激光熔覆 Fe 基高 Cr 涂层组织
及抗 CO_2 腐蚀性能研究

孙丽丽[1]　朱闯[2]　王勇[1]　张旭昀[1]　毕凤琴[1]

(1. 东北石油大学机械科学与工程学院；2. 中国石油大庆油田有限责任公司)

摘　要：采用激光熔覆方法制备了 Fe-20Cr、Fe-30Cr、Fe-40Cr 共 3 种不同 Cr 含量的 Fe 基涂层。用扫描电子显微镜（SEM）和 X 射线衍射（XRD）分析了涂层的微观组织和物相，借助电化学工作站测试了涂层在含 CO_2 饱和地层水中的腐蚀行为，并与 P110 套管钢进行了对比。结果表明：激光熔覆 Fe 基含 Cr 涂层皆为冶金结合，熔覆层质量良好。在地层水中通入 CO_2 后 3 种涂层腐蚀趋势变大，但由于表面疏松 $CaCO_3$ 和 $MgCO_3$ 产物的形成，抑制了涂层的腐蚀。低 Cr 含量的 Fe-20Cr 涂层具有较高的耐蚀性。Cr 含量增加时，由于 $Cr_{23}C_6$、Cr_7C_3 等碳化物在晶界处析出，产生晶体贫铬区，进而降低了高 Cr 含量涂层的耐蚀性，但碳化物析出提高了涂层的硬度。

关键词：激光熔覆；Fe 基高 Cr；CO_2 腐蚀；饱和地层水

0　引言

CO_2 驱油作为一种提高油田采收率的三次采油技术，以其适用范围大、驱油效率高及成本较低等优势受到世界各国的广泛重视。通过将 CO_2 注入到枯竭或开采后期的油气田中，可以把残留的油气驱替出，达到提高采收率的目的。这种驱油技术既可储存 CO_2，又可提高油气的开采量，不仅可以取得丰富的经济效益，而且具备社会效益，具有很好的发展和应用前景[1]。但 CO_2 溶入水后对部分金属材料有极强的腐蚀性，CO_2 腐蚀可能使油气井管柱的寿命大大低于设计寿命。CO_2 腐蚀作为现代油气工业中一种常见的腐蚀行为，还会造成严重的经济损失和社会后果。激光熔覆是材料表面改性技术的一种重要方法，利用高能激光束（$10^4 \sim 10^6 W/cm^2$）在金属表面辐照，通过迅速熔化、扩展和凝固，在基材表面熔覆一层具有特殊物理、化学或力学性能的材料，以弥补基体所缺少的高性能，能充分发挥两种材料的优势。Fe 基合金材料具有较高的耐磨、耐蚀性能，适用于要求局部耐磨且容易变形的零件，主要由于其组分与碳钢铸铁接近，与基体具有良好的相容性，且成本低廉。

笔者采用激光熔覆方法制备 3 种不同 Cr 含量的 Fe 基激光熔覆层，测试了 3 种涂层抗 CO_2 腐蚀性能和硬度，对涂层进行性能评价并分析机理，为油田 CO_2 驱油伴随产生腐蚀及磨损问题提供理论依据。

1　实验过程

1.1　实验设备及材料

采用 HGL9450-000 多功能数控激光熔覆设备，选择同步送粉法，将合金粉末在 ZYH-

20 自控远红外电焊条烘干炉中 100℃烘干 2h，基体材料预热后采用同步送粉氩气保护。所选激光熔覆参数为：输出功率 3.2kW、焦距 395mm、扫描速率 9mm/s、光斑直径 5mm。

熔覆基体选 15#钢，预置涂层前，试样的熔覆表面经喷砂处理再用无水乙醇和丙酮清洗。熔覆粉末以华工 325 自熔性 Fe 基合金粉末为基（颗粒粒度 38~45μm），3 种不同 Cr 含量的涂层成分见表 1（质量分数）。

表 1　3 种不同 Cr 含量粉末化学成分表　　　　　　（单位：%）

涂层	Cr	Mn	Ni	Si	Fe
Fe-20Cr	20.0	3.93	3.71	3.68	余量
Fe-30Cr	30.0	3.41	3.13	3.02	余量
Fe-40Cr	40.0	2.93	2.68	2.6	余量

1.2　实验介质

实验介质为模拟大庆油田饱和地层水，母液组分包括 2.294g/L 的 NaCl、1.86g/L 的 NaHCO$_3$、0.075g/L 的 Na$_2$SO$_4$、0.042g/L 的 CaCl$_2$、0.172g/L 的 MgCl$_2$·6H$_2$O。在母液中加入 NaCl，配制成含 Cl$^-$ 浓度为 0.5mol/L 的模拟地层水溶液。测试前在配制液中通入氩气以排除空气，随后通入 CO$_2$ 至饱和，再进行涂层和 P110 钢腐蚀性能对比研究，P110 钢的化学成分见表 2。

表 2　P110 钢化学成分表　　　　　　（单位：%）

元素	C	Mn	Si	Mo	P	Cr	S
含量	0.35	0.80	0.35	0.80	0.01	1.80	0.01

1.3　实验方法

腐蚀性能测试在 Gamry PCI4 工作站上进行，采用工作电极、辅助电极（Pt 片）和参比电极（饱和甘汞电极，SCE）三电极测试系统进行电化学阻抗谱和动电位极化曲线测试。涂层形貌和组织在日立 S-3400II 扫描电子显微镜、牛津 350 能谱仪上进行测试分析。采用日本理学 D/max-2200 X-射线衍射仪（XRD）进行物相分析，使用 Cu 靶，以 2θ 角进行分析得到涂层 X-射线衍射谱图。采用沃伯特 HV-401 型显微维氏硬度计测量涂层显微硬度（载荷 1.96 N，加载时间 10s），测试间隔为 0.25mm，测试结果取 3 次平均值。

2　实验结果与分析

2.1　涂层稀释率

在激光熔覆过程中，由于基材的熔入而引起的熔覆合金成分变化的程度，通常用基材合金在熔覆层中所占的百分率表示稀释率[2]：

$$\eta = \frac{A_1}{A_1 + A_2}$$

式中，A_1 为基体平面以上的涂层横截面积；A_2 为基体熔化区面积；η 为涂层的稀释率。

采用熔覆道横截面的面积测量计算几何稀释率（图1）。熔覆层的几何形貌可以用熔覆层宽度 W、熔覆层厚度 H、熔覆深度 h 以及横截面面积 S 等参数来描述。

利用线切割将各试样沿扫描方向切开，然后用砂纸打磨熔覆层的横截面，用王水轻微腐蚀，能较明显地区分熔覆层的各个区域即可。利用刻度的光学显微镜测量熔覆层的几何参数，3 种涂层的稀释率分别以 8%、6%、7% 为宜。

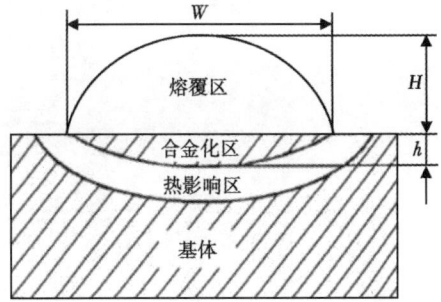

图 1　激光熔覆层的截面积示意图

2.2　熔覆层形貌及组织

3 种激光熔覆表面宏观形貌如图 2 所示，该熔覆层是多道搭接而得到的。从图 2 中可以看出 3 种熔覆层表面有褶皱，搭接处表面有少许熔体，均未出现宏观裂纹、孔洞等缺陷，且基体未出现大的变形，打磨后能够得到平整的熔覆层。

(a) Fe–20Cr　　(b) Fe–30Cr　　(c) Fe–40Cr

图 2　激光熔覆层宏观形貌

由图 1 可知，熔覆层横截面显微组织可分为熔覆层、合金化区和热影响 3 个区域。图 3 为 3 种涂层的合金化区和熔覆区的组织图。由图 3 知，3 种涂层组织均具有鲜明的枝晶和胞状晶生长特征，这是因为激光熔覆是以高能量密度的激光束为热源（温度升高速度快），扫描至金属粉末表面后使粉末快速熔化并凝固形成的，熔覆层凝固组织就是由于这种严重偏离平衡条件下的快速冷却所决定的。图 3 还反映出，熔覆层是由合金粉末迅速熔化、凝固后而形成的；合金化区是由熔化的合金粉末与基体材料部分微熔混合凝固后形成的。激光扫描时，固—液界面以下基体受高温热影响而发生固态相变，这部分为热影响区，熔覆过程中温度达到奥氏体温度以上，此时珠光体转变成奥氏体。当激光束扫描过后，由于快速急冷，奥氏体转变为马氏体，故该区域为马氏体+残余奥氏体+铁素体组织组成的粗大的马氏体。

激光熔覆过程中，基体与熔覆层的结合区是靠近基体在熔覆层底部的白亮带为平面晶，是典型的平面外延生长组织，宽约 $10\mu m$。3 种不同 Cr 含量熔覆层底部、中部大多为胞晶组织，与表层区晶粒相比较为粗大、方向性强、垂直于结合面生长，熔覆层表层区多为枝晶组织和粒状组织，晶粒较为细小，生长方向紊乱[3]。由此，3 种 Cr 含量的整个熔覆层组织为平面晶、胞状晶和树枝晶的混合结构。结合区平面晶的存在表明，3 种 Cr 含量涂层基体与熔覆层之间已形成良好的冶金结合，结合强度比较高。

随着 Cr 含量的增加，液态熔池中的碳首先与 Cr、Fe 等合金元素形成相应的碳化物颗粒[2]。伴随着温度不断降低，过多的 Cr 原子固溶于 γ-Fe 基体中形成大量的残余奥氏体，大量的残余奥氏体与渗碳体结合而形成高温莱氏体，随着温度进一步降低转变为低温莱氏体。

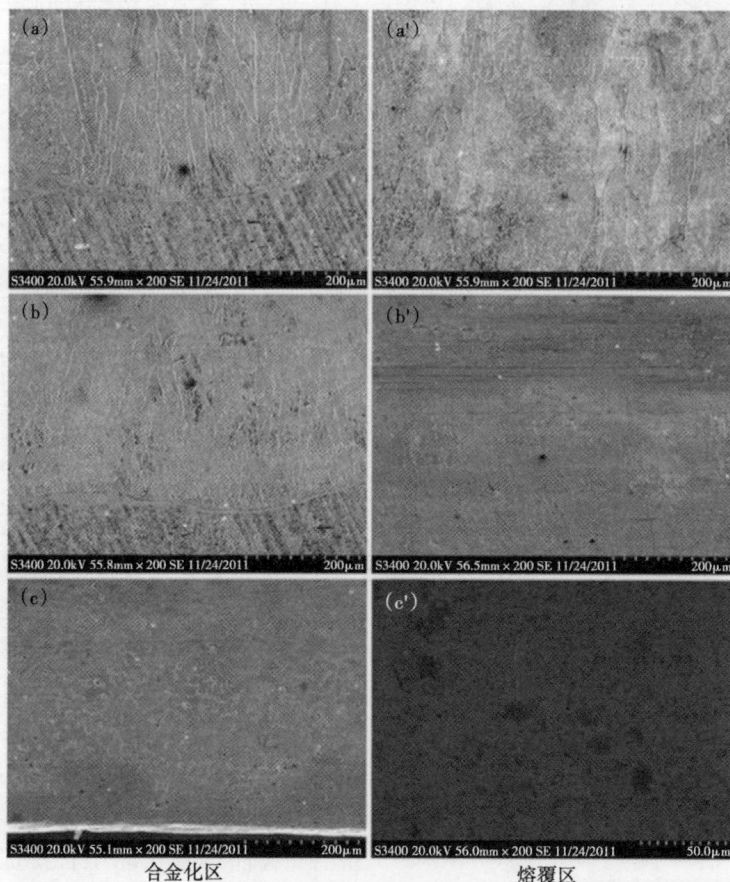

图 3 3 种涂层不同区组织图

(a)、(a′) 为 Fe-20Cr；(b)、(b′) 为 Fe-30Cr；(c)、(c′) 为 Fe-40Cr

2.3 CO$_2$ 腐蚀电化学行为

图 4[（a）~（d）]是激光熔覆涂层和 P110 钢在饱和 CO$_2$ 溶液与空白溶液中的动电位极化曲线。由图 4 可以得出，P110 钢在饱和地层水通入 CO$_2$ 后，极化曲线 Tafel 区阴阳极斜率（b_a、b_c）明显变大，拟合后所得到的腐蚀电流 I_{corr} 明显大于空白溶液中的。对于 3 种涂层，在通入 CO$_2$ 后，自腐蚀电位（E）均负移，其钝化电流密度也随着 CO$_2$ 的通入而降低。另外，涂层通入 CO$_2$ 后，腐蚀趋势变的越容易，而腐蚀速率变小，因为在空白溶液中，对腐蚀起主要作用的是饱和地层水，它主要含有 Ca^{2+}、Mg^{2+}、HCO$_3^-$，但当通入 CO$_2$ 后，大量的 CO$_2$ 溶于饱和地层水中，发生电离反应：

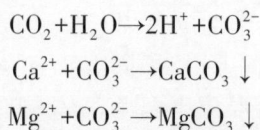

$$CO_2 + H_2O \rightarrow 2H^+ + CO_3^{2-}$$
$$Ca^{2+} + CO_3^{2-} \rightarrow CaCO_3 \downarrow$$
$$Mg^{2+} + CO_3^{2-} \rightarrow MgCO_3 \downarrow$$

常温下，CO$_2$ 与溶液中的 Ca^{2+}、Mg^{2+} 离子生成的难溶性的 CaCO$_3$、MgCO$_3$ 沉淀附着在涂层电极表面，阻碍了电极表面物质传递过程，造成腐蚀电流的降低。

326

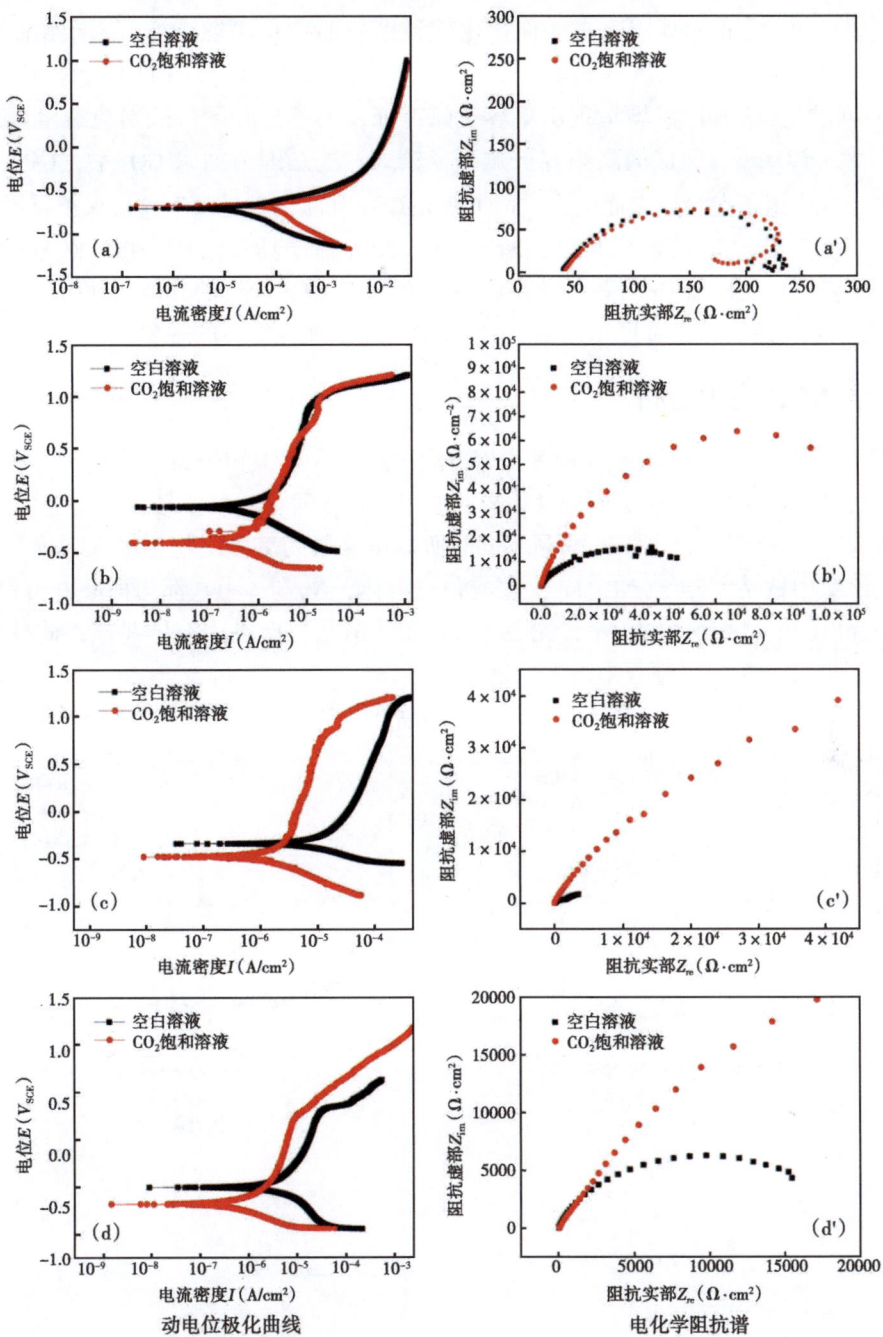

（a）、（a′）为P110钢；（b）、（b′）为Fe-20Cr；（c）、（c′）为Fe-30Cr；（d）、（d′）为Fe-40Cr

图4　激光熔覆涂层和P110钢的对比

图4[（a′）~（d′）]是激光熔覆涂层和P110钢在饱和CO_2溶液与空白溶液中的电化学阻抗谱。从Nyquist阻抗图谱可以看出，P110钢在高频区容抗弧大小基本一致，在通CO_2体系中的阻抗谱低频区出现了感抗弧，这是因为P110钢有比较高的腐蚀电流密度，随时间的延长，$CaCO_3$、$MgCO_3$沉淀与腐蚀产物晶粒的大小和晶向各不相同，结合能力较弱，生成的腐

蚀产物 $CaCO_3$、$MgCO_3$ 膜不致密，很容易剥落，造成局部腐蚀[4]。在空白溶液中，P110 钢腐蚀起主要作用的是饱和地层水在表面产生较致密的 $FeCO_3$ 腐蚀产物，该产物晶粒堆积均匀、致密，阻碍了阳离子的溶解和阴离子的还原，因此通入 CO_2 后 P110 钢耐蚀性下降。与 P110 钢不同的是，3 种涂层均呈现出单容抗弧特征，由于高频区一般对应着溶液反应电阻的大小，通入 CO_2 后，涂层的高频容抗弧半径较大，这是因为通入 CO_2 后，CO_2 与地层水中的 Ca^{2+}、Mg^{2+} 离子在涂层表面生成 $CaCO_3$、$MgCO_3$ 沉淀，正是由于这层保护膜提高了涂层的耐蚀性。从图 4[(b′) ~ (d′)] 可以判断出，Fe-20Cr 涂层具有比 Fe-30Cr 和 Fe-40Cr 涂层更高的容抗弧，反映出低 Cr 含量的涂层耐蚀性要高于高 Cr 含量的其他两种涂层，说明经过激光熔覆的 Fe 基涂层并没有展示出随着 Cr 含量增加耐蚀性提高的现象。

2.4　涂层 XRD 物相分析

由图 3 的组织图可以看出，3 种激光熔覆涂层熔覆区组织和分布大体相同。为了进一步分析低 Cr 含量的 Fe-20Cr 涂层耐蚀性优异的原因，对 3 种涂层的熔覆区分别进行了 XRD 测试（图 5）。由图 5 可以看出，Cr 含量不同时所制备涂层的熔覆层物相组成有较大区别，激光熔覆层组织中枝晶为 γ-（Ni，Fe）过饱和固溶体，Ni 与 γ-Fe 都为面心立方晶格结构，可以无限互溶，所以熔池中的 Ni 首先在 γ-Fe 表面结晶，形成固溶体晶核，此外还存在纯 Cr 物相，说明激光在熔覆过程中没有被氧化。

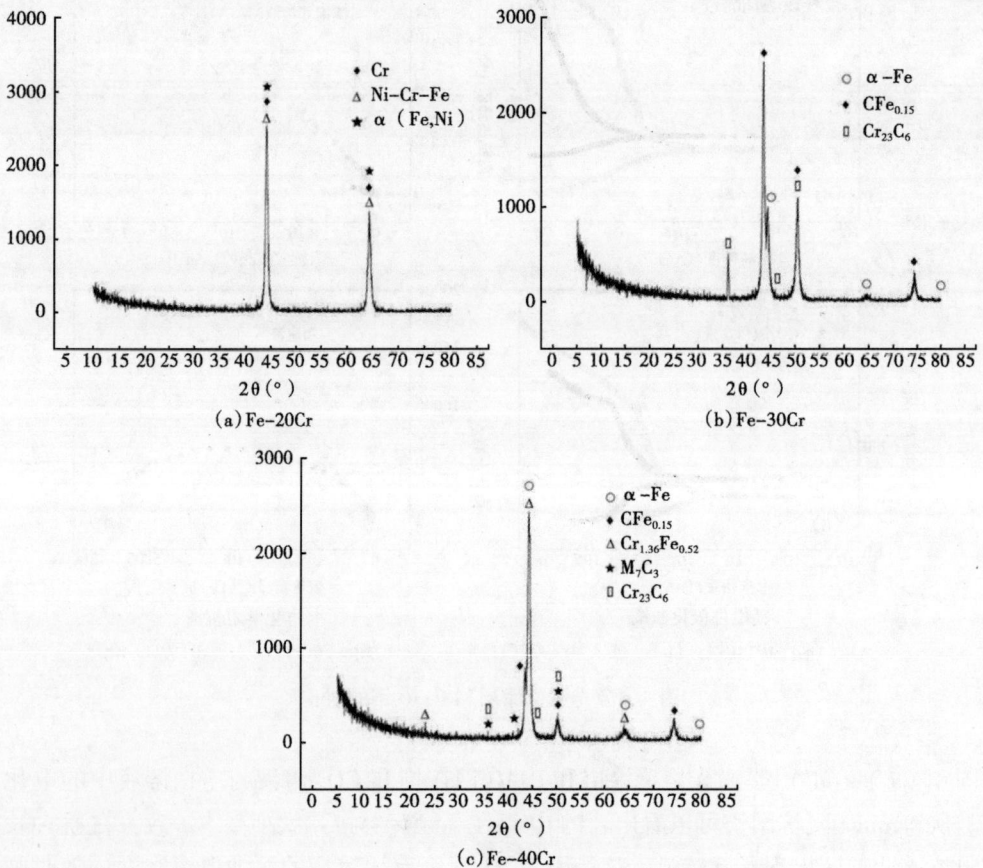

图 5　3 种涂层熔覆区 XRD 图谱

Fe-20Cr 涂层熔覆区物相由 α-Fe、Ni-Cr-Fe 及 α（Fe，Ni）等组成；Fe-30Cr 涂层熔覆区含 α-Fe、$CFe_{0.15}$、$Cr_{23}C_6$ 等物相组成；而 Fe-40Cr 涂层熔覆区除 α-Fe、$CFe_{0.15}$、$Cr_{23}C_6$ 相外还由 $Cr_{1.36}Fe_{0.52}$ 以及 M_7C_3 等相组成。可见，Fe-30Cr 和 Fe-40Cr 高 Cr 含量的涂层耐蚀性差是由于当 Cr 含量过多时，过多的 Cr 与 C 等元素具有一定的亲和力，逐渐形成了 Cr_7C_3、$Cr_{23}C_6$ 等相，这些物相在其熔覆区晶界上析出，导致晶界周围基体产生贫铬区，造成熔覆区耐蚀性大大降低。

2.5 涂层的硬度

图 6 是在不同 Cr 含量涂层激光熔覆层截面的显微硬度分布。硬度沿着熔覆层厚度方向向上呈现较明显的上升趋势，沿熔覆层表面到基体逐渐增大，3 种涂层得平均硬度分别是基体的 1.7 倍、2.1 倍、2.5 倍，熔覆区硬度均高于 P110 钢。随着 Cr 含量的增加，熔覆区硬度呈现逐渐增大的趋势，Fe-40Cr 涂层硬度高达 $520HV_{0.2}$ 左右。

熔覆层的硬度值与其试样表面显微组织是密切相关的。一方面，熔覆过程中，

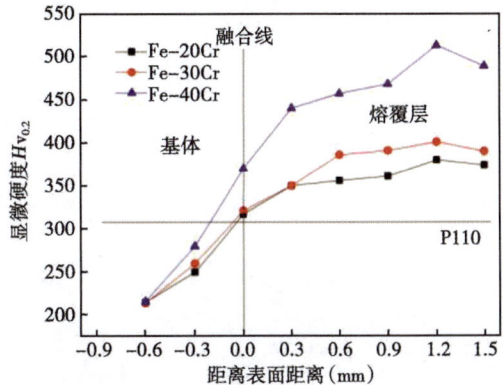

图 6　3 种涂层熔覆层和 P110 钢表面硬度

基体表层被加热产生相变硬化，组织转变成马氏体硬度增大。另一方面，M_7C_3、$Cr_{23}C_6$ 等多元共晶化合物引起的弥散强化，激光快速加热、快速凝固引起的细晶强化等因素，导致该区具有较大的硬度。整个熔覆层的硬度分布比较均匀，这是由于激光的快速加热、快速凝固特性，保证了熔覆层组织具有较高的耐磨性。

3　结论

（1）激光熔覆层组织由典型的熔覆层、合金化区和热影响区组成。Fe-20Cr Fe-30Cr 和 Fe-40Cr 这 3 种涂层皆为冶金结合，熔覆层质量良好，组织具有鲜明的枝晶和胞状晶生长特征。

（2）与饱和地层水相比，在通入 CO_2 后，3 种激光熔覆 Fe 基涂层腐蚀趋势变大，但由于其表面形成了疏松的 $CaCO_3$ 和 $MgCO_3$ 沉淀，抑制了涂层腐蚀的进行，降低了腐蚀速率。P110 钢的腐蚀过程则由未通入 CO_2 表面形成致密 $FeCO_3$ 产物转变为疏松的 $CaCO_3$ 和 $MgCO_3$ 腐蚀产物，膜层的脱落加剧了腐蚀过程，进而增大了腐蚀速率。

（3）低 Cr 含量的 Fe-20Cr 涂层具有较高的耐蚀性，Cr 含量增加时，Fe-30Cr、Fe-40Cr 熔覆层过多的 Cr 与 C 等元素具有一定的亲和力，逐渐形成了 Cr_7C_3、$Cr_{23}C_6$ 等相，在晶界处形成产生贫铬区，耐蚀性大大降低。但此类碳化物提高了涂层的硬度，有利于增强耐磨性。

参 考 文 献

[1] Marhohn AL, M A. Rahman S S. Treatment of Drilling Fluid to Common Drill Pipe Corrosion [J]. Corrosion, 1992 16（9）：721-726.

[2] 唐英, 杨杰. 激光熔覆镍基粉末涂层的研究 [J]. 热加工工艺, 2004（2）：16-17.

[3] 吕海燕,李双明,钟宏等. 激光表面快凝下 $Nd_{13.5}Fe_{79.75}B_{6.75}$ 过包晶合金的组织演化 [J]. 中国有色金属学报, 2009, 19 (2): 322-327.

[4] 吕祥鸿, 樊治海, 赵国仙, 等. 阳离子对 P110 钢高温高压 CO_2 腐蚀反应过程的影响 [J]. 腐蚀科学与防护技术 2005, 17 (2) 70-73.

Fe 基高 Cr 激光熔覆涂层 CO₂ 腐蚀行为研究

孙丽丽[1]　朱　闯[2]　王　勇[1]　张旭昀[1]　毕凤琴[1]

(1 东北石油大学；2 中国石油大庆油田有限责任公司)

摘　要：采用电化学工作站测试了激光熔覆 Fe-20Cr、Fe-30Cr、Fe-40Cr 涂层在含 CO₂ 饱和地层水中的腐蚀行为，研究了温度和 Cl⁻ 浓度对其腐蚀的影响规律，并与 P110 套管钢进行了对比。通过对涂层进行 X 射线衍射（XRD）分析，解释了涂层 Cr 含量不同时耐蚀性的差异。结果表明：随温度升高，涂层钝化区间变窄，耐蚀性随温度升高而降低；当温度为 85℃ 时，涂层具有比 P110 钢更低的腐蚀倾向和更优异的抗点蚀性；Cl⁻ 浓度与耐蚀性存在一临界值，当 Cl⁻ 浓度达 1mol/L 时，腐蚀性最强。低 Cr 含量的 Fe-20Cr 涂层具有较高的耐蚀性，随着 Cr 含量的增加，$Cr_{23}C_6$、Cr_7C_3 等碳化物在晶界处析出，产生晶体贫铬区，进而降低了高 Cr 含量涂层的耐蚀性。

关键词：Fe 基高 Cr；激光熔覆；CO₂ 腐蚀；饱和地层水

0　引言

CO₂ 在水和油中溶解度极高，当其大量溶解在原油中时，可使原油体积膨胀、黏度降低，同时可降低油水间的界面张力，溶于水后形成的碳酸还可以起到酸化作用，进而提高采收率。CO₂ 驱油作为一种提高采收率的三次采油技术，以其适用范围大、驱油效率高、成本较低等优势，受到了世界各国的广泛重视。这种驱油技术既储存 CO₂ 又提高了油气开采量，具有环保作用的同时也极大地提高了经济效益[1]。但由于 CO₂ 溶入水后对部分金属材料有极强的腐蚀性，也是困扰各国油气工业发展的一个极为突出的问题。激光熔覆是利用高能激光束（$10^4 \sim 10^6 W/cm^2$）在金属表面辐照，通过迅速熔化、扩展和迅速凝固在基材表面熔覆一层具有特殊物理、化学或力学特性的涂层材料。Fe 基合金材料具有较高的耐磨、耐蚀性能，适用于要求局部耐磨且容易变形的零件，主要由于其组分与碳钢铸铁接近，与基体具有良好的相容性，且成本低廉。

采用激光熔覆方法制备 3 种不同 Cr 含量的 Fe 基激光熔覆层，测试 3 种涂层在含 CO₂ 饱和地层水中的腐蚀性能，分析温度和 Cl⁻ 浓度对其腐蚀行为的影响，为油田 CO₂ 驱油伴随产生腐蚀问题提供防腐技术参考。

1　实验过程

1.1　设备及材料

采用 HGL9450-000 多功能数控激光熔覆设备，选择同步送粉法，将合金粉末在 ZYH-

20 自控远红外电焊条烘干炉中 100℃烘干 2h，基体材料预热后采用同步送粉氩气保护。所选激光熔覆参数为：输出功率 3.2 kW，焦距 395mm，扫描速率 9mm/s，光斑直径 5mm。

熔覆基体选择 15[#]钢，预置涂层前，试样的熔覆表面经喷砂处理再用无水乙醇和丙酮清洗。熔覆粉末以华工 325 自熔性 Fe 基合金粉末为基体（颗粒粒度 38~75μm），配制 3 种不同 Cr 含量的涂层成分（表 1）。

对比材料为套管 P110 钢，其成分的质量分数为：0.35% C、0.80% Mn、0.35% Si、0.80% Mo、0.01% P、1.80% Cr、0.01% S。

表 1   3 种不同 Cr 含量粉末化学成分表（质量分数，%）

涂层	Cr	Mn	Ni	Si	Fe
Fe-20Cr	20.0	3.93	3.71	3.68	余量
Fe-30Cr	30.0	3.41	3.13	3.02	余量
Fe-40Cr	40.0	2.93	2.68	2.6	余量

1.2 实验介质

实验介质为模拟大庆油田饱和地层水，母液组分为：2.294g/L NaCl，1.86g/L NaHCO$_3$，0.075g/L Na$_2$SO$_4$，0.042g/L CaCl$_2$，0.172g/L MgCl$_2$·6H$_2$O。分别在母液中加入 NaCl，配制成含 Cl$^-$浓度为：0.04mol/L、0.5mol/L、1mol/L、2mol/L 的模拟地层水溶液。测试温度分别为：25℃、45℃、65℃、85℃。测试前在配制液中通入氩气以排除空气，随后通入 CO$_2$ 至饱和后，进行涂层和 P110 钢腐蚀性能对比研究。

1.3 实验方法

腐蚀性能测试在 Gamry PCI4 工作站上进行，采用工作电极、辅助电极（Pt 片）和参比电极（饱和甘汞电极，SCE）三电极测试系统，测试前先测试试样的自腐蚀电位，稳定 1h 后再进行线性极化、电化学阻抗谱和动电位极化曲线的测试。

采用日本理学 D/max-2200 X—射线衍射仪（XRD）进行物相分析，使用 Cu 靶，以 2θ 角进行分析得到涂层 X—射线衍射谱图。

2 结果与分析

2.1 温度对电化学腐蚀性能的影响

线性极化法是在自腐蚀电位±10mV 附近，腐蚀金属上的外加极化电流和极化电位之间存在着近似的线性关系，即腐蚀电位（E）与腐蚀电流（I_{corr}）呈正比关系[2]。线性极化电阻（R_p）可表述为：

$$R_P = \frac{\Delta E}{\Delta I} \qquad (1)$$

332

根据斯特恩方程：

$$I_{corr} = \frac{b_a \cdot b_c}{2.3 \times (b_a + b_c)} \cdot \frac{1}{R_p} \tag{2}$$

式中，b_a 为阳极 Tafel 曲线斜率；b_c 为阴极 Tafel 曲线斜率。

式（1）表明材料的腐蚀电流与 R_p 呈反比，由此，可以通过 R_p 的大小判断金属的耐蚀性能。3 种涂层和 P110 钢在含 CO_2 饱和地层水中不同温度时的 R_p（图 1）。

随温度升高，P110 钢与 3 种涂层的 R_p 值均降低，这是因为温度升高后，离子穿过双电层的能力增大，即电荷传递电阻减小，同时也会使阳极反应电阻降低，腐蚀过程加剧。

图 2（a～d）是涂层和 P110 钢在含 CO_2 饱和地层水中的动电位极化曲线。可知，P110 钢在所有温度条件下均发生活性

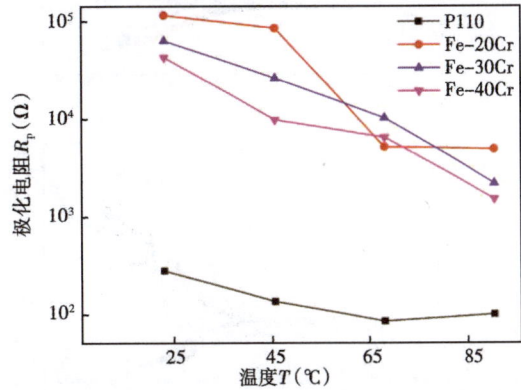

图 1　涂层和 P110 钢在不同温度条件下的 R_p 图

溶解，无钝化现象发生，而涂层则发生了明显的钝化。随着温度的升高，P110 钢自腐蚀电位均负移，腐蚀电流密度 I_{corr} 呈现增大的趋势，表明随温度升高 P110 钢腐蚀均加剧。这是由于温度较低时，生成的腐蚀产物膜对基体的保护作用好，随着温度的升高，离子在膜内的扩散作用增强，腐蚀过程加快。

Fe-20Cr、Fe-30Cr 和 Fe-40Cr 涂层在所有温度条件下，曲线均呈现出典型的钝化区间，说明激光熔覆 Fe 基涂层具有较稳定的钝化性能。随温度的升高，所有涂层的钝化电流均增加，说明温度升高涂层的钝化稳定性降低，耐蚀性恶化。从钝化区间看，Fe-20Cr 涂层在 4 种温度条件下均呈现出较宽的钝化区间（高于 1V）；Fe-30Cr 涂层在温度高于 45℃ 后，钝化区间急剧下降（低至 0.5V），钝化性能下降；Fe-40Cr 涂层的钝化区间则在所有温度条件下均较窄，钝化稳定性明显低于 Fe-20Cr 和 Fe-30Cr 涂层。这些结果说明经过激光熔覆的 Fe 基涂层并没有展示出随着 Cr 含量增加耐蚀性提高的现象。

图 2（a′～d′）是涂层和 P110 钢在含 CO_2 饱和地层水中电化学阻抗谱图。可知，P110 钢阻抗谱在 25℃ 时，出现 3 个时间常数，即高频容抗弧、中频感抗弧和低频的容抗弧，其中高频容抗弧与双电层电容、传递电阻有关，中低频感抗弧与基体溶解、吸附中间产物有关，说明试样表面部分区域被腐蚀产物膜覆盖，活化区与腐蚀产物膜并存，基体处于腐蚀状态。随着温度的增加，高频容抗弧半径逐渐减小，当温度增加到 65℃、85℃，中低频感抗弧消失，腐蚀产物膜 $FeCO_3$ 完全覆盖基体表面使阻抗谱呈双容抗特征，阳极反应过程受活化控制[3]。而 3 种涂层在 25℃、45℃、65℃ 时均呈现出单容抗弧特征，仅在 85℃ 时，低频区容抗弧逐渐出现一个"扩散尾"，表示出韦伯阻抗的性质，出现了容抗弧特征，低频区的容抗弧会大大增加试样表面的反应电阻。这是因为在 85℃ 表面产生了比较致密稳定的非晶态 $Cr(OH)_3$，表明试样电极表面反应由电化学控制变为扩散控制。

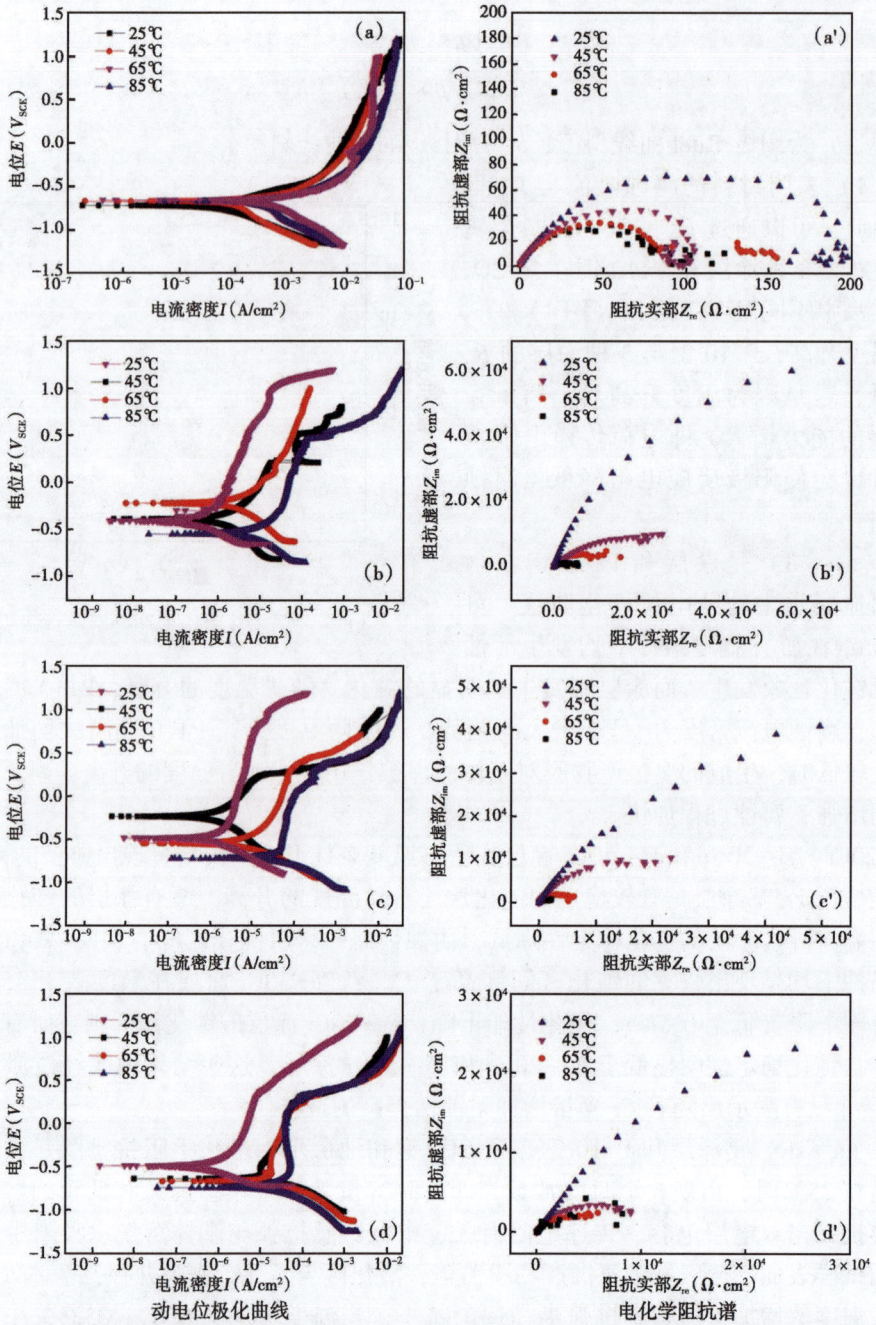

（a）、（a'）为P110钢 （b）、（b'）为Fe—20Cr （c）、（c'）为Fe—30Cr （d）、（d'）为Fe—40Cr

图 2　涂层和 P110 钢在含 CO_2 饱和地层水中不同温度下的动电位极化曲线和电化学阻抗谱图

2.2　Cl⁻浓度对电化学腐蚀性能的影响

图 3（a~d）是涂层和 P110 钢在含 CO_2 饱和地层水中不同 Cl⁻条件下的动电位极化曲线。可知，对于 P110 钢，Cl⁻浓度为 0.04mol/L 的腐蚀电流密度最小，Cl⁻浓度为 1mol/L 的

腐蚀电流密度最大。随着 Cl⁻ 浓度增加，自腐蚀电位逐渐负移。在 0.04~0.5mol/L，自腐蚀电位负移幅度越大，腐蚀电流密度增幅也越大，但当 Cl⁻ 浓度增加至 2mol/L 时，腐蚀电流密度反而降低。

对涂层来说，在 Cl⁻ 浓度为 1mol/L 时，3 种涂层的钝化电流密度也出现最小临界值，进一步增加 Cl⁻ 浓度加剧了钝化电流密度的值，耐蚀性下降。一方面 Cl⁻ 降低了试样表面腐蚀产物膜形成的可能性。另一方面 Cl⁻ 加速试样表面产物膜的破坏，在阳极区域发生活化溶解反应。但是当 Cl⁻ 浓度增大到 2mol/L，CO_2 在地层水溶液中的溶解度降低，pH 值增大，难溶物质 $CaCO_3$ 沉淀的倾向变大，抑制了均匀腐蚀反应速度。随 Cl⁻ 浓度的增大，溶液离子硬度变大，会抑制 H_2CO_3 的电离，抑制了阳极溶解反应，降低试样的反应速度[4]。

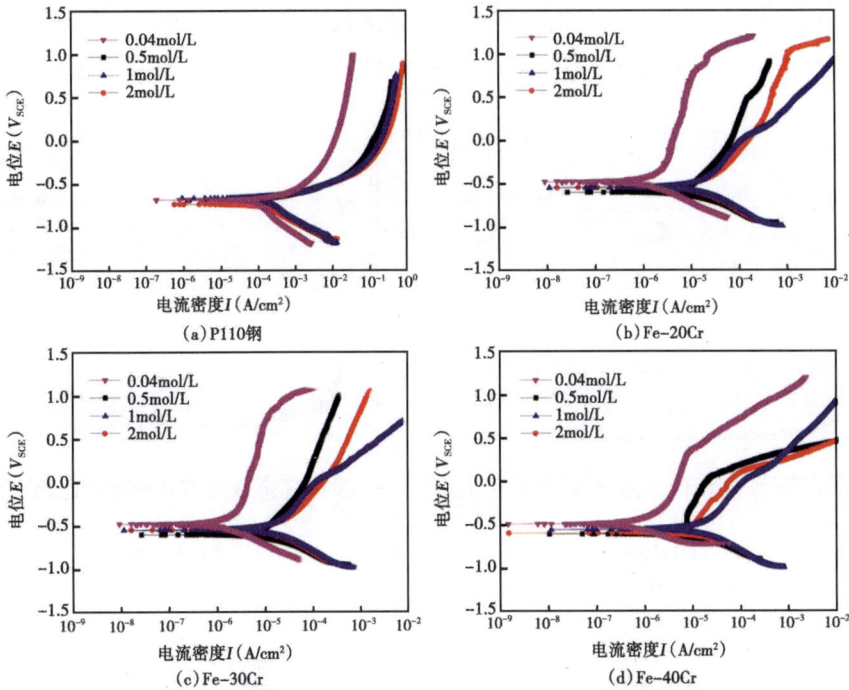

图 3　涂层和 P110 钢在含 CO_2 饱和地层水中不同 Cl⁻ 条件下的动电位极化曲线图

2.3　Cr 元素对于涂层耐蚀性的作用

以 85℃时 3 种涂层和 P110 钢在含 CO_2 饱和地层水中极化曲线图为例（图 4），P110 钢的自腐蚀电位远低于 3 种涂层，而腐蚀电流远高于 3 种涂层，说明激光熔覆涂层具有比 P110 钢更优异的耐点蚀性。因为涂层中 Cr 与 OH⁻ 有较强亲和力，容易优先生成 $Cr(OH)_3$ 吸附在表面，而 $Cr(OH)_3$ 阻碍离子在电极表面浸透，使界面外处的 Cl⁻、CO_3^{2-}、HCO_3^- 浓度低于介质中的含量，电极外 Cl⁻ 浓度减小，减弱对成膜的破坏性，点蚀阻力提高，

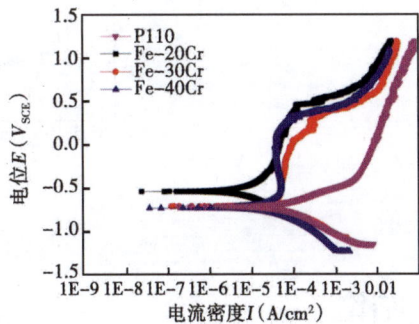

图 4　P110 和 3 种涂层在 85℃时含 CO_2 饱和地层水中极化曲线图

这也是涂层极耐点蚀的原因。

对于涂层来说，Fe-20Cr 涂层的自腐蚀电位高于 Fe-30Cr 和 Fe-40Cr 涂层，而钝化电流密度则远低于 Fe-30Cr 和 Fe-40Cr 涂层，说明涂层的钝化能力并没有随 Cr 含量的增加而增强，低 Cr 含量的 Fe-20Cr 涂层具有最优异的耐蚀性。

为了进一步分析低 Cr 含量的 Fe-20Cr 涂层耐蚀性优异的原因，对 3 种涂层的熔覆区分别进行了 XRD 测试（图 5）。由图 5 可以看出，Cr 含量不同所制备涂层的熔覆层物相组成有较大区别：Fe-20Cr 涂层熔覆区主要由 α-（Fe，Ni）、Cr、Ni-Cr-Fe 物相构成。Fe-30Cr 涂层熔覆区含 α-Fe、$CFe_{0.15}$、$Cr_{23}C_6$ 等物相，而 Fe-40Cr 涂层熔覆区除 α-Fe、$CFe_{0.15}$、$Cr_{23}C_6$ 相，还有 $Cr_{1.36}Fe_{0.52}$、M_7C_3 等相。

可见，Fe-30Cr 和 Fe-40Cr 高 Cr 含量的涂层耐蚀性差的原因，是由于当 Cr 含量过多时，过多的 Cr 与 C 等元素具有一定的亲和力，逐渐形成了 Cr_7C_3、$Cr_{23}C_6$ 等相，这些物相在其熔覆区晶界上析出，导致晶界周围基体产生贫铬区，因此造成熔覆区耐蚀性大大降低。

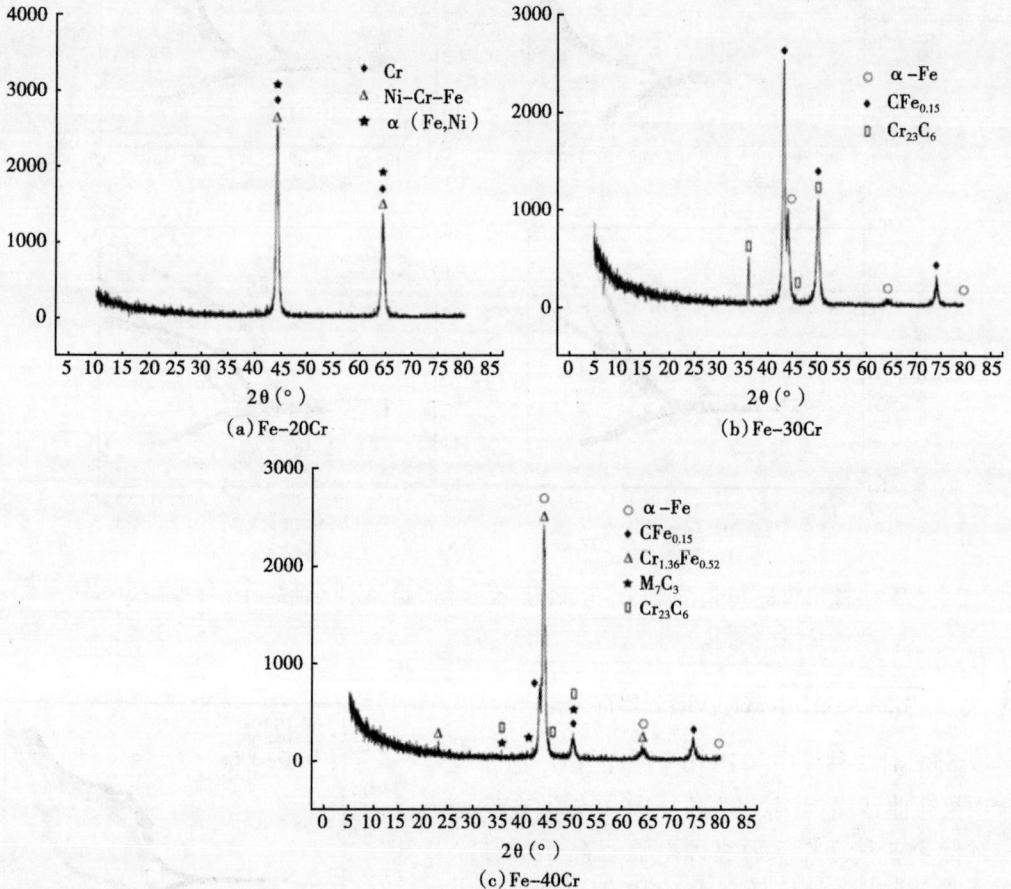

图 5　3 种涂层熔覆区的 XRD 图谱

3　结论

（1）P110 钢在所有温度条件下均发生活性溶解，腐蚀速率均随温度升高而提高。激光熔覆涂层则发生了明显的钝化，温度升高钝化区间变窄，钝化稳定性下降。

（2）P110 钢和涂层耐蚀性均随 Cl⁻浓度增加而降低，Cl⁻浓度与耐蚀性存在一临界值，当 Cl⁻浓度达到 1mol/L 时，腐蚀性最强。

（3）85℃时，涂层具有比 P110 钢更低的腐蚀倾向和更优异的抗点蚀性。低 Cr 含量的 Fe-20Cr 涂层具有较高的耐蚀性，Cr 含量增加时，由于 $Cr_{23}C_6$、Cr_7C_3 等碳化物在晶界处析出，产生晶体贫铬区，进而降低了高 Cr 含量涂层的耐蚀性。

参 考 文 献

[1] Marhohn AL, M A. Rahman S S. Treatment of Drilling Fluid to Common Drill Pipe Corrosion [J]. Corrosion, 1992 16 (9): 721-726.

[2] M. Stern, A. L. Geary. Electrochemical polarization I. A theoretical analysis of the shape polarization curves [J]. Journal of Electrochem. Society. 1957 (104): 56-63.

[3] 韩燕，赵雪会，白真权，等. 不同温度下超级 13Cr 在 Cl⁻/CO₂ 环境中的腐蚀行为 [J]. 腐蚀与防护，2011, 3 (5): 366-369.

[4] 韩燕，李道德，林冠发，等. Cl⁻、CO₂ 和微量 H₂S 共存时 13Cr 不锈钢的腐蚀性能 [J]. 理化检验，2010, 46 (3): 145-150.

高效固体缓释型阻垢剂的研制及性能分析

毕凤琴[1]　韩嘉平[1]　李会星[1]　赵红梅[2]　王　勇[1]

(1 东北石油大学；2. 中国石油大庆测试技术服务分公司检测中心)

摘　要：针对大庆油田集输管道腐蚀结垢现象，以主剂、分散剂、特殊缓蚀阻垢剂和载体配制固体缓蚀阻垢剂，通过复配试验筛选出适合其采出液水质的药剂，并对其缓蚀和阻垢性能进行分析。结果表明：加注药剂后的管道钢腐蚀速率为 0.070mm/a，远低于未加注前的 0.310mm/a；阻垢效果 15 天内稳定，阻垢率约为 91%，阻垢性能良好；缓释速率约为每 10 天溶解 8%，加注周期约为 3 个月。所制备药剂性价比高，缓蚀阻垢综合性能良好，可作为油田水驱采出液专用缓蚀阻垢药剂。

关键词：缓蚀阻垢剂；固体；稳定；性能分析

0　引言

大庆油田已进入开采中后期，油田管道结垢和腐蚀问题变得越来越突出，解决管道结垢和腐蚀问题已迫在眉睫，添加缓蚀阻垢药剂性价比高、针对性强，已成为目前各大油田比较可行且通用的方法[1,2]。目前通用的缓蚀阻垢剂多为液体型，由于随流体流失较大，因此添加量较大，作用周期和效果也不够理想，造成很大浪费，而固体缓蚀剂在缓慢溶解的过程中起到防腐和阻垢的作用，作用时间长，效果理想，经济性好，具有良好的性价比[3]。笔者针对大庆油田五厂区块水驱输油管道腐蚀结垢现象和油田采出液的成分特点，以缓蚀剂、阻垢剂、分散剂和载体为组分，配制出新型固体缓释型缓蚀阻垢剂，并对药剂性能进行评价。

1　实验

1.1　实验仪器和药品

实验仪器和药品有：美国 Gamry PCI4/7500 电化学工作站，恒温水浴锅，复配试剂（表1），大庆五厂杏-13 计量间油田采出液（成分见表2），20#钢试样以及实验室基本仪器等。

表 1　复配试剂表

名称	组分
主剂	ATMP（氨基三甲叉膦酸）[1]
分散剂	HEDP（羟基乙叉二膦酸）[1]
	EDTMP（乙二四甲叉膦酸）[1]
	PAA（聚丙烯酸）[1]
	AEC（聚烷基环氧羧酸）[1]

338

名称	组分
缓蚀	HAK（腐植酸钠）[2]
阻垢	HPMA（水解聚马来酸酐）[1]
	EDTMPS（乙二胺四亚甲基膦酸钠）[1]
载体	PVA（聚乙烯醇）[3]

注：1-山东省泰和水处理有限公司；2-天津大茂化学试剂厂；3-山西三维集团公司。

表 2　水驱管道采出液成分表　　　　　　　（单位：mg/L）

分析内容	含量
CO_3^{2-}	66.52
HCO_3^-	2705.63
OH^-	0.00
Cl^-	1035.61
Ca^{2+}	3.29
Mg^{2+}	3.81
K^+	1.25
Na^+	1712.87
Fe	0.90
矿化度	6370.61

1.2　缓蚀阻垢剂的制备

缓蚀阻垢剂复配由主剂（有机磷酸或磷酸盐，一种或多种）、分散剂、特殊缓蚀或阻垢剂和载体组成。

药剂复配方案：以一种或多种主剂分别和 3 种分散剂复配，制备至少 32 种缓蚀阻垢药剂，确定性能较稳定的缓蚀阻垢药剂。缓蚀阻垢剂的复配试验采用正交试验法，按复配方案依次编号为 CL01~CL32。

制备时先测试添加药剂水质成分，根据水质成分，确定腐蚀和结垢性。有针对性选取药剂成分，制备过程是在主剂中加入辅剂，加入聚丙烯醇后用恒温水浴锅加热，充分分散、均匀混合后，通过黏合剂进行黏合冲模，加工成特定形状，再经过脱模干燥，即为成品。

1.3　性能测试

采用中国石油天然气总公司行业标准 SY/5273—1991 评定缓蚀性能，阻垢性能采用 ED-TA 络合滴定法测定[4]。

模拟现场工况，采用失重法对添加的定量缓蚀阻垢剂进行动态模拟检测，根据单位时间溶解的质量确定缓释速率。

2 结果与分析

2.1 缓蚀性能分析

对 20# 管道钢作电化学实验，以油田采出液为介质，分别作空白和加注实验，得到极化曲线和阻抗谱如图 1 所示。

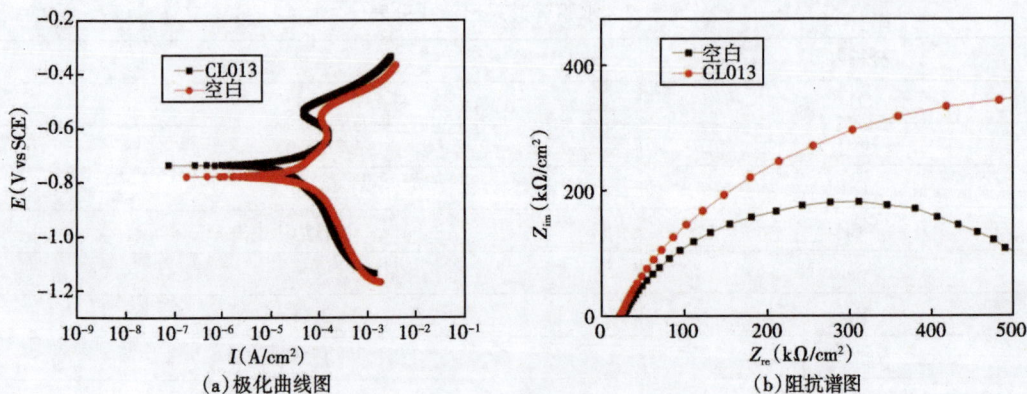

(a) 极化曲线图   (b) 阻抗谱图

图 1　加注药剂前后 20# 钢极化曲线和阻抗谱

如图 1（a）所示，加注药剂后自腐蚀电位上升，由于药剂溶解抑制腐蚀能力逐渐加强，有利于阻碍腐蚀的发生。腐蚀电流为 $22\mu A$，小于空白试验的 $34\mu A$，腐蚀电流变小，而极化电阻则增加到 $1.104k\Omega$，不利于腐蚀的发生。通过失重法测得腐蚀速率，空白和加药后试样的腐蚀速率分别为 $0.310mm/a$ 和 $0.070mm/a$，缓蚀效果明显，加药后试片的腐蚀速率低于 $0.076mm/a$，达到了油田集输管道的腐蚀标准。

电化学阻抗谱如图 1（b）所示，加药后容抗增大，耐腐蚀能力提高。经过 Zview 软件拟合后，加药后溶液电阻基本没有变化，说明加入药剂之后对溶液的导电性没有影响，对其腐蚀性影响较小。20# 钢膜层电阻和电荷传递电阻明显上升，耐腐蚀能力增强。

由于缓蚀主剂为有机磷酸，其结构中的 P、O 等原子能向金属原子提供孤对电子，在金属表面吸附成膜，抑制腐蚀电化学过程的发生，从而起到缓蚀作用。同时有机磷酸盐等的 O、P 原子均含有未成键电子对，可进入金属结构的空轨道形成配位体，在金属表面形成缓蚀剂分子的吸附层，抑制金属的腐蚀。

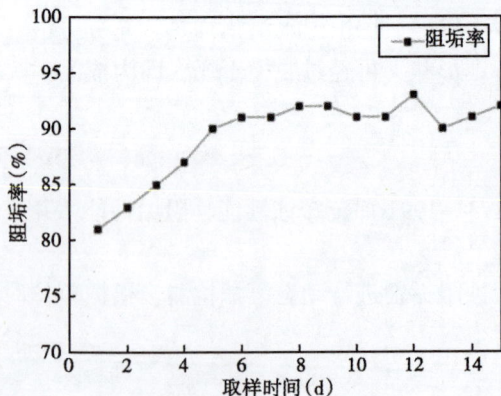

图 2　阻垢率随时间变化曲线

2.2 阻垢性能分析

在采出液中加入缓蚀阻垢剂，每天采样，持续 15d，分别对采出液样品做滴定实验，得到阻垢率如图 2 所示。阻垢作用主要由膦酸盐与共聚物阻垢剂复配而成，前者具有强螯合作用，与 Ca^{2+}、Mg^{2+} 等金属离子螯合生成易溶于水的螯合物，阻止垢的形成；后者为水溶性聚电解质，电荷密度高，产生离子间的斥力或空间位阻，阻止成垢晶核微粒相互接触，使垢不能聚结长大。此外，阻垢剂还可吸附在刚析出的晶核上，破坏和干扰晶核的正常生长，起到阻垢作用[5-6]。

从图2可知，在加入缓蚀阻垢剂的前几天里，阻垢率分别为81%和83%，这说明药剂有效成分尚未完全发挥阻垢性能；在第四、五天时药剂有效成分完全发挥作用，阻垢率也逐渐升高，可达到90%；在接下来的时间内阻垢率趋于稳定，在91%左右，波动不大，表明本产品的阻垢性能良好且稳定。

2.3 最佳用量确定

针对采出液水质设计5组加注用量实验，加注量分别为5mg/L、10mg/L、15mg/L、20mg/L、25mg/L，进行电化学实验，测试结果如图3所示。

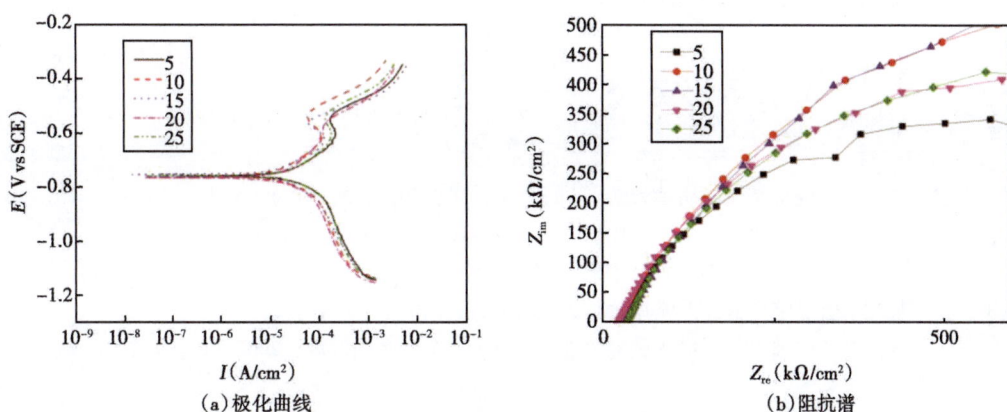

(a)极化曲线 　　(b)阻抗谱

图3　不同用量时极化曲线和阻抗谱

由图3（a）可知，当加注量为10mg/L时，腐蚀电流较小，为7.5μA；自腐蚀电位较高，采出液极化电阻变大，不利于腐蚀的发生。当加注量为10mg/L时，20#钢在采出液中的腐蚀速率为0.069mm/a，低于国家管道腐蚀标准。由图3（b）可知，当加注量为10mg/L和15mg/L时，溶液的容抗较大，耐腐蚀性较好，综合实验和经济性，选择加注量为10mg/L。

2.4 缓释性能分析

为了保证药剂的效果在采出液中持续作用，可以通过失重法对加入到采出液中定量的药剂进行检测，测得缓释速率，从而确定加药周期。固体缓蚀阻垢剂的黏合剂为聚乙烯醇，其综合性能良好，能有效固化缓蚀阻垢剂而不影响其缓蚀阻垢性能。由图4可知，固体药剂溶解和时间呈线性关系，约每10天溶解8%，由此确定加注周期约为3个月。根据采出液质量和加注药剂量计算，采出液中药剂含量在10～

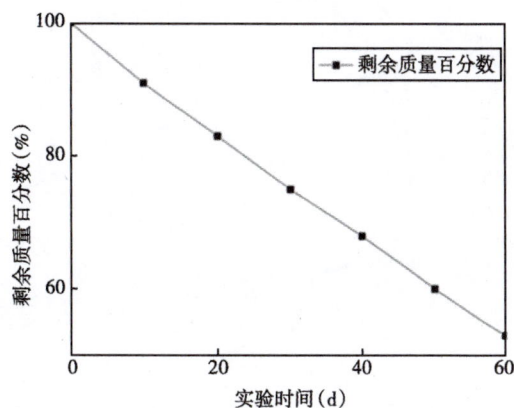

图4　剩余质量百分数随取样时间变化曲线图

15mg/L 之间，满足最佳加注用量要求。

3 结论

（1）加注药剂后，自腐蚀电位上升，腐蚀电流下降，极化电阻升高，腐蚀速率为 0.070mm/a，研制固体缓蚀阻垢药剂缓蚀性能良好。

（2）在加注后四至五天药剂可达到理想阻垢效果，阻垢率可达 90%；在加注 15d 内阻垢率波动不大，稳定在 91% 左右，阻垢效果稳定。

（3）聚乙烯醇为黏合剂缓释性能良好，缓释速率大约为每 10 天 8%，确定加注周期约为 3 个月。

（4）固体缓释型缓蚀阻垢剂加注用量为 10mg/L 时，缓蚀性能良好，可用作大庆油田五厂区块水驱管道的专用缓蚀阻垢剂。

参 考 文 献

[1] 杜春安，赵修太，邱广敏．缓释型固体缓蚀阻垢剂的研制与应用 [J]．石油与天然气化工，2005，34（2）：128-131.
[2] 汪双喜，谢小聪．油田结垢的防治方法 [J]．清洗世界，2010，26（12）：8-10.
[3] 张银．油田缓蚀阻垢剂筛选及复配室内研究 [J]．全面腐蚀控制，2008，22（3）：15-17.
[4] 上海化工厂．复合缓蚀阻垢剂的水质中稳定中试研究 [J]．化肥工业，1978，（1）：42-47.
[5] 王兵，李长俊．管道结垢原因及常用分析方法 [J]．油气储运，2008，27（2）：59-61.
[6] 王晏山，张洪林．缓蚀阻垢剂的筛选及工业应用 [J]．全面腐蚀控制，2005，19（4）：22-27.

不同类型缓蚀剂在含饱和 CO_2 油田采出液中对 P110 钢的缓蚀性能研究

毕凤琴[1] 张成果[1] 徐子怡[2] 张旭昀[1] 王 勇[1]

(1. 东北石油大学；2. 中国石油大学（北京）)

摘 要：采用电化学腐蚀法测试咪唑啉类 IMC-80-BH、有机胺类 ZK682 和有机磷类 N582 的缓蚀剂在常温下饱和 CO_2 油田采出液对 P110 钢的缓蚀行为，用高温高压反应釜借助失重法和腐蚀形貌观察研究了其在高温（80℃）条件下的缓蚀性能。结果表明：3 种缓蚀剂均为阳极抑制型，作用方式为"负催化效应"；3 种缓蚀剂均存在浓度极值现象，吸附膜稳定程度随浓度变化差别较大；在油田真实高温高压条件下，有机胺缓蚀剂 ZK682 缓蚀性能最优。

关键词：CO_2 腐蚀；缓蚀剂；电化学腐蚀；高温高压

0 引言

CO_2 是一种高效驱油溶剂，采用 CO_2 驱油在提高采收率的同时还可解决伴生 CO_2 出路问题，实现环保开发和有效减排。但在大量 CO_2 注入地层后，地面管线、井口、管柱、套管、井下工具等都引起严重的 CO_2 腐蚀。CO_2 腐蚀已成为油气田的主要腐蚀类型，其附加的阴极反应 H_2CO_3 的直接还原，是导致在相同的 pH 值下 CO_2 对钢铁的腐蚀比盐酸还严重的主要原因。目前控制油管 CO_2 腐蚀的方法主要有选用耐蚀材料（如 Cr13 钢）、涂覆有机或无机涂层、金属镀层或渗透层，水介质处理和加缓蚀剂等。实验和现场应用表明，Cr13 钢成本较高，管道内壁涂覆层实施有一定困难，水介质处理不太现实，而加注缓蚀剂是最为经济有效的措施，在控制油管的腐蚀中已普遍采用[1]。为此国内外学者开展了大量研究工作，并取得很多成果。尽管如此，有关缓蚀剂在真实油田环境下的缓蚀机理研究仍很薄弱，因此，深入研究不同类型缓蚀剂在真实油田环境下的缓蚀性能和机理具有重要的理论和实际意义。钢在饱和 CO_2 水溶液中的腐蚀是一个比较复杂的电化学过程，应用极化和交流阻抗法研究钢在饱和 CO_2 水溶液中的腐蚀是较为常用的方法。

以大庆油田采油四厂油田采出液为实验介质，用电化学工作站研究咪唑啉类 IMC-80-BH、有机胺类 ZK682 和有机磷类 N582 3 种不同类型缓蚀剂在常温时对套管钢 P110 在含有饱和 CO_2 的油田采出液中的腐蚀行为和缓蚀性能，借助高温高压反应釜用失重法研究其在高温（80℃）条件下模拟油田真实环境中的缓蚀性能，通过腐蚀形貌观察，确定最佳缓蚀剂种类，以期为油田 CO_2 腐蚀缓蚀剂的筛选和应用提供参考。

1 实验

材料为油田用套管 P110 钢（26CrMo4）。具体化学成分（质量分数）经德国 Spctrolab-

M10 直读光谱仪测量为 0.253% C、0.836% Cr、0.151% Mo、0.2517% Si、0.868% Mn、0.056%Ni、0.005%S、0.009%P、0.009%N。

实验介质为大庆油田采油四厂油田采出液，在其中通入饱和 CO_2，分压保持在 1MPa。缓蚀剂分别为咪唑啉类 IMC-80-BH、有机胺类 ZK682 和有机磷类 N582。

电化学测试在美国 Gamry PCI4/750 工作站上进行，采用 3 电极系统（辅助电极为铂片，参比电极为饱和甘汞电极，工作电极为有效面积为 $1cm^2$ 的 P110 钢），主要测试试样的电化学阻抗谱（EIS）和极化曲线，测试前将工作电极打磨露出光亮均匀的金属面，丙酮除油无水乙醇脱脂后，用去离子水冲洗，滤纸吸干。先测试试样的（腐蚀电位）E_{corr}-t 曲线，当 100s 内 E_{corr} 值的变化小于 1mV 时，可认为 E_{corr} 稳定。EIS（电化学阻抗谱）扫描频率范围为 10mHz~1kHz。极化曲线扫描范围为 -1.2~0.5V，扫描速度设为 0.33mV/s。电化学测试时，3 种缓蚀剂分别选择质量浓度为 50mg/L、100mg/L、150mg/L 进行。

质量损失法测试在高温高压反应釜中进行，试样尺寸为 50（mm）×20（mm）×3（mm），实验温度为 80℃，转速为 300r/min，实验前测试试样的质量和表面积，浸泡腐蚀 72h 后清除腐蚀产物，计算平均腐蚀速率：

$$B = 8.76(W_0 - W_1 - W_2)/\rho At \tag{1}$$

式中，B 代表按深度计算的腐蚀速率，mm/a；A 代表试样表面积，m^2；t 代表实验周期，h；W_0 代表试样初始重量，g；W_1 代表除去腐蚀产物后试样重量，g；W_2 代表清除腐蚀产物后试样的校正失质量，g；ρ 代表材料密度，g/cm^3。

腐蚀后采用日本 Hitachi S-3400N 扫描电镜对表面形貌进行观察分析。

2 结果与讨论

2.1 极化曲线测试

图 1 为 3 种缓蚀剂在不同质量浓度条件下的极化曲线。可以看出，在采出液中加入缓蚀剂后，3 种体系的自腐蚀电位均显著正移。阴阳极 Tafel 斜率变化不大，而自腐蚀电流密度大幅度减小，这主要是由于缓蚀剂吸附在电极表面使电极反应的活化能位垒升高，进而抑制电极反应所致。由此可见，3 种缓蚀剂均为阳极型缓蚀剂，作用方式为"负催化效应"[2]。

由图 1（a）可知，随着缓蚀剂 IMC-80-BH 浓度的增大，腐蚀电位逐渐正移，腐蚀电流密度逐渐减小，缓蚀效果越来越好，说明吸附膜随着浓度增大越致密。同理，随缓蚀剂 ZK682 和 N582 浓度的增大缓蚀效果变差，这是因为缓蚀剂分子中电子云密度较高的氮、硫等原子吸附于金属表面，而长链烷基等非极性基则位于离开金属的方向，排列在金属表面形成疏水薄膜，保护金属表面不受侵蚀[3]。使用 ZK682 和 N582 两种缓蚀剂吸附时，优先在阳极区的反应活性点吸附，随浓度增加，金属表面活性吸附中心被占完后，多余的缓蚀剂分子吸附于非活性点和阴极区。此时，再增加缓蚀剂的浓度，由于缓蚀剂分子间的相互作用和水的竞争吸附，使得处于活性区之外的缓蚀剂分子开始脱附，造成缓蚀性能的下降。当浓度达到一定值时，会达到吸脱附的平衡。缓蚀剂的这种阳极脱附行为，除与吸附在电极表面缓蚀剂分子的热运动有关外，在阳极极化足够大时，被吸附离子所覆盖金属表面的部分溶解，对

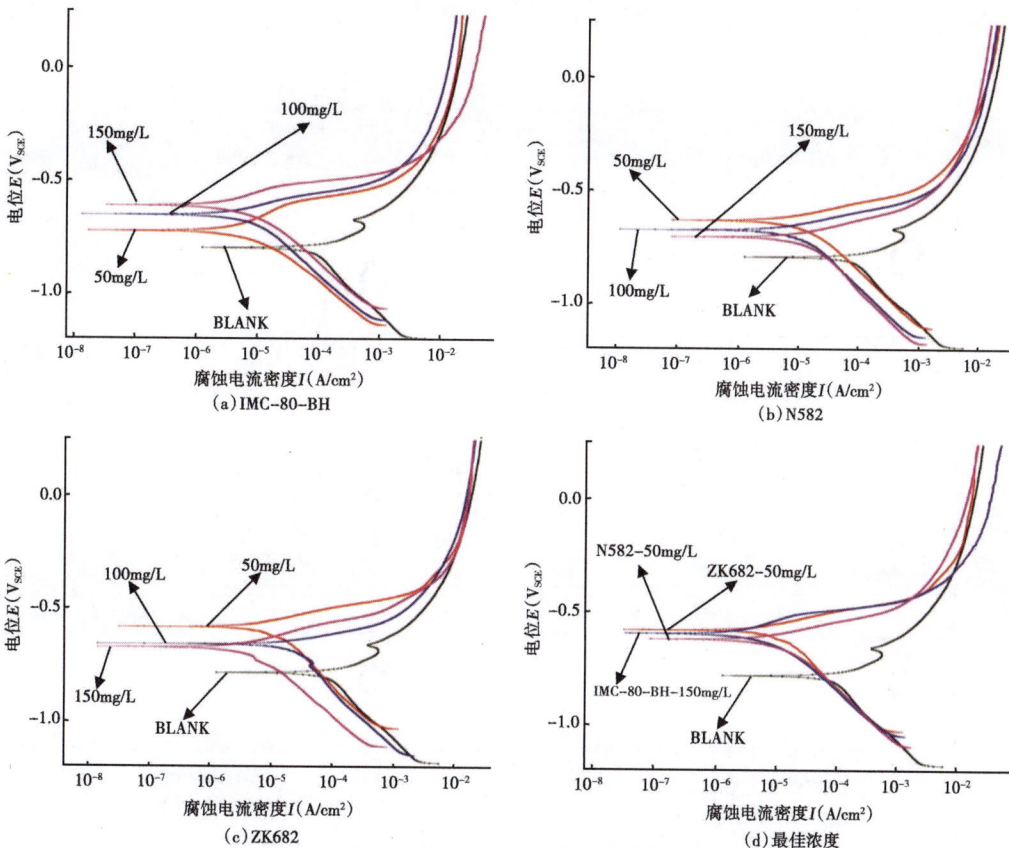

图1 3种缓蚀剂在不同质量浓度条件下的极化曲线图

吸附的缓蚀剂分子从电极表面离开也起到一定的牵动作用，阳极溶解电流密度越大，这种作用就越强[4]。

图1（d）为3种缓蚀剂在各自最佳质量浓度下（IMC-80-BH为150mg/L，ZK682为50mg/L，N582为50mg/L）的极化曲线。可看出，3种缓蚀剂在最佳浓度下其缓蚀效果存在一定差别，其中有机胺类ZK682（50mg/L）最优，有机磷类N582（50mg/L）最差。

2.2 电化学阻抗谱测试

图2为3种缓蚀剂在不同浓度条件下的电化学阻抗谱测试结果。图2（a）中缓蚀剂体系的高频容抗弧直径较空白体系均有较大的增幅，表明缓蚀剂显著降低了腐蚀速率。根据高频容抗弧直径的大小可以读出，缓蚀剂的缓蚀性能从高到底排序为150>100>50mg/L，图2（b）、（c）中为50 >150>100mg/L。由图2可知，3种缓蚀剂在一些浓度低频端有感抗出现，通常电化学活性反应产物或反映中间产物的吸、脱附过程会在阻抗谱上产生附加的容抗或感抗。根据图3的等效电路进行拟合计算。由于阻抗谱均出现弥散效应[5]，因此用常相位角元件描述电容性质。无感抗用图3（a）拟合，有感抗用图3（b）拟合。其中，R_s为溶液电阻，R_{ct}为电荷转移电阻，R_f为膜电阻，CPE为常相位角元件，CPE1由双电层电容C_d和弥散指数n_1组成，CPE2由膜电容C_f和弥散指数n_2组成，R_1为极化电阻，L为电感，$R_{ct}=R_1$ $+R_2$。电阻R_2和电感L可能与低频的中间反映过程有关[6]。电化学阻抗谱拟合结果见表1。

345

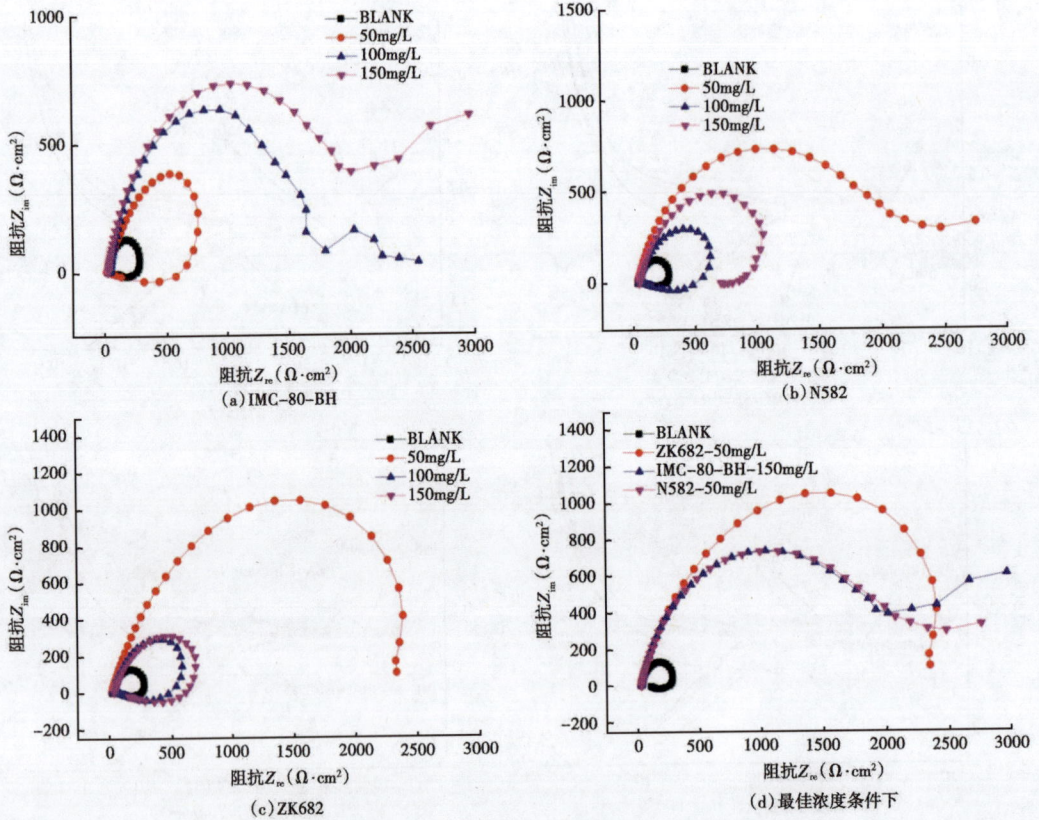

图 2　3 种缓蚀剂在不同质量浓度条件下的电化学阻抗谱

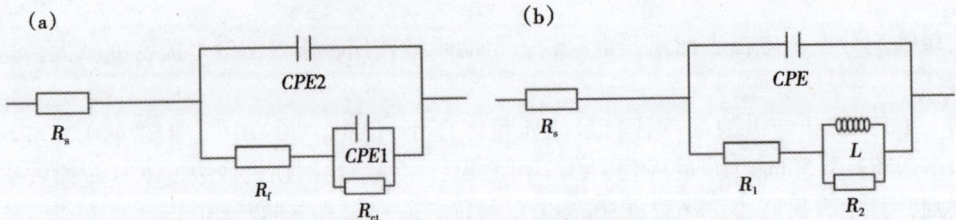

图 3　等效电路模型

表 1　电化学阻抗谱拟合结果表

缓蚀剂	质量浓度 (mg/L)	R_{ct} (k$\Omega \cdot$cm^2)	C_d [μF$\cdot$cm^2 (n_1)]	R_f ($\Omega \cdot$cm^2)	C_f (μF.cm^2 (n_2))	R_s ($\Omega \cdot$cm^2)
ZK682	50	1.85	379.19 (0.83)	750.6	239.3 (0.90)	18.55
	100	0.60	467.24 (0.79)	145.3 (R_1)	523.9KH (L)	26.09
	150	0.74	480.43 (0.76)	260.0 (R_1)	966.4KH (L)	36.46
N582	50	1.25	408.24 (0.89)	2065	938.42 (0.35)	26.37
	100	0.63	424.63 (0.78)	137.4 (R_1)	520.3KH (L)	27.21
	150	1.12	414.13 (0.82)	807.7 (R_1)	1002.0KH (L)	38.46

346

缓蚀剂	质量浓度 （mg/L）	R_{ct} （kΩ·cm²）	C_d [μF·cm² （n_1）]	R_f （Ω·cm²）	C_f （μF.cm² （n_2））	R_s （Ω·cm²）
IMC-80-BH	50	1.21	471.23 （0.65）	414.3 （R_1）	1571.0KH （L）	9.84
	100	1.30	416.46 （0.71）	432 （R_1）	1632.4KH （L）	11.42
	150	1.34	382.76 （0.78）	475 （R_1）	1638.8KH （L）	10.65

由表 1 的拟合结果可知，加入缓蚀剂 ZK682 的 R_{ct} 值随浓度的增大呈现出先减小后增大的趋势，而 C_d 值逐渐增大。由于吸附 H_2O 的介电常数比其他吸附物质的介电常数大很多，而且缓蚀剂吸附层厚度比 H_2O 吸附层厚，因此电极表面的吸附 H_2O 被缓蚀剂分子替代后 C_d 值会降低。缓蚀剂 ZK682 在 50mg/L 时 R_{ct} 值最大，C_d 值最小，因此有最好的缓蚀性能。C_d 值随浓度增加而增大，说明膜结构变得疏松，空隙率增大，缓蚀效果变差，这主要与缓蚀剂分子的脱附有关。

加入缓蚀剂 N582 的 R_{ct} 值随浓度的增大呈现出先减小后增大趋势。C_d 值基本不变，说明吸附膜致密度变化不大。从拟合结果可知 N582 在 50mg/L 时缓蚀效果最好。

加入缓蚀剂 IMC-80-BH 的 R_{ct} 值随质量浓度的增大而增大。而 C_d 值随浓度增大而减小，说明膜致密度不断增加。从拟合结果可知在 3 个质量浓度中，150mg/L 缓蚀效果最好，吸附膜最致密。

从拟合结果可知，3 种缓蚀剂在最佳质量浓度下 ZK682（50mg/L）的 R_{ct} 值最大，C_d 最小，有最好的缓蚀效果；而 N582（50mg/L）R_{ct} 值最小，C_d 值最大，缓蚀效果最差。图 2（d）为 3 种缓蚀剂在各自最佳浓度下的缓蚀效果，从拟合结果可知与极化曲线测试结果一致。

2.3 失重法

从失重法实验结果看（表 2），在模拟油田真实高温高压环境中，有机胺类缓蚀剂 ZK682 对 P110 钢缓蚀效果最好，有机磷类 N582 缓蚀效果最差。

表 2 失重法实验结果表

缓蚀剂	腐蚀速率（mm/a）	缓蚀率（%）
BLANK	0.3255	—
IMC-80-H	0.0420	84.55
ZK682	0.0094	93.95
N582	0.1043	61.66

图 4 为高温高压下添加不同缓蚀剂后 P110 钢的表面腐蚀形貌。可以看出，未加缓蚀剂时 P110 钢表面发生了严重的全面腐蚀（图 4a），有部分腐蚀产物膜破损且剥落。缓蚀剂 IMC-80-BH 质量浓度为 150mg/L 时金属表面（图 4b）发生轻微的局部腐蚀。缓蚀剂 ZK682 质量浓度为 50mg/L 时表面（图 4c）只有很少的腐蚀产物，偶见局部腐蚀。缓蚀剂 N582 质量浓度为 50mg/L 时表面（图 4d）覆盖着疏松的腐蚀产物，P110 钢发生了比较严重的局部腐蚀，且腐蚀产物疏松、多孔，这主要是因为有机磷类缓蚀剂在 P110 钢表面吸附作用不均匀。从腐蚀形貌分析，缓蚀剂 ZK682 在高温高压下对 P110 钢有较好的缓蚀作用。

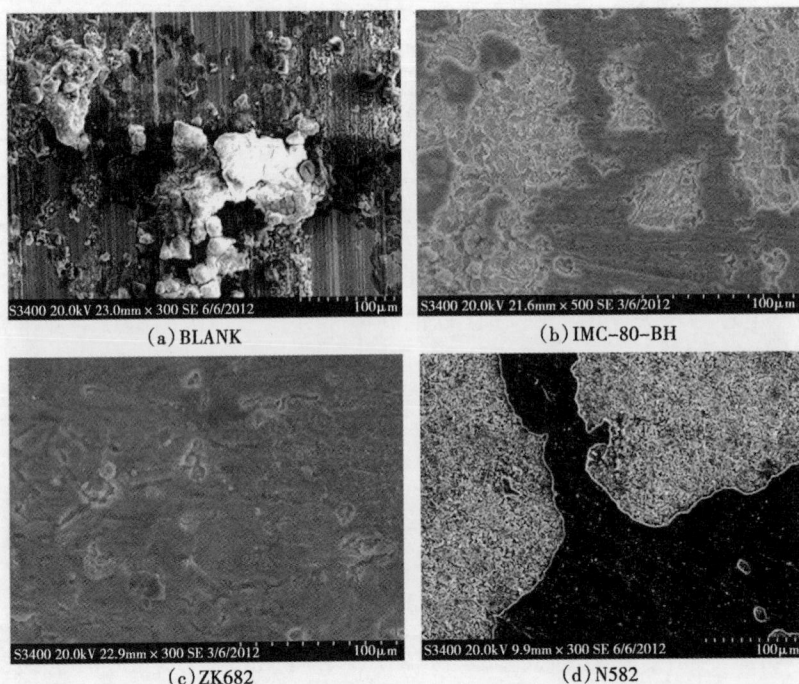

(a) BLANK   (b) IMC-80-BH

(c) ZK682   (d) N582

图4　高温高压下添加不同缓蚀剂后 P110 钢表面腐蚀形貌

3　结论

（1）咪唑啉类 IMC-80-BH、有机胺 ZK682 和有机磷 N582 3 种类型缓蚀剂均为阳极抑制型缓蚀剂，缓蚀机理为"负催化效应"。

（2）在实验条件下，3 种缓蚀剂均存在浓度极值。有机胺缓蚀剂 ZK682 随浓度的增大，存在脱附现象；有机磷缓蚀剂 N582 吸附膜随浓度改变变化不大；咪唑啉缓蚀剂 IMC-80-BH 在高浓度的缓蚀效果更优。

（3）在模拟油田真实高温高压条件下，加入有机胺类缓蚀剂 ZK682 后，P110 钢表面致密、腐蚀产物少，缓蚀效果最好，咪唑啉类缓蚀剂 IMC-80-BH 次之，有机磷类缓蚀剂 N582 缓蚀效果最差。

参 考 文 献

[1] 油气田腐蚀与防护技术手册编委会. 油气田腐蚀与防护技术手册 [M]. 北京：石油工业出版社，1999：30-35.

[2] 曹楚南. 腐蚀电化学原理 [M]. 北京：化学工业出版社，2004：201-206.

[3] 刘福国. 油田钻具、管道系统腐蚀规律及缓蚀剂缓蚀性能和机制研究 [D]. 青岛：中国海洋大学，2008.

[4] 杨怀玉，曹殿珍，陈家坚，等. CO₂ 饱和溶液中缓蚀剂的电化学行为及缓蚀性能 [J]. 腐蚀科学与防护技术，2000，12（4）：27-30.

[5] 王佳. 有机缓蚀剂作用机理和脱附行为的研究 [D]. 沈阳：中国科学院金属腐蚀与防护研究所，1990.

[6] 刘福国，杜敏，张静，等. 咪唑啉衍生物缓蚀剂对碳钢在 CO₂ 盐水中的缓蚀机理 [J]. 物理化学学报，2008，24（1）：146-150.

20 钢和 P110 钢 CO_2 腐蚀静态及动态缓蚀性能研究

王 勇[1] 李 洋[1] 孙世斌[2] 张旭昀[1] 徐子怡[3] 毕凤琴[1]

（1. 东北石油大学；2. 中国海洋石油工程股份有限公司；3. 中国石油大学（北京））

摘 要：借助高温高压腐蚀反应釜，测试了咪唑啉和有机胺类缓蚀剂在饱和 CO_2 油田采出液对 20 和 P110 钢的静态和动态缓蚀行为，并对表面腐蚀形貌进行分析。结果表明：20 钢静态电化学行为呈现活性溶解而 P110 钢发生钝化，缓蚀剂加入后均降低了腐蚀电流。缓蚀剂浓度是影响 20 钢和 P110 钢动态缓蚀行为的显著因素，材料表面均匀致密的产物膜层直接影响其缓蚀效果。

关键词：CO_2 腐蚀；缓蚀剂；电化学；动态；20 钢；P110

0 引言

CO_2 腐蚀一直是困扰油气工业安全生产的重要因素之一，随着 CO_2 驱油工艺的广泛应用和富含 CO_2 油气井的开采，如何抑制或减缓 CO_2 腐蚀成为油气生产管理部门亟待解决的关键问题[1]。目前，加注缓蚀剂是解决 CO_2 腐蚀的一种经济有效且切实可行的方法之一。普遍认为应用效果较好的缓蚀剂是吸附型，如咪唑啉型、有机胺类、季铵盐类或松香胺衍生物等[2-4]。

油田实际工况中，管材多处在高温、高压条件下及一定流速的环境中，再加上 CO_2 腐蚀性离子作用，腐蚀环境相当恶劣。多数的研究都集中在常温常压 CO_2 腐蚀条件下不同油管用钢缓蚀剂的筛选和缓蚀性能研究，其研究结果虽具有一定的借鉴意义，但与实际工况还存在相当大的差距。因此常温、常压下进行研究远远不够，很有必要结合油田实际工况，研究其在真实油田环境中的动态缓蚀性能，相应的研究结果在为 CO_2 腐蚀缓蚀剂现场应用提供参考的同时，更具有现实意义和工程指导价值。

首先以电化学测试两种典型的油管（20 钢）和套管（P110 钢）用钢，常温常压条件下，评价其静态腐蚀和缓蚀行为。借助高温高压腐蚀反应釜，以正交实验和失重法为手段，研究两种钢在真实油田采出液中的动态缓蚀性能。最后，用扫描电子显微镜分析腐蚀后的表面形貌，为确定腐蚀机理提供依据。

1 实验

1.1 材料

静态电化学测试仪器采用武汉 CorrTest 电化学工作站 CS310。腐蚀测试系统由三电极体

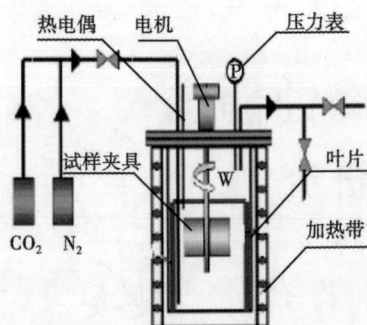

图 1　高温高压腐蚀反应釜

系组成，参比电极为甘汞电极（SCE），辅助电极为铂片，工作电极为暴露面积 1cm² 的 20 钢和 P110 钢（26CrMo4）。

动态缓蚀性能测试在高温高压腐蚀反应釜中以腐蚀挂片 [尺寸为 5（cm）×1（cm）×0.3（cm）] 形式进行，以失重法作为评价标准（图 1）。

实验介质选用油田采出液，测试前通入 N_2 以除去溶液中 O_2。再通入 CO_2 至少 2h 直至饱和。实验过程中，持续通入 CO_2 以确保 CO_2 饱和。

缓蚀剂为 20 钢选用 IMC-80ZS 咪唑啉型，沈阳中科防腐蚀工程技术有限公司生产；P110 钢选用有机胺类化合物 ZK682，广州市振科科技有限公司生产。

1.2　过程

（1）测试常温下 20 钢和 P110 钢在油田采出液中添加缓蚀剂时极化行为。

测量前先将工作面依次经 200#、400#、800# 水砂纸打磨，用无水乙醇、丙酮擦拭干净，去离子水冲洗晾干后备用。测试前先测试试样 E_{corr}-t 曲线，E_{corr} 稳定后进行动电位极化曲线测试。扫描范围为 -1.2~0.0V，扫描速度设为 0.33mV/s。

（2）采用高温高压腐蚀反应釜，以正交实验为手段，借助失重法，研究 20 钢和 P110 钢在油田采出液中动态缓蚀性能。

正交实验考虑 3 个因素：缓蚀剂浓度、介质流速和温度。CO_2 分压均为 1MPa，实验周期为 120h。每个因素选取 3 个水平，具体值参考油田实际工况，采用正交表 L9（3⁴），见表 1。

表 1　正交实验因素水平表

因素	缓蚀剂质量浓度（mg/L）	介质流速（m/s）	温度（℃）
1	50	1	50
2	100	2	70
3	150	3	90

失重法是根据腐蚀前后试件质量的变化来测定金属腐蚀速率，能反映出一段时间内平均的腐蚀情况。通常采用单位时间内单位面积上的质量变化来表征平均腐蚀速率 [g/（m²·h）]。计算公式：

$$v = \frac{W_o - W_1}{A \times t} \tag{1}$$

式中，v 为试样腐蚀速率，g/（m²·h）；A 为试样表面积，m²；t 为实验周期，h；W_o 为试样原始质量，g；W_1 为除去腐蚀产物后试样的质量，g。

根据腐蚀速率可计算各缓蚀剂的缓蚀效率：

$$\eta = \frac{v_o - v}{v_o} \times 100\% \tag{2}$$

式中，η 为缓蚀率，%；v_0 为未加缓蚀剂腐蚀速率，g/（$m^2 \cdot h$）；v 为加缓蚀剂腐蚀速率 g/（$m^2 \cdot h$）。

实验前用电子天平准确称量（精确至 0.0001g），测量各试片表面积。实验后，将试样浸于50℃除锈液（500mL盐酸+六次甲基四胺+500mL蒸馏水）中直至锈除尽，取出后用酒精清洗、滤纸擦干、丙酮除油，自然晾干后用电子天平准确称其质量。计算腐蚀速率及缓蚀率。

（3）用日本日立S-3400型扫描电子显微镜对浸泡腐蚀后试样表面形貌进行观察，为腐蚀机理提供参考。

2 结果与分析

2.1 静态电化学行为

20钢和P110钢在添加缓蚀剂条件下的极化曲线如图2所示。由图2可知，20钢和P110钢在油田采出液中呈现出不同的极化行为，20钢阳极呈现出活性溶解特征，而P110钢阳极则出现明显的钝化行为，这主要与P110钢中的合金元素所形成的钝化膜有关，致密钝化膜的形成可以抑制表面腐蚀的进一步发生。相对空白试验而言，添加缓蚀剂后，无论对活性溶解的20钢，还是发生钝化的P110钢，体系的自腐蚀电位均显著正移，而相应腐蚀电流则明显下降。

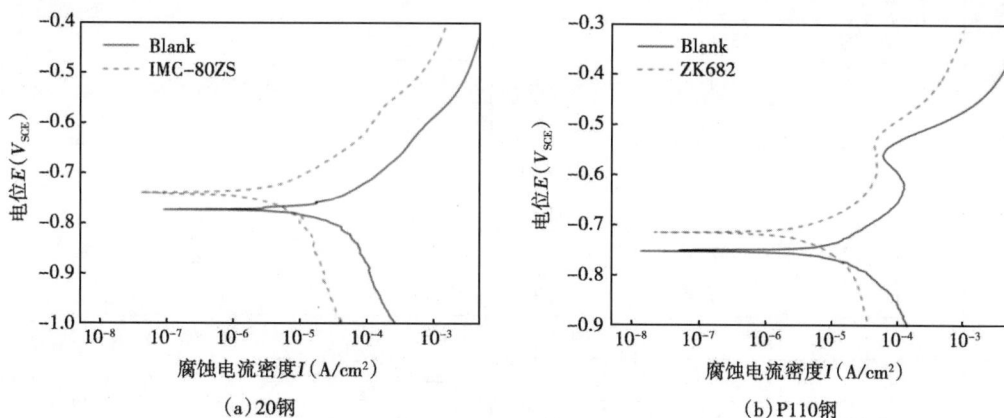

（a）20钢 　　　　　　　　（b）P110钢

图2　20和P110钢添加缓蚀剂条件下极化曲线图

腐蚀电位的升高说明钢的腐蚀倾向降低，而腐蚀电流密度的降低说明其腐蚀速率在添加缓蚀剂后受到抑制。

采用CView数据分析软件拟合分析极化曲线数据，所得腐蚀速率并计算缓蚀率，缓蚀剂IMC-80ZS对于20钢的缓蚀率超过94%，而缓蚀剂ZK682对于P110钢的缓蚀率则达到92%。

2.2 动态腐蚀和缓蚀性能

20钢和P110钢在高温高压反应釜中正交实验结果见表2。

表 2　正交实验结果表

	缓蚀剂质量浓度 A (mg/L)	介质流速 B (m/s)	温度 (℃)	空列	腐蚀速率 x_i (mm/a)	
					20 钢	P110 钢
1	1 (50)	1 (1)	1 (50)	1	0.07686	0.06549
2	1 (50)	2 (2)	2 (70)	2	0.07221	0.05385
3	1 (50)	3 (3)	3 (90)	3	0.05538	0.05668
4	2 (100)	1 (1)	2 (70)	3	0.04513	0.04728
5	2 (100)	2 (2)	3 (90)	1	0.03529	0.03894
6	2 (100)	3 (3)	1 (50)	2	0.04268	0.04328
7	3 (150)	1 (1)	3 (90)	2	0.03795	0.03567
8	3 (150)	2 (2)	1 (50)	3	0.04898	0.04486
9	3 (150)	3 (3)	2 (70)	1	0.03894	0.03689
20 钢计算结果						
K_{1j}	204.45	159.94	168.52	151.09	$T = \sum_{i=1}^{9} x_i = 453.42$	
K_{2j}	123.10	156.48	156.28	152.84		
K_{3j}	125.87	137.00	128.62	149.49		
K_1	68.15	53.31	56.17	50.36	$CT = T^2/9 = 22843.30$	
K_2	40.03	52.16	52.09	50.95		
K_3	41.96	45.67	42.87	49.83		
R	28.12	7.64	13.30	1.12		
K_{1j}^2	41799.80	25580.80	28398.99	22828.19	$S_j = 1/3 \ (K_{1j}^2 + K_{2j}^2 + K_{3j}^2) \ - T^2/9$	
K_{2j}^2	15153.61	24485.99	24423.44	23360.07		
K_{3j}^2	15843.26	18769.00	16543.10	22347.26		
P110 钢计算结果						
K_{1j}	176.02	148.44	153.63	141.32	$T = \sum_{i=1}^{9} x_i = 422.94$	
K_{2j}	129.50	137.65	138.02	132.80		
K_{3j}	117.42	136.85	131.29	148.82		
K_1	58.67	49.48	51.21	47.11	$CT = T^2/9 = 19875.36$	
K_2	43.17	45.88	46.01	44.27		
K_3	39.14	45.62	43.76	49.61		
R	19.53	3.86	7.45	5.34		
K_{1j}^2	30983.04	22034.43	23602.18	19971.34	$S_j = 1/3 \ (K_{1j}^2 + K_{2j}^2 + K_{3j}^2) \ - T^2/9$	
K_{2j}^2	16770.25	18947.52	19049.52	17635.84		
K_{3j}^2	13787.46	18727.92	17237.06	22147.39		

用 K_{1j}、K_{2j}、K_{3j} 表示各因素取第 1、2、3 水平时相应的试验结果之和，用 K_1、K_2、K_3 为各因素相应水平的综合平均值。R 为极差(各因素最大 K 值与最小 K 值之差)，$K_{1j} + K_{2j} + K_{3j} = T$。

对于 20 钢，根据表 3 实验结果，直接分析最优组合为 $A_2B_2C_3$，此条件在 5 组实验中腐蚀速率最小。通过计算最优组合为 $A_2B_3C_3$，以 $A_2B_3C_3$ 组合重做一次实验，得到腐蚀速率为 0.03748mm/a，高于 $A_2B_2C_3$ 组合。因此，可以确定最优组合为 $A_2B_2C_3$，即缓蚀剂 IMC-

80ZS 在质量浓度为 100mg/L、介质流速为 2m/s、90℃时 20 钢缓蚀效果最好。

表 3　20 钢方差分析结果表

方差来源	偏差平方和 S	自由度 f	平均偏差平方和 V（S/f）	F 值 （V/V_D）	显著性
A	$S_A = 1422.26$	$f_A = 2$	$V_A = 711.13$	$F_A = 756.52$	＊＊
B	$S_B = 101.96$	$f_B = 2$	$V_B = 50.98$	$F_B = 52.02$	＊
C	$S_C = 278.54$	$f_C = 2$	$V_C = 139.27$	$F_C = 148.16$	＊＊
误差 D	$S_D = 1.87$	$f_D = 2$	$V_D = 0.94$		
总和 T	$S_T = 1802.76$	$f_T = 8$			

用显著水平 a 表示某种判断时犯错误的概率，a 值不同，与其对应的 F 值分布表也不同，将实际的 F 值与临界值 Fa 作比较以评价其对指标影响的显著性。

评价方案如下：

（1）当 $F > F0.01$（f_1，f_2）时，表明此因素高度显著，记作"＊＊"；

（2）当 $F0.01$（f_1，f_2）$> F ≥ F0.05$（f_1，f_2）时，表明此因素显著，记作"＊"；

（3）当 $F0.05$（f_1，f_2）$> F ≥ F0.10$（f_1，f_2）时，表明此因素有一定影响，记作"＊"；

（4）当 $F0.10$（f_1，f_2）$> F$ 时，表明此因素无显著影响，即认为因素各水平的效应为零。

当 $a = 0.01$、0.05、0.10 时查 F 分布表，得 $F0.01$（2，2）= 99、$F0.05$（2，2）= 19、$F0.1$（2，2）= 9。可知 F_A、F_C 均大于 $F0.01$（2，2）= 99，所以 A、C 均为高度显著因素。

$F0.01$（2，2）大于 $F_B ≥ F0.05$（2，2）可知 B 为显著因素，均为高度显著因素。

通过方差分析（表 3）可知缓蚀剂的浓度和温度变化是影响缓蚀率的主要因素，介质流速次之，说明缓蚀剂吸附性较好，表面覆盖膜随流速的改变变化不大。因此，可以确定因素 20 钢缓蚀性能的主次顺序为：A > C > B，即缓蚀剂浓度 > 温度 > 流速。

对于 P110 钢，由表 3 直接分析最优组合为 $A_3B_1C_3$，此条件在 7 组实验中腐蚀速率最小。通过计算最优组合为 $A_3B_3C_3$，以 $A_3B_3C_3$ 组合再做一次实验，得到腐蚀速率为 0.03125mm/a，小于 $A_3B_1C_3$ 组合。可以确定最优组合为 $A_3B_3C_3$，即缓蚀剂 ZK682 在质量浓度为 150mg/L、介质流速在 3m/s、90℃时 P110 钢缓蚀效果最好。

同样，根据方差分析，由表 4 分析结果可知，$F0.05$（2，4）= 6.94、$F0.01$（2，4）= 18.0、$F0.10$（2，4）= 4.32。由于 $F_A ≥ F0.01$（2，4），A 为高度显著因素，因此缓蚀剂浓度对缓蚀剂的缓蚀效率影响较大。由 $F0.10$（2，4）$> F_C$ 可知 C 因素无显著影响。因此，缓蚀剂浓度是影响 P110 钢缓蚀率的主要因素，温度和流速影响较小。

表 4　P110 钢方差分析结果表

方差来源	偏差平方和 S	自由度 f	平均偏差平方和 V（S/f）	F 值 （V/V_D）	显著性
A	$S_A = 638.22$	$f_A = 2$	$V_A = 319.11$	$F_A = 18.04$	＊＊
B^Δ	$S_B = 27.93$	$f_B = 2$	$V_B = 13.97$		
C	$S_C = 87.56$	$f_C = 2$	$V_C = 43.78$	$F_C = 2.47$	
误差 D	$S_D = 42.83$	$f_D = 2$	$V_D = 21.42$		
误差 D^Δ	$S_D^\Delta = 70.76$	$f_D^\Delta = 4$	$V_D^\Delta = 17.69$		
总和 T	$S_T = 796.54$	$f_T = 8$			

2.3 腐蚀形貌观察

对于 20 钢和 P110 钢，9 组实验中分别选取腐蚀速度最大和腐蚀速度最小的表面进行分析。根据表 3 的结果，20 钢选取第 1 组（$A_1B_1C_1$）和第 5 组（最优组合 $A_2B_2C_3$），如图 3 所示；而 P110 钢选取第 1 组（$A_1B_1C_1$）和最优组合（$A_3B_3C_3$），如图 4 所示。

| （a）第1组（$A_1B_1C_1$） | （b）第5组（$A_2B_2C_3$） |

图 3　20 钢腐蚀形貌

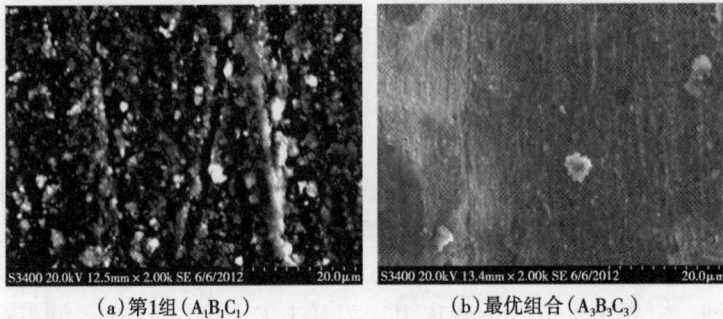

| （a）第1组（$A_1B_1C_1$） | （b）最优组合（$A_3B_3C_3$） |

图 4　P110 钢腐蚀形貌

由图 3 可看出，20 钢第 5 组（最优组合 $A_2B_2C_3$）的表面腐蚀较轻微，腐蚀产物膜层较均匀致密，可以有效保护基体进一步腐蚀；而第 1 组（$A_1B_1C_1$）腐蚀产物疏松，出现了大块或颗粒状的腐蚀产物，疏松的表面腐蚀产物易脱落，对金属表面保护性差，腐蚀较严重。

同理，由图 4 也可分析出，最优组合 $A_3B_3C_3$ 表面的腐蚀产物较致密、腐蚀较轻微。而第 1 组（$A_1B_1C_1$）表面腐蚀严重，腐蚀产物疏松且多为小颗粒状。另外，在膜层局部还出现明显较深的裂纹，这主要是由于疏松的腐蚀产物膜随着时间的增加，表面厚度不均引起应力集中，导致膜层开裂。一旦膜层开裂，腐蚀介质会迅速进入狭窄的裂缝，使缝内金属与缝外金属构成短路原电池。由于缝隙内部贫氧，氧的还原反应主要在缝隙外部的金属表面上进行，缝隙内部金属溶解产生了过多的正离子（Fe^{2+}），为维持电平衡，Cl^- 等有害离子从外部迁入缝内，使缝内的金属氯化物浓度增加，金属氯化物水解产生游离酸（H+），从而加速了金属的溶解速率，形成一个自催化过程，最终导致缝隙内发生强烈的腐蚀[5]。因此，腐蚀速率要远高于表面膜层致密的金属。

2.4 缓蚀机理分析

从图 3 和图 4 可以看出，加入缓蚀剂后，当缓蚀剂浓度和环境参数达到最佳时，可以实现金属表面腐蚀膜层均匀致密，大大降低腐蚀程度，这主要与缓蚀机理相关。缓蚀剂 IMC-80-ZS 主要成分为炔氧甲基季铵盐，缓蚀剂 ZK682 主要成分为有机胺类化合物，两者均为吸附型缓蚀剂。这两种缓蚀剂的中心原子同为氮原子。由于金属表面存在晶格缺陷使得金属表面存在大量的未耦合的 Fe 原子的 d 轨道。而 N 原子上有较多的孤对电子，两者接触时这些孤对电子会进入金属表面 d 轨道中，形成相对稳定的配位键。缓蚀剂吸附在金属表面从而提高了金属腐蚀液中的阳极反应的活化能，增大了金属腐蚀反应的能量壁垒，因此能降低金属阳极反应的腐蚀速率❶。当金属表面吸附了此类缓蚀剂后，可使金属表面能量状态趋于稳定。此外缓蚀剂中的非极性基排列在金属表面形成疏水薄膜，可以阻挡电荷的移动，从而使得腐蚀反应受到抑制。

3 结论

（1）20 钢和 P110 钢呈现出不同的静态电化学行为，20 钢呈现出活性溶解，而 P110 钢则发生钝化。缓蚀率均高于 92%。

（2）影响 20 钢和 P110 钢动态缓蚀效果的因素为缓蚀剂浓度、温度、介质流速（按影响力大小排序）。缓蚀剂浓度和介质流速为影响 20 钢显著因素，而浓度影响 P110 钢较为显著。

（3）两种缓蚀剂均为吸附型，配位键吸附增大了腐蚀反应能量壁垒，极大降低了阳极反应的腐蚀速率。

参 考 文 献

[1] 刘忠运，李莉娜. CO_2 驱油机理及应用现状 [J]. 技术创新，2009, 6 (10)：12-19.

[2] 王彬，杜敏，张静. 油气田中抑制 CO_2 腐蚀缓蚀剂的应用及其研究进展. 腐蚀与防护，2010, 31 (7)：503-507

[3] Jiang X, Zheng Y, Ke W. Corrosion inhibitor performances for carbon dioxide corrosion of N80 steel under static and flowing conditios [J]. Corrosion, 2005, 61 (4)：326~334.

[4] 毕凤琴，张成果，徐子怡，等. 不同类型缓蚀剂在含饱和 CO_2 油田采出液中对 P110 钢的缓蚀性能研究 [J]. 兵器材料科学与工程，2013, 36 (2)：9-12.

[5] 龚敏. 金属腐蚀理论及腐蚀控制 [M]. 北京：化学工业出版社，2012：113-116.

---

❶ 李芳. CO_2 驱油集输系统高效缓蚀剂的筛选及缓蚀性能研究，东北石油大学硕士学位论文，2012.

Sn 对 Q125 级 ERW 套管力学性能及腐蚀行为的影响

王立东　唐　荻　武会宾　蔡正旭

（北京科技大学高效轧制国家工程研究中心）

摘　要：为了分析 Sn 元素对 Q125 级 ERW 石油套管用钢力学性能和抗 CO_2 腐蚀性能的影响，采用扫描电子显微镜、透射电子显微镜、电子背散射衍射和电化学方法等手段对其进行了研究。结果表明：加入 0.05% 的 Sn 使得实验钢的耐腐蚀性能提高，但延伸率和冲击功下降显著，强度变化不明显. 少量 Sn 的加入改变了钢的活性，引起自腐蚀电位负移，腐蚀电流密度减小，同时实验钢的电极反应的极化电阻增大，腐蚀反应变慢，耐腐蚀性能得到提高；Sn 的加入使钢中小角度晶界的比例有所增加，提高了实验钢的耐腐蚀性，相反，大角度晶界比例减少，这使得延伸率和冲击功下降显著；但 Sn 的加入对钢的显微结构、析出物形态和尺寸均没有产生影响，因此强度变化不明显。

关键词：ERW 套管；腐蚀行为；大角度晶界；极化曲线

0　引言

套管是开采石油天然气必需的工程用具。随着"西部大开发"的深入，采油条件越来越恶劣，深井、超深井的开发量加大，对石油套管的性能提出了更高的要求[1]。随着 CO_2 驱油技术大量应用于深井、超深井的开发，CO_2 腐蚀给石油天然气工业造成了巨大损失[2]。因此，要求石油套管钢不仅具有优良的力学性能，而且还应具备一定的耐腐蚀性。目前，在研究 CO_2 腐蚀方面已经进行了大量的工作，但大多研究都集中于各项环境参数对腐蚀速率的影响以及合金元素对腐蚀产物膜的构成的影响[3-5]，而对于钢材本身抗 CO_2 腐蚀的文献相对较少。在钢铁冶炼过程中，Sn 为残留元素，不易去除，且会在晶界偏聚从而对钢的力学性能不利。但 Sn 元素可以削弱晶界结合力，减小晶界能，并使晶界扩散系数降低[6]❶，从而影响钢的腐蚀性能，但这方面的研究还未见报道。Q125 级是 API 5L 标准中强度级别最高的，其中对 Q125 钢性能的要求：屈服强度（$R_{t0.65}$）= 862～1034 MPa，抗拉强度（R_m）$\geqslant$ 931 MPa，延伸率（A）$\geqslant$14%，横向冲击功（0℃）$\geqslant$20J，纵向冲击功（0℃）$\geqslant$41J。目前，关于 Q125 级别石油套管的相关文献研究极少[7]，而在系统地研究 ERW 用 Q125 级石油套管钢的组织、力学性能及其耐腐蚀性能方面尚属空白。

文中选用自行研发的 ERW 用 Q125 级石油套管钢，通过透射电子显微镜（TEM）、电子背散射衍射（EBSD）、电化学等手段分析 Sn 元素对 Q125 级石油套管钢力学性能和抗 CO_2 腐蚀性能的影响。本研究可以为高强耐腐蚀钢种的开发以及 CO_2 驱油工程中深井、超深井

---

❶ 贾涓：《Sn 的晶界偏聚行为与低碳钢的热塑性》，学位论文，武汉科技大学，2003 年。

石油套管钢的腐蚀与防护提供理论依据。

1　实验材料和方法

实验材料的化学成分见表 1。实验用材料在 22kg 的真空感应电炉中冶炼，将其锻造成 80（mm）×80（mm）×90（mm）的尺寸，然后在加热炉中加热至 1250℃并保温 2h，取出在热轧机上经过粗轧和精轧两阶段控轧，轧至 9mm。热轧工艺参数：开轧温度 1150℃，终轧温度≥850℃，轧后的板材经过层流冷却至 620℃，放入加热炉中模拟卷曲，保温 2h 后空冷到室温。调质工艺：870℃淬火+520℃回火。为了满足性能的均匀性，淬火、回火的保温时间均为 1h。之后，从板材上切取金相试样进行打磨抛光，并采用 4%的硝酸酒精溶液进行浸蚀，在 ZEISS ULTRA 55 热场发射扫描电子显微镜（SEM）下进行组织观察及 EBSD 研究。切取尺寸为 10（mm）×10（mm）×0.3（mm）的试样，用砂纸打磨至 50μm。用酒精清洗后，在打孔器上打取 φ3mm 的样品，用 5%的高氯酸酒精溶液对样品进行电解双喷。萃取复型实验是将试样抛光后利用 4%硝酸酒精溶液浸蚀，然后在真空喷碳仪中喷一层碳膜，用 5%高氯酸酒精溶液进行电解剥离。采用 JEM-2000FX 型透射电镜对双喷后的样品和复型的碳膜进行形貌、析出物观察。

表 1　实验用 Q125 钢的化学成分表 *

钢	C	Si	Mn	Als	Cr	Nb	Ti	Ni+Cu+Mo	Sn
A 钢	0.19	0.27	1.66	0.03	0.28	0.06	0.02	≤0.5	—
B 钢	0.19	0.26	1.65	0.03	0.29	0.06	0.021	≤0.5	0.05

*：表中数据表示化学成分的质量分数。

腐蚀试样规格为直径 87mm、弧长 30mm、面宽 12mm、厚度 4mm 的弧形试样，如图 1 所示。经过 800# 砂纸打磨后，用丙酮除掉试样表面的油污，并用酒精清洗，然后用精度 0.1 mg 的电子分析天平称量试样的质量。实验溶液在实验前用 CO_2 除氧 10h，倒入反应釜后再除氧 2h 后升温，并调整 CO_2 压力和流速。每组实验采用 4 个平行试样，实验结束后取出试样，经清水酒精冲洗吹干拍照；然后将其中 3 个试样放入除锈液中，去除腐蚀产物膜后清洗干燥并称重。每种材料保留 1 个带腐蚀产物膜的试样，用扫描电子显微镜对试样表面形貌进行观察。

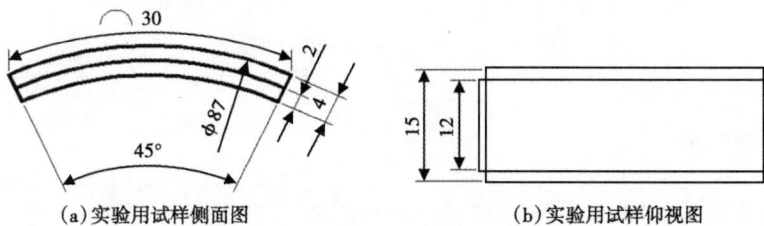

（a）实验用试样侧面图　　　　　　　（b）实验用试样仰视图

图 1　高温高压模拟实验用试样示意图（单位：mm）

抗 CO_2 腐蚀实验在 3L 的高温高压反应釜中进行，模拟采出液离子包括：Cl^-、SO_4^{2-}、HCO_3^-、Mg^{2+}、Ca^{2+} 和 Na^+，质量浓度分别为 1000mg/L、5450mg/L、600mg/L、1050mg/L、880mg/L 和 6320mg/L。环境参数设置如下：CO_2 分压为 1MPa，温度为 30℃，流速为 1m/s，

实验周期为 7d，通过失重法测得实验钢的腐蚀速率。采用 PAR2273 电化学测试仪在模拟采出液中测试极化曲线，辅助电极为铂电极，参比电极为饱和甘汞电极，测试面积为 $1cm^2$。测试电位范围为 $-500\sim600mV$，扫描速率为 $0.5mV/s$。

2 实验结果

2.1 力学性能

实验钢是以 Nb、Ti 和 Mo 为典型合金成分的 ERW 用石油套管钢，通过控轧控冷工艺（TMCP）+调质处理来实现其较好的强韧性配合。拉伸实验每种钢做 3 根，强度和延伸率取 3 根钢拉伸后所得数值的平均值。横向、纵向冲击功各做 5 个，去掉最大值和最小值而保留其余 3 个数值，所得力学性能见表 2。由表 2 可知，实验钢屈服强度、抗拉强度、延伸率、横/纵向冲击功，达到了 API 5L 标准中对 Q125 钢级的要求，具有较好的综合力学性能。与 A 钢相比，加入 Sn 元素的 B 钢强度没有产生明显变化，但延伸率和冲击韧性下降明显。

2.2 显微组织

图 2 为两种实验钢显微组织的 SEM 照片。由图 2 可知，实验钢组织均为典型回火马氏体组织。实验钢显微组织中板条特征比较明显，析出的细小碳化物主要在马氏体晶界或原奥氏体晶界聚集。在回火过程中，过饱和的碳原子从固溶体中不断析出，在马氏体板条内部、板条边缘和奥氏体晶界上大量弥散着析出碳化物质点。这些碳化物质点对位错运动阻碍作用非常明显，会使钢的强度大幅度提高[8]。

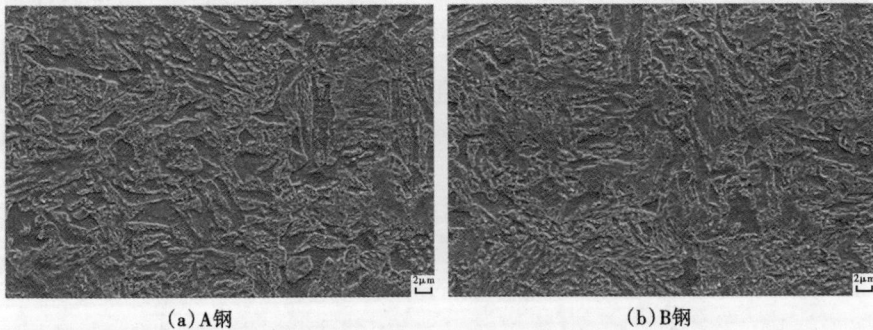

(a) A钢   (b) B钢

图 2　实验钢显微组织 SEM 照片

2.3 腐蚀实验结果

A 钢在采出液环境下的平均腐蚀速率和点蚀速率分别为 4.6348mm/a 和 4.8528 mm/a，而 B 钢的平均腐蚀速率和点蚀速率分别为 3.4439mm/a 和 4.4641mm/a。由此看出，加入 0.05% 的 Sn 后，实验钢的平均腐蚀速率和点蚀速率降低都比较明显，平均腐蚀速率由 4.6348mm/a 降到 3.4439mm/a，降幅为 25.7%；点蚀速率由 4.8528mm/a 降至 4.4641mm/a，降幅为 8.0%。图 3（a）、3（b）分别为腐蚀产物膜去除前后试样的宏观腐蚀形貌。从图 3（a）可以看出，两种实验钢腐蚀产物膜均比较致密，并且和基体的附着性好，局部腐蚀产物膜未发生脱落。从图3（b）上还可观察到去除腐蚀产物后，实验钢表面没有明显的点蚀坑，腐蚀较均匀。图4为腐蚀产物膜表面的微观形貌。由图4可知，两种实验钢表面腐蚀产物颗粒尺寸大小相同，腐蚀产物膜结构没有太大差异。腐蚀环境相同，腐蚀产物膜并没有太大差异，因此可以断定：实验钢腐蚀速率的降低只跟钢种本身有关，即Sn的加入使得腐蚀速率降低。

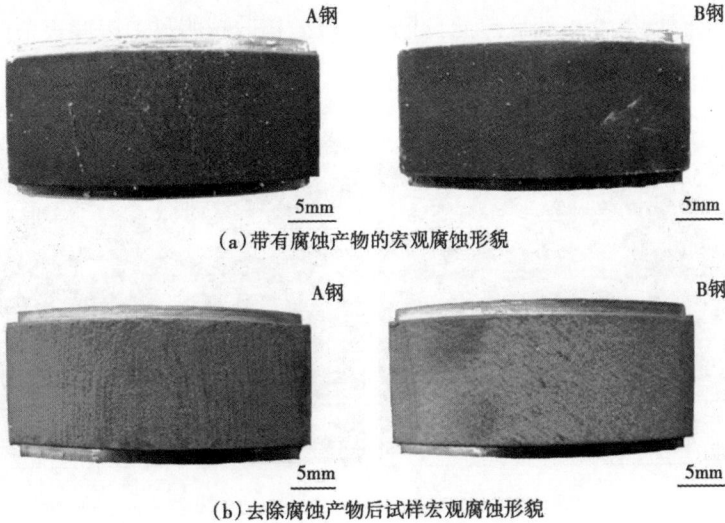

A钢   B钢

5mm   5mm

（a）带有腐蚀产物的宏观腐蚀形貌

A钢   B钢

5mm   5mm

（b）去除腐蚀产物后试样宏观腐蚀形貌

图3　去除腐蚀产物前后试样宏观腐蚀形貌

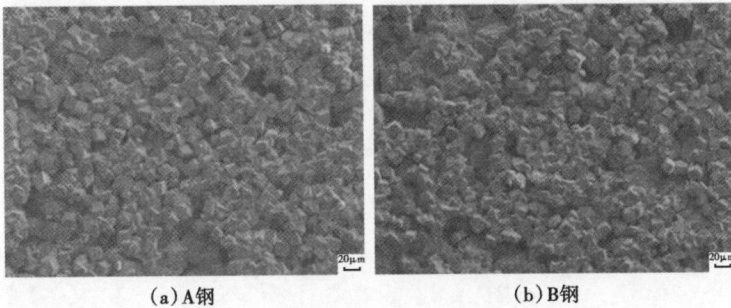

（a）A钢   （b）B钢

图4　腐蚀产物膜表面形貌

3　实验结果分析

3.1　力学性能分析

回火马氏体钢晶粒大小一般以原奥氏体晶粒的尺寸为准。在原奥氏体晶粒中，马氏体亚结构共分为3个层次[9]：晶区、板条束和板条。马氏体亚结构的尺寸由原奥氏体晶粒大小控制的。奥氏体晶粒越小，则晶区、板条束和板条的尺寸也就越小。马氏体内亚结构的大小直接影响回火马氏体钢的性能，晶区和板条束的尺寸是决定马氏体钢的强度和韧性直接因素[10-11]。有相关研究表明，晶区和板条束的尺寸对马氏体钢的强度的影响也呈Hall-Petch关系。

对两种实验钢的 EBSD 分析表明（图 5），A 钢、B 钢的晶区尺寸约为 2~5μm，二者差别不明显．图 7 为两种实验钢显微组织的 TEM 照片，可见板条状形貌非常明显，高度发达的位错相互缠结在一起，保证了钢的强度水平。由图 7 还可以看出，两种实验钢的板条束宽度大约为 70nm，尺寸相当。可见，Sn 的加入对实验钢微观组织影响不明显。但从图 7 上可以看出，与 A 钢相比，加入 Sn 元素的 B 钢板条界限变得有些模糊。通过测定静态相变曲线，A 钢的相变温度（Ac$_3$）温度为 837℃，而 B 钢的为 820℃。可见 Sn 的加入使相变点降低。在相同的回火温度下，相变点越低，板条界限处扩散越明显，因此板条形貌越不容易被保持。

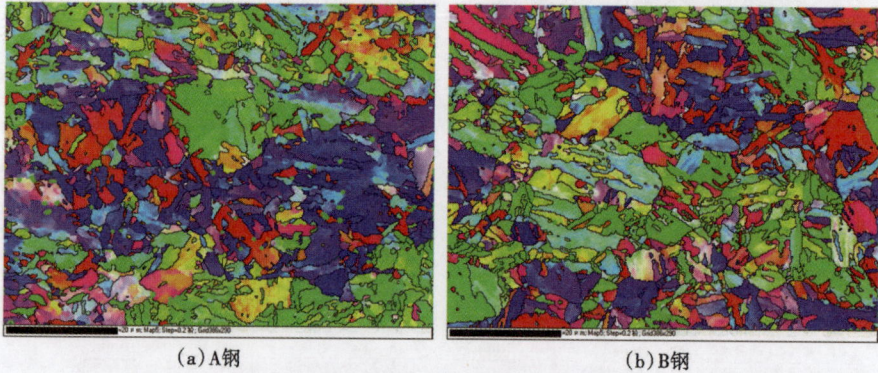

(a)A 钢　　　　　　　　(b)B 钢

图 5　实验钢取向成像显微图

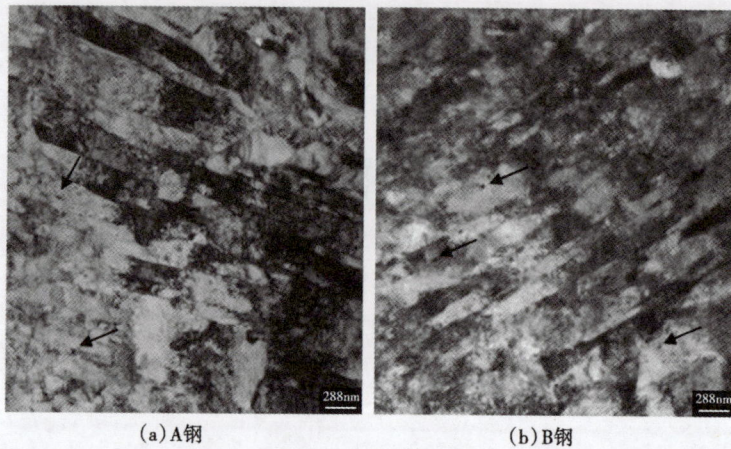

(a)A 钢　　　　　　　　(b)B 钢

图 6　实验钢显微组织的 TEM 照片

在两种钢的 TEM 照片上还可以发现许多细小的析出物（图 6 箭头所指）。图 7 是在透射电子显微镜下对实验钢的析出物进行萃取析出的结果。由图 7 可以看出，Sn 加入前后，析出物尺寸和形态没有发生明显的变化。析出物形态主要为椭圆形（红色箭头所指）和细小圆形（黑色箭头所指），棱角不分明．析出物较均匀地分布在晶粒内部，尺寸大部分小于30nm。通过对析出物能谱分析（图 8），发现析出物主要以（Nb，Ti）（C，N）、Nb（C，N）为主。在抗 CO_2 腐蚀钢中，通过添加 Nb、Ti 等合金元素可以促进晶粒的细化和细小碳氮化物的形成，以提高钢的强度和韧性[12]。在 TMCP 过程中，随着变形温度的降低和变形程度的增加，因微合金元素在钢板中的固溶度逐渐下降，所以应变能逐渐增加，在位错线或应变带析出的碳氮化物粒子不断增多，呈细小弥散状态，能够起到有效的强化作用并改善钢的冲击韧性[13]。实验钢在淬火极冷的条件下，较大的相变体积差产生了大量的位错，当位错在运动时，这些细小的椭圆形和细小圆形析出物可以起到钉扎位错的作用，阻止位错移动，对强度的提高贡献很大。EDS 能谱分析发现，A、B 钢中均未发现 Sn 的析出物。Sn 在钢中的存在形式有两种：一种是以第二相的形式存在；另一种则是以固溶的形式存在于钢中。通过对析出物的分析发现，析出物中不存在 Sn 元素，从而证明其是以固溶的形式存在于钢中的。

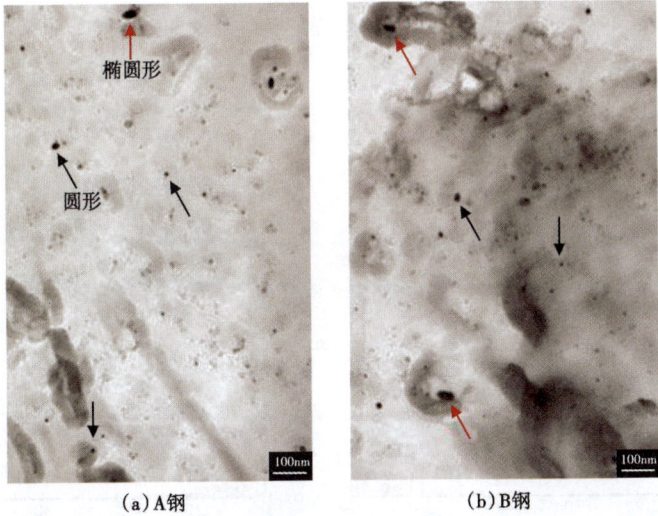

图 7　实验钢析出物 TEM 照片

（a）A钢　　　　（b）B钢

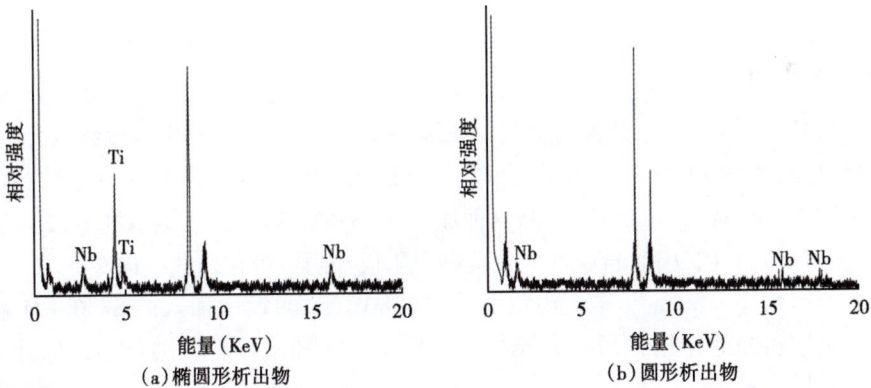

图 8　析出物 EDS 能谱图

（a）椭圆形析出物　　　　（b）圆形析出物

图 9 为两种实验钢的取向差分布图。一般而言，板条界面大多为小角度晶界，而相邻板条束之间以及晶区间的界面为大角度晶界（>15°）。从断裂角度分析，小角度晶界对裂纹扩展的阻力较小，大角度晶界与原奥氏体晶界类似，能有效改变裂纹扩展的方向，大角度晶界密度越高的组织也就能更有效地抑制裂纹扩展。从图 9 的统计结果中可看出，A 钢的大角度

361

晶界的比例约为 35.6%，而 B 钢的大角度晶界的比例约占 24.2%。这是因为：Sn 的存在可以降低钢的层错能。在上述分析中，析出物中未见 Sn，即 Sn 全部固溶于基体中，这使得钢保持了较低的层错能。而 Sn 在高温回火过程中又在晶界处偏聚，进而造成钢大角度晶界降低。Sn 的加入引起的实验钢大角度晶界比例的降低，使实验钢在断裂过程中的裂纹扩展得不到更多大角度晶界的阻碍，因此表现为 B 钢的延伸率和冲击功与 A 钢相比有大幅度的下降；同时，腐蚀一般容易在晶界、位错等表面结构的不均匀处发生。而在各种晶界中，大角度晶界的耐腐蚀性最差。大角度晶界能量较高，此处原子活性较大，反应速度常数较大，使基体金属反应速度增加。小角度晶界结构有序度高，自由体积小，界面能量低，具有较强的晶界失效抗力，能打断大角度晶界网络的连通性，可以有效地阻断材料沿大角度晶界腐蚀行为的连续扩展。因此，Sn 元素的加入使小角度晶界比例增加，是 B 钢腐蚀速率降低的一个原因。

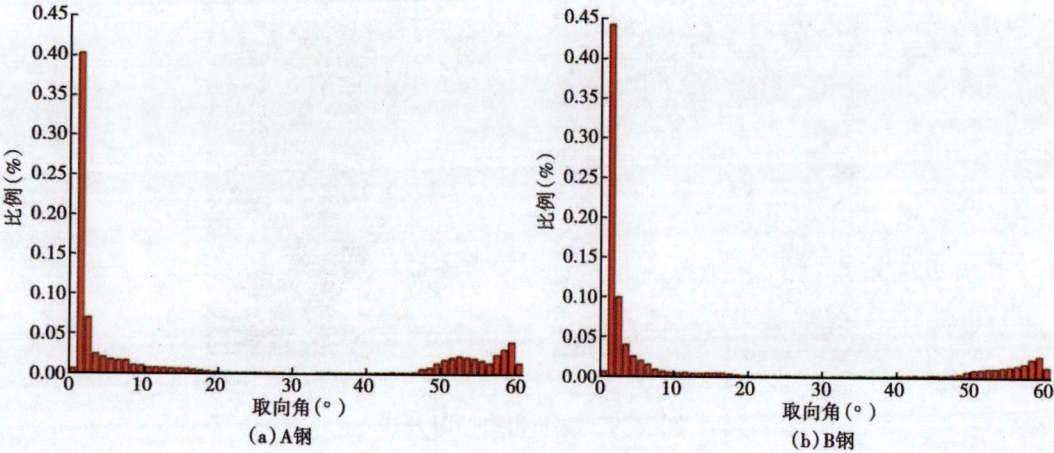

图 9　实验钢取向差分布图

3.2　电化学分析

A 钢、B 钢在溶液中的极化曲线测量结果如图 10 所示，通过电化学参数拟合结果得出：A 钢在溶液中的腐蚀电流密度和电流电位分别为 $20.1\mu A/cm^2$ 和 $-736mV$，B 钢的为 $20.1\mu A/cm^2$ 和 $-740mV$。可以看出，当钢种加入 0.05% 的 Sn 后，B 钢腐蚀电位由于少量 Sn 的加入改变了钢的活性，引起自腐蚀电位负移，阳极和阴极极化曲线都向左移动，腐蚀电流密度急剧减小至 $8.3\ \mu A/cm^2$，电流电位变为 $-740mV$。说明钢中加入的 Sn 在 CO_2 腐蚀中一方面起到抑制钢腐蚀的作用，另一方面起到了抑制阴极和阳极反应的作用，降低了腐蚀速率。大角度晶界能量较高，腐蚀容易在此处发生。随着腐蚀时间的延长，腐蚀失效行为将沿着大角度晶界不断地扩展，直至相互连通、整个晶粒被腐蚀。Sn 的加入，使得大角度晶界比例减小，这样就降低了腐蚀先发生的几率，同时也延长了失效的大角度晶界连通的时间，延缓了腐蚀。因此，Sn 的加入会对腐蚀反应的进行以及腐蚀路径的扩展有所影响，进而影响实验钢的腐蚀速率以及腐蚀产物的沉积速度。

A 钢、B 钢在溶液中的电化学交流阻抗谱 Nyquist 图和等效电路图如图 11 所示。Z_{re} 代表阻抗实部，Z_{im} 代表阻抗虚部。R_s 是溶液电阻，R_t 是电荷转移电阻，R_l 是电感元件的电阻，

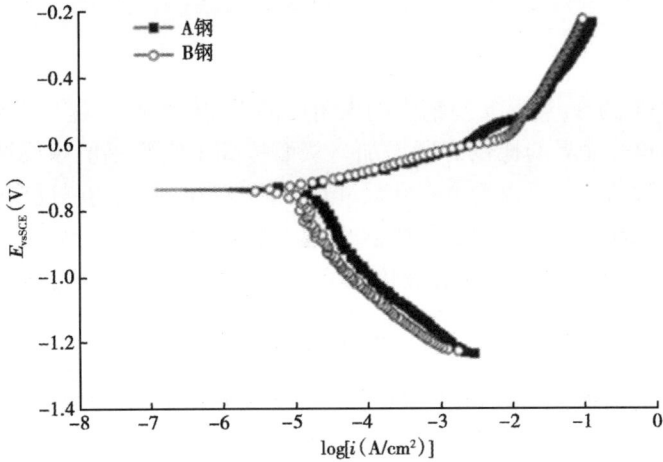

图 10　实验钢在采出液中的极化曲线图

L 是感抗，Q_{dl} 代表双电层电容。由 Nyquist 图可知，A 钢、B 钢在腐蚀环境下均呈现 3 个时间常数，在测试初期都形成较大的容抗弧，在中频区形成很小的感抗弧，又在低频区形成不明显的容抗弧，即高频容抗弧、中频感抗弧和低频容抗弧。高频容抗弧与双层电层电容和电荷传递电阻有关，中频感抗弧与吸附中间产物有关，而低频容抗弧与表面腐蚀产物膜沉积有关。与 A 钢相比，B 钢的高频容抗弧半径逐渐增大，表明双层电容 Q_{dl} 逐渐减小，电荷传递电阻 R_t 增大。根据等效电路对阻抗谱的拟合，所得 A 钢、B 钢的极化电阻分别为 720、815 $\Omega \cdot cm^2$，表明 A 钢的电极反应的极化阻力最小，腐蚀反应易于发生；而 B 钢的电极反应的极化电阻较大，腐蚀反应不易进行，因此具有较好的耐 CO_2 腐蚀性能，这与高温高压反应釜所得得实验结果及极化曲线分析结果一致。

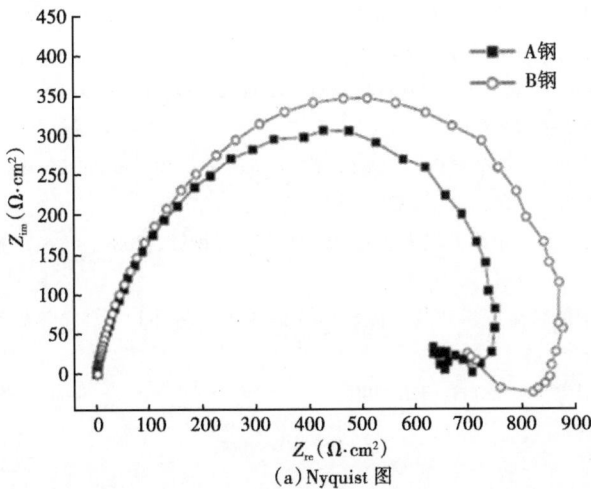

（a）Nyquist 图

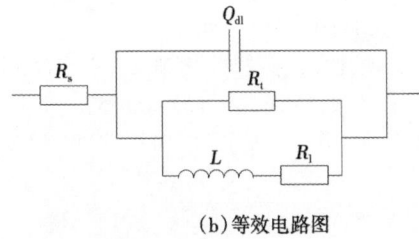

（b）等效电路图

图 11　实验钢在采出液中的交流阻抗谱和等效电路图

4 结论

(1) 加入 0.05% 的 Sn 后，实验钢中的大角度晶界比例减少，这使得延伸率和冲击功下降显著；相反，小角度晶界的比例有所增加。从而提高了实验钢的耐腐蚀性。然而，Sn 的加入对实验钢的显微结构、析出物形态和尺寸均未产生影响，钢的强度变化不明显。

(2) 少量 Sn 的加入改变了钢的活性，引起自腐蚀电位负移，腐蚀电流密度减小；同时，加入 Sn 后的实验钢的电极反应的极化电阻较大，腐蚀反应不易进行。

参 考 文 献

[1] 李亚欣，刘雅政，赵金锋，等. P110 级 25MnV 钢石油套管热处理工艺的优化 [J]. 特殊钢, 2009, 30 (6): 36-38.

[2] 李春福，王斌，代加林，等. P110 钢高温高压下 CO_2 腐蚀产物组织结构及电化学研究 [J]. 材料热处理学报, 2006, 27 (5): 73-78.

[3] Zhao Guo-xian, Lu Xiang-hong, Xiang Jian-min, et al. Formation characteristic of CO_2 corrosion product layer of P110 steel investigated by SEM and electrochemical Techniques [J]. Journal of Iron and Steel Research International, 2009, 16 (4): 89-94.

[4] 孙建波，柳伟，常炜，等. 低铬 X65 管线钢 CO_2 腐蚀产物膜的特征及形成机制 [J]. 金属学报, 2009, 45 (1): 84-90.

[5] 柳伟，陈东，路民旭. 不同 CO_2 分压下形成的 N80 钢腐蚀产物膜特征 [J]. 北京科技大学学报. 2010, 32 (2): 213-218.

[6] Yuan Z X, Jia J, Guo A M., et al. Cooling-induced tin segregation to grain boundaries in a low-carbon steel [J]. Scripta Materialia, 2003, 48: 203-206.

[7] 马爱清，郭兆成，贺景春. 30CrMnMo 钢调质 Q125 钢级套管的工艺研究 [J]. 包钢科技, 2009, 35 (1): 25-29.

[8] Qiu J, Ju X, Xin Y, et al. Effect of direct and reheated quenching on microstructure and mechanical properties of CLAM steel [J]. Journal of Nuclear Materials, 2010, 407: 189-194.

[9] Hiromoto Kitahara, Rintaro Ueji, Nobuhiro Tsuji, et al. Crystallographic features of lath martensite in low-carbon steel [J]. Acta Materialia, 2006, 54: 1279-1288.

[10] Morito S, Yoshida H, Maki T, et al. Effect of block size on the strength of lath martensite in low carbon steels [J]. Materials Science and Engineering A, 2006, 438-440: 237-240.

[11] 王春芳，王毛球，时捷，等. 低碳马氏体钢的微观组织及其对强度的影响 [J]. 钢铁, 2007, 42 (11): 57-60.

[12] 曹建春，雍岐龙，刘清友，等. 含铌钼钢中微合金碳氮化物沉淀析出及其强化机制 [J]. 材料热处理学报, 2006, 27 (5): 51-55.

[13] 翁宇庆. 超细晶钢 [M]. 北京：冶金工业出版社, 2003: 376-390.

温度对 X80 管线钢 CO_2 腐蚀行为的影响

张均生[1,2]　武会宾[1]　王立东[1]　蔡庆伍[1]

(1. 北京科技大学冶金工程研究院；2. 新余钢铁股份有限公司)

摘　要：在模拟实际工况下，利用高温高压反应釜对 X80 管线的 CO_2 腐蚀行为进行了研究，通过质量损失法、扫描电子显微镜（SEM）和 X—射线衍射（XRD）等分析手段，研究了温度对 X80 管线钢腐蚀性能的影响。结果表明：随着温度的升高，实验钢的平均腐蚀速率和点蚀速率均是先增大后减小。在温度为 30℃时，试样表面未形成完整的腐蚀产物膜，此时环境温度较低，平均腐蚀速率和点蚀速率也最小；在温度为 60℃时，平均腐蚀速率达到最大值，此时腐蚀产物膜脱落严重，但点蚀现象并不明显；当温度达到 90℃时，实验钢的点蚀速率达到最大值，并且点蚀速率与平均腐蚀速率相差程度最大；在温度为 120℃时，腐蚀产物膜与基体以及内外层之间的结合最为紧密，对基体的保护作用增强，所以此时的腐蚀速率比 60℃、90℃的腐蚀速率均低。

关键词：X80 管线钢；CO_2 腐蚀；点蚀速率；腐蚀速率

0　引言

随着社会进步及世界经济的发展，世界对石油资源的需求越来越大因此对石油开发、运输等所需的管线钢及配套资源的需求也日益增长。CO_2 作为伴生气体在石油、天然气的勘探和开采过程中同时产生，造成的 CO_2 腐蚀成为石油和天然气工业安全生产的主要问题[1-3]。特别是随着深层含油气层开发和"CO_2 驱油"增产技术的广泛应用，CO_2 腐蚀给石油天然气工业造成了巨大损失已成为制约管线钢在油气工业中应用的一个重要障碍[4]❶。这样不仅要求油气集输用管线钢具有优良的力学性能，而且还应具备一定的耐腐蚀性。

温度是管线钢 CO_2 腐蚀重要的影响因素之一，它对反应速度和溶液中溶解的 $FeCO_3$ 溶解度产生直接影响从而影响管线钢的腐蚀速率[5]。鉴于此，本文在模拟某油田现场采出液的环境中，研究了在不同温度下，管线钢腐蚀速率、腐蚀产物膜的结构、厚度的变化情况，研究结果对管线钢在 CO_2 驱油气田中的腐蚀防护具有理论意义和实际参考价值。

1　实验材料和方法

实验用钢选用 X80 管线钢，具体成分见表 1。将实验用钢加工成曲率 $\phi87mm$，弧长 30mm，面宽 12mm，厚度 4mm 的弧形试样；用 SiC 金相砂纸逐级打磨至 800#，清洗、除油、吹干；然后用精度为 0.1mg 的电子分析天平称量试样的重量，用精度 0.02mm 的游标卡尺测

---

❶ 张伟卫，《高级别管线钢抗 CO_2 腐蚀机理及生产工艺研究》，学位论文北京科技大学，2007 年。

量试样的尺寸，进行数据记录。而后将试样封装到试样夹具上。每种工况下准备4个平行试样，其中3个用失重法测量其腐蚀速率，剩余1个做表面分析与检测。

表1 实验用钢的化学成分表（质量分数,%）

钢号	C	Si	Mn	P	S	Ti	V+Nb+ Mo	Cr	Fe
含量	0.045	0.22	1.40	0.006	0.004	0.015	0.3	0.3	Bal

实验溶液介质的离子浓度见表2。根据管线钢的实际使用条件，本文中采用4种温度，分别为30℃、60℃、90℃、120℃，介质的相对流速均为2m/s，CO_2分压为1MPa。实验在高温高压反应釜中进行，实验溶液（其离子浓度见表2）在实验前用CO_2除氧10h后，倒入反应釜后再除氧2h后升温，到温后调整CO_2压力，开始计时，腐蚀实验周期为7天。

表2 实验溶液介质的离子浓度表（单位：mg/L）

种类	Cl^-	SO_4^{2-}	HCO_3^-	Mg^{2+}	Ca^{2+}	Na^+
含量	15000	5450	600	1050	880	9560

腐蚀实验结束后，取出试样，清洗吹干待用。将用于计算腐蚀速率的试样用除锈液清洗，脱水吹干之后称重，利用失重法计算出腐蚀速率，并对腐蚀产物进行观察。利用数码相机对腐蚀产物膜进行宏观观察，利用 ZEISS ULTRA 55 热场发射扫描电子显微镜对腐蚀产物膜表面进行微观形貌观察，利用日本理学（Rigaku）D/MAX—RB12KW 旋转阳极 X—射线衍射仪分析试样表面腐蚀产物的成分。

2 实验结果

实验钢分别在温度为30℃、60℃、90℃、120℃条件下的均匀腐蚀速率和点蚀速率见表3。可看出，随着温度的升高，平均腐蚀速率先增大后减小，在30℃时腐蚀速率最小，只有0.06mm/a，在60℃时平均腐蚀速率达到最大值。随着温度的增加，点蚀速率同样是呈现先增大后减小的趋势，最大值是在温度为90℃时，点蚀速率高达12.8310mm/a。图1为实验钢在不同温度下腐蚀产物的宏观形貌。从图1可以看出，在温度为30℃时，表面没有形成明显的腐蚀产物膜，试样表面呈现出金属的光泽，也没有明显的点蚀现象发生；温度为60℃时，局部腐蚀严重，腐蚀产物脱落较严重，但在脱落部分裸露的基体金属上没发现有明显的点蚀坑。Palacios 等[6]认为，出现腐蚀产物膜的剥落现象之后，在未脱落和脱落的地方会形成一个大阴极、小阳极的电偶腐蚀，最终会在腐蚀产物膜剥落的位置形成局部腐蚀，甚至点蚀，所以此温度下的腐蚀产物的保护作用较弱。而从温度为90℃时的实验钢宏观形貌上看出，虽然腐蚀脱落现象得到缓解，但脱落部分裸露的基体金属上发现有明显的点蚀坑，腐蚀产物膜颜色较深，并且较为致密和紧实，局部会有一些小的蚀坑；温度为120℃时，腐蚀产物较为均匀完整，表面未有腐蚀产物膜的明显剥落和蚀坑存在，最为致密。

表3 各个温度下实验钢平均腐蚀速率和点蚀速率（单位：mm/a）

温度	30℃	60℃	90℃	120℃
腐蚀速率	0.0671	5.0188	4.4784	2.2706
点蚀速率	0.4271	5.4038	12.8310	4.8782

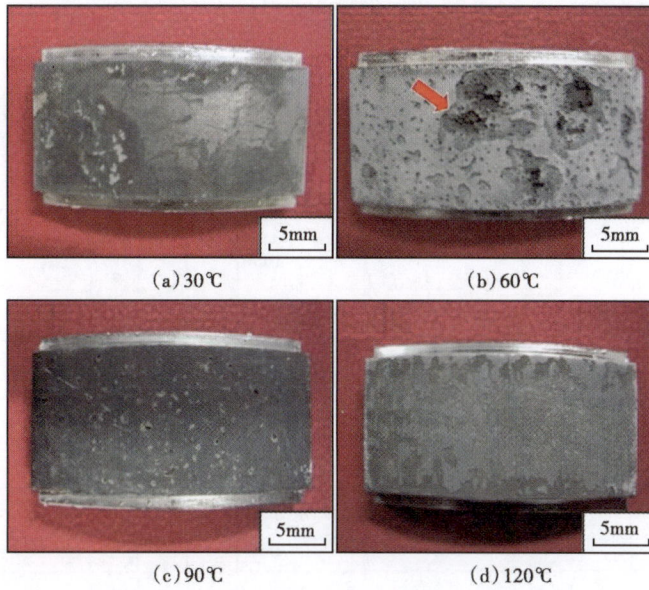

(a) 30℃   (b) 60℃

(c) 90℃   (d) 120℃

图 1　不同温度下实验钢腐蚀宏观形貌

3　分析与讨论

图 2、图 3 为实验钢在不同温度下腐蚀产物的微观形貌以及 XRD 分析结果。在温度为 30℃时，未形成晶粒状的腐蚀产物，腐蚀产物膜表面呈现了很大裂纹，这些裂纹的产生可能

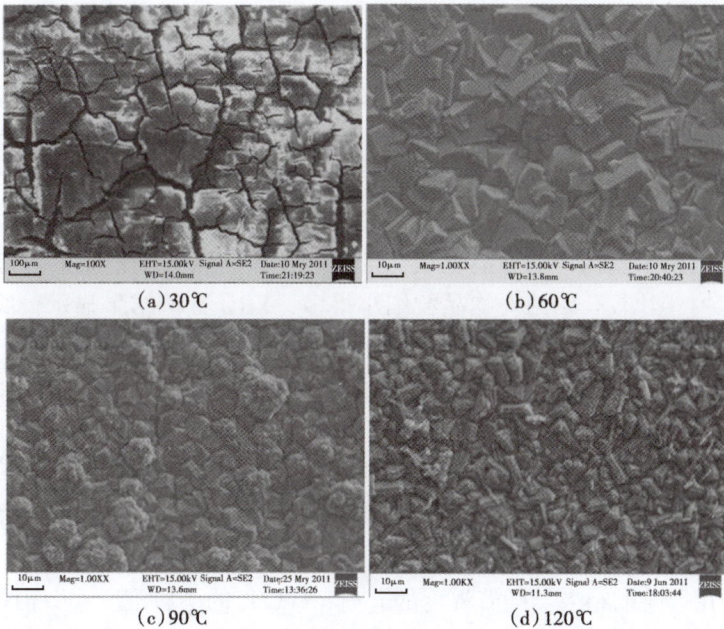

(a) 30℃   (b) 60℃

(c) 90℃   (d) 120℃

图 2　实验钢在不同温度下腐蚀产物的微观形貌

是由于腐蚀产物膜的具有内应力或者是腐蚀产物膜脱水的结果；XRD 分析只出现了 Fe 峰，是由于腐蚀产物膜比较薄，射线打到金属基体造成的。

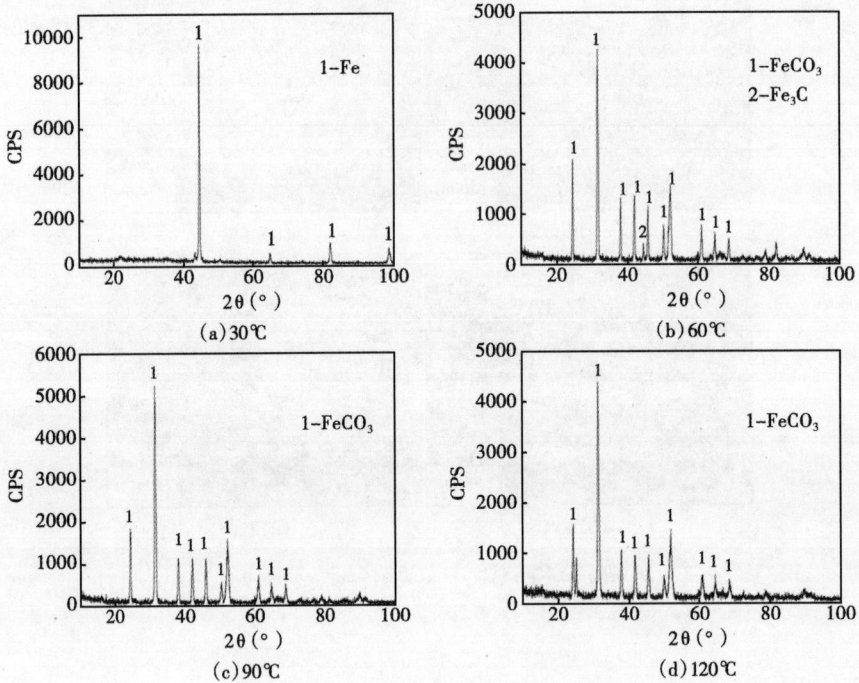

图 3　钢在不同温度下的 XRD 分析结果

在温度为 60℃时，腐蚀产物呈现规则的晶粒状排列，但是晶粒的尺寸较大；XRD 分析表明：产物主要为 $FeCO_3$，同时存在少量的 Fe_3C。Fe_3C 膜通过电偶腐蚀和局部酸化作用加速腐蚀速率[7]。电偶腐蚀表现在，相对于基体 Fe 而言，Fe_3C 的电位要低，两者之间的电偶作用可以加速铁的溶解；局部酸化表现在，阳极反应优先发生于 Fe_3C 上，从而在阴极和阳极腐蚀反应之间形成物理屏障，这将改变溶液成分，导致阴极区趋于碱性，阳极区趋于酸性[8-9]。但是 Fe_3C 同时可以组织 Fe^+ 在表面的扩散，促进 $FeCO_3$ 膜的形成并形成一定的保护性，切通过均匀混入 $FeCO_3$ 膜增强其性能和保护性。所以 Fe_3C 对腐蚀行为起对立的作用，这取决于其形成方式及其在膜的组织和形成过程中的控制作用。

在温度为 90℃时，腐蚀产物呈晶粒状，且在晶粒状的腐蚀产物膜上还有一些球状的腐蚀产物，与 60℃时相比较，腐蚀产物晶粒细小得多；XRD 分析结果显示，腐蚀产物主要为 $FeCO_3$。

在温度为 120℃时，腐蚀产物也成晶粒状排列，但是与 60℃、90℃条件下相比，晶粒的形状发生了很大的变化，60℃、90℃条件下晶粒的立体感比 120℃要强，发生这种情况的原因是因为 $FeCO_3$ 的溶解度具有负的温度效应和其分解作用引起的。在 CO_2 分压一定时，腐蚀速率和 $FeCO_3$ 的溶解度共同决定溶液中 Fe^+ 的浓度。腐蚀速率和 $FeCO_3$ 的溶解度越大，溶液的 Fe^+ 的浓度越大。可以看到 60℃之后，随着温度的升高，腐蚀速率越小，这是由于腐蚀产物膜的缓蚀作用影响的结果。随着温度的升高，$FeCO_3$ 的溶解度变小，但是可能是 $FeCO_3$ 的溶解度减小的幅度比腐蚀速率减小的幅度大，所以这个时候 $FeCO_3$ 晶体沉积过程就不是在溶解与沉积的动态平衡中进行，而是沉积的速率大于溶解的速率。温度越高，两者的差异越大，晶粒就越不规整，晶粒越小[10]。

由图 2 可以看出，随着温度的变化，腐蚀产物晶粒的大小和堆垛情况也各不相同，30℃时，未形成规则的腐蚀产物，且存在很大的空隙，60℃之后，随着温度的升高，晶粒尺寸越来越细小。晶粒的尺寸、形状和堆垛的方式决定这腐蚀产物膜的空隙率，晶粒越细小，堆垛越紧密，则空隙率越小，相反，空隙率则越大。腐蚀介质就是通过穿过空隙进入腐蚀产物膜内层甚至是金属基体，从而产生局部腐蚀的[11]。

图 4 为实验钢在不同温度下腐蚀产物膜的侧面形貌。由图可知，在 30℃ 时，腐蚀产物膜非常薄，此时主要发生金属的活性溶解，几乎没有腐蚀产物膜的形成。在 60℃ 时，腐蚀产物膜厚度最大，因为在这个温度情况之下，腐蚀速率最大，所以产生的腐蚀产物较多，所以厚度较大，且腐蚀产物内层与基体以及内外两层之间的存在这空隙结合不紧密，腐蚀产物膜会随着冲刷作用而剥落（图 4b）。在 90℃ 时，由于温度上升，腐蚀产物膜对金属的保护作用增强，腐蚀速率下降，所以腐蚀产物的沉积也下降，所以此时，腐蚀产物的厚度也有所下降，但是内外层腐蚀产物膜之间还是存在一些空隙。在 120℃ 时，腐蚀产物膜的与基体以及内外层之间的结合最为紧密，对基体的保护作用增强，所以腐蚀速率也为最低，此时的腐蚀产物膜也比 60℃、90℃ 的时候减薄。

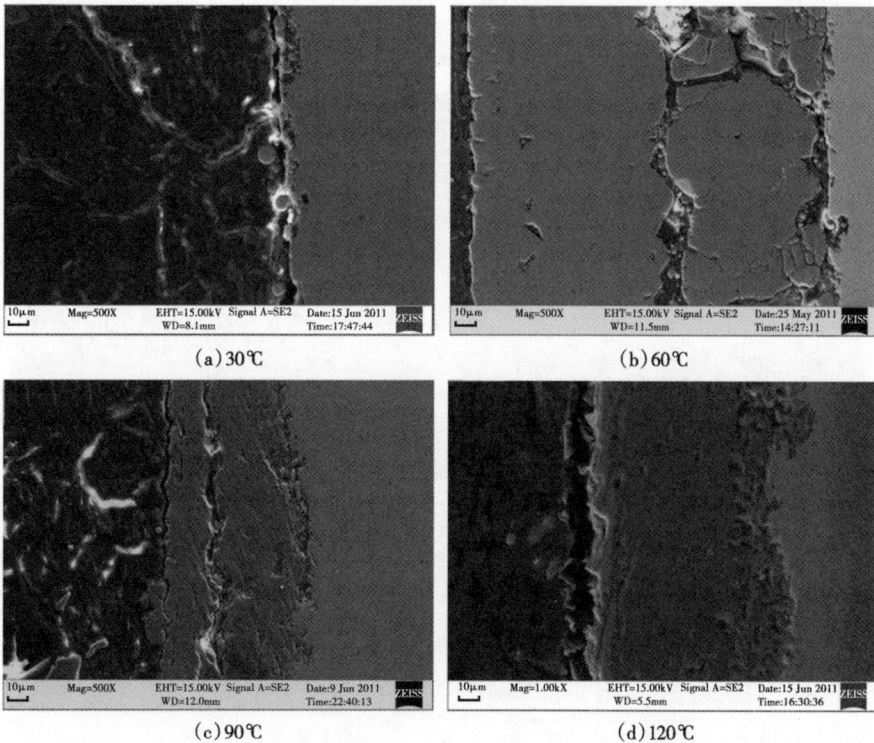

（a）30℃   （b）60℃

（c）90℃   （d）120℃

图 4　实验钢在不同温度下腐蚀产物的侧面形貌

4　结论

（1）在模拟实际工况的条件下，随着温度的升高，X80 管线钢的平均腐蚀速率和点蚀速率均呈现先增大后减小的趋势。

（2）在30℃时，试样表面未形成完整的腐蚀产物膜，腐蚀速率和点蚀速率最小；在60℃时，平均腐蚀速率达到最大值，但实验钢的点蚀现象并不严重；当温度到达90℃时，实验钢平均腐蚀速率开始下降，点蚀速率达到最大值；温度升高到120℃后，腐蚀速率呈现下降趋势，低于60℃、90℃时的腐蚀速率。

参 考 文 献

[1] 李士伦，张正卿．注气提高石油采收率技术 [M]．成都：四川科学技术出版社，2001．

[2] 张学元，邸超，雷良才．二氧化碳腐蚀与控制 [M]．北京：化学工业出版社，2000．

[3] WINNING I G, BRETHERTON N, MCMAHON A, et al. Evaluation of weld corrosion behavior and the application of corrosion inhibitors and combined scale, corrosion inhibitors [A]. The 59th NACE Annual Conference [C]. Houston: Omnipress, 2004.

[4] Kermani M B, Morshed A. Carbon dioxide corrosion in oil and gas production a compendium [J]. Corrosion, 2003, 59 (8): 559-683.

[5] VARELA F E, KURATAT Y, SANADA N. The influence of temperature on the galvanic corrosion of a cast iron-stainless steel couple [J]. Corrosion Science, 1997, 39 (4): 775-788.

[6] Palacios C A, Shadley J R. Characteristics of Corrosion Scales on Steels in CO_2-Sturated NaCl Brine [J]. Corrosion, 1991, 47 (2): 122-127.

[7] MORA-MENDOZA J L, TURGOOSE S. Fe_3C Influence on the Corrosion Rate of Mild Steel in Aqueous CO_2 Systems under Turbulent Flow Conditions [J]. Corrosion Science, 2002, 44 (6): 1223-1246.

[8] 何庆龙，孟惠民，俞宏英，等．N80油套管钢 CO_2 腐蚀的研究进展 [J]．中国腐蚀与防护学报，2007，27 (3): 186-192.

[9] 周琦，贾建刚，南雪丽，等．CO_2 环境介质下16Mn钢的高温高压腐蚀性能 [J]．兰州理工大学学报，2008，34 (1): 14-18.

[10] 周琦，贾建刚，南雪丽，等．高温高压 CO_2 环境介质中X60钢的腐蚀 [J]．腐蚀与防护，2008，29 (12): 720-723.

[11] 林冠发，白真权，赵新伟，等．温度对二氧化碳腐蚀产物膜形貌特征的影响 [J]．石油学报，2004，25 (3): 101-109.

低铬 X70 管线钢组织及其抗 CO_2 腐蚀性能

武会宾[1]　孙锐艳[2]　王立东[1]　王世刚[2]　唐荻[1]　梁金明[1]

(1. 北京科技大学 冶金工程研究院；2. 中国石油吉林油田公司 勘察设计院)

摘　要：为了明确 Cr 的质量分数高低对埋地用输油输气管线钢性能的影响，设计了 4 种 Cr 质量分数的 X70 管线钢，研究了不同 Cr 的质量分数下管线钢组织及其力学性能，并采用高温高压反应釜进行了实验钢的 CO_2 腐蚀试验。结果表明：钢中加入 0.1%～0.8% 的 Cr 后，其组织均由针状铁素体和准多边形铁素体构成，Cr 元素均呈现明显的沿晶界分布状态；随着 Cr 的质量分数的增加，钢板强度随之升高，晶界中 Cr 分布密度随之增大，$Cr(OH)_3$ 在腐蚀产物膜中的富集量增加，促使钢板的平均腐蚀速率降低；同时由于 $Cr(OH)_3$ 可以有效阻碍阴离子穿透腐蚀产物膜，因而大大减少了 Cl^- 的催化作用导致的点蚀，使得钢板点蚀速率明显降低。

关键词：X70 管线钢；Cr 的质量分数；CO_2 腐蚀；晶粒细化；固溶分布；腐蚀产物膜；平均腐蚀速率；点蚀速率

0　引言

近年来，在油气的开采过程中常采用 CO_2 的回注技术提高石油采收率。石油天然气中的 CO_2 的体积分数越来越高，从而使 CO_2 腐蚀成为油气采集输运过程中管道主要腐蚀原因之一[1-2]。CO_2 溶入水后对钢铁有极强的腐蚀性，在 pH 值相同的情况下，由于 CO_2 的总酸度比盐酸高，因此，它对钢铁的腐蚀比盐酸还严重，由此造成的油气管线腐蚀事故很多，甚至导致重大事故频发[3]。

普通碳钢管材不能抵抗 CO_2 腐蚀，而利用不锈钢代替碳钢成本较高，因此开发具有一定抗 CO_2 腐蚀能力的低 Cr 输油输气用钢已成为目前研究的热点[4-6]。作为一种新研发的抗 CO_2 腐蚀的钢材，低 Cr 合金钢具有力学性能好和生产成本低等优点，在输油输气管道方面具有广阔的应用前景[7]。目前针对低 Cr 合金钢耐蚀性的研究表明[8-9]，Cr 的质量分数为 1%～5% 的合金钢具有十分优异的抗 CO_2 腐蚀性能。地面输油输气管道中环境温度较低，腐蚀性较小，进一步降低管线钢中 Cr 的质量分数，均可以满足管线钢在此腐蚀环境中的应用，同时可以降低成本。然而，目前针对此方面的研究较少。本文设计了 4 种 Cr 的质量分数为 0.1%～0.8% 的 X70 管线钢，研究了 Cr 的质量分数对 X70 管线钢组织和力学性能的影响，并在模拟东北某油田采出流体（油—矿化水—CO_2 多相流环境）的实际工况下，采用高温高压反应釜开展了低 Cr X70 管线钢的抗 CO_2 腐蚀性能试验。本研究可以为低成本 X70 管线钢生产及其在地面输油输气工程中的应用提供理论依据。

1　实验材料和方法

实验用钢成分见表 1。Cr 的质量分数分别定为 0.1%、0.3%、0.6% 和 0.8%，对应的编号为 1#钢、2#钢、3#钢和 4#钢。设计成分经冶炼、锻造后，再经过控轧控冷工艺（TMCP）轧制成 12mm 厚板材。在 1#钢、2#钢、3#钢和 4#钢分别切取金相、透射样品。金相试样经打磨、抛光及 4%硝酸酒精溶液侵蚀后，在显微镜上观察金相组织，而后在扫描电子显微镜（SEM）下分析元素分布。透射试样经过磨光和 5%高氯酸酒精溶液双喷后，在透射电子显微镜上对析出物形貌和组成进行观察分析。

腐蚀试样规格为直径 87mm、弧长 30mm、面宽 12mm 和厚度 4mm 的弧形试样。经过 800#砂纸打磨后，丙酮除掉试样表面的油污，酒精清洗，然后用精度 0.1mg 的电子分析天平称量试样的重量。实验溶液在试验前用 CO_2 除氧 10h，倒入反应釜后再除氧 2h 后升温，并调整 CO_2 压力和流速。每组试验采用 4 个平行试样，试验结束后取出试样，经清水酒精冲洗吹干拍照；然后将其中 3 个试样放入除锈液中，去除腐蚀产物膜后清洗干燥并称重。每种材料保留一个带腐蚀产物膜的试样，用扫描电子显微镜对试样表面形貌进行观察和分析，运用 X-射线衍射（XRD）分析腐蚀产物的物质构成。

表 1   实验用 X70 管线钢的化学成分表（质量分数,%）

钢号	C	Si	Mn	P	S	Nb	V+Ti+ Mo	Cr	Fe
1#	0.04	0.20	1.42	0.007	0.004	0.03	≤0.4	0.1	余量
2#	0.04	0.23	1.40	0.007	0.004	0.03	≤0.4	0.3	余量
3#	0.04	0.22	1.41	0.007	0.003	0.03	≤0.4	0.6	余量
4#	0.04	0.23	1.40	0.007	0.004	0.03	≤0.4	0.8	余量

2   实验结果

2.1   组织和力学性能

表 2 是 4 种设计成分控轧控冷后板材的力学性能。由表 2 看出，随着 Cr 的质量分数增加，屈服强度由 540MPa 增至 570 MPa，抗拉强度由 635MPa 增至 675MPa，同时冲击功 198J 增至 232J，伸长率由 23.2%提高至 24%。可见，随着 Cr 质量分数的增加，钢的强度得到了提高，但塑韧性提高不明显。此外，Cr 的质量分数在 0.3%~0.6%之间变化时，其各项力学性能指标变化幅度较大；Cr 的质量分数由 0.6%增至 0.8%时，各项力学性能指标变化甚微。

表 2   实验钢力学性能

钢号	−20℃冲击功（J）	屈服强度（MPa）	抗拉强度（MPa）	屈强比	伸长率（%）
1#	198	540	635	0.85	23.2
2#	209	540	645	0.84	23.35
3#	225	570	670	0.85	23.9
4#	232	570	675	0.84	24.0

图 1 给出了性能达到 X70 要求的 4 块钢板的金相组织照片。由图 1 可见，随着 Cr 的质量分数从 0.1%增加至 0.8%时，实验钢组织主要由针状铁素体和边界呈锯齿状的准多边形铁素体构成且组织大小变化不大。实验钢强度提高的原因是 Cr 的质量分数增加使得固溶强化增高所致。

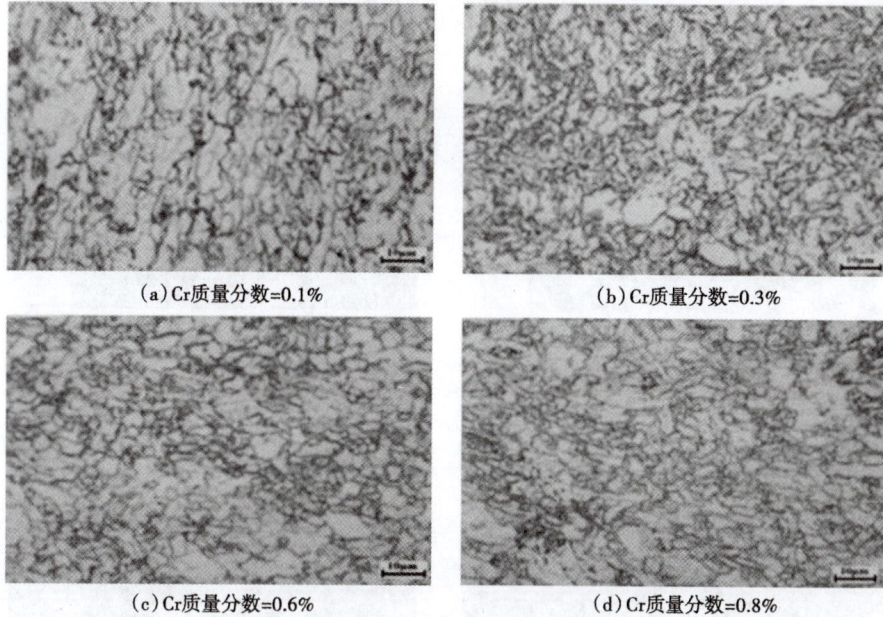

(a) Cr质量分数=0.1%　　　　　　　(b) Cr质量分数=0.3%

(c) Cr质量分数=0.6%　　　　　　　(d) Cr质量分数=0.8%

图 1　实验室设计 X70 管线钢金相照片

2.2　腐蚀实验结果

抗 CO_2 腐蚀实验在 3L 高温高压反应釜中进行，模拟采出液离子浓度见表 3。环境参数设置如下：CO_2 分压为 2MPa，温度为 30℃，流速为 1m/s，实验周期为 7d，通过失重法测得实验钢的腐蚀速率。X70 管线钢平均腐蚀速率和点蚀速率随 Cr 的质量分数变化曲线如图 2 所示。

表 3　油气田采出液的离子质量浓度表

离子种类	Cl^-	SO_4^{2-}	HCO_3^-	Mg^{2+}	Ca^{2+}	Na^+
质量浓度（mg/L）	15000	5450	600	1050	880	9560

由图 2 可以看出，4 种实验钢的平均腐蚀速率和点蚀速率均随着 Cr 的质量分数增加而降低，并且随着 Cr 的质量分数增加，平均腐蚀速率与点蚀速率的差值越来越小。当 Cr 的质量分数达到 0.8% 时，平均腐蚀速率为 0.0671mm/a，点蚀速率为 0.081mm/a，二者相差甚微。可见，在此腐蚀条件下，点蚀已经很难发生。

图 3、图 4 为腐蚀产物膜去除前后试样宏观腐蚀形貌，图 5、图 6 为腐蚀产物膜去除前后试样表面的 SEM 图。从图 3 可以看出，30℃时实验室设计 4 种成分管线钢腐蚀产物膜相对较致密，并且和基体附着性好，局部腐蚀产物膜未发生脱落，腐

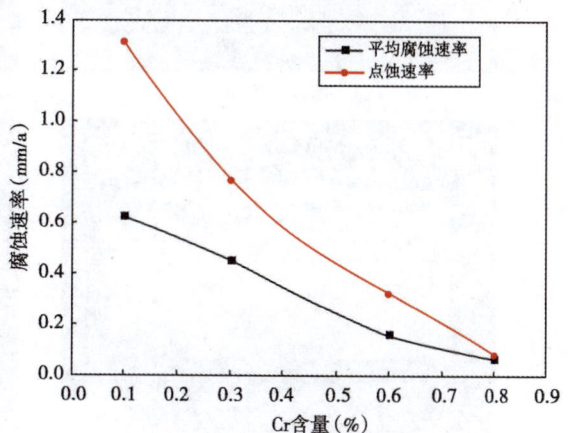

图 2　实验钢平均腐蚀速率和点蚀速率与
Cr 质量分数的关系

373

(a)Cr质量分数=0.1%   (b)Cr质量分数=0.3%

(c)Cr质量分数=0.6%   (d)Cr质量分数=0.8%

图 3　带有腐蚀产物的宏观腐蚀形貌

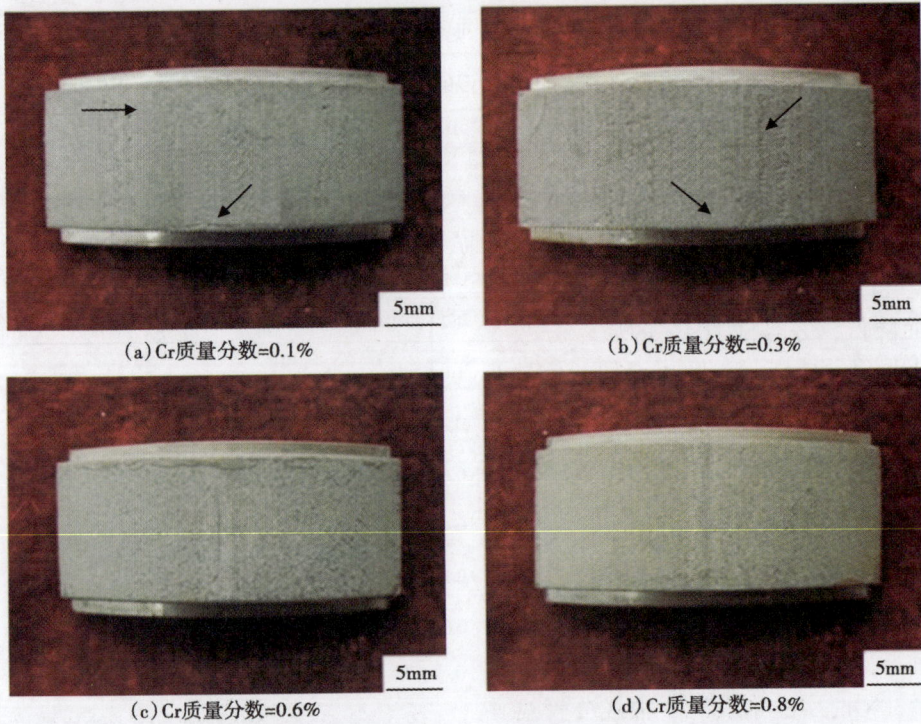

(a)Cr质量分数=0.1%   (b)Cr质量分数=0.3%

(c)Cr质量分数=0.6%   (d)Cr质量分数=0.8%

图 4　去除腐蚀产物的宏观腐蚀形貌

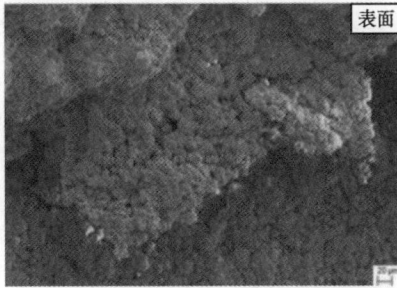

(a) Cr质量分数=0.1%时腐蚀产物膜的表面形貌图　　(b) Cr质量分数=0.1%时腐蚀产物膜的截面形貌图

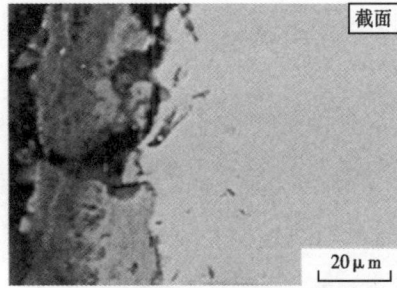

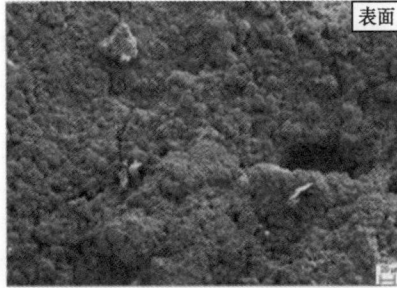

(c) Cr质量分数=0.3%时腐蚀产物膜的表面形貌图　　(d) Cr质量分数=0.3%时腐蚀产物膜的截面形貌图

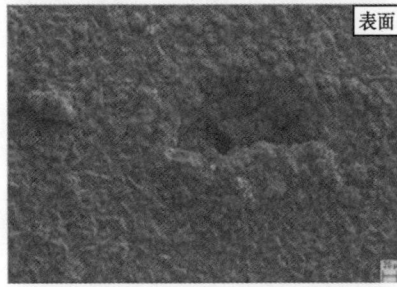

(e) Cr质量分数=0.6%时腐蚀产物膜的表面形貌图　　(f) Cr质量分数=0.6%时腐蚀产物膜的截面形貌图

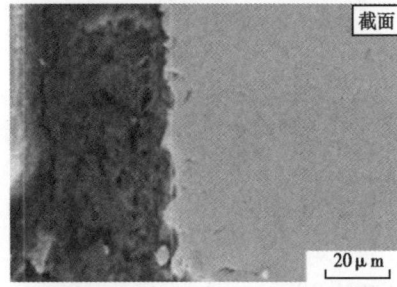

(g) Cr质量分数=0.8%时腐蚀产物膜的表面形貌图　　(h) Cr质量分数=0.8%时腐蚀产物膜的截面形貌图

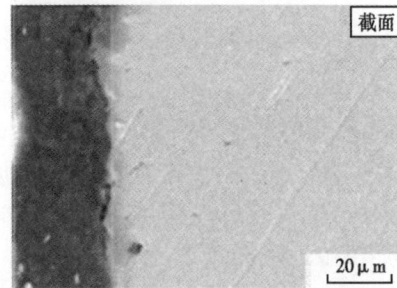

图 5　腐蚀产物膜表面及截面形貌

蚀产物膜对基体的保护性较好，腐蚀速率较小。从如图 4、图 6 上观察到 1# 钢点腐蚀坑较明显，数量较多且尺寸较大（箭头所指）；2# 钢仍然有可见的点蚀坑；3# 和 4# 钢已经观察不到明显的腐蚀坑；特别是 4# 钢，去除腐蚀产物的试样表面致密度较高（图 6）。图 5 为腐蚀产物膜放大后的形貌，由图 5 可见：4 种钢表面腐蚀产物颗粒都很细小；1# 钢腐蚀表面粗糙度较高，腐蚀产物膜致密度在 4 种钢中最低且呈层状，这些部位很可能成为产物膜脱离的开始点，进而产生局部腐蚀；2# 钢腐蚀产物膜的粗糙度较 1# 钢明显得到改善，颗粒依然细小，但

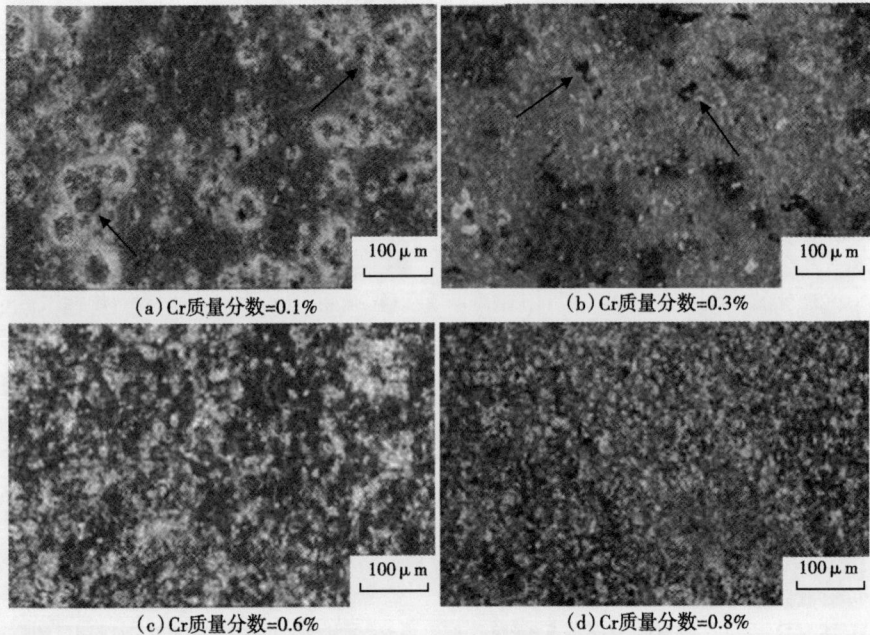

(a) Cr质量分数=0.1%　　　　　　　　(b) Cr质量分数=0.3%

(c) Cr质量分数=0.6%　　　　　　　　(d) Cr质量分数=0.8%

图 6　去除腐蚀产物膜后试样表面 SEM 图

在某些局部位置腐蚀产物膜并不均匀，有较深的腐蚀坑，根据图 4、图 6，这些腐蚀坑深度已经大于腐蚀产物膜的厚度；3#钢和 4#钢腐蚀产物表面粗糙度均较低，虽然 3#钢仍然有部分脱落，但深度较浅并有腐蚀产物继续覆盖表面，因此将在很大程度上缓解点蚀。可见，Cr 的质量分数增加可以有效地降低平均腐蚀速率，并且很大程度上可以抑制点蚀的发生。

3　分析与讨论

3.1　析出物分析

在透射电子显微镜下对各实验钢板的析出物进行分析，大量尺寸各异的析出物在针状铁素体晶内析出（图 7）。析出物大部分为椭圆形或圆形，棱角不分明，也有少数以方形存在，它们较均匀地分布在晶粒内部，尺寸大部分在 30nm 以内。通过对析出能谱分析（图 8），发现析出物主要以（Nb，Ti）（C，N）和 Nb（C，N）为主。

在抗 CO_2 腐蚀管线钢中，通过添加 Nb、V、Ti 和 Mo 等合金元素，可以促进针状铁素体和细小碳氮化物的形成，以提高管线钢的强度和韧性[10]。在 TMCP 过程中，随着变形温度的降低和变形程度的增加，因微合金元素在钢板中的固溶度逐渐下降，应变能逐渐增加，在位错线或应变带析出的碳氮化物粒子不断增多，呈细小弥散状态，能够起到有效的强化作用并改善钢的冲击韧性[11]。实验室设计的 1#、2#、3#和 4#成分钢板，C 的质量分数均控制在0.04%，但增加了 Nb、Mo 的质量分数来细化晶粒、弥补强度损失。微合金 Nb 的加入，在冷却和变形过程中沉淀析出，即能扩大形变奥氏体未再结晶区的温度范围，有利于增加奥氏体未再结晶区的轧制变形量，又能抑制晶粒长大，起到细化晶粒和沉淀强化作用。同时降低C 的质量分数，增加 Nb、Ti 的质量分数，有利于首先形成（Nb，Ti）（C，N）和 Nb（C，

376

N）析出，降低 Cr 的碳氮析出物，提高 Cr 的固溶，充分发挥 Cr 在耐 CO_2 腐蚀中的作用。经 EDS 能谱分析，各钢板均未发现 Cr 的析出物。

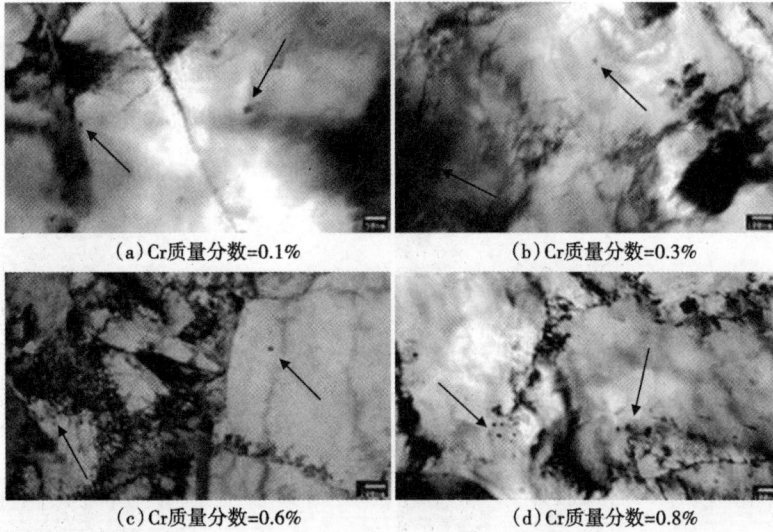

(a) Cr质量分数=0.1%　　　　(b) Cr质量分数=0.3%

(c) Cr质量分数=0.6%　　　　(d) Cr质量分数=0.8%

图 7　析出物 TEM 照片

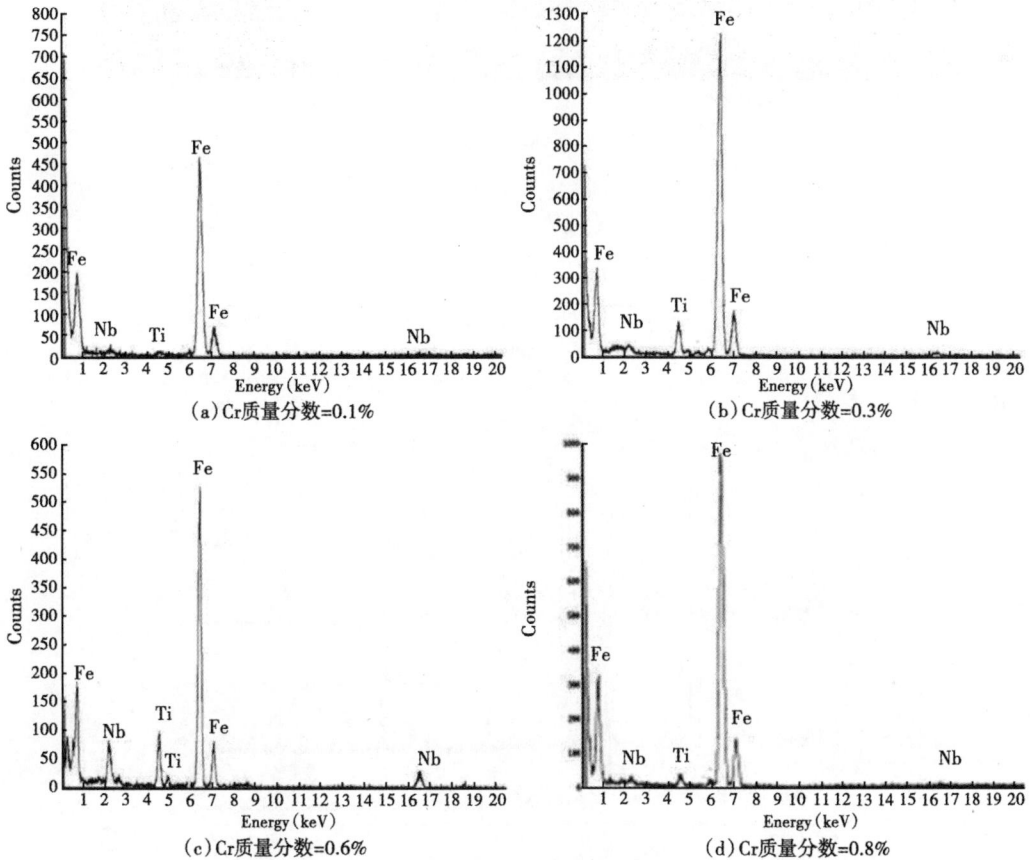

(a) Cr质量分数=0.1%　　　　(b) Cr质量分数=0.3%

(c) Cr质量分数=0.6%　　　　(d) Cr质量分数=0.8%

图 8　析出物能谱分析图

3.2 腐蚀性能分析

腐蚀一般容易在晶界、位错等表面结构的不均匀处发生，如果增加 Cr 在该处的固溶偏聚，有利于在此处首先形成耐蚀性保护膜，从而减缓腐蚀。实验中通过 SEM 扫描观察发现（图 9），本次实验用 X70 管线钢中的 Cr 均明显沿晶界分布。并且随基体中 Cr 的质量分数增加，晶界中 Cr 分布密度增大。通过对析出物分析发现，析出物主要是 Nb、Ti 的 C、N 化物，并没发现 Cr 的析出物。由此可以推断，沿晶界分布的 Cr 主要以固溶形式存在。

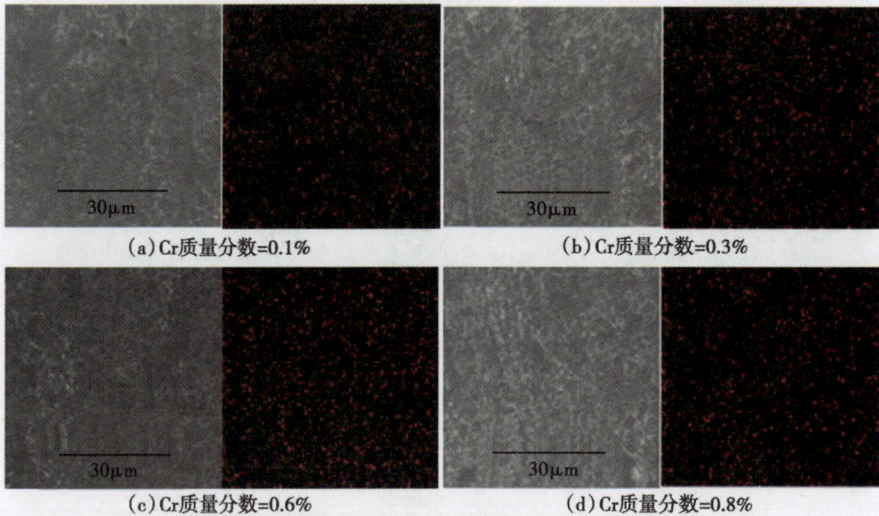

(a) Cr质量分数=0.1% (b) Cr质量分数=0.3%

(c) Cr质量分数=0.6% (d) Cr质量分数=0.8%

图 9 实验钢的 Cr 面分布图

图 10 为不同 Cr 的质量分数管线钢腐蚀后腐蚀产物膜的 X—射线衍射图。通过腐蚀产物膜的 XRD 图谱发现，腐蚀产物膜主要是 Fe、$FeCO_3$、Cr_2O_3 及少量的 Fe_2O_3。Fe_2O_3 可能是试样取出后在空气中氧化所致，基本上没有形成腐蚀产物膜。Fe_3C 是钢基体腐蚀后残留下来的物质，腐蚀过程中会暴露在钢铁表面充当腐蚀的阴极而加速钢铁的腐蚀。Cr_2O_3 是由于

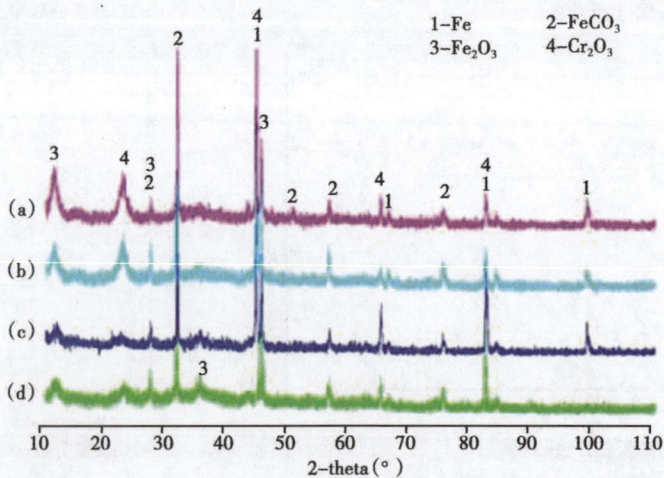

1-Fe   2-$FeCO_3$
3-Fe_2O_3 4-Cr_2O_3

图 10 实验钢表面腐蚀产物膜 XRD 图谱

Cr（OH）₃发生脱水反应而生成的。

在腐蚀介质中和一定的环境条件下，基体中含有 Cr 元素管线钢的 CO_2 腐蚀过程中存在以下 4 种阳极反应[4]：

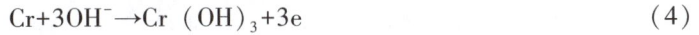

$$Fe \rightarrow Fe^{2+}+2e \qquad (1)$$

$$Fe+HCO_3^- \rightarrow FeCO_3+2e+H^+ \qquad (2)$$

$$Fe+CO_3^{2-} \rightarrow FeCO_3+2e \qquad (3)$$

$$Cr+3OH^- \rightarrow Cr（OH）_3+3e \qquad (4)$$

基体中的 Cr 与介质中的 OH^- 有较强的电子亲和力，容易优先生成 Cr（OH）₃（式 4），其化学性质比较稳定并将在金属表面沉积。在腐蚀过程中，随着腐蚀产物膜 $FeCO_3$ 的形成和溶解，Cr 元素会在腐蚀产物膜中富集。通过对腐蚀产物的能谱分析发现（图 11 和表 4），随着基体钢中 Cr 的质量分数的增加，Cr 的富集量也越多，4# 钢 Cr 元素的质量分数高达 13.15%。普遍认为 Cr 的富集是低 Cr 合金钢提高抗 CO_2 腐蚀能力的主要原因[12]。由于 Cr 元素会在腐蚀产物膜中富集，腐蚀产物膜对基体保护作用增强的原因可以归结为以下两个方面：一是腐蚀产物膜的致密度增大；二是膜的导电性降低，同时腐蚀产物膜具有了阴离子选择性，对介质中的腐蚀性阴离子阻碍作用较大，因此对金属基体的保护性大大增强。对比 4 种成分钢发现，随着基体含 Cr 量 0.1% 增至 0.8%，Cr（OH）₃ 在腐蚀产物膜中的富集量也明显增加，导致腐蚀产物膜结构发生显著变化，耐腐蚀性能有所提高，从而符合实验所得结果。而含有 Cr（OH）₃ 的腐蚀产物膜具有一定的阴离子选择性，即它可以有效阻碍阴离子穿透腐蚀产物膜到达金属表面，这样就降低了膜与金属界面处的阴离子浓度，界面处 Cl^- 浓度的降低会大大减少 Cl^- 的催化作用导致的点蚀[13]，这就是实验室设计 4 种成分管线钢随着 Cr 的质量分数增加，点蚀速率降低的原因。

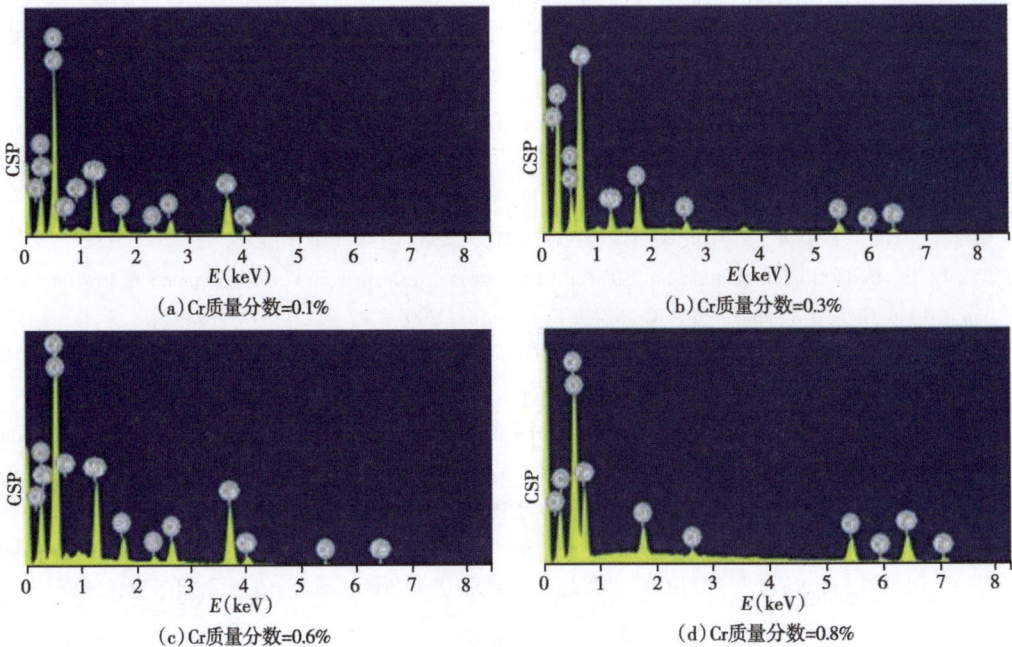

(a) Cr质量分数=0.1%
(b) Cr质量分数=0.3%
(c) Cr质量分数=0.6%
(d) Cr质量分数=0.8%

图 11　腐蚀产物膜能谱分析图

表4 对应图11的EDS分析结果表

元素种类	a		b		c		d	
	质量分数（%）	原子分数（%）	质量分数（%）	原子分数（%）	质量分数（%）	原子分数（%）	质量分数（%）	原子分数（%）
Fe	34.43	9.29	25.6	7.77	26.09	7.36	30.31	9.02
C	16.35	28.96	21.15	35.79	11.05	19.9	17.3	28.78
O	30.76	46.61	26.06	26.68	29.01	42.72	27.29	43.19
Cr	0.61	0.54	2.66	1.19	8.66	3.32	13.15	5.47
Cl	2.73	1.31	2.12	1.06	3.91	2.39	0.94	0.82
其他	15.12	13.29	22.41	27.51	21.28	24.31	11.01	12.72

4 结论

（1）对于不同Cr的质量分数X70管线钢而言，随着Cr加入量的增加，实验钢的强度有所提高，组织主要由针状铁素体和边界呈锯齿状的准多边形铁素体构成且组织大小变化不大。实验钢强度提高的原因是Cr的质量分数的增加使得固溶强化增高所致。同时，实验用钢的Cr元素均明显沿晶界固溶分布，并且随基体中Cr的质量分数增加，晶界中Cr分布密度增大。

（2）随着基体含Cr量0.1%增加至0.8%，$Cr(OH)_3$在腐蚀产物膜中的富集量增加，导致腐蚀产物膜结构发生显著变化，耐腐蚀性能得到提高。此外，$Cr(OH)_3$在腐蚀产物膜中富集加剧，可以有效阻碍阴离子穿透腐蚀产物膜到达金属表面，大大减少Cl^-的催化作用导致的点蚀，这使得X70管线钢点蚀速率降低。

参 考 文 献

[1] 郭万奎，廖广志，邵振波. 注气提高采收率技术 [M]. 北京：石油工业出版社，2003：53-66.

[2] Kermani M B, Morshed A. Carbon dioxide corrosion in oil and gas production a compendium [J]. Corrosion, 2003, 59 (8): 659-683.

[3] 张学元，邸超，雷良才. 二氧化碳腐蚀与控制 [M]. 北京：化学工业出版社，2000：19-27.

[4] Takabe H, Ueda M. The relationship between CO_2 corrosion resistance and corrosion products structure on carbon and low Cr bearing steels [J]. Corrosion Engineering, 2007, 56 (11): 514-520.

[5] 孙建波，柳伟，常炜，等. 低铬X65管线钢CO_2腐蚀产物膜的特征及形成机制 [J]. 金属学报，2009，45 (1): 84-90.

[6] Chen C F, Lu M X, Sun D B, et al. Effect of chromium on the pitting resistance of oil tube steel in a carbon dioxide corrosion system [J]. Corrosion, 2005, 61 (6): 594-601.

[7] 胡丽华，路民旭，常炜，等. Cr含量和组织对含1%Cr管线钢焊接接头抗CO_2腐蚀性能影响 [J]. 材料工程，2010 (7): 82-86.

[8] 张雷，胡丽华，孙建波，等. 抗CO_2腐蚀低Cr管线钢组织和性能研究 [J]. 材料工程，2009 (5): 6-10.

[9] Carvalho D S, Joia C J B, Mattos O R. Corrosion rate of iron and iron-chromium alloys in CO_2 medium [J]. Corrosion Science, 2005, 47 (12): 2974-2986.

[10] 曹建春, 雍岐龙, 刘清友, 等 . 含铌钼钢中微合金碳氮化物沉淀析出及其强化机制 [J]. 材料热处理学报, 2006, 27 (5)：51-55.

[11] 翁宇庆 . 超细晶钢 [M]. 北京：冶金工业出版社, 2003：76-390.

[12] 胡丽华, 张雷, 许立宁, 等 . 3Cr 低合金管线钢及焊接接头的 CO_2 腐蚀行为 [J]. 北京科技大学学报, 2010, 32 (3)：345-350.

[13] 张亮, 李晓刚, 杜翠薇, 等 . X70 管线钢在含 CO_2 库尔勒土壤模拟溶液中的腐蚀行为 [J]. 金属学报, 2008, 44 (12)：1439-1444.

CFD Validation of Scaling Rules for Reduced-Scale Field Releases of Carbon Dioxide

Ji Xing[1]   Zhenyi Liu[1]   Ping Huang[1]   Changgen Feng[1]
Yi Zhou[1]   Ruiyan Sun[2]   Shigang Wang[2]

(1. State Key Laboratory of Explosion Science and Technology, Beijing
Institute of Technology; 2. Oil Field Reconnaissance Designing Institute
of Jilin, Petro China Jilin Oilfield Company)

Abstract: Carbon Dioxide-Enhanced Oil Recovery (CO_2-EOR) has the potential risk that the blowout accident happens to cause casualties and environmental influence. To assess the accidental consequence, a reduced-scale field experiment of CO_2 release was performed based on the scaling rules instead of a full-size field test that cost too much without feasibility. A series of scaling rules was introduced to enlarge the reduced-scale field experiment to the full-size one like a real scenario. To validate the scaling rules, the numerical simulation was carried out based on the $k-\varepsilon$ turbulence model that was proofed a effective way to predict the concentration field on heavy gas dispersion. For concentration variation, the general tendencies of two methods kept identical except nearby the jet nozzle where the measured CO_2 concentration from the experiment was obviously higher than that in the simulation. The statistical performance indicators were introduced to verify the consistency between the scaled results and the simulated ones, and the results showed that using the scaling rules to scale the field experiment exhibited the acceptable accuracy at small flow rates and the scaling rules were applicable for the field experiment of accidental releases.

Keywords: Heavy gas dispersion; CFD simulation; $k-\varepsilon$ model; Carbon Dioxide; experiment

0   Introduction

Carbon capture and storage (CCS) technology has seen as one of the most promising solutions to the increasingly serious problem of global warming. Carbon Dioxide-Enhanced Oil Recovery (CO_2-EOR) technology, not directly targeting CO_2 sequestration, can help improve oil recovery and solve the problem of CO_2 sequestration, and it has been widely used in recent years[1-3]. CO_2 is an asphyxiating gas but not toxic, so suffocation phenomenon may occur during drilling and production when accidents of release happen[4]. The concentration of CO_2 over 7% in volume will cause death in a short time. If it occurs leaks of CO_2, over 3% in volume will be dangerous to the human body. Accidents of blowout and large-scale leakage during the CO_2-EOR process will cause great dangers to the surrounding staff[5]. According to the China's national standards GBZ 2.1-2007 "Occupational Exposure Limits for Industrial Workplace", the acceptable upper limit of CO_2 concentration for a short period of time (15min) exposure is 18000mg/m^3, equivalent to 1% in volume.

CO$_2$ has a greater density than air and its dispersion in the atmosphere complies with the heavy gas diffusion law[6]. The main experimental methods studying heavy gas diffusion include the full-size field experiment, the reduced-scale field experiment, the wind tunnel experiment and the brine experiment[7-10]. Although the full-size field experiment can reproduce the scenes of accidents, such as the weather condition, the topography and the characteristics of the leaking source, it costs too much labor and time. Furthermore, it needs high requirements of environmental conditions, especially for some specific weather conditions and terrain conditions, so it still lacks the repeatability for test. The wind tunnel experiment demands the high accordance of the dimensionless parameters between the model and the prototype to simulate the flow of the atmospheric boundary layer, such as Reynolds number (Re), Bulk Richardson number (Ri), Rossby number (Ro), Prandtl number (Pr) and Eckert number (Ec) [11-15]. However, it is unrealistic to keep all the dimensionless parameters identical. Moreover, the scaling of the wind tunnel experiment is too small usually between 1/100 to 1/1000. Hence, such a small scaling is difficult to reproduce the real environment with the humid atmosphere[16-18]. The brine experiment can use the liquid to simulate the gas diffusion for visual effect, but it hardly gets the quantitative results for gas concentration. The reduced-scale field experiment is conducted in the real terrain and weather conditions, and the parameters for the terrain and the weather conditions are scaled based on the similarity criteria. Absolutely it has to ensure the geometry, mechanics and kinematics for the characteristics of the source scaled correctly according to the similarity criteria. Consequently, the concentration information for the real scenario can be deduced based on the measured values of concentration in the reduced-scale field. In contrast with the full-size field test, the reduced-scale field experiment can greatly reduce the amount of hazardous materials to save the cost, and decrease the potential risk from abundant hazardous materials. Compared with the wind tunnel experiment, it has a better scale from 1/10 to 1/100 that is closer to a more realistic atmospheric environment.

Many gas diffusion models exist like the Gaussian model, the BM model, the Sutton model and the three-dimensional (3D) CFD model etc[19-22]. The 3D CFD model has the ability to accurately simulate the flow field based on the calculation principle. University of California, Lawrence Livermore National Laboratory (LLNL) in the 1980s developed the FEM3 model based on the finite element division and single-equation k theory turbulence model and simulated the series of instantaneous release tests of Thorney Island with it, and the compared results showed a good agreement. Now the model has been developed as the two-equation k-ε model FEM3C model. The Greek Spyros Sklavaounos used the software CFX 5. 6 to simulate the flow of the heavy gas over the obstacles with the k-ε and the SST equations, and he found the simulated results had a good agreement with the experimental data[23]. Steven R. Hanna and his group used the data from Kit Fox, MUST, Prairie Grass, and EMU test to verify the FLACS CFD model based on the k-ε turbulence equations and proved it effective[24]. Hence, the k-ε turbulence equations are able to be reliably used in the calculation for heavy gas dispersion[25].

This paper applied the similarity theory and described the scaling rules for the reduced-scale field releases of carbon dioxide[26], and it introduced the reduced−scale field experiment to simulate CO_2 blowouts, which was conducted by Beijing Institute of Technology. On the basis of the scaling rule, the reduced−scale field was enlarged back to the full−size scenario. The CO_2 concentration field in the full−size was simulated by the $k-\varepsilon$ model to verify the scaling rules that were used to scale the real CO_2 releases.

1　Scaling Rules

The reduced−scale field experiment is a type of field test scaled based on the similarity criteria and is conducted on the real terrain and real weather conditions. All the capital letters are used on behalf of the parameters in the full size while all the lower cases represent those in the reduced scale.

1. 1　Length Scales

The characteristic length (l, L) is described on the base of the volume of gas releasing into the ambient atmosphere, as

$$l = v^{1/3} \tag{1}$$

In the reduced−scale model and

$$L = V^{1/3} \tag{2}$$

In the full−size scale.
Define S_c as the reduced−scale/full−size length scaling factor. The same gas is released into the atmosphere both in the model and in the full scale, so that

$$\frac{v}{V} = \frac{m}{M} = \left(\frac{l}{L}\right)^3 = S_c^{\ 3} \tag{3}$$

It is apparent that the released required gas is decreased in abundance in the reduced−scale model. For instance, if $S_c = 0. 1$, it means that the releases of 10kg materials in the reduced−scale model are equivalent to the releases of 10t materials.

1. 2　Velocity Scales

As to the gas dispersion, both the density ratio

$$\frac{\rho_g}{\rho_a} \tag{4}$$

and the Froude number

$$F_\Gamma = \frac{U^2}{gL} \tag{5}$$

is supposed to keep identical at the reduced−scale model and the full−size scale considering the similarity theory.

384

There is another scaling parameter in the dispersion modeling defined as the Richardson number

$$R_i = g \frac{\rho_g - \rho_a}{\rho_a} \frac{L}{U^2} \tag{6}$$

According to the similarity theory, the Richardson number should also retain identical in the different scales. However, R_i will automatically be scaled right if Eqs. (4) and (5) keep identical at the reduced-scale model and the full-size scale.

Due to the same material released in different scales, the velocity rate should be defined as follows to satisfy Eqs. (4) and (5)

$$\frac{u_h}{U_H} = \left(\frac{l}{L}\right)^{0.5} = S_c^{0.5}$$

Thus, the inlet velocity and the wind speed are much higher in the full-size scale than those in the reduced-scale model. For example, if $S_c = 0.1$, it means that the velocity at 100m height in the full-size scale is three times more than that at 10m height in the reduced-scale model.

1.3 Time Scales

The dimensionless rate

$$\frac{UT}{L} \tag{7}$$

should keep identical in the full-size scale and in the reduced-scale model, so that

$$\frac{UT}{L} = \frac{ut}{l} \tag{8}$$

Due to $\frac{u}{U} = S_c^{0.5}$, it is easy to get the following result.

$$\frac{t}{T} = \frac{U}{u} \frac{l}{L} = S_c^{0.5}$$

1.4 Continuous Release Scales

The dimensionless rate

$$\frac{Q}{UL^2} \tag{9}$$

should keep identical in the full-size scale and in the reduced-scale model, thus

$$\frac{Q}{UL^2} = \frac{q}{ul^2} \tag{10}$$

$$\frac{q}{Q} = \frac{u}{U} \left(\frac{l}{L}\right)^2 = S_c^{2.5} \tag{11}$$

385

For example, if $S_c = 0.1$, it means that the release rate in the full-size scale should be three hundred times more than more than that in the reduced-scale model.

2   Description of the CO_2 Release Test

The CO_2 release field test was performed by Beijing Institute of Technology in Zhangjiakou. An artificial tunnel was built without completed limitation keeping the front and the rear open. The cube with 3m width and 3m height and 11m length was chosen as the experimental area from one small part of the whole tunnel, as shown in Fig. 1. The bottom was the ground and the other 5 faces were open. The CO_2 was released vertically upwards from the hole with the radius of 1 cm on the ground, and the initial CO_2 concentration released kept 99.9% in volume and the flow rates for ejecting CO_2 were limited from 0 to 20m³/h. In our trials, four different flow rates for ejecting CO_2 were 10m³/h, 12m³/h, 15m³/h and 18m³/h. Due to the diameter and flow rates fixed, the CO_2 were ejected from the source at the speeds of approximately 8.8m/s, 10.6m/s, 13.3m/s and 15.9m/s. Sensors were arranged on the central line across the CO_2 source on the ground along the downwind direction. The distances from the source to the ten sensors were 0.5m, 1m, 1.5m, 2m, 2.5m, 3m, 4m, 6m, 8m, 10m on the central line.

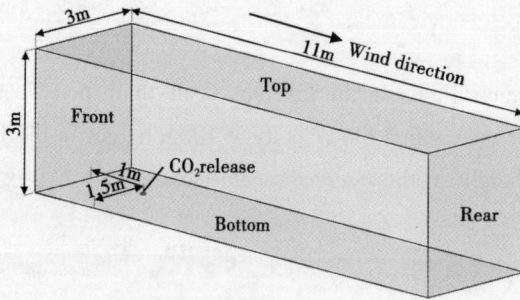

Fig. 1   Experimental Domain and Wind Direction

3   Turbulence Modeling

The $k-\varepsilon$ model is one of the most effective models evaluated from the Reynolds averaged Navier-Stokes (RANS) equations[27]. The $k-\varepsilon$ model includes two main parameters, and one is k, the turbulent kinetic energy, and the other is ε, the turbulence dissipation rate[28]. Launder and Spalding firstly proposed the classical model in 1972[29].

Continuity equation:

$$\frac{\partial \rho}{\partial t} + \frac{\partial}{\partial x_i}(\rho u_i) = 0 \tag{12}$$

Momentum equation:

$$\frac{\partial}{\partial t}(\rho u_i) + \frac{\partial}{\partial x_i}(\rho u_i u_j) = -\frac{\partial p}{\partial x_i} + \frac{\partial}{\partial x_j}\left(\mu \frac{\partial u_i}{\partial x_j} - \rho \overline{u_i' u_j'}\right) \tag{13}$$

The definitions for k and ε are presented as follows.

$$k = \frac{\overline{u_i' u_i'}}{2} = \frac{1}{2}(\overline{u'^2} + \overline{v'^2} + \overline{w'^2}) \tag{14}$$

$$\varepsilon = \frac{\mu}{\rho} \overline{\left(\frac{\partial u_i{'}}{\partial x_k}\right)\left(\frac{\partial u_i{'}}{\partial x_k}\right)} \tag{15}$$

The $k-\varepsilon$ equation defines that the function of the turbulence viscosity is determined by both the turbulence kinetic energy and the turbulence dissipation rate.

$$\mu_t = \frac{C_\mu \rho k^2}{\varepsilon} \tag{16}$$

As to the values of k and ε, they are both able to be determined from the following equations below.

$$\frac{\partial(\rho k)}{\partial t} + \frac{\partial(\rho k u_i)}{\partial x_i} = \frac{\partial}{\partial x_j}\left[\left(\mu + \frac{\mu_t}{\sigma_k}\right)\frac{\partial k}{\partial x_j}\right] + G_k - \rho\varepsilon \tag{17}$$

$$\frac{\partial(\rho\varepsilon)}{\partial t} + \frac{\partial(\rho\varepsilon u_i)}{\partial x_i} = \frac{\partial}{\partial x_j}\left[\left(\mu + \frac{\mu_t}{\sigma_\varepsilon}\right)\frac{\partial\varepsilon}{\partial x_j}\right] + C_{1\varepsilon}\frac{\varepsilon}{k}G_k - C_{2\varepsilon}\rho\frac{\varepsilon^2}{k} \tag{18}$$

where G_k is able be obtained from the given equation as follow.

$$G_k = \mu_t\left\{2\left[\left(\frac{\partial u}{\partial x}\right)^2 + \left(\frac{\partial v}{\partial y}\right)^2 + \left(\frac{\partial w}{\partial z}\right)^2\right] + \left(\frac{\partial u}{\partial y} + \frac{\partial v}{\partial x}\right)^2 + \left(\frac{\partial u}{\partial z} + \frac{\partial w}{\partial x}\right)^2 + \left(\frac{\partial v}{\partial z} + \frac{\partial w}{\partial y}\right)\right\} \tag{19}$$

where $C_{1\varepsilon}$ and $C_{2\varepsilon}$ are the determined constants empirically, and σ_k and σ_ε are separate corresponding Prandtl values for k and ε. These empirical constants could be defined as

$$C_{1\varepsilon} = 1.44, \quad C_{2\varepsilon} = 1.92, \quad C_\mu = 0.09, \quad \sigma_k = 1.0, \quad \sigma_\varepsilon = 1.3 \tag{20}$$

4 Statistical Performance Indicators

Statistical performance indicators were introduced to assess the effects of the models for simulation. Some typical statistical performance approaches were suggested by Hanna et al[30], including the geometric mean bias (MG), the geometric mean variance (VG), and the fraction of predictions within a factor of two of observations (FAC2). MG is the exponential function of the difference between means of logarithmic function of predicted values less the logarithmic function of observed value. VG is the exponential function of mean of square of the difference between logarithmic function of predicted values less the logarithmic function of observed values. FAC2 is defined as the percentage of predictions within a factor of two of the observed values. The definitions are shown as follows.

$$MG = \exp(\overline{\ln C_o} - \overline{\ln C_p}) \tag{21}$$

$$VG = \exp[\overline{(\ln C_o - \ln C_p)^2}] \tag{22}$$

$$FAC2 = \text{fraction of data that satisfy } 0.5 \leqslant \frac{C_p}{C_o} \leqslant 2.0 \tag{23}$$

In general, a perfect model should produce MG, VG and FAC = 1.0. To verify the predicted results, the acceptable range of the evaluation should be fixed in advance. According to Chang and

Hanna, $0.7 < MG < 1.3$, $VG < 4$ and $0.5 < FAC2 < 1$ mean the acceptable criteria for heavy gas dispersion models. Furthermore, Chang and Hanna presented a bottom limitation for the observed and predicted concentrations about the calculating MG and VG, because these two indicators were seriously influenced by tiny values. Consequently, the concentration values predicted of less 0.1% did not participate in calculations for MG and VG.

5 Simulation Results

The simulation was carried out in the FLUENT 6.3 code to calculate CO_2 concentration at a function of time, and species transport modeling was used in a transient form[31]. At the beginning of the transient simulation, the calculation was solved in steady state. The steady state was running for approximately $200 \sim 260$ iterations to achieve an acceptable convergence. The factor of 10^{-4} was chosen as a convergence residual criterion. The whole releasing continued 200s ($20 \times 0.25s + 50 \times 0.5s + 50 \times 1s + 60 \times 2s$). The entire simulation lasted approximately 4 hours on an 2.5 GHz Intel® Pentium Duo-Core with 3 GB RAM.

The CO_2 concentration field was simulated by species transport modeling, and the concentration field at the central line on the ground along the wind direction was the most significant concern, because the CO_2 concentration might maintains further along the downwind direction, and the CO_2 concentration field was directly related to personal safety for workers around the well outlet. For example, the CO_2 concentration permission in workplace is less than 1% in volume according to the National Standards of China. That central line represents the rough forward direction of the CO_2 cloud. Table 1 exhibits CO_2 concentration at different positions on this line predicted by the $k-\varepsilon$ model. These concentration values were recorded at 632s after CO_2 release.

Table 1 CO_2 Concentration at the Central Line on the Ground in the Simulation

Flow rate (m^3/h)	CO_2 concentration of different coordinates at 632s after CO_2 release (%)									
	(15, 1.5, 0)	(20, 1.5, 0)	(25, 1.5, 0)	(30, 1.5, 0)	(35, 1.5, 0)	(40, 1.5, 0)	(50, 1.5, 0)	(70, 1.5, 0)	(90, 1.5, 0)	(110, 1.5, 0)
10	0.4	0.7	1.8	2.0	1.5	1.1	0.7	0.5	0.2	0.1
12	0.2	0.3	1.1	1.9	1.7	1.3	0.8	0.5	0.3	0.1
15	0.1	0.1	0.3	1.2	1.7	1.5	0.9	0.6	0.3	0.2
18	0.1	0.1	0.1	0.3	1.2	1.5	1.0	0.7	0.5	0.3

6 Comparisons between Reduced-scale Experimental Results and Simulated Values

The scale of 1/10 was used in the field experiment on CO_2 release, so all the variables need be scaled in the original model for simulation based on the scaling rules described above. In the original model, it can be deduced that the distances between the source and sampling points were 5m, 10m,

388

15m, 20m, 25m, 30m, 40m, 60m, 80m, 100m in the central line; and CO_2 flow speeds were 27.8m/s, 33.5m/s, 42.1m/s and 50.3m/s after scaling instead of 8.8m/s, 10.6m/s, 13.3m/s and 15.9m/s in the field experiment; and the release time was prolonged approximately by three times and became 632s; and CO_2 flow rates became 3162m³/h, 3795m³/h, 4743m³/h and 5692m³/h instead of 10m³/h, 12m³/h, 15m³/h and 18m³/h. Gravity kept identical in the reduced -scale experiment and in the original model. In addition, wind speed was also scaled from 0.6 m/s at 10m height to 1.9m/s at 100m height. Table 2 shows CO_2 concentration at different positions on the central line after scaling in the original model.

Table 2　CO_2 Concentration at the Central Line on the Ground in the Original Model after Scaling

Flow rate (m³/h)	CO_2 concentration of different coordinates at 632s after CO_2 release (%)									
	(15, 1.5, 0)	(20, 1.5, 0)	(25, 1.5, 0)	(30, 1.5, 0)	(35, 1.5, 0)	(40, 1.5, 0)	(50, 1.5, 0)	(70, 1.5, 0)	(90, 1.5, 0)	(110, 1.5, 0)
10	0.8	1.3	1.9	2.2	1.4	1.0	0.6	0.4	0.2	0.1
12	0.6	0.5	0.8	1.4	1.6	1.2	0.7	0.4	0.3	0.2
15	0.6	0.4	0.3	0.8	1.4	1.5	1.0	0.5	0.4	0.2
18	0.4	0.2	0.5	0.6	0.9	1.4	1.3	0.7	0.5	0.2

CO_2 concentrations of at the central line on the ground were compared between the simulation and the scaled results from the field experiment at different flow rates, as Fig. 2 shown. It was ap-

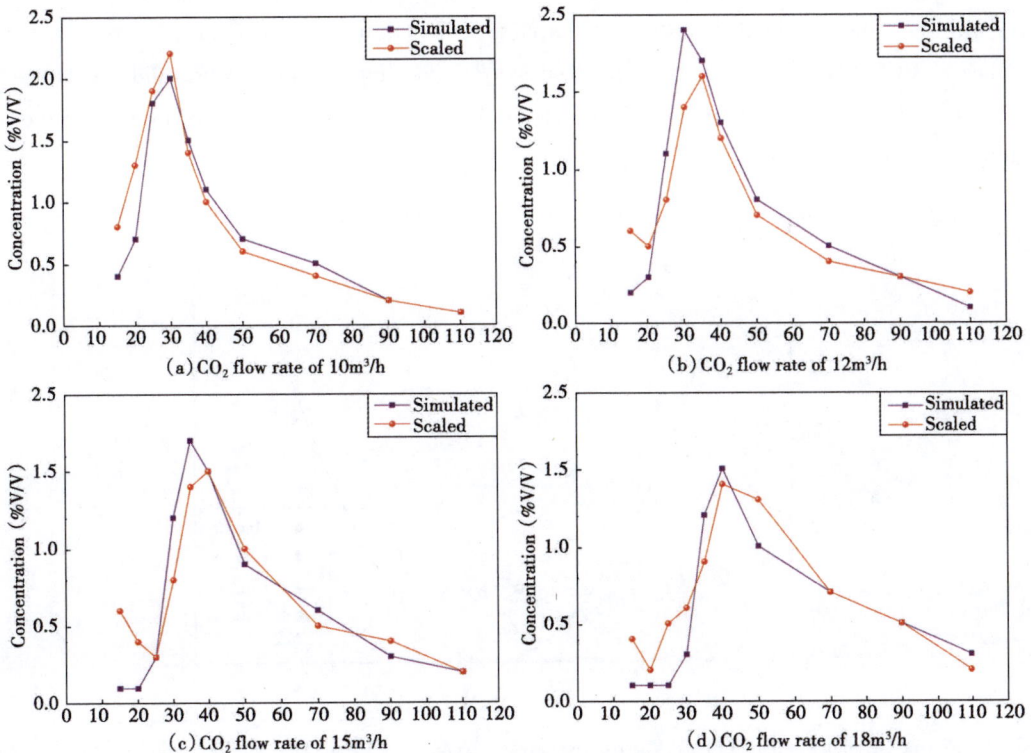

Fig. 2　CO_2 Concentration Comparisons between the Simulation and the Scaled Results from the Field Experiment

parent that the simulation and the scaled results exhibited a good accordance for concentration tendency except near the jet nozzle. With increasing the flow rate, the concentration difference near the jet nozzle became obvious. For the flow rate of $10m^3/h$, the concentration trend kept identical even near the jet nozzle, but the trend discrepancy came out before 20m for the flow rates of $12m^3/h$ and $15m^3/h$. The inconformity for the flow rates of $12m^3/h$ continued until 30m. The scaled results had higher CO_2 concentration than the simulation. It is because that the CO_2 diffusion near the jet nozzle in the real scenario performances more intensely than in the simulation. Therefore, the difference between the simulation and the scaling methods mainly focused on the near area by the CO_2 outlet, and the scaling method could really show the tendency of CO_2 concentration variation.

Table 3 Summary of the Statistical Performance for the Original Model after Scaling Comparing to the Simulated CO_2 Concentration Values in the Central Ground Line Parallel to the Wind Velocity; the Acceptable Ranges are 0.7<MG<1.3, VG<4, 0.5 < FAC2<2, and an Ideal Model Produces MG = VG = FAC2 = 1.0

Flow rate (m^3/h)	MG	VG	FAC2
10	0.912	1.10	1.0
12	0.889	1.25	0.9
15	0.876	2.14	0.7
18	0.678	1.78	0.8

To verify the scaling rules, concentration values after scaling in the field experiment were compared with those in the simulation through the statistical performance indicators. Through the performance of MG, VG and FAC2, the effectiveness of the scaling rules could be validated. Except a low MG=0.678 for the flow rate of $18m^3/h$ outside the acceptable range, others all arise in the reasonable region according to Chang's criteria. The FAC2 was the most robust performance indicator

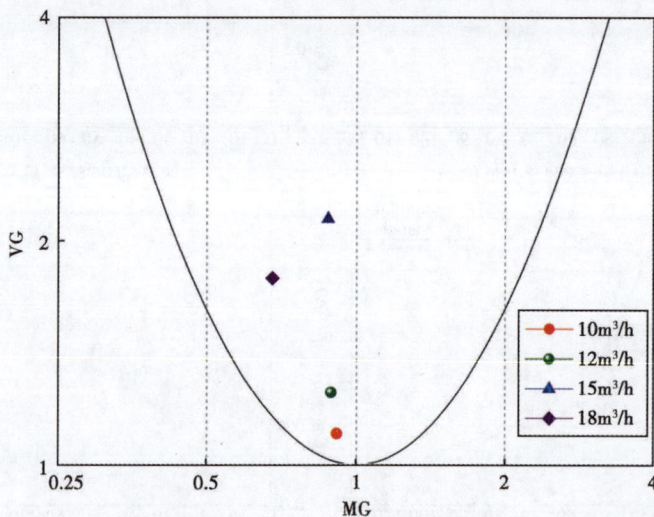

Fig. 3 Plot of Geometric Mean (MG) Versus Geometric Variance (VG) for CO_2 Concentration for the Original Model after Scaling the Field Experiment. The Solid Parabola Represents the Minimum VG for a Given Value of MG.

because it was not considerably influenced by exceeding high or too low values. The FAC2 = 1 for the flow rate of $10m^3/h$ shows the best performance followed by that for $12m^3/h$. The FAC2 = 0.7 for the flow rate of 15 m^3/h presented the worst in all the cases but still in the acceptable arrange of FAC2 >0.5. For MG and VG, the value nearer to MG = 1 and VG = 1 is better. As Fig. 2 Shown, it is easily seen that the MG and VG of the flow rate of 10 m^3/h is the most satisfied due to the nearest distance from the point of MG = 1 and VG = 1, however, the MG and VG for 15 m^3/h and 18 m^3/h is further from the point (1, 1). Hence, we can get the conclusion that the scaling rules bring more accuracy when flow rates are not too high. It is because that increasing the flow rate has to improve the CO_2 outlet speed that means the height of CO_2 spilling comes higher. It will take more time for spilled gas to return the ground and the distribution of CO_2 on the ground becomes harder for prediction. So the accuracy of the scaling rules for concentration prediction descends with the flow rate increasing.

7 Conclusions

A series of scaling rules was introduced and used to scale the field experiment on CO_2 dispersion, which was conducted by Beijing Institute of Technology. The length scaling factor 1/10 was applied for scaling this field experiment. To validate the scaling rules, the successful $k-\varepsilon$ turbulence model was used to simulate the CO_2 concentration distribution of the scaled original model for comparison. The whole tendencies of two methods kept identical except nearby the jet nozzle where the measured CO_2 concentration from the experiment was obviously higher than that in the simulation. The results of comparison were evaluated using statistical performance indicators, including MG, VG and FAC2. It was shown that all the VG and FAC2 results were satisfactory at different flow rates according to the Chang's criteria. The values of MG represented the descending trend for accuracy with the flow rate increasing, and even the MG = 0.678 for $18m^3/h$ was outside the acceptable range. Hence, it was concluded that using the scaling rules to scale the field experiment exhibited the acceptable accuracy at small flow rates. Through the comparison with the $k-\varepsilon$ model and the statistical performance indicators, we concluded that the scaling rules were useful for the field experiment of accidental releases.

References

[1] Jiang X. A review of physical modelling and numerical simulation of long−term geological storage of CO_2 [J]. Appl Energy 2011, 88: 66−3557.

[2] Zhu L, Fan Y. A real options−based CCS investment evaluation model: Case study of China's power generation sector [J]. Appl Energy 2011, 88: 33−4320.

[3] Liang X, Reiner D, Li J. Perceptions of opinion leaders towards CCS demonstration [J]. Appl Energy 2011, 88: 85−1873.

[4] Wee J. A review on carbon dioxide capture and storage technology using coal fly ash [J]. Appl Energy 2013, 106: 51−143.

[5] Uddin M, Jafari A, Perkins E. Effects of mechanical dispersion on CO_2 storage in Weyburn CO_2−EOR field— Numerical history match and prediction [J]. Int J Greenhouse Gas Control 2013, 16: 35−49.

[6] Witlox H W M, Harper M, Oke A. Modelling of discharge and atmospheric dispersion for carbon dioxide releases [J]. J. Loss Prev. Process Ind. 2009, 22: 795-802.

[7] Kumar A, Mahurkar A, Joshi A. Study of the spread of a cold instantaneous heavy gas release with surface heat
transfer and variable entrainment [J]. J. Hazard. Mater. 2003, 101: 77-157.

[8] Viebahn P, Daniel V, Samuel H. Integrated assessment of carbon capture and storage (CCS) in the German
power sector and comparison with the deployment of renewable energies [J]. Appl Energy 2012, 97: 48-238.

[9] Damao L A, Eboibi O, Howell R. An experimental investigation into the influence of unsteady wind on the performance of a vertical axis wind turbine [J]. Appl Energy 2013, 107: 11-403.

[10] Zhou Y, Lu Z, Tang S, et al. Research advances of heavy gas dispersion [J]. J. Saf. Eviron. 2012, 12:
47-242.

[11] Hachicha A A, Rodriguez I, Castro J, et al. Numerical simulation of wind flow around a parabolic trough solar
collector [J]. Appl Energy 2013, 107: 37-426.

[12] Wu Z, Caliot C, Bai F et al. Experimental and numerical studies of the pressure drop in ceramic foams for volumetric solar receiver applications [J]. Appl Energy 2010, 87: 13-504.

[13] Wang D. Effects of the earth's rotation on convection: Turbulent statistics, scaling laws and Lagrangian diffusion Dyn [J]. Atmos. Oceans 2006, 41: 20-103.

[14] Suga K. Computation of high Prandtl number turbulent thermal fields by the analytical wall-function [J]. Int.
J. Heat Mass Transfer 2007, 50: 74-4967.

[15] Abel M S, Mahesha N, Tawade J. Heat transfer in a liquid film over an unsteady stretching surface with viscous dissipation in presence of external magnetic field [J] . Appl. Math. Modell. 2009, 33: 41-3430.

[16] Koopman R P, Ermak D L. Lessons learned from LNG safety research [J]. J. Hazard. Mater. 2007, 140:
28-412.

[17] Anay L H. A review of large-scale LNG spills: Experiments and modeling [J]. J. Hazard. Mater. 2006,
A132: 40-119.

[18] Pitblado R, Baik J, Ranhunathan V. LNG decision making approaches compared [J]. J. Hazard. Mater.
2006, 130: 54-148.

[19] Mazzoldi A, Hill T, Colls J J. CFD and Gaussian atmospheric dispersion models: A comparison for leak from
carbon dioxide transportation and storage facilities [J]. Atmos. Environ. 2008, 42: 54-8046.

[20] Hanna S R, Hansen O R, Ichard M, Strimaitis D. CFD model simulation of dispersion from chlorine railcar releases in industrial and urban areas [J]. Atmos. Environ. 2009, 43: 70-262.

[21] Sun B, Utikar R P, Pareek V K, et al. Computational fluid dynamics analysis of liquefied natural gas dispersion for risk assessment strategies [J] . J. Loss Prev. Process Ind. 2013, 26: 28-117.

[22] Khatir Z, Paton J, Thompson H, et al. Computational fluid dynamics (CFD) investigation of air flow and temperature distribution in a small scale bread-baking oven [J]. Appl Energy 2012, 89: 89-96.

[23] Sklavounos S, Rigas F. Validation of turbulence models in heavy gas dispersion over obstacles [J]. J. Hazard. Mater. 2004, 108: 9-20.

[24] Hanna S R, Hanseh O R, Dharmavaram S. FLACS CFD air quality model performance evaluation with Kit
Fox, MUST, Prairie Grass, and EMU observations [J]. Atmos. Environ. 2004, 38: 87-4675.

[25] Tauseef S M, Rashtchian D, Abbasi S A. CFD-based simulation of dense gas dispersion in presence of obstacles [J]. J. Loss Prev. Process Ind. 2011, 24: 76-371.

[26] Hanna S R, Chang J C. Use of the Kit Fox field data to analyze dense gas dispersion modeling issues [J]. Atmos. Environ. 2001, 35: 42-2231.

[27] Weydahl T, Jamaluddin J, Seljeskog M, et al. Pursuing the pre-combustion CCS route in oil refineries-The
impact on fired heaters [J]. Appl Energy 2013, 102: 39-833.

[28] Raj A R G S, Mallikarjuna J M, Ganesan V. Energy efficient piston configuration for effective air motion-A CFD study [J]. Appl Energy 2013, 102: 54-347.

[29] Launder B E, Spalding D B. Lectures in Mathematical Models of Turbulence [M]. London: Academic Press. 1972.

[30] Chang J C, Hanna S R. Technical description and User's guide for the BOOT statistical model evaluation software package, Version 2. 0 [M]; 2005.

[31] Bansal V, Misra R, Agarwal GD, et al. 'Derating Factor' new concept for evaluating thermal performance of earth air tunnel heat exchanger: A transient CFD analysis [J]. Appl Energy 2013, 102: 26-418.

CO$_2$ 管线泄漏扩散小尺度实验研究

刘振翼[1]　周　轶[1]　黄　平[1]　孙瑞艳[2]　王世刚[2]　马晓红[2]

(1. 北京理工大学爆炸科学与技术国家重点实验室;

2. 中国石油吉林油田分公司勘察设计院)

摘　要：CO$_2$ 气体综合开发和利用过程中涉及很多安全问题, 为了研究 CO$_2$ 作为典型重质气体从输送管线泄漏后的扩散规律, 在综述重气扩散实验的基础上设计了 CO$_2$ 管线泄漏扩散缩比例实验; 通过幂函数拟合得到监测点峰值浓度曲线, 并据此确定了生产现场的安全距离; 根据危险气体非正常排放模型对实验进行了理论预测, 结果与实验值进行了对比。结果表明小风条件下, 当以 950 m^3/h 初始流量释放 20 min 左右的 CO$_2$ 时, 现场安全距离可设为 46m, 而当初始流量为 6300 m^3/h 时, 现场安全距离为 160m; 实验值和理论预测值的偏差最小为-3.0%, 最大为 28.1%。经分析可知 CO$_2$ 管线泄漏近源区一定距离内主要靠射流动能扩散, 其后范围主要受浓度梯度的影响, 危险气体非正常排放模型可较准确的预测近源区的扩散情况; 安全距离与泄放流量基本成指数关系。

关键词：CO$_2$ 扩散; 管线泄漏; 小尺度实验; 安全距离

0 引言

近年来, 温室气体排放引起全球气候变暖的问题日益凸显, 其中 CO$_2$ 的作用占 63%。为缓解全球气候变化的影响, 一些发达国家大力研究 CO$_2$ 捕集与封存 (CO$_2$ capture and storage , CCS) 技术。CCS 技术包括三个方面[1]：一为捕集, 即收集并浓缩排放的 CO$_2$; 二为运输, 即将捕集的 CO$_2$ 输运到合适的封存地点; 三为封存, 即将 CO$_2$ 注入地下地质构造中, 或通过工业流程使之固化为碳酸盐。现阶段 CO$_2$ 地质储存方法主要有：注入油气井提高油气采收率[2] (CO$_2$ enhanced oil recovery, CO$_2$-EOR), 注入煤层获得甲烷, 注入废弃油气田、地下咸水层和海底储存。在油田, 利用 CO$_2$ 提高油田采收率 (EOR) 已经获得使用, 主要是利用注入 CO$_2$ 来溶解孔隙当中的剩余油[3]。该方法既能够一定程度上提高油田采收率, 又能较有效地存储 CO$_2$。

CO$_2$ 主要通过管线实现大规模运输。在世界范围内有多条 CO$_2$ 长输管线正在运营, 随着 CCS 技术的深入应用, 更多的 CO$_2$ 管网将会投入使用。由于 CO$_2$ 具有弱极性, 易溶于水并反应生成 H$_2$CO$_3$, 长期作用会导致管线腐蚀发生泄漏, 会对人体和环境造成重大危害[4]。常温条件下, CO$_2$ 气体对空气的相对密度约为 1.5, 大量的 CO$_2$ 气体突然释放与扩散具有典型的重气扩散特征, 易在地势较低的区域形成高浓度区。绝热瞬时泄漏情况下 (如从高压运输设施的泄漏), 温降超过 100℃ (焦耳—汤姆逊效应) 使泄漏源附近气体密度增大到约为 2.8kg/m^3。这增加了气体贴近地面的趋势, 且低温易造成人体冻伤。在复杂地形和低风速时这种情况更为严重[5,6]。

CO_2 是一种能使脑血管扩张的窒息性气体，在空气中的浓度高于 7% 时，会造成循环功能障碍从而导致昏迷甚至死亡。CO_2 有三种浓度阈值[7]：1%、3% 和 10%。当体积浓度达到 3% 时，人会出现头痛、弥漫性出汗和呼吸困难等症状。我国国家标准 GBZ2.1—2007《工作场所工业有害因素职业接触限值》中规定 CO_2 的时间加权平均容许浓度 PC-TWA 为 9000mg/m³，短时间（15min）接触容许浓度 PC-STEL 为 18000mg/m，相当于 1%（体积分数），本文以此值作为划分安全距离的临界浓度值。

目前，国内外研究人员对 CO_2 管线运输的安全性研究主要包括过程风险分析[2-3]、CO_2 泄漏扩散后浓度场和速度场变化的数值模拟[8]等方面，对 CO_2 管线泄漏扩散的实验研究鲜有报道。在含 CO_2 天然气田的发现和开发中，CO_2 泄漏扩散安全防护距离的确定是一个重要的安全问题。本文通过对 CO_2 管线泄漏进行缩比例实验研究的方法来确定生产现场的安全距离，对生产实际中制定应急预案、采取有效现场救护措施并最终减少事故伤亡损失具有重要意义。

1　缩比例原则

缩比现场实验即参照相似准则将实际情形缩小的野外实地实验，选择真实、适合的地形和气象条件，所用释放源的几何、力学和运动学特性等都根据相似原理加以缩小❶，测量得到的浓度场、流场数据根据相似原理[8]再换算到实际释放的情形。与全尺寸现场实验相比，缩比实验[9]可以大大减少实验物质的消耗，节省成本，降低实验风险。本文对 CO_2 以不同泄放流量的管线泄漏进行了缩比例现场实验，研究了 CO_2 气体泄漏后的扩散规律。

1.1　长度比例

以大气环境下的气体体积代替（以小写字母代表小型实验，大写字母代表全尺寸实验）危险气体的长度尺度：$L = V^{1/3}$；$l = v^{1/3}$。假设大气边界层的湍流强度仅受空气动力学粗糙 z_0 度控制，根据几何相似准则，有：

$$\frac{l}{L} = \frac{z_0}{Z_0} = \frac{h}{H} = S_c \tag{1}$$

由于实验中使用同一种危险气体，因此：

$$\frac{v}{V} = \frac{m}{M} = \left(\frac{l}{L}\right)^3 = S_c^3 \tag{2}$$

1.2　速度比例

小型实验中，要求危险气体与大气密度比 ρ_g/ρ_a、Froude 数 F_r、Richardson 数 R_i 分别相等。只要前两个量满足，第三个量自动相等。故当速度满足下面条件时可达到要求：

$$\frac{u}{U} = \left(\frac{l}{L}\right)^{0.5} = S_c^{0.5} \tag{3}$$

---

❶　郑远攀，《含二氧化碳天然气燃爆特性及扩散危险性研究》，学位论文，北京理工大学，2009 年.

1.3 释放流量

对于连续释放，体积流量的量纲 1 释放速率是 $\dfrac{Q}{UL^2}$，无论在缩比试验还是在全尺寸实验中这个量纲 1 数值必须相同，因此有：

$$\frac{q}{Q} = \frac{u}{U}\left(\frac{l}{L}\right)^2 = S_c^{2.5} \tag{4}$$

1.4 现场浓度的时间对应

在小型现场实验与全规模实验中量纲 1 量 $\dfrac{UT}{L}$ 应当保持不变，即：

$$\frac{UT}{L} = \frac{ut}{l} \tag{5}$$

由于 $\dfrac{u}{U} = S_c^{0.5}$，于是：

$$\frac{t}{T} = \frac{U}{u} \cdot \frac{l}{L} = S_c^{0.5} \tag{6}$$

1.5 浓度场的实验与现场对应

对于关心区域的任意量纲 1 空间点 (x, y, z)，在给定风场和给定的空间点源排放条件下，其量纲 1 浓度 C^* 可表示为[●]：

$$C^* = \frac{CU\sigma_Y\sigma_Z}{C_0Q} = \frac{cu\sigma_y\sigma_z}{c_0q} \tag{7}$$

式中，C^*、c（小型实验）和 C（全尺寸实验）为无量纲量，采用体积浓度，即气体在大气环境中所占体积比，单位为 1；x 方向为顺风向下游；y 方向为水平面垂直于 x 轴方向；z 方向为垂直于地面向上；坐标原点为释放源地表；σ_y 和 σ_z 为下风向距离 x 处的水平和垂直方向的扩散参数，它们都是 x 的函数，m。

根据高斯烟羽模型，有：

$$\sigma_x = \sigma_y = \gamma_1 x^{\alpha_1}, \quad \sigma_z = \gamma_2 x^{\alpha_2} \tag{8}$$

其中，α_1，α_2 和 γ_1，γ_2 在不同的大气稳定度条件下有不同的值。

这样从模型浓度到原型浓度的转换关系为[10]：

$$C = \frac{u}{U} \times \frac{C_0Q}{c_0q} \times \frac{\sigma_y\sigma_z}{\sigma_Y\sigma_Z}c \tag{9}$$

● 吴庆善，《含碳天然气田的钻井井喷风险研究》，学位论文，北京理工大学，2009 年.

2 CO_2 泄漏缩比例实验

本文共进行了 6 组 CO_2 气体泄漏扩散实验，每次实验泄放持续时间约为 20min，泄漏强度分别为 $3m^3/h$、$8m^3/h$、$10m^3/h$、$12m^3/h$、$14m^3/h$、$20m^3/h$。

2.1 气象测量

气象条件是影响气体扩散的重要因素，因此实验中测量了环境温度、相对湿度以及风向风速几个气象参数。其中温度和相对湿度由温湿表测量，采用多次测量取平均值的方法，仪器置于距地面 1.5m 高处。温湿表的温度测量范围为 $-20 \sim 50℃$，相对湿度测量范围为 $0 \sim 100\%$。风向风速由深圳弗兰德电子有限公司的 FYF-1 型电接风向风速仪测量，仪器置于实验场地地面。

2.2 气体泄放装置

实验物质是经过高压压缩到 40L 钢瓶中的 CO_2 气体，纯度为 99.9%，充装压力为 5MPa。实验时气瓶出口连接软管形成地面释放源，泄放流量则通过减压阀控制，数值由玻璃转子流量计读出。

2.3 浓度测量及测点布置

CO_2 浓度由量程 $0 \sim 100\%$（体积），分辨率 0.1%（体积）的 CO_2 红外传感器测量，采用非色散红外检测原理。浓度数据由 THY-SL 气体报警控制器采集为 RS485 信号，通过 RS485 转 USB 转接头转化为 RS232 信号传输到计算机中；通过与气体报警控制仪配套的上位机软件实时显示测量浓度值和变化曲线。实验仪器各参数见表 1。

表 1 实验仪器列表

用途	仪器名称	型号	相关参数	功能	测量原理/方法	布置方式
气象测量	风向风速仪	FYF-1	—	测量风向、风速	多次测量取平均值	实验场地地面
	温湿表		温度 $-20 \sim 50℃$，相对湿度 $0 \sim 100\%$	测量温度、相对湿度		悬挂于 1.5m 高处
浓度检测	CO_2 传感器	TB10	量程 $0 \sim 100\%$（体积分数）精度 0.1%（体积分数）	测量 CO_2 浓度	红外	各取样点地面
	气体报警控制器	THY-SL	40 通道	采集浓度信号	—	—
	上位机软件		—	将采集器信号传输到计算机		

当流量为 $3m^3/h$、$8m^3/h$、$10m^3/h$ 时（记为情况 1）监测点布置如图 1 所示。沿下风向轴线方向共设置 10 个 CO_2 监测点，近源区稠密，较远处稀疏。由于 CO_2 扩散具有典型的重气扩散特征，泄漏后因重力作用沉降并沿地表扩散，故 CO_2 红外传感器放置于监测点地面，最远监测点距释放源 10m。

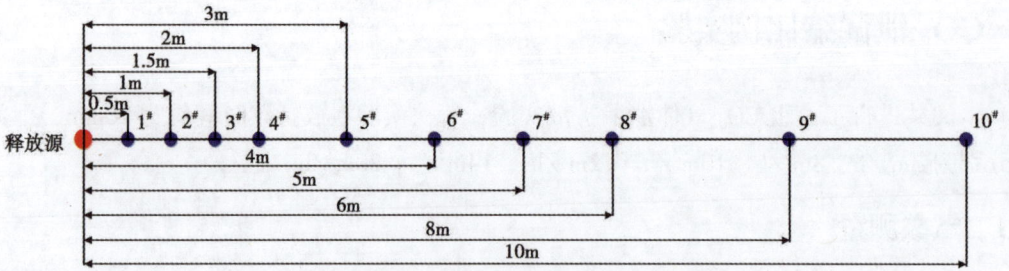

图 1　情况 1 测点布置图

流量为 12m³/h、14m³/h 和 20m³/h 时（记为情况 2）测点布置方式如图 2 所示。与图 1 不同的是，相邻监测点的距离增大，最远监测点距释放源 20m。

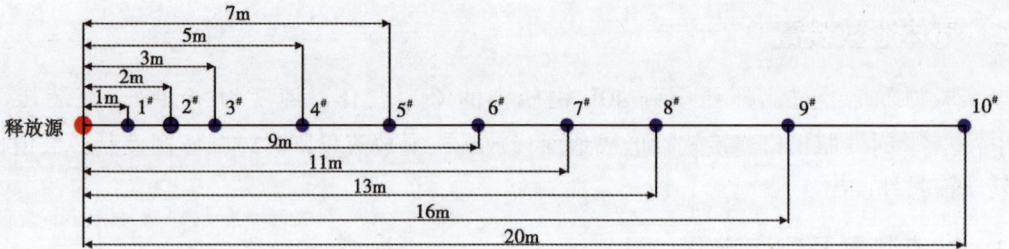

图 2　情况 2 测点布置图

2.4　实验步骤

CO_2 属典型重气，泄放后沿地表扩散，浓度场分析为实验的主要研究目标。

2.4.1　实验准备阶段

（1）在气瓶瓶口安装减压阀，读取瓶内气体初始压强。通过软管将减压阀出气口与流量计进气口连接，流量计的出气口连接软管作为气体泄放口，即完成气体泄放装置的组装。

（2）每次实验前，必须检查气体泄放装置的气密性。

（3）将 CO_2 传感器按监测点布置图（图 1）中位置排放好，并通过电源线路和 24V/220V 稳压开关为传感器供电，通过信号线路将传感器与气体报警控制仪连接。后通过 RS485/RS232 转接头转换信号，继而采集到计算机中，通过配套上位机软件读取数据和变化曲线。

（4）将风向风速仪置于实验场地地面，温湿表置于距地面 1.5m 高处，检查准确性。

2.4.2　试验阶段

（1）打开 CO_2 气瓶气阀，通过减压阀控制泄放流量为 3m³/h、8m³/h、10m³/h，12m³/h、14m³/h、20m³/h，气体开始泄漏扩散，气体报警控制仪开始记录数据，时间间隔为 1s。

（2）实验中每隔 15min 记录一次风向、风速、相对湿度数值。

（3）25min 后关闭减压阀停止泄放气体，流量计转子回到零点，数据采集系统继续工作，直到上位机软件中显示各传感器浓度降到零值，本次实验结束。

3 实验结果分析

取缩放比例为 1:10，即 $S_c = \dfrac{1}{10}$。实验结果包括气象测量条件、泄放参数以及根据缩比原则换算得到的原型实验参数汇总情况见表 2。

表 2　实验条件表（气象条件和泄放参数）

		实验 1	实验 2	实验 3	实验 4	实验 5	实验 6
泄放流量（m³/h）	模型	3	8	10	12	14	20
	原型	950	2500	3200	3800	4400	6300
温度（℃）		32	16	35	21	15	17
相对湿度（%RH）		33	52	30	35	57	32
风速（m/s）	模型	0.5	0.5	0.4	0.5	0.6	0.6
	原型	1.6	1.6	1.3	1.6	1.9	1.9
风向		东北风向					
泄放时间（min）	模型	20	20	20	20	20	20
	原型	65	65	65	65	65	65

国家推荐标准"环境影响评价技术导则/大气环境"HJ/T2.2-93[11]中建议采用帕斯奎尔稳定度分类方法，该方法根据五类地面风速、三类日间的日射和两类夜间云量把扩散天气分成六类，即 A 强不稳定、B 不稳定、C 弱不稳定、D 中性、E 较稳定和 F 稳定。综合实验环境情况，我们认为符合 D 类大气稳定度条件，故满足：

$$\gamma_1 = 0.110726, \quad \alpha_1 = 0.929481;$$
$$\gamma_2 = 0.104634, \quad \alpha_2 = 0.826212$$

根据公式（9）将实验浓度换算到现场原型。由前文内容知，风场满足 $\dfrac{u}{U} = S_c^{0.5} = 10^{-0.5}$，

$\dfrac{\sigma_y \sigma_z}{\sigma_Y \sigma_Z} = (S_c)^{\alpha_1 + \alpha_2} = 10^{-1.76}$，$\dfrac{Q}{q} = S_c^{-2.5} = 10^{2.5}$，$C_0 = 99.99\%$，$c_0 = 99.9\%$，换算得 $\eta = \dfrac{u}{U} \cdot \dfrac{C_0 Q}{c_0 q} \cdot$

$\dfrac{\sigma_y \sigma_z}{\sigma_Y \sigma_Z} = 1.74$，故有 $C = \eta c = 1.74 c$。

3.1　情况 1：CO_2 浓度变化情况

情况 1 按 1:10 的缩放比例转换到现场原型，泄放流量分别为 950m³/h、2500m³/h、3200m³/h，该种情况下平均风速为 1.5m/s。现场原型各相应测点浓度随时间变化如图 3 所示，气体开始泄放后，各监测点传感器按由近及远的顺序逐次响应，并且浓度迅速升高。流量为 950m³/h 时，5m 处监测点 1# 在 26 s 时刻达到峰值浓度 20%，10m 处监测点 2# 在 64 s 时刻上升到最大值 7.5%；流量为 3200m³/h 时，5m 处监测点 1# 在 15s 时刻达到峰值浓度 45%，10m 处监测点 2# 在 25s 时刻上升到最大值 10%。由此可见，距释放源越近，峰值浓度

出现得越早，且峰值浓度越高；泄放流量越大，响应时间越短，峰值浓度越高。同时由图3可知，距离释放源越近峰值浓度持续时间越长。

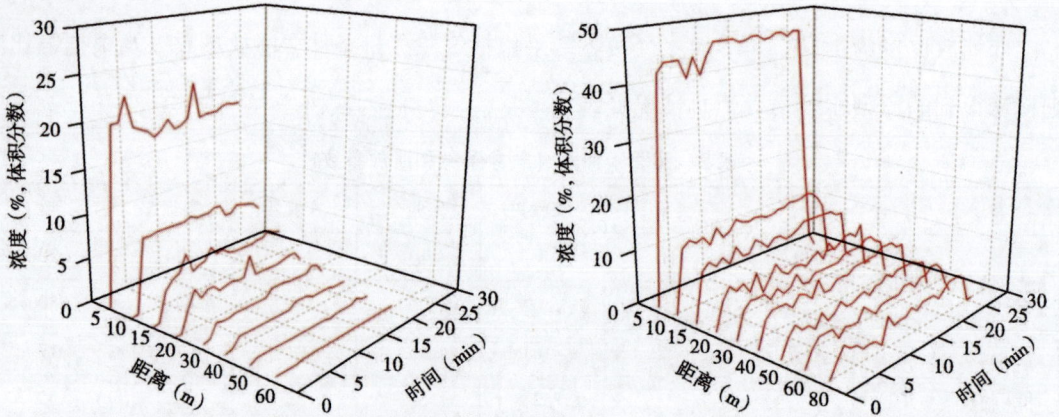

图3　原型流量为 $950m^3/h$ 和 $3200m^3/h$ 时气体浓度随下风向距离、时间变化三维示意图

3.2　情况2：CO_2 浓度变化情况

情况2按1:10的缩放比例转换到现场原型，泄放流量分别为 $3800m^3/h$、$4400m^3/h$、$6300m^3/h$，平均风速为 $1.8m/s$。

由前面分析可知，地面管线泄漏后 CO_2 很快扩散形成稳定流场，且浓度分布呈随距释放源距离增加而逐渐减小的基本规律。如图4所示，流量为 $3800m^3/h$ 时，10m处监测点 $1^\#$ 在16s时刻达到峰值浓度16%，20m处监测点 $2^\#$ 在24s时刻上升到最大值6%；流量为 $6300m^3/h$ 时，10m处监测点 $1^\#$ 在16s时刻达到峰值浓度33%，20m处监测点 $2^\#$ 在21s时刻上升到最大值10%。情况2遵循情况1的基本规律，与情况1不同的是，当流量由 $3800m^3/h$ 增大到 $6300m^3/h$ 时，峰值浓度到达时间缩短不是非常明显，这可能是由于传感器响应时间的限制等。

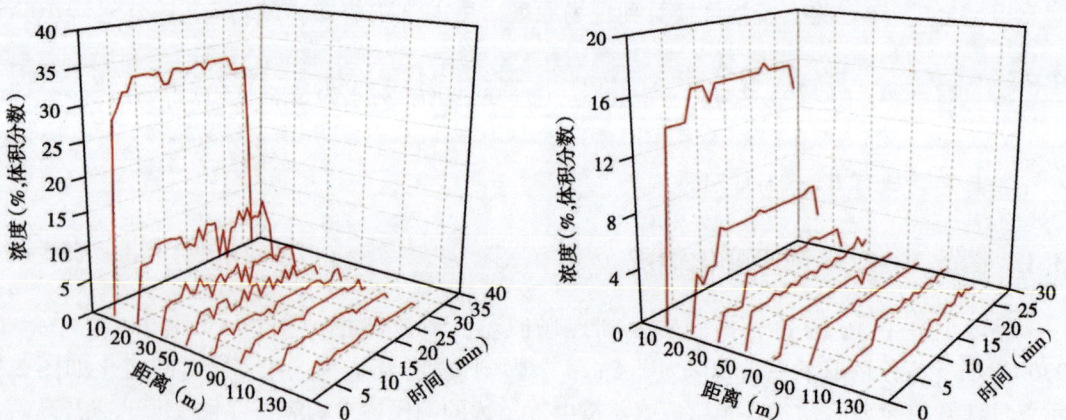

图4　原型流量为 $3800m^3/h$ 和 $6300m^3/h$ 时气体浓度随下风向距离、时间变化三维示意图

3.3 峰值浓度及安全距离的确定

根据监测点数据可以得到轴线峰值浓度随下风向距离的分布。经分析可知，下风向近源区峰值浓度随距离的变化呈幂函数关系，如式 $C_{max}(x) = K \cdot x^{-\alpha_E}$，（式中，$K$ 是指前因子，与初始释放量有关；α_E 是指数因子，决定气体浓度扩散的快慢程度）[8]。图5、图6给出了各释放流量下峰值浓度与下风向距离的拟合关系。

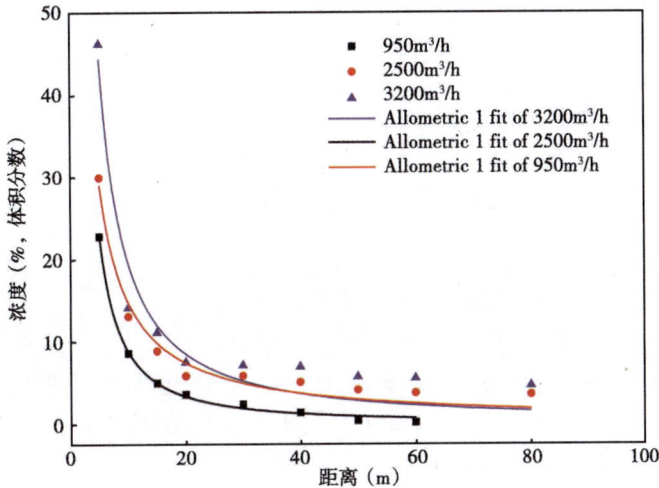

图5 原型现场泄放流量为950m³/h、2500m³/h、3200m³/h
时峰值浓度与下风向距离拟合关系

3.3.1 情况1中3种流量峰值浓度与下风向距离拟合关系

情况1三种流量条件下，下风向轴线峰值浓度与下风向距离均符合幂函数关系，各拟合关系式见表3。从图5中可清楚观察到，距泄放源最近的测点 CO_2 气体浓度很高，当流量为3200m³/h 时浓度可达46%。该测点附近有大量 CO_2 气体连续泄放出来，由于距泄放源较近，气体自由扩散速度低于射流速度且有新鲜的气体连读补充，故而可维持较高浓度。随着距离的增加，空气卷吸作用累积浓度迅速下降，并且下降速度逐渐趋于平缓。由图5可知，扩散到一定距离时，各流量下峰值浓度趋于一致，这是因为较远距离处 CO_2 的浓度梯度逐渐降低，最终归于零。

3.3.2 情况2中3种流量峰值浓度与下风向距离拟合关系

情况2三种流量下，下风向轴线峰值浓度与下风向距离的拟合关系见表3。该种情况下监测点位置有所调整，故测点1（10m）相比情况1的测点1（5m）流量虽然增大但气体浓度反而减小，但情况二测点1（10m）与情况一测点2（10m）相比，浓度明显有增大，故图6中曲线是合理的。与情况1相比，情况2具有相同的趋势，距泄放口最近的测点浓度很高，随后迅速下降并逐渐趋于稳定，在一定距离处，各流量浓度趋于一致，由于泄放流量的增大，该一致点有所远移。

生产安全距离是在一定区域范围内，在特定风险源的作用下，假设不存在任何人为影响或风险防护措施等情况下，为保证不损害周围受体而需要在风险源与周围受体之间设定的最小空间水平距离。我国国家标准 GBZ2.1-2007[12] 规定，CO_2 的短时接触浓度阈值为

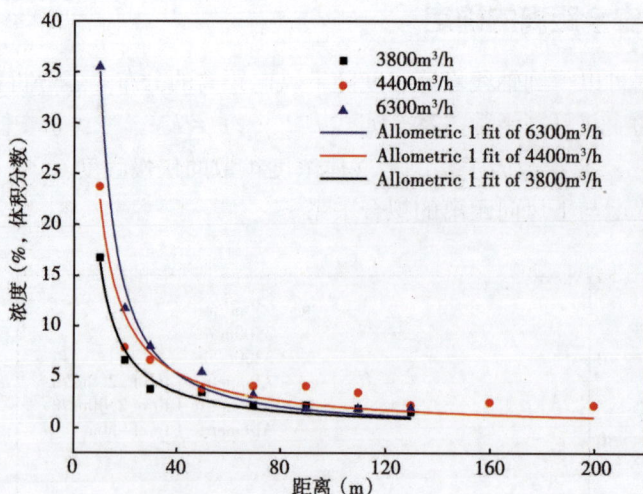

图6 原型现场泄放流量为 3800m³/h、4400m³/h、6300m³/h
时峰值浓度—下风向距离拟合关系

18000mg/m³，相当于 1%（体积），将此值作为 CO_2 浓度的临界值，根据拟合得到的函数关系式可以求得环境风速较小时，各泄放流量下气体扩散的安全距离（表3）。

表3 各泄放流量下气体扩散的安全距离表（平均风速 $u = 1.65m/s$）

泄放流量（m³/h）	拟合关系	相关系数	安全距离（m）
950	$C_{max}(x) = 210.3x^{-1.4}$	0.99	46
2500	$C_{max}(x) = 141.3x^{-1.1}$	0.97	89
3200	$C_{max}(x) = 301.1x^{-1.2}$	0.94	116
3800	$C_{max}(x) = 209.1x^{-1.1}$	0.97	129
4400	$C_{max}(x) = 237.7x^{-1.1}$	0.93	145
6300	$C_{max}(x) = 649.3x^{-1.28}$	0.97	160

经分析在实验设定气象条件下，泄放流量与安全距离大致符合指数关系，安全距离随泄放流量的增大而逐渐增大，到达一定程度后趋于平稳，拟合图形如图7所示，拟合函数为 $sd = 203.4 - 203.4\exp(-0.00026q)$，据此可计算当生产现场流量为 10000m³/h 时，需要的安全距离大致为 190m。

3.4 理论模型预测

根据中华人民共和国环境保护行业标准环境影响评价技术导则大气环境，危险气体非正常排放模式[13]下，t 时刻地面任一点的浓度可按式（10）计算

$$C(x, y) = \frac{Q}{\pi u \sigma_y(x) \sigma_z(x)} \exp\left[-\frac{y^2}{2\sigma_y^2(x)}\right] \tag{10}$$

则下风向地面轴线浓度为：

$$C(x, 0) = \frac{Q}{\pi u \sigma_y(x) \sigma_z(x)} \tag{11}$$

402

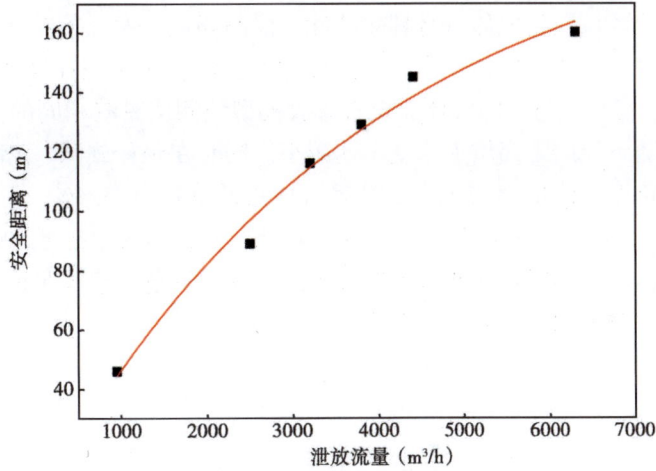

图 7 安全距离与泄放流量的拟合关系曲线图

$$\sigma_x = \sigma_y = \gamma_1 x^{\alpha_1}, \quad \sigma_z = \gamma_2 x^{\alpha_2} \tag{12}$$

D 类大气稳定度条件下，有：

$$\gamma_1 = 0.110726, \quad \alpha_1 = 0.929481;$$
$$\gamma_2 = 0.104634, \quad \alpha_2 = 0.826212$$

根据式（11），对以 950m³/h、6300m³/h 的流量泄放 15min 时，下风向轴线各监测点浓度实验值与理论预测值分别进行比较，见表 4、表 5。

表 4 流量为 950m³/h 时实验值与理论预测值对比表

下风向距离（m）	下风向地面轴线浓度（%，体积分数）		偏差
	实验值	模型预测值	
5	22.8	27.075	−18.7%
10	7.6	8.016	−5.4%
15	4.6	3.935	14.4%
20	2.8	2.375	15.1%

表 5 流量为 6300m³/h 时实验值与理论预测值对比表

下风向距离（m）	下风向地面轴线浓度（%，体积分数）		偏差
	实验值	模型预测值	
10	35.076	44.7	−27.4%
20	10.744	13.2	−22.8%
30	6.320	6.51	−3.0%
50	5.060	3.64	28.1%

重气扩散是一个比较复杂的过程，目前人们对扩散机理的认识有限，现有的理论模型都是基于对扩散现象本身做一定假设后得到的理想化情形，故而实验值和理论预测值必然存在

偏差。危险气体非正常排放模型一般用于非重气扩散，但由于发生泄漏事故后，气体扩散的过程较为复杂，危险区域受到气体自身理化特性、泄漏环境、大气条件、地形和泄漏强度等多种因素的影响，对于某些重气体在小风速条件下的扩散，采用该计算模型也是可行的[5]。由表4、表5可知，实验值与危险气体非正常排放模型预测值具有相同的变化趋势，距释放源越近浓度越高，随下风向距离增加浓度迅速减小。同时在距释放源一定距离范围内，二者结果具有较高的相符度，流量为950m³/h时20m内最低偏差为−5.4%，最高偏差为18.7%；流量为6300m³/h时50m内最低偏差为−3.0%，而最高为28.1%。在更远距离处，由于受模型以及实验值多次换算等得影响，相符度明显下降。这说明危险气体非正常排放模型可以比较准确的预测CO_2气体从管线中泄漏后近源区的扩散情况。

4 结论

（1）本文对CO_2气体管线泄漏扩散进行了缩比例实验研究，得到了CO_2管线泄漏扩散的基本规律：由于泄放口射流的影响，在近源区一定距离内CO_2扩散主要靠射流动能，在其后范围主要受浓度梯度的影响；气体开始泄放后浓度迅速增加到一定峰值，随后保持稳定，直至停止泄放，浓度再次迅速下降；距释放源越近，峰值浓度出现得越早，且峰值浓度越高，峰值浓度持续时间越长；泄放流量越大，响应时间越短，峰值浓度越高。

（2）分析了微风条件下，下风向不同距离处监测点的峰值浓度，并经过幂函数拟合得到气体浓度与下风向距离的关系，如释放流量为950m³/h时，拟合关系式为$C_{\max}(x)=210.3x^{-1.4}$，流量为6300m³/h时，拟合关系式为$C_{\max}(x)=649.3x^{-1.28}$，相关系数最高可达0.99。以1%作为$CO_2$的临界浓度值，计算得到流量为950m³/h时，安全距离可设为46m，而当流量为6300m³/h时，安全距离为160m；综合分析泄漏流量与安全距离的关系，指数拟合得到关系式$sd=203.4-203.4\exp(-0.00026q)$，该关系式对油田$CO_2$管线输送安全距离的确定具有一定指导意义。

（3）实验值与危险气体非正常排放模型理论预测值进行了对比，结果表明在近源区依靠射流动能扩散阶段，实验值与理论预测值相比较保守，而在较远处靠浓度梯度变化阶段实验值稍高于理论预测值。在距释放源一定距离范围内二者具有较高的符合性，偏差最低可达−3.0%，最高为28.1%，危险气体非正常排放模型可以比较准确的预测CO_2管线泄漏近源区的扩散情况。

符号说明

C_0——释放口体积浓度；

F_r——佛罗德数，$F_r=U^2/gL$；

H——原型风速的参考高度，m；

h——实验模型风速的参考高度，m；

Q——现场原型泄放流量，g/s；

q——缩比例实验泄放流量，g/s；

R_i——理查森（Richardson）数，$R_i=g\dfrac{\rho_g-\rho_a}{\rho_a}\cdot\dfrac{L}{U^2}$；

S_c—— 缩比实验与现场原型的缩放比例；

U—— 原型实验和缩比实验环境风速，m/s；

u—— 缩比例实验环境风速，m/s；

z_0—— 紊流的空气动力学粗糙度；

σ_y—— 下风向距离 x 处水平扩散参数，m；

σ_z—— 下风向距离 x 处垂直扩散参数，m。

参 考 文 献

［1］杜磊，湛哲，徐发龙，等大规模管道长输 CO_2 技术发展现状［J］. 油气储运 . 2010，29（2）：86-89.

［2］Gale J，Davison J. Transmission of CO_2——safety and economic considerations［J］. *Energy*. 2004，29（9-10）：1319-1328.

［3］Duncan I J，Nicot J P，Choi J W. Risk Assessment for future CO_2 Sequestration Projects Based CO_2 Enhanced Oil Recovery in the U. S.［J］. *Energy Peocedia*，2009，1（1）：2037-2042.

［4］Molag M，Dam C. Modeling of accidental releases from a high pressure CO_2 pipelines［J］. *Energy Procedia*. 2011，4：2301-2307.

［5］Mazzoldi A，Hill T. A Consideration of the jet-mixing effect when modeling CO_2 emissions from high pressure CO_2 transportation facilities［J］. *Energy Procedia*. 2009，1（1）：1571-1578.

［6］Witlox H W M，Harper M. Modeling of discharge and atmospheric dispersion for carbon dioxide releases［J］. *Journal of Loss Prevention in the Process Industries*. 2009，22（6）：795-802.

［7］Chow F K，Granvold P W. Modeling the effects of topography and wind on atmospheric dispersion of CO_2 surface leakage at geologic carbon sequestration sites［J］. *Energy Procedia*. 2009，1（1）：1925-1932.

［8］Dandrieux A，Dusserre G. Small scale field experiments of chlorine dispersion［J］. *Journal of Loss Prevention in the Process Industries*. 2002，15（1）：5-10.

［9］Dimbour J P，Dandrieux A，Dusserre G. Reduction of chlorine concentrations by using a greenbelt.［J］. *Journal of loss prevention in the process industries*. 2002，15（5）：329-334.

［10］宣捷 . 大气扩散的物理模拟［M］. 北京：中国气象出版社，2000：215.

［11］HJ/T2. 2-93 中华人民共和国环境保护行业标准环境影响评价技术导则大气环境［S］，1993.

［12］GBZ2. 1—2007《工作场所工业有害因素职业接触限值》［S］. 2007.

［13］罗艾民，魏利军 . 有毒重气泄漏安全距离数值方法 .［J］. 中国安全科学学报 . 2005，15（8）：98-108.

405

Study on Safety in Treatment, Transportation and Injection of Carbon Dioxide in Jilin Oilfield

Shuren Liu[1]   Wanli Kang[1]   Shaoran Ren[1]   Ruiyan Sun[2]   Sheng Yu[2]

(1. China University of Petroleum (East China); 2. Jilin Oilfield Company, Ltd.)

Abstract: Carbon dioxide geological storage may be one of the effective measures to mitigating greenhouse gas emissions that is being studied worldwide. Carbon dioxide enhanced oil recovery is a relatively mature technology in petroleum industry by which the gas is injected into an oil reservoir to achieve miscible displacement. Carbon dioxide miscible displacement in Jilin oilfield involves production, purification, liquefaction, transportation and injection to oil wells and carbon dioxide undertakes phase variation in these processes. This paper analyses the safety problems and the causes in dehydration and purification, compression and liquefaction, transportation and injection according to the present situation of these processes. Some well-targeted, technically feasible and economically reasonable safety countermeasure suggestions are put forward to enhance safety management, safety operation and potential treatment. This study has some practical significance to safely implement carbon dioxide miscible displacement in Jilin oilfield.

Keywords: carbon dioxide; miscible displacement; transportation and treatment; injection; safety

0   Introduction

Carbon dioxide is generally acknowledged as a kind of greenhouse gas emissions, which is being studied worldwide on its mitigation. Carbon dioxide geological storage may be one of the effective measures to mitigating greenhouse gas emissions. Carbon dioxide enhanced oil recovery is a relatively mature technology in petroleum industry by which the gas is injected into an oil reservoir to achieve miscible or miscible displacement[1-3]. In South America, carbon dioxide displacement has been one of main enhanced oil recovery technologies owing to its abundant natural carbon dioxide resources. Based on OGJ analysis, 80 out of 94 developed displacement projects worldwide in 2004 belong to America, whose total injection rate reaches about 33×10^6t. At present, a carbon dioxide pipe network with long distance has been constructed to transport natural carbon dioxide to nearby oilfields for enhanced oil recovery in American target zones. Now Weyburn demonstration project of carbon dioxide storage in North America is being developed, which is the first project of carbon dioxide storage and EOR using artificial carbon dioxide resources in a large scale. Since 2000, Canada has transported carbon dioxide from coalification gas factories in northern part of Dakota in America to Weyburn oilfield by a 325km pipe, where carbon dioxide is injected into 19 wells in a rate of

$1.8 \times 10^6 t/a$ with a anticipation of increase of $155 \times 10^6 bbl$ of crude oil and anticipation of $19 \times 10^6 t$ of carbon dioxide storage during the project. With the rising of carbon dioxide geological storage, technologies such as trapping, transportation and injection of carbon dioxide has been hot issues in domestic and abroad studies.

Technology of carbon dioxide displacement in Jilin oilfield has been studied for many years. In view of the reservoir conditions in Jilin oilfield, all the technologies that carbon dioxide miscible displacement involves have been systematically studied. After laboratory investigation and field test have been finished, the matching technologies of carbon dioxide miscible displacement are basically formed. At present, the program using natural associated carbon dioxide for enhanced oil recovery in Jilin oilfield will significantly be the first demonstration project using carbon dioxide resources in large scale in China. Carbon dioxide miscible displacement in Jilin oilfield involves production, purification, liquefaction, transportation and injection to oil wells and carbon dioxide undertakes phase variation in these processes. This paper analyses the safety problems and the causes in dehydration and purification, compression and liquefaction, transportation, and injection according to the present situation of these processes. Some well-targeted, technically feasible and economically reasonable safety countermeasure suggestions are put forward to enhance safety management, safety operation and potential treatment. This study has some practical significance to safely implement carbon dioxide miscible displacement in Jilin oilfield.

1 Current Situation of CO_2 Flooding in Jilin Oilfield

The CO_2 resource of Jilin oilfield is Changshen gas field, which locates in Ganhua town of Qianguo county in Jilin province. This gas field is in the south of Daqingzijing well, and belongs to Yingcheng Formation. Its top buried depth is $3547 \sim 3809m$. There are 3 wells, respectively Changshen 2, 4 and 6 well. The gas temperature of the well head is below to 25°C. The gas is mainly CO_2, accounting for $97.23\% \sim 98.31\%$. There is also $0.877\% \sim 2.47\%$ hydrocarbon gas.

The procedure of the CO_2 miscible displacement is as follows: the gas from the Changshen 4 well is transmitted to the gas-gathering station through the gathering pipeline network, and then is liquefied after purification and dehydration. The liquid CO_2 gained is stored in the storage tank. After that, the liquid CO_2 is transported to the Hei 59 gas injection station through the pressurized pipelines and the pressure of the gas is increased to $25MPa$ there. Finally the gas is injected to the bottom of a well by the gas-injection network. The whole procedure is shown in Fig. 1.

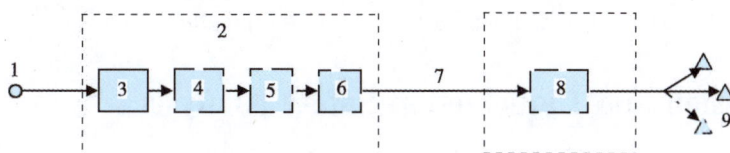

Fig. 1　Procedure of the Surface CO_2 Gathering and Transporting System

1—Changshen 4 gas well; 2—Gas gathering station; 3—Dehydration unit; 4—Liquefaction unit;
5—Storage tank; 6—Booster pump; 7—8km pipeline; 8—Injection station; 9—Injection wells

According to the conditions of the reservoir, wells and surface, Hei 59 block of the Daqingzi-jing field is selected as the test zone. The producing oil-bearing area is $2km^2$, the producing geologic reserve is $86 \times 10^4 t$ in this area.

2　Dehydration and Purification Safety Analysis

Molecular sieve dehydration technology is used to dehydrate CO_2 in Jilin oilfield, which is a procedure of physical adsorption[4-5]. The physical adsorption is caused by Vander Waals attraction or the diffusion force. The gas adsorption is like the gas condensation, which is non-selective and reversible. And at the same time the adsorption heat is small and the activation energy adsorption needed is also small. So the adsorption velocity is large and the reaction can reach balance easily. The dehydration absorption tower used now is fixed bed absorption tower, which is multi-tower process.

As the presence of water in the feed gas, the corrosion and the formation of hydrate need to be controlled. The corrosion of absorption tower is mainly electrochemical corrosion which is caused by carbonic acid produced when carbon dioxide dissolves in water. When CO_2 and water coexist, there are several corrosion types of steel caused by the carbon dioxide aqueous solution. Besides, the corrosion rate is different. The possible corrosion types are: uniform corrosion without carbonate membrane covered; uniform corrosion with carbonate membrane covered; table-shaped corrosion caused by flowing; localized corrosion in the non-membrane area etc[6-7]. Corrosion can cause gas leakage in the absorption tower, even the explosion or other accidents. Corrosion can be prevented by adding corrosion inhibitors into the adsorption tower or coating on the adsorption tower.

Under the condition of a certain temperature and pressure, CO_2 and liquid water can form a kind of white crystalline solid hydrate[8-10]. And the local accumulation of hydrate will reduce the area that the airflow gets through in the gas transmission and absorption tower, limit the flow of gas and reduce the processing volume of gas, even blocking the gas pipelines and other processing facilities in severe cases, which will cause great difficulties in gas dehydration and purification. There are several methods to prevent the formation of hydrate: adding various inhibitors to prevent the formation of hydrate, increasing gas temperature and so on[9].

In prophase of the production, a small amount of welding slag and other debris will be left in the pipeline. It will cause the sealing surface of the ball valve of programmable valve wear and tear to degrade the valve sealing performance and to cause gas leakage. Therefore, when conducting maintenance, the internal leakage valve should be removed and the valve core should be wiped with the water emery cloth to ensure good sealing performance of the programmable valve[11].

3　Compression and Liquefaction Safety Analysis

The low temperature and low pressure liquefaction technology is used in the CO_2 liquefaction in Jilin oilfield. The advantage of this method is using low liquefied pressure under the condition of low temperature, reducing the pressure-resistance requirements of equipments and investment costs, increasing the production capacity significantly and facilitating the transportation[12]. Cryogenic refrigeration equipments are needed in the low-temperature and low-pressure liquefied technology[12-13], Jilin oilfield adopts the liquid ammonia refrigerating machine system, which is designed according to the endothermic property of liquid ammonia when it is vaporizing in the condition of low pressure and low temperature. This system is a closed circulatory system; the ammonia gas is absorbed by refrigeration compressor in the evaporator and then is undertaken compression when it enters into the compressor after passing through the liquid ammonia separator (to prevent the compressor from liquid impact damage) ; compressed high pressure gas goes into the liquid ammonia storage tank with water cooler (high pressure and temperature gas cooled into liquid) ; then the liquid ammonia in the tank enters into the liquid ammonia evaporator after passing through the expansion valve (liquid ammonia vaporized rapidly in the low pressure area) absorbing low temperature heat from the external environment to achieve the purpose of cooling.

The safety problem in this procedure is the leakage of liquid ammonia. Ammonia belongs to the IV light harmful gas according to the "Classification of health hazard levels from occupational Exposure to toxic substance" . People who is exposed to certain concentration of NH_3 may become poisoning. On the other hand, the mixture of air and ammonia may get burning or explosion when the concentration is suitable and there is open fire. This will be more dangerous if there are fuel and other combustible matter around. The explosive limits of mixture of air and ammonia is 16% ~ 25% (the most optimal concentration is 17%)[14-15]. Therefore, inflammable gas detector should be used in the leakage of liquid ammonia and placed where the leakage may occur. When the concentration of ammonia is higher than the limit, the equipment can give alarm signal. The operator can open the emergency switching-off system in time, and then the ammonia resource and the equipment will be shut off.

If the operators will contact with cold gas or cold liquid, they must wear a protective mask, put on leather gloves, and wear no bag trousers and high boots, long-sleeved clothes and so on. When frostbite occurs, frostbited skin should be washed with plenty of warm water. That the injured should not be cured by heat drying method and should be moved to a warm place (about 22). If the injured can not get immediate treatment, he should be sent to the hospital without delay[16]. In case of poisoning, the injured should be sent to the hospital in time.

4　Transportation Safety Analysis

The statistics on the natural gas pipeline safety and hazardous liquid pipeline accidents, including CO_2 pipeline accident data made by Office of the U. S. transportation sector show that from 1990 to 2001, the frequency of CO_2 pipeline accident is lower than that of natural gas pipeline accidents. Jilin oilfield CO_2 pipeline can have the potential safety hazards to the people because CO_2 must be transported through 8km pipeline to the gas injection station after the dehydration liquefaction and the pipeline goes through residential areas, farmland and other areas of intensive human activities. The main causes of the CO_2 pipeline accidents are failure of pressure relief valve, bad sealing performance of welds, gaskets and valves, corrosion, external forces and so on [17-18].

Because of the dehydration treatment of CO_2, corrosion will not be considered here. The external forces including train derailment, vandalism, agricultural activities and other factors by contractors, farmers and workers will cause the pipeline rupture and the leakage of carbon dioxide. The failure of pressure relief valve, bad sealing performance of welds, gaskets and valves, corrosion, will also cause CO_2 leakage.

Automatic control system can be used to monitor the volume flow velocity and pressure fluctuation. Moreover, stop valves should be set every few hundred files along the pipeline to ensure that it can be closed in time when pipeline incidents occur. In the case of pipeline rupture, the sharp drop of the pressure can be quickly monitored by the automatic control system, and the rupture segments should be isolated within a few minutes. Thus, the harm caused by this accident should be reduced to a minimum. CO_2 is a colorless and odorless gas, therefore, the leakage and accumulation of CO_2 can not be easily detected by human and animals. If the mercaptan and other additives are added into CO_2, the odors of which are familiar to people, the potential dangers of CO_2 leakage can be realized when the odors of these additives are smelt, and then the measures will be taken to remove this danger.

5 Injection Safety Analysis

Carbon dioxide is transported to the Hei 59 by pipeline, pressurized by the booster pump and then injected into the reservoir. The sate of CO_2 during injection is liquid state. In the carbon dioxide injection, the leakage of CO_2 through the strings and the corrosion of the strings exist and can cause security problems.

Leakage of carbon dioxide through injection well strings is partly because of the bad bonding performance of the cement and the formation, the cement and the casing in cementing. And CO_2 can leak through the crack between the cement sheath and formation, that between the cement sheath and the casing, and the self-crack of cement; CO_2 leakage is also cause by the injection strings, such as packers, tubing, fittings, valves etc. Cement bond logging can be used to test the quality between cement and cement casing[19]; before a production well changes into a CO_2 injection well, its leakage history should be checked and the detailed workover measures should be checked to ensure the well integrity if this well has leakage history. In addition, tracer can be used to detect the leakage location of the casing. Ultrasonic detector can also be used to detect pipe leakage because the leakage wave frequency is in connection with the pressure drop, leakage location and the gas leakage amount.

In CO_2 injection, corrosion of injection strings will occur[20-22]. The simplest method to detect underground corrosion and erosion is to use multi-arm caliper. Electromagnetic thickness measurement tools can be used to test the underground corrosion and erosion, too. In addition, underground imaging device can also be used to observe underground corrosion. Currently, there are many anti-corrosion measures such as adding corrosion inhibition, using anticorrosion steel pipe and coating of tubing, etc.

410

6 Conclusions

The safety problems caused by CO_2 phase change, such as CO_2 corrosion, hydrate plugging, gas leakage etc; exist during the operation of purification, liquefaction, storage and transportation, injection in CO_2 miscible displacement to improve oil recovery in Jilin oilfield. In order to avoid accidents and to ensure CO_2 miscible displacement implementation safely, effective measures should be taken to strengthen the safety of pipeline, station equipment, automatic control system and other aspects of management.

References

[1] Gozalpour F, Ren S R, Tohidi B. CO_2 EOR and Storage in Oil Reservoirs. Oil & Gas Science and Technology [J]. 2005, 60 (3): 537-546.

[2] IEA Energy Technology Essentials. CO_2 Capture &Storage [M]. 2006.

[3] Lain W, Wright. The In Salah Gas CO_2 Storage Project [J]. IPTC 11326. Presented at the International Petroleum Technology Conference held in Dubai, U. A. E. , 2007.

[4] Hu X M, Lu Y K, Zeng L Q. Description of Molecular Sieve Dehydration Process [J]. Natural Gas and Oil. 2008, 26 (1): 39-41.

[5] Guo Z, Zeng S B, Chen W F. The Application of Molecular Sieve Dehydrator in LNG Project [M]. Chemical Engineering of Oil and Gas. 2008, 37 (2): 138-140.

[6] Kou J, Liang F C, Chen J. Corrosion and Protection of Oil and Gas Pipeline [M]. China Petrochemical Press, Beijing, 2008.

[7] Zhou H L, Cao H X. Research on Tube Corrosion of Natural Gases [J]. Inner Mongolia Petrochemical Industry. 2009, 13: 5-6.

[8] Chen J X. Long Distance Delivery Gas Hydrate Preventive Measures and Dehydration Process [J]. Offshore Oil. 2001, 4: 56-60.

[9] Gao Z H and Yang H W. The Analysis of Carbon Dioxide Hydrate Forming in Gas Pipeline [J]. Low Temperature and Specialty Gases. 2006, 24 (6): 36-38.

[10] Xiao Y, Liu D P, Xie Y M, et al. Current Research Advancement of Carbon Dioxide Hydrates [J]. Natural Gas Geoscience. 2007, 18 (4): 607-610.

[11] Zhao J B, Ai G S and Chen Q H. Optimization of Molecular Sieve Dehydration Technology in the Yengimahalla Condensate Gas Field [J]. Natural Gas Industry. 2008, 28 (10): 113-115.

[12] Yu X C, Li Z J, Zheng X P, et al. Carbon Dioxide Ground Processing, Storage and Transportation [J]. Natural Gas Industry. 2008, 28 (8): 99-101.

[13] Qu T F, Zhang Z X, Wang R Z, et al. Analysis on Energy Conservation of Improved CO_2 Liquefaction Measures [J]. Compressor Technology. 2001, 1: 15-20.

[14] Luo A M, Shi L C, Duo Y Q, et al. Comparative Studies on Accident Mode of Anhydrous Ammonia Leakage [J]. Journal of Safety Science and Technology. 2007, 3 (3): 21-24.

[15] Peng L. Discussion on Environmental Risk Assessment of Accidental Leaking of Liquid Ammonia Storage Tank [J]. Guangdong Chemical Industry. 2009, 36 (4): 110-113.

[16] Liu Y. Dangers and Safeguards of LNG [J]. Natural Gas Industry. 2004, 24 (7): 105-107.

[17] Li G X, Liu Y. Safety Issues and Countermeasures on Long Distance Gas Transmission Pipeline [J]. Oil & Gas Storage and Transportation. 2006, 25 (7): 52-56.

411

［18］ Discussion on the Analysis and its Countermeasure of the Elements on NG Pipeline's Security Harmful ［J］. Shanghai Gas. 2008, 3: 38-41.

［19］ Wei Z L. Zhou C C. Geophysical Log ［M］. Geological Publishing House, Beijing, 2005.

［20］ Yang G, Gong P B, Wang W, et al. Study on CO_2 Corrosion in Oil & Gas Wells ［J］. Corrosion & Protection in Petrochemical Industry. 2009, 26 (1): 11-13.

［21］ Wu M J. Corrosion Study and Management of Surface System of CO_2 Flooding Tertiary Recovery ［J］. Oil-gasfield Surface Engineering. 2004, 23 (1): 16-18.

［22］ Zhang X Y, Wang F P, Yu H Y, et al. Research on the Protection of Carbon Dioxide Corrosion ［J］. Corrosion and Protection. 1997, 18 (3): 8-11.

地质封存过程中 CO_2 泄漏途径及风险分析

任韶然　李德祥　张　亮　黄海东

（中国石油大学（华东））

摘　要：CO_2 的捕集和地质封存可望成为减少温室气体排放的重要且有效的方法，但其安全性一直受到广泛关注。在研究各种 CO_2 地质封存体及其圈闭机理的基础上，根据不同封存体中潜在的泄漏途径及主控因素，结合实例对可能的 CO_2 泄露进行了描述和相应的风险分析。研究结果表明，枯竭油气藏的地质封闭性已得到证实，并的实效将是 CO_2 泄露的主要途径。通过与提高油气采收率技术相结合，在油气藏中封存 CO_2 具有一定的经济性，但由于油气田分布不广，封存潜力有限，仅适合于中短期的 CO_2 处置。深部盐水层分布较广，可选择的圈闭和封存机理较多，泄漏途径和未知因素也较多，泄漏风险较高，但封存潜力巨大，是目前最具前景的 CO_2 封存体。煤层通过吸附可达到封存 CO_2 的目的，但其圈闭机理单一，对煤层压力的依附性较大，且影响未来煤资源的利用，其安全性和经济性相对较差。海底水合物封存方案具有热力学可行性，但海底水合物层埋深浅，地质圈闭性差，CO_2 泄漏风险较高，其封存和泄露机理以及 CO_2 注入方法有待进一步研究和关注。

关键词：CO_2 地质封存；深部盐水层；圈闭机理；泄漏途径；风险分析

0　引言

全球气候变化所引起的环境问题使各国政府和科学家面临着前所未有的挑战。各种极端天气的频繁发生，可能与以 CO_2 为代表的温室气体大量排放有关。CO_2 处置技术引起了国内外研究者越来越多的关注，期望能在保持经济持续发展的前提下，采用多种方法控制温室气体的排放。CO_2 捕集与埋存（CCS）作为一项新兴的、具有大规模应用潜力的 CO_2 处置技术，为化石能源使用的 CO_2 近零排放提供了一种可能[1-2]。CCS 技术可将工业中产生的 CO_2 捕集并安全地存储于特定地质构造中，以减少其向大气中的排放量，从而缓解全球气候变化。其中用于 CO_2 封存的地质体包括深部盐水层、枯竭的油气藏及不可采深部煤层等。据不完全统计，目前世界范围内完成或正在实施的大型 CO_2 捕集和地质封存示范工程项目达二十多个，同时有近百个 CO_2 地质封存研究项目在进行中。据估计 2015 年全球发电厂示范项目将达到 1000×10^4t 的 CO_2 捕集能力，每年至少有 10 个 100×10^4t 规模的新封存项目[3]。近年来，中国也相继开展了 CO_2-EOR 和地质封存技术的研究。"十一五"（2006—2010 年）期间，中国石油天然气集团公司所属的吉林油田公司利用分离的天然气伴生 CO_2 进行驱油和封存示范工程试验，2016 年计划年注入量将超过 100×10^4t/a。中国石油化工集团公司所属的胜利油田分公司也在进行电厂 CO_2 的捕集和注入油藏提高采收率的工程试验，在"十二五"（2011—2015 年）期间要达到 CO_2 年处置量（80~100）$\times 10^4$t 的规模。CO_2 在煤层中的封存也具有一定的吸引力，一方面可以封存 CO_2，另一方面通过置换甲烷提高煤层气采收率。"十一五"期间，中联煤层气有限责任公司与加拿大合作在沁水盆地开展了注 CO_2 提高煤层气采收率（CO_2—ECBM）的现场试验[4]。此外，一些能源集团公司也在积极研发 CO_2 捕集和封存技术，如神华集团研发了在燃煤电厂捕集 CO_2 的新工艺，并在鄂尔多斯盆地进行低渗透地层的 CO_2 注入和封存试验[5]。

从技术层面上来说，虽然 CO_2 捕集和地质封存技术已趋成熟，但制约其大规模推广应用主要在于其经济性和封存的安全性。CCS 工程的经济性可以通过不断的技术创新和实施碳税得到改善，但是对可能发生的 CO_2 从地下封存体的渗漏或泄漏一直存有担心和疑问（即封存的安全性问题）。因而如何有效预防、监测和控制 CO_2 泄漏，确保 CO_2 封存的安全性，已成为 CO_2 封存技术研究的一项重要内容。在 CO_2 地质封存示范项目中，CO_2 泄漏风险分析、泄漏监测技术和封堵技术等的研究，就受到了越来越多的重视[6-9]。

笔者通过分析各种 CO_2 地质封存方案和圈闭机理的基础上，总结了不同封存地质体中 CO_2 泄漏或渗漏的途径，结合 CO_2 和天然气藏的泄漏和渗漏的实例，分析论证了地质封存过程中 CO_2 的泄漏机理和可能性，并进行了相应的泄漏风险分析，为 CO_2 地质封存体的筛选和评价提供了一定的理论基础，为建立温室气体资源化利用和永久封存技术体系，提高公众对 CCS 技术的信心及消除对其安全性的疑虑提供技术依据，为制定 CO_2 地质封存的法律法规提供决策支持。

1 CO_2 地质封存形式及圈闭机理

根据目前的研究，可以考虑用于 CO_2 封存的地质体包括油气藏、盐水层、不可开采煤层和水合物层等。在地质封存过程中，CO_2 的有效封存依赖于一系列的圈闭机理或其相应的组合，不同地质封存体中封存 CO_2 的主控因素存在差异，这就导致了 CO_2 的泄漏途径、机理及风险水平有所差别。

1.1 构造圈闭

有效的 CO_2 构造圈闭依赖于封存地质体的构造特征，包括断层、储盖层组合、岩性等形式。对于油气藏或盐水层封存而言，因为 CO_2 的密度低于油和水，在浮力的作用下，大部分 CO_2 具有向上运移的趋势，当遇到密封性良好的盖层或断层时，由于不渗透层中的毛细管力远大于 CO_2 的浮力，从而使 CO_2 停止向上运移，并在不渗透层下聚集起来，从而起到了对上移 CO_2 的水力屏障作用。构造圈闭就可以作为 CO_2 注入埋存体后长期和永久的主要封存机理[10-12]，石油和天然气在地下的储存也主要依赖于构造圈闭。

1.2 残余气圈闭

当 CO_2 在封存体中运移的过程中遇见较小岩石孔隙时，由于毛细管压力存在而被束缚起来，这种过程称为残余气圈闭，这可使一定量的 CO_2 束缚在岩石孔隙和小的裂隙中，起到对 CO_2 的封存作用，其为实现 CO_2 有效圈闭的一种重要机理[10-11]❶。

---

❶ 张亮，《CO_2 盐水层封存机理及南海西部天然气田伴生 CO_2 埋存方案设计》，学位论文，中国石油大学（华东），2011 年。

1.3 吸附圈闭

通过物理吸附作用，CO_2 分子会被束缚在煤、油母岩质及矿物岩石等的表面，地层内的液体静压力控制着整个吸附解吸过程，当孔隙压力高于临界解吸压力时，就会实现对 CO_2 气体的有效束缚，达到封存 CO_2 的目的；反之，当压力低于临界解吸压力时，将会失去圈闭效果。煤层和页岩层等类型的地质体是典型依靠物理吸附圈闭机理的埋存体[12-13]。

1.4 溶解圈闭

将 CO_2 注入油气藏及盐水层后，会有部分 CO_2 溶解于地层水或油中。在油层内，CO_2 在驱油的同时，会溶解于油中，进而使部分 CO_2 圈闭于残余油中[10-11]。在盐水层中，饱和 CO_2 后水的密度将有所提高，例如，在 15MPa、60℃的条件下，饱和 CO_2 后纯水的密度可提高到 1000.6kg/m^3，同样溶有 CO_2 的盐水密度比原始的盐水密度要高[14-15]，这就能促使溶于盐水中的 CO_2 随盐水向下运移，有利于 CO_2 的进一步溶解和扩散。在较高的地层压力下，以及较大体积的盐水层内，注入的 CO_2 可能完全溶解于盐水中，类似于未饱和油藏中天然气在原油中的溶解。因而，溶解圈闭也是一种可靠的 CO_2 封存形式。

1.5 矿化作用圈闭

CO_2 溶于水中后产生 HCO_3^- 和 H^+ 离子。在地层中，CO_2 的衍生物会与含有 Ca、Mg 及 Fe 的硅酸盐矿物发生化学反应，生成一些固相的碳酸盐（如 $CaCO_3$ 和 $MgCO_3$）或生成可溶解的含水复合盐（如在碳酸盐岩储层内）。这些盐类化合物将会沉淀在地层内或保留在地层水中，达到圈闭部分 CO_2 的目的，也是一种较为永久和安全的 CO_2 封存形式[16-18]。目前已有不少有关 CO_2—岩石—水的相互作用研究[19-20]。赵仁宝等利用长庆油田含碳酸盐岩心，研究了在与 CO_2 作用过程中对其矿物组成、孔隙结构及力学性能的影响，结果表明了 CO_2 水溶液对岩心的溶蚀和离子化作用非常明显[21]。Kelzer 等进行了盐水层中 CO_2 的矿物圈闭机制研究，表明了 CO_2 在地层内既有溶解作用，又有化学沉淀作用[22]。然而 CO_2 的矿化作用很缓慢，反应的量有限，且与地层水和矿物的组成有很大关系，因此可以作为一种 CO_2 封存的有效辅助形式。

1.6 水合物圈闭

CO_2 水合物在压力高于 5MPa、温度低于 10℃就可能生成❶。CH_4 水合物生成的条件是温度低于 10℃、压力高于 10MPa❷，即 CO_2 水合物比 CH_4 水合物更容易生成。将 CO_2 以水合物的形式封存于已有 CH_4 水合物的海底沉积物或永久冻土带中也是一种埋存形式，因而提出了一种用 CO_2 置换甲烷水合物的机理[23-25]。

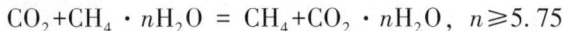

$$CO_2 + CH_4 \cdot nH_2O = CH_4 + CO_2 \cdot nH_2O, \quad n \geqslant 5.75$$

通过将 CO_2 注入到甲烷水合物层，达到置换和开采甲烷气体，并将 CO_2 封存在地层内的目的。

---

❶ 王在明，《超临界二氧化碳钻井液特性研究》，学位论文，中国石油大学（华东），2008 年。

❷ 孙晓杰，《天然气水合物地质物理力学性质实验研究》，学位论文，中国石油大学(华东)，2008 年。

2 不同封存地质体的 CO_2 泄漏途径

通过对各种 CO_2 封存及圈闭机理的分析表明，将 CO_2 注入到枯竭油气藏、煤层和天然气水合物层，结合油气开采和置换等技术，可以实现封存 CO_2 及增产油气的双重目的，因而目前容易受到广泛青睐。利用深部盐水层封存 CO_2 涉及的圈闭机理较多，其潜在的泄漏途径及风险大小也大不相同。

由于自然或人为的地质活动，在油气藏、盐水层和煤层中不可避免地存在或产生一些 CO_2 逃逸途径[8]。如图 1 所示[26]，在长期 CO_2 注入和封存过程中，可能发生 CO_2 逃逸和泄漏途径主要有 3 个：（1）通过注入井或废弃井；（2）通过未被发现的断层、断裂带或裂隙[9]；（3）通过盖层的渗漏。CO_2 从封存构造可泄漏到大气、海洋或其他地质构造中，使大气中局部 CO_2 浓度过高、或使海水、地下水或土壤被污染，将对人类健康和生态系统产生不利影响[3]。因而，对 CO_2 不同封存构造的气密性及潜在泄漏途径进行分析，对比其风险水平，为后续的环境控制技术的选择提供决策依据，是确保 CO_2 封存安全性的关键因素。

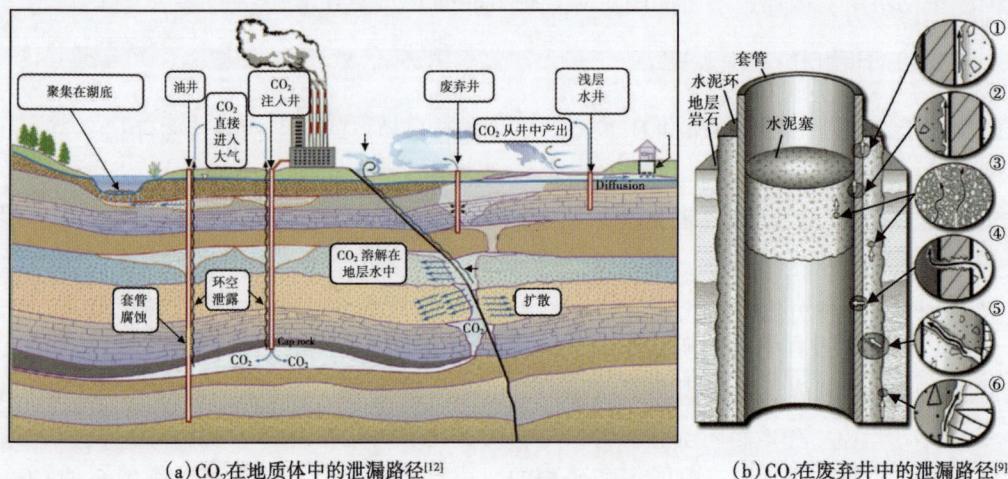

(a) CO_2 在地质体中的泄漏路径[12]　　　　(b) CO_2 在废弃井中的泄漏路径[9]

图 1　发生 CO_2 泄漏的潜在途径

2.1　枯竭油气藏封存

油气藏经过多年的勘探开发，储层性质及盖层情况相对明确，具有很好的气密性，是一种安全性和操作性较强的封存方案。注入的 CO_2 通过占据原来在较长地质时期存储油气的构造，达到有效的封存目的。CO_2 可以使原油黏度降低、体积膨胀，进而提高原油采收率。注 CO_2 提高原油采收率被认为是目前较为安全和经济可行的封存方案。通过 CO_2 的循环注入，可将大量的 CO_2 封存于地层的孔隙体积内及溶于残余油和地层水中，目前得到了国际社会的广泛认可，中国也特别推荐了一种 CO_2 资源化利用和处置 CO_2 的有效方案。正在进行的大型示范工程包括加拿大 Weyburn 油田及中国吉林油田等 CO_2 驱工程[6]。枯竭的气藏也是 CO_2 封存的很好选择。一般情况下，气藏通过衰竭式开采就可达到较高的采收率，虽然注入 CO_2 提高气体采收率（EGR）的经济性不高，但随着 CO_2 封存技术的推广及气体分离技术成本的降低，CO_2 用于气藏 EGR 和埋存会受到越来越多关注。

无论是枯竭油气藏封存还是以提高采收率为目的将 CO_2 注入油气藏中，虽然 CO_2 会溶于残余油、地层水和注入水中，溶解圈闭和残余圈闭机理也会起一定作用，但是大部分 CO_2 被注入后在相当长的时间内是以游离状态存在的，浮力会导致 CO_2 向构造上部运移，这会增大封存有效性对盖层的依赖，此时构造圈闭机理是主控因素。对于枯竭油气藏来说，油气藏圈闭构造在很长地质时期内能够储存油气，其气密性已经被证实，但在 CO_2 注入过程中局部压力过高在盖层产生新的裂隙或者导致部分井密封失效，使 CO_2 从构造中泄漏出来。而且枯竭油气藏有很多废弃的生产井和注水井，年久失修，其水泥胶结强度降低及套管的腐蚀，也是潜在和主要的泄漏通道（图 1b）。

CO_2 在地层条件下，表现出较好的传质性能（尤其是超临界状态下），很容易溶于水中形成碳酸，进而导致较低 pH 值的酸性环境，这种酸性环境的形成会使矿物溶解，削弱圈闭的地层，损害井的套管和水泥环，导致新的泄漏通道的产生。其中井的密封失效引起的泄漏途径有：（1）套管与水泥环胶结变差出现的裂隙与胶结缺陷；（2）水泥环的缝隙或裂缝；（3）套管缺陷；（4）水泥环与岩石胶结失效。由于开发过的油气田都有相当数量的生产和注入井，埋存体范围内井的数目及完整性程度决定着 CO_2 泄漏风险的水平[6,8,9]。

在油气田开发过程中，石油工业界积累了大量的处理油气井井喷、井涌和泄漏事故的经验，包括天然气和有毒气体的井喷和泄漏事故。目前也正在进行 CO_2 对油气井套管和固井水泥石腐蚀机理和防腐技术的研究[27-29]。地下天然气储集库工程的经验表明，通过注入井和废弃井的泄漏是主要的天然气泄漏途径。在美国总储存量为 1.6×10^8 t 的 470 个天然气储气库工程中，发生的 9 个泄漏事故中有 5 个跟井的完整性有关，如在 Kansas 发生的一次严重泄漏事故中，尽管泄漏量只占总储量的 0.002%[26]，可是总量达到 3000t，井筒密封失效就是事故主要原因。其他 4 个泄漏的案例是由于盖层泄漏和库的选择及储存设计不合理造成的，即与储气库的构造圈闭失效有关。虽然目前 CO_2 在油气藏实际封存过程中通过井或其他途径引起的大量泄漏鲜有报道，但国内外现正在进行的几个大型的 CO_2 提高采收率工程中，都加强了 CO_2 泄露的监测工作，以取得预测和防止泄漏的经验[30-32]。

由于在油气藏勘探和开发过程中收集了大量的地质及周围环境的信息，有利于对 CO_2 泄漏的风险进行评估，因而与其他封存体和封存方案相比，对利用枯竭油气藏埋存 CO_2 的安全性应更有信心。

2.2 深部盐水层封存

目前世界范围内进行和计划进行的大型 CO_2 地质封存示范工程都集中在盐水层封存体上，如北海挪威海域的 Sleipner、阿尔及利亚的 In Salah、挪威巴伦支海的 Sn hvit 和澳大利亚的 Gorgon 等。政府间气候变化专门委员会（IPCC）封存潜力评估研究表明，盐水层封存的潜力巨大[10]（表 1），可以满足减排和长远封存的需要。笔者前期通对中国南海莺歌海盆地深部盐水层的评估研究，仅 DF-1-1 气田附近就有很多与气藏类似的盐水层构造，满足封存与南海气藏伴生 CO_2 的封存要求[2]。选择 CO_2 源附近的盐水层进行就地封存，比通过长距离输送到油气田或采用其他方法处置更具经济性和可操作性。

与油气田相比，盐水层的分布相对较广，储层较厚，涉及的区域较大，其封存 CO_2 的潜力巨大。通常在油气田勘探开发中，会发现油层上下部大都分布着具有很好圈闭结构的盐水层，但在人类油气勘察活动中，往往忽视了盐水层的地质结构和圈闭完整性，造成对其地质信息了解不够，整个区域内可能存在贯穿盐水层的断层、裂隙及不完整盖层等泄漏隐患，而且盐水层往往与地面露头联通，地层和地表水文活跃，这些都是影响 CO_2 圈闭有效性的因素。盐水层内用于 CO_2 注入的钻井数量较少，且设计的井身质量较高，所以井的泄漏不会是主要的途径[2,33-35]。

表 1　IPCC 全球 CO_2 地质封存潜力评估表

埋存体类型	已评估证实的 CO_2 封存能力（下限）（10^9t）	潜在的 CO_2 封存能力上限（10^9t）
油气藏	675	900
不可开采煤层	3～15	200
深部盐水层	1000	远大于 10000

在注入盐水层过程中，CO_2 会进入充满盐水的岩石孔隙，驱替部分盐水，占据其空间，并扩散和溶解于盐水中。CO_2 在盐水中的主要迁移过程如图 2 所示[12]。在注入初期，由于 CO_2 大部分以自由气状态存在，构造圈闭机理依然是主控因素，如果 CO_2 封存过程中注入压力不高于岩层的破裂压力，盖层密封性没被破坏，盐水层相对独立（即不与地面露头等连通，水文相对静态），随着时间的推移，将有更多的 CO_2 溶于盐水中，逐步降低储层压力，减小对盖层的依赖。同时，在运移过程中会有残余的 CO_2 气体滞留在较小孔隙中（残余气圈闭）；CO_2 也可与岩石和地层水中离子发生反应（矿物圈闭）。如图 3 所示，在 CO_2 运移过程中，溶解和矿物圈闭作用越来越大，随着时间的推移，封存的安全性也越来越高，泄漏风险水平会逐步降低[●]。

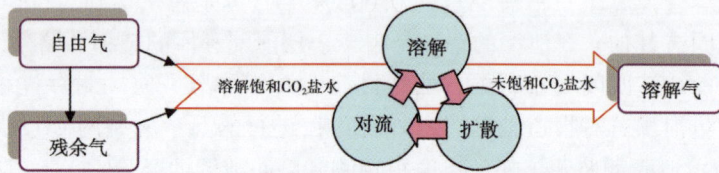

图 2　盐水层中各种状态 CO_2 之间的转化关系示意图

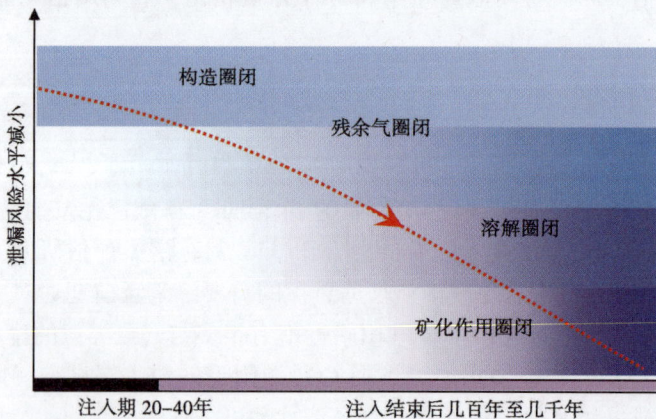

图 3　盐水层封存中 CO_2 泄漏风险水平及圈闭机理随时间的变化关系图

---

❶　张亮，《CO_2 盐水层封存机理及南海西部天然气田伴生 CO_2 埋存方案设计》，学位论文，中国石油大学（华东），2011 年。

如果 CO_2 注入的盐水层水文条件活跃，并且与其他地面露头连通，CO_2 泄漏的风险将随时间的推移逐渐增大。在盖层完好的情况下，注入初期，CO_2 由于浮力作用聚集于盖层下方可以形成有效封存。随着时间的推移，溶于盐水的 CO_2 量会越来越大，将慢慢开始横向运移扩散，在压力较低的部位，或在盐水层与地面露头处和与地表水联通处，可能产生 CO_2 的渗漏。因而一定的构造圈闭是保证 CO_2 封存安全的必要条件，溶解和矿物圈闭机理将起到辅助作用，降低 CO_2 泄漏的风险。

国内外盐水层封存示范工程项目中尚无关于 CO_2 泄漏的报道，但位于美国怀俄明尤因塔县 Leroy 盐水层天然气储存工程中已经出现了天然气泄漏到地面的现象，其可为 CO_2 盐水层封存的安全性分析提供很好的借鉴。

Leroy 盐水层是背斜构造，在其西部方向有断层存在，天然气被注入到深度约为 900m 的 Thaynes 小层中，储层由中粗粒、高孔隙度、高渗透率砂岩构成。Leroy 盐水层上方主要由页岩及砂、泥岩组成的盖层。储气库区域内有几口老井及新钻的井。1973 年，当累计天然气注入量为 $110×10^6 m^3$，压力达到 12MPa 时，Leroy 层的 3 号井表层套管周围出现泄漏。检测表明，天然气泄漏产生于 Leroy 的 4 号井 415m 处，由腐蚀的套管所致。气体通过石灰岩层迁移到了邻近的井中，引起了 3 号井泄漏。这主要是由于井的腐蚀，以及连通层的存在。盐水层封存中，通常需要气体注入压力大于原始地层压力，以压缩及驱替孔隙中的盐水。地层压力的升高增大了气体向其他层位渗漏的趋势。Thaynes 盐水层的原始地层压力为 10.3MPa，截至 1978 年，当天然气注入量为 $246×10^6 m^3$，地层压力达到 12.6MPa 时，地面监测显示，储气库上方及附近的小溪及水塘里出现了气体泄漏。通过采用示踪剂方法，证实了气体可能是从储气库直接泄漏到地面的。另一种情况是气体先渗流到浅层构造，再通过浅层构造缓慢渗漏到地面露头。另外还发现，当储气库内的气体被完全采出时，储层将被水淹，由储层到地面的直接泄漏通道由于失去气源而停止泄漏，但还是有一部分露头存在渗漏，这些渗漏气体可能是在水淹之前聚集在浅层构造中，缓慢渗漏到地面的[36]。

位于美国犹他州中部 Paradox 盆地北端溶有 CO_2 的天然盐水层的泄漏，也表明在盐水层封存中存在潜在的 CO_2 泄漏机理。Little Grand Wash 断层和 Salt Wash 地堑带北部的断层为地层流体提供了横向圈闭，但贯穿盖层的断层的封闭性较差，形成了 CO_2 迁移到地面的泄漏通道。研究发现，溶有 CO_2 的地下水通过断层面不断从泉水和间歇泉中泄漏出来，这表明即使溶解圈闭机理起作用，如果与地面露头连通或有断层裂隙，泄漏风险依然存在[37]。

对于盐水层封存，需要进行充分的地质勘查和钻井探查工作，深入了解盐水层位和周围地层的地质特征，掌握地下不同层位的连通特性，充分利用不同的圈闭机理，降低 CO_2 泄漏或渗漏的风险。

2.3 深部煤层埋存

美国作为世界上最早、开发规模最大也最成功的煤层气开采国家，2007 年煤层气的销售总量占天然气销售总量的 8.7%。中国山西沁水盆地等几个大型盆地内也有较广泛的煤层气分布。CO_2 在煤层中封存主要通过吸附圈闭机理来实现。煤层表面对 CO_2 的吸附量约是甲烷的两倍[38]，因而很多深层不易开采的煤层，会成为潜在的 CO_2 封存空间。煤层吸附主要是以范德华力为主的可逆物理吸附，煤岩的气体吸附能力大小依次为 CO_2、CH_4、N_2。煤层内大都吸附着大量甲烷气体，由于煤表面对 CO_2 的吸附能力大于对 CH_4 的吸附能力，因此可以向煤层注入 CO_2，通过竞争吸附将 CH_4 置换下来，以达到封存 CO_2 和提高煤层气采收率的目的。近年来，国内外相继进行了一些 CO_2 煤层封存的现场试验及示范性工程。美国最早进行了 CO_2 注入煤层现场试验，并在 San Juan 盆地建立了示范工程[39]；加拿大、荷兰和日本等国家的研究人员也相继对深部煤层进行了 CO_2 处置能力评估；中国在山西沁水盆地也进行了先导性试验，显示了 CO_2 在煤层中的封存和提高煤层气采收率的双重作用[40-43]。对于埋深较浅的煤层来说，如果其地质圈闭能力不足，吸附圈闭机理相对单一，CO_2 泄漏风险分析就显得尤为重要[14]。

CO_2 和 CH_4 等气体在煤表面吸附后，都存在临界解吸压力，当煤层压力等于此压力时，吸附和解吸达到平衡，而低于此压力时气体就会解吸出来，由于没有相应的地质圈闭构造，解吸下来的气体很容易发生向上或周围（非煤层）地层迁移，产生泄漏风险。此外，当注入压力过高时，CO_2 有可能进入微裂隙，由于气体分子的楔开作用可能导致煤基质间的胶结强度降低，引起煤层宏观上渗透率和力学性能的转变，增大泄漏的风险水平[13-14]。

目前，对于 CO_2 在煤层中的封存安全性的研究比较少，其潜在的泄漏风险还未引起关注。因此借鉴美国 Black Warrior 盆地、Alabama 和 Powder River 盆地、Wyoming 煤层气的泄漏情况，来分析 CO_2 在煤层中潜在的泄漏途径及机理。

煤层气需要通过一定的煤层压力被吸附或储存在地下煤层中，但某些诱因可导致压力扰动和变化，引起气体解吸和流动。在 Black Warrior 盆地中，发现有些区域内煤层中 CH_4 含量很低，而其他煤层中 CH_4 含量很高。对这种低 CH_4 含量的异常现象地质解释表明，CH_4 气体由于煤层孔隙内液体静压降低，通过断层或其他裂隙泄漏，导致气体含量降低[44]。在 Powder River 盆地的 Gillette 区块内，地面的开矿活动导致了区域内液体静压力的降低，引起了煤层内 CH_4 气体解吸，并通过一系列断层及裂隙运移到土壤中，表层土壤中高浓度 CH_4 严重损害了当地居民的生活环境。如果泄漏气体为 CO_2，其后果将会是非常严重的[45]。从此特例中可以看出，泄漏可能是由于煤层液体静压力的降低导致气体解吸，吸附圈闭机理失效造成的，所以煤层封存方案中，需要控制煤层液体静压力，减少煤层及煤层区域内的开采活动等，使其不低于 CO_2 临界解吸压力。

另一方面，煤层内除有可开采的煤层气外，还存在大量的煤炭资源，虽然目前还不具备深部煤层中的煤炭开发技术，但考虑到未来新技术的发展，如果现在将 CO_2 埋存于煤层内，可能影响到将来煤炭资源的利用，因而选用地下煤层封存 CO_2，必须深入研究和认证其埋存的安全性及煤炭资源的可利用等问题。

2.4 深海水合物层封存

利用海底水合物层封存 CO_2 的方案目前还处于概念研究阶段，许多实验表明了其热力学可行性[46]。从世界范围内来看，具备天然气水合物形成的温度、压力条件的区域是十分广泛的，其中27%的陆地和90%以上的海域都具备甲烷水合物稳定的条件。CO_2 水合物形成所需的压力和温度条件比甲烷水合物的要低，所以其潜在的封存区域和位置会很多。

近年来，不少国家相继开展了天然气水合物资源的勘探开发研究。2013 年 3 月，日本宣布在爱知县渥美半岛及三重县志摩半岛海域从天然气水合物层内提取了天然气，这在全球尚属首例[47]。中国也有广阔海域和永久冻土带适合水合物形成的区域，在 2007 年，南海北部神狐海域成功钻探获得了天然气水合物的样品[48]。目前提出的水合物开采方法包括热激法、减压法和注入化学剂法等，即通过外部扰动，使水合物分解成自由气，通过气井产出。但也有学者担心，水合物的分解可能破坏原水合物地层的胶结状态，易引起海底滑坡现象及其他的自然灾害[49]。与 CO_2 置换煤层中甲烷类似，Ebinuma[50] 和 Ohgaki[51] 等提出了以 CO_2 置换水合物层中 CH_4 的设想，既能维护水合物沉积层的稳定性，又可实现 CO_2 的有效封存。

研究发现，直接注入 CO_2 置换 CH_4 效率不高，需要通过其他方法强化置换过程，使注入 CO_2 能形成水合物，从而被迅速封存[23]。利用海底水合物层封存 CO_2 的圈闭机理也比较单一，虽然水合物本身具有一定"自保性"，但压力和温度出现波动时，会引起水合物的分解，导致 CO_2 的泄漏。同时对这种封存方案的关键问题的研究还很欠缺，如 CO_2 的注入部位及注入 CO_2 的迁移途径还不明确。如图 4 所示，如将 CO_2 注入到甲烷水合物的稳定区内，CO_2 将容易形成水合物，且不论其能否置换 CH_4，形成的水合物都将影响或阻止后续 CO_2 的注入；虽然在 CO_2 水合物稳定区下部的注入可行性高，CO_2 开始会向上部水合物稳定区迁移，但产生的 CO_2 水合物可能导致地层渗透率降低，引起后续注入的 CO_2 向侧向或水平方向流动，不但影响封存效率，而且如果注入层位与大陆架露头联通，CO_2 可能沿地层在浮力作用下上移，产生泄漏。

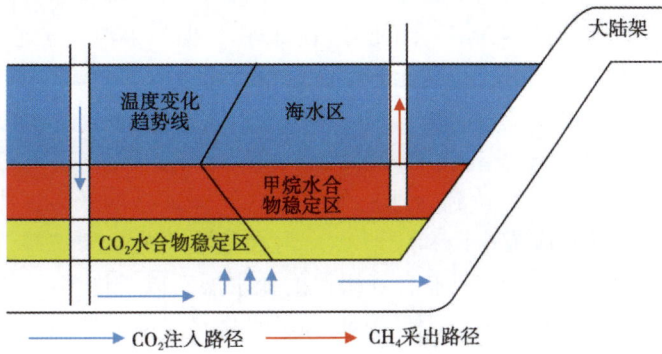

图 4　水合物封存方案示意图

对于海底水合物层来说，其上部海水的压力和海底低温，有利于 CO_2 水合物的稳定，但海底浅层的地质圈闭能力较差，其圈闭机理单一，所以泄漏风险也较高，CO_2 封存的可行性较差。对于陆上冻土层来说，由于埋深较浅，且处于开放的地质系统，且受环境温度的影响很大，从安全角度来说，不具备 CO_2 封存的可行性。

3　不同封存方案 CO_2 泄漏风险及安全性

CO_2 地质封存的泄漏风险主要取决于封存体的选择及不同圈闭或封存机理的联合利用。通过原油和天然气的地质储存及开发经验表明，构造圈闭是有效天然气封存机理，对于 CO_2 的封存也同样有效。近几十年来，注 CO_2 用于油藏提高采收率工程的经验也提供了有利的证据。对于利用油气藏埋存 CO_2 来说，由于其地质特性的认识较为深入，井的风险通过相应的监测和封堵措施也会降低，所以枯竭油气藏是最为安全的 CO_2 封存体。但从解决减排问题的角度来讲，油气藏分布不是太广泛，其封存量不足，不能从根本上缓解全球的温室效应气体问题。研究结果表明，油藏封存 CO_2 的潜力大约是其原始地质储量的 30%（以 t 为计量单位）。中国目前探明的原油地质储量约为 $940×10^8$ t（即 CO_2 的总封存潜力约为 $280×10^8$ t）[52]，而现在每年的 CO_2 排放量约为 $80×10^8$ t（国际能源署 2013 年 6 月 10 日的报告中指出全球 2012 年 CO_2 排放量超过 $316×10^8$ t，中国和美国各占 1/4）[53]，国内学者预测，若中国维持目前的经济发展水平，2020 年中国的 CO_2 排放量将达到 $151×10^8$ t[54]。因而现探明油藏的 CO_2 封存潜力明显不足。但对当前来说，由于油藏注入和封存 CO_2 的技术比较成熟，可以通过这种封存方案，取得一定的减排效果和经验，同时结合 CO_2 提高采收率，可以得到一定的经济效益，因此其是 CO_2 封存工程实施初期最优，也是最容易接受的封存方案。在 CO_2 驱与地质封存工程中[55]，注入的 CO_2 约有 50%~60% 会随原油开采出来，通过循环注入，最终 CO_2 将通过各种圈闭机理被封存起来。值得注意的是，在油田封存 CO_2 的过程中，除了具有通过井的泄漏风险外，CO_2 的注入也可能诱发新的泄漏通道，即可能通过未知的裂隙、裂缝及断层等产生泄漏。因此需要采用针对 CO_2 泄漏的油藏和环境监测方法，确保封存的有效性及安全性。

利用盐水层封存 CO_2 涉及圈闭机理较多，其泄漏风险因封存体不同也大相径庭。由于对盐水层地质特性的认识不像油气藏那样充分，虽然通过油气藏勘探方法，可以找到具有一定地质圈闭的盐水层，但其盖层的密封性能及构造圈闭的可靠性不能得到有效的证实，即使其气密性较好，随着时间的推移，溶解圈闭开始起作用，但如果与其连通的地层水文活跃并与地面露头相联，就可能出现封存初期无泄漏而封存后期会出现泄漏的情况。总体而言，盐水层的泄漏风险相比枯竭油气藏要高。

目前通过盐水层封存 CO_2 没有任何经济效益，但盐水层分布较广，其封存潜力巨大，考虑未来埋存 CO_2 的规模以及要达到全球范围内温室气体减排的效果，深部盐水层会是极具吸引力的封存方案，其将成为中期和长期 CO_2 封存地质体的主要选择。在油气田地质勘探和开发过程中，已经发现了很多具有良好构造的盐水层，以及许多与油气藏连通的水层，这些都可以作为 CO_2 封存体的首选。对于其他区域的盐水层，需要通过地质普查及钻井详查的方法，对盐水层的封存潜力和安全性进行评估，通过合理利用各种圈闭机理的联合协同作用，如一定构造圈闭与溶解圈闭机理相结合，也能实现 CO_2 在盐水层内的有效及安全封存。近年来，国外对盐水层封存 CO_2 的研究比较重视，特别是欧洲和澳大利亚，已经完成了区域性和场地性的普查和详查工作，为将来利用盐水层大规模的封存 CO_2 提供了技术依据。中国有关盐水层的研究工作刚刚起步，建议应将深部盐水层当作潜在的地质资源，尽快开展相应的地质普查和详查工作。

对于煤层封存方案，由于其圈闭形式相对单一，CO_2 在现阶段不可开采煤层中的封存是利用煤层的吸附圈闭机理，但是任何可能扰动地层压力的地面或地下活动（如钻井、层位连通及地面开矿活动等），都有可能导致煤层中被吸附 CO_2 的解吸。此外煤层构造一般处于"开放"状态，解吸的 CO_2 容易泄漏到浅层构造或地面。另一方面，随着能源需求的增长及煤炭开采技术的提高，现阶段不可动用的煤层，将来有可能变为可动用资源，一旦开采，将导致 CO_2 再次释放。因此，煤层封存方案的泄漏风险较高，不是有效的 CO_2 处置方式。

海底水合物层封存 CO_2 还处在概念设计及机理研究阶段，其泄漏风险和封存还不清楚，其圈闭机理也相对单一。虽然水合物分解有一定"自保性"，但可认为水合物封存中的 CO_2 泄漏风险与煤层封存相当（风险水平如图 5 所示），水合物封存 CO_2 在热力学上的可行性已被证明，但在实际操作上还存在问题，现场应用也较为困难。因此，在现阶段是一种不可取的封存方案，有许多技术难点需要攻克，理论基础及配套工艺也有待完善。

通过对 CO_2 圈闭机理及泄漏风险的分析，可以得到不同地质封存体中封存 CO_2 的潜力及泄漏风险的定性描述（图5）。从图5可以看出，枯竭油气藏封存的安全性最高，泄漏风险最低，但其封存潜力有限，可以作为近期封存 CO_2 的试验场地，解决部分减排问题；盐水层的封存潜力巨大，封存风险也相对较小，通过有效的地质筛选和评估，可以达到安全有效的封存及全球减排的目的，可以作为中长期 CO_2 封存的主要选择。目前对于煤层和海底水合物层封存 CO_2 的机理及可行性的认识尚不足，需要进行深入研究，建议不作为有效的选择（表2）。

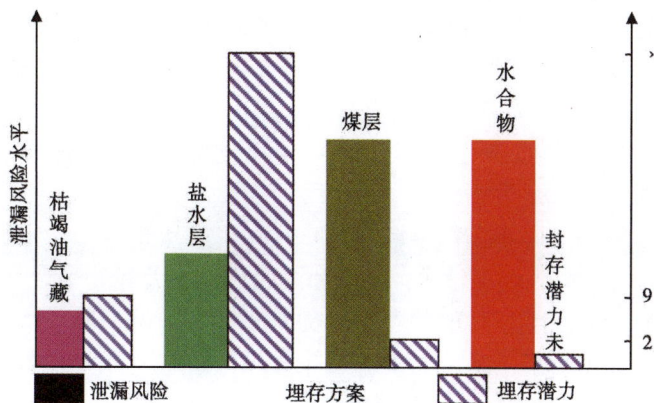

图 5　世界范围内不同 CO_2 封存体的泄漏风险水平和封存能力

表 2　不同封存体泄漏风险定性分析表

埋存类型	主要封存机理	圈闭条件认可度	主要泄漏机理和通道	泄漏风险水平	应用前景
油气藏	构造、溶解和残余气圈闭	较好	注入井和废弃井失效	较低	中短期处置温室气体
盐水层	构造、溶解、残余气和矿物圈闭	构造圈闭的完整性认识不够	与地面露头和浅层的连通层及断层裂隙	中	长远处置方案
煤层	吸附圈闭	不够	吸附圈闭失效后，煤层处于开放状态，泄漏途径多	高	有争议的处置方式
海底水合物沉积层	水合物圈闭	不够	埋深浅，地层处于开放状态，泄漏途径多	高	理论及技术有待完善

4　结论与展望

（1） CO_2 地质封存的主要封存机理包括构造圈闭、残余气圈闭、吸附气圈闭、溶解圈闭、矿物圈闭和水合物圈闭等，由于每种封存体和封存方案的主控圈闭机理的差异，将导致其潜在的泄漏途径及泄漏风险水平不同。

（2）对枯竭油气藏封存来说，由于长时间对油气资源的有效圈闭，其封存安全性已被证实，潜在的泄漏主要是由于生产井和注入井的失效和不完整性引起。井的泄漏风险需要通过相应的封堵来降低，因而枯竭油气藏是比较安全的封存体。中国油气藏分布虽然广泛，但是相对于中国的 CO_2 排放量而言，其封存潜力仍然有限，不能满足减排和缓解温室效应的要求。但是在中短期（10~30a）内，可以通过油气藏封存方案，结合 CO_2 驱提高油气采收率技术，取得一定的经济效益，因而 CO_2 驱和封存是目前中国最佳和容易接受的封存方案。

（3）地下盐水层分布较广，封存潜力巨大，能够满足全球减排的需要，是长远解决 CO_2 封存问题的主要选择。CO_2 在深部盐水层封存涉及的圈闭机理较多，但对其地质结构圈闭的认识没有油气藏那么深入，所以可能存在潜在的泄漏通道，有一定的泄漏风险，需要进行认真评估，选择有效的地质构造及溶解、矿物封存机理的组合，降低泄漏风险。目前，国际上多个盐水层封存示范工程尚无泄漏的报道，这表明通过合理利用不同的圈闭机理，可以确保盐水层封存的安全性。建议中国尽快开展一定规模盐水层的地质普查和 CO_2 封存潜力和安全性的评估研究。

（4）对于利用煤层和海底水合物层封存 CO_2 来说，虽然煤层对 CO_2 的吸附及 CO_2 水合物都是有效封存形式，但由于埋深较浅，煤层和海底浅地层大都处于"开放"的地质状态，缺少一定的地质圈闭结构，且 CO_2 的状态对压力很敏感，泄漏风险较大。同时由于煤层和海底水合物层内都有潜在的煤、煤层气和甲烷等资源，虽然可将 CO_2 封存其中并与部分甲烷气体置换，但可能影响将来和长远的能源开发和利用，因而其封存的安全性和长远经济可行性还需要深入研究和评估。

参 考 文 献

[1] 任韶然，张莉，张亮 . CO_2 地质埋存：国外示范工程及其对中国的启示 [J]. 中国石油大学学报：自然科学版，2010，34（1）：93-98.

[2] 张亮，任韶然，王瑞和，等 . 东方 1-1 气田伴生 CO_2 盐水层埋存可行性研究 [J]. 中国石油大学学报：自然科学版，2010，34（3）：89-93.

[3] 朱跃钊，廖传华，王重庆，等 . 二氧化碳的减排与资源化利用 [M]. 北京：化学工业出版社，2011：110-116.

[4] 李小春，方志明 . 中国 CO_2 地质埋存关联技术的现状 [J]. 岩土力学，2007，28（10）：2229-2233.

[5] 吴秀章，崔永君 . 神华 10 万 t/a CO_2 盐水层封存研究 [J]. 石油学报（石油加工），2010，26（增刊1）：236-239.

[6] 任韶然，任博，李永钊，等 . CO_2 地质埋存监测技术及其应用分析 [J]. 中国石油大学学报：自然科学版，2012，36（1）：106-111.

[7] 谷丽冰，李治平，侯秀林 . 二氧化碳地质埋存研究进展 [J]. 地质科技情报，2008，27（4）：80-84.

[8] 许志刚，陈代钊，曾荣树 . CO_2 地质埋存渗漏风险及补救对策 [J]. 地质论评，2008，54（3）：373-386.

[9] 张森琦，刁玉杰，程旭学，等 . 二氧化碳地质储存逃逸通道及环境监测研究 [J]. 冰川冻土，2010，32（6）：1251-1260.

[10] 蔡博峰，格雷格·利蒙，刘兰翠 . 二氧化碳地质封存和环境监测 [M]. 北京：化学工业出版社，2013：1-9.

[11] 师春元，黄黎明，陈赓良 . 机遇与挑战—二氧化碳资源开发与利用 [M]. 北京：石油工业出版社，2006：42-61.

［12］王烽，汤达祯，刘洪林，等．利用 CO_2-ECBM 技术在沁水盆地开采煤层气和埋藏 CO_2 的潜力［J］．天然气工业，2009，29（4）：1-4.

［13］王晓峰，朱卫平．注 CO_2 提高煤层气采收率技术研究现状［J］．资源与产业，2010，12（6）：125-129.

［14］Garcia J E. Density of aqueous solutions of CO_2［EB/OL］. 2013-04-02. http：//escholarship. org/uc/item/6dn022hb.

［15］Song Y，Chen B，Nishio M，et al. The study on density change of carbon dioxide seawater solution at high pressure and low temperature［J］. Energy，2005，30（11/12）：2298-2307.

［16］Bachu S，Gunter W D，Perkins E H. Aquifer disposal of CO_2：hydrodynamic and mineral trapping［J］. Energy Conversion and Management，1994，35（4）：269-279.

［17］Gunter W D，Perkins E H，McCann T J. Aquifer disposal of CO_2-Rich gases：reaction design for added capacity［J］. Energy Conversion and Management，1993，34（9/11）：941-948.

［18］Xu T F，Apps J A，Pruess K. Reactive geochemical transport simulation to study mineral trapping for CO_2 disposal in deep arenaceous formations［J］. Journal of Geophysical Research，2003，108（B2）：2071-2084.

［19］Czernichowski-Lauriol I，Sanjuan B，Rochelle C，et al. Analysis of the geochemical aspects of the underground disposal of CO_2［M］. Salt Lake City：Academic Press，1996：565-585.

［20］Gunter W D，Wiwchar B，Perkins E H. Aquifer disposal of CO_2-Rich greenhouse gases：extension of the time scale of experiment for CO_2-sequestering reactions by geochemical modeling［J］. Mineralogy and Petrology，1997，59（1/2）：121-140.

［21］赵仁宝，孙海涛，吴亚生，等．二氧化碳埋存对地层岩石影响的室内研究［J］．中国科学：科学技术，2010，40（4）：378-384.

［22］Ketzer J M，Iglesias R，Einloft S，et al. Water-Rock-CO_2 interactions in saline aquifers aimed for carbon dioxide storage：experimental and numerical modeling studies of the Rio Bonito formation（Permian），Southern Brazil［J］. Applied Geochemistry，2009，24（5）：760-767.

［23］李臻，王欣．绿色开采天然气水合物技术研究［J］．钻采工艺，2010，33（6）：71-74.

［24］郭平，刘士鑫，杜建芬．天然气水合物气藏开发［M］．北京：石油工业出版社，2006：14-17.

［25］Komui T，Sakamoto Y，Tanaka A. Enhanced CO_2 geological storage system using gas hydrates and environmental risk assessment：proceedings of the Twentieth International Offshore and Polar Engineering Conference，Beijing，June 20-25，2010［C］. Beijing：ISOPE，2010：115-118.

［26］Bert M，Ogunlade D，Manuela L. IPCC special report on carbon dioxide capture and storage［M］. New York：Cambridge University Press，2005.

［27］张学元，邸超，雷良才．二氧化碳腐蚀与控制［M］．北京：化学工业出版社，2000：1-60.

［28］林永学，陈雷，王立志．CO_2 腐蚀环境下油管防腐技术方法初探［J］．石油钻采技术，1999，27（3）：34-36.

［29］宋伟，冯小波，李娅，等．重庆气井井下油管腐蚀及缓蚀剂的应用［J］．腐蚀与防护，2005，26（8）：347-350.

［30］Ren S，Niu B L，Ren B，et al. Monitoring on CO_2 EOR and storage in a CCS demonstration project of Jinlin oilfield China［R］. SPE 145440，2011.

［31］Talman S J，Perkins E H. Pembina cardium CO_2 monitoring project，Alberta，Canada-geochemical interpretation of produced fluid compositions［J］. Energy Procedia，2009，1（1）：2151-2159.

［32］Emberleya S，Hutcheonb I，Shevalierb M，et al. Geochemical monitoring of fluid rock interaction and CO_2 storage at the Weyburn CO_2-injection enhanced oil recovery site，Saskatchewan，Canada［J］. Energy，2004，29（9/10）：1393-1401.

［33］张亮，任韶然，王瑞和，等．南海西部盐水层 CO_2 埋存潜力评估［J］．岩土力学，2010，31（4）：

1238-1242.

[34] R6veillfire A, Rohmer J. Managing the risk of CO_2 leakage from deep saline aquifer reservoirs through the creation of a hydraulic barrier [J]. Energy Procedia, 2011 (4): 3187-3194.

[35] Nakayama K, Takahashi T. New concept for mechanism of CO_2 leakage from aquifer [J]. Energy Procedia, 2009, 1 (1): 3345-3350.

[36] Araktingi R E, Benefield M E, Bessenyei Z, et al. Leroy storage facility, Uinta County, Wyoming-a case history of attempted gas migration control [J]. Journal of Petroleum Technology, 1984, 36 (1): 132-140.

[37] Doekrill B, Shipton Z K. Structural controls on leakage from a natural CO_2 geologic storage site: Central Utah, U. S. A [J]. Journal of Structural Geology, 2010, 32 (11): 1768-1782.

[38] Kelemen S R, Kwiatek L M. Physical properties of selected block Argonne Premium bituminous coal related to CO_2, CH_4, and N_2 adsorption [J]. International Journal of Coal Geology, 2009, 77 (1/2): 2-9.

[39] Reeves S R. Assessment of CO_2 sequestration and ECBM potential of U. S. coalbeds [R]. Houston: US Department of Energy, 2003.

[40] Gentzis T. Subsurface sequestration of carbon dioxide: an overview from an Albetra (Canada) perspective [J]. International Journal of Coal Geology, 2000, 43 (1/4): 287-305.

[41] Hamelinck C, Faaij A, Tukenburg W, et al. CO_2 enhanced coalbed methane production in the Netherlands [J]. Energy, 2002, 27 (7): 647-674.

[42] Yamazaki T, Aso K, Chinju J. Japanese potential of CO_2 sequestration in coal seams [J]. Applied Energy, 2006, 83 (9): 911-920.

[43] Wong S, Law D, Deng X H, et al. Enhanced combed methane and CO_2 storage in anthracitic coals: micro-pilot test at south Qinshui, Shanxi, China [J]. International Journal of Greenhouse Gas Control, 2007, 1 (2): 215-222.

[44] Malone P G, Briscoe F H, Camp B S, et al. Discovery and explanation of low gas contents encountered in coalbeds at the GRI/USSC Big Indiana Greek site, Warrior Basin, Alabama: proceedings of the Coalbed Methane Symposium, Tuscaloosa, Alabama, November 1987 [C]. Tuscaloosa: International Coalbed Methane Symposium, 1987.

[45] Clayton J L, Leventhal J S, Rice D D, et al. Atmospheric methane flux from coals-preliminary investigation of coal mines and geologic structures in the Black Warrior Basin, Alabama [R] //Howell D G. The future of energy gases: Geological Survey Professional Paper 1570, United States Geological Survey, 1994: 471-492.

[46] 颜克凤, 李小森, 陈朝阳, 等. 二氧化碳置换开采天然气水合物研究 [J]. 现代化工, 2012, 32 (8): 42-49.

[47] 日经能源环境网. 日本在全球首次利用海底甲烷水合物产出天然气 [EB/OL]. [2013-04-02]. http://finance.people.com.cn/n/2013/0320/c348883-20848867.html.

[48] 国土资源部网站. 再钻南海可燃冰-聚焦中国海洋天然气水合物勘探 [EB/OL]. [2013-05-02]. http://www.gov.cn/gzdt/2012-07/24/content_2190366.htm.

[49] 徐文世, 于兴河, 刘妮娜, 等. 天然气水合物开发前景和环境问题 [J]. 天然气地球科学, 2005, 16 (5): 680-683.

[50] Ebinuma T. Method for dumping and disposing of carbon dioxide gas and apparatus therefor [P]. US 5261490A, 1993.

[51] Ohgaki K, Takano K, Sangawa H, et al. Methane exploitation by carbon dioxide from gas hydrates-phase equilibria for $CO_2 - CH_4$ mixed hydrate system [J]. Journal of Chemical Engineering of Japan, 1996, 29 (3): 478-483.

[52] 沈平平, 赵文智, 窦立荣. 中国石油资源前景与未来10年储量增长趋势预测 [J]. 石油学报, 2000, 21 (4): 1-6.

［53］驻旧金山总领馆经商室. 国际能源署：2012 年全球 CO_2 排放增长 1.4% ［EB/OL］. ［2013-06-07］. http：//www. mofcom. gov. cn/article/i/jyj1/1/201306/20130600163368. shtml.

［54］刘燕华，葛全胜，何凡能，等. 应对国际 CO_2 减排压力的途径及我国减排潜力分析 ［J］. 地理学报，2008，63（7）：675-682.

［55］Enick R M，Olsen D，Ammer J，et al. Mobility and conformance control for CO_2 EOR via thickeners，foams，and Gels-A literature review of 40 years of research and pilot tests ［R］. SPE 154122，2012.

第四篇
二氧化碳驱油与埋存动态监测及潜力评价技术

CO_2 驱试验区试井测试资料分析及应用

王敬瑶

(中国石油勘探开发研究院提高采收率国家重点实验室)

摘　要：二氧化碳驱试井测试资料表现复杂渗流特性。解释分析某注二氧化碳试验区同井不同时间段和同时间段不同井的试井测试资料应用复合区试井模型，得到表皮因数、流度和外推地层压力等油藏参数；比较二氧化碳驱前后地层流度和外推地层压力的变化特征，获得有效的二氧化碳驱开发参数；对试验区试井测试资料综合分析，反映二氧化碳驱受效情况。结果表明：与生产动态相吻合，准确反映油井生产状况。

关键词：二氧化碳驱；试井；测试资料分析；开发参数；受效情况

0　引言

注二氧化碳驱油能够提高石油采收率，具有适用范围广、驱油效率高、成本低等优点[1]。试井分析是对试井测取的资料进行解释分析，取得与测试井相关的油藏及井的特性参数，如油藏外推地层压力、平均渗透率、流动系数、表皮因数等，对开发评价油藏具有重要作用[2]。从二氧化碳驱试井测试资料中获得开发动态信息对二氧化碳驱理论研究和生产具有重要意义。Tang R W 等通过对二氧化碳注入井不稳定压力资料的研究发现，在流体流动与相态变化的交互作用下，压力的瞬时变化表现为 3 个区的复合油藏[3]；MacAllister D J 对二氧化碳注入与生产井的不稳定压力进行分析，阐述基于拟压力的压力不稳定分析的理论基础[4]；雷友忠应用均质模型与裂缝模型对国内某油田二氧化碳注入井试井资料进行解释，对得到的地层参数进行对比，但是并未提出完善的试井模型与参数计算公式[5]。目前针对二氧化碳驱试井测试资料解释分析的研究比较少[6]。对吉林油田某二氧化碳驱示范区的试井测试资料进行解释分析，应用复合区试井模型及现代试井分析方法获得表皮因数、储层平均渗透率、流度和外推地层压力等参数；通过对同时段不同井与同井不同时段试井测试资料综合分析，准确反映二氧化碳驱油提高采收率效果。

1　分析原理及方法

二氧化碳驱的渗流区域可划分为内区、过渡区与外区[3-11]。内区为二氧化碳超临界流体，其渗流特征与气体类似；过渡区是由油气相互接触作用形成的区域，依据驱替方式不同，过渡区流体可分为单相流或油气两相流；外区即未驱替的原油。二氧化碳驱油物理模型见图 1。对油井进行试井测试，数据常表现为过渡区和未驱替区部分。

图 1 二氧化碳驱油物理模型

由图 1 可建立相应试井数学模型，用渗流方程组表示。

（1）无因次量渗流方程。考虑二氧化碳驱中 3 个区域渗流过程满足达西定律，应用渗流理论建立无因次拟压力方程：

$$\begin{cases} \dfrac{\partial^2 m_{1D}}{\partial r_D^2} + \dfrac{1}{r_D}\dfrac{\partial m_{1D}}{\partial r_D} = \dfrac{1}{C_D e^{2s}}\dfrac{\partial m_{1D}}{\partial(t_D/C_D)} \\[2ex] \dfrac{\partial^2 m_{2D}}{\partial r_D^2} + \dfrac{1}{r_D}\dfrac{\partial m_{2D}}{\partial r_D} = \dfrac{1}{C_D e^{2s}}\dfrac{M_{13}}{\omega_{12}}\dfrac{\partial m_{2D}}{\partial(t_D/C_D)} \\[2ex] \dfrac{\partial^2 m_{3D}}{\partial r_D^2} + \dfrac{1}{r_D}\dfrac{\partial m_{3D}}{\partial r_D} = \dfrac{1}{C_D e^{2s}}\dfrac{M_{13}}{\omega_{13}}\dfrac{\partial m_{3D}}{\partial(t_D/C_D)} \end{cases} \tag{1}$$

式中，m_{1D} 为无量纲内区拟压力函数；m_{2D} 为无量纲过渡区拟压力函数；m_{3D} 为无量纲外区拟压力函数；r_D 为无量纲半径；M_{12} 为内区和过渡区的流度比；M_{13} 为内区和外区的流度比；ω_{12} 为内区和过渡区的储容比；ω_{13} 为内区和外区的储容比；C_D 为无量纲井筒存储因数；s 为表皮因数；t_D 为无量纲时间。

（2）考虑在原始时刻整个地层压力为原始地层压力，则初始条件为

$$m_{1D}(r_D,\ 0) = m_{2D}(r_D,\ 0) = m_{3D}(r_D,\ 0) = 0 \tag{2}$$

（3）考虑井筒存在井筒存储效应，并以定流量注入，因此内边界条件为

$$\begin{cases} \dfrac{\partial m_{wD}}{\partial(t_D/C_D)} - \dfrac{\partial m_{1D}}{\partial r_D}\bigg|_{r_D=1} = 1 \\[2ex] m_{wD} = m_{1D}\big|_{r_D=1} \end{cases} \tag{3}$$

式中，m_{wD} 为无量纲井筒拟压力函数。

（4）考虑整个地层为无限大储层，则外边界条件为

$$m_{3D}(\infty,\ t_D) = 0 \tag{4}$$

（5）考虑各流动区交接面不存在压力损失，则连接面条件为

$$\begin{cases} M_{1D}(r_{1D},\ t_D) = m_{2D}(r_{1D},\ t_D) \\[2ex] \dfrac{\partial m_{1D}}{\partial r_D}\bigg|_{r_D=r_{1D}} = \dfrac{\partial m_{2D}}{\partial r_D}\bigg|_{r_D=r_{1D}} \\[2ex] M_{2D}(r_{1D},\ t_D) = m_{3D}(r_{1D},\ t_D) \\[2ex] \dfrac{\partial m_{2D}}{\partial r_D}\bigg|_{r_D=r_{2D}} = \dfrac{\partial m_{3D}}{\partial r_D}\bigg|_{r_D=r_{2D}} \end{cases} \tag{5}$$

式中，r_{1D}为无量纲内区半径；r_{2D}为无量纲过渡区半径。

对式（1）至式（5），应用 Laplace 积分变换方法可以求得无因次井底压力的解，可获得双对数试井特征曲线（图 2），由图 2 可以看出内外区流度不同，在无因次时间较大处拟压力导数曲线偏离水平线。通过曲线拟合技术对实际测试的二氧化碳驱试井测试资料应用该模型进行井底压力及其导数拟合分析，可以确定地层平均渗透率、井筒存储系数、表皮因数、内区半径、流度比、外推地层压力等相关参数[12-15]。

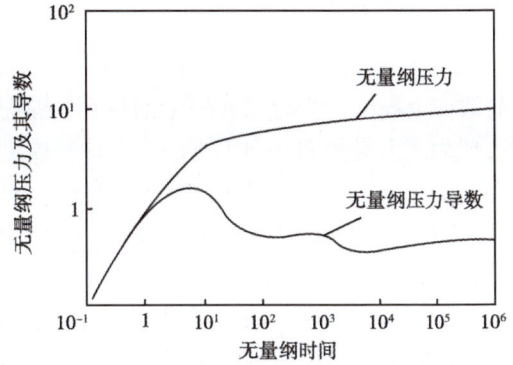

图 2　二氧化碳驱试井特征曲线

2　测试资料解释分析

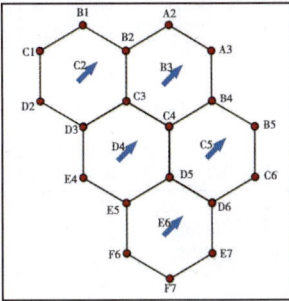

图 3　试验区块井位分布

二氧化碳驱试验区的井位分布见图 3，各测试井的孔隙度为 0.18，井半径为 0.1m，黏度为 5.4mPa·s，体积系数为 1.113，综合压缩系数为 0.0013MPa^{-1}，测试井其他基本数据见表 1。在该测试井区中 2008—2009 年对 E7、F6、D6、E4、B4、C3、B2 等井进行试井测试，取得 9 个井次有效测试资料，其中 F6 井和 B2 井分别在 2008 年与 2009 年进行 2 次不同时间段的试井测试。在这些测试井中 F6 井与 D6 井为 1 个井组、E4 井与 C3 井为 1 个井组、B4 井与 C3 井及 B2 井为 1 个井组。

在 9 个井次的有效试井测试资料中有同时间段不同井的，如 E4 井、C3 井与 B2 井；有同井不同时间段的，如 F6 井和 B2 井，这些不同井次试井测试资料富含油藏信息，通过分析测试资料，评价二氧化碳驱油效果。

表 1　二氧化碳驱试验区测试井基本数据

测试井名	测试时间	流量（m³/d）	储层厚度（m）
E7	2009-08	4.0	26
F6	2008-11	5.0	26
	2009-06	8.2	13
D6	2009-05	11.5	9
E4	2009-03	11.1	11.6
B4	2009-09	7.2	16.4
C3	2009-03	13.7	7.8
B2	2009-03	18	9
	2008-06	6.7	9

2.1 同井不同时间段

2.1.1 B2井

B2井试井测试资料拟合结果见图4，由图4可以看出，2008年测试资料表现为压力导数曲线向上翘起的复合试井模型的特征，是外区流度比内区流度小的表现；2009年测试资料表现为压力导数曲线向下探的复合试井模型的特征，是外区流度比内区流度大的表现。应用现代试井分析方法对测试资料进行分析，解释结果见表2。由表2可以看出，从2008年6月—2009年3月该井井筒附近表皮因数和内区流度变化不大，外推地层压力有较大程度的上升，上升3MPa以上，外区流度也变化较大，上升近6倍。内区的流度反映压裂后近井地带的流度，外区流度反映压裂后该井实际地层流度，2008年水驱时地层流度明显低于2009年二氧化碳驱时地层流度，原因是二氧化碳驱后地层油和二氧化碳混合，实现混相驱油，降低流体的黏度。

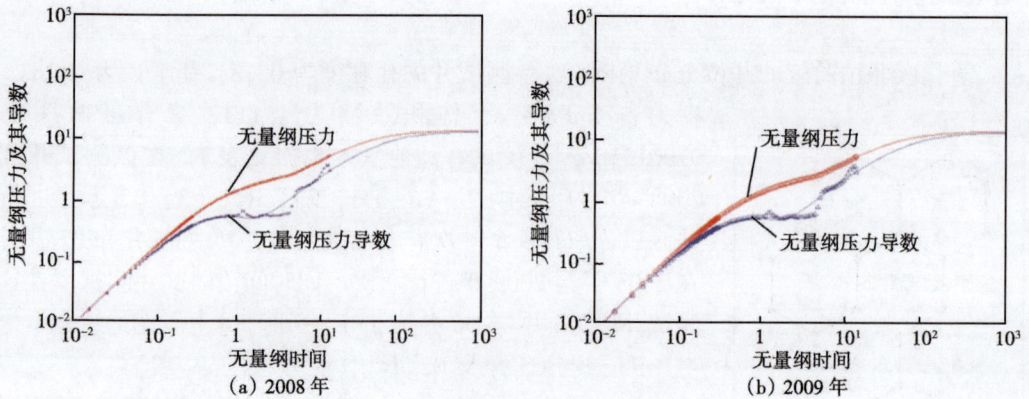

图4　B2井试井测试资料拟合结果

表2　B2井试井测试资料解释结果

测试时间	表皮因数	内区流度 ($10^{-3}\mu m^2/$ (mPa·s))	外区流度 ($10^{-3}\mu m^2/$ (mPa·s))	内区半径 (m)	外推地层压力 (MPa)
2009-03	-4.21	0.4375	0.7432	53.97	27.27
2008-06	-4.57	0.4646	0.1043	105.05	23.91

2.1.2 F6井

F6井试井测试资料拟合结果见图5，由图5可以看出，2次测试表现为均质试井模型的特征。应用现代试井分析方法对测试资料进行分析，解释结果见表3。由表3可以看出，2008年11月—2009年6月该井井筒附近表皮因数有一定程度的升高，外推地层压力也有所上升，相对其他井上升幅度较小，储层的流度略有下降，这是水驱的结果，二氧化碳驱在该井未产生作用。

表3　F6井试井测试资料解释结果

测试时间	表皮因子	流度 ($10^{-3}\mu m^2/$ (mPa·s))	地层压力 (MPa)
2008-11	-2.08	0.1607	22.65
2009-06	1.13	0.1512	23.24

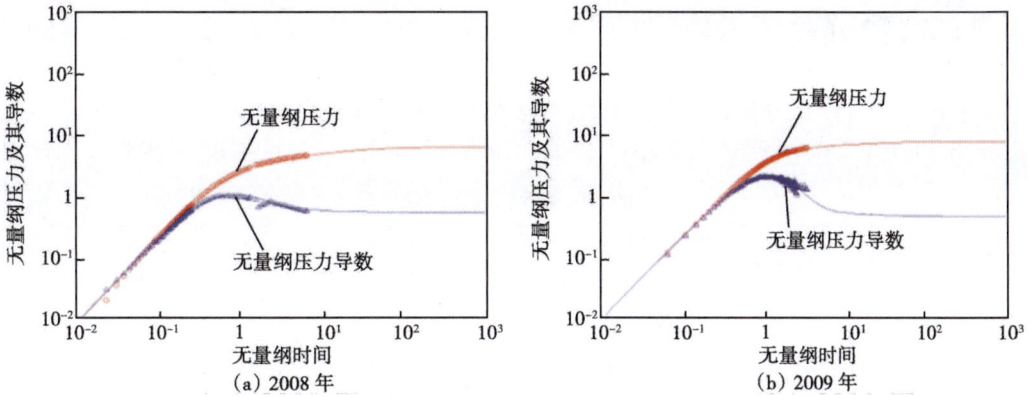

图 5　F6 井试井测试资料拟合图

2.2　同时间段不同井

　　同时间段 E4、C3 井二氧化碳驱后试井测试资料拟合结果见图 6。E4、C3 井及 C3、B4 井分别为菱形反 9 点井组中 2 口生产井。由图 6 可以看出，2 次测试结果表现为压力导数曲线向上翘起的复合试井模型的特征，是外区流度比内区流度小的表现。

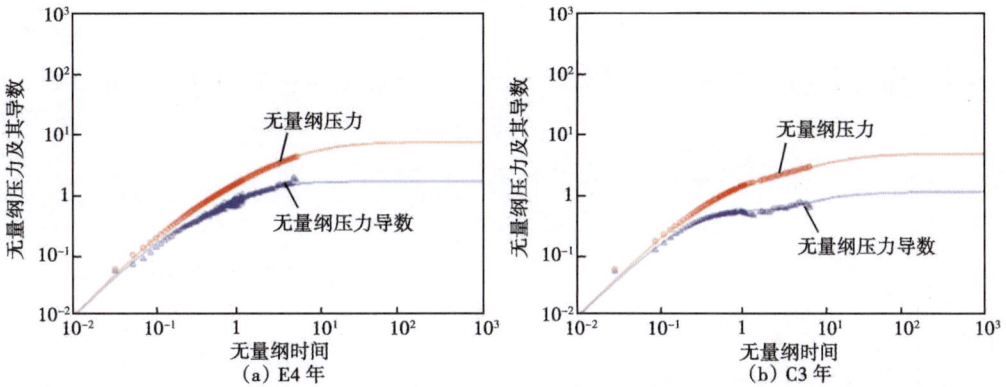

图 6　E4 和 C3 井试井测试资料拟合结果

　　应用现代试井分析方法分析测试资料，解释结果见表 4。由表 4 可以看出，E4 和 C3 井的外推地层压力较高，达到 29 MPa 以上，储层的流度明显升高，比水驱时流度（$0.1×10^{-3}$ $\mu m^2/$（mPa·s））升高 1~2 倍；B4 井的地层压力明显低于同井组其他井，储层的流度与水驱时比较没有变化。因此，该井在二氧化碳驱时没有受效，或者注入井 B3 注入的二氧化碳往该井推进较少。

表 4　E4、C3 和 B4 井试井测试资料解释结果

测试井名	测试时间	表皮因数	内区流度 （$10^{-3}\mu m^2/$（mPa·s））	外区流度 （$10^{-3}\mu m^2/$（mPa·s））	内区半径 （m）	外推地层压力 （MPa）
E4	2009-03	-4.76	0.5225	0.1907	31.78	29.19
C3	2009-03	-4.79	0.6933	0.3404	99.87	29.59
B4	2009-09	-3.97	0.6141	0.1018	40.64	22.13

3 测试资料解释结果评价

试验区 9 个井次试井测试资料解释结果见表 5。气驱前平均地层压力为 22MPa，由表 5 可以看出，F6 井、B4 井受效程度小；E7 井、D6 井、E4 井、C3 井和 B2 井受效程度较大。

表 5 试验区试井测试资料解释结果

测试井名	测试时间	表皮因数	内区流度 ($10^{-3}\mu m^2/$ (mPa·s))	外区流度 ($10^{-3}\mu m^2/$ (mPa·s))	内区半径 (m)	外推地层压力 (MPa)
E7	2009-08	-4.15	0.7307	0.1711	34.64	29.71
F6	2008-11	-2.08	0.1607	0.1607	—	22.65
	2009-06	1.13	0.1512	0.1512	—	23.24
D6	2009-05	-4.39	0.8965	0.3417	31.87	26.95
E4	2009-03	-4.76	0.5225	0.1907	31.78	29.19
B4	2009-09	-3.97	0.6141	0.1018	40.64	22.13
C3	2009-03	-4.79	0.6933	0.3404	99.87	29.59
B2	2009-03	-4.21	0.4375	0.7432	53.97	27.27
	2008-06	-4.57	0.4646	0.1043	105.05	23.91

B4 井生产动态曲线见图 7。由图 7 可以看出，B4 井日产液量、日产油量持续下降，动液面基本保持较低水平，二氧化碳体积分数较低。未见到混相驱油二氧化碳体积分数上升的

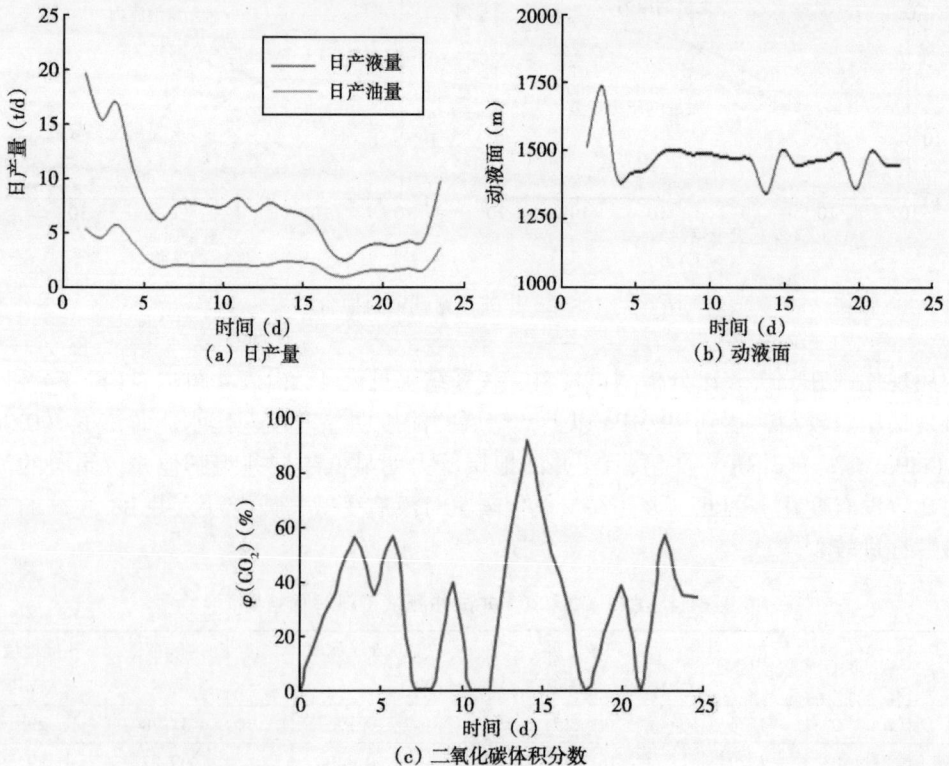

(a) 日产量

(b) 动液面

(c) 二氧化碳体积分数

图 7 B4 井生产动态曲线

动态特征，说明该井为二氧化碳驱未见效井，这与试井测试资料一致。

B2 井生产动态曲线见图 8。由图 8 可以看出，由于注二氧化碳后地层压力上升至混相压力（22.4 MPa）之上，地层能量得到很大程度补充，目前日产油量为 4.8t，并且日产液量、日产油量处于上升趋势。这与该井试井解释结果表现出地层压力上升一致。

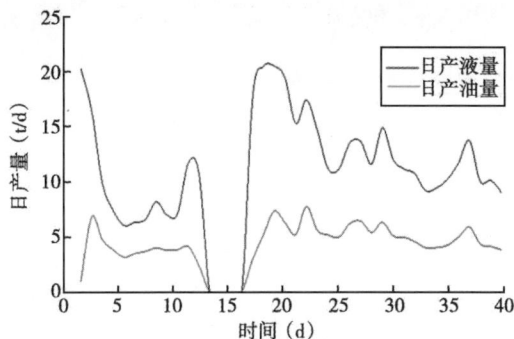

图 8　B2 井生产动态曲线

4　结论

（1）二氧化碳驱试井测试资料表现为 3 部分，对其驱前后特征进行解释和对比，可以反映二氧化碳驱受效情况。

（2）对同井不同时段和同时间段不同井试井测试资料解释结果对比，结果为：二氧化碳驱过程中 F6 井和 B4 井受效程度小，E7 井、D6 井、E4 井、C3 井和 B2 井受效程度较大。

（3）将前述解释结论与实际井动态资料进行对比，具有较好一致性，解释结果基本满足实际要求，具有一定意义。

参 考 文 献

[1] 廖新维，沈平平. 现代试井分析 [M]. 北京：石油工业出版社，2002.

[2] 沈平平，廖新维. 二氧化碳地质埋存与提高石油采收率技术 [M]. 北京：石油工业出版社，2009.

[3] Tang R W，Ambastha A K. Analysis of CO_2 pressure transient data with two and three region analytical radial composite models [J]. SPE 18275，1988.

[4] MacAllister D J. Pressure transient analysis of CO_2 and enriched-gas injection and production wells [J]. SPE 16225，1987.

[5] 张川如，虞绍永. 二氧化碳气井测试与评价方法 [M]. 北京：石油工业出版社，1999.

[6] 邓才. 二氧化碳驱试井分析方法研究 [D]. 北京：中国石油大学（北京），2010.

[7] 李卓. 芳 48 试验区注气开发效果研究 [D]. 大庆：大庆石油学院，2005.

[8] 钟立国，韩大匡，李莉，等. 特低渗透油藏二氧化碳吞吐模拟 [J]. 大庆石油学院学报，2009，33（4）：120-124.

[9] 陈方方，贾永禄，霍进，等. 三孔均质径向复合油藏模型与试井样版曲线 [J]. 大庆石油学院学报，2008，32（6）：67-75.

[10] 吕秀凤，刘振宇，张瞳阳. 数值试井分析的有限元法 [J]. 大庆石油学院学报，2005，29（5）：12-17.

[11] 侯健. 一种基于流线方法的 CO_2 混相驱数学模型 [J]. 应用数学和力学，2004，25（6）：635-641.

［12］王海涛，张烈辉，冀秀香，等. 基于边界元法的含局部不渗透区域任意形状气藏渗流问题［J］. 大庆石油学院学报，2009，33（2）：62-67.

［13］宁正福，廖新维，高旺来，等. 应力敏感裂缝性双区复合气藏压力动态特征［J］. 大庆石油学院学报，2004，28（2）：34-36.

［14］张怀文，张翠林，多力坤. CO_2 吞吐采油工艺技术研究［J］. 新疆石油科技，2006，16（4）：19-21.

［15］李孟涛. 低渗透油田注气驱油实验和渗流机理研究［D］. 北京：中国科学院研究生院（渗流流体力学研究所），2006.

适合 CO_2 驱油开发特点的油藏动态监测技术

陈国利[1]　胡永乐[2]　刘运成[1]　张　辉[1]　张云海[1]

(1. 中国石油吉林油田公司；2. 中国石油勘探开发研究院)

摘　要： CO_2 混相驱油是一种重要的提高原油采收率的方法，可以大幅度提高单井产量和采收率，但 CO_2 驱油存在混相不稳定、流体运移难控制、腐蚀问题突出、安全环保要求高等问题。为了解决这些问题，在油藏监测方面相对应地需要增加一些特殊项目，主要有吸气剖面监测、直读压力监测、井流物分析、气相示踪剂、腐蚀监测和环境监测等。这些监测项目在吉林黑 59 区块 CO_2 驱油先导试验的实际应用中取得了较好的效果，明确了试验区动态变化的特点和趋势，为保混相、防气窜、防腐蚀、防泄漏提供了技术支撑，已经初步形成了适合 CO_2 驱油开发特点的油藏动态监测技术。在下一步矿场试验中，需要进一步对油藏监测项目进行优化，继续实施多项监测技术组合应用，重点完善注气前缘监测技术，并适时开展驱替效果监测。CO_2 驱油藏动态监测技术将向系统化、高精度、多学科交叉融合和实时监测与调控发展。

关键词： CO_2 混相驱油；油藏监测；吸气剖面；压力监测；井流物分析；气相示踪剂

0　引言

2008 年以来，陆续在吉林油区大情字井油田黑 59 和黑 79 南区块开展了 CO_2 驱油先导试验和扩大试验。在 CO_2 驱油开发试验过程中，对油藏监测技术进行了探索。在水驱油藏监测的基础上，针对 CO_2 驱油的开发特点，初步研发形成了适合 CO_2 驱油开发特点的油藏动态监测技术。这些监测技术在 CO_2 驱油矿场试验中发挥了重要的作用，明确了试验区动态变化的特点和趋势，为保混相、防气窜、防腐蚀、防泄漏提供了技术支撑。

1　CO_2 驱油开发特点及监测需求

1.1　混相状况识别及监测需求

在 CO_2 混相驱油开发过程中，对混相状况的识别至关重要。国内外 CO_2 驱油试验表明，混相驱替效果要远远好于非混相或近混相[1]，油藏只有达到混相，才能大幅度提高采收率。由于低渗透油藏一般混压差小，混相状况不稳定，因此，需要对混相状况进行定期监测。对混相状况的识别需应用地层压力、生产动态、井流物分析、试井等多种方法进行综合分析。

1.2　流体运移控制及监测需求

当用 CO_2 驱替原油时，由于流度比和油藏非均质性的影响，尤其是在未混相的情况下，

CO_2 的运移难以控制，CO_2 的波及系数较低。因此，需要应用气相示踪剂、井间地震、电位法井间监测等技术对 CO_2 的运移方位、速度等进行监测，分析流体运移特点，为开发调整提供依据。

1.3 腐蚀防护及监测需求

CO_2 为酸性注入剂，由于液态水的存在，会发生强烈的腐蚀。CO_2 驱注入井、采油井及地面设备的腐蚀问题非常突出，需要长期对井下、井口和地面设施进行腐蚀防护和腐蚀监测，确保注采和地面系统安全运行。CO_2 腐蚀物被带入地层后，还会造成储层孔隙堵塞[2]。

1.4 预防泄漏及监测需求

在 CO_2 驱油与埋存过程中，存在 CO_2 泄漏的风险。CO_2 可能从断层、裂缝、井筒、管线泄漏，影响安全生产，对周边环境造成损害[3]。为了预防和及时处理可能发生的 CO_2 泄漏，需要系统地进行环境监测，使试验区及其附近的大气、土壤和地表水中的 CO_2 气体含量达到安全标准。

2 几项特色油藏监测技术

2.1 吸气剖面监测

吸气剖面测试是注气开发油田必不可少的监测项目，能够确定油层的吸气剖面及吸气特点[4,5]。

2.1.1 测试目的

通过对注 CO_2 井实施吸气剖面测试，获得井下测试段温度、压力和流量数据，通过软件解释获得井下测试层位的吸气剖面。

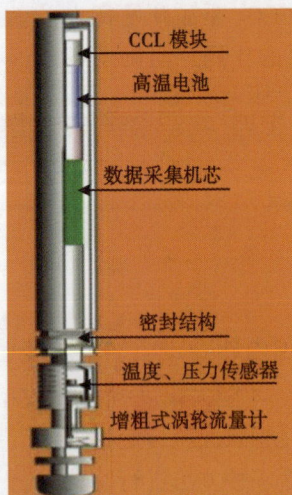

图 1 井下仪器结构及
测试原理图

2.1.2 测试原理及测试工艺

测试仪器采用存储式测试工艺，采用铂电阻测温，硅蓝宝石传感器测压，增粗式高精度涡轮流量计测流量，磁定位技术校深，由高效电池供电实现对井下温度、压力、流量和深度等参数测试，并通过软件解释得到吸气剖面（图1）。

2.1.3 应用情况

黑 59 区块共测试 CO_2 吸气剖面 5 口井 17 井次，明确了低渗透油藏的吸气剖面及吸气特点。

通过与吸水剖面对比，明确了各层吸气状况与吸水状况的差异；明确了未注水井组与注水井组在注气后剖面变化的不同特点。低渗透油藏吸气能力比较强，未注水井组注气后各层吸气比较均匀，可以有效驱替低渗层，改善层间矛盾。注水井组在注气后吸气剖面与吸水剖面相比变化不大。

2.2 直读压力监测

井下直读式压力监测系统可以长期实时监测采油井井底流压和地层压力，为确定混相状况和采油井的合理工作制度提供依据。

2.2.1 地面直读监测目的

通过压力和温度的直读监测，及时调整对应注气井的配注量，合理调整采油井的工作制度和参数。

2.2.2 地面直读监测技术工艺

采用直读式监测方法。地面仪表箱通过电缆供电，由地面仪表系统采集和显示测试的数据。管缆泵下部分走管内，泵上部分走管外（图2）。

2.2.3 应用情况

黑59区块直读监测实施4口井，长期实时监测了生产井井底流压，并取得了大量的地层压力数据，指导了现场注采调控方案的制订。

通过直读式测压，为流压控制、抑制气体突破及生产井工作制度的及时调整提供了依据。在产气量大影响液面的情况下，分析了气体突破井液面和压力的关系。通过对压力的及时监测，对水气交替过程中的注水能力和注气能力的变化情况进行了分析，并及时优化了水气交替方案。

2.3 井流物分析

通过对产出气和原油组分的分析，为确定采油井的混相状况提供依据。

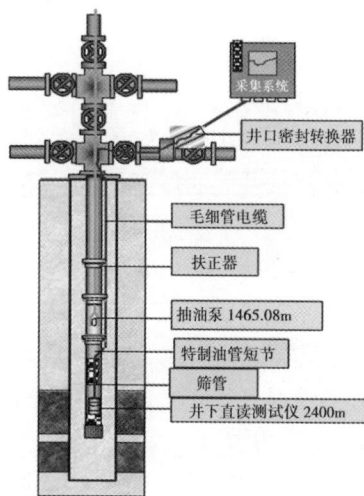

图2 地面直读监测技术工艺图

2.3.1 产出气组分分析

分析轻组分萃取情况及 CO_2 含量变化情况。根据产出气中 CO_2 含量变化规律，确定混相状况。

2.3.2 原油组分分析

分析原油各组分及平均相对分子质量，测定样品原油密度、黏度、凝固点、含蜡量。观察产出物组分变化，建立原油组分与地层压力、饱和压力及混相状态之间的关系。

2.3.3 高压物性取样分析

分析地层条件下 CO_2 驱替过程中的原油组分及性质的变化规律，判断混相状态。

2.3.4 应用情况

目前已化验分析了大量的样品，为油井混相状况的判断、效果分析及方案调整提供了依据。长岩心驱替实验表明，混相驱 CO_2 突破后产出气中 CO_2 含量快速升高。试验区几口井产出气中 CO_2 含量呈现快速大幅度增高的态势，因此判断为混相驱（图3）。

2.4 气相示踪剂

气体示踪剂在注气混相驱中有较好的应用[6,7]。采用气体示踪剂可以更好地了解 CO_2 在油层中的运移特点。

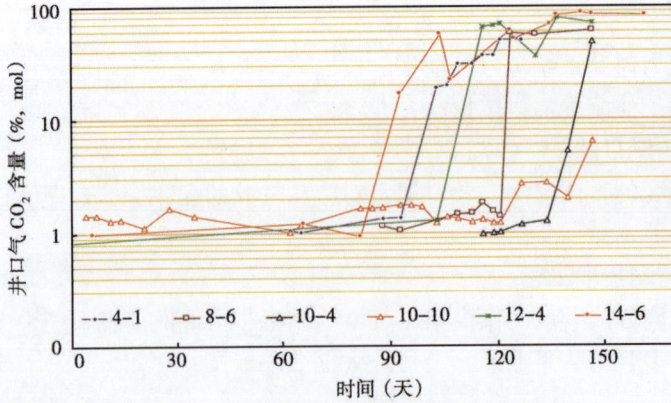

图 3 CO_2 突破后产出 CO_2 含量变化曲线

2.4.1   目的

在注气井中注入气体示踪剂，在周边观察井中进行取样分析，可以了解注入气体的波及情况、注采井间的连通情况、注入气体的推进方向及速度、地层压力场分布和裂缝发育等情况。

2.4.2   原理

井间气体示踪技术所依据的理论是层析理论。利用气体示踪剂可以跟踪注入气体（CO_2 气）的流动速度、流动方向。通过软件对示踪剂产出曲线进行拟合处理，可认识注采井间的连通情况、注入气体的波及情况以及气体示踪剂波及层的非均质性。

2.4.3   应用情况

黑 59 区块实施气相示踪剂监测 6 口井，初步明确了注入气体运移方向及速度。气相示踪剂的推进速度远远大于液相示踪剂的推进速度。平面上示踪剂的运移速度差异较大，波及不均匀，非均质性较强。一线井示踪剂的浓度大，但运移速度较慢；二、三线井浓度小，但速度较快。一线井中东西方向井示踪剂的浓度大，运移较快，南北方向井示踪剂的浓度小，运移较慢（图 4）。

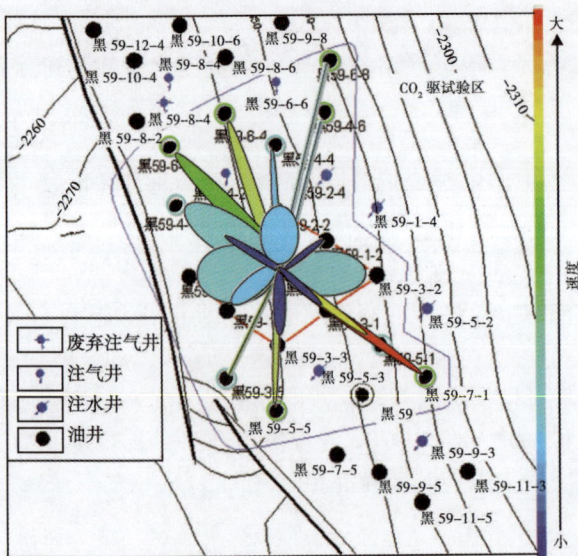

图 4   监测井中示踪剂浓度和速度示意图
宽窄表示相对浓度

2.5   腐蚀监测

2.5.1   井口腐蚀监测技术系列

井口腐蚀监测技术系列有：（1）箱体式井口评价技术；（2）井口缓蚀剂残余浓度检测技术；（3）井口在线探针监测技术；（4）超声波测厚技术。

2.5.2   井下腐蚀监测技术系列

包括（1）井下挂片、挂环腐蚀监

测技术；（2）井下实时存储腐蚀监测技术；（3）井下弱极化腐蚀监测技术；（4）井下在线直读腐蚀监测技术（图5）。

2.5.3 应用情况

系列化腐蚀监测技术现场实施40余井次，认识了注气井和采油井的腐蚀剖面，准确评价了防腐效果。腐蚀速率总体低于行业标准。黑59区块油管、杆、泵解剖无明显腐蚀结垢特征，注采系统安全运行。

2.6 环境监测

2.6.1 目的及意义

环境监测作为CO_2地质埋存监测的重要组成部分，是对储层监测的有益补充，具有不可替代的价值和作用。其目的是保证人居环境不受CO_2埋存的影响，确保所

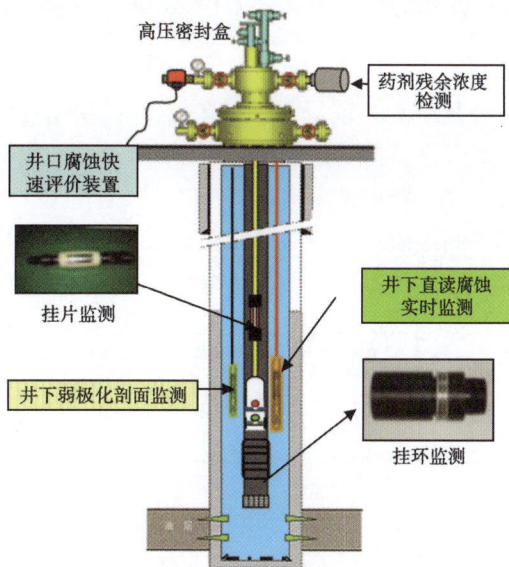

图5 CO_2腐蚀监测技术系列

埋存的CO_2不会泄漏到大气、土壤及淡水层，并保障安全的作业环境和有效的地下埋存。

2.6.2 环境监测内容和方法

（1）大气中CO_2含量监测。CO_2从埋存地点发生泄漏后可能会导致大气中CO_2通量和浓度发生明显变化，因此可以使用便携式CO_2红外探测器进行大气中CO_2含量测试。

（2）土壤气体浓度监测。储层中CO_2气体的逸出会导致土壤气体成分（CH_4、N_2、C_2/C_3、CO_2、O_2）的变化，需要监测土壤气体成分的浓度变化，了解地球化学反应和气体可能的迁移途径，评估逸出量。

（3）地表水及湖泊水监测。CO_2从埋存地点发生泄漏后可能会渗入到地表水中，引起地表水或临近湖泊、湿地等水的性质（pH和电导性）和成分以及其中的溶解气含量（CO_2等气体）的变化。可以通过监测浅层地表水和湖泊水的变化，从而确定地表浅层泄漏的可能性及泄漏的地点和区域。

3 下步发展方向

在下一步矿场试验中，需要进一步对油藏监测项目进行优化，继续实施多项监测技术组合应用，完善注气前缘监测技术，并适时开展驱替效果监测。CO_2驱油藏动态监测将向系统化、高精度、多学科交叉融合和实时监测与调控发展。

3.1 继续实施多项监测技术组合应用

为满足CO_2驱油试验动态监测的目的及要求，实施多项监测技术组合应用比较适合。例如通过"井间示踪"与"不稳定试井"监测，相互印证，明确注采井间的连通状况及渗流能力。

3.2　完善注气前缘监测技术

利用井间地震、时移地震、微地震以及电位法监测等资料分析储层中 CO_2 饱和度的变化，可判断 CO_2 前缘[8-10]。在黑 59 区块已实施井间地震监测 6 条测线。

通过电位法井间监测，可了解 CO_2 驱油试验注入的 CO_2 在油层中的推进方位、推进范围[11]。在黑 59 区块实施了 5 口井，对于 CO_2 驱油推进方位的探测基本可行。

井间地震和电位法井间监测取得了一定的成果。下步应充分考虑油藏地质条件、资料录取、处理解释等因素，优化组合，完善注气前缘监测技术系列。

3.3　适时开展驱替效果监测

CO_2 驱油效果评价对试验及后期推广意义重大。通过脉冲中子测井可确定 CO_2 驱油过程中含油饱和度变化，确定 CO_2 驱油方向及残余油分布。分阶段钻取心检查井，可通过对储层特征及流体性质分析，确定驱替程度、残余油分布及性质，进一步明确 CO_2 驱油机理和驱替效果。

3.4　不断探索新的监测技术和监测方法

20 世纪 90 年代以来，油藏监测技术向系统化、高精度、多学科交叉融合和实时监测与调控发展，在油藏管理中发挥了重要作用。美国 SACROC 油田注 CO_2 项目已进行了油藏动态实时监测与调控试验[12]。多传感器、多参数监测将成为未来油藏动态监测的主要发展方向。随着 CO_2 驱油矿场试验的不断推进，应根据油藏的特点和试验的进程，不断探索新的监测技术和监测方法。

4　结论

（1） CO_2 驱油开发存在混相不稳定、流体运移难控制、腐蚀问题突出、安全环保要求高等问题，与水驱相比，在油藏监测方面需要增加一些特殊项目，主要有吸气剖面监测、直读压力监测、井流物分析、气相示踪剂、腐蚀监测和环境监测等。

（2） CO_2 驱油藏监测技术在吉林黑 59 区块先导试验的实际应用中取得了较好的效果，明确了试验区动态变化的特点和趋势，为保混相、防气窜、防腐蚀、防泄漏提供了技术支撑。

（3）在下一步矿场试验中，需要继续实施多项监测技术组合应用，完善注气前缘监测技术，并适时开展驱替效果监测。 CO_2 驱油藏动态监测技术将向系统化、高精度、多学科交叉融合和实时监测与调控发展。

参 考 文 献

[1] 仵元兵，胡丹丹，常毓文 . CO_2 驱提高低渗透油藏采收率的应用现状 [J]. 新疆石油天然气，2010，6（1）：36-39，54.

[2] 杨彪，于永，李爱山 . CO_2 驱对油藏的伤害及其保护措施 [J]. 石油钻采工艺，2002，24（4）：42-44.

[3] 许志刚，陈代钊，曾荣树 . CO_2 地质埋存渗漏风险及补救对策 [J]. 地质论评，2008，54（3）：373-386.

［4］吴志良．复杂断块油田 CO_2 驱油动态监测技术应用与分析［J］．石油实验地质，2009，31（5）：542-546.

［5］程杰成，雷友忠，朱维耀．大庆长垣外围特低渗透扶余油层 CO_2 驱油试验研究［J］．天然气地球科学，2008，19（3）：402-409.

［6］曹雅萍，龙华，王运萍．井间气体示踪监测技术在齐 40 块蒸汽驱中的应用［J］．石油钻采工艺，2004，26：12-14.

［7］鲜波，熊钰，孙良田．油藏多孔介质中气体示踪剂运移特征研究［J］．海洋石油，2010，26（4）：52-55.

［8］张进铎．井间地震技术在油气藏开发中的应用［J］．中国石油勘探，2007，（4）：42-46.

［9］陈小宏，易维启．时移地震油藏监测技术研究［J］．勘探地球物理进展，2003，26（1）：1-6.

［10］吴伟忠．微地震监测技术在油藏注水开发中的应用［J］．特种油气藏，2009，16（增）：175-177.

［11］刘新茹，张向林．油藏监测技术概述［J］．石油仪器，2008，22（3）：31-33.

［12］张凯，姚军，刘均荣．油藏动态实时监测与调控［J］．石油矿场机械，2010，39（4）：4-8.

Monitoring on CO_2 EOR and Storage in a CCS Demonstration Project of Jilin Oilfield China

Shaoran Ren[1]   Baolun Niu[1]   Bo Ren[1]   Yongzhao Li[1]
Wanli Kang[1]   Guoli Chen[2]   Hui Zhang[2]   Hua Zhang[2]

(1. China University of Petroleum; 2. Jilin Oilfield Company, PetroChina)

Abstract: Jilin oilfield, located in Jilin Province of Northeast China, is conducting the first large scale demonstration project on CO_2 EOR and storage. During the first stage, CO_2 separated from a nearby natural gas reservoir ($20\% \sim 97\%$ CO_2 content) is injected into the northern part of Hei$-$59 oil block. Currently, the targeted oil reservoir, with an area of 1.7 km^2 and oil reserve of $0.78 \times 10^6 t$, has six CO_2 injectors and twenty$-$five production wells with an inverted seven$-$spot injector$-$producer pattern. Up to now, nearly 150000 t of CO_2 has been injected for a miscible flooding, and an expected incremental oil recovery of 10% would be achieved. In order to better evaluate the reservoir performance and ensure a safe injection and long$-$term storage of CO_2, a monitoring program has been designed and deployed to monitor the EOR performance and the state of CO_2 in the reservoir.

The monitoring techniques used include CO_2 tracers to monitor gas breakthrough in production wells and flow of CO_2 across the reservoir, and bottom$-$hole pressure survey of producers to evaluate the miscibility effect on EOR combined with production data analysis. Geophysical techniques are also employed to detect possible fractures and their orientation and dimensions, which include electric spontaneous potential measurement and micro$-$seismic detections. Cross$-$well seismic is intended to monitor the flow and track the front of CO_2 between wells. Data of geochemical monitoring from produced water can be used to estimate the amount of CO_2 dissolved in reservoir fluids. CO_2 dissolution in formation water and residual oil is one of the main storage mechanisms, which can greatly enhance the safety and effectiveness of subsurface storage. In this paper, the results of reservoir monitoring applied in the oilfield are present and analyzed, and forward work will be focused on optimizing EOR performance and verification of the geo$-$capacity storage in the targeted reservoir.

Keywords: CO_2 EOR; Storage; Monitoring; CCS demonstration

0   Introduction

The CO_2 EOR and Storage Project in Jilin oilfield is the first large CCS (carbon capture and storage) demonstration project in China for CO_2 geological storage into depleting oil reservoirs. It aims at enhancing the understanding of CO_2 EOR mechanisms, movement of CO_2 in the reservoir and relevant physical$-$chemical reactions involved in the storage process, meanwhile gaining practical experience of monitoring and verification of CO_2 storage technology in tight oil reservoirs. During the first phase of the project, Block Hei-59 was selected as the target reservoir for the pilot test after an extensive geological characterization and laboratory assessments, and the project will be extended to other oil blocks nearby in the future.

The CO_2 pilot was carried out in the northern part of the Block Hei-59. It consists of 31 wells with a spacing of 440m×140m; 25 production wells and 6 CO_2 injectors. They are divided into 6 well groups, each with a central injector, as shown in Fig. 1. In April, 2008, CO_2 was firstly injected into wells 6-6 and 12-6, and then during the following six months, CO_2 injection was subsequently started with other wells. Injection of CO_2 can maintain and increase reservoir pressure in terms of EOR operations. In the first phase, the CO_2 used has been from a natural gas reservoir, which contains up to 97% of CO_2. Later on, CO_2 will be supplied from a separation plant for processing CO_2 associated natural gas. The injection pressure is currently of 11~14 MPa, and at -22 °C (liquid CO_2). From June, 2009, water alternative CO_2 injection was initiated in well 12-6

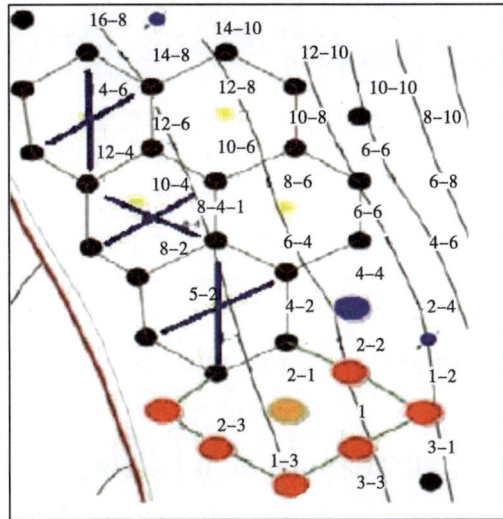

Fig. 1   Diagram of Well location and Surface Layout of Cross-well Seismic lines: Yellow Dots Are CO_2 Injectors, and the Seismic Lines Are in Deep Blue Color. Well 1 (large orange dot) was converted to a CO_2 injector in later September, 2009

to improve sweeping efficiency. Up to now, approximately 0. 18 HCPV (hydrocarbon pore volume) of CO_2 has been injected into the oil reservoir, with an injection rate of 100~170 t/d. Water flooding in the southern part of the Block is simultaneously underway in comparison with CO_2 flooding.

In order to understand the EOR performance and monitor the movement of CO_2 inside the reservoir, a monitoring program has been designed and conducted during the injection process. In this paper, the preliminary results are analyzed and presented corresponding to various monitoring techniques applied, including gas tracer monitoring, production fluid sampling, electric spontaneous potential measurement, cross-well seismic, micro seismic measurement and daily productivity analysis. These results can be used for studying the effects of CO_2 injection on EOR, mapping CO_2 distribution in the reservoir and verifying geological storage capacity.

1   Reservoir Description

Block Hei-59 is composed of several producing zones from the Qing shankou Group formed in the age of Cretaceous. It was deposited in a delta front environment and controlled by two antithetic down faults. The oil-bearing formations are manifested with good development of sand body, good connectivity, and well defined cap rocks, which are relatively independent with hydro-geological systems and have relatively maturely-developed fractures[1]. A relatively stable cap zone is promised

447

by thick mudstone layers with well−developed shale intercalations and minor fractures/faults. The target oil zones are buried at an averaged depth of 2400m, with a small dip angle of $2 \sim 4$ degrees from west to east. The gross sand layers range from $11.2 \sim 18.2$ m in thickness, which include subzone−7, 12, 14, 15 of the lower Qing−1 interval and subzone−23, 24 of the upper Qing−2 interval. The average porosity and permeability are of 12.7% and 3.5 mD, respectively. The reservoir temperature is of 98.9℃, and the original reservoir pressure is 24.15 MPa. The oil viscosity varies from $1.85 \sim 2.18$ mPa · s at the reservoir condition, and the averaged formation water salinity is about 14607 mg/L. The reservoir has been produced via primary depletion and water injection in the past four years, and the oil recovery factor is estimated to be 16% prior to CO_2 injection. In order to increase productivity of the tight reservoir, fracturing treatments of production wells have been conducted.

2 Designed Monitoring Program

2.1 Well Tests and Well Logging

Most CO_2 injectors were converted from previous water injectors or oil producers. In order to verify their integrity over CO_2 injection, a well−bore logging program was applied before and during the injection, which include pressure—temperature logs, Cement Bond Logs (CBL), ultra sound logs, electromagnetic inspection (EMI) and mechanical integrity test (MIT) using helium. A designed reservoir saturation test (RST), to measure CO_2 or gas saturation along the well bore region, will be deployed in the future.

During the injection, annulus pressure, well head temperature and pressure and gas rates were recorded for all the injectors, which are prime parameters needed for monitoring safe injection operation and for reservoir simulation study. In the production sides, well productivity measurement, bottom−hole pressure survey and fluid level measurement were conducted in order to monitor the EOR performance and movement of injected CO_2. The injection wells were profiled using radioactive surveys, and the radioactive profiles were also run on selected production wells to determine flow distribution or the location of gas breakthrough within the wellbore.

2.2 Fluid Sampling

Fluid sampling from the production wells were taken and analyzed since the injection started. Water, gas and oil samples were collected from well−head for the measurement of pH, major ion concentrations, and gas compositions, such as nitrogen, CO_2 and natural gas components. Well−head oil samples were also analyzed for viscosity, density and the contents of wax and asphaltene. In addition to the oil samples obtained at well heads, down−hole fluid samplings were also taken using high pressure sampler in selected wells.

2.3 Gas Tracer

Tracers are unique or highly indicative chemical species that can be used to "fingerprint" the fluid of interest and distinguish it from other sources. They can be used to trace the movement of injected CO_2, which are very effective to gain information on inter-well reservoir heterogeneity, relative fluid velocities, and sweeping efficiency of CO_2. From September, 2008 to January, 2009, five tracers were selected, and individually injected into five injection wells. Meanwhile, nineteen production wells in the five well groups were sampled once a day in order to adequately monitor the possible breakthrough of the gas tracers, using gas chromatograph method.

2.4 Crosswell Seismic

A 4D surface seismic program was recommended for active seismic monitoring on CO_2 storage[2], but the thin inter-bedded layers (<2 m) of the targeted reservoir pose a great challenge to the resolution of the 4D seismic method. On the other hand, crosswell seismic profiling can provide higher resolution, at a much lower cost than a 3D survey, which can be reliable and cost-effective. Reservoir properties (e. g. porosity) can be imaged and characterized using crosswell seismic survey[3], and in a time-elapse mode, it also can be used to monitor changes in the reservoir (e. g. saturation and pressure). In March, 2008, crosswell seismic surveys were undertaken in Hei-59 block. The program layout consists of six seismic lines, ranging around 600 m in length and centered in three CO_2 injection wells in the west side of the pilot region, as indicated in Fig. 1. The frequency band used ranges from 100~800 Hz, and a geophysical receiver string at 10 m spacing, were placed downhole over 200 m along the formation layers.

2.5 Microseismic

The stress on formation rocks around the fractures and pore pressure may undergo a significant increase during well fracturing treatment, which can affect the stability of the formation and cause shear slippage, forming possible seepage for CO_2 leakage. Micro-fracture may also be caused during CO_2 injection due to high injection pressure. Acoustic signals appearing in fracturing or micro-fracturing process can be detected using appropriate receivers and processed to determine the locations of these microseismic events[4,5]. Therefore, microseismic techniques can be used to map fractures, including their azimuth, length and height. In this study, microseismic fracture mapping was implemented in wells around the fracturing treatment wells and CO_2 injectors. An array of 12 three-component geophone receivers was placed in nearby well at 5~10 m spacing over 90 m across the oil layers.

2.6 ESP Measurement

When fluid flows within a porous media, it can produce an electrical potential due to the separation of ions across flow boundaries, which is called electric spontaneous potential (ESP). This offers a possible means of detecting CO_2 migration in a time-elapse mode[6]. Since May, 2009, ESP measurement was performed for fracturing treatment well 8-2 in order to monitor the fracture azimuth. Repeated measurement will be deployed and extended to track CO_2 migration.

3 Results and Discussions

3. 1 Injection and Production Responses

Good production response has been observed after about six months of CO_2 injection since April 2008, as shown in Fig. 2. Oil production in the whole pilot area has rapidly increased to around 100t/d from 20 t/d, and the present oil production rate has been maintained at 60t/d. Water cut has been kept stable with an approximately 6% reduction. However, it should be noted that GOR has greatly increased after CO_2 breakthrough, and the oil response of some wells are much below the average level. For oil wells located in the east side of the injectors and in the north part of the pilot area, oil production has been significantly enhanced, while wells in the west side responded poorly.

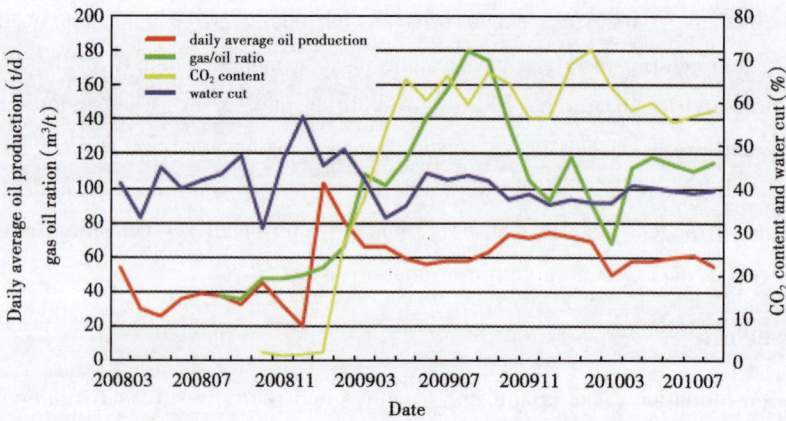

Fig. 2   Measured Oil Production, Water Cut, CO_2 Content and GOR in the CO_2 Miscible Pilot Area

The CO_2 injectivity of individual layers (injection profile) is much better than that of water. Fig. 3 shows typical injection profiles measured for water and CO_2 injection at different survey times.

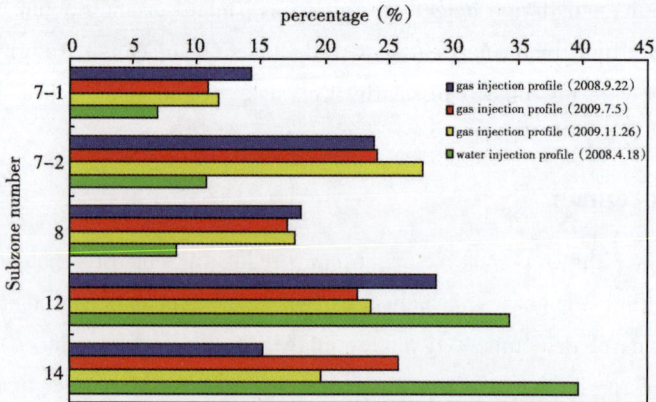

Fig. 3   Water and CO_2 Injection Profiles of Subzones along the Wellbore Taken at Different Survey Times in Well 10-8 (Water Injection Profiling Was Taken before CO_2 Injection)

It can be observed that, in comparison with water injection, the injection profiles of CO_2 are much uniformly distributed along the sub oil layers, which is very beneficial to mobilize oils in the upper formations where water injectivity was poor.

The production data also show that, compared to the simultaneous operation of water flooding in the southern part of the Block Hei−59, CO_2 flooding in the northern part is expected to achieve an incremental oil recovery of 8%~9% in 5 years.

3.2 CO_2—oil Miscibility Analysis

The MMP (minimum miscibility pressure) of the Block Hei−59 was determined as 22.3 MPa. In order to achieve effective miscible flooding, it is designed for the bottomhole pressure of the producers to be closely monitored and maintained. If any well exhibiting a GOR increase over 500 $m^3/$ t, or the liquid production less than 2t/d, the well would be shut in, waiting for the well pressure to rise.

High pressure sampling data indicate that dissolved gas/oil ratio has been largely increased, and almost 60% mole fraction of CO_2 was dissolved into the reservoir fluid at a pressure of 24.2MPa. After CO_2 breakthrough, the measured density of produced oil (degassed) in several wells has slightly decreased; this may be attributed to the extraction effect of light and intermediate components from oil by CO_2.

Compositions of the produced oil have been analyzed to evaluate the CO_2—oil miscibility effect. The degassed oil compositions sampled from different wells and at different times, are shown in Fig. 4. The data shows that, before CO_2 breakthrough, contents of light hydrocarbon components (C_2—C_{10}) were relatively stable, while they fluctuate after CO_2 breakthrough. Similar trends were observed for the content of heavy oil components (C_{20+}). This indicates the interactions of oil and CO_2 can change the compositions of the produced oils, while it might not be obvious to see the expected CO_2—oil miscibility effect: such as light component extraction etc. Detailed study in combination with pressure and production data analysis is needed.

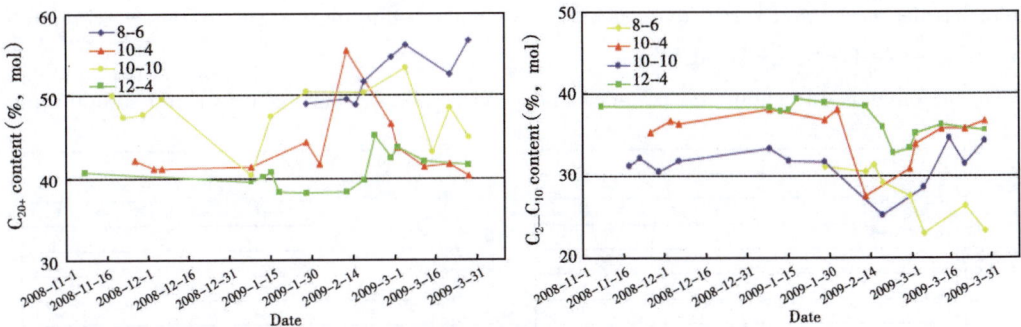

Fig. 4　Oil Analysis Results: C_{20+} and C_2—C_{10} Content
versus Sampling Date from Different Well Groups

3.3 Movement of CO$_2$

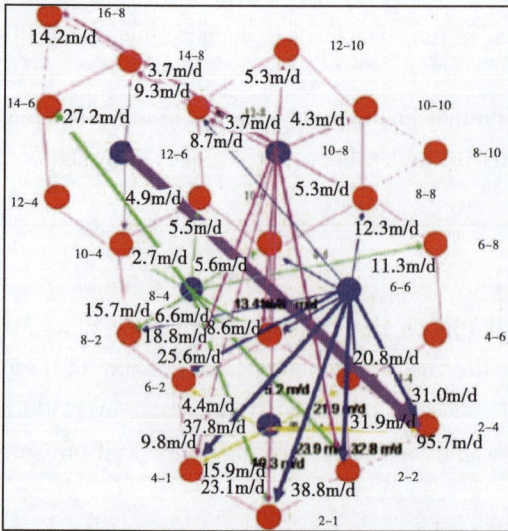

Fig. 5　Gas Tracer Results, the Value for Each Arrow Line Represents CO$_2$ Movement Rate between the Connected Two Wells

The methods that can be used to monitor the movement of CO$_2$ include geochemical and fluid sampling, tracer surveys, and a variety of geophysical surveys. Reservoir modeling and simulation can be used to verify the measurement results. The most effective response obtained so far was using CO$_2$ gas tracers, which has clearly tracked the spread of CO$_2$ in the reservoir and identified high permeability zones or channels. As indicated in Fig. 5, the movement of CO$_2$ from each injector and its speed toward relevant producers are plotted. The data indicate that reservoir heterogeneity has a significant effect on CO$_2$ migration and make the movement of CO$_2$ in the reservoir much more complex. The tracer measurements can be also used to quantify the CO$_2$ volume breakthrough in individual production wells and track back to its origins (injection wells). Gas tracers can be an effective tool to map high permeability channels in the formation.

The content of CO$_2$ in well-head gas samples measured at different times is graphed in Fig. 6 for different producers, which is another important parameter to study the CO$_2$—oil interaction and EOR performance. Early breakthrough and high CO$_2$ contents were observed in wells 14-6 and 2-4, which normally have a negative effect on oil production. Serious gas channeling can lead to early breakthrough and very high CO$_2$ content, and conformance control techniques, such as WAG (water alternative gas) and foam injection, are needed to improve EOR performance. Some wells (12-4 and 8-6) had a significant increase of CO$_2$ content shortly after CO$_2$ breakthrough, and then CO$_2$ content has decreased and varied dynamically. This may be due to the effect of the WAG operations in injection wells 12-6, 8-4 and 6-6.

On the other hand, short well distance and good response from gas tracer and sampling meas-

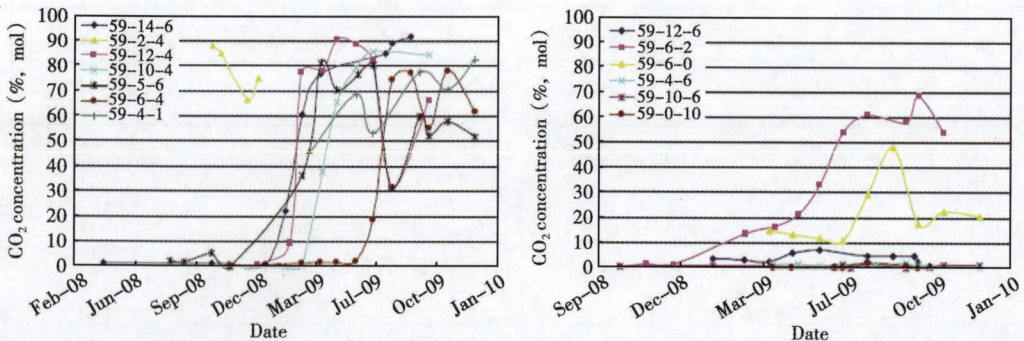

Fig. 6　Content of CO$_2$ from Well-head Gas Samples for Different Producers

urements might overshadow the needs to detect the CO_2 movement and fronts in the formation using other techniques, such as the crosswell seismic and electric spontaneous potentials methods as described above, though it has been found that the vertical resolution of crosswell seismic is satisfactory, nearly five times greater than that of conventional surface 3D seismic from the aspect of reservoir characterization.

3.4   Reservoir Heterogeneity and Its Effect on Well Performance

Natural fractures exist in the reservoir. Gas tracer results indicate that natural west—east fractures is well-developed between the well 8−41 and 8−6, and well 6−6 with 8−2 and 4−1. Fractures generated during fracturing treatment prevail in the direction of north—east between well 4−1 and 2−1 and 6−2.

Fig. 7 shows the results using the microseismic technique to map the fractures generated in a fracturing treatment well. The data analysis reveals that a fracture of approximately 190 m in length, with an azimuth at N85°E, has been generated during the process, which is in accordance with the expected fracturing design. The microseismic events recorded in the two wings of the fracture are almost symmetric. As shown in Fig. 7b, both upward and downward fractures growth was observed; the rectangular box that covers the majority of the microseisms is interpreted as the main fracture body, showing fracture heights of approximately 37 m and 43 m, respectively. Fig. 7c reveals that fractures appear to grow downward toward the east wing. This is consistent with the formation dip of approximately 2°~4°, indicating that fracture grows preferably along the formation dip in the reservoir.

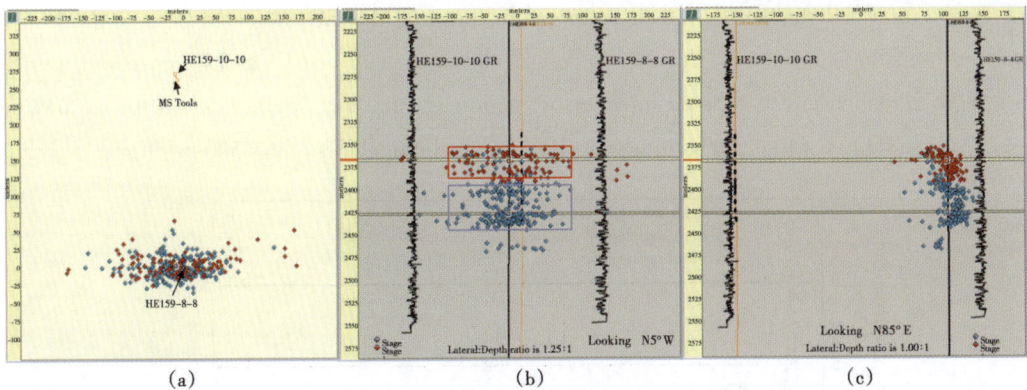

(a)   (b)   (c)

Fig. 7   Top Plan (a) and Side-section (b, c) Views of Microseismic
Signals Received from a Fracturing Treatment Well

The blue line in (a) represents the measured fracture, with approximately 190m in
length and an azimuth at N85°E

Electric spontaneous potential measurement in another fracturing treatment of well 8−2 is shown in Fig. 8. It can be observed that both fractures in the subzone−7 and 12 are almost in the same direction from west to east. This is in accordance with the expectation of hydraulic fracturing design.

Reservoir heterogeneity, in terms of fractures and high permeability channels, has seriously

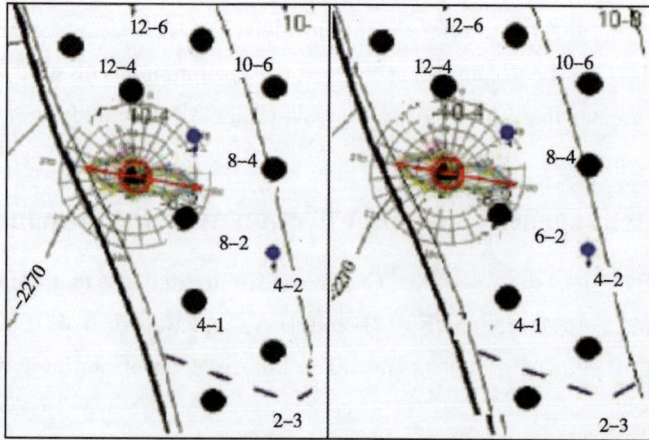

Fig. 8   Fracture Diagnosis in Well 8-2 of Jilin Oilfield Using Earth Potential

compromised production performance, especially for wells located in the upper part of the reservoir. Figs 9 and Fig 10 show the data of liquid and gas production for well 14-6 from 1/2007 to 10/2010. It has revealed that, from the tracer results of Fig. 5, a relatively high movement speed of CO_2 of 27 m/d from injector well 8-4 to 14-6 was observed. After gas breakthrough in February, 2009, high gas ratio and low oil production occurred, and a repeated well shut-in and reopen procedure was a-dopted to control CO_2 production. The produced CO_2 and gas oil ratio (GOR) increased rapidly dur-ing the well-reopen period, and CO_2 content reached over 95% with GOR as high as 1300 m^3/t. As for this, it was speculated that a high permeability channel may, exist between well 14-6 and 8-4. It has been confirmed that, from a well temperature logging, the CO_2 detected from well 14-6 was channeled from the subzone-7. In order to control the movement of CO_2 and improve well perform-ance, WAG injection and CO_2 foam injection were initiated in May 2010. Good response has been observed for well14-6 (Fig. 10), in which GOR has maintained less than 500 m^3/t and oil produc-

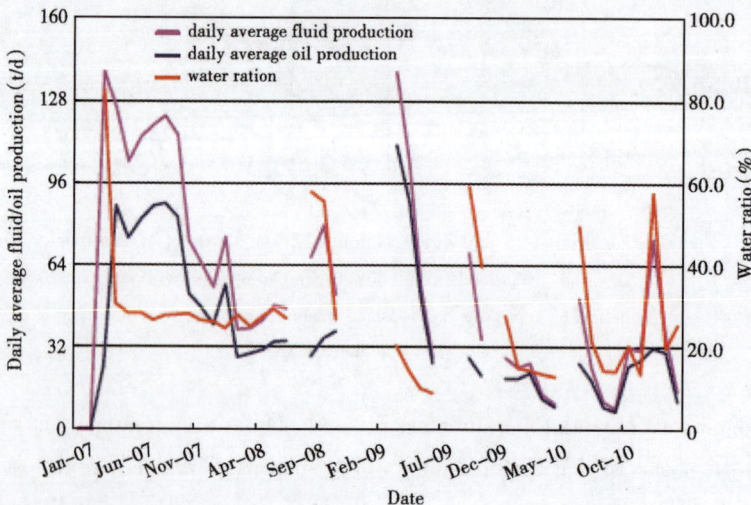

Fig. 9   Daily Fluid and Oil Production and Water Cut for Well 14-6 (2007—2010)

454

tion is of 4t/d in average so far.

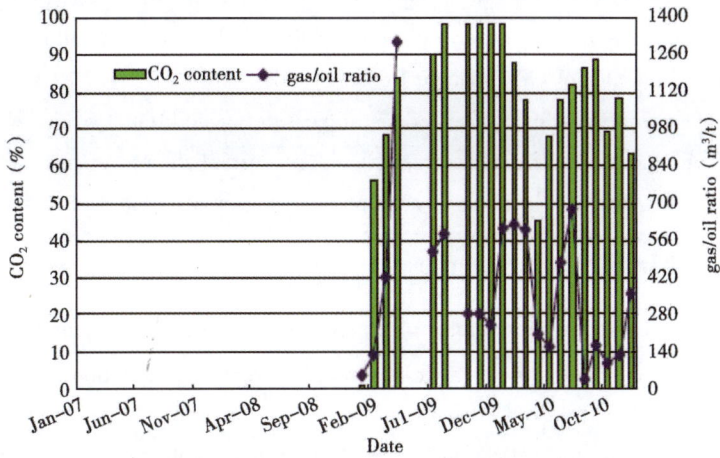

Fig. 10 CO$_2$ Content and Gas Oil Ratio in Well 14-6, Gas Breakthrough Was during February, 2009

3.5 Formation Water Sampling

Fig. 11 shows the concentration variations of Ca^{2+}, Mg^{2+} and HCO$_3^-$, sampled from the produced water in well 12-10, in which CO$_2$ has not significantly breakthroughed (see Fig. 6, the diagram in the right). It can be observed that the concentrations of Ca^{2+} and HCO$_3^-$ have slightly increased over the sampling period, while the concentration of Mg^{2+} has greatly increased after eight months of CO$_2$ injection. It may indicate that CO$_2$ has dissolved and reacted with the formation water, which may become another important factor to consider for the study of the CO$_2$—water interaction, CO$_2$ distribution in the formation and the storage potentials. Water sampling may not always be available due to low water cut and highly emulsified oil.

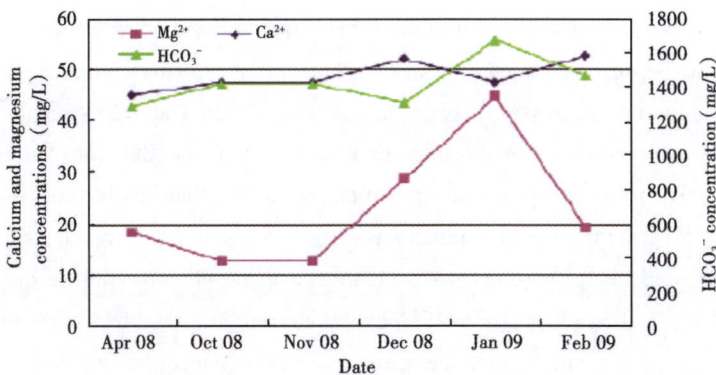

Fig. 11 Calcium, Magnesium and Biocarbonate versus Date in Well 12-10

Small amount of CO$_2$ were produced around Feburary, 2009 and the CO$_2$ content was kept less than 8% afterwards

3.6 CO$_2$ Leakage Monitoring

Reservoir characterization has demonstrated that the targeted reservoir has good natural barriers and cap rock integrity, therefore the main challenge for the storage safety or CO$_2$ leakage is the in-

455

tegrity of wells. For the old wells in the field, CO_2 corrosion can be a serious problem, which may lead to corrosions of casing and production tubes. Well head and annulus pressures have been monitored to ensure the casing is in good order and no leakage of CO_2 into the annulus. Up to now, there has been no obvious indicator of CO_2 leakage from the wells.

For the leakage monitoring and long-term storage safety, additional works have been carried out, including development of anti-CO_2 corrosion cement, establishment of risk evaluation methodology for well-bore integrity, deployment of cement bond logs (CBL) and ultra sound logs.

3.7　Follow-up Monitoring Schemes

Areas of further work in the next phase of the project would include determining CO_2 distribution and verifying the geological storage capacity through injection-production analysis, repeated interwell seismic, ESP and RST. Conformance control will be further implemented to better optimize the EOR performance. Also, the collection and analysis of the geochemical data will continue. Reservoir and geochemical simulations will be conducted to evaluate the flow and chemical processes in the reservoir in order to understand and determine the long-term fate of CO_2 sequestered in various forms. There are also substantial needs to conduct environmental monitoring and assessment, including gas sampling from soils, surface water and atmosphere, to further verify and provide assurance that CO_2 has not leaked back to the surface.

4　Conclusions

(1) For CO_2 EOR and storage projects, it is necessary to apply and maintain an effective monitoring program over the migration and distribution of CO_2, its final state in the reservoir and the integrity of reservoir and wells, in order to eliminate the risk of large CO_2 leakage. On the other side, it is beneficial to monitor the injection-production response and the level of miscibility between CO_2 and oil for the EOR performance to be optimized. The monitoring and verification program is site-specific, and should be designed according to the characteristics of individual reservoirs.

(2) The combination of crosswell seismic, gas tracers, fluid and geochemical sampling techniques can provide new insights into the migration/evolution of the CO_2 injected in oil reservoirs, which is more effective in terms of cost and technical reliability than an expensive and conventional 4D surface seismic program in oilfield applications. Fracture diagnosis using electric spontaneous potential measurement and micro-seismic has been proven to be effective during the fracturing treatments in the field application of the Hei-59 block. Bottom-hole pressure survey and fluid sampling in conjunction with injection-production data analysis can enhance our understanding on the performance of CO_2 EOR and storage capacity.

(3) Over three years of CO_2 EOR and storage project in the Hei-59 block, more than 150, 000 tons of CO_2 have been injected, and oil production has significantly increased, without obvious CO_2 leakage. Longer-term operation of the project continues to be guided by a comprehensive, cost effective and fit-for-purpose monitoring program in the second phase of the pilot in the region, which will include 15 more CO_2 injection wells.

456

Reference

[1] Xuan Z, He S L. Potential and early opportunity-analysis on CO_2 geo-sequestration in China [C]. Paper SPE 130358 presented at the 2010 SPE EUROPEC/EAGE Annual Conference and Exhibition held in Barcelona, Spain, 14-17 June, 2010.

[2] Wurdemann H, Moller F, Kuhn M, et al. CO_2 sink-from site characterization and risk assessment to monitoring and verification: One year of operational experience with the field laboratory for CO_2 storage at Ketzin, Germany [J]. International Journal of Greenhouse Gas Control, 2010, 4 (6): 938 - 951, doi: 10.1016/j.ijggc.2010.08.010.

[3] Harris J, Langan R. Crosswell seismic profiling: Principle to applications [J]. Search and Discovery, 2001.

[4] Warpinski N R, Wolhart S L, Wright C A. Analysis and prediction of microseismicity induced by hydraulic fracturing [C]. Paper SPE 71649 presented at the 2001 SPE Annual Technical Conference and Exhibition, New Orleans, Louisiana, USA, 30 September-3 October, 2001.

[5] Wang F, Liu X H, Liu C Y, et al. Fracture diagnostics and modeling help to understand the performance of horizontal wells in the Jilin Oilfield, China [C]. Paper SPE 122438 presented at the 2009 SPE Asia Pacific Oil and Gas Conference and Exhibition held in Jakarta, Indonesia, 4-6 August, 2009.

[6] Hoversten G M, Gasperikova E. Non-seismic geophysical approaches to monitoring [C]. In: Thomas D C & Benson S M. (ed.) Carbon dioxide Capture for Storage in Deep Geologic Formations-Results from the CO_2 Capture Project, 2005, Volume 2: Geological Storage of Carbon Dioxide with Monitoring and Verification. Elsevier, Oxford, 2005, 1071-1112.

CO_2 地质埋存监测技术及其应用分析

任韶然[1]　任　博[1]　李永钊[1]　张　亮[1]
康万利[1]　刘运成[2]　陈国利[2]　张　华[2]

(1. 中国石油大学石油工程学院；2. 中国石油吉林油田勘探开发研究院)

摘　要：对 CO_2 地质埋存监测目的和意义进行阐述，对地震、重力测试、井流体取样、示踪剂及 CO_2 泄漏等主要监测技术的原理、特点和应用进行分析，监测系统及相应的监测技术可对储层、盖层和周围环境进行描述，概括总结监测技术的研究进展和发展趋势。结果表明：示踪剂、井流体取样和测井技术是监测储层内流体运移和 CO_2 状态以及油藏和井完整性的有效手段；土壤气体分析和大气监测能够有效地监测 CO_2 泄漏和地表环境，监测系统的优化设计要结合监测目的和储层条件，根据现有的技术水平及应用经验合理筛选监测技术，可以经济有效安全地对注入 CO_2 的状态和泄漏进行监测。

关键词：CO_2；地质埋存；油藏监测；示踪剂；监测技术

0　引言

近年来，各国加强了 CO_2 捕集和埋存（CCS）相关技术的研究，世界上已有多个 CCS 示范工程相继实施，其中包括 Sleipner 和 In Salah 气田伴生 CO_2 的盐水层埋存与加拿大 Weyburn 油田的油藏埋存及提高原油采收率（EOR）工程[1]。中国近年来也开展了 CO_2 的捕集和资源化利用的研究，并于 2007 年在吉林油田进行先导试验。将 CO_2 注入油藏提高原油采收率是石油工业的一项成熟技术，已有 40 余年的工程实践历史。CO_2 注入油气藏不仅可以提高采收率，而且可以实现永久储存 CO_2 的目的[2]。但是，对于这一新技术，除有政治、政策问题及经济可行性考虑外，还存在许多技术问题及政府和公众的接受性问题。其中焦点之一是能否保证 CO_2 的安全永久埋存，这也是目前众多示范工程的主要目的。因此，需要系统、完整、有效地监测，以保证 CO_2 埋存的安全实施。笔者阐述 CO_2 地质埋存监测的目的、监测阶段及监测工具和技术的选择，对相关监测技术的原理及其应用现状进行分析，并介绍监测技术的研究进展和发展趋势。

1　监测目的及阶段划分

在 CO_2 地质埋存过程中，CO_2 可能从井筒、断层、裂缝泄漏，也有可能通过分子扩散从储层和盖层溢出。监测的目的是检验储层和盖层的完整性，分析 CO_2 的分布和埋存状态，发现可能的泄漏途径并采取补救措施，从而确保所埋存的 CO_2 不泄漏到大气、海洋及淡水层中。另外，通过开展监测项目，了解监测工具和技术的优势和局限性，可以改进和发展

CO_2 地质埋存监测的技术和方法[3]。

在油田实施注 CO_2 提高采收率和地质埋存工程时，需要考虑提高采收率的幅度，要进行油藏内 CO_2 和原油混相状况、驱替流体运移和注气前缘监测，注入井和生产井动态监测等，为模型的修正、CO_2 埋存潜力和安全性评估及油田注采方案的调整提供支持。

CO_2 地质埋存工程可以分为地质评估（预备）、注入、关井和关井后四个阶段。不同的工程阶段具有不同的监测目标和参数，需进行不同的监测项目设计。在预备工作阶段，需获得地质等基准状况，进行储层描述和工程设计，确定潜在的泄漏途径；在 CO_2 注入过程中，要进行储层（油气藏）、生产井、注入井的监测以及地面泄漏监测；在关井阶段，要进行井的封堵或者废弃处理，监测 CO_2 的泄漏；在关井后一段时间内，如果监测信息和预期的埋存状态相符，除非出现意外泄漏或者额外需要埋存状态的信息，可以停止监测。

2 监测技术

根据监测对象，可以分为地面环境、盖层和储层系统监测；根据所采用的方法与技术可以分为地下模拟、地球物理和地球化学方法等；根据监测技术的使用状况，分为主要技术、次要技术和潜在技术；根据获取地下信息的方式，可分为直接和间接监测技术。上述分类方法相互关联，如储层系统监测要结合使用模型模拟和地球物理等方法。

现有的监测技术大部分与石油工业相关。监测技术的组合取决于以下因素：现场条件、埋存深度、油藏组分特性和注入 CO_2 性质等，同时也取决于所用监测方法和技术的类型、使用期限、覆盖范围、分辨率、重复测量的必要性和成本等，因此需要联合使用以发挥每种技术的优势。

2.1 地球物理监测

2.1.1 地震监测

1) 四维地震

四维地震使用地表震源波，经地下储层反射回地表，从而对储层和上覆岩层进行全容积成像。在时移模式下，地震响应变化可以表征储层流体的变化，追踪注入流体前缘，确定 CO_2 的分布及可能通过上覆岩层的泄漏。在油气田开发中，利用四维地震可以认识油藏内油气水的分布，寻找剩余油气带。

四维地震技术具有覆盖面积广、检测通量下限低等优点。在有利的条件下，CO_2 注入量小于 1000t 时仍然可以监测到。West Pearl Queen[4] CO_2 注入矿场试验中，四维地震监测到了注入至 1.4km 深处 CO_2 的分布。但是，四维地震存在很大的局限性：垂向分辨率不够高（2~5m），地震成像的精确度很大程度上取决于 CO_2 聚集的性质、储层流体特点及压力特征[5]，而且四维地震只能对 CO_2 自由相成像，当 CO_2 饱和度低、储层较薄时，即使能对 CO_2 的分布成像，也很难精确计算其质量。

在 Sleipner[5]、CO_2 SINK[6]、Weyburn[7] 等 CO_2 地质埋存示范工程中应用了四维地震。Sleipner 地震监测结果清晰地显示了 CO_2 层状羽状体在盐水层内的逐步发展和运移状况，表明该技术监测较厚盐水层内 CO_2 的有效性。而在 Weyburn[7] 油田，研究表明其利用四维地震对 CO_2 羽状体成像要比 Sleipner 困难，主要可能由于油层厚度较薄，地震反射不够明显。在 Pembina Cardium CO_2 EOR[8] 工程中，注入 CO_2 前后的地震波特性没有发生明显的变化，因

此未能对储层内 CO_2 羽状体成像。

2）井间地震

井间地震将震源与检波器分别置入相邻井中进行观测，能够避开地表低速带对高频信号的吸收，获得分辨率较高的信号，可评价井间储层构造、流体分布、注气效果和微量 CO_2 的泄漏等[9,10]。美国得克萨斯州 Frio[10] 盐水层埋存矿场试验将约 1600t 的 CO_2 注入到一个相对均匀的砂岩盐水层中，并开展了井间地震监测，结果清楚地成像了 CO_2 羽状体的声速异常区域，展现了井间地震监测盐水层中 CO_2 的有效性。

3）垂直地震剖面（VSP）

VSP 使用井内检波器和地面震源进行数据采集，提供井眼附近的相关信息，包括盖层完整性、井附近 CO_2 的运移等。通过多变井源距方式实施 VSP，可以扩大其覆盖面积，为潜在 CO_2 泄漏的早期预警提供可能。与地表地震相比，VSP 有更好的分辨率。在 Weyburn[7]、Frio[10]、Pembina Cardium[11] 等工程中开展了 VSP 测试。在 Pembina Cardium CO_2 EOR 工程中，VSP 结果表明已对 1.6km 深处的油层内 CO_2 羽状体成功成像。在油气藏中进行 CO_2 埋存时，CO_2 沿井的泄漏可能性较大，VSP 或许是不错的选择。

4）微地震

微地震借助地面或井下检波器对 CO_2 注入可能引发的微地震事件进行检测，评估 CO_2 注入诱发裂缝的可能性。通过对微地震波形的分析，可了解储层中传导性裂缝和再活动断裂构造形态等信息以及 CO_2 沿裂缝运移的情况。在 Otway[12]、Weyburn[7]、In Salah[13]、Sleipner[14] 等工程中均使用了微地震，但尚未观察到因 CO_2 注入引起的显著的微地震。在吉林油田尝试使用井下微地震对压裂井进行裂缝方位监测，如图 1 所示。检波器安装在邻井 H-8-2，监测结果显示 H-6-2 主裂缝方向近东西向（N84°E），附近有一组近北东向（N55°E）的次裂缝，主裂缝西翼长为 135m，东翼长为 100m，裂缝的两翼基本对称。

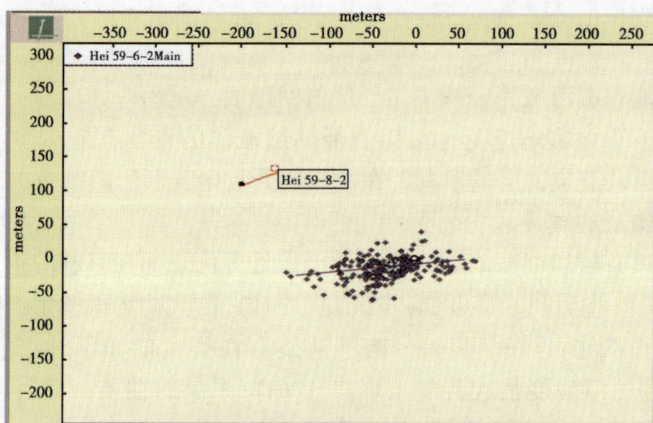

图 1　H-6-2 井微地震监测裂缝俯视图

2.1.2　电磁监测

1）重力测试

重力测试通过测量由物质分布改变所导致的储层重力变化来评估储层岩石和流体密度及其分布[15]。该技术包括地表和井下两种方式。地表重力测试具有探测范围广、重复周期短等优点，结合地震资料，可估计 CO_2 的埋存潜力，发现浅层低密度 CO_2 的聚集，但是地表重力测试获取的井附近饱和度的信息分辨率较低。如果注入的 CO_2 和油藏流体密度较为接近，地表重力测试的难度较大。井下重力测试方法为监测井底附近 CO_2 的运移提供了可能。该技术受井径和埋深的影响较小。在盐水层中，由于 CO_2 和盐水的性质相差较大，注入 CO_2 后可引起储层流体密度明显变化，宜考虑重力测试以扩大监测的覆盖范围。

2）电磁测量

电磁波方法利用电场或磁场的传播对地下电导率/电阻率的变化进行成像。探头可布置在井中和地面上。井间电磁波测量与井间地震类似，两种技术结合使用可相互补充，降低不确定性。井间和单井电磁成像技术已在美国应用于监测 CO_2 在 EOR 过程中的运移[16]。

电阻层析成像技术（ERT）[16] 依据 CO_2 注入后引起储层流体电阻率变化的原理对 CO_2 的运移分布进行成像，包括地表和井间 ERT 测试。地表 ERT 对土壤的电阻率成像，提供土壤类型、温度和饱和度信息，从而发现可能泄漏到地表的 CO_2。在油藏 CO_2 埋存中，考虑到其组分的多样和多变性，开展电磁测量难度相对较大。

3）大地电位

流体在多孔介质中的流动和大地电位相互耦合，通过测试后者的变化监测 CO_2 在多孔介质中的运移。大地电位技术相对简单，成本较低。在静态模式下，可以使用大地电位对水力压裂井进行裂缝方位的监测。图 2 为吉林油田采用大地电位监测反演后的结果，图中显示了 H59-8-2 经水力压裂后，上下两小层的裂缝方向基本一致（近东西向），与压裂设计一致。

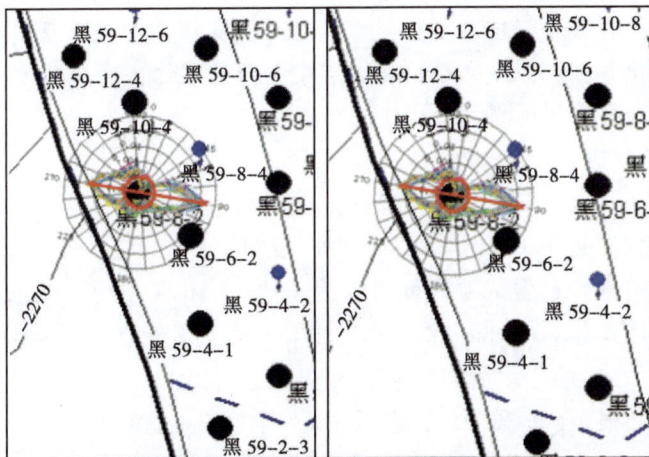

图 2　吉林油田 H59-8-2 重复压裂井借助大地电位监测裂缝方位

2.1.3　测井技术

井下测井的标准组合包括电阻率、中子、声波、密度、伽马射线、自然电位、温度和水泥胶结测试。这些技术可以提供 CO_2 饱和度的信息、监测 CO_2 的突破，从而降低地震和其他资料定性评价中的不确定性。在 CO_2 埋存过程中，CO_2 注入井的监测非常重要。地质埋存中井和油藏监测的相关测井技术总结见表 1。

表 1 地质埋存中井和油藏监测的相关测井技术

测试类型	测试方式	目的和作用
泄漏监测	温度、视频和超声波噪声测井	查找漏失点，确定维修位置
管柱诊断测井	磁通量漏失和超声波检测管材状况	评估管柱和潜在漏失
生产动态测井	注入/产出剖面、流量和流体识别及压力和温度测井	评估注入和产出能力，优化生产动态
流体和饱和度测试	中子、中子衰减和光谱测井	监测近井 CO_2 相态、位置和水的存在
水泥完整性测井	水泥胶结、声波速度和光记录测井	探测钻井液通道、可能胶结缺陷和微孔隙
机械完整性和环压监测	井口密封性、环压及井口井底压力和温度测试	确定管柱等的机械完整性

日本长冈 Nagaoka[17]盐水层 CO_2 埋存研究表明，中子孔隙度测井可用来推测 CO_2 的饱和度，但精度不高。考虑到横截面的热吸收即热中子的捕获速率对 CO_2 饱和度变化非常敏感，油藏饱和度测试（RST）也许是更有前景的技术，这在美国 Frio[10]盐水层 CO_2 埋存矿场试验中得到了验证。

在油藏 CO_2 埋存过程中，结合生产井饱和度测试和压力场以及油气组分资料，可以评价混相状况。考虑到注入井发生 CO_2 泄漏的可能性远远高于生产井，因此开展注入井饱和度测试，获取自由相 CO_2 分布的相关信息更有价值。

2.2 地球化学监测

地球化学监测是 CO_2 埋存监测中的重要组成部分，通过对注入 CO_2 后储层、地表土壤层和地下水的地球化学监测，评价储层储存 CO_2 的能力和安全性。其主要包括：通过对储层在 CO_2 注入前后的矿物化学成分变化，评估 CO_2 在储层内的化学反应；通过对注入井和生产井流体的地球化学分析，评价 CO_2 在储层中的运移规律；通过对地表土壤层气体和水进行地球化学监测，评价和预测储存 CO_2 的安全性。

2.2.1 井流体化学分析

通过井流体化学分析，了解 CO_2 的运移、溶解及与地层流体的反应。该技术的主要优点是以较低的成本获取地下 CO_2 分布的详细而敏感的数据。监测可以在生产井、监测井、注水井和浅层水井中进行。测试项目重点包括 pH 值、HCO_3^- 浓度、碱度、溶解气、烃类、阴阳离子和稳定同位素等。Weyburn 油田[7,18]注 CO_2 后油井流体取样结果分析得到了 HCO_3^- 和 pH 值变化情况，CO_2 溶于水后增加地层水的 HCO_3^- 浓度，降低其 pH 值。在油藏 CO_2 埋存中，可供采样的井很多，覆盖面广，进行井流体监测具有优势。

2.2.2 示踪剂技术

示踪剂由极微小的固体颗粒、可溶解气体或液体组成，作为注入 CO_2 的添加剂监测其运移，另外通过示踪剂测试，发现 CO_2 沿上覆岩层和下伏地层可能的迁移途径，估算 CO_2 的流速和体积。在美国 Frio[19]盐水层埋存矿场试验中通过对示踪剂的相分离获得了非自由相 CO_2 的质量，从而评价残余圈闭机制的作用。

对于油田 CO_2 埋存来说，油井数量多、分布广，非常有利于示踪剂的应用。图 3 显示吉林油田 CO_2 驱和埋存示范区块注 CO_2 过程中采用气体示踪剂监测 CO_2 在油层内的运移情况及其速度（示踪剂在注气井 H-8-4 投注）。结果清楚地表明，由于地层非均质性，CO_2 在地层内的运移及在井间的突破时间存在很大差异。

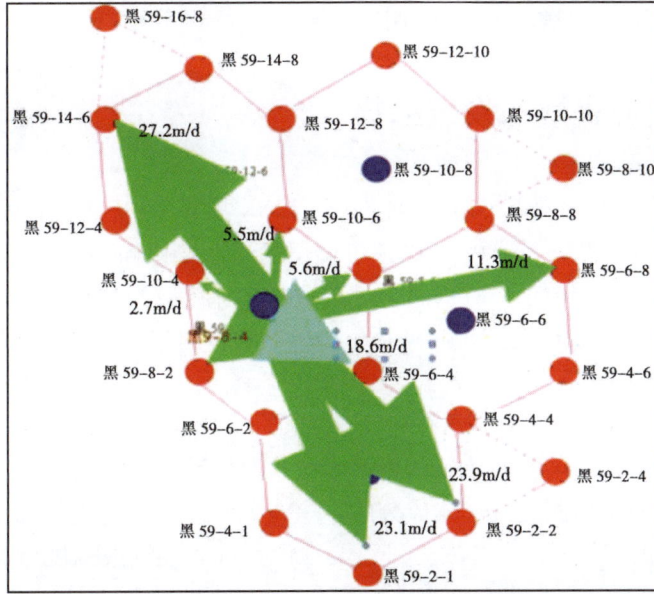

图 3　吉林油田 CO_2 驱和埋存示范区气相示踪剂监测结果

2.2.3　土壤气体分析

土壤气体分析为近地表 CO_2 监测提供了一种很好的方法。储层中 CO_2 的逃逸会导致土壤气体成分的变化，而且与油藏有关的物质（氢、氦、甲烷等）会伴随着 CO_2 向上迁移，因此土壤气体分析可示踪深层气体的流动，发现气体可能的迁移途径，评估地球化学反应和 CO_2 逃逸量。主要应用的技术包括浅层土壤气体（CH_4、C_2/C_3、CO_2、N_2、O_2 和微量元素等）分析和地表 CO_2 通量以及氡流速监测[7,20]。

油田注 CO_2 地质埋存工程中开展土壤气体分析时，首先要综合考虑井距、裂缝和断层分布以及地形地貌等相关因素，确定合理的浅层土壤气体采样网格分布，对比基准资料以及作业区和非作业区的相关数据。在此基础上，对 CO_2 浓度和气体变化异常区域以及可能发生泄漏的高风险区域，如裂缝、断层、废弃井和注入井周围等，进行连续监测和其他监测，如安装氡探测器，测定 $\delta^{13}CO_2$ 等，以确定 CO_2 来源，验证泄漏是否发生，寻找泄漏途径。

2.2.4　大气监测

CO_2 从埋存地点的泄漏可能产生地面可观的 CO_2 通量和浓度变化，因此可以采用涡度相关法和 CO_2 探测器进行监测。考虑到气体复杂的渗流通道和地表风密度差异驱动可能产生的大气扩散等各种因素，利用通量变化判别 CO_2 是否泄漏和评估泄漏量存在着较大的误差。便携式 CO_2 红外探测器是一个较好的选择，其检测下限低［体积分数（1~2）×10^{-6}］，易于操作，可连续进行，能够及时发现 CO_2 浓度的异常升高，适用于点检测。

2.3　地下模拟技术

地下数值模拟是 CO_2 埋存监测方案的重要组成部分。模拟工作和现场监测相辅相成，只有选择合适的模拟器，确定所需模拟参数，才能够优化监测技术和方法，确定相关的监测参数、测量时机、位置和分辨率，将模型结果和监测数据紧密有效地联系起来。在过去的十几年，CO_2 埋存模拟技术得到了广泛的研究[14]。主要包括 CO_2 的运移、CO_2 渗流特征和最终埋存状态、地质力学/机械/化学/流动耦合等模拟。

3 监测技术研究进展与发展趋势

在地球物理监测方面，德国 Ketzin 地区 CO_2 SINK 项目中[21]设计了新的电磁监测工具，部署了分布式热扰动传感器（DTPS）以评价储层的温度特性和 CO_2 在储层中的分布状况。在地震技术中，正开发高级的 AVO 分析、叠前成像和速度衰减层析成像等新技术，开展了永久性或者半永久性的井间地震[22]和 VSP[23]传感器部署的相关研究和现场测试，以便更加高效地监测 CO_2 的运移状况。

在地球化学和环境监测方面，正在研制深层地下水 CO_2 通量检测装置，开发了新的井底取样工具，包括 U 型取样系统[24]、多参数水化学取样和气取样装置[25]。在土壤和大气监测[26]方面，开发了自动监测土壤气体成分的半永久式装车站和低造价的激光设备以测量所需范围内的 CO_2 浓度，另外正在开发先进泄漏检测系统和连续波温度可调分布式半导体激光器等。在井监测方面，开发了全面整体监测系统，包括压力温度计、多功能永久传感器串和取样系统等，以便从地球化学和地球物理等各方面，全方位、高效率、较低成本进行连续的监测作业。

4 结束语

CO_2 埋存监测的目的是保证注入和埋存安全，验证物质平衡，了解和评估 CO_2 状态和迁移规律，确保 CO_2 不发生泄漏。埋存监测可以分为预操作、操作阶段、关井、关井后四个阶段，不同的监测阶段和监测系统有具体的监测目标和监测技术。监测技术应该联合使用，扬长避短，相互补充。四维地震技术可以有效地监测超临界 CO_2 羽状体在盐水层内的形成和迁移，但对于构造复杂、储层较薄的油藏而言，其有效性大大降低，井间地震、VSP 和井下重力测试也许更为有效；示踪剂、井内流体取样分析和测井技术是监测储层内流体运移和 CO_2 状态以及油藏和井完整性的有效手段；电磁监测是有前景的低成本 CO_2 地质埋存监测技术；土壤气体分析和大气监测能够对 CO_2 泄漏和地表环境进行有效的监测；地下 CO_2 模拟是评估地下 CO_2 运移和预测埋存状态的有效工具。

参 考 文 献

[1] Metz Bert, Davidson Ogunlade, De Coninck Heleen, et al. IPCC special report on carbon dioxide capture and storage [M]. New York: Cambridge University Press, 2005, 35-38.

[2] 任韶然，张莉，张亮. CO_2 地质埋存：国外示范工程及其对中国的启示 [J]. 中国石油大学学报（自然科学版），2010, 34 (1)：93-98.

[3] Litynski J T, Plaskynski S, Mcilvried H G, et al. The United States department of energy's regional carbon sequestration partnerships program validation phase [J]. Environment International, 2008, 34: 127-138.

[4] Pawar R, Lorenz J, Byrer C, et al. Sequestration of CO_2 in a depleted sandstone oil reservoir: results of a field demonstration test: 8th international conference on greenhouse gas control technologies, 2006 [C].

［5］Monea M, Knudsen R, Worth K, et al. Considerations for monitoring, mitigation, and verification for GS of CO₂ ［J］. Geophysical Monograph, 2009, 183: 303-316.

［6］S Yordkayhun, A Tryggvason, B Norden, et al. 3D seismic travel time tomography imaging of the shallow subsurface at the CO₂ SINK project site, Ketzin, Germany ［J］. Geophysics, 2009, 74 (1): 1-15.

［7］Wilson M, Monea M. IEA GHG Weyburn CO₂ monitoring & storage operation summary report 2000—2004 ［R］. Regina: Petroleum Technology Research centre, 2004.

［8］Gunter B, Grobe M, Chalaturnyk R, et al. Pembina cardium CO₂ monitoring pilot. A CO₂—EOR project, Alberta, Canada ［R］. ISBN 13: 9780968084458, 2009.

［9］Kikuta K, Hongo S, Tanase D, et al. Field test of CO₂ injection in Nagaoka, Japan ［M］// Rubin E S, Keith D W, Gilboy C F. Greenhouse Gas Control Technologies. Oxford: Elsevier Science Ltd, 2005, 1367-1372.

［10］Hovorka S D, Benson S M, Doughty C, et al. Measuring permanence of CO₂ storage in saline formations: the Frio experience ［J］. Environmental Geosciences, 2006, 13 (2): 105-121.

［11］Lawton D, Alshuhail A, Coueslan M, et al. Pembina cardium CO₂ monitoring project, Alberta, Canada: time -lapse seismic analysis-lessons learned ［J］. Energy Procedia, 2009, 1 (1): 2235-2242.

［12］Dodds K, Daley T, Friefeld B, et al. Developing a monitoring and verification plan with reference to the Australian Otway CO₂ pilot project ［J］. The Leading Edge, 2009, 28 (7): 812-818.

［13］Mathieson A, Midgley J, Dodds K, et al. CO₂ Sequestration monitoring and verification technologies applied at Krechba, Algeria ［J］. The Leading Edge, 2010, 29 (2): 216-222.

［14］Wildenborg T, Bentham M, Chadwick A, et al. Large-scale CO₂ injection demos for the development of monitoring and verification technology and guidelines (CO₂ Remove) ［J］. Energy Procedia, 2009, 1 (1): 2367-2374.

［15］Nooner S, Zumberge M, Eiken O, et al. Constraining the density of CO₂ within the Utsira Formation using time-lapse gravity measurements: Proceedings of the 8th international conference on greenhouse gas control technologies ［C］. Norway, 2006, 1-6.

［16］Tseng H W, Lee K H. Three dimensional interpretations of single-well electromagnetic data for geothermal applications: Proceedings of twenty-ninth workshop on geothermal reservoir engineering Stanford University ［C］. USA, 2004.

［17］Freifelda B M, Daleya T M, Hovorkab S D, et al. Recent advances in well-based monitoring of CO₂ sequestration ［J］. Energy Procedia, 2009, 1 (1): 2277-2284.

［18］Emberley S, Hutcheon I, Shevalief M, et al. Monitoring of fluid-rock interaction and CO₂ storage through produced fluid sampling at the Weyburn CO₂-injection enhanced oil recovery site, Saskatchewan, Canada ［J］. Applied Geochemistry, 2005, 20 (6): 1131-1157.

［19］Mcallum S D, Phelps T J, Riestenberg D E, et al. Interpretation of per-fluorocarbon tracer data collected during the Frio CO₂ sequestration test: Poster presented at AGU Fall Meeting ［C］. USA, 2005.

［20］Klusman R W. Rate measurements and detection of gas micro-seepage to the atmosphere from an enhanced oil recovery/sequestration project: Rangely, Colorado, USA ［J］. Applied Geochemistry, 2003, 18 (12): 1825-1838.

［21］Lewicki J L, Birkholzer J, Tsang C. Natural and industrial analogues for release of CO₂ from storage reservoirs: Identification of features, events, and processes and lessons learned ［R］. DE-ACO₂-05CH11231, 2006.

［22］Daley T M, Solbau R D, Ajo-Franklin J B, et al. Continuous active-source monitoring of CO₂ injection in a brine aquifer ［J］. Geophysics, 2007, 72 (5): 57-61.

［23］Majer E L, Daley T M, Korneev V, et al. Cost-effective imaging of CO₂ injection with borehole seismic methods ［J］. The Leading Edge, 2006, 25 (10): 1290-1302.

［24］Freifeld B M, Trautz R C, Yousif K K, et al. The U-tube: A novel system for acquiring borehole fluid samples from a deep geologic CO_2 sequestration experiment [J]. J of Geophysical Research, 2005, 110 (B10): 203-229.

[25] Jones D G, Barlow T, Beaubien S E, et al. New and established techniques for surface gas monitoring at onshore CO_2 storage sites [J]. Energy Procedia, 2009, 1 (1): 2127-2134.

[26] Humphries S D, Nehrir A R, Keith C J, et al. Testing carbon sequestration site monitor instruments using a controlled carbon dioxide release facility [J]. Applied Optics, 2008, 47 (4): 548-555.

Assessment of CO_2 EOR and Its Geo-storage Potential in Mature Oil Reservoirs, Changqing Oil Field, China

Xinwei Liao[1]   Chunning Gao[2]   Pingcang Wu[2]
Kun Su[1]   Yangnan Shangguan[2]

(1. China University of Petroleum, Beijing; 2. Changqing Oil Field, Petrochina)

Abstract: Most oil reservoirs in the Changqing Oilfield area are low permeability and have entered middle development stages after several ten years of production, and they are suitable for applying CO_2 EOR and carbon storage techniques. This study is aimed at assessing the potential of CO_2 EOR and storage in Changqing oil fields based on the data of 261 mature oil reservoirs. The assessments include a regional geology assessment, storage site screening, and reservoir screening for CO_2 EOR and EOR potential and storage capacity calculations. Of 261 reservoirs, 113 are suitable both for miscible or near-miscible flooding EOR and storage while 148 reservoirs are found suitable for immiscible flooding EOR and storage. The total EOR potential could be 9836.03×10^4 t and the CO_2 storage potential could reach 23920.34×10^4 t. The average incremental oil recovery rate in reservoirs suitable for miscible or near-miscible flooding could be 12.19%. The average incremental oil recovery rate in reservoirs suitable for immiscible could be 6.63%. The greater OOIP the oil reservoir has, the greater potential for CO_2 EOR and storage it will have, and the more suitable for large-scale storage projects it will be. Those oil reservoirs suitable for CO_2 EOR with large OOIP will be the preferred sites for CO_2 storage.

Keywords: CO_2 EOR; Geo-storage potential; Mature oil reservoir

0 Introduction

The mitigation of green gas emission, especially CO_2, has drawn worldwide attention as the aggravation of global warming and climate change. CO_2 geological sequestration in oil reservoirs can not only decrease CO_2 concentration in atmosphere, but also enhance oil recovery by CO_2 flooding (CO_2 EOR). In North American countries, application of CO_2 EOR have been maturely developed for decades. For those in China, where reservoirs are mainly characterized as heterogeneous layers and viscous crude oil, proper evaluation criteria for CO_2 EOR and sequestration should be developed. Changqing Oilfield, which is also the second largest oilfield in China, is targeted to check newly established criteria and then recovery increment and sequestration potential can be researched. Characterized as low permeability, water flooding method benefits turned down during last decades in this field. Considering unique advantages of CO_2 solvent over water, including feasible injectivity and high displacing efficiency, CO_2 flooding in Changqing Oilfield is expected for extra oil production and sequestration as well. By then, an evaluation criterion of sequestration is developed based on 261 production layers and then potential benefits are predicted via that.

1 Evaluation Criteria of CO_2-EOR and Sequestration in China Oil Reservoirs

1.1 Criteria Establishment

J J Taber et al concluded screening criteria of CO_2 flooding on the analysis of successful field application[1]. Bradshaw J et al also suggested screening parameters over previous researches and ranked candidate reservoirs by setting optimum value and parametric weight, which proved ideal application in Alberta reservoirs[2]. In China, similar researches have also been processed which is helpful for reservoir screening[3-6]. Based on those above, screening criteria of CO_2 EOR and sequestration in Changqing Oilfield can be obtained as in Table 1.

Table 1   Screening Criteria for CO_2 Application

Screening parameters		Miscible	Immiscible	Evaluated target
Crude oil properties	Density (g/cm^3)	<0.9	<0.95	Miscibility
	Viscosity (mPa·s)	<10	<600	Miscible performance Inject ability
	Components	Rich in C_2—C_{10}	—	Miscibility
Reservoir characters	Depth (m)	900~3000	>900	Miscibility
	Permeability (mD)	<10mD	NC	Inject ability
	Temperature (℃)	<90	—	Miscibility
	Oil saturation (%)	>50	>50	EOR potential
	Variation coefficient	<0.75	<0.75	Sweeping efficiency
	K_v/K_h	<0.1	<0.1	Floating effect
	Kh (m^3)	>$10^{-13} \sim 10^{-14}$	>$10^{-13} \sim 10^{-14}$	Inject ability
	$S_o * \varphi$	>0.05	>0.05	Sequestration Potentials
	Pressure (MPa)	p>MMP	—	Miscible requirement
Cap formation characters	Seal	Cap fractures are undeveloped		Security
	Leakiness	Less than 0.05% in 300 thousand years		Security

1.2 Evaluation Method of CO_2 Application

CO_2 sequestration is widely evaluated using CO_2 utilization coefficient in American and European countries, which is defined as total sequestration amount divided by cumulative oil production[7-10]. Here to predict storage potentials, we defined it as below:

$$M_{CO_2} = R_{CO_2} \cdot N \qquad (1)$$

in which, M_{CO_2} is the CO_2 storage potentials, 10^4 t; R_{CO_2} is the sequestration coefficient, dimensionless; N is the OOIP, 10^4 t。

Sequestration coefficient can be calculated by either numerical simulation or streamline simulation. By such methods respectively, two typical pilots in Changqing Oilfield are predicted on EOR increment and sequestration coefficient.

1.2.1 Huang 116 Block

pVT test and slim tube observation has been taken for Chang 6_1^1 formation, the principal producing zone in Huang 116 block, and then fluid properties could be obtained in Table 2. Initial formation pressure (18MPa), which is less than minimum miscible pressure (19.5MPa), indicates immiscible performance by CO_2 flood.

Table 2 pVT and MMP Observation of Well H−1

Parameters		Value
Reservoir conditions	Initial formation pressure	18.0 MPa
	Initial formation temperature	71.56℃
Formation fluid properties	Fluid type	Black oil
	Saturation pressure (71.56℃)	5.60MPa
	Gas oil ratio (flashed gas/ flashed oil)	37.8m^3/m^3
	Formation volume factor (71.56℃, 18.0MPa)	1.1372
	Initial density (71.56℃, 18.0MPa)	0.7751g/cm^3
	viscosity (71.56℃, 18.0MPa)	1.64mPa·s
	Tank oil (0.101MPa, 20℃)	0.8414g/cm^3
	MMP (71.56℃)	19.5MPa
Fluid components mole fraction	C_1+N_2	18.79%
	CO_2+C_2—C_{10}	37.82%
	C_{11+}	43.39%

As illustrated in Fig. 1, well group pattern is inverted nine−point, of which well space and line space are 540m×130m respectively. Geological properties of each grid of the model are interpolated by Kriging method based on data including well deviation, porosity and permeability, etc. Then geological model dimensioned as 29m × 16m×3m is obtained, of which size of each cell is 50m (X) ×50m (Y) ×4.5m (Z) meters. Reservoir permeability varied from 0.1 ~ 0.8 mD, and 0.42 mD on average. For porosity, it ranges from 11.53% ~ 12.38%, and the average value is 13.72%. Immobile water saturation is 37.58%, and average thickness of sand body and production layer is 17.5/13.5 meters respectively.

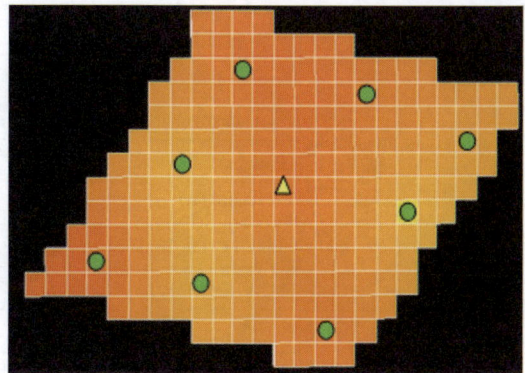

Fig. 1   Grid Model Illustration of
Group Pattern in Huang 116 Block

Solvents (water and solvent) injection rate are both 35 m^3 (under reservoir condition), and bottom pressure of production well is set to be above 8 MPa (others see Table 3). By two methods mentioned above, total amount of oil production and CO_2 sequestration during flooding are predicted after water flooding (Table 4).

Table 3　Scenario Design in Huang 116 Block

Scenario	Injection rate (m^3)	Cumulative injected volume (PV)	Production period (year)
Water flooding	35	0.6	20
CO_2 flooding	11000	0.6	20

Table 4　Prediction Results of Huang 116 Block

Methods	Recovery factor after water flooding (%)	Recovery factor after CO_2 flooding (%)	Recovery increment (%)	Sequestration coefficient
Numerical simulation	21.6	29.8	8.2	0.264
Stream tube simulation	22.7	31.9	9.2	0.232

As is shown in Table 4, another 9.2 percent of oil would be produced by CO_2 over water solvent, and sequestration coefficient is 0.232 meantime.

1.2.2　Muo-1 Block

As for another typical reservoir in Changqing Oilfield, Muo 1 Block, of which the main zone is Chang $8_1{}^1$ layer, was processed by indoor experiments to obtain fluid properties under reservoir condition (Table 5). CO_2 flooding performance in this block is miscible for lower minimum miscible pressure (19.8MPa) than initial pressure (22MPa).

Table 5　*pVT* and MMP Observation of Well M-1

Parameters		Value
Reservoir conditons	Initial formation pressure	22MPa
	Initial formation temperature	92.0℃
Formation fluid properties	Fluid type	Black oil
	Saturation pressure (92.0℃)	9.25 MPa
	Gas oil ratio (flashed gas/ flashed oil)	65.1m^3/m^3
	Formation volume factor (92.0℃, 22.0MPa)	1.2523
	Initial density (92.0℃, 22.0MPa)	0.7295 g/cm^3
	viscosity (92.0℃, 22.0MPa)	1.35 mPa·s
	Tank oil (0.101MPa, 20℃)	0.8311 g/cm^3
	MMP (92.0℃)	19.8MPa
Fluid components mole fraction	C_1+N_2	26.67%
	$CO_2+C_2—C_{10}$	38.99%
	C_{11+}	34.34%

The same group pattern with previous one, illustrated in Fig. 2, well and line space are changed to 480m×150 m. The dimension and size of geological model are 28m×39m×3m and 30m（X）×30m（Y）×3m（Z）for each grid. Reservoir permeability ranges from 0.1mD ~ 0.6mD and its average is 0.58mD. As to porosity, it ranges from 6% ~ 11% and average thickness of sand body and production layer are 13.7/10.5 meters.

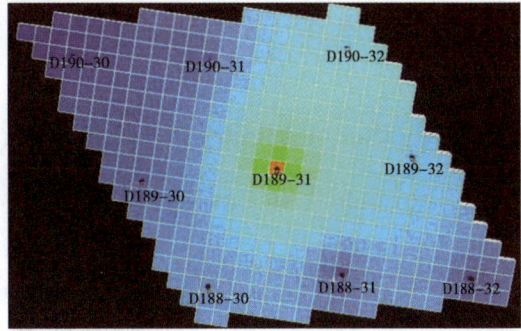

Fig. 2　Grid Model Illustration of Group Pattern in Muo 1 Block

In this block, the daily injection rate of each solvent is 40 m³（under reservoir condition）, and bottom pressure is set above 10 MPa（others is shown in Table 6）. So, oil production and sequestration amount can be predicted by numerical and stream tube simulation method（Table 7）.

Table 6　Scenario Design

Scenario	Injection rate（m³）	Cumulative injected volume（PV）	Production period（year）
Water flooding	40	0.6	20
CO_2 flooding	11000	0.6	20

Table 7　Prediction Results of Muo 1 Block

Methods	Recovery factor after water flooding（%）	Recovery factor after CO_2 flooding（%）	Recovery increment（%）	Sequestration coefficient
Numerical simulation	22.24	35.13	12.9	0.258
Stream tube simulation	22.78	36.58	13.8	0.237

As is shown in Table 7, recovery increment of CO_2 can be 13.8 percent over water flooding, and its sequestration coefficient is predicted as 0.237 in this block. As illustrated in Table 4 and Table 7, the error on oil recovery factor and sequestration coefficient between the two methods is tolerable. Considering far more details are need by former method comparing with stream tube, the latter one is chosen to evaluate CO_2 application.

Given OOIP of Huang 116 Block and Muo 1 Block, oil production increment by CO_2 over water is $172.5×10^4$ t and $765.4×10^4$ t respectively, during which their CO_2 sequestration amount could see $435.0×10^4$ t and $1338.2×10^4$ t.

1.3　Evaluation Procedure of CO_2—EOR and Sequestration

To evaluate CO_2 application in oil reservoirs, screening among candidate reservoirs should be taken following Table 1. And then, oil recovery increment as well as sequestration coefficient in target blocks can be calculated by stream tube simulation. After, predicted sequestration potentials in each block will be predicted via Formula（1）, which adds to the cumulative potentials of entire oilfield.

2 Evaluation of CO_2 Application in Changqing Oilfield

2.1 Prediction of Minimum Miscible Pressure

Compared with the slim tube observation on field oil, empirical method developed by NPC[11] is adjusted as below:

$$MMP = (- 329.558 + 7.727 \cdot MW \cdot 1.005^T - 4.377 \cdot MW)/145 \qquad (2)$$

$$MW = \left(\frac{8864.9}{G} \right)^{0.988} \qquad (3)$$

$$G = \frac{141.5}{\gamma_0} - 131.5 \qquad (4)$$

in which, MMP is the minimum miscible pressure, MPa; MW is the molar weight of C_{5+}, dimensionless; G is the gravity of tank oil, °API; γ_0 is the relative density of tank oil, dimensionless; T is the reservoir temperature, °F。

2.2 Evaluation Results of CO_2 Application in Changqing Oilfield

Evaluation on Changqing Oilfield indicates 26 out of 30 fields, which are constituted of 261 production layers, are proper for CO_2 application. Among them, 14 fields (113 production layers) could fulfill miscible flooding, and immiscible flooding in other 12 ones (148 production layers).

Followed by the procedure discussed above, evaluation on miscible areas, 14 fields (113 production layers), proved 22.42 percent oil would be produced by water flooding, comparing 34.61 percent by CO_2 flooding. Moreover, total oil production increment and CO_2 sequestration amount can reach about 48 million tons and about 103 million tons respectively, detail shown in Table 8.

Table 8 Evaluation Results of Miscible Areas in Changqing Oilfield

Oil field	OOIP (10^4 t)	Recovery by water flooding (%)	Recovery by CO_2 flooding (%)	Recovery increment (%)	sequestration coefficent (t/t)	Oil production increment (10^4 t)	Sequestration potentials (10^4 t)
DHZ	113.09	20.45	34.42	13.97	0.36	15.80	40.71
BYJ	478.14	23.71	36.67	12.96	0.24	61.97	114.75
DSK	398.20	21.36	30.49	9.11	0.24	36.28	93.58
FJC	1599.48	22.06	35.15	13.09	0.25	209.37	399.87
HJZ	1470.13	24.91	38.10	13.20	0.25	194.06	367.53
JY	7421.49	25.60	39.66	14.07	0.30	1044.20	2226.45
MF	1069.81	22.97	36.72	13.76	0.27	147.21	288.85
ML	8478.78	21.51	33.90	12.39	0.26	1050.52	2204.48
NL	2711.01	19.74	30.43	10.70	0.26	290.08	704.86
XF	12683.20	23.65	33.92	10.28	0.24	1303.83	3043.97

Oil field	OOIP (10^4 t)	Recovery by water flooding (%)	Recovery by CO_2 flooding (%)	Recovery increment (%)	sequestration coefficent (t/t)	Oil production increment (10^4t)	Sequestration potentials (10^4 t)
YW	219. 92	23. 02	36. 87	13. 85	0. 29	30. 46	63. 78
WJ	415. 91	19. 58	30. 42	10. 85	0. 20	45. 13	83. 18
YC	1819. 23	21. 64	32. 31	10. 67	0. 20	194. 11	363. 85
BB	1415. 14	23. 73	35. 47	11. 75	0. 23	166. 28	325. 48

For immiscible areas, 12 fields (148 production layers), CO_2 flooding (27. 63%) will obtain 6. 63 percent more oil than water solvent (20. 78%) . By average sequestration coefficient (0. 19), cumulative oil production increment is about 50 million tons and CO_2 sequestration potentials are about 136 million tons, detail shown in Table 9.

Table 9　Evaluation Results of Immiscible Areas in Changqing Oilfield

Oil field	OOIP (10^4 t)	Recovery by water flooding (%)	Recovery by CO_2 flooding (%)	Recovery increment (%)	Sequestration coefficient (t/t)	Oil production increment (10^4t)	Sequestration potentials (10^4t)
AS	26037. 8	21. 36	27. 61	6. 25	0. 18	1627. 36	4686. 80
CH	1107. 48	20. 69	30. 81	10. 12	0. 24	112. 08	265. 80
HJS	1664. 53	20. 51	27. 52	7. 01	0. 18	116. 68	299. 62
HC	4477. 83	21	27. 76	6. 75	0. 18	302. 25	806. 01
JA	29323. 3	21	29. 05	8. 06	0. 21	2363. 46	6157. 89
WWZ	355. 35	20. 65	29. 8	9. 15	0. 22	32. 51	78. 18
WQ	2332. 58	20. 1	27. 31	7. 21	0. 2	168. 18	466. 52
LZZ	149. 59	19. 73	23. 55	3. 81	0. 12	5. 70	17. 95
YFZ	2334. 26	20. 02	28. 72	8. 7	0. 19	203. 08	443. 51
SJ	1526. 48	21. 89	25. 97	4. 09	0. 18	62. 43	274. 77
ZL	1528. 98	21. 21	25. 19	3. 36	0. 06	51. 37	91. 74
MW	31. 91	21. 2	28. 3	5. 1	0. 32	1. 63	10. 21

3　Conclusions

In this paper, a reliable screening criterion is firstly established and quick evaluation method is discussed. By such, 30 fields in Changqing Oilfield are checked for CO_2 application, and 26 ones among them are proved to be feasible for CO_2—EOR and sequestration. Furthermore, after predicting flooding performance according adjusted empirical method from NPC, miscible blocks (113 layers in 14 fields) can expect about 48×10^6t of oil production increment and about 103×10^6t of CO_2 sequestration during that application, comparing about 50×10^6t of oil increment and about 136×10^6t of sequestration in immiscible blocks (148 layers in 12 fields). Considering complication by developed faults in Changqing Oilfield, CO_2 sequestration during CO_2—EOR is strong recommended to meet environmental and economic benefits to prevent risk especially for reservoirs which are small in sequestration potential and far away from emission source.

References

[1] Taber J J, Martin F D, Seright R S, EOR screening criteria revisited—Part 1: introduction, to screening criteria and enhanced recovery field projects [J]. SPE Reservoir Engineering, 1997, Vol. 12, No. 3: 189-198.

[2] Bradshaw J, Bachu S. Screening, evaluation, and ranking of oil reservoirs suitable for CO_2-flood EOR and carbon dioxide sequestration [J]. Journal of Canadian Petroleum Technology, 2002, Vol. 41, No. 9: 51-61.

[3] Zheng Yun-Chuan, Xiong Yu, Hou Tian-Jiang. Screening method based on fuzzy optimum for gas injection in candidate reservoir [J]. Journal of Southwest Petroleum Institute, 2005, Vol. 27, No. 1: 44-47. (ISSN1000-2634. In Chinese).

[4] Lei Huai-Yan, Gong Cheng-Lin, Guan Bao-Cong. New screening method for reservoir by CO_2 injection miscible flooding [J]. Journal of China University of Petroleum, 2008, Vol. 32, No. 1: 72-76, (in Chinese).

[5] Zeng Shun-Peng, Yang Xiu-Wen, et al. Fuzzy hierarchy analysis-based selection of oil reservoirs for gas storage and gas injection [J]. Henan Petroleum, 2005, Vol. 19, No. 4: 40-46, (in Chinese).

[6] Zhang Liang, Wang Shu, Zhang Li, et al. Assessment of CO_2 EOR and its geo-storage potential in mature oil reservoirs, Shengli Oil field, China [J]. Petroleum Exploration and Development, 2009, 36 (6): 737-741.

[7] Bradshaw J, Bachu S, Bonijoly D, et al. A taskforce for review and development of standards with regards to storage capacity measurement [DB/OL]. http: //www. cslforum. org/documents/Taskforce_ Storage_ Capacity_ Estimation_ Version_ 2. pdf, 2005.

[8] Bradshaw J, Bachu S, Bonijoly D, et al. Estimation of CO_2 storage capacity in geological media [DB/OL]. http: //www. cslforum. org/documents/PhaseIIReportStorageCapacityMeasurementTaskForce. pdf, 2007.

[9] Bradshaw J, Bachu S, Bonijoly D, et al. Comparison between methodologies recommended for estimation of CO_2 storage capacity in geological media [DB/OL]. http: //www. cslforum. org/documents/ Phase III Report Storage Capacity Estimation Task Force0408. pdf, 2008.

[10] Shen Pingping, Liao Xinwei, Liu Qiujie. Methodology for estimation of CO_2 storage capacity in reservoirs [J]. Petroleum Exploration and Development, 2009, 36 (2): 216-220.

[11] Robl F W, Emanuel A S, Van Meter Jr, O E. The 1984 natl. petroleum council estimate of potential EOR for miscible processes [J]. Journal of Petroleum Technology, August 1986, 875-882.

Evaluation of CO_2 Enhanced Oil Recovery and Sequestration Potential in Low Permeability Reservoirs, Yanchang Oilfield, China

D F Zhao[1]   X W Liao[1]   D D Yin[2]

(1. MOE Key laboratory of Petroleum Engineering, China University of Petroleum , Beijing;
2. EOR Research Institute, China University of Petroleum , Beijing)

Abstract: Sequestrating CO_2 in reservoirs can substantially enhance oil recovery and effectively reduce greenhouse gas emission. To evaluate the potential of CO_2 enhanced oil recovery (EOR) and sequestration for Yanchang Oilfield in China, a screening standard which was suitable for CO_2 EOR and sequestration in Yanchang oilfield was proposed based on its characteristics of strong heterogeneity, high water content and severe fluid channeling after water flooding. In addition, an efficient calculation method stream tube simulation method was presented to figure out CO_2 sequestration coefficient and oil recovery factor. After screening and evaluating, it turned out that 148 out of 176 blocks in 22 oilfields were suitable for CO_2 EOR and sequestration. CO_2 flooding after water flooding can produce 180.21×10^6t more crude oil and sequestrate 223.38×10^6t CO_2. The average incremental oil recovery rate of miscible reservoirs was 12.49% and the average CO_2 sequestration coefficient was 0.27t/t while the two values were 6.83% and 0.18t/t for immiscible reservoirs. There are comparatively more reservoirs that are suitable for CO_2 EOR and sequestration in Yanchang Oilfield than normal, which can obviously enhance oil recovery and means a great potential for CO_2 sequestration. CO_2 EOR and sequestration in Yanchang Oilfield has a bright application prospect.

Keywords: CO_2 sequestration; Enhanced oil recovery; Capacity estimation; CO_2 EOR screening; China oil reservoirs

0　Introduction

Greenhouse gas reduction and energy demand are two significant problems confronting the economic development in China. Sequestrating CO_2 in reservoirs can not only enhance oil recovery drastically, but also be able to reduce CO_2 emission effectively[1]. Sequestrating CO_2 in reservoirs is one of the most effective ways to sequestrate CO_2. Because compared with saline layer, reservoir reserve has been known, a large amount of reservoir data has been acquired and the injection equipment has already been built[2, 3].

During the evaluation of CO_2 EOR and sequestration, it is of paramount importance to screen the reservoirs. Daniel D and Taber J J et al concluded screening criteria of CO_2 flooding on the analysis of successful field application[4,5]. Bradshaw Jet al also suggested screening parameters over previous researches and ranked candidate reservoirs by setting optimum value and parametric weight, which proved ideal application in Alberta reservoirs[1,6-8]. For most of the oilfields in China, reservoir forming materials are continental sedimentation with their crude oil to be high viscosity, high paraffin content, high freezing point and reservoirs having complex structure and strong heterogeneity. Besides, among all these reservoirs, low permeability reservoirs (or tight reservoirs) and reservoirs with high temperature and high salinity account for a sizeable percentage, and most oilfields have fluid channeling after a long term water flooding[9]. Zhang Liang et al conducted a research into the screening standard for enhancing oil recovery through CO_2 flooding and put forward a screening standard suitable for CO_2 EOR and sequestration in China oilfields[10-12]. Based on this research and combined with the characteristics of Yanchang oilfield, this paper presented a screening standard for CO_2 EOR and sequestration for Yanchang oilfield.

There are four kinds of CO_2 sequestration capacity: theoretical sequestration capacity, effective sequestration capacity, practical sequestration capacity and matched sequestration capacity[2,3], among which theoretical sequestration capacity and effective sequestration capacity are two concerning values. Theoretical sequestration capacity presented by Carbon Sequestration Leadership Forum (CSLF) suits the theory well[6], but it had not taken CO_2 resolved in the water and oil into consideration. As most reservoirs in China are high water-cut reservoirs, theoretical sequestration capacity considering CO_2 resolved in water and oil was presented by Shen Pingping and Liao Xinwei[4, 13].

Effective sequestration capacity is a subset of theoretical storage capacity and it considers factors like reservoir property, sealing capacity of reservoir, burial depth, reservoir pressure system, pore volume, etc[1]. CSLF presented a calculation method for effective sequestration capacity based on material equilibrium method, and this method has taken a lot of factors into account, such as buoyancy, gravity override, mobility ratio, heterogeneity, water saturation, aquifer strength, etc[2]. These effective storage coefficients are hard to determine, and only through numerical simulation method can these data be obtained.

Personnel from related organizations and countries like USA and EU raised analogy method to calculate CO_2 sequestration capacity in the CO_2 EOR program. This method obtained effective sequestration capacity through introducing CO_2 utilization coefficient[2, 3, 14, 15]. However, CO_2 utilization coefficient fluctuated strongly, so the sequestration coefficient was introduced in this paper to better reflect the variation tendency of CO_2 sequestration capacity.

To figure out the recovery factor and sequestration coefficient is the key to study the assessment of CO_2 EOR and sequestration potential, and the project data of EOR or numerical simulation are usually used. However, it is obviously unsuitable to carry out preliminary CO_2 EOR and sequestration appraisal in large scale, so a reliable and efficient stream tube simulation method based on fractional flow theory was proposed in this paper.

This paper put forward a screening standard that was suitable for CO_2 EOR and sequestration in Yanchang oilfield, and introduced sequestration coefficient to better reflect the variation tendency of CO_2 sequestration capacity. Besides, based on fractional flow theory, stream tube simulation method was proposed to efficiently calculate CO_2 sequestration coefficient and recovery coefficient, and numerical simulation method was used to verify the values so as to ensure that the deviation was within

476

acceptable range. Finally, stream tube simulation method was used to evaluate the potential of CO_2 EOR and geological sequestration through miscible flooding and immiscible flooding respectively. The results states that there is a great potential of conducting CO_2 EOR and sequestration in Yanchang oilfield and the technique has a bright application prospect there.

1 Procedure

Based on some former researches and through the utilization of the reservoir data of Yanchang oilfield, typical models were established to analyze the effect law of each parameter on the oil recovery and storage capacity as well as the sensibility of each parameter. A comprehensive appraisal standard for CO_2 EOR and sequestration in Yanchang oilfield was proposed by using fuzzy comprehensive evaluation and analytic hierarchy process methods in this paper[16, 17], as shown in Table 1.

Table 1   Screening Standard for CO_2 EOR and Sequestration in Yanchang Oilfield

First class index	Weight coefficient of first class index	Second class index	Weight coefficient of second class index	The worst value	The optimal value
Reservoir characters	0.65	Sedimentary rhythm	0.16	reverse	positive
		The initial oil saturation	0.16	0.3%	0.65%
		Reservoir effective thickness	0.16	30m	1m
		Depth of the reservoir pressure	0.16	300m	2500m
		Average reservoir permeability	0.10	0.1mD	1.5mD
		Interlayer development situation (K_V/K_H)	0.10	0.5	0.001
		Heterogeneity (coefficient of variation)	0.06	0.7	0.1
		Formation dip angle	0.04	5°	35°
		Directional permeability (K_Y/K_X)	0.04	100	5
		The temperature	0.02	75℃	19℃
Fluid properties	0.25	Fluid density	0.5	1.000g/cm^3	0.7g/cm^3
		Fluid viscosity	0.5	1000mPa·s	1mPa·s
Development factors	0.1	Well pattern	0.5	Seven spot	Inverted nine spot
		Well spacing	0.5	400m	100m

2 Calculation Method of Theoretical Sequestration Capacity and Effective Sequestration Capacity

2.1 Calculation Method of Theoretical Sequestration Capacity

In high water cut reservoir, CO_2 theoretical sequestration capacity includes three parts, theoretical sequestration capacity in free space of oil reservoir, theoretical sequestration capacity dissolving in water and theoretical sequestration capacity dissolving in oil[2, 3].

$$M_{CO_2t} = M_{CO_2displace} + M_{CO_2inoil} + M_{CO_2inwater} \tag{1}$$

$$M_{CO_2displace} = \rho_{CO_2r}(R_f \times POIP - V_{iw} + V_{pw}) \tag{2}$$

$$M_{CO_2inwater} = E_f \times \rho_{CO_2r} \times (PWIP + V_{iw} - V_{pw}) \times m_{CO_2inwater} \tag{3}$$

$$M_{CO_2inoil} = E_f \times \rho_{CO_2r} \times POIP \times (1 - R_f) \times m_{CO_2inoil} \tag{4}$$

in which, M_{CO_2t} is the theoretical sequestration capacity, 10^6 t; $M_{CO_2displace}$ is sequestration capacity in the process of CO_2 flooding, 10^6 t; M_{CO_2inoil} is sequestration capacity of CO_2 dissolved in the reservoir water, 10^6 t; $M_{CO_2inwater}$ is sequestration capacity of CO_2 dissolved in crude oil, 10^6 t; POIP is the amount of oil in the reservoir after water flooding, 10^6 t; PWIP is the amount of water in the reservoir after water flooding, 10^6 t; E_f is sweep efficiency of CO_2 displacement; $m_{CO_2inwater}$ is the solubility of CO_2 in water; m_{CO_2inoil} is the solubility of CO_2 in oil.

As CO_2 can only dissolve in oil and water by directly contacting, CO_2 swept coefficient was introduced when the storage volumes that CO_2 dissolved in water and oil were calculated.

2. 2 Calculation Method of Effective Sequestration Capacity

Effective sequestration capacity is a subset of theoretical sequestration capacity. So based on theoretical sequestration capacity, CSLF proposed a calculation method to work out effective sequestration capacity with factors like buoyancy, gravity override, mobility ratio, heterogeneity, water saturation, aquifer strength been taken into consideration[2, 3].

$$M_{CO_2e} = C_e \times M_{CO_2t} = C_m \times C_b \times C_h \times C_w \times C_a \times M_{CO_2t} \tag{5}$$

in which, M_{CO_2e} is the effective sequestration capacity, 10^6 t; C_e is comprehensive influence coefficient of the effective sequestration of various factors; C_m is the effective sequestration coefficient caused by mobility influence; C_b is the effective sequestration coefficient caused by buoyancy influence; C_h is the effective sequestration coefficient caused by heterogeneity influence; C_w is the effective sequestration coefficient caused by water saturation influence; C_a is the effective sequestration coefficient caused by aquifer strength influence.

The calculation coefficients of this method are hard to determine, and effective sequestration coefficients are usually determined through numerical simulation. CO_2 utilization coefficient (utilization coefficient = total storage capacity/ accumulated oil production) is wildly used in USA and European countries to evaluate CO_2 sequestration capacity[2, 3]. However, the CO_2 utilization coefficient (R_{CO_2}) has relative large amplitude (about 0. 1 ~ 0. 8t/bbl), so it cannot better illustrate the variation tendency of CO_2 sequestration. To solve this problem, a concept of CO_2 sequestration coefficient (= total storage capacity/ oil geological reserve) was introduced. As shown in Fig. 1 and Fig. 2, along with the increase of planar heterogeneity, the sequestration coefficient S_{CO_2} was on a declining curve, and along with the increase of diffusion coefficient, sequestration coefficient showed an increase trend. Therefore, here came a conclusion—as the variation range can be known directly from

478

the value of S_{CO_2}, CO_2 sequestration coefficient S_{CO_2} could better convey the variation tendency of CO_2 sequestration capacity.

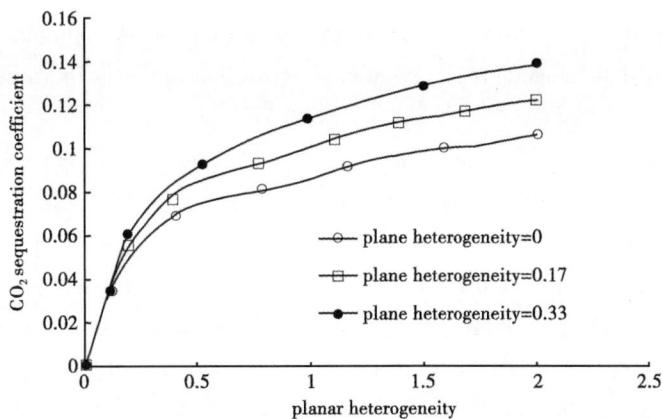

Fig. 1   Impact of Plane Heterogeneity on Sequestration Coefficient

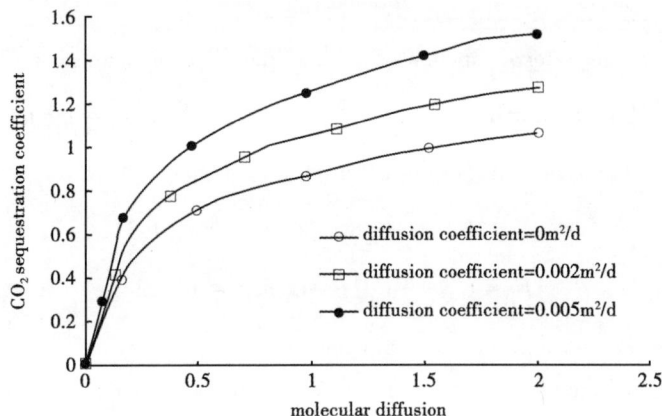

Fig. 2   Impact of Molecular Diffusion on Sequestration Coefficient

By applying equation (1) to equation (5), the efficient sequestration capacity can be given by

$$S_{CO_2} = C_e \times [(1 - S_w) \times R_f + S_w \times R_w + E_f \times S_{pw} \times (1 - R_w)] \times m_{CO_2 inwater}$$
$$+ E_f \times (1 - S_{pw}) \times (1 - R_w) \times m_{CO_2 inoil} \qquad (6)$$

in which, S_{CO_2} is the sequestration factor; S_w is the initial water saturation, %; S_{pw} is the present water saturation, %; R_w is the water recovery factor, %。

2.3   Calculation of Recovery Factor and Sequestration Coefficient

Two key parameters were involved in CO_2 EOR and sequestration appraisal: recovery factor (R) and sequestration coefficient (S_{CO_2}). The two parameters must be determined before the evaluation, and numerical simulation is the best way to do this job. However, for now, commercial software like Eclipse and CMG all need detailed data of the oilfield for real calculation and take a long

479

time, which is obviously unsuitable to apply for a large number of oilfields during a short time. Therefore, efficient stream tube simulation method was presented in this paper to determine R and S_{CO_2}.

Stream tube simulation method is based on fractional flow theory and with key factors like viscous fingering, vertical heterogeneity, gravitational separation and miscible effect being taken into account.

The assumption conditions for the model are: (1) oil and gas will not volatilize to gas-phase during the flow process; (2) displacement is a constant temperature process; (3) Koval coefficient is used to depict viscous fingering; (4) when the injection mode is water—gas alternate, water and CO_2 will be injected with a certain WAG ratio simultaneously; (5) there's no free gas in miscible flooding; (6) there's no big fracture in the reservoir and no leak during CO_2 injection.

According to the law of conservation of mass, the relation equation of component concentration and fractional flux was obtained:

$$\frac{\partial C_i}{\partial t_D} + \frac{\partial F_i}{\partial X_D} = 0 \tag{7}$$

in which, $i = 1$ is the water components; $i = 2$ is the oil components; $i = 3$ is the injected gas components; $X_D = X/L$ is the dimensionless distance of the system; $t_D = \int_0^t q \mathrm{d}t / \mathrm{d}V_p$ is dimensionless time in the pores; C_i is the overall concentration of component.

$$C_i = C_{i1}S_1 + C_{i2}S_2 + C_{i3}S_3, i = 1,2,3 \tag{8}$$

$$F_i = C_{i1}f_1 + C_{i2}f_2 + C_{i3}f_3, i = 1,2,3 \tag{9}$$

in which, C_{ij} is the concentration of component i in the j phase.

According to the assumptions that the oil and water do not evaporate into the gas phase, So, in the above formula, $C_{33} = 1$, $C_{13} = C_{23} = 0$. $j = 1$ is the water phase; $j = 2$ is the oil phase; $j = 3$ is the gas phase; S_j is the saturation of the j phase; f_j is the fractional flow of the j phase.

The model above is immiscible model, for miscible model, the $C_{i3}S_3$ in equation (8) will be 0, $C_{i3}f_3$ in equation (9) will be 0, and $i = 1$, 2.

In the model, the breakthrough time of injected gas and oil recovery were calculated by revised fractional flow theory, which had considered the effect of factors like viscous fingering, volume swept coefficient, vertical heterogeneity and gravity segregation. Finally, characteristics method was used to solve the model.

Koval coefficient was used in the model to depict the effects that factors like viscous fingering and vertical heterogeneity have on fractional flow. As the density of injected gas was lower than oil and water, injected gas had a trend to move toward the top of oil layer and override or slip stream at relative lower part in the oil layer. According to that, Koval enlargement method was used in the model to amend the effect of gravity. The model transformed two-dimension planar displacement calculation into one-dimension displacement calculation through stream tube simulation based on fractional flow theory so as to obtain relative precise and efficient seepage calculation model. This model

480

can be used to evaluate the breakthrough time of oil-gas front, oil recovery, geological sequestration coefficient, etc.

The parameters need to be inputted for fractional flow theory includes porosity, permeability, saturation, oil viscosity, oil density, relative permeability, well pattern type, injection mode, injection time, etc. This method can evaluate the sequestration coefficient and oil recovery reliably and quickly.

2. 4   Determination of the Minimum Miscible Pressure

Before the appraisal of CO_2 EOR and geological sequestration, the minimum miscible pressure (MMP) of the target reservoir needs to be determined first to make sure that whether miscible state is realizable or not. This paper combined the experimental results of slim tube test with the NPC empirical formulas method of United States Department of Energy (USDOE) —reservoir temperature and molecular weight C_{5+} in the crude oil were used to evaluate MMP so as to meet the demand of being both accurate and efficient. Below is MMP calculation model suitable for Yanchang oilfield.

$$MMP = -329.558 + (7.727 \times MW \times 1.005^T - 4.377 \times MW) \qquad (10)$$

$$MW = \left(\frac{8864.9}{G}\right)^{\frac{1}{1.012}} \qquad (11)$$

in which, MW is molecular weight of C_{5+}; MMP is minimum miscible pressure, psi; T is reservoir temperature, ℉; G is gravity of tank oil, °API.

2. 5   Results and Conclusion

The method was tested through its application to two synthetic fields and then to a real field in Yanchang oilfield. The results of these applications are presented and discussed in this section.

3   Synthetic Fields

3. 1   Chuan 72 Block

pVT test and slime tube observation have been taken for Chang 6^1 in Chuang 72 block, and the fluid properties are shown in Table 2. The initial formation pressure was 17.1MPa, which was lower than MMP (18.4MPa), and the CO_2 displacement was immiscible flooding. The model was inverted nine-point well network, and the well spacing and array pitch were 340m×150m. Geological parameters of each grid like permeability and porosity were obtained through Kriging interpolation method according to the values around the well. The main parameters are as follows: the gridding dimension of geological model was 27m×16m×3m; the gridding size was 30m (X) ×30m (Y) × 4.5m (Z); the variation range of reservoir permeability was 0.1~0.9mD with its average value to be 0.54mD; the porosity variation range was 11.31% ~ 13.25% with its average value to be 12.42%; the immobile water saturation was 36.86%; the average thickness of sand body and production layer were16. 8m and 13. 6m respectively.

The injection velocities of water and CO_2 were both $35m^3/d$ (under reservoir condition), and the flowing bottom hole pressure was about 9MPa (Table 3). Through the two methods above, the total volume of produced oil and CO_2 sequestration capacity during CO_2 flooding after water flooding were predicted, as shown in Table 4. Through CO_2 flooding, 8.7% more oil will be produced and the sequestration coefficient was 0.228.

Table 2　*pVT* and MMP Observation of Well C-1

Parameters		Value
Reservoir conditions	Initial formation pressure	17. 1MPa
	Initial formation temperature	70. 26℃
Formation fluid properties	Fluid type	Black oil
	Saturation pressure（70.26℃）	5. 59 MPa
	Formation volume factor（70.26℃，17.1MPa）	1. 1365
	Initial density（70.26℃，17.1MPa）	0. 7762 g/cm^3
	Viscosity（70.26℃，17.1MPa）	1. 62mPa · s
	Tank oil（0.101MPa，20℃）	0. 8325g/cm^3
	MMP（70.26℃）	18. 4MPa
Fluid components mole fraction	C_1+N_2	18. 53%
	$CO_2+C_2—C_{10}$	38. 01%
	C_{11+}	43. 46%

Table 3　Scenario Design in Chuan 72 Block

Scenario	Injection rate（m^3）	Cumulative injected volume（PV）	Production period（year）
Water flooding	35	0. 6	20
CO_2 flooding	12000	0. 6	20

Table 4　Prediction Results of Chuan 72 Block

Methods	Recovery factor after water flooding（%）	Recovery factor after CO_2 flooding（%）	Recovery increment（%）	Sequestration coefficient
Numerical simulation	20. 7	28. 8	8. 1	0. 261
Stream tube simulation	21. 9	30. 6	8. 7	0. 228

3. 2　Fang 5 Block

For another typical block, *pVT* test and slime tube observation have been taken for Chang 8 in Fang 5 block, and the fluid properties are shown in Table 5. The initial formation pressure was 20. 6MPa, which was lower than MMP (18. 9MPa), and the CO_2 displacement was immiscible flooding.

The model was inverted nine-point well network, and the well spacing and array pitch were 420m×140m. Geological parameters of each grid like permeability and porosity were obtained through Kriging interpolation method according to the values around the well. The main parameters are as follows: the gridding dimension of geological model was 29m×38m×3m; the gridding size was 30m (X) ×30m (Y) ×3m (Z); the variation range of reservoir permeability was 0.1~0.7mD with its average value to be 0.41mD; the porosity variation range was 6.3%~11.4% with its average value to be 10.2%; the immobile water saturation was 37.61%; the average thickness of sand body and production layer were14.2m and 11.2m respectively.

Table 5 *pVT* and MMP Observation of Well F-5

Parameters		Value
Reservoir conditions	Initial formation pressure	20.6MPa
	Initial formation temperature	84.52℃
Formation fluid properties	Fluid type	Black oil
	Saturation pressure (84.52℃)	9.21MPa
	Formation volume factor (84.52℃, 18.0MPa)	1.1286
	Initial density (84.52℃, 18.0MPa)	0.7332g/cm³
	Viscosity (84.52℃, 18.0MPa)	1.31mPa·s
	Tank oil (0.101MPa, 20℃)	0.8303g/cm³
	MMP (84.52℃)	18.9MPa
Fluid components mole fraction	C_1+N_2	25.26%
	$CO_2+C_2-C_{10}$	39.35%
	C_{11+}	35.39%

The injection velocities of water and CO_2 were both 38m³/d (under reservoir condition), and the flowing bottom hole pressure was about 10MPa (Table 6). Through the two methods above, the total volume of produced oil and CO_2 sequestration capacity during CO_2 flooding after water flooding were predicted, as shown in Table 7. Through CO_2 flooding, 12.6% more oil will be produced and the sequestration coefficient was 0.235.

Table 6 Scenario Design in Fang 5 Block

Scenario	Injection rate (m³)	Cumulative injected volume (PV)	Production period (year)
Water flooding	38	0.6	20
CO_2 flooding	12000	0.6	20

Table 7 Prediction Results of Fang 5 Block

Methods	Recovery factor after water flooding (%)	Recovery factor after CO_2 flooding (%)	Recovery increment (%)	Sequestration coefficient
Numerical simulation	22.3	34.0	11.7	0.256
Stream tube simulation	22.8	35.4	12.6	0.235

A comparison between the results of stream tube simulation appraisal method and numerical simulation method stated that the oil recovery and geological sequestration coefficient calculated by stream tube simulation method were a little bigger. This is because that stream tube simulation method has left factors like capillary force out of consideration. However, the deviation was acceptable, so the stream tube simulation method could still be applied for the evaluation of CO_2 EOR and geological sequestration potential.

3.3 Real Field in Yanchang Oilfield

Yanchang oilfield reservoir has an 2100m average depth, 15.1 MPa average formation pressure, and 70.6℃ average formation temperature. It is a normal warm-pressing system. Viscosity of the formation oil is 0.32~11.5mPa $\cdot$ s, and the range of oil saturation is 46%~72%.

Screening standard for CO_2 EOR and sequestration mentioned before in this paper was adopted to carry out appraisal calculation for the reservoir suitability of CO_2 flooding in 174 blocks from Yanchang oilfield, and then all the appraisal values of all the blocks were ranked. 0.53 was used as the boundary for the evaluation of Yanchang oilfield. The detailed statistics are exhibited in Table 8 and 9. Among the 176 blocks, there are 148 blocks that were suitable for CO_2 EOR and sequestration, which accounted for 84% of all the oilfield blocks and possessed 80% geological reserves of Yanchang oilfield. Miscible flooding and immiscible flooding potential evaluations were also conducted for some appropriate reservoirs using the CO_2 EOR and sequestration appraisal method studied above, and the results were shown in Table 10 and 11.

Table 8   Ranking of Low Permeability Reservoirs CO_2 Flooding Appraisal Value in Yanchang Oilfield

Oil field	Block	Horizon	Geological reserves value (10^4t)	Value
DB	FX	Chang6	207	0.858
WQ	WWZ	Chang4+5	2200	0.832
CK	BB	Chang 8	1864	0.533
NNW	XY	Chang 6	875	0.530
WYB	YMH	Chang 8	213	0.528
GGY	T114	Chang 6	1008.98	0.526
GGY	DBLQ	Chang 6	4502.8	0.457
NNW	SSL	Chang 4+5	1225	0.447

Table 9   Potential Estimation of Low Permeability Reservoirs of CO_2 Flooding in Yanchang Oilfield

Oil field	Total number of reservoir in block	Suitable for CO_2 flooding reservoir	Suitable for CO_2 flooding reservoir number proportion	Geological reserves value (10^4t)	Suitable for CO_2 flooding reserves (10^4t)	Suitable for CO_2 flooding reservoir reserves proportion
XZC	13	12	0.92	10685	10495	0.98
JB	15	14	0.9333	12340.5	12071	0.97816
YN	5	4	0.8	25315.5	24727.33	0.97677
XB	3	3	1.00	12011.9	12011.87	1.00
DB	20	20	1.00	14431.6	14431.64	1.00

484

Oil field	Total number of reservoir in block	Suitable for CO_2 flooding reservoir	Suitable for CO_2 flooding reservoir number proportion	Geological reserves value (10^4t)	Suitable for CO_2 flooding reserves (10^4t)	Suitable for CO_2 flooding reservoir reserves proportion
WQ	34	34	1.00	37080	37080	1.00
CK	8	8	1.00	7004.23	7004.23	1.00
QLC	8	1	0.13	12691	1934	0.15
GGY	6	1	0.17	9624.31	1670.31	0.17
WJC	1	0	0.00	8111	0	0.00
QPC	1	1	1.00	4305.2	4305.2	1.00
QHZ	3	2	0.67	9894.63	4321.63	0.44
PL	4	4	1.00	1987	1987	1.00
NNW	10	5	0.50	7595	2065	0.27
NQ	4	4	1.00	1139.2	1139.2	1.00
XSW	8	8	1.00	11601	11601	1.00
YW	2	2	1.00	647.84	647.84	1.00
ZL	4	4	1.00	4080	4080	1.00
HS	4	4	1.00	2068.16	2068.16	1.00
ZC	5	5	1.00	7619.05	7619.05	1.00
WYB	16	11	0.6875	13923.9	11141.7	0.80019
ZB	4	3	0.75	4981.9	3872.9	0.78
Total	178	150	0.84	219138	176274.1	0.80

Table 10　the Results of Assessment of Miscible Flooding of Yanchang Oilfield

Oil field	Geological reserves value (10^4t)	Recovery by water flooding (%)	Recovery by CO_2 flooding (%)	Recovery increment (%)	Sequestration coefficient (t/t)	Oil production increment (10^4t)	Sequestration potentials (10^4t)
XZC	10495	20.45	34.42	13.97	0.36	1466.152	40.71
JB	12071	23.71	36.67	12.96	0.24	1564.402	114.75
XB	12011.87	21.36	30.49	9.13	0.24	1096.684	93.58
WQ	37080	22.06	35.15	13.09	0.25	4853.772	399.87
QLC	1934	24.91	38.1	13.19	0.25	255.0946	367.53
QHZ	4321.63	25.6	39.66	14.06	0.3	607.6212	2226.45
PL	1987	22.97	36.72	13.75	0.27	273.2125	288.85
XS	—	—	—	—	—	—	—
W	11601	21.51	33.9	12.39	0.26	1437.364	2204.48
YW	647.84	19.74	30.43	10.69	0.26	69.2541	704.86
ZC	7619.05	23.65	33.92	10.27	0.24	782.4764	3043.97
ZB	3872.9	23.02	36.87	13.85	0.29	536.3967	63.78

Table 11　the Results of Assessment of Immiscible Flooding of Yanchang Oilfield

Oil field	Geological reserves value (10^4 t)	Recovery by water flooding (%)	Recovery by CO_2 flooding (%)	Recovery increment (%)	Sequestration coefficient (t/t)	Oil production increment (10^4 t)	Sequestration potentials (10^4 t)
YN	24727.33	21.36	27.61	6.25	0.18	1545.45	4450.91
DB	14431.64	20.69	30.81	10.12	0.24	1460.48	3463.59
CK	7004.23	20.51	27.52	7.01	0.18	490.99	1260.76
GGY	1670.31	21	27.76	6.76	0.18	112.91	300.65
WJC	0	21	29.05	8.05	0.21	0	0
QPC	4305.2	20.65	29.8	9.15	0.22	393.92	947.14
NN	—	—	—	—	—	—	—
W	2065	20.1	27.31	7.21	0.2	148.88	413
NQ	1139.2	19.73	23.55	3.82	0.12	43.51	136.70
ZL	4080	20.02	28.72	8.7	0.19	354.96	775.2
HS	2068.16	21.89	25.97	4.08	0.18	84.38	372.26
WY	—	—	—	—	—	—	—
B	11141.7	21.21	25.19	3.98	0.06	443.43	668.50

As shown in Table 10 and Table 11, 148 blocks in Yanchang oilfield were suitable for CO_2 EOR and sequestration. CO_2 flooding after water flooding can produce 180.21×10^6 t more crude oil and sequestrate 223.38×10^6 t CO_2. The average incremental oil recovery rate of miscible reservoirs (65 blocks) was 12.49% and the average CO_2 sequestration coefficient was 0.27 t/t while the two values were 6.83% and 0.18 t/t for immiscible reservoirs (83 blocks). Crude oil increased 129.42×10^6 t and sequestrated 95.48×10^6 t CO_2 in miscible reservoirs while the crude oil increased 50.78×10^6 t and sequestrated 127.88×106 t CO_2 in immiscible reservoirs. The sequestration coefficient of CO_2 miscible reservoirs was larger than immiscible reservoirs. That is mainly because that when CO_2 reaches miscible state with crude oil, it can displace more oil and further spare more space for CO_2 accumulation.

4　Conclusions

(1) Based on the characteristics of Yanchang oilfield, this paper presented a reservoir screening standard that was suitable for CO_2 EOR and geological sequestration in Yanchang oilfield in China for the first time. CO_2 sequestration coefficient R_{CO_2} was introduced to better reflect the variation tendency of CO_2 storage capacity, and stream tube simulation method was proposed to figure out CO_2 sequestration coefficient and oil recovery more precisely and efficiently. This method is more convenient than numerical simulation method, and the data for this method is easier to obtain. In addition, this appraisal method can also be used for other analogous potential evaluation in other oilfields.

(2) 148 blocks in Yanchang oilfield were suitable for CO_2 EOR and sequestration. CO_2 flood-

ing after water flooding can produce 180.21×10^6t more crude oil and sequestrate 223.38×10^6t CO_2. The average incremental oil recovery rate of miscible flooding was 12.49% and the average CO_2 sequestration coefficient was 0.27 t/t while the two values were 6.83% and 0.18t/t for immiscible flooding.

References

[1] S. Bachu, J Shaw. Evaluation of the CO_2 sequestration capacity in Alberta's oil and gas reservoirs at depletion and the effect of underlying aquifers [J]. Journal of Canadian Petroleum Technology, 2003, 42 (9): 51-61.

[2] S Bachu, D Bonijoly, J Bradshaw, et al. CO_2 storage capacity estimation: Methodology and gaps [J]. International Journal of Greenhouse Gas Control, 2007, 1 (4): 430-443.

[3] J Bradshaw, S Bachu, D Bonijoly, et al. CO_2 storage capacity estimation: Issues and development of standards [J]. International Journal of Greenhouse Gas Control, 2007, 1 (1): 62-68.

[4] J J Taber, F D Martin, R S Seright. EOR screening criteria revisited—Part 1: Introduction to screening criteria and enhanced recovery field projects [J]. SPE Reservoir Engineering, 1997, 12 (3): 189-198.

[5] D Diaz, Z Bassiouni, W Kimbrell et al. Screening criteria for application of carbon dioxide miscible displacement in waterflooded reservoirs containing light oil [J]. SPE/DOE Improved Oil Recovery Symposium, 1996.

[6] J Shaw, S Bachu. Screening, evaluation, and ranking of oil reservoirs suitable for CO_2-flood EOR and carbon dioxide sequestration [J]. Journal of Canadian Petroleum Technology, 2002, 41 (9).

[7] F Gozalpour, S R Ren, B Tohidi. CO_2 EOR and storage in oil reservoir [J]. Oil & Gas Science and Technology, 2005, 60 (3): 537-546.

[8] T Babadagli. Optimization of CO_2 injection for sequestration/enhanced oil recovery and current status in Canada [J]. 2006, 261-270.

[9] G Moritis. CO_2 sequestration adds new dimension to oil, gas production [J]. Oil & Gas Journal, 2003, 101 (9): 39-44.

[10] Z Yunchuan, X Yu, H Tianjiang. Screening method based on fuzzy optimum for gas injection in candidate reservoir [J]. Journal of Southwest Petroleum Institute, 2005, 27 (1): 44-47, (ISSN1000-2634. in Chinese).

[11] L Huaiyan, G Chenglin, G. Baocong. New screening method for reservoir by CO_2 injection miscible flooding [J]. Journal of China University of Petroleum, 2008, 32 (1): 72-76, (in Chinese).

[12] Z Shunpeng, Y Xiuwen. Fuzzy hierarchy analysis-based selection of oil reservoirs for gas storage and gas injection [J]. Henan Petroleum, 2009, 19 (4): 40-46, (in Chinese).

[13] S Pingping, L Xinwei, L Qiujie. Methodology for estimation of CO_2 storage capacity in reservoirs [J]. Petroleum Exploration and Development, 2009, 36 (2): 216-220.

[14] S Bachu. Comparison between methodologies recommended for estimation of CO_2 storage capacity in geological media [R]. Report by the CSLF task force on CO_2 storage capacity estimation and the US DOE capacity and fairways subgroup of the regional carbon sequestration partnerships program-phase III report, 2008.

[15] S Bachu, D Bonijoly, J Bradshaw, et al. Estimation of CO_2 storage capacity in geological media, phase 2 [R]. Prepared for the task force on CO_2 storage capacity estimation for the technical group of the carbon sequestration leadership forum, 2007.

[16] S Wang, H Jiang. Determine level of thief zone using fuzzy ISODATA clustering method [N]. Transport in porous media, 2011, 86 (2): 483-490.

[17] W Peixi, Z Jing. Application and design of fuzzy intelligent evaluation software for sand production and steam channeling prediction of steam injection well [J]. Procedia Engineering, 2011, 24: 546-550.

Potential Evaluation of CO_2 Sequestration and Enhanced Oil Recovery of Low Permeability Reservoir in Junggar Basin, China

Huan Wang   Xinwei Liao   Xiangji Dou   Baobing Shang   Heng Ye
Dongfeng Zhao   Changlin Liao   Xiaoming Chen

(Petroleum Engineering Department, China University of Petroleum, Beijing)

Abstract: The Chinese government is seeking CO_2 gas emission reduction measures. CO_2 capture and geological sequestration is one of the main measures. Injecting CO_2 into oil reservoirs can not only achieve the environmental protection purpose of CO_2 geological sequestration but also improve oil recovery and realize economic benefits, which helps to offset the cost of CO_2 sequestration. Therefore, the oil reservoir is one of the best sites for CO_2 sequestration. As for the reservoir of CO_2 flooding after water flooding, there are two methods for evaluating the potential of CO_2 enhanced oil recovery (EOR) and sequestration capacity, which are the mass balance method and analogy method. Through a combination of these two methods, this paper presents a new method, which can be reasonably used to evaluate these potentials. Besides, the screening criteria of CO_2 sequestration and EOR in the Junggar Basin are also proposed. On the basis of the guidelines of CO_2 source matching, reservoir characteristics, and fluid characteristic, four typical low permeability reservoirs (Caiman oil reservoir, Karamay oil reservoir, Beisantai oil reservoir, and Luliang oil reservoir) of the Junggar Basin are selected to study their potential of CO_2 EOR and sequestration. And then the potential of CO_2 EOR and sequestration capacity for the Junggar Basin oil reservoirs of CO_2 flooding after water flooding is studied by applying the method mentioned above. For 275 development blocks of 24 oil fields in the Junggar Basin, 139 development blocks are suitable for CO_2 miscible flooding EOR and sequestration, whereas 136 development blocks are suitable for CO_2 immiscible flooding EOR and sequestration. The total EOR potential could be 18407.76×10^4t and the CO_2 sequestration potential could amount to 47486.0×10^4t. The evaluation results show that the Junggar Basin's oil reservoirs are suitable sites for CO_2 EOR and sequestration and have great potentials. It can provide the decision basis for the future implementation of CO_2 emission reduction projects in western China.

Keywords: CO_2 flooding; Enhanced oil recovery; Effective storage; Analogy method; Basin; Potential evaluation

0 Introduction

There is a broad scientific consensus that global warming result primarily from increased concentrations of atmospheric greenhouse gases[1]. The main greenhouse gases that cause climate change are CO_2, CH_4, N_2O, HFC_S, PFC_S and SF_6. Among these gases, CO_2 is the main culprit for the earth warming[2]. Along with economic development and all-round social progress, China has become the world's second largest economy and second largest CO_2 gas emitter[3,4]. As a result, China is seeking CO_2 gas emission reduction measures. CO_2 capture and geological storage is one of the main measures, and reservoir, deep saline aquifers, and coal beds are the main places of CO_2 geological storage. [5-14] Oil reservoirs can provide safe geologic traps for CO_2 storage. Injecting CO_2 into oil reservoirs can not only achieve the goal of emission reduction, but also improve oil recovery and ensure the satisfactory economic benefits, which can reduce the cost of CO_2 storage. Therefore, storing CO_2 in a reservoir is a preferred method. The Chinese government attaches great importance to CO_2 capture and storage research and supports many national basic research programs of China (973 Program) and the Chinese national major science and technology (863 Program), which have conducted a lot of mechanism and laboratory experiments and carried out field tests in Jilin, Shengli, Daqing and other oilfields. Hao et al used slim tubule tests and PVT experiments to investigate the mechanism of CO_2 flooding[15]. Zhang et al used a mass balance method, which considers different trapping states of CO_2 in oil reservoirs and aquifers, to assess the potential of CO_2 storage in oil reservoirs associated with large aquifers[16]. Zhou et al conducted preliminary assessments on the effective CO_2 storage capacity in the Pearl River Mouth Basin (PRMB) offshore Guangdong[17]. Vincent et al used published methods to evaluate the CO_2 storage potential for Dagang oilfield and Shengli oil field[18]. Su et al introduced China's first field−scale reservoir demonstration project in Jilin oil field and evaluated its performance with respect to both EOR and carbon storage[19]. The content of this study is supported by the 973 Program "Carbon Dioxide Emission Reduction, Storage and Resource Utilization".

The Junggar Basin is an important oil−bearing basin in China. Good geologic trap and complete injection−production facilities enable it to be a promising location for CO_2 geological storage in China. Thus, evaluating CO_2 effective storage and EOR capacity in the Junggar Basin is of practical significance. On the basis of current capacity calculation methods for CO_2 storage in oil reservoir, this paper proposed a new method, which is appropriate for evaluating CO_2 effective storage capacity and EOR in China. The calculation method used in the paper is different from the mass balance method and analogy method. Through a combination of the two methods, this paper presents a new method to evaluate the EOR effects and sequestration potential of CO_2. The new method is elaborated, and the background of the proposal is presented. Furthermore, the paper conducts a completed case study, which is based on a lot of laboratory experiments, geological data, geologic modeling, and numerical simulation. Considering low permeability oil reservoirs in the Junggar Basin are mainly developed by water flooding, and the development effects of flooding would become worse in the high water cut stage. CO_2 flooding is an effective method to slow down the decline of development effects, so it would be used to improve oil recovery after water flooding. On the basis of the factors mentioned above, this paper focuses on the potential evaluation of CO_2 storage and EOR of low permeability oil reservoirs after water flooding in the Junggar Basin.

1 Calculation Models for CO_2 Effective Storage Capacity in Oil Reservoirs

Classification for CO_2 storage capacity can be described by the resource pyramid proposed by

McCabe in 1998. The CO_2 storage capacity includes theoretical storage capacity, effective storage capacity, practical storage capacity, and matched storage capacity (Fig. 1)[20]. Theoretical storage capacity, which occupies the whole of resource pyramid, represents the physical limit of what the geological system can accept. The effective storage capacity, which is a subset of theoretical capacity, is obtained by applying a range of technical cutoff limits to a storage capacity assessment, including the consideration of the part of theoretical storage capacity that can actually be physically assessed. Effective storage capacity is the CO_2 storage capacity considering reservoir properties, sealing of reservoir, storage depth, pressure system of reservoir, and pore volume. The capacity, which is influenced by fluidity, gravitational differentiation effect, reservoir heterogeneity, and formation water body, is always smaller than the theoretical one. Practical storage capacity, a subset of the effective capacity, is determined by considering technical, legal, and regulatory conditions of certain country or area, infrastructure, and general economic limit to CO_2 geological storage. Matched storage capacity is the subset of practical capacity, the value is obtained by the detailed matching of large stationary CO_2 sources with geological storage sites that are adequate in terms of capacity, injectivity, and supply rate. When calculating the effective storage capacity, some factors are taken into account, which are buoyancy, gravity override, mobility contrast, heterogeneity, water saturation, and so on. The effective storage capacity is close to the practical storage capacity, and it is meaningful for the primary evaluation of low permeability oil reservoirs in a whole basin. This paper focuses on the evaluation of CO_2 effective storage capacity in the Junggar Basin, thus the effective storage capacity evaluation models are mainly discussed.

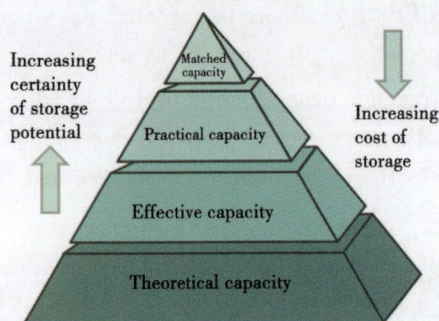

Fig. 1   Techno-economic Resource—Reserve Pyramid for CO_2 Sequestration Capacity (CSLF, 2005; Bradshaw et al, 2007)

Many researchers in different countries and organizations have studied the calculation method for CO_2 effective storage capacity. All the research methods can be divided into two categories, the mass balance method and the analogy method. On the basis of the calculation method of theoretical storage capacity, the Carbon Sequestration Leadership Forum (CSLF) proposed a calculation method based on mass balance, which considers buoyancy, gravity override, mobility contrast, heterogeneity, water saturation, and aquifer strength[21].

$$M_{CO_2e} = C_m \times C_b \times C_h \times C_w \times C_a \times M_{CO_2t} = C_e \times M_{CO_2t} \tag{1}$$

in which, M_{CO_2e} is the CO_2 effective storage capacity in an oil reservoir, t; M_{CO_2t} is the CO_2 theoretical storage capacity in an oil reservoir, t; C_e is the effective storage coefficient influenced by comprehensive factors, subscripts m, b, h, w and a stand for mobility contrast, buoyancy, heterogeneity, water saturation, and aquifer strength, respectively. The advantage of this method is to consider relative comprehensive influence factors of CO_2 storage, whereas the disadvantage is that each effective storage coefficient cannot be easily obtained. Each of the coefficients can be calculat-

490

ed by numerical simulation, which is heavy and complicated work. Equation 1 can be used to calculate the CO_2 effective storage capacity in both depleted oil reservoirs and CO_2 flooding oil reservoirs.

The analogy method is another calculation method based on the actual data of CO_2 EOR projects, thus this method is mainly used for CO_2 flooding reservoirs. The United States and European Union have conducted in-depth research for this method. This method calculates the effective capacity by introducing a CO_2 utilization factor[22].

$$M_{CO_{2e}} = N_P \times R_{CO_2} \tag{2}$$

in which, $M_{CO_{2e}}$ is the CO_2 effective storage capacity in an oil reservoir, t; N_P is the extra oil obtained by CO_2 injection, m^3; R_{CO_2} is the CO_2 utilization factor, which equals to the ratio of the net CO_2 injection amount to oil production amount, t/m^3.

The method has the advantage of a simple calculation. Although, the disadvantage is that the CO_2 utilization coefficient has little correlation with crude oil and reservoir properties, and cannot form the corresponding laws.

By applying the CO_2 flooding data of seven Permian sedimentary basins, Stevens proposes an empirical relationship between oil gravity and extra oil recovery due to the injection of CO_2[23]. The empirical relationship shows that original oil in place (OOIP) can be determined with the following formula if the final recoverable reserves and oil gravity were known.

$$OOIP = \frac{URR \times 100}{API + 5} \tag{3}$$

in which, URR is the final recoverable reserves, $10^9 m^3$; API is the crude oil gravity, °API.

2　Evaluation Method for CO_2 Effective Storage Capacity in Oil Reservoirs

Great differences of oilfields are exist between China and other countries due to specific sedimentary environments, reservoir forming conditions and oil properties. The low permeability reservoirs of China are continental deposits, which is different from marine deposits of abroad[24]. Generally speaking, they have complicated structures and strong anisotropy. The physical property of low permeability reservoirs is poor with low-porosity and extra-low-permeability. In China, the low permeability reservoirs are usually developed by water flooding, most of them have reached at the high water cut stage, and the crude oil is close to the black oil. So calculation models that are suitable for oil reservoirs in some literature cannot be applied to the Junggar Basin directly[21, 22]. Collection difficulty and accuracy of key parameters for capacity calculation are very critical to determine the reasonable evaluating model. OOIP, a key parameter, can be obtained easily and accurately for each reservoir in Chinese oil fields. Then on the basis of the mass balance method and analogy method, a new method to calculate the CO_2 effective storage capacity is proposed in this paper. On the basis of Equation 2 ($M_{CO_{2e}} = N_P \times R_{CO_2}$), where, $N_P = N \times R_{oil}$, N is the OOIP and R_{oil} is the oil recovery by CO_2 injection, so $M_{CO_{2e}} = N \times R_{oil} \times R_{CO_2}$. As mentioned above, considering the

491

weakness of Equation 2, the oil recovery factor and CO_2 utilization factor are calculated, respectively, which is in a complicated manner. Furthermore, in this study, we found that merging the oil recovery factor with the CO_2 geological storage coefficient can show the good relationship with oil properties and reservoir properties[25]. By introducing CO_2 storage coefficient R_s, which considers some influence factors (such as buoyancy, gravity override, mobility contrast, heterogeneity, and water saturation) by using the idea of mass balance as the same as Equation 1, the model is as follows[26, 27].

$$M_{CO_2e} = N \times R_s \tag{4}$$

in which, M_{CO_2e} is the CO_2 effective storage coefficient in the oil reservoir, t; N is the OOIP for the oil reservoir, t; R_s is the CO_2 geological storage coefficient, which equals to the dimensionless ratio of CO_2 storage amount to oil production amount.

How to determine R_s is the key to calculate the CO_2 effective storage coefficient. Generally, this coefficient can be determined empirically from completed CO_2 EOR and storage projects or numerical simulation. Considering few of those projects have been carried out in China as well as large differences between the geological characteristics and fluid properties of China and abroad, this paper mainly determines R_s with numerical simulation based on some experiments.

This method can be used for a small reservoir as well as a whole basin to evaluate the CO_2 effective storage capacity and EOR. When a certain reservoir is evaluated, an average storage coefficient can be obtained from numerical simulation on several typical well groups in different locations of the reservoir, and then the average value can be applied to the whole reservoir. When the capacity for a whole basin is calculated, it is impossible to build geological models and conduct numerical simulations for every reservoir in this basin, thus several typical reservoirs should be selected (Fig. 2). By conducting simulation on models of these typical reservoirs, important parameters such as the CO_2 effective storage coefficient and oil recovery factor can be obtained. Then, by using the analogy method, the capacity of other reservoirs can be determined by "borrowing" these important parameters from the typical reservoir based on the law of similarity. Finally, the potential evaluation of CO_2 effective storage capacity and EOR for the whole basin can be completed.

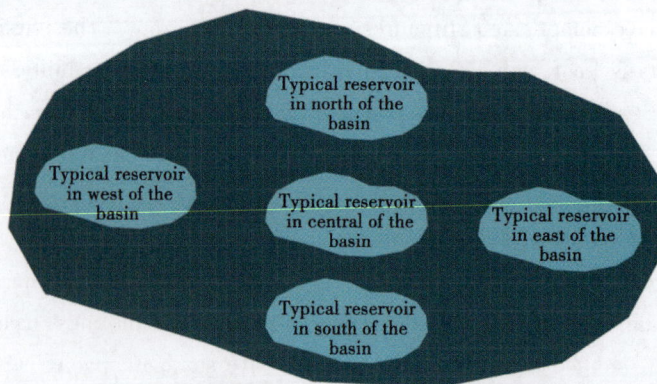

Fig. 2 Selecting Schematic Diagram of the Typical Reservoir in a Basin

492

3 CO₂ EOR and Storage Potential Evaluation of Typical Reservoirs

3.1 Geological Properties of the Junggar Basin

Junggar Basin is located in north of the Xinjiang Uygur Autonomous Region, China, shown in Fig. 3[28]. The area of the basin is $13.487 \times 10^4 km^2$. The Junggar Basin is a large superposition basin, which is mainly composed of later Paleozoic, Mesozoic and Cenozoic continental deposits based on Precambrian crystal basement, late Proterozoic era, and Paleozoic platform sedimentary face. Then, a composite and superimposed basin is formed by tectonic activities. The basin has experienced Hercynian, Indo-Chinese, Yanshan, and Himalayan orogeny periods. The multiple types of structural combination and sedimentary system are the results of the polycyclic sedimentary tectonic evolution. Sedimentary lithofacies of this basin include diluvia facies, fluvial facies, delta facies, lake facies, and volcanic facies. Integrated sedimentary layers of the Junggar Basin are developed from middle-late period of the Permian to Quaternary, and the maximum thickness of which could be $1.6 \times 10^4 m$. Currently, six oil-bearing series and formations including Carboniferous, Permian, Triassic, Jurassic, Cretaceous, and Tertiary have been found. The main oil-bearing series are Triassic, Jurassic, Cretaceous Permian, and Carboniferous.

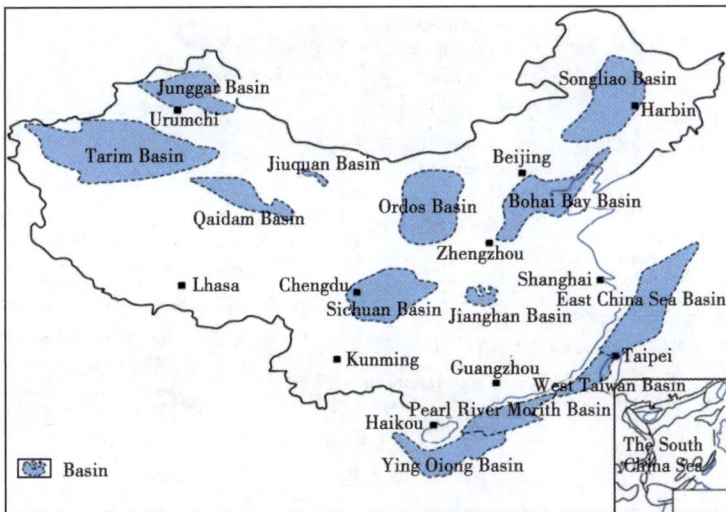

Fig. 3 Location of the Junggar Basin (Modified from Jia et al, 2012)

Tectonics of some oil fields are monoclines controlled by faults (e. g. Karamay and Baikouquan oil fields), anticlines (Cainan, Huoshaoshan, Beisantai, and Dushanzi oil fields) that are intact or simply cut by faults, complex fault blocks (Hongshanzui, Xiazijie oil fields), etc. The reservoir buried depth varies from about 300~4500 m (Shinan oil field). The pore structure of the reservoir is complicated, and the porosity and permeability are low, most of which are less than 20% and 50 mD, respectively. Besides, the plane and vertical heterogeneities of the reservoirs are strong. In the Junggar basin, the oil and water distribution of a handful of reservoirs (Huoshaoshan, Xiazijie,

493

etc.) is relatively complex due to the influence of structure, lithology, and fracture. And these reservoirs have edge-bottom waters and gas caps. Although, in most of the oil fields, the oil and water distributions are simple and the fluid properties are much better (Karamay, Baikouquan, and Cainan oil fields, etc.), which have no edge-bottom waters and gas caps. Therefore, in this paper, only the amount of CO_2 that is buried into the reservoir, including the CO_2 that exists as free gas and dissolves into the remaining oil and reservoir water during CO_2 flooding, is taken into consideration to evaluate the potential of CO_2 effective storage capacity, while the paper does not consider the CO_2 that dissolves into the edge-bottom water of the reservoir.

3.2　Typical Oil Reservoirs Selection in the Junggar Basin

The exploration and development of the Junggar Basin date back to the 1930s. By the end of 2013, there was a total of 25 proven oil fields in the basin (Fig. 4), including 9 oil fields in the northwest (Karamay, Honshanzui, Baikouquan, Wuerhe, Fengcheng, Xiazijie, Mabei, Xiaoguai, and Chepaizi), 7 oil fields in the east (Huoshaoshan, Beisantai, Santai, Ganhezi, Shanan, Shabei, and Dishuiquan), 3 oil fields in the south (Dushanzi, Qigu and Kayindike), and 6 oil fields in the central basin (Shixi, Cainan, Luliang, Mobei, Mosuowan, and Shinan).

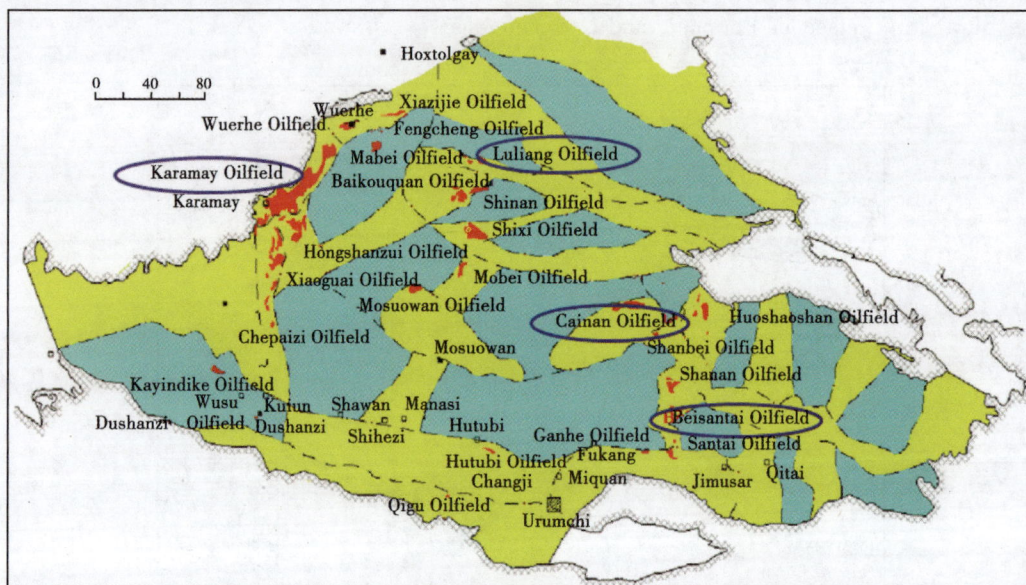

Fig. 4　Typical Reservoirs and CO_2 Emission Sources Map in Junggar Basin

On the basis of the deposition, reservoir and fluid characteristics of the Junggar Basin, and the matching relation of the location between the target oil reservoir and the source gas, the selection standards of typical reservoirs are determined. (1) When CO_2 flooding is conducted in an oil reservoir, fewer heavy components of crude oil that means a higher possible EOR. So, when the candidate oil reservoir is selected, the amount of heavy components, especially resin and asphaltene of the crude oil, should be as little as possible. (2) The candidate oil reservoir should be near the gas source to reduce the CO_2 transport costs. (3) The candidate oil reservoir should have good sealing

494

with no fracture development, or else the CO_2 will leak to cause greater environmental pollution.

According to the above selection standards, Cai 9 Xishanyao reservoir of the Cainan oil field, Ke8 Upper Karamay reservoir of the Karamay oil field, Bei 16 Wutonggou reservoir of the Beisantai oil field, and the Lu 9 Xishanyao reservoir of the Luliang oil field are chosen as the typical oil reservoirs in the Junggar Basin (Fig. 4). The four candidate reservoirs are typical low permeability reservoirs. The oil and gas have been sealed in these reservoirs for more than hundreds of millions of years, which have demonstrated that they have good sealing cap rocks. They have relative uniform distributions in the basin. Also, they can represent the characteristic of low permeability reservoirs located in different places of the basin. Source-sink matching is indispensable for CO_2 EOR and storage, so the selection also considers the matching relation of the location between the target oil reservoir and the CO_2 emission source. The statistical data of CO_2 emission sources near the typical oil reservoirs is shown in Table 1. For the Cainan, Karamay and Beisantai oil fields, their distances to the CO_2 emission source are all within 25 km. But the Luliang oil field is relatively far from the CO_2 emission source, whose distance is about 100km.

Table 1 Statistics of CO_2 Emission Sources Surrounding a Typical Reservoir in the Junggar Basin

Distance (km)	Oil field	Company category	CO_2 emission (10^4t)	Remarks
25	Cainan	Coal industrial park	—	Under construction
	Karamay	Methanol plant, refinery	141	
	Beisantai	Power plant, industrial park, ethylene plant, coal chemical industry	136	
	Luliang	—	—	Without statistics
50	Cainan	Power plant, ethylene plant, coal chemical industry	1870	Predict
	Karamay	Methanol plant, refinery	141	
	Beisantai	Power plant, industrial park, ethylene plant, coal chemical industry	136	
	Luliang	—	—	Without statistics
100	Cainan	Power plant, ethylene plant, coal chemical industry	230/1870	Have/Predict
	Karamay	Methanol plant, refinery	141	
	Beisantai	Power plant, industrial park, ethylene plant, coal chemical industry	1774	
	Luliang	—	—	Without statistics

3.3 Geological Characteristics of the Typical Oil Reservoirs

3.3.1 Xishanyao Reservoir of the Cainan Oil Field

The Xishanyao reservoir of the Cainan oil field belongs to lacustrine delta front subfacies deposition, including the J_2x_1 and J_2x_2 sections (Table 2). The main sand body belongs to the J_2x_2 section, which is a sandy distributary channel deposition. This sand body is an overlay to a depositional massive reservoir body, and it stably distributes in the Cai 9 block. The sedimentary thickness of the sand body is 33~60m with an average of 47.2m. The lithology of the reservoir is mainly sandstone and fine sandstone, and the interstitial material is mainly composed of argillaceous and kaolinite. According to the core petrophysical analysis data, the average porosity and permeability of the Cai 9 Xishanyao reservoir are 16% and 10.05mD, respectively. The pore development degree of the reservoir is medium. The dominated pore type is intergranular dissolved pores. The pore throat radius is not very big and the maximum connected throat radius lies between 1.527~20.83μm with an average of 4.78μm. The horizontal heterogeneity of the reservoir is controlled by the degree of sand body development and its distribution direction. The planar permeability contrast is 50~150 and the average vertical permeability contrast is 13.43.

Table 2   Stratigraphic Sequence of Typical Reservoirs

Typical reservoir	Erathem	System	Series	Formation	Section	Age (Ma)
Cai 9 Xishanyao	Mesozoic	Jurassic	Middle	Xishanyao	J_2x_1, J_2x_2	166.1~178.0
Lu 9 Xishanyao	Mesozoic	Jurassic	Middle	Xishanyao	J_2x_4	166.1~178.0
Ke 8 Upper Karamay	Mesozoic	Triassic	Middle	Upper Karamay	T_2k_2	235.0~241.1
Bei 16 Wutonggou	Palaeozoic	Triassic	Upper	Wutonggou	P_3wt	245.0~256.0

3.3.2   Upper Karamay Reservoir of the Karamay Oil Field

The Upper Karamay reservoir is a uniclinal structure inclining to the southeast with the dip angle of 5°~7°. The Nanbaijiantan fault that traverses the north part of the district is an overthrust fault and the fault surface inclines to the northwest. The Jurassic Sangonghe formation and Triassic Baijiantan formation are at the top of the fault. The dip angles of these two formations are about 70°~80° and 25°~30°, respectively. The fault throw of the Upper Karamay reservoir is 400~500m, and the horizontal fault throw is about 200~450m. The Upper Karamay reservoir conformably deposits on the Lower Karamay reservoir with the average deposition thickness of 218m. Fan delta deposition formed when the braided river flowed into the lake. The average porosity and permeability of the reservoir are 14.9% and 14.2mD, respectively.

3.3.3   Wutonggou Reservoir of the Beisantai Oil Field

The north edge of the Beisantai embossment was cut by the north fault of the Beisantai formation, and thus formed the Bei 16 well block, which is a nose structure. The top of the Bei 16 well block is a nose uplift with gentle slope. Its east and west wings become steeper. And the nose uplift pitches to the north. The tectonic axis lies exactly on the connecting line of well B1041 and well B1079. The east and west parts of the fault nose are constrained by a nearly north-south trending fault. In this area, the Wutonggou reservoir is a alluvial fan-fan delta facies deposition. $P_3wt_1^{2-1}$—$P_3wt_1^{3-2}$ is the main oil-bearing sand layer, and it belongs to fan delta front sub-facies depositon. Its deposition thickness is about 110 m. The main lithology of the reservoir is small conglomerate, pebbly medium-coarse sandstone and medium sandstone. The average porosity of the reservoir is

496

17% and the average permeability is 28. 5mD.

3. 3. 4   Xishanyao Reservoir of the Luliang Oil Field

The Lu 9 well block is located on three fountain bugles and it is a secondary structural unit of the Luliang uplift. The three fountain bugles and Luliang uplift have undergone the whole process of three uplift movements, which happened at the end of Early Permian, the end of Jurassic and the end of Cretaceous to Tertiary, respectively. The Xishanyao reservoir J_2x_4 section of the Lu 9 well block is a sedimentary association, which is composed of thick sandstones and thin mudstones. The reservoir thickness lies between 27. 4~39. 7 m with the average of 32. 4 m. The Xishanyao reservoir J_2x_4 section is vertically divided into three subsections $J_2x_4^1$, $J_2x_4^2$ and $J_2x_4^3$. Laterally, the reservoir of the whole area distributes stably. The lithology of the upside of the reservoir is mudstone and argillaceous siltstone, whereas the lithology of middle and bottom parts is mainly medium sandstone. The average porosity and permeability of the reservoir are 18. 9% and 18. 2 mD, respectively.

3. 4   Models of the Typical Oil Reservoirs

3. 4. 1   Geological Models of the Typical Oil Reservoirs

To build the geological models of the four typical reservoirs, different kinds of data were collected, including the coordinate data of the production and injection wells, stratified data, logging data, fault data, deposition facies diagrams, core analysis data, producing test, and various kinds of well test data. Then, facies−controlled models of the four typical oil reservoirs were built based on the collected data. When the geological models were built, first of all, the deposition facies and formation structure models were established. Then, interwell interpolation and stochastic simulation for different facies were accomplished based on the distribution law of the formation parameters. As a result, the formation parameter distribution models were built. The geological models of the four typical reservoirs are shown in Fig. 5.

3. 4. 2   Oil Pressure, Volume, Temperature Test of the Typical Oil Reservoirs

To study the influence of the injected gas on the physicochemical properties of the crude oil, pVT tests were conducted. First, bottom hole oil samples of typical wells were taken from the four typical oil reservoirs. Then these oil samples' components were analyzed and several pVT tests were conducted, including the saturation pressure test, flash separation experiment, constant composition expansion experiment, differential liberation experiment, and formation oil viscosity test. The main equipment used for the experiments is the mercury−free transparent piston high−pressure pVT device produced by French ST Company (Fig. 6). This device is mainly composed of pVT container, constant temperature air bath, pressure and temperature sensors, sample bucket, high−pressure metric pump, operation control system, and the observation and recording system. The autoclave is a piston device and its volume is variable. The volumetric change of the autoclave can be controlled by the piston that is driven by the precise motor.

On the basis of the above experiments, pVT module of the numerical simulation software Eclipse was applied to match the experimental data. The critical parameters of the heavy components and equation of state (EOS) parameters were corrected to obtain the pVT data that can be used in the numerical simulation. PR3 was chosen as the appropriate EOS. When the pseudocomponents of the

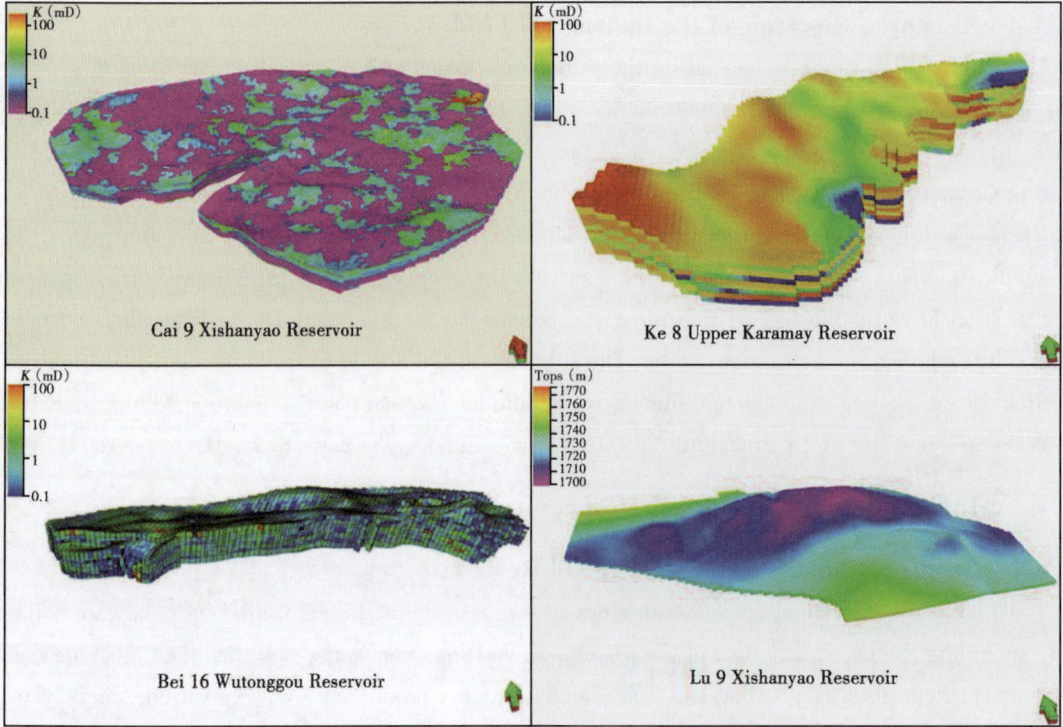

Fig. 5   Geological Models of Four Typical Reservoirs

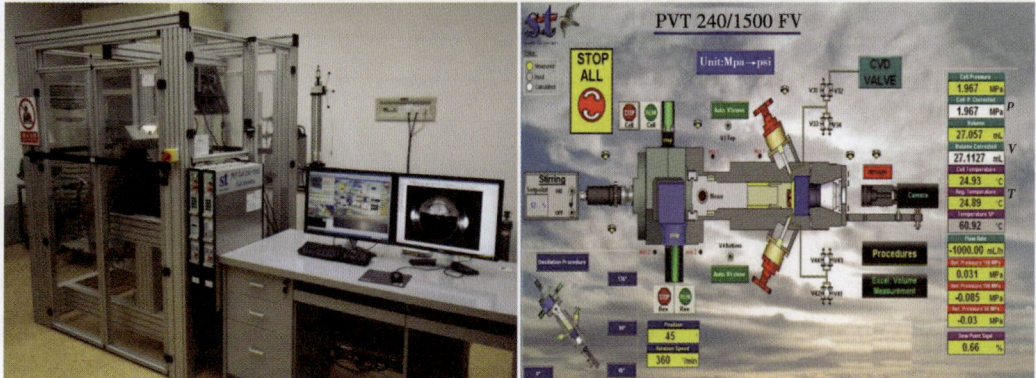

Fig. 6   ST-PVT Experimental Set-up

reservoir fluid were determined (Table 3), the data of the above experiments was regressed to obtain the characteristic parameters of the reservoir fluid, such as saturation pressure, gas-oil ratio, surface oil density, formation oil density, viscosity, and the change of the saturation pressure after gas injection. The data calculated by EOS are applied to match the experimental data. When the curves of calculated data are approximate to the curves of experimental data, the good matching results could be obtained. The matching results, which are the critical parameters of the oil simple from the Cai 9 Xishanyao reservoir, are shown in Table 4 (these critical parameters of the oil simples from the other three typical oil reservoirs are not shown here).

Table 3 Pseudocomponents of Crude Oil Sample

Pseudocomponent	CO_2	N_2+C_1	C_2—nC_4	C_5—C_6	C_7—C_{10}	C_{11}—C_{17}	C_{18}—C_{27}	C_{28+}
Mole fraction (%)	0.1	30.35	4.23	3.15	14.49	21.67	13.98	12.03

Table 4 Fluid Critical Parameters of EOS

Components	MW (g/mol)	Ωa	Ωb	p_c (bar)	T_c (K)	V_c ($m^3 kg^{-1} mol^{-1}$)	Z_c	AF
CO_2	44.010	0.457	0.078	73.866	304.700	0.094	0.225	0.225
CH_4+N_2	16.185	0.515	0.020	12.598	94.405	0.098	0.013	0.013
C_2H_6	33.877	0.547	0.037	30.859	280.513	0.162	0.113	0.113
C_{5+}	71.988	0.457	0.078	12.014	468.983	0.308	0.247	0.247
C_{7+}	114.288	0.457	0.078	27.269	587.887	0.460	0.335	0.335
C_{11+}	187.666	0.457	0.078	18.998	700.025	0.721	0.510	0.510
C_{18+}	298.750	0.457	0.078	13.762	730.429	1.052	0.731	0.731
C_{28+}	518.632	0.910	0.109	2.763	787.548	1.999	4.511	4.511

3.4.3 CO_2-crude Oil Expansion Experiment of Typical Oil Reservoirs

A gas-injection expansion experiment was conducted to study the effect of different proportions of injected gas on the fluid phase. This experiment could also help to determine the oil-displacement mechanism by injecting gas and to provide fundamental parameters for phase regression. Under the current formation pressure, a 10% mole fraction of CO_2 was added into the oil. After gas was injected, the pressure of the system increased gradually until all of the injected gas dissolved into the oil. Then the saturation pressure, PV relation, and viscosity of the new system were tested. After that, above procedure was repeated several times until the percentage of injected gas reached the designed amount (Table 5).

Table 5 Influence of Fluid Phase by Injecting CO_2

Mole ratio of CO_2 (mol/mol)	Saturation pressure (MPa)	Oil density (g/cm^3)	Volume factor B_o	Solution gas-oil ratio (m^3/t)	Swell factor V/V
0	11.79	0.8150	1.1220	40.6	1.0000
10%	14.75	0.8109	1.1968	56	1.0379
20.0%	17.8	0.8060	1.2521	76	1.0858
30.0%	20.3	0.8026	1.3190	101	1.1438
40.0%	23.65	0.8014	1.4035	135	1.2171
50.0%	30.52	0.8026	1.5166	182	1.3151
60.0%	42.84	0.8079	1.6782	252	1.4553

In this paper, the crude oil of the Bei16 Wutonggou oil reservoir was used to conduct the CO_2-injection expansion experiment. The main physical characteristic changes of the crude oil under bubble point pressure after injecting CO_2 are shown in Table 5. The saturation pressure of the crude oil increases greatly when the molar volume ratio of injected CO_2 increases. When the ratio reaches to

499

60%, the oil saturation pressure rises to 42.84 MPa, the state of the CO_2 and crude oil have not yet reached a critical point, which indicates that the first contact minimum miscible pressure (MMP) of them is higher than 42.84 MPa. The crude oil volume and its volume factor are increasing with the increased amount of CO_2, and the effects of swell and gas flooding are better. The solution gas–oil ratio is also increasing with the increase of injected CO_2, but crude oil density decreases at first and then increases with the increase of injected CO_2.

3.4.4　MMP Experiments of CO_2 and Crude Oil of Typical Reservoirs

For the technology of gas flooding to improve oil recovery, it can be divided into miscible flooding and immiscible flooding. The miscible displacement mechanism is that, under reservoir conditions, the oil and gas can be mixed without the presence of interfacial tension due to the diffusion and mass transfer between two fluids. As a result, the oil displacement efficiency could be greatly improved and the residual oil saturation could be minimized. The flowchart of the slim tube experiment is shown in Fig. 7, and the main parameters of the slim tube model are shown in Table 6.

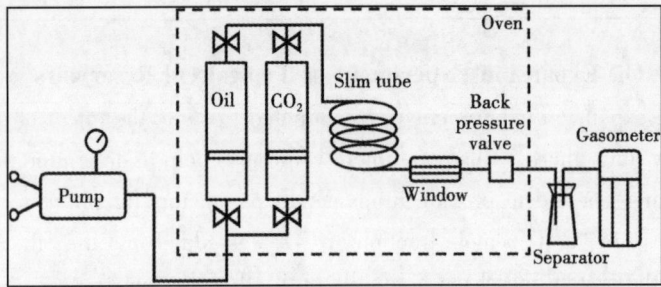

Fig. 7　Flowchart of Slim Tube Experiment

Table 6　Main Parameters of Slim Tube Experiment

Main parameters	Value
Maximum temperature (℃)	150
Maximum pressure (MPa)	50
Length (m)	20
Inner diameter (mm)	3.86
Outside diameter (mm)	6.35
Filler (quartz sand) (mesh)	170~325
Porosity (%)	39
Air permeability (μm^2)	5.43

The miscible flooding efficiency is much higher than that of immiscible flooding proved by theory and practice[29-31]. The oil displacement efficiency of CO_2 injection depends largely on the displacement pressure. It can achieve miscible flooding when the displacement pressure is higher than the MMP, whereas it can't reach mixed phase and achieve high oil recovery when the displacement pressure is less than the MMP. But the displacement pressure should not be too high, because a-

500

chieving high pressure conditions needs greater investment and costs. MMP is a key parameter to i-dentify if those four typical reservoirs can achieve miscible flooding, to identify CO_2 miscible / im-miscible displacement design and prediction, as well as to identify feasible studies, development plan designs and economic evaluations of CO_2 miscible flooding. So, the MMP experiments are an indispensable part of the study of CO_2 flooding and storage of typical reservoirs in the Junggar Basin. Fig. 8 shows the MMP of CO_2 and crude oil system of the four typical reservoirs.

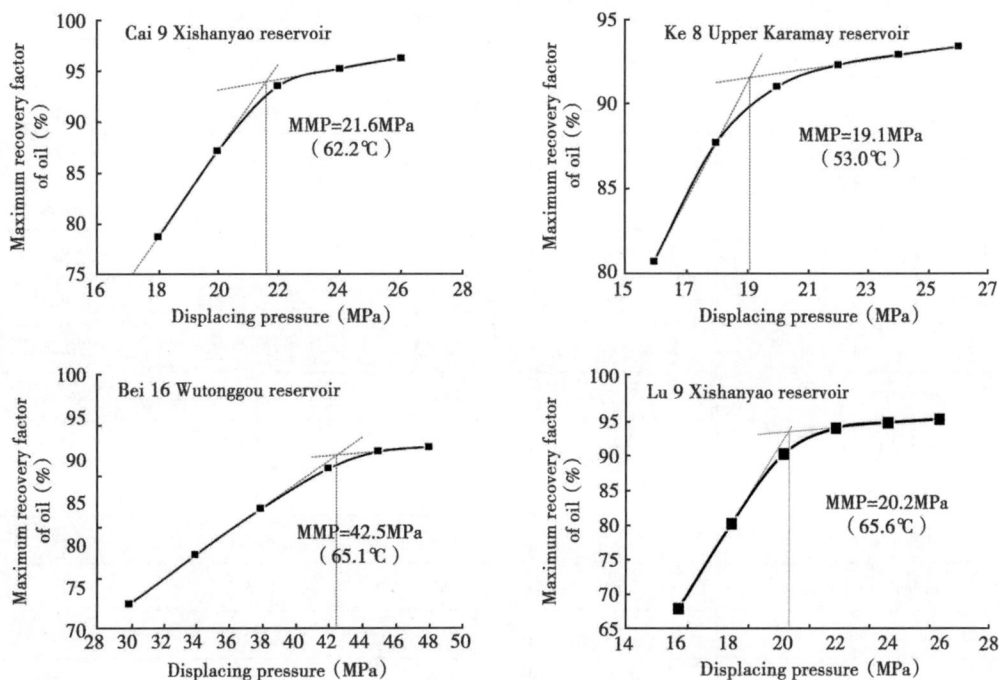

Fig. 8   MMP of CO_2 and Crude Oil System of the Four Typical Reservoirs

3. 4. 5   Numerical Simulation Models of Typical Reservoirs

The oil—water relative permeability curve was obtained through core displacement experiments and oil—gas relative permeability curve was obtained through Corey model calculation. The numeri-cal simulation models were established by applying the static field parameters of the geological mod-el, the pVT parameters, oil—water and oil—gas relative permeability data, the basic parameters of the rock, and the historical production data of wells. After the four numerical simulation models were established, then history matching was conducted to revise those geological models and to ob-tain the current residual oil saturation distribution and the reservoir parameters, whose aim is to pro-vide accurate models for the following study of CO_2 EOR and storage.

3. 5 CO_2 EOR and Storage Capacity Calculation of Typical Reservoirs

Currently the reservoirs in the Junggar Basin are usually developed by water flooding, so the po-tential evaluation of CO_2 EOR and storage focuses on the reservoirs after water flooding. The basic parameters of four typical reservoirs are shown in Table 7.

Table 7 Parameters of Four Typical Reservoirs

Parameters	Cai 9 Xishanyao reservoir	Ke 8 Upper Karamay reservoir	Bei 16 Wutonggou reservoir	Lu 9 Xishanyao reservoir
Geologic reserves (10^4 t)	2655	5711	714	9873. 9
Temperature (°C)	62. 2	53	65. 1	65. 6
Reservoir depth (m)	2230	2060	2235	2175
Effective thickness (m)	30. 2	19. 5	15. 7	16. 7
Reservoir type	Structure	Monoclinic	Structure	Structure
Porosity (%)	16	14. 9	17	18. 9
Permeability (mD)	10. 05	14. 2	28. 5	18. 2
Initial oil saturation (%)	56	60	52	50
Pressure factor	0. 93	1. 25	1. 12	0. 926
Initial reservoir pressure (MPa)	21. 27	25. 65	25. 3	20. 6
Saturation pressure (MPa)	8. 4	14. 35	19. 05	4. 19
Initial gas/oil ratio (m^3/m^3)	40	141	67	17
Oil volume factor	1. 235	1. 281	1. 185	1. 128
Density of surface oil (kg/m^3)	825. 2	854. 1	898. 8	853
Density of formation oil (kg/m^3)	776. 2	739	833. 7	829
Viscosity of surface oil (cp)	2. 78	25. 3	15. 4	10. 85
Viscosity of formation oil (cp)	1. 18	2. 01	6. 61	4. 9
Formation water salinity (mg/L)	10210	18392	14589	13380
MMP (MPa)	21. 6	19. 1	42. 5	20. 2
Miscible or immiscible	immiscible	miscible	immiscible	miscible

To obtain the correct numerical simulation models, the history matching was conducted. When the history matching is carried out, the production system is set to constant a single well production rate. To judge the accuracy of history matching, main indicators are original oil in place (OOIP), reservoir pressure, water cut, gas-oil ratio, and the production index. According to these indicators, finally good history matching results of the four models are realized, and their qualified rates of matching are all more than 90%. Thus, the four models meet the accuracy requirements of CO_2 EOR and storage potential evaluation.

It is necessary to determine a reasonable gas injection volume before applying CO_2 EOR and geological sequestration in a reservoir. From the numerical simulation result of Fig. 9, we can see that there is an inflection point near 0. 7 HCPV of CO_2. The oil recovery factor increases greatly with the increasing of HCPV of CO_2 before the inflection point. But after that point, the oil recovery factor increases slightly with the increasing of CO_2 HCPV. Therefore, the 0. 7 HCPV of CO_2 is a reasonable value.

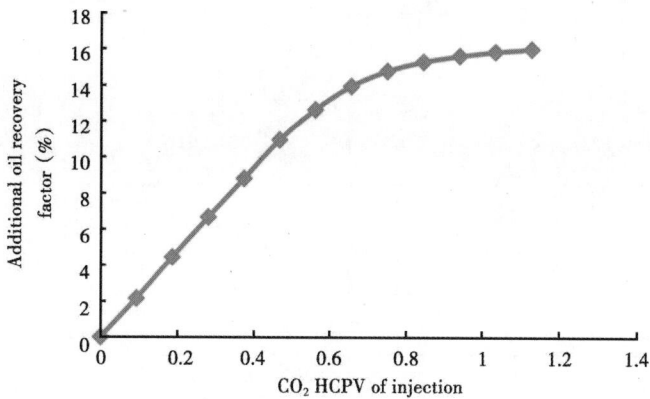

Fig. 9   Oil Recovery Factor with Different Volumes of CO_2 Injection

To obtain the oil recovery factor and CO_2 storage coefficient, we injected 0.7 HCPV of CO_2 after the water cut rose to 95%. The gas injection rate is $2 \times 10^4 m^3/d$. The production well was controlled by constant flowing bottom hole pressure (FBHP), and the economic limit indicator is set for the produced gas-oil ratio to $1000 m^3/m^3$. Then the recovery factor and CO_2 storage coefficient is calculated. The evaluation results of the four typical reservoirs are shown in Table 8.

Table 8   EOR and CO_2 Storage Evaluation Results of the Four Typical Reservoirs

Reservoir	Geologic reserves (10^4t)	EOR (10^4t)	CO_2 storage (10^4t)	Enhance recovery factor (%)	CO_2 storage coefficient (10^4t)	Miscible or immiscible
Cai 9	2148.000	176.351	508.345	8.210	0.237	immiscible
Ke 8	1445.000	204.323	474.784	14.140	0.329	miscible
Bei 16	714.300	53.358	187.861	7.470	0.225	immiscible
Lu 9	6755.000	971.369	2411.535	14.380	0.357	miscible

4 CO_2 EOR and Storage Potential Evaluation of the Junggar Basin

4.1   Screening Criteria for CO_2 EOR and Storage

In recent years, the reservoir screening criteria that are suitable for CO_2 miscible flooding are proposed by domestic and foreign experts. Although each experts' screening criteria are different at various periods, the value of their screening criteria have the same trend and have similar value boundaries. The screening criteria for CO_2 miscible flooding that are suitable for light oil reservoir in the Junggar Basin are proposed by taking into account other experts' screening criteria and reservoir characteristics of the Junggar Basin (Table 9) [32, 33].

Table 9   Screening Criteria for Application of CO_2 Miscible Flooding

Reservoir parameter	Carcoana (1982)	Taber (1983)	Klins (1984)	Ren S (2008)	Zhao F (2001)	S Bach (2004)	Ninth five-year plan (1998)	The paper (2010)
Crude oil density (g/cm^3)	<0.8227	<0.8948	<0.8762	0.9218~0.7587	<0.9218	0.8924~0.7883	<0.9042	<0.9218
Oil gravity (°API)	>40	>26	>30	22~55	>22	27~48	>25	>22
Depth (m)	<3000	>700	>914	600~3500	>762	—	—	>600
Original pressure (MPa)	>8.3	—	>10.3	—	—	>7.5	≥MMP	>7.5
Temperature (℃)	<90	—	—	<120	—	32~120	—	32~120
Viscosity (mPa·s)	<2	<15	<12	<188	<10	—	—	<188
Permeability (mD)	>1	—	—	>5	—	—	—	>1
Oil saturation	>0.30	>0.30	>0.25	0.28~0.64	>0.20	>0.25	—	>0.20

According to the reservoir screening criteria suitable for CO_2 immiscible flooding of foreign heavy oil reservoirs, for the heavy oil reservoirs whose crude oil density is greater than 920 kg/cm^3 in the Junggar Basin, the screening criteria for CO_2 immiscible flooding (or near miscible flooding) is shown in Table 10.

Table 10   Screening Criteria for Application of CO_2 Immiscible Flooding in the Junggar Basin

Reservoir parameter	Value
Crude oil density (g/cm^3)	0.92~0.98
Reservoir depth (m)	>550
Oil viscosity (mPa·s)	<600

4.2 CO_2 EOR and Storage Potential Evaluation

By the end of 2013, there were 25 proven oil fields in the Junggar Basin. According to the screening criteria for CO_2 EOR and storage, which are suitable for Chinese geological features and fluids characteristics (Table 9 and Table 10), the reservoirs in the Junggar Basin were screened, of which there are 275 development blocks of 24 oil fields suitable for CO_2 EOR and storage. There are 139 development blocks with 87495.24×10^{4}t OOIP that are suitable for CO_2 miscible flooding, and 136 development blocks with 75649.78×10^{4}t OOIP that are suitable for CO_2 immiscible flooding.

The oil recovery factors of two typical CO_2 miscible flooding reservoirs (Ke 8 Upper Karamay reservoir and Lu 9 Xishanyao reservoir) are 14.14% and 14.38%, respectively, whereas the CO_2 storage coefficients are 0.329 and 0.357, respectively. The oil recovery factors of two typical CO_2 immiscible flooding reservoirs (Cai 9 Xishanyao reservoir and Bei16 Wutonggou reservoir) are 8.21% and 7.47%, respectively, whereas the CO_2 storage coefficients are 0.237 and 0.225, respectively. The CO_2 EOR and storage potential evaluation of water flooding oil reservoirs in the Junggar Basin are conducted by "borrowing" the oil recovery factor and storage coefficients of typical reservoirs. For two typical CO_2 miscible flooding reservoirs, the average additional oil recovery factor is 14.26%, and the average CO_2 storage coefficient is 0.343. For two typical CO_2 immiscible flooding reservoirs, the average additional oil recovery factor is 7.84% and the average CO_2 storage coefficient is 0.231. For the 275 development blocks in the Junggar Basin, the CO_2 miscible flooding reservoirs can improve additional oil for $12476.82 \times 10^4 t$, and the CO_2 immiscible flooding reservoirs can improve additional oil for $5930.94 \times 10^4 t$. The total improved additional oil is $18407.76 \times 10^4 t$. The CO_2 storage capacity of miscible flooding reservoirs is $30010.87 \times 10^4 t$, and the CO_2 storage capacity of immiscible flooding reservoirs is $17475.1 \times 10^4 t$. The total CO_2 storage capacity is $47486.0 \times 10^4 t$ (Fig. 10).

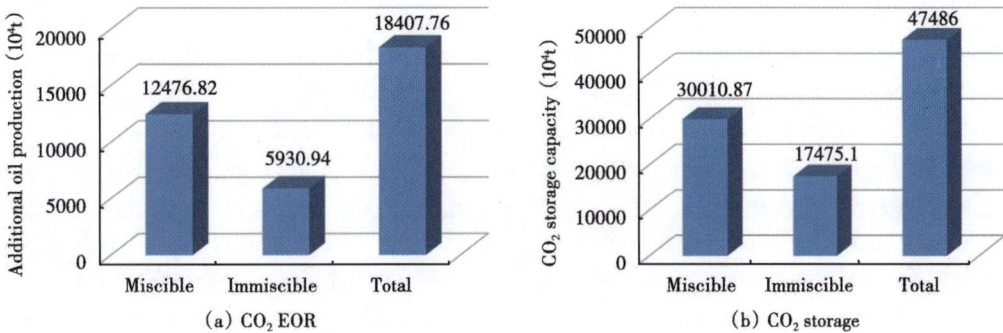

Fig. 10   Results of CO_2 EOR and Storage

5   Conclusions

(1) The CO_2 EOR and storage calculation method proposed in the paper has combined the advantages of the mass balance method and analogy method. By introducing the CO_2 storage coefficient to calculate CO_2 storage capacity, it can not only be calculate easily but also can consider many influence factors, such as buoyancy, gravity override, mobility contrast, heterogeneity, and water saturation, when using the numerical method to calculate the CO_2 storage coefficient. However, the weakness of the method is that it needs plenty of geological and fluid data to build geological and numerical models. And the influence factors of the CO_2 storage coefficient are composed in one integrated parameter, thus it is difficult to clearly recognize which factor is important.

(2) For two typical miscible flooding reservoirs in the Junggar Basin, the average additional oil recovery factor is 14.26% and the average CO_2 storage coefficient is 0.343. For two typical immiscible flooding reservoirs in the Junggar Basin, the average additional oil recovery factor is 7.84% and the CO_2 storage coefficient is 0.231.

(3) For the 275 development blocks in the Junggar Basin, the CO_2 miscible flooding reservoirs can improve additional oil for $12476.82 \times 10^4 t$, while the CO_2 immiscible flooding reservoirs can improve additional oil for $5930.94 \times 10^4 t$, and the total improved additional oil is $18407.76 \times 10^4 t$. The CO_2 storage capacity of miscible flooding reservoirs is $30010.87 \times 10^4 t$, the CO_2 storage capacity of

immiscible flooding reservoirs is 17475.1×10^4 t, and the total CO_2 storage capacity is 47486.0× 10^4 t.

(4) The evaluation results show that the low permeability oil reservoirs in the Junggar Basin are suitable locations for CO_2 EOR and storage and have huge potential. The results can provide the decision basis for the future implementation of CO_2 emission reduction projects in western China. Furthermore, the method used in the paper can be used in other basin for evaluating CO_2 EOR and storage at home and abroad.

References

[1] Shen P, Jiang H. Utilization of greenhouse gas as resource in EOR and storage it underground [J]. Engineering Sciences. 2009, 11 (5): 54-59.

[2] Jiang H, Shen P, Song X, et al. Global warming and current status and prospect of CO_2 underground storage [J]. Palaeogeography, 2008, 10 (3): 323-328.

[3] Barboza D. China passes Japan as second-largest economy [N]. New York Times, 2010, 15.

[4] Zhang M, Mu H, Ning Y. Accounting for energy-related CO_2 emission in China, 1991—2006 [J]. Energy Policy, 2009, 37 (3): 767-773.

[5] Zhao X, Liao X. Evaluation method of CO_2 sequestration and enhanced oil recovery in an oil reservoir, as applied to the Changqing Oilfields, China [J]. Energy & Fuels, 2012, 26 (8): 5350-5354.

[6] Jahangiri H R, Zhang D. Ensemble based co-optimization of carbon dioxide sequestration and enhanced oil recovery [J]. Greenhouse Gas Control, 2012, 6 (8): 22-33.

[7] Hughes T J, Honari A, Graham B F, et al. CO_2 sequestration for enhanced gas recovery: New measurements of supercritical CO_2—CH_4 dispersion in porous media and a review of recent research [J]. Greenhouse Gas Control, 2012, 6 (9): 457-468.

[8] Underschultz J, Boreham C, Dance T, et al. CO_2 storage in a depleted gas field: An overview of the CO_2CRC Otway Project and initial results [J]. Greenhouse Gas Control, 2011, 5 (4): 922-932.

[9] Hatzignatiou D G, Riis F, Berenblyum R, et al. Screening and evaluation of a saline aquifer for CO_2 storage: Central Bohemian Basin, Czech Republic [J]. Greenhouse Gas Control, 2011, 5 (6): 1429-1442.

[10] Bachu S, Pooladi-Darvish M, Hong H. Chromatographic partitioning of impurities (H_2S) contained in a CO_2 stream injected into a deep saline aquifer: Part 2. Effects of flow conditions [J]. Greenhouse Gas Control, 2009, 3 (4): 458-467.

[11] Ogawa T, Nakanishi S, Shidahara T, et al. Saline-aquifer CO_2 sequestration in Japan-methodology of storage capacity assessment [J]. Greenhouse Gas Control, 2011, 5 (2): 318-326.

[12] Taku Ide S, Jessen K, Orr jr F M. Storage of CO_2 in saline aquifers: Effects of gravity, viscous, and capillary forces on amount and timing of trapping [J]. Greenhouse Gas Control, 2007, 1 (4): 481-491.

[13] Ren S, Zhang L, Zhang L. Geological storage of CO_2: Overseas demonstration projects and its implications to China [J]. China University of Petroleum, 2010, 34 (1): 93-98.

[14] Shi J Q, Durucan S, Fujioka M. A reservoir simulation study of CO_2 injection and N_2 flooding at the Ishikari coalfield CO_2 storage pilot project, Japan [J]. Greenhouse Gas Control. 2008, 2 (1): 47-57.

[15] Hao Y, Bo Q, Chen Y. Laboratory investigation of CO_2 flooding [J]. Explor. Dev, 2005, 32 (2): 110-112.

[16] Zhang L, Ren S, Ren B, et al. Assessment of CO_2 storage capacity in oil reservoirs associated with large lateral/underlying aquifers: Case studies from China [J]. Greenhouse Gas Control, 2011, 5 (4): 1016-1021.

[17] Zhou D, Zhao Z, Liao J, et al. A preliminary assessment on CO_2 storage capacity in the Pearl River Mouth

Basin offshore Guangdong, China [J]. Greenhouse Gas Control, 2011, 5 (2): 308-317.

[18] Vincent C J, Poulsen N E, Zeng R, et al. Evaluation of carbon dioxide storage potential for the Bohai Basin, north-east China [J]. Greenhouse Gas Control, 2011, 5 (3): 598-603.

[19] Su K, Liao X, Zhao X, et al. Coupled CO_2 enhanced oil recovery and sequestration in China's demonstration project: Case study and parameter optimization [J]. Energy & Fuels, 2013, 27 (1): 378-386.

[20] Bachu S, Bonijoly D, Bradshaw J, et al. CO_2 storage capacity estimation: Methodology and gaps. Int. J. Greenhouse Gas Control. 2007, 1 (4), 430-443.

[21] Bradshaw J, Bachu S, Bonijoly D, et al. A taskforce for review and development of standards with regards to sequestration capacity measurement [J]. Carbon Sequestration Leadership Forum (CSLF), 2005, 8: 6-8, http://www.cslforum.org/documents/Taskforce_Sequestration_Capacity_Estimation_Version_2.pdf (accessed Aug 23, 2007).

[22] Hendriks C, Graus, W, Van Bergen F. Global carbon dioxide storage potential and costs [J]. 2004, 1-71. http://www.langtoninfo.com/web_content/9780521685511_frontmatter.pdf.

[23] Stevens S H, Kuuskraa V A, Taber J J. Sequestration of CO_2 in depleted oil and gas fields: Barriers to overcome in implementation of CO_2 capture and sequestration [J]. International Energy Agency Greenhouse Gas R&D Programme, International Energy Agency (IEA): Cheltenham, U K , 1999, Report IEA/CON/98/31.

[24] Zhao H, Liao X. Key problems analysis on CO_2 displacement and geological storage in Low Permeability oil reservoir, China [J]. Journal of Shaanxi University of Science & Technology, 2011, 29 (1): 1-6.

[25] Liao X, Chen Y, Zhao H, et al. Sensitivity analysis of CO_2 storage coefficient and CO_2-EOR [C]. Proceedings of the Power and Energy Engineering Conference (APPEEC), Chengdu, Mar 28-31, 2010.

[26] Chen Y, Liao X, Zhao H, et al. Determination two key dissolution coefficients in calculation of CO_2 storage capacity [J]. Science and Technology Review, 2010, 28 (1): 98-101.

[27] Shen P, Liao X, Liu Q. Methodology for estimation of CO_2 storage capacity in reservoirs [J]. Petroleum Exploration and Development, 2009, 36 (2): 216-220.

[28] Jia C, Zheng M, Zhang Y. Unconventional hydrocarbon resources in China and the prospect of exploration and development [J]. Petroleum Exploration and Development, 2012, 39 (2): 139-146.

[29] Gao H, He Y, Zhou X. Research progress on CO_2 EOR technology [J]. Special Oil and Gas Reservoirs, 2009, 16 (1): 6-12.

[30] Koottungal L. 2012 worldwide EOR survey [J]. Oil & Gas Journal, 2012, 110 (4): 57-69.

[31] Wang H, Liao X, Zhao X, et al. Potential evaluation of CO_2 flooding enhanced oil recovery and geological sequestration in Xinjiang Oilfield [J]. Shanxi University of Science & Technology, 2013, 31 (2): 74-79.

[32] Zhang L, Wang S, Zhang L, et al. Assessment of CO_2 EOR and its geo-storage potential in mature oil reservoirs, Shengli Oilfield , China [J]. Explor. Dev, 2009, 36 (6): 737-742.

[33] Shaw J, Bachu S. Screening, evaluation, and ranking of oil reservoirs suitable for CO_2 flood EOR and carbon dioxide sequestration [J]. Petrol. Technol, 2002, 41 (9): 51-61.

吉林低渗透油藏气驱开发潜力

张　辉　于孝玉　马立文　陈国利　郑雄杰　赵世新

（中国石油吉林油田公司勘探开发研究院）

摘　要： 在老区稳产和新区上产难度越来越大的形势下，为有效提高低渗透油藏采收率和储量动用率，吉林油田提出转变开发方式，全面推进气驱开发的发展战略。从油田资源特点出发，对气驱开发的技术需求、可行性和潜力进行分析，对矿场试验和工业化推广作出部署。吉林油田拥有丰富的 CO_2 资源，可动用储量具备形成年产 $200×10^4$ t CO_2 的产能规模。适合气驱的油藏储量资源规模较大，混相和近混相驱石油资源占 39.9%。气驱开发将以 CO_2 混相驱技术为引领，逐步拓宽气驱类型。在长岭气田周边优先开展 CO_2 驱，在距离 CO_2 气源较远的油田开展空气驱，规划 2020 年气驱年产油达到 $80×10^4$ t。吉林油田气驱开发具有资源优势和技术基础，潜力很大。

关键词： 低渗透油藏；含 CO_2 气藏；混相；CO_2 驱；空气驱；矿场试验；潜力

0　引言

近年来，吉林油区已开发老油田进入双高开发阶段，但水驱采收率较低；未动用和新探明的储量常规水驱开发经济效益差，难以动用。为有效提高油田采收率和储量动用率，需要转变开发方式，全面推进气驱开发。因此，当前应对气驱开发的潜力进行深入评价，对矿场试验和工业化推广作出部署安排。

气驱开发的主要类型有烃类气驱、CO_2 驱、氮气驱和空气驱。根据油藏地质开发特点、气源情况、矿场试验效果和国内外气驱开发的趋势，认为吉林油田目前应重点攻关 CO_2 驱和空气驱技术。

1　气驱开发技术需求及可行性

1.1　气驱开发技术的发展与应用情况

1.1.1　CO_2 驱开发技术

CO_2 驱油的机理是，注入 CO_2 与地层油之间不断发生传质（溶解、蒸发、凝析）作用，进行组分交换，直至完全消除界面张力，形成混相。CO_2 混相驱通过改变原油性质、消除界面张力，可大幅提高驱油效率。CO_2 非混相驱通过溶解、膨胀和降黏作用提高驱油效率[1-4]。

CO_2 驱技术于 20 世纪 80 年代在美国工业化推广，目前已在世界各国广泛应用。其中 CO_2 混相驱项目多，产量规模大，技术成熟。2012 年世界有 CO_2 混相驱项目 135 个，CO_2

非混相驱项目 6 个。CO_2 驱技术应用主要在美国，2012 年 CO_2 驱项目有 121 个，年产油达 1687×10^4 t，较水驱可提高采收率 7.5% ~ 17%。CO_2 驱典型油田为美国的兰奇利油田（Rangely）和加拿大的韦本油田（Weyburn）[5,6]。

国内 CO_2 驱技术应用起步较晚，总体上规模小、经验少、技术不配套。2008 年开始在吉林油田开展了具有一定规模的矿场试验，基本形成了 CO_2 驱油与埋存配套技术，目前正处于工业化推广过程中。

1.1.2　空气驱开发技术

空气驱油的机理是，空气注入油藏后，氧气与原油发生低温氧化反应，氧气被消耗，产生的热效应可以使原油降黏和热膨胀，并使得油藏压力升高。生成碳的氧化物 CO、CO_2 以及由 N_2 和轻烃组分等组成的烟道气，可实现氮气驱或间接的烟道气驱。部分油藏采用低温氧化重力驱，效果相对较好[7-9]。

空气驱技术在国外开展相对较晚，实施项目少。低温氧化空气驱项目世界有 8 个，年产油 28×10^4 m^3。试验较水驱提高采收率 4% ~ 12%。国外空气驱典型油田为美国水牛油田（Buffalo）[10,11]。国内百色、长庆、延长、中原、大庆等油田开展了空气驱的研究和试验，但均没有规模实施。

1.1.3　注气重力稳定驱

注气重力稳定驱是针对注气介质特性，采用高注低采模式，提高气体波及效率的一种气驱开发方式，适用的油藏条件比较宽泛。注气重力稳定驱典型油田为 Hawkins 油田。

1.1.4　CO_2 吞吐

CO_2 单井吞吐已成为提高轻质油采收率和增产的主要方法。对油藏条件要求不苛刻，技术要求不高，经济上可行，可作为常规措施应用[12,13]。

注气作为一种提高采收率的常用方法，受到普遍关注。在美国注气采油法在提高采收率中仅次于热力采油，排第二位。注气开发特别适合低渗透油田[14]。

1.2　吉林油田气驱开发技术需求

1.2.1　已开发低渗透老油田和特低渗透油藏

已开发油田含水 85.7%，可采储量采出程度 63.9%。目前标定采收率 23.5%，整体进入中高采出程度阶段，稳产难度大。

已开发油田可分为三类油藏：

（1）中高渗老油田，例如扶余、红岗。

（2）低渗透老油田，例如新立、新民。

（3）特低渗新开发油田，例如大情字井、大安—海坨子。

针对 3 类已开发油田，需要大力攻关改善水驱配套技术，积极探索气驱有效技术。

已开发低渗透老油田逐步进入中高含水开发阶段，常规井网调整采收率增加幅度减缓，应用常规水驱开发技术，进一步提高采收率空间有限。

"十五"以来开发的特低渗透油藏单纯依靠水驱保证原油产量稳定难度大，对气驱提高采收率技术需求迫切。例如英台八面台油田储层物性差（渗透率为 0.5 ~ 5mD），水驱难以建立有效驱替关系；注入压力高，压力保持水平低，仅为 65.9%；开发效果差，水驱采收率为 12.7%。

1.2.2　待开发低渗透薄油层和超低渗油藏

待开发储量以中部组合低渗透薄油层和扶余超低渗油藏为主，油层厚度薄，储量丰度低，储层致密、含油饱和度低，单井产量低，无法实现效益动用。

中部组合低渗透薄油层，主要分布在乾安地区。油层厚度一般 1~3m，储层发育不连续，采用常规直井开发，效益油层钻遇率低、单井控制可采储量少，应用常规技术已无法实现效益勘探和规模开发。

扶余超低渗油藏，主要分布在中央凹陷周边地区。储层致密，渗透率一般小于 0.5mD，可动流体饱和度低，油层平面和纵向非均质性强，采用常规注水方式开发，单井产能低、采收率低，效益开发难。

1.3　全面推进气驱开发可行性分析

1.3.1　气驱开发的优势

气驱开发方式具有较大优势：

（1）注入能力明显好于水驱，吸气指数是吸水指数的 4~6 倍。

（2）地层压力保持水平高，可以保持在原始地层压力 90%~110%。

（3）驱油效率高，可以达到 70%以上。

（4）油层动用程度提高，油层动用下限更低，渗透率下限可达到 0.2mD，可以有效动用小孔喉内的残余油，实现低渗透差油层的有效驱替。

（5）提高单井产量 30%~60%，提高采收率 4%~15%。

1.3.2　全面推进气驱开发的可行性

（1）全面推进气驱开发具有资源和技术基础。吉林油田拥有丰富的 CO_2 资源，适合气驱的油藏储量资源规模较大，国外具有成熟的气驱开发技术可供借鉴。

（2）吉林油田基本形成 CO_2 混相驱油技术，矿场试验效果较好，经济上可行。目前黑 59 等 3 个试验区注气井达到 34 个，年产量规模达到 10×10^4t。大情字井油田 CO_2 混相驱矿场试验证实，陆相低渗透油藏 CO_2 混相驱油效果明显好于水驱，矿场试验采油速度持续保持在 2%以上，单井产量与水驱对比提高 25%，利用数值模拟预测采收率将提高 10%以上。通过重大专项形成了 10 项 CO_2 驱油与埋存配套技术。

（3）吉林油田全面推进气驱开发的态势正在形成。CO_2 吞吐技术在老油田应用效果明显，大安黑帝庙油层单井组空气驱试验初步见到效果。目前正在实施大情字井油田 CO_2 混相驱工业化推广，英台油田方 118 区块空气驱先导试验，以及大情字井油田黑 168 区块水平井 CO_2 吞吐试验。

2　气驱开发技术应用潜力

2.1　适合气驱开发油藏潜力

2.1.1　CO_2 驱开发油藏潜力

吉林油区有 3 个油田可以实现混相驱，分别为大情字井、长春和莫里青油田。有 3 个油田可以实现近混相驱。混相和近混相驱石油资源占 39.9%。混相驱资源主要集中在长岭凹陷和红岗阶地。另外，一些非混相驱资源技术上有 CO_2 驱开发需求（图 1）。

大情字井油田是具有较强非均质性的复杂岩性油藏，具有得天独厚的气源优势，根据油层的发育情况，已开发区大多适合实施CO_2驱。

莫里青油田构造倾角大，扇体沉积储层复杂多变，储层敏感性强，地层压力与混相压力接近，在解决气源的情况下可有效应用重力稳定驱。

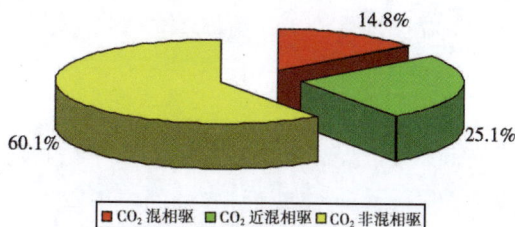

图 1 吉林油区探明储量混相驱资源状况

对于近混相的油田，可以利用高注低采的稳定重力驱的方式实施CO_2驱，以解决乾安老油田采收率低，大安—海坨子开发效果差的实际问题。

2.1.2 空气驱开发油藏潜力

根据空气驱的驱油机理[15-20]，在注气过程中需要发生稳定的低温氧化反应，才能取得较好的效果，因此对油藏温度要求较为严格。根据这一条件空气驱最好在埋深大于1850m的油藏实施，但目前在研助燃剂可以将埋藏深度拓宽到1500m。

对比国外空气驱技术应用实例（油藏温度均大于80℃），吉林油田适合空气驱的资源主要分布在乾安、大安—海坨子和英台八面台地区。

2.2 CO_2资源状况和可利用潜力

国内含CO_2气藏和纯CO_2气藏储量资源十分有限，仅少数地区具有比较丰富的储量资源[21-23]。

松辽盆地CO_2资源主要受控于后期岩浆活动的深大断裂带，CO_2天然气具有点状、带状分布，局部富集的特点。初步分析具有工业价值的CO_2气藏主要是火山岩储层。

通过对探井试气、试采资料分析，目前初步认为具备形成$10×10^8m^3$（$200×10^4t$）CO_2产能规模，其中含CO_2气藏开发可形成年产$2.9×10^8m^3$（$58×10^4t$）产能规模。

地面可外供气源总量少，运费高，无法满足规模实施CO_2驱开发的气源供应。必须动用纯CO_2气藏，才能满足CO_2驱开发需求。

含CO_2气藏开发在提供清洁能源甲烷的同时，还产出伴生的CO_2气，成本较低。纯CO_2气藏开发产出的CO_2气不用分离，成本更低。

3 气驱开发整体部署及矿场试验安排

3.1 气驱开发规划部署原则

（1）气驱开发以CO_2混相驱技术为引领，逐步拓宽气驱类型。根据技术成熟度，近期以CO_2混相驱为主体，以顶部注气CO_2近混相驱和空气驱作为技术储备，以周期注气采油和CO_2吞吐作为新区水平井能量补充的重要方式。

（2）优化资源配置，提高经济效益。在长岭气田周边优先考虑开展CO_2驱，距离CO_2气源较远或CO_2气源不足时，考虑开展空气驱。在西南大情字井和乾安等油田形成CO_2驱开发区，在西北英台和大安等油田形成空气驱开发区。

（3）根据 CO_2 注采平衡，统筹安排规模进度。气驱开发要在矿场试验的基础上逐步扩大规模，确保平稳推进，规模应用。

3.2 气驱开发规划部署

气驱开发以提高低渗油田开发水平为核心，以 CO_2 混相驱技术为引领，超前试验空气驱技术，形成 2 种气驱优势互补的应用模式。当前矿场试验的总体部署安排是推广大情字井油田，加快伊通油田，试验英台油田。

乾安地区：为特低渗透油藏，采收率为 20%，主体为混相、近混相资源，距 CO_2 气源较近，适合 CO_2 驱。为提高已开发区采收率和动用低丰度层储量，处理好碳平衡，应加快 CO_2 驱推广。

大安—海坨子：为致密油，采收率为 14%，主体为近混相资源，距 CO_2 气源较远。重点开展 CO_2 驱，辅以空气驱。从水平井开发区块能量补充出发，推进新区水平井 CO_2 吞吐和周期注气采油试验。

莫里青油田：为水敏性储层，主体为混相资源，距 CO_2 气源很远。适合 CO_2 驱和空气驱。从为水敏性储层补充能量和动用外围储量出发，寻找气源、开展试验；优化方案，整体实施。

扶新地区：为低渗透油藏，水驱矛盾突出。非混相驱，不适合空气驱。立足 CO_2 驱试验，优选气驱方式。

英台地区：为特低渗油藏，采收率仅 15%。非混相驱，适合空气驱。加快试验，明确空气驱应用领域。

气驱开发产量规模：预计到 2020 年注入井组达到 300 个，气驱开发覆盖地质储量 $6900 \times 10^4 t$，气驱年产油 $80.8 \times 10^4 t$，其中 CO_2 混相驱 $58.5 \times 10^4 t$，CO_2 近混相驱 $16.6 \times 10^4 t$，CO_2 非混相驱 $2.2 \times 10^4 t$，空气驱 $3.5 \times 10^4 t$。

2020 年最高日注 CO_2 7380t，年注 $244 \times 10^4 t$，部分循环利用，年需 CO_2 $220 \times 10^4 t$。

3.3 重点矿场试验安排

立足吉林油田油藏特性，围绕 2 大气驱技术，宏观规划部署，有序推进矿场试验。

试验要解决 4 个关键性的问题：

（1）裂缝、非均质、薄互层、低构造倾角油藏合理注采井网设计技术（重力驱技术的应用）。

（2）非均质油藏注采参数优化设计、注采调控、剖面调整等扩大波及体积技术。

（3）复杂井况气驱安全注采关键技术。

（4）空气驱、CO_2 吞吐注入、集输及处理技术。

重点矿场试验安排如下：

（1）乾安油田 CO_2 重力稳定近混相驱矿场试验。

（2）大安油田扶余致密油藏水驱转 CO_2 驱矿场试验。

（3）大安油田致密油藏水平井 CO_2 周期注气采油能量补充矿场试验。

（4）英台油田空气重力驱矿场试验。

（5）新民油田 CO_2 重力稳定非混相驱矿场试验。

（6）莫里青油田 CO_2 重力稳定混相驱矿场试验。

（7）大情字井黑帝庙油藏水平井 CO_2 吞吐能量补充矿场试验。

4 结论

（1）吉林油田拥有丰富的 CO_2 资源，可动储量具备形成年产 CO_2 200×10^4 t 的产能规模。适合气驱的油藏储量资源规模较大，混相和近混相驱石油资源占39.9%。

（2）气驱开发应以 CO_2 混相驱技术为引领，逐步拓宽气驱类型。要优化资源配置，在长岭气田周边优先开展 CO_2 驱，在距离 CO_2 气源较远的油田开展空气驱。当前矿场试验的总体部署安排是推广大情字井油田，加快伊通油田，试验英台油田。

（3）要根据 CO_2 注采平衡，统筹安排规模进度。规划2020年气驱年产油达到 80×10^4 t。在近几年安排重点矿场试验7项。吉林油田对气驱开发需求迫切，具有资源优势和技术基础，气驱开发的潜力很大，应用前景广阔。

参 考 文 献

[1] 李士伦，张正卿，冉新权，等. 注气提高石油采收率技术 [M]. 成都：四川科学技术出版社，2001.
[2] 郭平，苑志旺，廖广志. 注气驱油技术发展现状与启示 [J]. 天然气工业，2009，29（8）：92-96.
[3] 计秉玉，王凤兰，何应付. 对 CO_2 驱油过程中油气混相特征的再认识 [J]. 大庆石油地质与开发，2009，28（3）：103-109.
[4] 杨永智，沈平平，张云海，等. 中国 CO_2 提高石油采收率与地质埋存技术研究 [J]. 大庆石油地质与开发，2009，28（6）：262-267.
[5] 沈平平，廖新维. 二氧化碳地质埋存与提高石油采收率技术 [M]. 北京：石油工业出版社，2009.
[6] 李士伦，孙雷，郭平，等. 再论我国发展注气提高采收率技术 [J]. 天然气工业，2006，26（12）：30-34.
[7] 张旭，刘建仪，易洋. 注气提高采收率技术的挑战与发展—注空气低温氧化技术 [J]. 特种油气藏，2006，13（1）：6-9.
[8] 张旭，刘建仪，孙良田，等. 注空气低温氧化提高轻质油气藏采收率研究 [J]. 天然气工业，2004，24（4）：78-80.
[9] 曹维政，罗琳，张丽平，等. 特低渗透油藏注空气、N_2 室内实验研究 [J]. 大庆石油地质与开发，2008，27（2）：113-117.
[10] 王杰祥，徐国瑞，付志军，等. 注空气低温氧化驱油室内实验与油藏筛选标准 [J]. 油气地质与采收率，2008，15（1）：69-71.
[11] 张霞林，张义堂，吴永彬. 油藏注空气提高采收率开采技术 [J]. 西南石油学院学报，2007，29（6）：80-84.
[12] 李士伦，周守信，杜建芬，等. 国内外注气提高石油采收率技术回顾与展望 [J]. 油气地质与采收率，2002，9（2）：1-5.
[13] 赵彬彬，郭平，李闽，等. CO_2 吞吐增油机理及数值模拟研究，大庆石油地质与开发，2009，28（2）：17-120.
[14] 曹学良，郭平，杨学峰. 低渗透油藏注气提高采收率前景分析 [J]. 天然气工业，2006，26（3）：100-102.
[15] 李延军. 海拉尔油田注空气和 CO_2 驱试验效果及应用潜力 [J]. 大庆石油地质与开发，2013，32（5）：132-136.
[16] 刘尚奇，王伯军，孔凡忠. 低渗透油藏注空气提高采收率物理模拟实验 [J]. 大庆石油地质与开发，

2011, 30 (6)：143-147.

[17] 秦佳，周亚玲，王清华，等．注空气轻质原油低温氧化油气组分变化研究 [J]．大庆石油地质与开发，2008，27 (5)：111-113.

[18] 杨宝泉，任韶然，王杰祥，等．基于模糊评判的轻质油藏注空气筛选评价方法 [J]．大庆石油地质与开发，2010，29 (1)：119-123.

[19] 赵国忠，杨清彦，唐文锋，等．大庆长垣外围低渗透油田开发机理 [J]．大庆石油地质与开发，2009，28 (5)：126-133.

[20] 王粤川，魏刚，王昕．秦南凹陷秦皇岛 29-2 高含 CO_2 气藏天然气成因与成藏过程 [J]．大庆石油地质与开发，2013，32 (2)：22-26.

[21] 杜灵通．无机成因二氧化碳气藏研究进展 [J]．大庆石油地质与开发，2005，24 (2)：1-4.

[22] 李士伦，郭平，戴磊，等．发展注气提高采收率技术 [J]．西南石油学院学报，2000，22 (3)：41-45.

[23] 侯启军，邵明礼，李晶秋，等．松辽盆地南部深层天然气分布规律 [J]．大庆石油地质与开发，2009，28 (3)：1-5.

CO$_2$ 混相驱的可行性评价

汤 勇[1] 尹 鹏[1] 汪 勇[2] 孙 博[1] 侯大力[1]

(1. 西南石油大学"油气藏地质及开发工程"国家重点实验室;
2. 中国石油大学(北京)提高采收率研究院)

摘 要:CO$_2$ 驱是提高原油采收率和实现温室气体埋存的双赢举措,在中国具有广泛的应用前景。开展 CO$_2$ 混相驱可行性评价是 CO$_2$ 驱矿场实施的重要基础。基于文献和矿场经验,分析了 CO$_2$ 混相驱的不利因素,指出了 CO$_2$ 混相驱过程中需要重点考虑的问题以及发展趋势。提出 CO$_2$ 混相驱可行性评价主要包括:油藏筛选、室内实验评价、油藏数值模拟、经济风险评价、先导试验等环节。并对各个环节进行了分析。同时认为,可行性评价还需要考虑 CO$_2$ 注入过程中对储层、扰流或气窜、腐蚀、钙质与沥青质沉积等的影响。指出 CO$_2$ 混相驱成功的重点是保证油藏中较高的混相程度和 CO$_2$ 波及效率。在油藏方案和工程方案基础上,通过经济效益论证才能保障 CO$_2$ 混相驱的可行性。

关键词:CO$_2$ 混相驱;CO$_2$ 排放;温室效应;可行性评价;提高采收率

0 引言

注 CO$_2$ 提高采收率已成为世界范围内提高原油产量的重要手段之一[1,2],具有广泛的应用前景,比水驱具有更明显的技术优势[3-6]。利用 CO$_2$ 驱技术可以在常规技术的基础之上进一步提高石油采收率 10%~20%[1-4]。

中国适合注 CO$_2$ 混相驱开发的原油地质储量在 10.57 ×10^{8}t 以上,而新发现的原油地质储量大多是低渗透油藏,动用难度大,从国内外 EOR(Enhanced Oil Recovery)技术应用趋势来看,CO$_2$ 混相驱在提高低渗透油藏采收率上很有前景[5-8]。Weyburn 油田实施了目前世界上最大的 CO$_2$ 驱和埋存项目[1],目前我国吉林油田,草舍油田和中原濮城油田沙一下亚段特高含水油藏开展了 CO$_2$ 驱矿场试验,取得了较好经济效益和社会效益。更多的高含水后期和低渗透油田都在开展相关的可行性论证[9-11]。

基于 CO$_2$ 实际矿场经验[11-22],从油藏筛选、室内实验、油藏数值模拟方案优化、经济和风险评估和先导试验及其再评估等方面探讨了 CO$_2$ 混相驱可行性论证方法和必须考虑的因素。

1 CO$_2$ 混相驱可行性评价流程

1.1 油藏筛选

油藏筛选可以参考大量的文献和已有的矿场经验,主要集中在几个关键参数的筛选。

CO_2 混相驱油藏筛选的指导原则[1,2,7]如下。

（1）达到混相原则。一个油藏是否适合 CO_2 混相驱，取决于油藏的地层压力是否大于混相压力。CO_2 驱所需要的混相压力要比天然气或氮气所需的混相压力低很多，这是注 CO_2 的一个主要优点。

（2）比较有利的流度比原则。原油黏度较高的油藏不适合 CO_2 混相驱，一般建议将 $10\sim12$ mPa·s 作为粗略的筛选标准。

（3）避免严重非均质性原则。严重非均质油藏中应用 CO_2 混相驱可能导致 CO_2 过早气窜。因此，严重的层状非均质性或裂缝性油藏应避免采用 CO_2 混相驱。油藏的非均质性可以根据注水动态史以及地质、测井和油井不稳定试井等资料得到。

（4）最低含油饱和度原则。为了确保 CO_2 混相驱提高采收率的经济效益，在实施注 CO_2 之前剩余油饱和度应大于 25%。

根据矿场经验，主要考虑 5 个方面：油藏埋深、原油密度、油藏压力、油藏温度和原油组成[7,8]。CO_2 混相驱油藏筛选的准则主要有以下几点：

（1）水驱效果较好的油藏一般都适合 CO_2 驱。

（2）水驱采收率大于 20% 且小于 50% 的油藏较适宜。

（3）油藏埋深宜大于 762m，以满足油藏压力大于最小混相压力。

（4）油藏条件下，原油黏度小于 10mPa·s，密度小于 0.8895g/cm^3 的油藏较适宜。

（5）储层孔隙度大于 12%，且有效渗透率大于 10mD 时较理想。

油藏要尽可能地满足筛选准则，另外，油藏资料必须要充足，以满足各个环节的研究需要。在对油藏进行 CO_2 混相驱适应性研究时，可以结合油田特有的地质、油藏特征，提出针对整个油田的筛选准则。

1.2 混相能力及混相机理

混相能力评价主要是通过室内实验测试 CO_2 与原油之间的最小混相压力（Minimum Miscibility Pressure，MMP）评价 CO_2 的溶解、膨胀、降黏、油气界面张力降低以及 CO_2 与原油之间的组分传质作用[13,14,20]。

1.2.1 最小混相压力

最小混相压力的确定可分为理论法和实验室测试两种，理论计算主要有相平衡计算和经验公式预测；实验室测试方法主要有 PVT 仪测试法、升泡仪法和细管实验法[2]。目前普遍采用的是细管实验测试。大量研究发现，油藏温度、原油性质和注入气组成是影响 CO_2—原油体系 MMP 的 3 个主要因素[1,2,16]。油藏温度与 MMP 之间存在一定的关系：MMP 随着油藏温度的增加而增加。原油相对分子质量越大，MMP 越大。一些组分对 CO_2—原油的 MMP 有一定影响，H_2S、C_2—C_4 的临界温度要高于 CO_2，这会增加 CO_2 在油中的溶解性，从而降低 MMP。然而，N_2、O_2、C_1 的存在会使 MMP 变大，因为这些气体的临界温度高于 CO_2，会降低 CO_2 在油中的溶解性。相关研究表明，注入气的拟临界温度会影响 MMP，因此，可以将注入气拟临界温度作为混相能力评价的参数[16]。

1.2.2 CO_2 驱提高采收率机理

CO_2 驱提高采收率机理可以概括为：降低原油黏度，膨胀原油体积，蒸发原油中间烃组分（CO_2 是很强的蒸发剂），利用混相效应降低油气界面张力，溶解气驱，溶于水形成碳酸水，改善储层渗透率，溶于原油形成溶解气驱。可通过室内实验测试或相态软件包模拟计算进行评价[12-20]。

1）膨胀降黏作用

可通过 PVT 仪测试地层油注入 CO_2 体系的饱和压力、溶解气油比、饱和压力下的黏度和密度变化。从而分析 CO_2 与地层油的增容配伍性。饱和压力增加越平缓，说明配伍性越好。大量实验显示[1-2]，油混相时，CO_2 在原油中的溶解度较高（物质的量比例约 60%），使原油体积膨胀（约 30%），进而降低原油黏度（降低到原始状态的 20%～30%），改善原油的流动性。

2）多次接触混相的组分传质

通常在一定条件（温度、压力、原油组成等）下，CO_2 与原油多次接触后才能够达到混相，即多次接触混相。地层流体通过注 CO_2 不断抽提地层油的中间组分，CO_2 被加富，富化的 CO_2 混合气不断向前继续与地层油接触，不断被加富，最后形成富含中间烃的 CO_2 气，在前缘与地层油达到混相（向前接触混相或蒸发混相）。多次接触混相过程中，CO_2 与原油之间出现组分传质作用，形成一个驱替相过渡带[13,14]。多次接触混相可通过 PVT 仪和油气色谱仪测试向前接触和先后接触混相过程的组分传质作用，特别是 CO_2 抽提作用。也可通过相平衡模拟计算多次接触过程。

3）动态混相过程

在 CO_2—原油体系相态模拟和一维细管模拟基础上，可通过组分模型模拟多孔介质中注入 CO_2 后油气之间的组分变化、油气性质变化（黏度、密度及界面张力等）、饱和度变化。通过分析不同压力、注入烃孔隙体积条件下过渡带变化特征，可以更直观地认识到流动过程中油气蒸发—凝析传质过程和混相机理[1,2,14]。

1.3 驱油效率及渗流特征

油藏 CO_2 驱可行性研究中长岩心驱替实验是非常重要的基础资料[2,20]。一方面可以确定某具体油藏条件下的 CO_2 驱油效率；另一方面可以获取高温高压条件下的油气相渗曲线（常规方法难以准确测试，该资料在注气数值模拟中非常重要）。长岩心实验能反映实际的地层温度、流体组成、渗透率、孔隙结构和润湿性等性质，可以较真实地对比水驱后剩余油饱和度、注气压力、注气方式等对 CO_2 驱油效果的影响。当然实际应用时还需结合宏观的非均质性和井网。通过对 CO_2 驱长岩心实验的拟合，可为三维油藏数值模拟获取油气相对渗透率曲线，也可分析高温高压下油气渗流特征[21]。

1.4 注气油藏工程方案及注气参数优化

基于 CO_2—原油相态、最小混相压力和驱油效率室内实验，结合精细地质描述成果和油藏工程认识，利用组分油藏数值模拟技术，对一些重要的参数开展敏感性研究[1,2]。数值模型中能够反映出 CO_2 混相驱的一些特征：如油水中溶解性、混相性、膨胀、降黏以及相渗滞后作用等。

敏感性评价主要开展：注气层位、注气井网的选择、注气井距、注气井型的组合、注气井位置等。另外有必要对油藏非均质性、韵律、裂缝等进行敏感性研究，以评价 CO_2 的气窜和突破。CO_2 驱可以放大注采井距，解决低丰度油藏经济与技术井距的矛盾。

在完成主要参数的敏感性研究后，开展注气参数的优化，包括注气保持压力水平、注气速度、注气量、采油强度和注采比等。最后得到推荐的注气方案及指标预测结果。

1.5 经济和风险评估

注气技术方案完成后，需要开展经济和风险评估。大量的油田应用实例已经证明，CO_2混相驱是一种有效提高油田采收率的方法，但是其经济性仍然备受争议[1-8]。当今，高油价和低采收率增加了人们对 CO_2 驱油的关注，特别是大部分油田已经进入水驱开发后期并且是注水难以开发的低渗透油藏。CO_2 混相驱的经济性与工程的有效时间、注采井网井距、渗透率、油层厚度、埋深、成本（包括气源、注入设备、流体处理等）、税费、油价等有关。经济性评价基于客观油藏评价、可靠的实验结果、准确的油藏模型，贯穿于整个流程的各个环节。在前期，经济性评价可能就是简单地对比 CO_2 混相驱与水驱开发效果，在后期，风险评估能够定量研究 CO_2 混相驱过程存在的经济风险。风险定量化评估主要包括两个方面：可能的损失量与损失发生的可能性。损失量与损失发生的可能性同等重要，但是实际上在资料较少情况下，很难对风险作出准确的评估。

2 CO_2 混相驱评价需要考虑的问题

在可行性评价过程中，还需要考虑 CO_2 混相驱过程中的一些不利因素，例如绕流、钙质与沥青质沉积、腐蚀、水化物等。

2.1 绕流或气窜

向油藏中注气后，注入的气体不可能与所有原油都接触，这就是绕流或气窜。大部分绕流的发生归功于油藏的非均质性或重力分异作用。波及效率可以理解为评价驱替过程中可动油的动用程度的一个指标。在 CO_2 混相驱过程中，一方面，CO_2—原油的混相效应会增加微观驱替效率。另一方面，黏性指进和重力分异作用会降低体积波及效率，油藏非均质性、裂缝、层理等会加重绕流效应，降低驱替效率。为了消除或减弱气体的突进，可以加入其他物质来控制流度，如水、聚合物、表面活性剂等[10,11,21]。此外，在 CO_2 驱替过程中，当 CO_2 与原油接触，原油性质会发生改变。原油的黏度降低，同时，地层水的黏度增加，这将大大提高油相的流动能力。

2.1.1 水气交替注入

为了降低 CO_2 混相驱过程的流度比，水气交替在 CO_2 驱过程中已经被广泛应用。由于水的注入增加了储层的含水饱和度，注入溶剂的相对渗透率会降低，因此注入剂的流度会降低，注入溶剂的驱替效率会有所增加。相比只注 CO_2 而言，水气交替驱替过程更趋稳定。但是，水气交替驱仍然有一些缺点，例如，在油藏中水和 CO_2 分配不均，某些区域水相饱和度太高，导致微观驱替效率降低。由于 CO_2 无法波及到含水饱和度高的区域，因此这些区域会有大量剩余油存在。

依据矿场经验和文献调研[1-11]，大部分水气交替驱均应用于陆上油田。影响水气交替驱可行性的主要参数有：油藏非均质性、岩石类型、流体饱和度和性质、注入气、水气交替注入比、混相能力和重力。

近几年，国外将三相相渗滞后理论的研究成果应用到气水交替驱动态预测中并取得了重要进展[15]。也有学者[19]研究了 CO_2 在地层水中溶解对驱油的影响，发现注气初期，考虑 CO_2 溶解时由于有效 CO_2 的损失，原油采出程度低，气突破时间更晚。

2.1.2 化学剂

CO_2 驱替过程中加入化学剂可以控制注入气的流度。化学流度控制剂必须满足几个条件，首先，化学流度控制剂不能损害环境，不能与其他工艺材料发生反应，不能被岩石吸附。其次，其还要承受住一定的压力和温度，承受压力一般在 1000 ~ 5000psi（1psi = 6.895kPa）之间，温度在 20~120℃ 之间。再次，油藏原油中含有各种有机表面活性剂，地层水中含有大量的钙、镁、钠、碳酸根和氯离子等，这些都需要考虑。最后，化学流度控制剂在经济上也必须可行。尽管化学流度控制剂的要求很多，但是，其对 CO_2 驱效果可能有较大的影响，因此，化学流度控制剂可以降低 CO_2 混相驱的风险。

2.2 腐蚀

干燥的纯 CO_2 对钢材基本上没有腐蚀性，但如果处于 CO_2/H_2O 的系统中，钢材的腐蚀问题是不可忽视的。当 CO_2 气体遇到水时，部分 CO_2 溶解到水中并形成碳酸，CO_2 在水中的溶解度主要取决于 CO_2 分压和温度。影响 CO_2 腐蚀的主要因素为：温度、CO_2 分压、流速等。详细的方案设计、合理的工艺措施和即时的腐蚀检测和监测可以有效地控制 CO_2 腐蚀[1,2]。在防腐方面，可以借鉴矿场经验，其中最常用的方法就是将 CO_2 和水严格分隔开来。另外，为管线和设备加上防护层和采用不锈钢材质的管线也是很有必要的，也可以在生产井或注入井中添加缓蚀剂。

2.3 钙质沉积

CO_2 驱替过程中，大量的 CO_2 存在于气相、油相以及水中，这可能会引起钙质沉积，进而结垢，在产出的水中钙质沉积尤为明显[19]。钙质沉积有几方面的原因，首先，由于压力太大导致 CO_2 在水中的溶解度增大，溶解的 CO_2 部分与水反应生成碳酸氢盐，且地层水的 pH 值会降低 2~3，酸性的地层水溶解了石灰岩中的钙，使得水中钙的浓度增加。其次，当地层水从地下流到井口，进入地面管线、分离器，最终进入注水站，期间压力会逐级降低，压力降低会减小钙的溶解度，进而引起钙质沉积。同样地，在该过程中由于 CO_2 的逸出，水的 pH 值会有所增加，这也会促使钙质沉积。

2.4 沥青质沉积

在大部分注 CO_2 驱过程中，沥青质沉积是一个必须要考虑的问题[17,18]。沥青质沉积可以改变储层的润湿性，影响注水能力，还可能引起储层伤害和井筒堵塞。沥青质是极性的、多芳基、高分子的碳氢化合物，且不溶于低级正构烷烃（nC_5—nC_8）。沥青质是以比较稳定的分散胶体形式存在于原油中的，胶质附着于沥青质的表面。如果原油中胶质含量不足，则沥青质会进一步缔合形成更大的分子团，从而产生沥青质的絮凝和沉淀。沥青质与胶质含量的比决定了是否会出现沥青质沉积。在 CO_2 驱替过程中，原油的沥青质与胶质的比被改变了，因而出现沥青质沉积。在静态 pVT 实验中，可以观测到沥青质的絮凝，因此在岩心驱替实验中可能会发生沥青质沉积。

2.5 水驱对油藏的影响

对于油藏后期 EOR 而言，前期的水驱效果是决定 EOR 成功与否的关键因素，水驱效果差的油藏通常 CO_2 混相驱效果也较差，由于水驱效果差通常是由于储层非均质性和分布的不连续性导致的，而这些因素也同样会影响 CO_2 驱过程。然而，没有经历水驱开发的油藏也可能适合 CO_2 混相驱，只不过其不确定性会有所增加。在非均质油藏中，先前水驱形成的窜流通道会加剧 CO_2 的窜流和指进。长期水驱过后，油藏性质会发生变化，高的含水率、窜流通道、低含油饱和度等是其主要特征，油藏非均质性不单受静态地质因素影响，更多的与动态过程有关[19-22]。

3 结论

（1）由于 CO_2 混相驱的复杂性，针对特定油藏，需要经过充分细致的研究和先导试验来评价 CO_2 混相驱可行性。主要包括油藏筛选、室内实验评价、油藏数值模拟、经济风险评价、先导试验。

（2）油藏中的混相程度和流度控制对 CO_2 混相驱有着重要的影响。同时，可行性评价时还需要考虑 CO_2 对储层、绕流、腐蚀、钙质与沥青质沉积等的影响。

（3）CO_2 混相驱重点是保持油藏中较高的混相程度和波及效率。在油藏方案和工程方案基础上，通过经济效益论证才能保障 CO_2 混相驱的可行性。

参 考 文 献

[1] 沈平平，廖新维. 二氧化碳地质埋存与提高石油采收率技术 [M]. 北京：石油工业出版社，2009，128-165.

[2] 李士伦，张正卿，冉新权，等. 注气提高石油采收率技术 [M]. 成都：四川科学技术出版社，2001，85-99.

[3] Gao P, Towler B, Jiang H. Feasibility investigation of CO_2 miscible flooding in south slattery minnelusa reservoir, Wyoming [J]. SPE133598, 2010, 27-29.

[4] 孙扬，杜志敏，孙雷，等. CO_2 的埋存与提高天然气采收率的相行为 [J]. 天然气工业，2012，32 (5)：39-42.

[5] 胡滨，胡文瑞，李秀生，等. 老油田二次开发与 CO_2 驱油技术研究 [J]. 新疆石油地质，2013，34 (4)：436-440.

[6] 李中超，杜利，王进安. 水驱废弃油藏注二氧化碳驱室内试验研究 [J]. 石油天然气学报，2012，34 (4)：131-135.

[7] 郑云川. 注气提高采收率候选油藏筛选方法及其应用研究 [D]. 南充：西南石油学院，2003.

[8] 曹学良，郭平，杨学峰，等. 低渗透油藏注气提高采收率前景分析 [J]. 天然气工业，2006，26 (3)：100-102.

[9] 秦积舜，张可，陈兴隆. 高含水后 CO_2 驱油机理的探讨. 石油学报，2010，31 (5)：797~800.

[10] 关振良，谢丛姣，齐冉，等. 二氧化碳驱提高石油采收率数值模拟研究 [J]. 天然气工业，2007，27 (4)：142-144.

[11] Yu Hongwei, Song Xinmin, Yang Siyu, et al. Experimental and numerical simulation study on single layer injectivity for CO_2 flooding in low permeability oil reservoir [C]. SPE 144042, 2011.

[12] 孙扬，杜志敏，孙雷，等. 注 CO_2 前置段塞+N_2 顶替提高采收率机理 [J]. 西南石油大学学报（自

然科学版），2012，34（3）：89-97.

[13] 苏畅，孙雷，李士伦. CO_2混相驱多级接触过程机理研究［J］. 西南石油学院学报，2001，23（2）：33-36.

[14] 汤勇，孙雷，周涌沂，等. 注气混相驱机理评价方法［J］. 新疆石油地质，2004，25（4）：414-417.

[15] Ayirala S, Rao D. Miscibility determination from gas-oil interfacial tension and P-R equation of state［J］. The Canadian Journal of Chemical Engineering Volume 85, 2007, 302-312.

[16] Emera M, Sarma H. A reliable correlation to predict the change in minimum miscibility pressure when CO_2 is diluted with other gases［J］. Journal SPE Reservoir Evaluation & Engineering, 2006, 366-373.

[17] 黄磊，贾英，全一平，等. CO_2驱替沥青质原油长岩芯实验及数值模拟［J］. 西南石油大学学报：自然科学版，2012，34（4）：135-140.

[18] Choiri M, A A Hamouda. Study of CO_2 effect on asphaltene precipitation and compositional simulation of asphaltenic oil reservoir［C］. SPE 141329, 2011.

[19] 汤勇，杜志敏，孙雷. CO_2 在地层水中溶解对驱油过程的影响［J］. 石油学报，2011，32（2）：311-314.

[20] Ghedan S. Global laboratory experience of CO_2-EOR flooding［C］. SPE 125581, 2009.

[21] Sohrabi M, Danesh A, Tehrani D H, et al. Microscopic mechanisms of oil recovery by near-miscible gas injection［J］. Transport in Porous Media, 2008, 72（3）：351-367.

[22] Wei L. Sequential coupling of geochemical reactions with reservoir simulations for waterflood and EOR studies［C］. SPE 138037, 2012.

Assessment of Oil Reservoirs Suitable for CO₂ Flooding in Mature Oil Reservoirs, Changqing Oilfield, China

Assessment of Oil Reservoirs Suitable for CO_2 Flooding in Mature Oil Reservoirs, Changqing Oilfield, China

iel Diaz put forward 8 parameters[9]. Before 1990s, countries compare the parameters according to these standards, but the disadvantage is voting the reservoir which has the potential to inject gas[3,10]. In 1992 the Department of Energy entrusted Louisiana State University to study the comprehensive screening method of the reservoir which had finished water flooding and began to CO_2 flooding. The influence of the various parameters on the reservoir recovery is integral. So in order to quantitative evaluation, the study should be based on the quantification of various parameters and determine the influence of the parameters[11-14]. So adopt the Fuzzy mathematics and the recent developing Fuzzy Analytic Hierarchy Process, quantize the evaluation parameters and put forward a comprehensive theory and method to screen all the reservoirs in a oildom for CO_2 flooding.

1　The Design of The Typical Model for CO_2 Flooding

After analyzing and screening the parameters, choose 12 parameters for the typical model in Table 1. Every parameter adopts single factor analysis means. There are 69 projects.

Table 1　Parameter Designing of Typical Model

No	Factor	The level of the parameter	Remark
1	Stratigraphic dip (°)	5, 10, 15, 25, 35	Single factor analysis
2	Formation pressure (MPa)	3, 5, 7, 9, 12, 18, 25	Single factor analysis
3	Effective pay thickness (m)	5, 7.5, 10, 13, 15, 20, 30	Single factor analysis
4	Average reservoir permeability (mD)	0.1, 0.5, 1.5, 10, 50	Single factor analysis
5	Interlayer growth, K_V/K_H	0.001, 0.01, 0.1, 0.3, 0.5	Single factor analysis
6	The initial oil saturation (%)	30, 40, 50, 60, 65	Single factor analysis
7	Permeability direction development, K_Y/K_X	5, 10, 20, 50, 100	Single factor analysis
8	Well spacing (m)	100, 150, 250, 350, 500	Single factor analysis
9	Well pattern	Five spot, seven spot, inverted seven spot, nine spot, inverted nine spot	Single factor analysis
10	Viscosity and density of the fluid	By adjusting the five components to complete	Single factor analysis
11	Variable coefficient	0.1, 0.2, 0.35, 0.5, 0.7	Single factor analysis
12	Variable coefficient	Positive rhythm, reverse rhythm	Single factor analysis

The typical model is nine-spot pattern. Oil wells and water well is in the same horizon, Oil wells locate in the four sides and four corners of the square, while water wells locate in the center of the square, while contains 8 oil wells and 1 gas injection well.

The default parameters of the typical model are as following: (1) Horizontal permeability is 1.5mD, variation coefficient is 0.5 and reverse rhythm. The horizontal permeability of different horizons is respectively 0.50, 0.88, 0.98, 1.40, 1.43, 1.54, 1.59, 1.72, 2.04, 2.14, 2.30, 2.72, 3.14; (2) the permeability near wellbore (the distance to wellbore is about 50m) is 10mD; (3) vertical permeability is the same as the horizontal permeability; (4) the effective thickness of the horizon is 12m; (5) the porosity is 11%; (6) the well spacing is 250m; (7) the initial

523

oil saturation is 60%, and the reservoir temperature is 49.8℃; (8) the flowing bottomhole pressure of oil well is 3.5MPa, while the liquid production of all the reservoir is 5.0m³/d; (9) the maximum pressure of gas injection wells is less than 1.8 times of the reservoir pressure; (10) the depth of the reservoir is 1500m and the corresponding pressure is 15MPa (Table 1). Choose the component model of CMG as the simulation software.

2　The Establishment of The Sensitivity Model for CO_2 and The Analysis of The Reference model

2.1　The Model Selection

When select the model, the following factors should be considered: reservoir complexity (anisotropy, hierarchy intercalation and so on), well placement, driving mechanism of liquid, phase number, phase behavior and the reliability and integrity of the data.

Based on the development of Changqing Oilfield, choose the GEM compositional model of CMG to simulate the reservoir. According to the feature of sedimentary rhythm and early geological research, the model can be divided 13 horizons vertically and the thickness of every horizon is 1m. The model is homogeneous and equal thickness. The model adopts orthogonal grid system. The space of the direction X and Y is same and every space is 15m. The direction X contains 41 grids and the direction Y contains 35 grids. The total spots of the model: 41×35×13=18655. The geological reserves are 25.5×10⁴t (Fig.1).

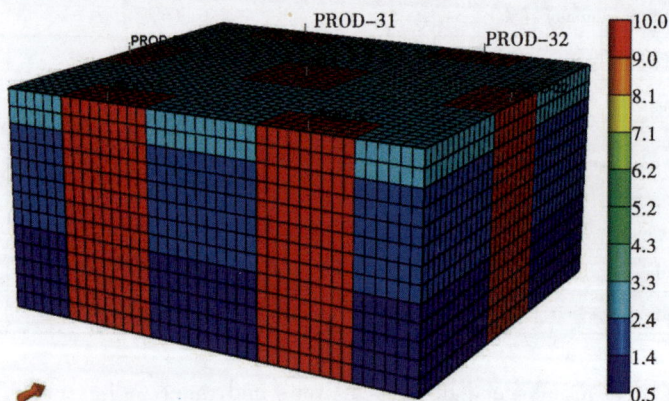

Fig. 1　Permeability of the Model

2.2　Physical Parameters

(1) The physical properties of the reservoir and liquid.

The oil density at surface condition is 0.78g/cm³, the initial rock compressibility is 5.351× 10^{-4}MPa^{-1}.

(2) The oil—water relative permeability (Fig. 2).

524

(3) Relative permeability of oil and gas (Fig. 3).

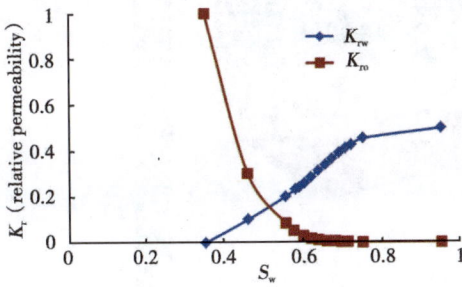

Fig. 2　Oil—Water Relative Permeability Curve

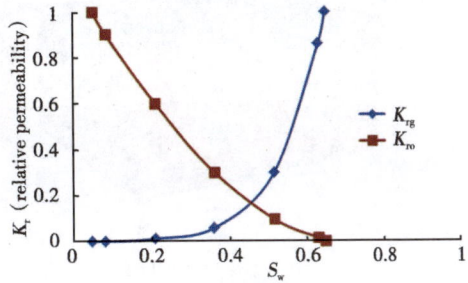

Fig. 3　Oil and Gas Relative Permeability Curve

2.3　Initial Model

The porosity of the model is 11%, average oil saturation is 60%, and the initial temperature is 49.8℃. The molar fraction of every component in oil is as following: CO_2 is 0.0082, C_1 is 0.1990, C_{2+} is 0.1954, C_{4+} is 0.1281, C_{9+} is 0.2760, C_{20+} is 0.1934.

2.4　Dynamic Model

There are 13 horizons perforated. The well spacing is 250m. The maximum gas injection pressure is less than 1.8 times of reservoir pressure and the flow pressure of oil well is 3MPa.

3　Factor Sensitivity Analysis

The factor sensitivity analysis bases on the model with default parameters. Then change the parameters, calculate and analyze the results. There are 12 simulate analysis of sensitivity parameters. Here take an example of interlayer growth to introduce the method.

Change the ratio between vertical permeability and horizontal permeability, which is the value of K_V/K_H. They are respectively 0.001, 0.01, 0.1, 0.3, 0.5. The results are as following.

3.1　Viscosity Distribution

The three—dimensional diagram of viscosity (the time when the gas—oil ratio of all the oil wells is more than $1000m^3/m^3$) is as in Fig. 4.

The oil saturation distribution at the end of displacement (as shown in Fig. 5).

It is observed from the two above diagram that the more degree of interlayer development can prevent CO_2 from moving upward staightly and not reaching the middle and lower horizon that make the oil saturation decreases sharply. Therefore the interlayer development can enhance the swept volume of CO_2, then the sweep efficiency and recovery raise.

3.2　Producing Gas—Oil Ratio and Recovery Ratio

According to the time when the gas—oil ratio of every producing well reach $1000m^3/m^3$, find

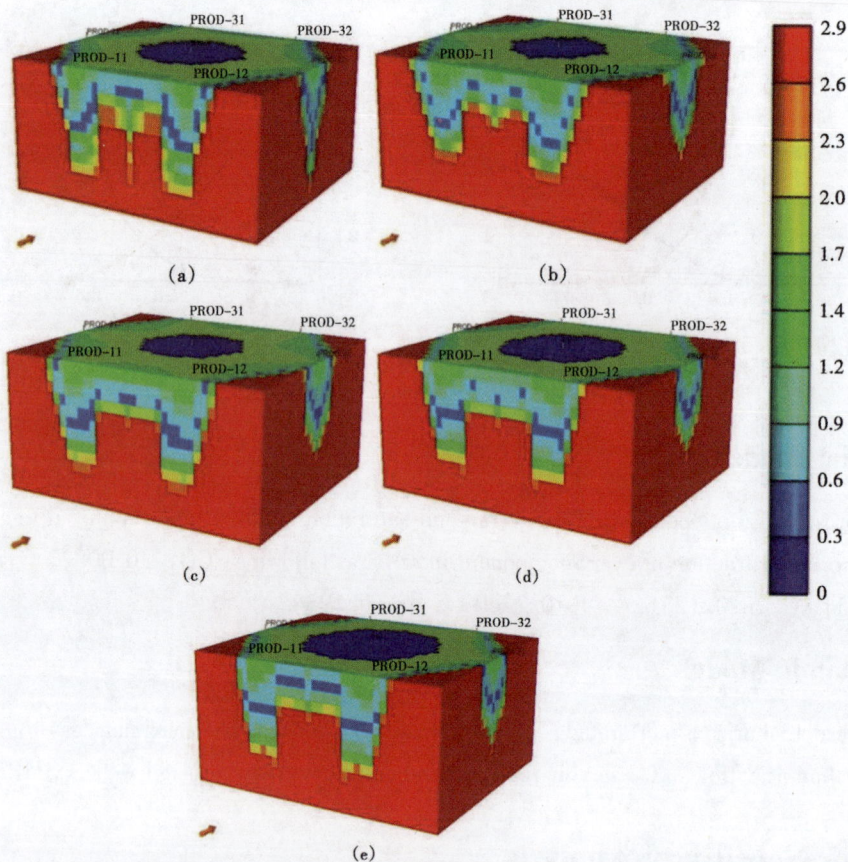

Fig. 4   the Three-dimensional Distribution of Viscosity When Interlayer with Different Development Degree
(a) Three-dimensional distribution of viscosity when K_V/K_H is 0. 001; (b) three-dimensional distribution of viscosity when K_V/K_H is 0. 01 ; (c) three-dimensional distribution of viscosity when K_V/K_H is 0. 1; (d) three-dimensional distribution of viscosity when K_V/K_H is 0. 3; (e) three-dimensional distribution of viscosity when K_V/K_H is 0. 5

out the the corresponding cumulative prodution of every component. In the same way, record the cumulative prodution of eight wells when the gas—oil ratio reaches $1000m^3/m^3$, the datum is as in Table 2.

Table 2   The Summary Table of Production Data under Different Interlayer Conditions

Cumulative output	C_1 (10^4t)	C_{2+} (10^4t)	C_{4+} (10^4t)	C_{9+} (10^4t)	C_{20+} (10^4t)	Total (10^4t)
$K_V/K_H = 0.001$	0. 182	0. 750	0. 610	2. 502	4. 607	8. 651
$K_V/K_H = 0.01$	0. 151	0. 622	0. 505	2. 076	3. 809	7. 163
$K_V/K_H = 0.1$	0. 147	0. 608	0. 497	2. 028	3. 786	7. 066
$K_V/K_H = 0.3$	0. 133	0. 554	0. 455	1. 844	3. 495	6. 480
$K_V/K_H = 0.5$	0. 125	0. 523	0. 431	1. 741	3. 320	6. 139

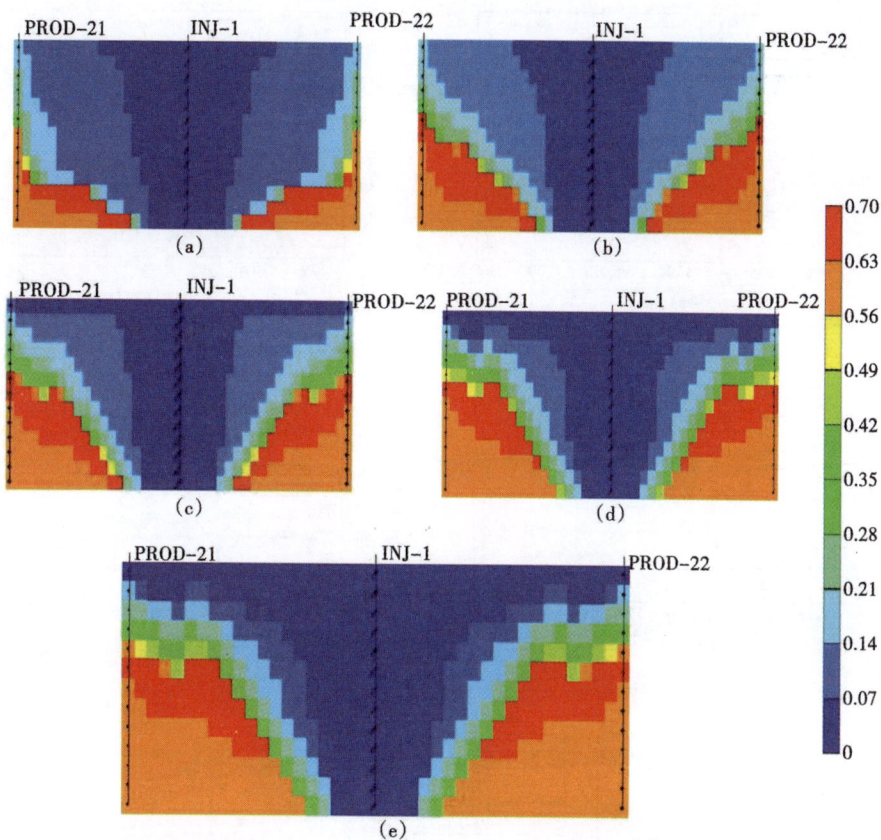

Fig. 5   the Two-dimensional Section of Saturation When Interlayer with Different Development Degree
(a) Two-dimensional section of saturation when K_V/K_H is 0.001; (b) two-dimensional section of saturation when K_V/K_H is 0.001; (c) two-dimensional section of saturation when K_V/K_H is 0.001; (d) two-dimensional section of saturation when K_V/K_H is 0.001; (e) two-dimensional section of saturation when K_V/K_H is 0.001

The geological reserves under the condition of different interlayer can be found out in calculated file. Hence the recovery can be calculated. The results are as in Fig. 7 and Table 3.

Table 3   The Summary Table of Production Data under Different Interlayer Conditions

K_V/K_H	Cumulative output (10^4t)	Geologic reserve (10^4t)	Recovery	The cumulative amount of injected CO_2 (10^4t)	The cumulative gas—oil ratio (10^4t/10^4t)
0.001	8.651	21.875	0.3953	10.224	0.8461
0.01	7.163	21.875	0.3274	8.461	0.8466
0.1	7.066	21.875	0.3230	8.121	0.8701
0.3	6.480	21.875	0.2962	7.384	0.8776
0.5	6.139	21.875	0.2806	6.968	0.8810

527

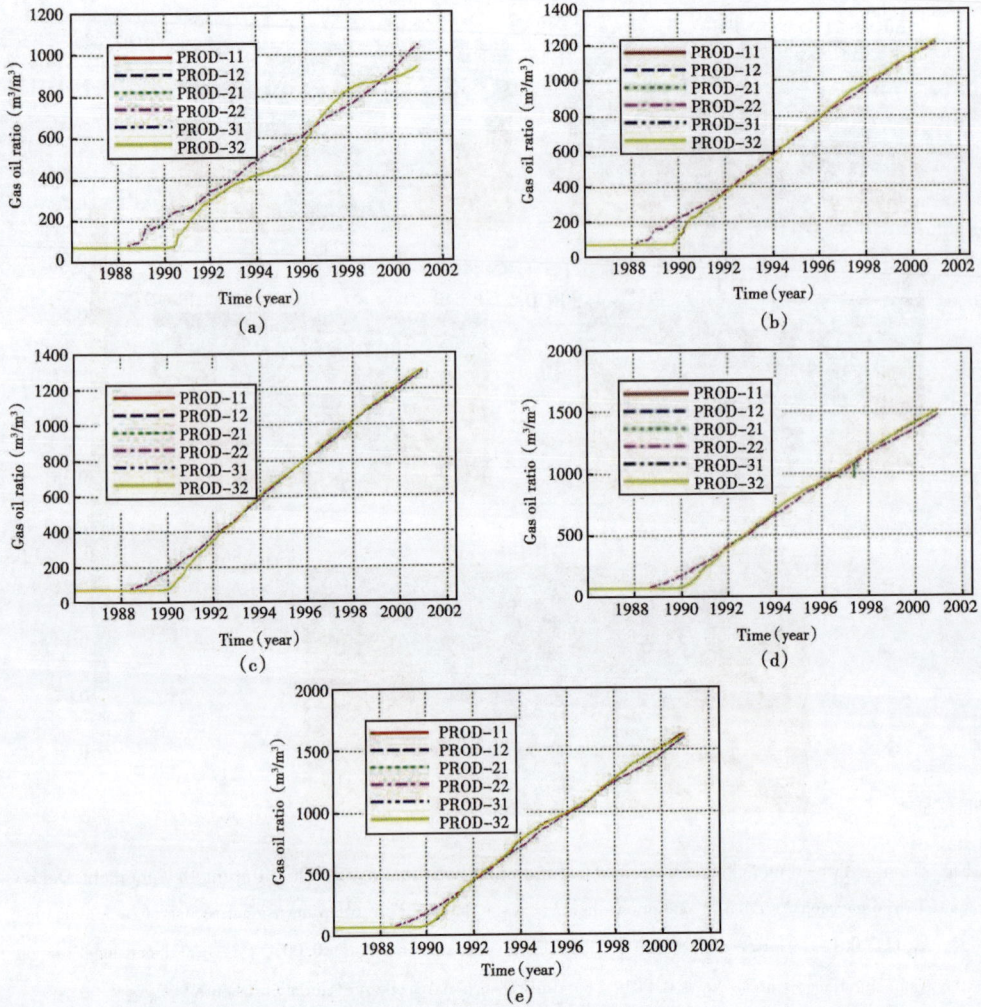

Fig. 6 the Curves of the Producing Gas—Oil Ratio When with Different K_V/K_H

(a) The curves of the producing gas–oil ratio when K_V/K_H is 0. 001; (b) the curves of the producing gas–oil ratio when K_V/K_H is 0. 01; (c) the curves of the producing gas–oil ratio when K_V/K_H is 0. 001; (d) The curves of the producing gas–oil ratio when K_V/K_H is 0. 001; (e) the curves of the producing gas–oil ratio when K_V/K_H is 0. 5

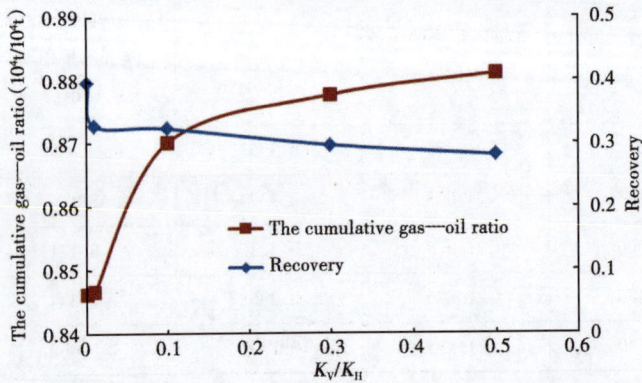

Fig. 7 the Curves of the Cumulative Gas—Oil Ratio under Different K_V/K_H

4 Sensitivity Analysis

According to the above results from reservoir numerical simulation, consider every factor's influence on the recovery and gas-oil ratio and find out whether the factors are sensitive as Table 4.

Table 4   Factors Sensitivity and Screening Index of CO$_2$ Flooding of Low Permeability Reservoir in Changqing Oilfield

No	Factors	Whether or not sensitive factors	Comprehend
1	Stratigraphic dip (°)	Relatively insensitive	Stratigraphic dip is conducive to recovery
2	Formation pressure (MPa)	Very sensitive	The Increase of pressure is conducive to recovery
3	Effective pay thickness (m)	Very sensitive	The increase of the thickness is not conducive to oil recovery
4	Average reservoir permeability (mD)	Sensitive	The average permeability increased first is conducive to recovery then against
5	Interlayer growth, K_V/K_H	Sensitive	Interlayer development is conducive to recovery
6	Oil saturation (%)	Very sensitive	The increase of initial oil saturation is conducive to recovery
7	Permeability direction development, K_Y/K_X	Whether or not sensitive factors	Recovery is not sensitive, the cumulative gas—oil ratio is sensitive: permeability direction development is slightly conducive to recovery
8	Well spacing (m)	Sensitive	The increase of Well spacing is not conducive to recovery
9	Well pattern	Sensitive	The recovery of five spot, inverted seven spot and inverted nine spot well pattern is higher
10	Viscosity and density of the fluid	Sensitive	The increases of Viscosity and density are not conducive to recovery
11	Variable coefficient	Whether or not sensitive factors	Recovery is sensitive, I the cumulative gas—oil ratio is not sensitive: The variation coefficient increased is not conducive to recovery, and it has Little effect on the cumulative gas—oil ratio
12	Sedimentary rhythm	Very sensitive	Positive rhythm is more favorable to the reverse rhythm of CO$_2$ flooding recovery efficiency

5 Screening Standard for CO$_2$ Flooding of Changqing Oilfield

There are many factors influencing CO$_2$ flooding of a reservoir. If a reservior has too severe fracture or active edge and bottom water, it is not fit for CO$_2$ flooding. Based on the results from sensitivity analysis, recovery trend, cumulative gas-oil ratio and other factors, some screening standard for CO$_2$ flooding in low permeability reservoir can be obtained as Table 4.

6 Suitability Rank for CO_2 Flooding in Changqing Oilfield

Sort all the parameters that influence CO_2 flooding as geological factors, liquid property and development indexes. Rank the parameters following importance and establish evaluation index system, the hierarchical structure of construction is in Table 5.

Table 5　Hierarchy Structure of Screening Index for CO_2 Flooding Reservoir in Changqing Oilfield

	First class index	Second class index
Changqing oilfield CO_2 drive reservoir comprehensive evaluation	Geological factors	Sedimentary rhythm
		Oil saturation
		Effective thickness of oil reservoir
		Formation pressure
		Average reservoir permeability
		Interlayer growth
		Variable coefficient
		Stratigraphic dip
		Permeability direction development
	Fluid property	Density of the fluid
		Viscosity of the fluid
	Developmental factors	Well pattern
		Well spacing

Adopt the principle of hierarchical analysis and calculate the weight of every parameter in Table 6.

Table 6　Screening Index Weights of CO_2-flooding Reservoir in Changqing Oilfield

First class index	The weigh of first class index	Second class index	The weigh of second class index
Geological factor	0.65	Sedimentary rhythm	0.16
		Oil saturation	0.16
		Effective thickness of oil reservoir	0.16
		Formation pressure	0.16
		Average reservoir permeability	0.10
		Interlayer growth	0.10
		Variable coefficient	0.08
		Stratigraphic dip	0.04
		Permeability direction development	0.04
Fluid property	0.25	Density of the fluid	0.5
		Viscosity of the fluid	0.5
Developmental factors	0.1	Well pattern	0.5
		Well spacing	0.5

As the effective thickness, the pressure in reservoir depth, average permeability and liquid viscosity and density given by Changqing Oilfield is relatively perfect, index weights should be distributed to these 6 indexes. The final weights after distribution is in Table 7. When evaluate other reservoirs, weight can be distributed in this way according to the practical datum of oilfield.

Table 7 Screening Index Weights of CO_2-flooding Reservoir in Changqing Oilfield

Index	The index weight
Effective thickness in oil reservoir	0.219
Formation pressure	0.190
Average reservoir permeability	0.157
The initial oil saturation	0.193
Density of the fluid	0.120
Viscosity of the fluid	0.120

Normalize the geological data. If there are l reservoirs, every reservoir has m indexes. Item j of reservoir k is $X'_{k,j}$.

$$ X_{k,j} = \frac{|X'_{k,j} - X^*_{w,j}|}{|X^*_{o,j} - X^*_{w,j}|} \tag{1} $$

where, the optimal value $Y^*_{o,j} = Y^*_{n,j}$, the corresponding index value is $X^*_{o,j} = X^*_{n,j}$; the worst value $Y^*_{w,j} = Y^*_{1,j}$, the corresponding index value is $X^*_{w,j} = X^*_{1,j}$.

Evaluate the recovery and accumuatice gas—water ratio. The optimal value and the worse value of every parameter adopted by normalization are in Table 8.

Table 8 The Optimal Value and Worst Value Adopted to Normalized Index of Each Reservoir

Index	The worst value	The optimal value
Effective thickness of oil reservoir (m)	30	1
Formation pressure (MPa)	3	25
Average reservoir permeability (mD)	0.1	1.5
	100	1.5
Oil saturation (%)	30	65
Density of the fluid (kg/m³)	1000	700
Viscosity of the fluid (mPa·s)	1000	1

As the distance between the optimal value and the worse value of liquid viscosity is too far, adopt the logarithmic formula when normalize this index and label it as item 6. Item 6 of k reservoir is $X'_{k,6}$.

$$ X'_{k,6} = \frac{|\lg X'_{k,6} - \lg X^*_{w,6}|}{|\lg X'_{o,6} - \lg X^*_{w,6}|} \tag{2} $$

where, the optimal value $Y^*_{o,j} = Y^*_{n,j}$, the corresponding index value is $X^*_{o,j} = X^*_{n,j}$; the worst value $Y^*_{w,j}$

$= Y_{1,j}^*$, the corresponding index value is $X_{w,j}^* = Y_{1,j}^*$.

The evaluating value k is：

$$T_k = \Sigma X_{k,j} \times P_j \tag{3}$$

where, $X_{k,j}$ is geological data by normalizing；P_j is the weighted value of item j calculated by analytic hierarchy process.

Adopt the above theory and results and evaluate the reservoir suitability of all the blocks for CO_2 flooding in Changqing Oilfield. Rank the evaluating value calculated of all the blocks and evaluate the protential for CO_2 flooding by limit the value as 0. 52. The evaluation results are in Table 9 and Table 10.

Table 9　the Value of CO_2 Flooding in Low Permeability Reservoir in Changqing Oilfield

Oil field name	The name of the area	Horizon	Geologic reserves (10^4t)	Evaluation value
DB	FX	Chang 6	207	0. 858
WQ	WWZ	Chang 4+5	2200	0. 832
WQ	XZ	Chang 8	160	0. 818
GGY	DBLQ	Chang 6	4502. 8	0. 457
NNW	SSL	Chang 4+5	1225	0. 447

Table 10　Potential Evaluation of CO_2 Flooding in Low Permeability Reservoir in Changqing Oilfield

Oil field name	The total number of reservoirs in the area	Number of reservoir suitable for CO_2 flooding reservoir	Number of reservoir not suitable for CO_2 flooding reservoir	Number proportion of reservoir suitable for CO_2 flooding	Total geological reserves (10^4t)	Reserves suitable for CO_2 flooding (10^4t)	Reserves not suitable for CO_2 flooding (10^4t)	Number proportion of reserves suitable for CO_2 flooding
XZC	13	12	1	0. 92	10685	10495	190	0. 98
JB	15	14	1	0. 9333	12340. 5	12071	269. 5	0. 97816
WYB	16	11	5	0. 6875	13923. 9	11141. 7	2782	0. 80019
ZB	4	3	1	0. 75	4981. 9	3872. 9	1109	0. 78
total	178	150	28	0. 84	219138	176274. 1	42864	0. 80

From the above table it can be seen that 86 percent blocks are suitful for CO_2 flooding in Changqing Oilfield, whose reserves occupy 82 percent geological reserves. Therefore CO_2 flooding have a good prospect in Changqing Oilfield and may bring immeasurable economic benefits.

7　Conclusions

（1）Analysis the twelve parameters' influence on the viscosity distribution with the numerical simulation, final gas—oil ratio and oil recovery during the CO_2 flooding. Hence the varying sensitivity to CO_2 flooding of every parameter can be arrived, and put forward the evaluation system of CO_2

flooding.

(2) Based on the theory of the fuzzy comprehensive evaluation, set up the grading evaluation of the influence parameters, determine the weights of every parameter, propose the concrete steps to calculate the screening parameters of CO_2 flooding.

(3) Filter and evaluate the parameters of 253 reservoirs in Changqing Oilfield during CO_2 flooding, which can guide the development of CO_2 flooding in Changqing Oilfield.

References

[1] Wood D J, L W Lake. A screening model for CO_2 flooding and storage in Gulf Coast reservoirs based on dimensionless groups [C]. SPE/DOE Symposium on Improved Oil Recovery, Tulsa, Oklahoma, USA, Society of Petroleum Engineers, 2006.

[2] Wo S, P Yin. Simulation evaluation of gravity stable CO_2 flooding in the muddy reservoir at Grieve Field, Wyoming [C]. SPE/DOE Symposium on Improved Oil Recovery, Tulsa, Oklahoma, USA, Society of Petroleum Engineers, 2008.

[3] Wood D, L Lake. A screening model for CO_2 flooding and storage in Gulf Coast reservoirs based on dimensionless groups [J]. SPE Reservoir Evaluation & Engineering, 2008, 11 (3): 513-520.

[4] Wo S, L D Whitman. Estimates of potential CO_2 demand for CO_2 EOR in Wyoming basins [C]. SPE Rocky Mountain Petroleum Technology Conference, Denver, Colorado, Society of Petroleum Engineers, 2009.

[5] Bang V. A New Screening Model for Gas and Water Based EOR Processes [C]. 2013 SPE Enhanced Oil Recovery Conference, Kuala Lumpur, Malaysia, Society of Petroleum Engineers.

[6] Rivas O, S Embid. Ranking reservoirs for carbon dioxide flooding processes [C]. SPE Advanced Technology Series 2, 1994, (1): 95-103.

[7] Babadagli T. Optimization of CO_2 injection for sequestration/enhanced oil recovery and current status in Canada [J]. 2006, 65: 261-270.

[8] Núñez-López, V, M H Holtz. Quick-look assessments to identify optimal CO_2 EOR storage sites [J]. Environmental Geology, 2008, 54 (8): 1695-1706.

[9] Dickson J, A Leahy-Dios. Development of improved hydrocarbon recovery screening methodologies [R]. SPE Improved Oil Recovery Symposium, 2010.

[10] Patil S B, S L Patil. Screening of oil pools on Alaskan North Slope and phase behavior study of viscous oil and CO_2 system in conjunction with CO_2 sequestration [J]. Petroleum Science and Technology, 2008, 26 (7-8): 844-855.

[11] Zhang L, S R Ren. Assessment of CO_2 storage capacity in oil reservoirs associated with large lateral/underlying aquifers: Case studies from China [J]. International Journal of Greenhouse Gas Control, 2011, 5 (4): 1016-1021.

[12] Liao X, C GAO. Assessment of CO_2 EOR and its geo-storage potential in mature oil reservoirs, Changqing Oil Field, China [C]. Carbon Management Technology Conference, 2012.

[13] Ran, X, Y Zhao. An assessment of a CO_2 flood for EOR and sequestration benefits in the Ordos Basin, northwest China [C]. Carbon Management Technology Conference, 2012.

[14] Zhao X, X Liao. Evaluation method of CO_2 sequestration and enhanced oil recovery in an oil reservoir, as applied to the Changqing Oilfields, China [J]. Energy & Fuels, 2012, 26 (8): 5350-5354.

Integrated Assessment of CO_2 Enhanced Oil Recovery and Storage Capacity

Y. Zhang   L. Zhang   B. Niu   S. Ren

(China University of Petroleum)

Abstract: CO_2 enhanced oil recovery (EOR) has been used as a commercial process for enhancing oil recovery since the 1970s, whereas limited field applications of CO_2 storage were undertaken only recently. In practice, considerable reservoir engineering design effort was made to reduce the total amount of CO_2 required to recover each barrel of oil in a CO_2 EOR project. For CO_2 storage, however, the objective is to increase the amount of CO_2 left behind at the end of the injection process; therefore, the approach to the design question changes. Consequently, optimization of CO_2 EOR and CO_2 storage processes differs significantly from the current CO_2 injection practices. In this paper, techniques were developed to systematically assess CO_2 EOR and storage capacity in a hydrocarbon reservoir selected for a demonstration project. More specifically, oil recovery was assessed and determined under miscible conditions, while CO_2 storage capacity was determined by using an estimation model improved in this study. In addition, economic analysis was conducted, assuming that CO_2 was captured from a chemical plant and transported 120 km to the oilfield. It is found that the geological framework is suitable for CO_2 storage in the selected reservoir and that, due to a favourable CO_2 miscible displacement mechanism, high oil recovery and storage capacity can be achieved, which leads the demonstration project to be economically profitable if prices of crude oil and CO_2 remain above certain values.

Keywords: CO_2 EOR; Storage capacity; Assessment

0 Introduction

The concentration of CO_2 in atmosphere has risen significantly since the beginning of the industrial revolution, primarily as a consequence of fossil fuel combustion for energy production and other industrial activities. Anthropological CO_2 emissions to the atmosphere are identified as a major drive for the global warming effect[1]. Therefore, governments and energy producers are interested in reducing the intensity of CO_2 emissions into atmosphere. A major mitigation method for reducing the amount of CO_2 emissions into atmosphere is carbon dioxide capture and storage (CCS), of which geological storage is a major component [2]. Some CCS demonstration projects have already been underway; however, in order to prove the validity of CO_2 sequestration, more demonstrations are needed in various geological formations worldwide. Generally, separating and capturing CO_2 from power plants is costly; but there are some low-cost CO_2 sources in China that produce NH_3 from coal via gasification, which are attractive candidates for storage demonstrations if there are suitable storage sites nearby [3]. Although the CO_2 capture and storage has been proven technologically, the CO_2

534

storage capacity should be assessed carefully in order to guarantee economic viability of the CCS project prior to large scale implementation.

The following types of storage reservoirs were distinguished: depleted oil fields, depleted natural gas fields, CO_2 EOR reservoirs, such as the IEA GHG Weyburn CO_2 monitoring and storage project[2], unmineable coal layers to enhance coal bed methane recovery using CO_2 (CO_2 ECBM), and deep aquifer, such as Sleipner project [4]. Among these storage methods, although CO_2 storage during EOR process had the lowest storage capacity, it was most likely to be carried out firstly, as the additional economic benefit from the incremental oil production can offset the cost of CO_2 sequestration[5]. In contradiction to the commercial CO_2 EOR projects whose main purpose was to maximize oil recovery with a minimum amount of CO_2, CO_2 EOR and storage aimed at maximizing both oil production and storage potentials of CO_2. Therefore, different approaches are needed to optimize the CO_2 EOR and storage projects. Prior to implementing of CO_2 EOR and storage project, the reservoir should be assessed for its suitability for miscible or immiscible CO_2 flooding first since the CO_2 EOR capacity was related closely with flooding model. The screening criteria of CO_2 miscible flooding were summarized as follows[6-9]: (1) The ratio of CO_2 pressure in the reservoir to minimum miscibility pressure (MMP) should be greater than 0.95; (2) The viscosity of crude oil was less than 12 cp; (3) Reservoir temperature was between 32℃ and 121℃; (4) Specific gravity of crude oil was less than 0.9; and (5) Minimal remaining oil saturation was greater than 25%. In general, miscible CO_2 flooding should be avoided in stratified and highly heterogeneous reservoirs, and in fractured reservoirs. Then, the CO_2 EOR and storage capacity should be calculated carefully.

Currently, several methods have been developed to calculate CO_2 EOR and storage capacity, such as, (1) assessment methods developed by Shaw and Baschu which were suitable for oil reservoirs under CO_2 EOR, and depleted oil reservoirs after water flooding[10, 11]; (2) assessment method provided by Stevens which was suitable for oil reservoirs through CO_2 EOR[12]; (3) assessment method established by ECL Technology Company under the operation models of water−alternating−gas and gravity stable gas injection[13]. These methods can be applied easily and require few calculation parameters; and how to determine CO_2 utilization coefficient and incremental oil production is the key to calculate CO_2 storage capacity. Generally, the incremental oil production can be calculated using the empirical formula or reservoir simulation. More credible CO_2 storage potential can be obtained if CO_2 utilization coefficient was determined through case analysis, but it was not practical due to limited CCS projects available. The CO_2 utilization coefficient was usually taken from literatures, but the value varied greatly by regions, so this would increase the uncertainties about the CO_2 storage capacity.

In this study, an improved methodology was proposed to assess the CO_2 EOR and storage capacity in an oil reservoir which was selected to carry out CO_2 EOR and storage demonstration project in China. More specifically, the CO_2 EOR potential was determined by the empirical curve between oil gravity and additional oil recovery proposed by Stevens. The calculation method for CO_2 storage capacity, which considers CO_2 dissolution, according to the features of water−flooding reservoir development in China, depends on storage mechanism, and the reservoir status. Not only can this method be suit-

535

able for CO_2 EOR reservoir, but also it can be suitable for depleted reservoirs and aquifers. Based on CO_2 EOR and storage capacity assessment, economic benefits of the project was then evaluated.

1   Methodology

Generally, the additional oil recovery due to CO_2 injection can be calculated from reservoir numerical simulation or empirical formula, although the reservoir simulation is relative accuracy and also much more complex. The Stevens's empirical relationship between oil gravity (API) and additional oil recovery was used to calculate the CO_2 EOR potential, as shown in Fig. 1, which considered oil gravity as the determinant parameter on additional oil recovery due to CO_2 EOR, though other parameters may affect the additional oil recovery of original oil in place (OOIP); and oil gravity was in this method considered representing the composition of the oil, since the composition of the oil, its density and viscosity were related closely [12].

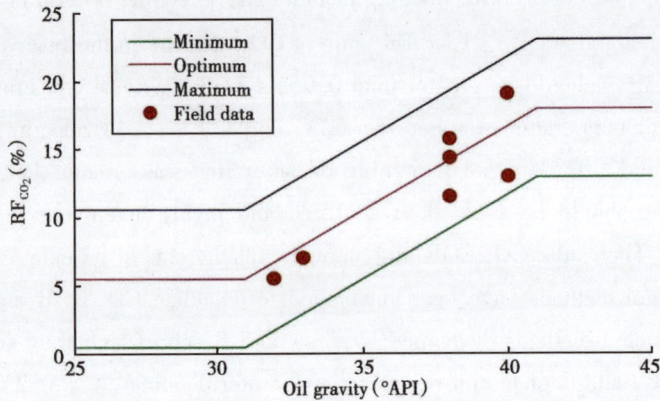

Fig. 1   Relationship between API Gravity and Percentage of Additional Oil due to EOR

In this study, the calculation method of CO_2 storage capacity in reservoirs is improved on the basis of storage mechanism, reservoir status, material balance equation and the original work of Tanaka[14] and Shaw[11], with the assumption of the volume previously occupied by the produced oil becomes available for CO_2 sequestration. According to CO_2 storage mechanism, CO_2 storage in reservoirs is realized through structural trapping, dissolution trapping and mineral trapping, and the first one is the key storage model. But with storage duration increases, the dissolution in crude oil and water is not negligible, which should be considered in the calculation method. Therefore, the theoretical capacity of CO_2 storage (M_{CO_2}) in oil reservoir can be divided into four parts:

$$M_{CO_2} = M_1 + M_2 + M_3 + M_4 \tag{1}$$

where M_1 is the storage capacity of CO_2 taking the volume previously occupied by produced oil; M_2 is the storage capacity of CO_2 dissolved in residual oil; M_3 is the storage capacity of CO_2 dissolved in water contained in reservoir; M_4 is the storage capacity of CO_2 reacting with reservoir rock.

According to the assumption that CO_2 would occupy all the volumn of produced oil, the storage capacity of M_1 can be calculated. And the assumption is usually valid for reservoirs that do not contact with an aquifer; otherwise, water will invade into the reservoirs as the reservoir pressure declines during production. The CO_2 storage capacity will decrease, so effect of aquifer influx (C_{aq}) should then be considered, and M_1 can be written as:

$$M_1 = RF_{ultimate} \times OOIP \times B_o \times C_{aq} \times \rho_{CO_2res}/\rho_{CO_2sc} \qquad (2)$$

where $RF_{ultimate}$ is the ultimate oil recovery factor of the reservoir; B_o is the oil formation volume factor; C_{aq} is the effective efficient of aquifer influx if a weak aquifer is present, $C_{aq} = 0.97$ and for a strong aquifer, $C_{aq} = 0.50$[11]; ρ_{CO_2res} is the CO_2 density at reservoir conditions, kg/m^3; ρ_{CO_2sc} is the CO_2 density at standard conditions, 1.977 kg/m^3.

Generally, the injected CO_2 will not contact all the remaining oil in place in a reservoir during CO_2 EOR process[15], so sweep efficiency (E_f) should be considered while calculating M_2; and M_2 can be written as:

$$M_2 = OOIP \times B_o \times (1 - RF_{ultimate}) \times E_f \times R_{o(CO_2)} \times \rho_{CO_2res}/\rho_{CO_2sc} \qquad (3)$$

where E_f is the sweep efficiency, %; $R_{o(CO_2)}$ is the CO_2 solubility in oil, %.

In an oil reservoir, the water content (M_w) includes two parts: initial water content and influx water content. The injected CO_2 will not contact all the water in place, sweep efficiency (E_f) should be considered while calculating M_3, so M_3 can be written as:

$$M_3 = M_w \times E_f \times R_{w(CO_2)} \times \rho_{CO_2res}/\rho_{CO_2sc} \qquad (4)$$

where M_w is the amount of water in the oil reservoir; $R_{w(CO_2)}$ is the CO_2 solubility in formation water, %.

In addition, premature breakthrough of CO_2 at production wells needs to be anticipated as the produced CO_2 is usually recycled and re-injected to increase CO_2 storage efficiency in the reservoir[16]. Moreover, the method assumes reservoir pressure is to be boosted up to the original reservoir pressure after the CO_2 EOR process terminates. And the CO_2 solubility in oil ($R_{o(CO_2)}$) can be determined from equations developed by Xue[17], the CO_2 solubility in water ($R_{w(CO_2)}$) can be calculated from the correlations developed by Enick and Klara[18].

Since the reaction rate between the injected CO_2 and reservoir rocks is very slow and the shortage capacity of CO_2 reacting with reservoir rock (M_4) is small, it is usually neglected. Finally, Equation 1 can be rewritten as:

$$M_{CO_2} = M_1 + M_2 + M_3 = [RF_{ultimate} \times OOIP \times B_o \times C_{aq} + OOIP \times B_o \times (1 - RF_{ultimate})$$
$$\times E_f \times R_{o(CO_2)} + M_w \times E_f \times R_{w(CO_2)})] \times \rho_{CO_2res}/\rho_{CO_2sc} \qquad (5)$$

2 Case Study

2.1 Field Background

Caoshe Oilfield is located in Subei basin, Jiangsu Oilfield Company Ltd. SINOPEC. The geological map of Subei basin and the location of Caoshe Oilfield are shown in Fig. 2. The Caoshe Oilfield was discovered in 1970s, with an area of 5 km². Taizhou formation is its main oil-bearing formation, which is a relatively small reservoir with oil trapped within structures formed by faults. The area of Taizhou formation is 0. 703 km², and its OOIP is $1.67×10^6 m^3$. The structure of the Taizhou formation is an anticline, where there are main fine sandstones, and the oil reservoirs are mainly distributed in the lower sandstones which range between 50m and 70m in thickness [19]. The geological structure map and oil reservoirs cross-section diagram of Taizhou formation are shown in Fig. 3 and 4, respectively[20]. Oil production of Taizhou formation reservoirs began in 1981, and there were three main stages in past thirty year's development history: the primary production stage (5/ 1981—8/1990); the secondary production stage through water flood (9/1990—12/1994); and the improved oil production stage via adjusting well pattern (1995—present). The oil recovery factor was 19. 4%, the remaining oil saturation was 30%~50%, and water cut reached 50. 2% in 2006. The current reservoir pressure is 28-30 MPa and reservoir temperature is about 110℃. During the first three development periods, the oilfield has developed a complete well pattern of injectors and producers with good well connection, the water cut at the producers was relative low, and reservoir pressure has been well maintained. The detailed reservoir parameters are shown in Table 1.

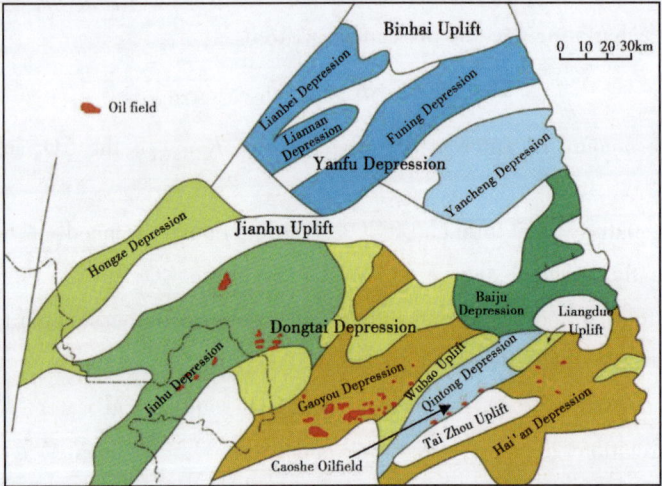

Fig. 2   Geotectonic Map Showing the Main Depression and Uplift Regions in the
Subei Basin (where the Caoshe Oilfield is located)

Table 1   Basic Geological and Reservoir Parameters of the Caoshe Oilfield

Parameter	Value	Parameter	Value
Oil area (km²)	0. 703	Initial gas/oil ratio (m³/ m³)	20. 6
OOIP (m³)	$1.67×10^6$	Formation volume factor	1. 13
Mean thickness (m)	17	Underground viscosity (mPa · s)	7. 02
Porosity (%)	14. 1	Underground oil specific gravity	0. 82
Permeability ($10^{-3}μm$)	46	Surface oil specific gravity	0. 85

538

Parameter	Value	Parameter	Value
Initial oil saturation (%)	51.9	Surface viscosity (mPa·s)	54.2
Reservoir depth (m)	3020	Freezing point (℃)	36.5
Initial reservoir pressure (MPa)	35.9	Wax content (%)	19.2
Saturation pressure (MPa)	4.9	Sulfur content (%)	0.58
Initial reservoir temperature (℃)	119	Cl^- (10^{-6}mg/L)	19123

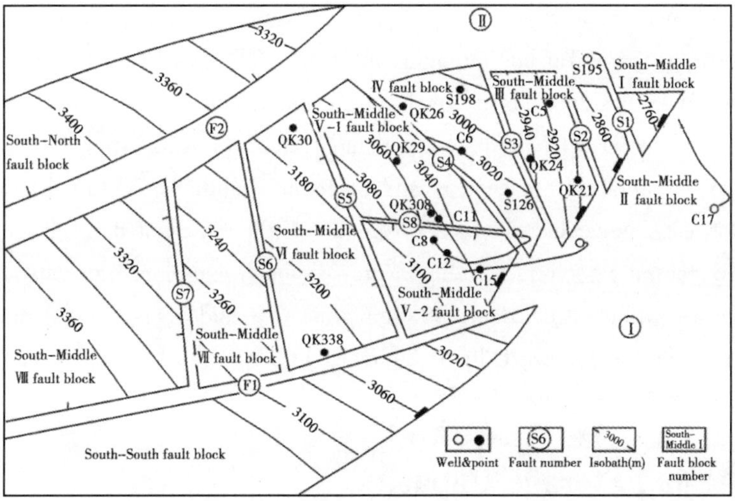

Fig. 3　Geological Structure Map of Taizhou Formation

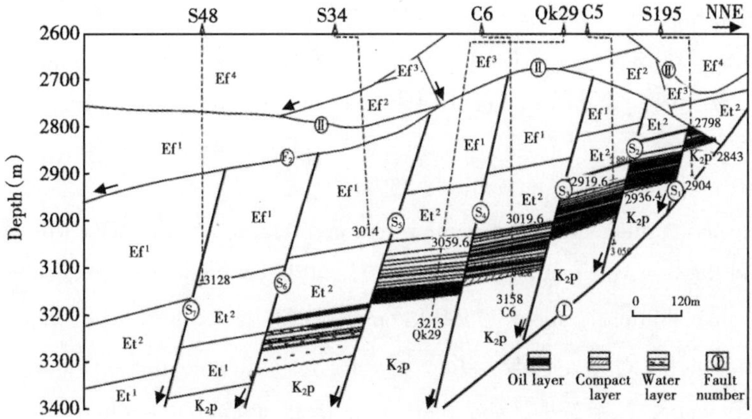

Fig. 4　Oil Reservoirs Cross-section Map of Taizhou Formation

2.2　Suitability of the Reservoir for CO_2 EOR and Storage

Although the Taizhou formation are trapped by fault, the boundary faults have large throws, greater than 100m, and extend for distance of more than 3 km, they therefore provide valid isolation between the reservoirs, which can be proven by the existence of gas and oil[20]. The overburden of the oil reservoirs in the Taizhou formation mainly comprises mudstones and gypsum, which represent an excellent regional caprock or a partial seal as they have proven capability to retain the hydrocarbons[20, 21], so the stratigraphic and structure traps of Taizhou formation reservoirs can provide safety in the CO_2 storage. The reservoirs of Taizhou formation have an average permeability of 46 mD and an average porosity of 14.1%, indicating fair infectivity potential. The crude oil in the Taizhou formation reservoirs in general is of medium density, and the oil viscosity underground is relative low. The reservoir temperature and pressure were adequate to make CO_2 and crude oil achieve miscibility[19]. Besides, the well formation test data and production data indicated that there were no obvious fractures developed in the reservoir[19]. According to the reservoir parameters and the development history, the reservoirs of Taizhou formation meet the conditions for a CO_2 miscible displacement process, which has been designed and is ready to be implemented. Therefore, Caoshe Oilfield has been selected to conduct CO_2 EOR and storage demonstration project.

In addition to the suitability of geology and reservoir conditions for CO_2 EOR and shorage, a CO_2 injection pilot test was carried out in 2006 in the southern part of the Taizhou formation using CO_2 from a nearby natural reservoir, which can provide much experience and many fundamental facilities for the Caoshe Oilfield CCS demonstration project. Besides, a detailed numerical reservoir model has been built by the Jiangsu Oilfield Company Ltd., SINOPEC, which can be used to provide reliable data for optimising the CO_2 EOR and storage project[19].

3 CO_2 EOR and Storage Capacity

The API gravity of crude oil in Taizhou formation reservoirs is 34.9. According to the Stevens's empirical relationship between oil gravity (API) and additional oil recovery, as can be seen from Fig. 1, the additional oil recovery of OOIP due to CO_2 EOR ranges from 5.2% to 15.2%, the minimum value is 5.2%, the optimal value is 10.4% and the maximum value is 15.2%. Correspondingly, the minimum incremental oil production for the Caoshe reservoires is $8.7 \times 10^4 m^3$, the optimal value is $17.4 \times 10^4 m^3$, and maximum value is $25.4 \times 10^4 m^3$.

According to assumption that the originl reservoir pressure of 35.9MPa will be achieved after CO_2 injection and the reservoir temperature of 110°C will be maintained, the density of the crude oil under this reservoir condition is about 735.0 kg/m$^{3[22]}$, CO_2 solubility in oil, $R_{o(CO_2)} = 36.6\%$ and CO_2 solubility in water, $R_{w(CO_2)} = 6\%$. In addition, according to the previous reservoir simulation[19], thep efficiency, $E_f = 22\%$, and there is a week aquifer under the reservoirs i.e. $C_{aq} = 0.97$[19]. Besides, the Caoshe Oilfield CO_2 EOR and storage project will be conducted when the water flooding recovery reaches 24%. Therefore, according to Equation 2 to Equation 5, the minimum CO_2 storage capacity at standard conditon is estimated to be $24.7 \times 10^7 m^3$, the optimal value is $27.9 \times 10^7 m^3$ and the maximum value is $30.9 \times 10^7 m^3$. The corresponding each part of the CO_2 storage capacity, M_1 M_2 and M_3, is shown in Table 2. And the optimal values will be used to do economic evaluation.

Table 2 CO$_2$ EOR and Storage Capacity Assessment Results

Type	Additional oil recovery (%)	Incremental oil production ($10^4 m^3$)	M_1 ($10^7 m^3$)	M_2 ($10^7 m^3$)	M_3 ($10^7 m^3$)	M_{CO_2} ($10^7 m^3$)
Minimum	5.2	8.7	19.9	4.0	0.8	24.7
Optimum	10.4	17.4	23.4	3.7	0.8	27.9
Maximum	15.2	25.4	26.7	3.4	0.8	30.9

4 Economic Evaluation

In addition to the technique issues, economic evaluation is another crucial measure to ensure the CO$_2$ EOR and storage project's feasibility. The economics of the project vary greatly, depending on the field size, CO$_2$ gas sources, location and the existing facilities. Thus, several issues should be designed carefully prior to the full scale implementation. In general, four economic elements (i. e. , CO$_2$ capture and compression cost, CO$_2$ transportation cost, CO$_2$ storage cost and revenue from incremental oil production) should be used to evaluate the economic efficiency[23].

As for the Caoshe Oilfield CCS demonstration project, Nanjing chemical plant, a synthetic ammonia plant 120 km away from the Caoshe Oilfield, was selected as a low-cost CO$_2$ source[3]. According to the study of Ecofys and TNO, the CO$_2$ capture and compression cost in a synthetic ammonia plant (pure stream) was about $28.7×10^{-3}/m$^{3[8]}$. The road transportation method was then selected for CO$_2$ because of small transportation amount, whose cost was about $40.3×10^{-3}/m$^{3[24]}$. The CO$_2$ storage cost includes two parts: injection cost and monitoring cost. According to the study of IPCC, the CO$_2$ injection cost of a typical onshore CO$_2$ EOR project was $20.8×10^{-3} ~ 47.4× 10^{-3}/m$^{3[25]}$, and the injection cost should be relative low for the Caoshe Oilfield because the CO$_2$ EOR pilot in this region can provide facilities and experience for the demonstration project, so an average value of $33.6×10^{-3}/m^3 was assumed; the cost of monitoring was comparatively small, ranging from $0.2×10^{-3} ~ 0.6×10^{-3}/m$^{3[26]}$. The total cost of the CCS demonstration project was about $103.2×10^{-3}/m^3. According to Table 2, the incremental oil production and CO$_2$ storage capacity are 17.4 × 10^4 m^3 and 27.9 × 10^7 m^3, respectively, provided that the world oil price averaged $471.8/m^3 recently. The total investment for the demonstration project is about $28.8 million, and the revenue from incremental oil production is about $82.1 million. Therefore, the revenue from the incremental oil production is significantly higher than the cost. This means the revenue not only can offset the cost of the CO$_2$ storage, but also can generate certain economic benefit to the Caoshe Oilfield. Based on the integrated assessment of CO$_2$ EOR and storage capacity, it is feasible to carry out CO$_2$ EOR and storage demonstration project in the Caoshe Oilfield.

5 Conclusions

(1) The addition oil recovery in CO$_2$ EOR and storage project was determined from the Ste-

vens's empirical relationship between oil gravity (API) and additional oil recovery via CO_2 EOR; the calculation method of CO_2 storage capacity in reservoirs is mainly improved on the basis of storage mechanism and material balance equation, assuming that all geometrical space of produced oil can be used to sequestrate CO_2, which also considers the water influx and sweep efficiency.

(2) The stratigraphic and structure traps of the Taizhou formation reservoirs can be used to safely sequestrate CO_2, and the reservoirs are suitable for CO_2 miscible flooding.

(3) The optimum incremental oil production and CO_2 storage capacity are calculated to be $17.4 \times 10^4 m^3$ and $27.9 \times 10^7 m^3$ for Taizhou formation reservoirs, respectively.

Reference

[1] Bachu S. CO_2 storage in geological media: Role, means, status and barriers to deployment [J]. Progress in Energy and Combustion Science, 2008, 34: 254-273.

[2] Sengul M. CO_2 sequestration—A safe transition technology [C]. International Health, Safety & Environment Conference, Abu Dhabi, UAE, 2-4 April 2006.

[3] Meng K C, Williams R H. Opportunities for low-cost CO_2 storage demonstration projects in China [J]. Energy Policy, 2007, 35 (4): 2368-2378.

[4] Korbul R, Kaddour A. Sleipner vest CO_2 disposal—injection of removed CO_2 into the Utsira Formation [J]. Energy Conversion and Management, 1995, 36 (6-9): 509-612.

[5] Holt T, Lindeberg E G B, Taber J J. Technologies and possibilities for larger-scale CO_2 separation and underground storage [C]. Annual Technical Conference and Exhibition, Dallas, TX, 1-4 October 2000.

[6] Kovscek A R. Screening criteria for CO_2 storage in oil reservoirs [J]. Petroleum Science and Technology, 2002, 39 (9): 841-866.

[7] Taber J J, Martin F D, Seright R S. EOR screening criteria revised—Part1: Introduction to screening criteria and enhanced recovery field projects [J]. SPE Reservoir Engineering, 1997, 12 (3): 189-198.

[8] Hendriks C, Bergen F V. Global carbon dioxide storage potential and costs [R]. Ecofys in Cooperation with TNO, Netherlands, 2004.

[9] Martin F D, Taber J J. Carbon dioxide flooding [J]. Journal of Petroleum Technology, 1992, 44 (4): 396-400.

[10] Shaw J C, Bachu S. Screening, evaluation, and ranking of oil reservoirs suitable for CO_2-flood EOR and carbon dioxide sequestration [J]. Journal of Canadian Petroleum Technology, 2002, 41 (9): 51-61.

[11] Bachu S, Shaw J C. Estimation of oil recovery and CO_2 storage capacity in CO_2 EOR incorporating the effect of underlying aquifers [C]. SPE/DOE Symposium on Improved Oil Recovery, Tulsa, OK, 17-21 April 2004.

[12] Stevens S H. Sequestration of CO_2 in depleted oil and gas field: Barriers to overcome in implementation of CO_2 capture and storage (disused oil and gas fields) [R]. Report for the IEA Greenhouse Gas R&D Programmer, 1999: PH3/22.

[13] Goodfield M, Woods C. Potential UKCS CO_2 retention capacity from IOR Project [C]. DTI's Improved Oil Recovery Research Dissemination Seminar, 25 June 2002.

[14] Shafeen A, Croiset E, Douglas P L. CO_2 sequestration in Ontario, Canada, Part I: Storage evaluation of potential reservoirs [J]. Energy Conversion and Management, 2004, 45 (17): 2645-2659.

[15] Hadlow R E. Update of industry experience with CO_2 injection [C]. SPE Annual Technical Conference and Exhibition, Washington, DC, 4-7 October 1992.

[16] Bondor P L. Applications of carbon dioxide in enhanced oil recovery [J]. Energy Conversion and Management, 1992, 33 (5-8): 579-586.

[17] Xue H T. Forecasting model of solubility of CH_4, CO_2 and N_2 in crude oil [J]. Oil and Gas Geology, 2005, 24 (4): 444-449.

[18] Metz B. Carbon dioxide capture and storage [J]. Intergovernmental Panel on Climate Change, 2002, 384-390.

[19] Zhu P, Shen Z H. The research on CO_2 flooding for enhanced oil recovery in complex fault block oil reservoirs [D]. PhD Dissertation, China University of Petroleum, China, October 2007.

[20] Yu K, Liu W. Study of CO_2 miscible flooding technique in the Caoshe oil field, Qingtong sag, Subei basin [J]. Petroleum Geology & Experiment, 2008, 30 (2): 312-612.

[21] Wong Q L. Structures and petroleum distribution in the Caoshe oil field of Qingtong depression in the Subei basin [J]. Marine Geology Letters, 2003, 19 (4): 26-29.

[22] Lu C, Han W S. Effects of density and mutual solubility of a CO_2-brine system on CO_2 storage in geological formation [J]. "Warm" vs. "Cold" Formations, Advances in Water Resources, 2009, 32 (12): 1685-1702.

[23] Algharaib M. Economic modeling of CO_2 capturing and storage project [C]. SPE Saudi Arabia Section Technical Symposium, Al-Khobar, Saudi Arabia, 12 May 2008.

[24] Yu X C, Li Z J. Carbon dioxide ground processing, storage and transportation [J]. Natural Gas Industry, 2008, 28 (8): 99-101.

[25] IPCC. IPCC special report on carbon dioxide capture and storage [M]. Prepared by Working Group III of the Intergovernmental Panel on Climate Change, Cambridge University Press, Cambridge, United Kingdom and New York, NY, USA, 2005, 442.

[26] Benson S M. Monitoring carbon dioxide sequestration in deep geological formations for inventory verification and carbon credits [C]. SPE Annual Technical Conference and Exhibition, San Antonio, TX, 24-27 September 2006.

The Comprehensive Evaluation on the Integral Development of Volcanic Gas Reserves & CO_2 Flooding in Jilin Oilfield

Xu Qing[1] Ran Qiquan[1] Song Wenli[2]
Chen Guoli[2] Xiang Dong[2] Li Yan[2]

(1. RIPED, CNPC; 2. Jilin Oil Field, CNPC)

Abstract: Pilot-CO_2 flooding in Jilin Oil Field has been got a first base in recent years in order to ensure CO_2 coming from the development of volcanic gas reserves is available for utilizing. The project has come into extended testing stage and will get into a carbon industry chain (CIC) production stage of integral development of volcanic gas reserves and CO_2 flooding. CIC will become a huge oil & gas development project, which composing volcanic gas reserves development, CO_2 flooding and CO_2 sequestration. The project has its own characteristics of long cycle construction, superior technical difficulty and high investment risk. Therefore, evaluating micro & macro-benefit of CIC becomes a crucial element task for management promoting, successful financing and governmental support. Maximizing ROI (return on investment) of CIC has become a core objective and theory of hierarchy analysis has been utilized in order to set up a CIC comprehensive evaluation index system which contains society, economy, environment, resource, science & technology and management. Analytical method of hierarchy has been used for establishing CIC evaluation model proceed to the evaluation of the multi-effect of the CIC based on SAS, MAPLAB etc. The study achieves a prophase comprehensive evaluation of CIC in Jilin Oil Field. The results indicate that the CIC has a certain economic benefit and preferable macro performance. It may promote the development & production of CIC, enhance the utilizing of oil & gas resource and carbon emission reduction especially under high oil price condition.

Keywords: Evaluation; Volcanic gas reserves; CO_2 flooding

0 Introduction

From the year of 2008, CNPC has begun pilot test for CO_2 enhanced oil recovery (EOR) in Jilin Oil Field. The main purpose has three points: (1) Realization of CO_2 emission reduction. According to China's Policies and Actions for Addressing Climate Change (2008), China's climate warming trend is basically consistent with the general trend globally. In the recent 50 years, the spatial distribution of precipitation has changed obviously in China, the sea level of coastal areas has risen by 90mm. As to ease the further aggravation of climate warming in future, we must find out the practical and effective methods to reduce emission of CO_2 into the atmosphere during production and consumption of fossil fuels. Existing research and application results at home and abroad have shown that hydrocarbon reservoirs are well closed underground gas storages, which can realize long-term CO_2 sequestration. (2) Improvement of crude oil recovery. Lots of successful experience from indoor and field tests of CO_2 flooding in America, Canada and Britain has shown that it has broad application prospects to CO_2 flooding technology in raising the use rate of low permeability oil reservoir, crude oil recovery, recovery of thickened oil reservoir, natural gas recovery and coal bed methane (CBM) recovery, etc. In these years, pretty much of the developed oil fields have come into the middle and late stages of water flooding, of which the average recovery was only 32% and more than 70% of new producing reserves were low permeability reserves, therefore, development technologies having advantages over water flooding are urgently needed. (3) Promotion of development of volcanic gas reservoirs. They have been successively discovered to CO_2-containing natural gas reservoirs and high CO_2-containing gas reserves at a large number since 2002 in Songliao Basin, to ensure safe and effective development of volcanic gas reservoirs, these works must be done well including separation, treatment, transportation, use and sequestration of CO_2 gas in production gas, in order that natural gas resources can be developed and domestic natural gas supply and energy security can be improved.

Development of volcanic gas reservoirs as well as CO_2 flooding and sequestration in Songliao Basin has formed a complementary CO_2 industry chain, microcosmically, the industry chain not only promotes the development and production of unconventional natural gas reserves, but also promotes the technical progress in EOR, as well as enables R & D on CO_2 geological sequestration technology. Macroscopically, it promotes the development of CO_2 emission reduction and environmental protection technologies in China, and improves the energy security of domestic natural gas and oil.

1 Background

1.1 Development of Volcanic Gas Reservoirs

There are lots of CO_2-containing gas reservoirs found in deep volcanic reservoirs in Song Liao Basin and Junggar Basin since 2002. The accumulated proved natural gas geological reserves have exceeded $4000 \times 10^8 m^3$ and the accumulated proved recoverable reserves have reached $2000 \times 10^8 m^3$ by the end of 2012, and the annual natural gas production of volcanic gas reservoirs has been close to $30 \times 10^8 m^3$.

The established natural gas production capacity is approximate $15 \times 10^8 m^3$ with an annual output of $13 \times 10^8 m^3$ of natural gas, its CO_2 gas output is approximate $2 \times 10^8 m^3$. All the CO_2 captured is used for CO_2 flooding EOR oil field development activities. At present, the production runs well and its benefit is considerable.

1.2 CO_2 Flooding and CO_2 Sequestration

Water flooding development of CO_2 pilot test block H-B in Jilin Oil Field was gradually put into action in August 2002, and the daily oil output of a single well was 9t/d with 6.4t/d verified in the initial stage. As the producing time went on, the output dropped gradually, which was basically in keeping with the index declining rule. In 2007, the average annual daily output of a single well was 4. 8t/d in oil, 3. 3t/d verified.

In 2008, it was designed injections well group of 22 in the development program in H−B block, i. e. 11 gas injection wells, 11 water injection wells and 63 oil production wells. The development of CO_2 flooding was in the way of alternate injection of water and gas. In 2010, CO_2 flooding of the pilot test area acquired a remarkable effect, which the annual oil increment of a single well in CO_2 flooding was $1. 6×10^4$t compared with water flooding as well as the estimated final EOR was approximate 10%. At present, Phase−II construction of the extended test area is completed on schedule, and the water−gas alternate development has acquired good flooding effect.

CO_2 flooding is able to realize CO_2 geological sequestration while enhancing oil recovery; Jilin Oil Field where the pilot test area is located, the study shows that its basin, oil field and oil reservoirs are agreeable in terms of CO_2 sequestration security. Pilot tests since 2008 also confirm that CO_2 EOR geological sequestration is agreeable and effective. The ratio of CO_2 geological sequestration is approximate 40% in the H−B test area so far.

2 Comprehensive evaluation of CO_2 industry chain

As industrialized application of CO_2 gas reservoir development and CO_2 flooding in Jilin Oil Field has basically formed a carbon industry chain (CIC), in order to comprehensively analyze the influence of the carbon industry chain on the society, environmental protection, resource development and utilization, as well as technical progress and management performance, an advanced analytic hierarchy process (AHP) is selected in this paper to evaluate the micro & macro benefits of the CIC, thus putting forward some alternative decision−making suggestions for oil companies and government.

Comprehensive evaluation refers to making an objective, fair, rational and comprehensive comment on objects to be evaluated. It is a method that uses multiple indexes and multiple ensembles for simultaneous quantitative evaluation and comparison. Generally speaking, a comprehensive evaluation modeling includes four steps below:

(1) Establishment of evaluation index system;

(2) Determination of weights;

(3) Standardization of indexes;

(4) Calculation of comprehensive evaluation indexes.

2. 1 Establishment of Evaluation Index System

2. 1. 1 Establishment of Objects

All is for guaranteeing continuous and effective operation of the CO_2 industry chain. The maximum return on investment is taken as a general objective in term of comprehensive utilization of the CO_2 industry chain of PetroChina of which those influence are fully taken into consideration in term of on the society, environmental protection, resource development and utilization as well as technical

546

progress and management performance. So three links involving CO_2-containing gas development, CO_2 flooding and CO_2 sequestration are put forward as sub-object hierarchies, under which rule hierarchies for six fields are designed, namely society, economy, environment, resource, science & technology, and management, under each rule hierarchy, sub-rule hierarchies reflecting comprehensive benefits are designed according to characteristics of all fields under all sub-objects, and then evaluation index hierarchies are established under every rule, so that completing establishment of the comprehensive benefit evaluation index system of the CO_2 industry chain, completing the comprehensive benefit evaluation based on this system, and putting forward measures and suggestions that promote benign development of the CO_2 industry chain.

2.1.2 Principle of Establishment

There are three principles for establishing the comprehensive benefit evaluation index system of the CO_2 industry chain. The first principle is the combination of comprehensiveness and emphasis to comprehensively reflect characteristics of sustainable development of the CO_2 industry chain and objectively reflect major factors that influence the CO_2 industry chain, so as to guarantee rationality of evaluation results. The second one is the combination of objectiveness and operability to reduce the influence of subjective factors on the establishment and application processes of the evaluation index system and to comprehensively consider availability of all data information required by evaluation, so as to guarantee operability of the evaluation index system. The third one is the combination of qualitative indexes and quantitative indexes to focus on the evaluation of quantitative factors involving the society, economy, environment and resources in the model and to establish qualitative evaluation factors of the six fields, so as to complete the establishment and analysis of the evaluation index system by combining qualitative indexes and quantitative indexes.

2.1.3 Establishment Method

With maximizing the comprehensive benefit of the CO_2 industry chain as the general object, using the hierarchy analysis principle, a system hierarchy structure of the CO_2 industry chain evaluation index system is established under the guide of sustainable development index system and relevant recycling economy theories, and the system includes general object hierarchy, sub-object hierarchy, rule hierarchy, sub-rule hierarchy and index hierarchy. To be specific, firstly, the maximum comprehensive benefit of the CO_2 industry chain is determined as the general object; secondly, the sub-object hierarchies including three links i. e. CO_2 gas field development, CO_2 flooding development and CO_2 sequestration are designed by combining the characteristics of the CO_2 industry chain, and by taking the sub-objects under the general optimal object of the whole industry chain as the optimal sub-objects in all links to achieve optimal comprehensive benefit of all links; thirdly, evaluation rule hierarchies that mainly evaluate all sub-objects in the fields of society, economy, resource, environment, science & technology, and management are established based on production and operation features of three links; fourthly, the sub-rule hierarchies under all sub-objects that reflects comprehensive benefits of all rules are designed based on different sub-object requirements; and finally, all sub-rules are broken down into evaluation indexes, which forms the index hierarchy. (Fig. 1)

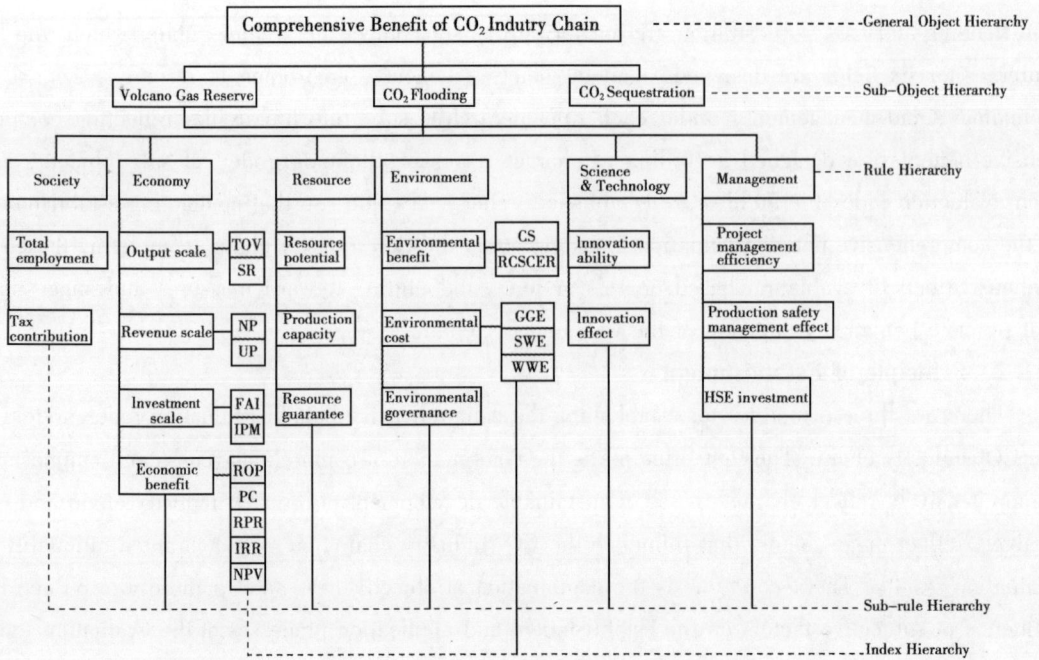

Fig. 1   Composition of Comprehensive Evaluation Index System of the CIC

2. 1. 4   Architecture of the Index System

The comprehensive benefit evaluation index system of the CO_2 industry chain is the basis and standard for comprehensive evaluation and study of harmonious development of the society, economy, resource, environment, technology and management systems in all links of the CIC, and is a hierarchically ordered set comprehensively reflecting different attribute indexes of all systems and composed based on subordination relations and hierarchical relationship. In the society field, employment opportunities for the society and tax contribution to the country made by three links of the industry chain are mainly considered; in the economy field, the output scale, profit scale, investment scale and economic benefit of the CO_2 industry chain are mainly investigated; in the resource field, resource potential, production capacity and economic development security are considered; in the environment field, environmental benefit, environmental cost, environmental governance are mainly investigated, and CO_2 emission, emission reduction and comprehensive environmental governance problems are comprehensively considered; in the technology field, technical innovation ability and effects of the project are mainly investigated; and in the management field, project management efficiency, production safety management effects and HSE investment are mainly investigated.

The comprehensive benefit evaluation index system of the CO_2 industry chain includes three links (sub−object hierarchies), namely, CO_2−containing gas development, CO_2 flooding and CO_2 sequestration, and comprehensive evaluation of each link includes six fields (rule hierarchies), namely, society, economy, resource, environment, technology and management. Based on theoretical research of CO_2−containing gas field development and CO_2 flooding development as well as oil and gas field development practices, the evaluation index study comprehensively analyzed typical

548

block development status of volcanic gas reservoirs in Song Liao Basin, studied the development and production, investment, cost and benefit of the CO_2 pilot test areas by trace research, proposed the main technical and economic characteristics of CO_2-containing gas field development, CO_2 flooding and sequestration in Song Liao Basin, and established the comprehensive benefit evaluation index system of the CIC.

2.2 Determination of Weights

2.2.1 Advanced-AHP

The advanced analytic hierarchy process (AHP) is applied to establish hierarchy index weights. When the AHP is applied to comparison and ranking of all indexes, a systematic hierarchical structure shall be established, and pairwise comparison and judgment matrixes shall be constructed; then, the judgment matrix will be used to calculate relative weights of compared indexes, finally, the synthetic relative weights of all hierarchies to the system object is calculated and ranked, specifically:

(1) Establishment of hierarchical structure;

(2) Establishment of judgment matrix;

(3) Importance ranking of single hierarchy factors or indexes;

(4) Hierarchy general ranking: comprehensive evaluation.

Weights shall be established in close combination with actual situation as well as comprehensive consideration of opinions and views of experts in the fields of economic society and oil and gas development, that is, when the judgment matrix is established, Delphi method (expert survey) shall be used to collect related expert advices, and then fuzzy comprehensive evaluation method shall be used to sort out and handle the expert advices. The main technical route is shown in Fig. 2.

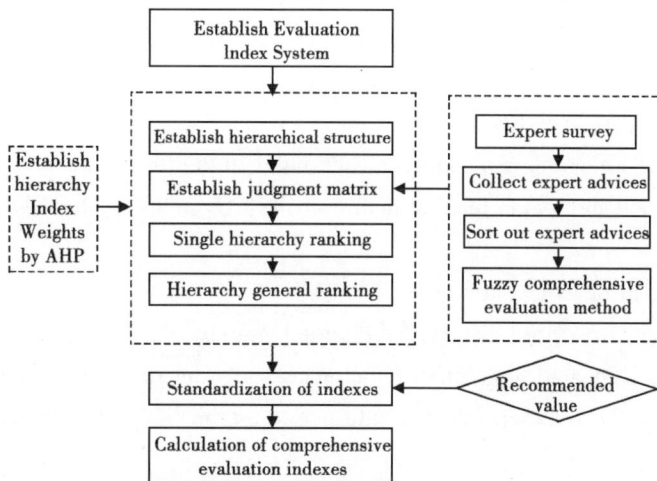

Fig. 2　Flow Chart of Establishing Hierarchy Index Weights by AHP

The AHP is applied to establishing the hierarchical structure and the judgment matrix, single rule ranking is used to approximately calculate the relative weights of all elements of a hierarchy under a certain rule, and then combined influence consistency check is performed to determine the effects of different elements on decision making. The advanced AHP is used to establish a four-scale matrix, which simplifies provision and solution process of the matrix, and finally establishes economic evaluation system model of the CIC, which has guiding significance for effective operation of the CO_2 industry chain.

The economic evaluation system of the CO_2 industry chain can be considered as a hierarchical structure of the AHP. The judgment matrix is established according to the basic steps of the AHP; weights of evaluation indexes of CO_2-containing gas field and CO_2 flooding oil field are determined in accordance with the domestic existing CO_2-containing gas field development and CO_2 flooding oil field development scheme and in combination with the actual data; and comprehensive evaluation is evaluation and analysis of Jilin CIC case are completed.

2.2.2 Establishment of Hierarchical Structure Model

Factors contained in problems and correlations between problems are analyzed to break down related factors into several hierarchies from top to bottom by different attributes. Factors at the same hierarchy are subject to the upper hierarchy or have effect on the upper level, and they also dominate the factors at the next hierarchy or are affected by them. Namely, targeted index evaluation system and tree index hierarchical structure are established through the research on the development and comprehensive utilization of CO_2 gas fields, so as to reflect the economic, social, environmental and resource benefits of the CIC. For example, the first hierarchy is the general object hierarchy, i. e. the comprehensive evaluation results of the CO_2 industry chain (comprehensive evaluation indexes), reflecting the development and comprehensive utilization of CO_2 gas fields; the second one includes economy, society, resource, environment, science & technology and management, reflecting various aspects influenced by the CIC; the third one includes evaluation indexes of corresponding fields, reflecting the relevance of these fields; and the fourth one and fifth one are similar, the indexes at the upper hierarchy are weighted for calculation after standardizing the values of the indexes of the corresponding next hierarchy.

2.2.3 Establishment of Weights

In this paper, the weight corresponding to each index in the tree structure is determined through the advanced AHP, and the hierarchical structure of the CO_2 industry chain has been established, as shown in Fig. 3.

(1) Establishment of judgment matrix.

In the hierarchical structure, factors at the same hierarchy which belong to factors at the upper hierarchy are pairwise compared to get their importance degree for rule, and the importance degree is quantified according to the scale specified in advance to establish a matrix form, i. e. the judgment matrix.

$$A = \begin{bmatrix} a_{11} & a_{12} & \cdots & A_{1m} \\ a_{21} & a_{22} & \cdots & A_{2m} \\ \vdots & \vdots & \ddots & \vdots \\ a_{m1} & a_{m2} & \cdots & A_{mm} \end{bmatrix}$$